OXFORD GRADUATE TEXTS IN MATHEMATICS

Series Editors

M. BRIDSON
G.G CHEN
S.K. DONALDSON
T.J. LYONS
M.J. TAYLOR

OXFORD GRADUATE TEXTS IN MATHEMATICS

Books in the series

1. Keith Hannabuss: *An Introduction to Quantum Theory*
2. Reinhold Meise and Dietmar Vogt: *Introduction to Functional Analysis*
3. James G. Oxley: *Matroid Theory*
4. N.J. Hitchin, G.B. Segal, and R.S. Ward: *Integrable Systems: Twistors, Loop Groups, and Riemann Surfaces*
5. Wulf Rossmann: *Lie Groups: An Introduction through Linear Groups*
6. Qing Liu: *Algebraic Geometry and Arithmetic Curves*
7. Martin R. Bridson and Simon M. Salamon (eds): *Invitations to Geometry and Topology*
8. Shmuel Kantorovitz: *Introduction to Modern Analysis*
9. Terry Lawson: *Topology: A Geometric Approach*
10. Meinolf Geck: *An Introduction to Algebraic Geometry and Algebraic Groups*
11. Alastair Fletcher and Vladimir Markovic: *Quasiconformal Maps and Teichmüller Theory*
12. Dominic Joyce: *Riemannian Holonomy Groups and Calibrated Geometry*
13. Fernando Villegas: *Experimental Number Theory*
14. Péter Medvegyev: *Stochastic Integration Theory*
15. Martin A. Guest: *From Quantum Cohomology to Integrable Systems*
16. Alan D. Rendall: *Partial Differential Equations in General Relativity*
17. Yves Félix, John Oprea, and Daniel Tanré: *Algebraic Models in Geometry*
18. Jie Xiong: *Introduction to Stochastic Filtering Theory*
19. Maciej Dunajski: *Solitons, Instantons, and Twistors*
20. Graham R. Allan: *Introduction to Banach Spaces and Algebras*
21. James Oxley: *Matroid Theory, Second Edition*
22. Simon Donaldson: *Riemann Surfaces*
23. Clifford Henry Taubes: *Differential Geometry: Bundles, Connections, Metrics and Curvature*
24. Gopinath Kallianpur and P. Sundar: *Stochastic Analysis and Diffusion Processes*
25. Selman Akbulut: *4-Manifolds*
26. Fon-Che Liu: *Real Analysis*
27. Dusa Mcduff and Dietmar Salamon: *Introduction to Symplectic Topology, Third Edition*
28. Chris Heunen and Jamie Vicary: *Categories for Quantum Theory: An Introduction*
29. Shmuel Kantorovitz and Ami Viselter: *Introduction to Modern Analysis, Second Edition*
30. Stephan Ramon Garcia, Javad Mashreghi, and William T. Ross: *Operator Theory by Example*
31. Maciej Dunajski: *Solitons, Instantons, and Twistors, Second Edition*

An Introduction to Module Theory

IBRAHIM ASSEM
FLÁVIO U. COELHO

OXFORD
UNIVERSITY PRESS

OXFORD
UNIVERSITY PRESS

Great Clarendon Street, Oxford, OX2 6DP,
United Kingdom

Oxford University Press is a department of the University of Oxford.
It furthers the University's objective of excellence in research, scholarship,
and education by publishing worldwide. Oxford is a registered trade mark of
Oxford University Press in the UK and in certain other countries

© Ibrahim Assem and Flávio U. Coelho 2024

The moral rights of the authors have been asserted

All rights reserved. No part of this publication may be reproduced, stored in
a retrieval system, or transmitted, in any form or by any means, without the
prior permission in writing of Oxford University Press, or as expressly permitted
by law, by licence or under terms agreed with the appropriate reprographics
rights organization. Enquiries concerning reproduction outside the scope of the
above should be sent to the Rights Department, Oxford University Press, at the
address above

You must not circulate this work in any other form
and you must impose this same condition on any acquirer

Published in the United States of America by Oxford University Press
198 Madison Avenue, New York, NY 10016, United States of America

British Library Cataloguing in Publication Data
Data available

Library of Congress Control Number: 2024939638

ISBN 9780198904908
ISBN 9780198904915 (pbk.)

DOI: 10.1093/9780198904939.001.0001

Printed and bound by
CPI Group (UK) Ltd, Croydon, CR0 4YY

Links to third party websites are provided by Oxford in good faith and
for information only. Oxford disclaims any responsibility for the materials
contained in any third party website referenced in this work.

Dedicated to Sonia and Marcia

Contents

Introduction   1

I. Rings and Algebras   5
    I.1 Introduction   5
    I.2 Rings and modules   5
    I.3 Algebras   17
    I.4 Algebra morphisms   30
    I.5 Principal ideal domains   40

II. Modules   49
    II.1 Introduction   49
    II.2 Modules and submodules   49
    II.3 Module morphisms   69
    II.4 The isomorphism theorems   80

III. Categories and functors   111
    III.1 Introduction   111
    III.2 Categories and functors   111
    III.3 Products and coproducts of modules   126
    III.4 Free modules   137

IV. Abelian categories   145
    IV.1 Introduction   145
    IV.2 Linear and abelian categories   145
    IV.3 Fibered products and amalgamated sums   168
    IV.4 Equivalences and dualities of categories   178

V. Modules over principal ideal domains   191
    V.1 Introduction   191
    V.2 Free modules and torsion   191
    V.3 The structure theorems   201
    V.4 An application: the Jordan form of a matrix   222

VI. Functors between modules   231
    VI.1 Introduction   231
    VI.2 The tensor product of modules   231
    VI.3 Exact functors   244
    VI.4 Projectives, injectives and flats   255

VII. The chain conditions   287
    VII.1 Introduction   287
    VII.2 Artinian and noetherian modules and algebras   287
    VII.3 Decompositions of algebras   300
    VII.4 Composition series   314
    VII.5 Semisimple modules and algebras   329

VIII. Radicals   339
    VIII.1 Introduction   339
    VIII.2 Radical and socle of a module   339
    VIII.3 Radicals of algebras   350
    VIII.4 Indecomposability   363
    VIII.5 The radical of a module category   376

IX. Projectives and quivers   387
    IX.1 Introduction   387
    IX.2 Projective modules over artinian algebras 387
    IX.3 Morita equivalence   406
    IX.4 Bound quiver algebras   424

X. Homology   449
    X.1 Introduction   449
    X.2 Homology and cohomology   449
    X.3 Derived functors   467

XI. Extension and torsion   479
    XI.1 Introduction   479
    XI.2 The extension and torsion functors   479
    XI.3 Exact sequences and extensions   505

XII. Homological dimensions   531
    XII.1 Introduction   531
    XII.2 Homological dimensions of modules   531
    XII.3 Homological dimensions of algebras   545
    XII.4 Classes of algebras   560

Bibliography   579
Index   581

Introduction

This book is a byproduct of graduate courses taught by the authors in their respective universities over the last thirty years. During this period, the area it addresses— Module theory—has experienced deep changes, to which teaching had to adapt. Module theory is one of the fundamental areas of algebra and has important applications in other parts of mathematics—such as in Representation Theory (of groups and algebras), Combinatorics, Geometry, Algebraic Topology, Mathematical Physics and Functional Analysis. It is being taught at the Master level to students in algebra, and also to students who need algebraic tools in their work. We believe our book to be an accessible and reasonably selfcontained account of Module theory, that can be used as a textbook for graduate students, advanced undergraduate students and also for selfstudying. In fact, we only assume that the reader knows some linear algebra and ring theory, at the level normally acquired in standard undergraduate courses of most universities.

Module theory can be thought of as an advanced form of linear algebra. Indeed, the definition of a module is the same as that of a vector space, except that the scalars, instead of being taken in a field, are taken in a ring, or, equivalently, in an algebra. In other words, scalars do not necessarily have multiplicative inverses. This makes their study very different from that of vector spaces—perhaps the most striking difference comes from the fact that, usually, a module does not have a basis. Another one is that, because our rings (or algebras) are not assumed to be commutative, one has to consider the two possibilities for multiplying scalars, namely, on the right or on the left. These, and other differences, make the setting of modules much richer than that of vector spaces and call for new methods and tools in order to study this algebraic structure.

Module theory is a standard area but it has undergone profound changes in recent years, the most evident one being the new emphasis on categorical and homological techniques. Our approach stresses the use of both from the start. Category theory, originally introduced as an elegant unifying language, furnishes a deep conceptual insight into the theory. It provides a new light in which to look at modules and, therefore, new research problems. From the practical point of view, category theory forces us to concentrate on maps rather than on sets and their elements, and this allows us to present simpler and shorter proofs of classical results. We do not assume any previous knowledge of category theory. Instead, we provide all relevant definitions and results. Similarly, homological algebra, which takes its origins in algebraic topology, is used to provide invariants for modules. Accordingly, our book features a selfcontained introduction to homological methods and their applications to module theory.

Our treatment is based on the strong conviction that the top priority should be given to clarity and intelligibility of the text. Thus we tried to provide the reader with

An Introduction to Module Theory. Ibrahim Assem and Flávio U. Coelho, Oxford University Press.
© Ibrahim Assem and Flávio U. Coelho (2024). DOI: 10.1093/9780198904939.003.0001

as many pictures and examples as possible. Mathematicians do a lot of examples and draw a lot of pictures on the blackboard to help visualising concepts and thus develop their students' intuition, but these pictures seldom make it into the books. We made it a point to include as many pictures as possible, explaining their advantages as well as their limitations. We included a large number of examples of all kinds worked out in detail. We focused on showing the student how to construct and analyse examples by himself, in a systematic way, leading to the proof of a statement or a counterexample to an incorrect one. Besides taking our examples from classical areas, such as abelian groups or modules over matrix rings, we introduce representations of quivers, by now a well-developed and established research area, which provides us with an inexhaustible source of examples, elementary enough for any beginner to handle (and to construct by himself). Finally, we try to motivate the introduction of each new concept by explaining, in a few words, why it is needed. Likewise, we present in several places the rationale for new results and the strategy followed in order to prove them. We did not hesitate to repeat ourselves when it seemed to us to enhance our exposition. As a consequence, the pace of our presentation is deliberately pedestrian and we try to give as many details as possible, especially in the proofs of the main results. We have included a large number of exercises of all levels of difficulty, both computational and theoretical, and provided hints for some of them.

The choice of the material presented here and our approach to it are consequences both of our teaching experience and our mathematical itinerary. The contents have nothing original: the concepts and results presented here have appeared in different forms in many other books or research papers. We acknowledge our debt to the texts cited in the bibliography, and to so many others we did not specifically mention. This book is conceived as a textbook for a beginner Master Student in Mathematics. Due to this constraint, we could not be encyclopedic in this work. Perhaps the most obvious omission is the study of limits and colimits in module categories, leading to the description of infinitely generated modules. We apologise to the reader for this and any other omission.

We explain the contents of the book. Chapter I recalls the necessary facts from undergraduate ring theory and presents the equivalent concept of an algebra. We start working with modules in Chapter II, treating them both from a classical perspective and from a categorical one. Chapters III and IV are devoted to the rudiments of category theory and direct applications in module theory. In Chapter V, we pause from theory to present a classical topic: the classification of finitely generated modules over a principal ideal domain, with its straightforward application to the Jordan form of a matrix. Chapter VI concentrates on the study of two functors, namely, the Hom and tensor functors, which allow to pass from one module category to another and to introduce projective, injective and flat modules. Chapters VII and VIII contain the classical results of module theory, such as the Jordan–Hölder, Wedderburn–Artin, Hopkins–Levitsky theorems, and finally, the Krull–Schmidt Unique Decomposition theorem, which shows how a module can be built starting from ones that cannot be decomposed into direct sums of smaller ones. In Chapter IX, we describe finitely generated projective modules over Artinian algebras, we prove the Morita and Eilenberg–Watts theorems, then return to representations of quivers, culminating in the computation of indecomposable finitely generated modules over the so-called

Nakayama algebras. Chapter X is meant as an introduction to the main tools of homological algebra, the (co)homology sequences and derived functors. These are applied in Chapter XI, where we present extension and torsion modules and the resulting functors. We conclude in Chapter XII by discussing homological dimensions of modules and algebras and showing how they can be used in the study of hereditary and selfinjective algebras.

This book has developed from graduate courses given at the universities of Sherbrooke and São Paulo. It is a pleasure for us to acknowledge our debt to our students and all the colleagues with whom we discussed the contents of our courses. Their questions, remarks and criticisms gave us an invaluable feedback. We thank in particular Shiping Liu, Marco Armenta and Juan Carlos Bustamante for their useful suggestions.

We also warmly thank Jean-Philippe Morin and Souheila Hassoun for their precious help in getting our manuscript into shape. We are also grateful to Dan Taber, Giulia Lipparini as well as the whole OUP team for their help during its preparation.

The authors gratefully acknowledge financial support from CNPq and FAPESP, Brazil, as well as NSERC, Canada.

Sherbrooke and São Paulo,
January 2024

Chapter I
Rings and Algebras

I.1 Introduction

The subject of this book is module theory. The easiest way to think of a module is as 'a vector space over a ring' or, equivalently, 'over an algebra'. More precisely, a *module* is defined by the same axioms as a vector space but the scalars, instead of being taken inside a field, are taken inside a ring, or an algebra. However, different techniques are needed to deal with modules. Because nonzero scalars do not usually have inverses, many basic properties of vector spaces do not hold true for modules in general. For instance, modules do not necessarily have bases. In this first chapter, we recall the basic properties of rings and introduce the equivalent notion of algebra, which is technically easier and more fruitful. We also present several examples.

I.2 Rings and modules

I.2.1 Rings

In elementary mathematics, most sets under consideration, such as integers, real numbers, polynomials and matrices, have (at least) two operations, 'addition' and 'multiplication', and many properties of these sets arise from the interplay between these operations. If X is a set, an **operation** or **internal operation** in X is a map from $X \times X$ to X. That is, it assigns to each pair of elements of X a uniquely determined element of X, called **result** of the operation.

Typical operations are the usual addition and multiplication of integers, real numbers, polynomials or matrices. There are differences between properties of these operations in these sets. For instance, multiplication of integers, real numbers or polynomials is commutative, but not multiplication of matrices. Further, any nonzero real number has a multiplicative inverse which belongs to the same set of nonzero real numbers, but this property does not hold true for integers, polynomials or matrices. Now, the set of real numbers is a typical example of a field, as defined in linear algebra courses. The other three sets mentioned satisfy all axioms of a field, except perhaps commutativity of multiplication, and existence of multiplicative inverses. Removing the latter two from the axioms of a field, we get the following definition.

I.2.1 Definition A *ring* $(K, +, \cdot, 1)$, or simply K, is a quadruple, where K is a set such that:

    (A) There is an operation $+$ on K, called **addition**, which associates to each pair $(\alpha, \beta) \in K \times K$ its **sum** $\alpha + \beta \in K$, and which makes the pair $(K, +)$ an **abelian group**, that is, it satisfies the following axioms:

An Introduction to Module Theory. Ibrahim Assem and Flávio U. Coelho, Oxford University Press.
© Ibrahim Assem and Flávio U. Coelho (2024). DOI: 10.1093/9780198904939.003.0002

(A1) It is commutative, that is, $\alpha + \beta = \beta + \alpha$, for any $\alpha, \beta \in K$.

(A2) It is associative, that is, $\alpha + (\beta + \gamma) = (\alpha + \beta) + \gamma$, for any $\alpha, \beta, \gamma \in K$.

(A3) There exists an element $0 \in K$, called the **zero**, such that $\alpha + 0 = 0 + \alpha = \alpha$, for any $\alpha \in K$.

(A4) To any $\alpha \in K$ is associated an element $(-\alpha) \in K$, called its **negative**, or **opposite**, such that $\alpha + (-\alpha) = (-\alpha) + \alpha = 0$.

(B) There is an operation $\cdot$ on K, called **multiplication**, which associates to each pair $(\alpha, \beta) \in K \times K$ its **product** $\alpha \cdot \beta \in K$, also denoted simply by $\alpha\beta$, and an element 1 of K, which make the triple $(K, \cdot, 1)$ a **monoid**, that is, it satisfies the following axioms:

(B1) It is associative, that is, $\alpha(\beta\gamma) = (\alpha\beta)\gamma$, for any $\alpha, \beta, \gamma \in K$.

(B2) The element $1 \in K$, called the **identity**, or **one**, is such that $\alpha 1 = 1\alpha = \alpha$, for any $\alpha \in K$.

(C) These operations are compatible, in the sense that multiplication is **left and right distributive** over addition:

(C1) $\alpha(\beta + \gamma) = \alpha\beta + \alpha\gamma$, for any $\alpha, \beta, \gamma \in K$.

(C2) $(\beta + \gamma)\alpha = \beta\alpha + \gamma\alpha$, for any $\alpha, \beta, \gamma \in K$.

Some authors consider rings without identity elements, that is, abelian groups with a multiplication satisfying axioms (B1), (C1) and (C2) (but not necessarily (B2)). In this text, we always consider rings with identities, sometimes called **unitary rings**. This assumption is usual when one deals with modules, and can be made without loss of generality, as explained after Definition I.3.1.

While we required addition of a ring to be commutative, we made no such assumption on multiplication. This is an additional axiom that a ring may or may not satisfy. A ring K is called **commutative** if $\alpha\beta = \beta\alpha$, for any $\alpha, \beta \in K$. In a commutative ring, the left distributivity axiom (C1) is equivalent to the right (C2).

I.2.2 Examples.

(a) One may ask what is the 'smallest' possible ring. Because of Definition I.2.1, any ring contains 0 and 1, so no ring is empty. But nothing says that 0 and 1 should be distinct. Indeed, let $K = \{0\}$ be a one-element set, and define operations by setting $0 + 0 = 0$ and $0 \cdot 0 = 0$. Then K becomes a (commutative) ring, called the **zero** or **trivial** ring, and denoted by $K = 0$.

(b) The sets $\mathbb{Z}$ of integers, $\mathbb{Q}$ of rationals, $\mathbb{R}$ of reals and $\mathbb{C}$ of complexes, with their usual addition and multiplication, are commutative rings.

(c) For any integer $m \geq 2$, the set $\mathbb{Z}/m\mathbb{Z}$ of integers modulo m is a commutative ring. Indeed, let $\bar{a} = a + m\mathbb{Z}$ denote the residual class of an integer a. It is easily verified that the addition and multiplication given respectively by $\bar{a} + \bar{b} = \overline{a + b}$ and $\bar{a} \cdot \bar{b} = \overline{ab}$, for $a, b \in \mathbb{Z}$, are unambiguously defined operations which satisfy the required axioms.

(d) Let K be a ring. The set $K[t]$ of polynomials in one indeterminate t with coefficients in K is a ring under ordinary addition and multiplication of polynomials. The ring $K[t]$ is commutative if (and only if) K is commutative.

(e) Let K be a ring and n a positive integer. The set $M_n(K)$ of $n \times n$-matrices with coefficients in K is a ring under ordinary addition and multiplication of matrices. If $n > 1$ and $1 \neq 0$ in K, then $M_n(K)$ is not commutative even if K is commutative. For instance, the matrices $\begin{pmatrix} 1 & 1 \\ 0 & 1 \end{pmatrix}$ and $\begin{pmatrix} 1 & 0 \\ 1 & 1 \end{pmatrix}$ do not commute.

(f) Let K be a ring. The **opposite ring** K^{op} has the same elements and the same addition as K, so that $K^{\mathrm{op}} = K$, *as abelian groups*, but multiplication in K^{op} is defined by $\alpha \cdot \beta = \beta\alpha$, for $\alpha, \beta \in K$, where the product on the right-hand side is the product in K. Thus, a ring K is commutative if and only if $K^{\mathrm{op}} = K$, *as rings*.

Rings satisfy most arithmetic properties of integers, as is shown by the following proposition.

I.2.3 Proposition. *Let K be a ring, then:*

(a) *The zero element of K is unique.*

(b) *The identity element of K is unique.*

(c) *The negative of any $\alpha \in K$ is unique.*

(d) *For any $\alpha \in K$, we have $\alpha \cdot 0 = 0 \cdot \alpha = 0$.*

(e) *For any $\alpha, \beta \in K$, we have $(-\alpha)\beta = \alpha(-\beta) = -(\alpha\beta)$.*

(f) *For any $\alpha, \beta_1, \ldots, \beta_m \in K$, we have $\alpha(\sum_{i=1}^{m} \beta_i) = \sum_{i=1}^{m}(\alpha\beta_i)$.*

(g) *For any $\alpha_1, \ldots, \alpha_n, \beta \in K$, we have $(\sum_{j=1}^{n} \alpha_j)\beta = \sum_{j=1}^{n}(\alpha_j\beta)$.*

Proof. (a) If both $0, 0'$ satisfy axiom (A3) then, because 0 is a zero, we have $0 + 0' = 0'$. But $0'$ is also a zero, so $0 + 0' = 0$. Hence, $0 = 0'$.

(b) This is similar to (a) and left to the reader.

(c) If both $(-\alpha)$ and $(*\alpha)$ satisfy (A4), then, thanks to (A2),

$$(-\alpha) = (-\alpha) + 0 = (-\alpha) + (\alpha + (*\alpha)) = ((-\alpha) + \alpha) + (*\alpha) = 0 + (*\alpha) = (*\alpha).$$

(d) Let $\alpha \in K$. Then $0 + 0 = 0$ gives, thanks to (C1),

$$\alpha \cdot 0 = \alpha \cdot (0 + 0) = \alpha \cdot 0 + \alpha \cdot 0.$$

Subtracting $\alpha \cdot 0$ from both sides, that is, adding $-(\alpha \cdot 0)$ to both sides, we get $\alpha \cdot 0 = 0$. Similarly, $0 \cdot \alpha = 0$.

(e) Let $\alpha, \beta \in K$. Because of (C2) and (d) above, we have:

$$(-\alpha)\beta + \alpha\beta = ((-\alpha) + \alpha) \cdot \beta = 0 \cdot \beta = 0.$$

Then (c) above gives $(-\alpha)\beta = -(\alpha\beta)$. Similarly, $\alpha(-\beta) = -(\alpha\beta)$.

(f) This is proved by induction on m. If $m = 1$, there is nothing to prove. If $m > 1$, then, using the induction hypothesis and (C1), we have:

$$\alpha \left(\sum_{i=1}^{m} \beta_i \right) = \alpha \left(\left(\sum_{i=1}^{m-1} \beta_i \right) + \beta_m \right) = \alpha \left(\sum_{i=1}^{m-1} \beta_i \right) + \alpha \beta_m$$

$$= \sum_{i=1}^{m-1} (\alpha \beta_i) + \alpha \beta_m = \sum_{i=1}^{m} \alpha \beta_i.$$

(g) This is similar to (f) and left to the reader.   ◻

Because K is an abelian group, we also have $-(-\alpha) = \alpha$ and $-(\alpha + \beta) = (-\alpha) + (-\beta)$, for $\alpha, \beta \in K$.

A nice consequence of (d) is that a ring equals the zero ring if and only if its identity equals zero. Clearly, $K = 0$ implies $1 = 0$. Conversely, if $1 = 0$ in K, then, for any $\alpha \in K$, we have $\alpha = \alpha \cdot 1 = \alpha \cdot 0 = 0$, so $K = 0$.

When we wrote that rings satisfy 'most' arithmetic properties of integers, we meant that they usually do not satisfy them all. For instance, the product of two nonzero integers is nonzero, but in the ring $\mathbb{Z}/6\mathbb{Z}$, the product of the nonzero elements $\bar{2}$ and $\bar{3}$ is $\bar{2} \cdot \bar{3} = \bar{6} = \bar{0}$. Such elements are called zero divisors.

I.2.4 Definition.   Let K be a nonzero commutative ring.
   (a) A nonzero element $\alpha \in K$ is a **zero divisor** if there exists a nonzero element $\beta \in K$ such that $\alpha \beta = 0$.
   (b) The ring K is an **integral domain** if it contains no zero divisor.

Thus, a nonzero commutative ring K is an integral domain if and only if it satisfies one of the equivalent conditions:

   (i) If $\alpha, \beta \in K$ are such that $\alpha \beta = 0$, then $\alpha = 0$ or $\beta = 0$.
   (ii) If $\alpha, \beta \in K$ are such that $\alpha \beta = 0$ and $\alpha \neq 0$, then $\beta = 0$.
   (iii) If $\alpha, \beta \in K$ are such that $\alpha \neq 0$ and $\beta \neq 0$, then $\alpha \beta \neq 0$.

A typical example of an integral domain is the set $\mathbb{Z}$ of integers. If K is an integral domain, then the ring of polynomials $K[t]$ is also an integral domain, because the product of the (nonzero) leading coefficients of two nonzero polynomials is itself nonzero.

In noncommutative rings, one distinguishes left and right zero divisors, but we shall not need that.

A nonzero commutative ring is an integral domain if and only if it satisfies the so-called **cancellation property**.

I.2.5 Lemma.   *A nonzero commutative ring K is an integral domain if and only if, for any $\alpha, \beta, \gamma \in K$ such that $\alpha \neq 0$ and $\alpha \beta = \alpha \gamma$, we have $\beta = \gamma$.*

Proof. *Necessity.* From $\alpha \beta = \alpha \gamma$ we infer that $\alpha(\beta - \gamma) = 0$. Because K is integral and $\alpha \neq 0$, this gives $\beta - \gamma = 0$, hence $\beta = \gamma$.
   Sufficiency. If $\alpha \neq 0$ and $\alpha \beta = 0$, then $\alpha \beta = \alpha 0$ yields $\beta = 0$, hence K is integral.   ◻

We now consider multiplicative inverses in rings. Let K be a ring. An element $\alpha \in K$ is called **invertible** if there exists $\alpha^{-1} \in K$ such that $\alpha \cdot \alpha^{-1} = \alpha^{-1} \cdot \alpha = 1$. If such an element α^{-1} exists, then it is easily shown, as in Proposition I.2.3(c), that it is unique. It is called the **inverse** of α. As expected, if $\alpha, \beta \in K$, then $(\alpha^{-1})^{-1} = \alpha$ and $(\alpha\beta)^{-1} = \beta^{-1}\alpha^{-1}$ whenever α^{-1}, β^{-1} exist. In the ring $\mathbb{Z}$ of integers, only the elements 1 and -1 have multiplicative inverses. It follows from Proposition I.2.3(d) that, in a nonzero ring, the zero element is not invertible.

I.2.6 Definition. Let K be a nonzero ring. If any nonzero element of K is invertible, then K is a **division ring** or a **skew field**. If, moreover, K is commutative, then it is a **field**.

The reader is certainly familiar with several examples of fields, such as $\mathbb{Q}, \mathbb{R}$ or $\mathbb{C}$. Before giving more examples, we observe the following.

I.2.7 Lemma. *Invertible elements in a commutative ring are not zero divisors. In particular, any field is an integral domain.*

Proof. Assume that $\alpha \neq 0$ and $\alpha\beta = 0$. If α^{-1} exists, then

$$\beta = 1 \cdot \beta = (\alpha^{-1}\alpha)\beta = \alpha^{-1}(\alpha\beta) = \alpha^{-1} \cdot 0 = 0.$$

This proves the first statement. The second is an obvious consequence. □

The converse of the lemma is false: $\mathbb{Z}$ is an integral domain but not a field. However, any *finite* integral domain is a field. For, let K be a finite integral domain and $\alpha \in K$ be nonzero. Define a map $f : K \to K$ by $x \mapsto \alpha x$, for $x \in K$. Because of Lemma I.2.5, f is injective. But K is a finite set, and a map from a finite set to itself is injective if and only if it is surjective. Hence, $f : K \to K$ is surjective. So there exists $x \in K$ such that $1 = f(x) = \alpha x$. Thus, α is invertible, as required.

I.2.8 Examples.

(a) The best known example of a noncommutative division ring is the set $\mathbb{H}$ of Hamilton's quaternions. A **quaternion** is a 2×2 matrix of the form

$$\begin{pmatrix} u & v \\ -\bar{v} & \bar{u} \end{pmatrix}$$

where u, v are complex numbers, and $\bar{u}, \bar{v}$ their respective conjugates. We leave to the reader the verification that $\mathbb{H}$ is a ring under addition and multiplication of matrices. We prove that $\mathbb{H}$ is a division ring. Let $\begin{pmatrix} u & v \\ -\bar{v} & \bar{u} \end{pmatrix} \in H$ be nonzero. Its determinant equals $u\bar{u} + v\bar{v} = |u|^2 + |v|^2 \neq 0$. Hence, this matrix has a multiplicative inverse, namely $\frac{1}{|u|^2+|v|^2}\begin{pmatrix} \bar{u} & -v \\ \bar{v} & u \end{pmatrix}$, which lies in $\mathbb{H}$. Therefore, $\mathbb{H}$ is a division ring. Multiplication in $\mathbb{H}$ is not commutative: for instance, the matrices $\begin{pmatrix} 0 & 1 \\ -1 & 0 \end{pmatrix}$ and $\begin{pmatrix} i & 0 \\ 0 & -i \end{pmatrix}$, both lying in $\mathbb{H}$, do not commute. Therefore, $\mathbb{H}$ is a noncommutative division ring.

(b) Let $m \geq 2$ be an integer. The commutative ring $\mathbb{Z}/m\mathbb{Z}$ is a field if and only if m is prime. Indeed, assume that m is prime, and $\bar{a} \in \mathbb{Z}/m\mathbb{Z}$ is nonzero. Then there exists a unique $k \in \mathbb{Z}$ such that $0 < k < m$ and $\bar{k} = \bar{a}$ (actually, k is the remainder of the division of a by m). But then k and m are coprime, so, because of the Bézout–Bachet theorem, there exist $s, t \in \mathbb{Z}$ such that $sk + tm = 1$. Passing modulo m, this gives $\bar{s}\bar{k} = \bar{1}$, because $\bar{m} = \bar{0}$. Hence, $\bar{s} = \bar{k}^{-1} = \bar{a}^{-1}$ and $\bar{a}$ is invertible, as required.

Conversely, if m is not prime, there exist a, b such that $1 < a, b < m$ and $m = ab$. But then $\bar{a}, \bar{b}$ are nonzero, while $\bar{a}\bar{b} = \overline{ab} = \bar{m} = \bar{0}$. So, $\mathbb{Z}/m\mathbb{Z}$ has zero divisors. Because of Lemma I.2.7, it is not a field.

To end this subsection, we prove that any integral domain can be embedded in a field. The prototype of this construction is that of the field $\mathbb{Q}$ of rationals, starting from the integral domain $\mathbb{Z}$. If K, K' are rings, then a map $f: K \to K'$ is called a **ring homomorphism** provided it preserves addition, multiplication and the multiplicative identity, that is: $f(\alpha + \beta) = f(\alpha) + f(\beta), f(\alpha\beta) = f(\alpha)f(\beta)$, for any $\alpha, \beta \in K$ and $f(1) = 1$. A **ring isomorphism** is a ring homomorphism that is bijective.

I.2.9 Definition. Let K be a commutative ring. A field, or a division ring, K' is an **extension** of K if there exists an injective ring homomorphism from K to K'.

That is, K' is an extension of K if and only if it is a field, or a division ring, containing (an isomorphic image of) K as a subring. For instance, the inclusions $\mathbb{Z} \subseteq \mathbb{Q} \subseteq \mathbb{R} \subseteq \mathbb{C}$ indicate three field extensions of the integers. The set of quaternions $\mathbb{H}$ is a (noncommutative) division ring extension of $\mathbb{C}$. The subfield $\mathbb{Q}[\sqrt{2}] = \{a + b\sqrt{2} : a, b \in \mathbb{Q}\}$ of $\mathbb{R}$ is a field extension of $\mathbb{Q}$.

Let K be an integral domain. Define an equivalence relation on the set $K \times K^* = \{(\alpha, \beta) : \alpha, \beta \in K, \beta \neq 0\}$ by setting:

$$(\alpha, \beta) \sim (\gamma, \delta) \text{ if and only if } \alpha \cdot \delta = \beta \cdot \gamma.$$

For instance, $(\alpha, \beta) \sim (0, \delta)$ if and only if $\alpha\delta = 0$. But $\delta \neq 0$ by definition, so $\alpha = 0$. Thus, $(\alpha, \beta) \sim (0, \delta)$ if and only if $\alpha = 0$.

We prove that $\sim$ is an equivalence. Reflexivity and symmetry are easy to verify. We prove transitivity. Assume that $(\alpha, \beta) \sim (\gamma, \delta)$ and $(\gamma, \delta) \sim (\lambda, \mu)$. If $\gamma = 0$, then the previous argument shows that $\alpha = \lambda = 0$ and there is nothing to show. So, suppose $\gamma \neq 0$. Then $\alpha\delta = \beta\gamma$ and $\gamma\mu = \delta\lambda$. Multiplying yields $\alpha\delta\gamma\mu = \beta\gamma\delta\lambda$. But $\delta \neq 0$, hence $\delta\gamma \neq 0$. Therefore, Lemma I.2.5 gives $\alpha\mu = \beta\lambda$, that is, $(\alpha, \beta) \sim (\lambda, \mu)$, as required.

We denote by α/β or $\frac{\alpha}{\beta}$ the equivalence class of a pair $(\alpha, \beta) \in K \times K^*$ and call it a **fraction**. Let $Q(K) = (K \times K^*)/\sim$ be the quotient set. Define operations on $Q(K)$ by setting:

$$\frac{\alpha}{\beta} + \frac{\gamma}{\delta} = \frac{\alpha\delta + \beta\gamma}{\beta\delta}$$

$$\frac{\alpha}{\beta} \cdot \frac{\gamma}{\delta} = \frac{\alpha\gamma}{\beta\delta}$$

for $\alpha/\beta, \gamma/\delta \in Q(K)$.

We must prove that these operations are unambiguously defined. Assume that $(\alpha, \beta) \sim (\alpha', \beta')$, $(\gamma, \delta) \sim (\gamma', \delta')$. Then $\alpha\beta' = \alpha'\beta$ and $\gamma\delta' = \gamma'\delta$. Hence,

$$(\alpha\delta + \beta\gamma)\beta'\delta' = \alpha\delta\beta'\delta' + \beta\gamma\beta'\delta' = \alpha\beta'\delta\delta' + \beta\beta'\gamma\delta' =$$
$$= \alpha'\beta\delta\delta' + \beta\beta'\gamma'\delta = \alpha'\delta'\beta\delta + \beta'\gamma'\beta\delta = (\alpha'\delta' + \beta'\gamma')\beta\delta$$

therefore $(\alpha\delta + \beta\gamma, \beta\delta) \sim (\alpha'\delta' + \beta'\gamma', \beta'\delta')$. Thus, addition of classes does not depend on the choice of the representatives. The proof for multiplication is similar (even easier).

I.2.10 Lemma. *With these operations, $Q(K)$ is a field extension of K.*

Proof. It consists in checking the axioms. The element $0/1$ is the zero, and $1/1$ the identity. The inverse of a nonzero element α/β is β/α (which makes sense because $\alpha \neq 0$). Finally, in order to consider K as embedded inside $Q(K)$, identify $\alpha \in K$ with the fraction $\alpha/1$.   □

The field $Q(K)$ is called the ***field of fractions*** of the integral domain K. For instance, if K is a field, then the field of fractions of the integral domain $K[t]$ of polynomials in one indeterminate t is the field $K(t) = \{p/q : p, q \in K[t], q \neq 0\}$ of rational fractions.

It is well-known that a system of homogeneous linear equations with a nonzero rational solution also has a nonzero integral solution. This generalises to the following proposition.

I.2.11 Proposition. *Let K be an integral domain and $Q(K)$ its field of fractions. If the system of homogeneous linear equations*

$$\begin{cases} \alpha_{11}x_1 & + & \cdots & + & \alpha_{1n}x_n & = & 0 \\ \vdots & & & & \vdots & & \\ \alpha_{m1}x_1 & + & \cdots & + & \alpha_{mn}x_n & = & 0 \end{cases} \qquad (*)$$

with coefficients $\alpha_{ij} \in K$ has a nonzero solution in $Q(K)^n$, then it has a nonzero solution in K^n.

Proof. Suppose $(\beta_1, \cdots, \beta_n) \in Q(K)^n$ is a nonzero solution of $(*)$. For each i, write β_i as γ_i/δ_i, where $\gamma_i, \delta_i \in K$ and $\delta_i \neq 0$. Let $\delta = \delta_1 \cdots \delta_n$. Because $\delta_i \neq 0$, for each i, we have $\delta \neq 0$. Now, for each i, we have $\delta(\gamma_i/\delta_i) = \gamma_i'/1$, where $\gamma_i' = (\prod_{j \neq i} \delta_j)\gamma_i \in K$. Then

$$\delta(\beta_1, \cdots, \beta_n) = (\delta\beta_1, \cdots, \delta\beta_n) = (\gamma_1'/1, \cdots, \gamma_n'/1)$$

is also a solution of $(*)$ which can be interpreted as $(\gamma_1', \cdots, \gamma_n') \in K^n$, as required.   □

I.2.2 Modules

As we said, modules are 'vector spaces over rings' by which we mean that they have an addition and a scalar multiplication by elements of rings. Rings are not assumed commutative. Therefore, we must define two types of modules, one with the scalar multiplied on the right of the vector, and the other with the scalar multiplied on its left. These are respectively called right and left modules.

I.2.12 Definition. Let K be a ring. A *right K-module* $(M, +, \cdot)$, or simply M, is a triple, where M is a set such that:

(A) There is an operation + on M, called **addition**, which associates to each pair $(x, y) \in M \times M$ an element $x + y \in M$, called its **sum**, and which makes the pair $(M, +)$ an abelian group, that is, it satisfies the following axioms:

(A1) It is commutative, that is, $x + y = y + x$, for any $x, y \in M$.

(A2) It is associative, that is, $x + (y + z) = (x + y) + z$, for any $x, y, z \in M$.

(A3) There exists an element $0 \in M$, called the **zero**, such that $x + 0 = 0 + x = x$, for any $x \in M$.

(A4) To any element $x \in M$ is associated its **negative**, or **opposite**, $(-x) \in M$ such that $x + (-x) = (-x) + x = 0$.

(B) There is a map $M \times K \to M$, called **external multiplication**, or **right K-action**, associating to each pair $(x, \alpha) \in M \times K$ an element $x \cdot \alpha$, or simply $x\alpha$, in M, called their **product**, satisfying the following axioms:

(B1) For any $x \in M, \alpha, \beta \in K$, we have $x(\alpha\beta) = (x\alpha)\beta$.

(B2) For any $x \in M$, we have $x \cdot 1 = x$.

(C) These operations are compatible: multiplication is left and right distributive over addition:

(C1) $x(\alpha + \beta) = x\alpha + x\beta$, for any $x \in M, \alpha, \beta \in K$.

(C2) $(x + y)\alpha = x\alpha + y\alpha$, for any $x, y \in M, \alpha \in K$.

The product on M is called external because elements of M are multiplied by elements outside of M (lying in K) to obtain elements of M. Axiom (B1) is called **mixed associativity** because it involves two products, the one inside K and the external one on M. In (B2), 1 represents the identity of the ring K.

Left *K-modules* are defined in the same way: in this case, the external multiplication, or left K-action, is a map $K \times M \to M, (\alpha, x) \mapsto \alpha x$, for $\alpha \in K, x \in M$. In order to emphasise the side on which external multiplication is defined, we denote a right K-module M as M_K, and a left K-module N as $_K N$.

Let M be a right K-module, with K not necessarily commutative. Then M has a left module structure over the opposite ring K^{op} : for any $x \in M, \alpha \in K$, set $\alpha \cdot x = x\alpha$, where the product on the right is calculated inside the right K-module M_K. It is easily verified that this action makes M a left K^{op}-module. So, right K-modules coincide with left K^{op}-modules. In the same way, left K-modules coincide with right K^{op}-modules. Because of that, it is equivalent to work with right or with left modules. If K is commutative, then $K = K^{op}$ as rings, and therefore left and right K-modules coincide.

In order not to repeat statements, we have to fix a side, so we agree to work with right modules. From now on, and unless otherwise specified, the word 'module' means a right module.

I.2.13 Examples.
- (a) If K is a field, then K-modules are just K-vector spaces. Indeed, the axioms for a module are the same as for a vector space. Thus, classical linear algebra is part of module theory.
- (b) By definition, a module is an abelian group. Conversely, an abelian group M has a natural $\mathbb{Z}$-module structure. We must define an external operation $M \times \mathbb{Z} \to M$. For nonnegative integers, define it inductively: for $x \in M$, set $x0 = 0$ and, for $n \geq 1$, set $xn = x(n-1) + x$. If $n < 0$, set $xn = -(x|n|)$. It is easily seen that this multiplication satisfies conditions (B) and (C) of the definition and thus makes M a $\mathbb{Z}$-module. So, abelian groups coincide with $\mathbb{Z}$-modules (and abelian group theory is part of module theory).
- (c) By definition, any module contains a zero element 0, so no module is empty. On the other hand, the one-element set $M = \{0\}$ is a K-module for the operations defined by $0 + 0 = 0$ and $0 \cdot \alpha = 0$, for any $\alpha \in K$. This is the **zero** or **trivial** module, denoted as $M = 0$.
- (d) Any ring K is itself a right (and a left) K-module, the right (or left) K-action on K is the multiplication of the ring.
- (e) The set $K[t]$ of polynomials in t with coefficients in a ring K is a right K-module under ordinary addition of polynomials and right multiplication of polynomials by elements of K. Also, $K[t]$ may be viewed as a left K-module. These two module structures coincide if K is commutative.
- (f) For an integer $n \geq 1$, the set $M_n(K)$ of $n \times n$-matrices with coefficients in a ring K is a right K-module under ordinary addition of matrices and right multiplication of matrices by elements of K. Again, $M_n(K)$ may be considered as left K-module, and these two structures coincide when K is commutative.

We refrain from giving more examples because they arise more naturally later on. We need a notation. When dealing with a K-module M, we deal with two zeros, the one of M and the one of K. Usually, both are denoted by 0. When we want to distinguish them, we denote the first by 0_M and the second by 0_K. Arithmetic properties of modules resemble those of vector spaces, as is seen from the following statement.

I.2.14 Proposition. *Let K be a ring and M_K a module, then:*
- (a) *The zero element of M is unique.*
- (b) *The negative of any $x \in M$ is unique.*
- (c) *For any $\alpha \in K$, we have $0_M \cdot \alpha = 0_M$.*
- (d) *For any $x \in M$, we have $x \cdot 0_K = 0_M$.*
- (e) *For any $\alpha \in K, x \in M$, we have $(-x)\alpha = x(-\alpha) = -(x\alpha)$.*
- (f) *For any $\alpha \in K, x_1, \ldots, x_m \in M$, we have $(\sum_{i=1}^{m} x_i)\alpha = \sum_{i=1}^{m}(x_i\alpha)$.*
- (g) *For any $\alpha_1, \ldots, \alpha_n \in K, x \in M$, we have $x(\sum_{j=1}^{n} \alpha_j) = \sum_{j=1}^{n}(x\alpha_j)$.*
- (h) *For any $\alpha \in K, x \in M, n \in \mathbb{Z}$, we have $x(n\alpha) = (x\alpha)n = (xn)\alpha$.*

Proof. The proofs of statements (a) to (g) are similar to those of the corresponding statements in Proposition I.2.3, so they are left to the reader. We just prove (h). If $n = 0$, then equality follows from (c) and (d). If $n > 0$, it follows from (f) and (g). Assume that $n < 0$. Because of (e), we have:

$$(xn)\alpha = (-x\,|n|)\alpha = -x(|n|\,\alpha) = x(-\,|n|\,\alpha) = x(n\alpha).$$

Similarly, $(xn)\alpha = (x\alpha)n$. □

The abelian group structure of a module M implies that we also have $-(-x) = x$ and $-(x + y) = (-x) + (-y)$, for $x, y \in M$.

Together with a mathematical object come its subobjects.

I.2.15 Definition. Let K be a ring and M_K be a module. A **submodule** L of M is a subset of M which is itself a module under the same operations as M.

Thus, if L is a submodule of a module M and $x, y \in L$, then their sum in L is the same as their sum in M and, if $\alpha \in K$, the product of x by α in L is the same as their product in M. This is expressed by writing $x + y \in L$ and $x\alpha \in L$, respectively. A consequence is that, if $x, y \in L$, and $\alpha, \beta \in K$, then $x\alpha + y\beta \in L$. We prove that, conversely, this last condition on a nonempty subset L of a module M implies that L is a submodule of M.

I.2.16 Proposition. *Let M be a K-module and L a nonempty subset of M. Then L is a submodule of M if and only if for any $x, y \in L$, and $\alpha, \beta \in K$, we have $x\alpha + y\beta \in L$.*

Proof. We have just shown necessity so we prove sufficiency. The hypothesis (setting first $\alpha = \beta = 1$, then $\beta = 0$) guarantees the existence of operations in L. We check that the module axioms are satisfied in L. Identities (A1), (A2), (B1), (B2), (C1) and (C2) are satisfied in L because they are satisfied in M, so we just have to check (A3) and (A4). Because L is nonempty, there exists $x \in L$. The hypothesis gives $x \cdot 0 = 0 \in L$, hence L contains the zero of M (which is thus a zero element in L). Similarly, if $x \in L$, then $x(-1) = -x \in L$. The proof is complete. □

In a module M, an expression of the form $x\alpha + y\beta$ as above is called a **linear combination** of x, y with coefficients $\alpha, \beta \in K$. The proposition says that a nonempty subset is a submodule if and only if it is **stable**, or **closed**, under linear combinations.

Again, examples of submodules will appear more naturally later on. For the time being, we content ourselves with obvious ones.

I.2.17 Examples.
  (a) Any module M has two submodules, itself and the set $\{0\}$. These coincide if and only if $M = 0$.
  (b) If K is a field, then a submodule of a K-module (=vector space) is a subspace. Indeed, the condition of Proposition I.2.16 is the well-known criterion allowing to verify whether a subset of a vector space is a subspace or not.

(c) If $K = \mathbb{Z}$, that is, the module is an abelian group, then a submodule is a subgroup. Actually, one proves easily that the condition in Proposition I.2.16 reformulates as follows: a nonempty subset L of an abelian group M is a submodule if and only if, for any $x, y \in L$, we have $x - y \in L$.

Exercises of Section I.2

1. (a) Let K be a ring. By developing $(\alpha + \beta)(1 + 1)$ in two different ways, show that commutativity of addition follows from the other axioms for rings.
   (b) State and prove the corresponding statement for K-modules.

2. Let K be a ring.
   (a) If $\alpha_1, \ldots, \alpha_m, \beta_1, \ldots, \beta_n$ are elements of K, prove that:

$$\left(\sum_{i=1}^{m} \alpha_i \right) \left(\sum_{j=1}^{n} \beta_j \right) = \sum_{i=1}^{m} \sum_{j=1}^{n} (\alpha_i \beta_j).$$

   (b) If $\alpha, \beta \in K$ satisfy $\alpha\beta = \beta\alpha$, prove that $\alpha^m \beta^n = \beta^n \alpha^m$, for any integers $m, n \geq 0$.
   (c) If $\alpha, \beta \in K$ satisfy $\alpha\beta = \beta\alpha$, prove that, for any integer $n > 0$, the **binomial theorem** holds true

$$(\alpha + \beta)^n = \sum_{i=0}^{n} \binom{n}{i} \alpha^i \beta^{n-i}.$$

3. A ring K in which $\alpha^2 = \alpha$, for any $\alpha \in K$, is called a **boolean ring**.
   (a) Let K be a boolean ring. Prove that $\alpha = -\alpha$, for any $\alpha \in K$.
   (b) Prove that any boolean ring is commutative.
   (c) Let E be a set and $A = \mathcal{P}(E)$ its power set (that is, the set of its subsets). For $X, Y \in A$, define $X + Y = X \triangle Y$, where $\triangle$ denotes the symmetric difference between the two sets, and $XY = X \cap Y$. Prove that A, together with these operations, is a boolean ring.

4. Let K, K' be rings. We define on the cartesian product $K \times K' = \{(\alpha, \alpha') : \alpha \in K, \alpha' \in K'\}$ an addition and a multiplication componentwise, that is, we set

$$(\alpha, \alpha') + (\beta, \beta') = (\alpha + \beta, \alpha' + \beta') \quad \text{and} \quad (\alpha, \alpha') \cdot (\beta, \beta') = (\alpha\beta, \alpha'\beta')$$

for $\alpha, \beta \in K$ and $\alpha', \beta' \in K'$. Show that:
   (a) $K \times K'$, endowed with these operations, is a ring, called the **product** of K and K'.
   (b) $K \times K'$ is commutative if K and K' are commutative.
   (c) The product of two nonzero integral domains is not an integral domain.

5. Let K be an integral domain. If there exists a nonzero integer n such that $n\alpha = 0$ for any $\alpha \in K$, the least such positive n is called the **characteristic** of K and

denoted by char$K$. If no such integer exists, we say that K has characteristic zero and write char$K = 0$. For example, char$\mathbb{R} = 0$.

  (a) Prove that $n\alpha = 0$, for any $\alpha \in K$, if and only if $n \cdot 1 = 0$.
  (b) Prove that the characteristic of an integral domain is either zero or a prime number.
  (c) Let p be a prime integer. Prove that char$(\mathbb{Z}/p\mathbb{Z}) = p$.
  (d) Assume that char$K = p$ is a prime integer. Prove that, for any $\alpha, \beta \in K$, we have $(\alpha + \beta)^p = \alpha^p + \beta^p$. Deduce that, in this case, the map $\varphi : K \to K$ given by $\varphi(\alpha) = \alpha^p$, for $\alpha \in K$, is a ring homomorphism.
  (e) Assume that char$K = p$ is a prime integer and let $\alpha \in K$ be such that there exists n, which is not a multiple of p, satisfying $n\alpha = 0$. Prove that $\alpha = 0$.

6. Prove that a ring in which any nonzero element has a right inverse is a division ring.

7. Using quaternions, prove that, if two integers m, n are sums of 4 squares (of integers), then so is their product mn.

8. Let $\mathbb{Z}[i] = \{\alpha + i\beta : \alpha, \beta \in \mathbb{Z}\}$ be the ring of **Gaussian integers**. Prove that $\mathbb{Z}[i]$ is an integral domain and that its field of fractions is $\mathbb{Q}[i] = \{\lambda + i\mu : \lambda, \mu \in \mathbb{Q}\}$.

9. Prove that the set of real matrices of the form

$$\begin{pmatrix} a & b & c & d \\ -b & a & -d & c \\ -c & d & a & -b \\ -d & -c & b & a \end{pmatrix}$$

equipped with the usual addition and multiplication of matrices is a noncommutative division ring.

10. Let K be a commutative ring, $n > 0$ an integer, $M_n(K)$ the ring of $n \times n$ matrices with coefficients in K and K^n the abelian group $K^n = \{(\alpha_1, \ldots, \alpha_n) : \alpha_i \in K\}$, where addition is defined componentwise.
  (a) Prove that ordinary matrix multiplication of an element of K^n by a matrix induces on K^n the structure of a right $M_n(K)$-module.
  (b) By considering column instead of row vectors in K^n, define a left $M_n(K)$-module structure on K^n.

11. Let $n > 1$ be an integer and G an additive abelian group. Prove that G has a natural $\mathbb{Z}/n\mathbb{Z}$-module structure defined by $x\bar{a} = xa$, for $x \in G$ and $\bar{a} = a + n\mathbb{Z} \in \mathbb{Z}/n\mathbb{Z}$, if and only if $nx = 0$, for any $x \in G$.

12. Let G be an abelian group and $K = \operatorname{End}_{\mathbb{Z}} G$ the set of all its endomorphisms (that is, group homomorphisms from G to itself). Prove that:
  (a) K is an abelian group under the addition defined by $(f+g)(x) = f(x) + g(x)$, for $f, g \in K, x \in G$.
  (b) K is a monoid under the usual composition of maps. Deduce that K is a ring.
  (c) G is a left K-module under the multiplication defined by $f \cdot x = f(x)$ for $f \in K, x \in G$.

13. Let K be a ring. Show that the set of invertible elements of K is a group under the multiplication of K. Is it true that this group is abelian if and only if K is commutative?

I.3 Algebras

I.3.1 The definition of an algebra

It is sometimes useful to look at a ring as having an enriched structure. It is then called algebra. Because algebras have a richer structure than rings, working with the former offers more technical possibilities than working with the latter.

From now on, unless otherwise specified, the letter K stands for a commutative ring. A K-algebra A is a K-module together with an operation $A \times A \rightarrow A$, called multiplication, which is right and left distributive over addition and compatible with the K-action on A, that is,

$$(ab)\alpha = a(b\alpha) = (a\alpha)b$$

for $a, b \in A$ and $\alpha \in K$. Because the ring K is commutative, the right K-module A is also a left K-module, that is, one can write the multiplication on A by K on either side. We simply say that A is a K-module. Listing the axioms for algebras yields the following definition.

I.3.1 Definition. Let K be a commutative ring. A set A is a **K-algebra** or **algebra over K** if:
- (A) There is an operation $+$ on A which associates to each pair $(a, b) \in A \times A$ its **sum** $a + b \in A$, and which makes the pair $(A, +)$ an abelian group, that is:
  - (A1) It is commutative: $a + b = b + a$, for any $a, b \in A$.
  - (A2) It is associative: $a + (b + c) = (a + b) + c$, for any $a, b, c \in A$.
  - (A3) There exists in A an element called the **zero**, and denoted by 0 or 0_A, such that $a + 0 = 0 + a = a$, for any $a \in A$.
  - (A4) To any element $a \in A$ is associated its **negative**, or **opposite**, $(-a)$ which is such that $(-a) + a = a + (-a) = 0$.
- (B) The set A is equipped with an external operation, or right K-action, associating to each pair $(a, \alpha) \in A \times K$ an element $a\alpha \in A$, called its **external product**, and making A a K-module, that is:
  - (B1) $a(\alpha\beta) = (a\alpha)\beta$, for any $a \in A$, $\alpha, \beta \in K$.
  - (B2) $a \cdot 1_K = a$, for any $a \in A$.
- (C) External multiplication is left and right distributive over addition:
  - (C1) $(a + b)\alpha = a\alpha + b\alpha$, for any $a, b \in A$, $\alpha \in K$.
  - (C2) $a(\alpha + \beta) = a\alpha + a\beta$, for any $a \in A$, $\alpha, \beta \in K$.
- (D) The set A is equipped with an **internal multiplication** associating to each pair $(a, b) \in A \times A$ an element $a \cdot b$, or simply ab, in A, called its **product**, and which is compatible with the two previous operations:

(D1) $(a + b)c = ac + bc$, for any $a, b, c \in A$.
(D2) $c(a + b) = ca + cb$, for any $a, b, c \in A$.
(D3) $(ab)\alpha = a(b\alpha) = (a\alpha)b$, for any $a, b \in A, \alpha \in K$.

In axiom (B2), 1_K denotes the (multiplicative) identity of the ring K. We did not assume that the multiplication $A \times A \to A$ has an identity. If this is the case, that is, if there exists an element 1_A, or 1, in A such that $a \cdot 1_A = 1_A \cdot a = a$, for any $a \in A$, then A is called a **unitary algebra**. Similarly, if the multiplication $A \times A \to A$ is associative, that is, if $a(bc) = (ab)c$, for any $a, b, c \in A$, then A is called an **associative algebra**. In this text, unless otherwise specified, the word 'algebra' means an associative and unitary algebra.

The important restriction here is associativity. The existence of an identity is more an apparent constraint than a real one, because any algebra can be embedded into a unitary one, see Exercises I.3.1 and I.4.2, and modules over these two algebras can be considered to be the same, see Exercises II.2.1 and II.3.20. Thus, an algebra is at the same time a ring, in the sense of Definition I.2.1, and a K-module, in the sense of Definition I.2.12, these two structures being compatible.

An algebra A is called a **commutative algebra** if the multiplication $A \times A \to A$ is commutative, that is, if $ab = ba$, for any $a, b \in A$ (but we do not assume it in general).

Given a K-algebra A, one defines its **opposite algebra** A^{op} in the same way as the opposite ring: it is the same set, with the same K-module structure, but the product of the elements a, b is defined as $a \cdot b = ba$, where the product on the right is computed inside A. Thus, $A = A^{\mathrm{op}}$, *as algebras*, if and only if A is commutative.

We shall see in Subsection I.3.3 several examples of algebras, but here are three immediate ones. The zero K-module $A = \{0\}$ becomes an algebra if one defines multiplication by $0 \cdot 0 = 0$. This is the **zero**, or **trivial** algebra, denoted (predictably) by $A = 0$. The set $K[t]$ of polynomials in one indeterminate t with coefficients in K has a K-module structure, see Example I.2.13(e), and also a ring structure, see Example I.2.2(d). Because these structures are compatible, that is, axiom (D3) is satisfied, $K[t]$ is an algebra, obviously commutative, called the **algebra of polynomials** in t over K. Given an integer $n > 0$, the set $M_n(K)$ of $n \times n$-matrices with coefficients in K has a K-module structure, see Example I.2.13(f), and also a ring structure, see Example I.2.2(e). Because these structures are compatible, $M_n(K)$ is an algebra, usually not commutative, called the **full matrix algebra** over K.

Because an algebra is at the same time a ring and a module, verifying the extra compatibility condition (D3), then it satisfies all arithmetic properties listed in Propositions I.2.3 and I.2.14.

As defined, algebras are particular types of rings. On the other hand, any ring can be viewed as an algebra over its centre $Z(A) = \{a \in A : xa = ax \text{ for any } x \in A\}$ or over the integers $\mathbb{Z}$. This follows from the next proposition which, given a commutative ring K, says that a ring is a K-algebra if and only if there exists a ring homomorphism from K to the centre of the ring.

I.3.2 Proposition. *Let K be a commutative ring.*
  (a) *If A is a K-algebra, then the map $\varphi : K \to A$ defined by $\alpha \mapsto 1_A \cdot \alpha$, for $\alpha \in K$, is a ring homomorphism whose image is contained in the centre $Z(A)$ of A.*

(b) *Let A be a ring and $\varphi : K \to Z(A)$ a ring homomorphism. The external multiplication $A \times K \to A, (a, \alpha) \mapsto a\varphi(\alpha)$, for $a \in A, \alpha \in K$, gives A a K-algebra structure.*

Proof. (a) It is easily seen that φ is a ring homomorphism. We prove that its image is contained in $Z(A)$. Let $\alpha \in K$ and $a \in A$, then we have:

$$a\varphi(\alpha) = a(1_A \cdot \alpha) = (a\alpha)1_A = a\alpha = 1_A(a\alpha) = (1_A \cdot \alpha)a = \varphi(\alpha)a$$

where we used axiom (D3) at the second and fifth equality.

(b) It is easily verified that the given K-action makes A a K-module. We prove compatibility of this module structure with the ring structure of A. Let $a, b \in A$ and $\alpha \in K$, then:

$$a(b\alpha) = a(b\varphi(\alpha)) = a(\varphi(\alpha)b) = (a\varphi(\alpha))b = (a\alpha)b$$

because $\varphi(\alpha) \in Z(A)$. Also, we have:

$$(ab)\alpha = (ab)\varphi(\alpha) = a(b\varphi(\alpha)) = a(b\alpha)$$

because multiplication is associative in the ring A. □

Taking in (b), $K = Z(A)$ and $\varphi = 1_{Z(A)}$, we deduce that any ring is an algebra over its centre. Also, taking $K = \mathbb{Z}$, and $\varphi : n \mapsto 1_A \cdot n$, we get that any ring is a $\mathbb{Z}$-algebra. Thus, the two concepts of ring and algebra may be considered equivalent. This allows us to use the terminology of algebras.

In the notation of the proposition, $\varphi(0_K) = 0_A$ and $\varphi(1_K) = 1_A$. We may thus identify the zeros of K and A, as well as their identities, and denote them simply by 0 and 1, respectively. In fact, from now on, we use the same symbols 0 and 1 for all rings and algebras under consideration. This will entail no confusion.

If K is a field and A is nonzero, then the nonzero map $\varphi : K \to A$ is necessarily injective (because of Proposition I.3.9 below), and so K may be identified to a subfield of A, contained inside its centre. Also, A is a K-vector space, so we may talk about its dimension $\dim_K A$, defined, as usual, as the cardinality of an arbitrary basis in A. The algebra A is said to be a ***finite-dimensional algebra*** if $\dim_K A < \infty$, and an ***infinite-dimensional algebra*** otherwise. For instance, the set $\mathbb{C}$ of complex numbers is an algebra over the set $\mathbb{R}$ of the reals. It is two-dimensional, an obvious basis being $\{1, i\}$.

I.3.2 Subalgebras and ideals

It is natural, when one defines an algebraic structure to study the corresponding substructures, as we did for modules and submodules in Subsection I.2.2.

I.3.3 Definition. Let A be a K-algebra. A subset B of A is a **subalgebra** if it is an algebra for the same operations as A.

In particular, a subalgebra is never empty, because it contains the zero and the identity of A. While any algebra A is a subalgebra of itself, one sees that, if $A \neq 0$, then the subset $\{0\}$ is *not* a subalgebra, because it does not contain the identity of A.

The next criterion allows us to verify whether a subset of an algebra is a subalgebra or not.

I.3.4 Lemma. *Let A be an algebra. A subset B of A is a subalgebra if and only if the following conditions are satisfied:*
   (a) *$1 \in B$,*
   (b) *B is stable under K-linear combinations, that is, if $a, b \in B, \alpha, \beta \in K$, then $a\alpha + b\beta \in B$,*
   (c) *B is stable under multiplication, that is, if $a, b \in B$, then $ab \in B$.*

Proof. Necessity is obvious. Sufficiency follows from the fact that the stated conditions guarantee that B is nonempty and the result of any operation in A performed between elements of B lies in B. Algebra axioms are satisfied in B because they are satisfied in A. $\qquad\square$

Clearly, if A is commutative, then so is any of its subalgebras.

A consequence of the previous lemma is that the intersection of any family of subalgebras is also a subalgebra. In particular, the intersection of all subalgebras of A containing a given subset $X \subseteq A$ is itself a subalgebra of A, called the subalgebra **generated** by X.

In practice, the notion of ideal is more important than that of subalgebra.

I.3.5 Definition. Let A be a K-algebra. An **ideal** or, more precisely, **two-sided ideal**, I of A is a K-submodule of A such that $x \in I, a \in A$ imply $xa \in I$ and $ax \in I$.

That is, ideals are additive subgroups which are right and left stable under multiplication by elements of the algebra. The fact that I is an ideal of A is expressed by the notation $I \trianglelefteq A$.

Considering a ring A as a $\mathbb{Z}$-algebra, we get the classical definition of ideal in the ring A. Conversely, if A is a K-algebra, the notion of ideal of A considered as a ring coincides with that of ideal of A as an algebra; indeed, let I be an ideal in A as a ring, in order to show that I is an ideal in A as an algebra, it suffices to show that I is a K-submodule of A, that is, it is stable under the right K-action. Now, for any $x \in I, \alpha \in K$, one has

$$x\alpha = (1 \cdot x)\alpha = (1 \cdot \alpha)x \in I$$

due to axiom (D3), as required. Any algebra has at least two ideals, namely itself and $\{0\}$. The only ideal containing the identity 1 of an algebra A is A itself: for, if $I \trianglelefteq A$ and $1 \in I$, then, for any $a \in A$, we have $a = a \cdot 1 \in I$ and so $I = A$.

Fundamental construction procedures for ideals starting from others are intersection and sum.

I.3.6 Lemma. *Let $(I_\lambda)_{\lambda \in \Lambda}$ be a family of ideals, then so is their intersection $\cap_\lambda I_\lambda$.*

Proof. Because $0 \in I_\lambda$, for any λ, we have $0 \in \cap_\lambda I_\lambda$, so $\cap_\lambda I_\lambda \neq \emptyset$. If $x, y \in \cap_\lambda I_\lambda$, and $\alpha, \beta \in K$, we have $x\alpha + y\beta \in I_\lambda$, for any λ, because each I_λ is a K-submodule of A. Therefore, $x\alpha + y\beta \in \cap_\lambda I_\lambda$. On the other hand, if $a \in A$, then $ax, xa \in I_\lambda$, for any λ, because each I_λ is an ideal of A. Therefore, $ax, xa \in \cap_\lambda I_\lambda$, and the proof is complete. $\square$

Consequently, the intersection of all ideals containing a given subset X of A is itself an ideal containing X. It is the smallest ideal containing X and for this reason is said to be **generated** by X. For instance, the ideal generated by the empty set is the zero ideal $\{0\}$. Given a nonempty set, one can describe explicitly the ideal it generates.

I.3.7 Lemma. *Let $X \subseteq A$ be nonempty. The ideal it generates is the set*

$$\langle X \rangle = \left\{ \sum_{i=1}^{n} a_i x_i b_i : n \geq 0, x_i \in X, a_i, b_i \in A \right\}.$$

Proof. We leave to the reader the verification that the given set $\langle X \rangle$ is an ideal. We prove it is the smallest one containing X. If $x \in X$, then $x = 1 \cdot x \cdot 1$ shows that $X \subseteq \langle X \rangle$. Let I be an ideal containing X, and $y \in \langle X \rangle$. Then there exist an integer $n \geq 0, x_i \in X$ and elements $a_i, b_i \in A$ such that $y = \sum_{i=1}^{n} a_i x_i b_i$. Because each x_i lies in I, which is an ideal, we have $a_i x_i b_i \in I$, for any i. Hence, $\sum_{i=1}^{n} a_i x_i b_i \in I$ and $\langle X \rangle \subseteq I$, as required. $\square$

If A is commutative, then the result of the lemma takes on a particularly nice form: in this case, $\langle X \rangle$ is the set of linear combinations $\sum_{i=1}^{n} c_i x_i$, with $n \geq 0, x_i \in X$ and $c_i \in A$. If, in particular, $X = \{x\}$ is a one-element set, then $\langle X \rangle = \langle x \rangle$ is the set $Ax = \{cx : c \in A\}$ of the multiples of x. Such an ideal Ax, generated by a single element, is called a **principal ideal**. For instance, it is well-known that any subgroup of $\mathbb{Z}$ is of the form $n\mathbb{Z}$, for some integer n. Because each $n\mathbb{Z}$ is an ideal when $\mathbb{Z}$ is considered as an algebra, any ideal of $\mathbb{Z}$ is of the form $n\mathbb{Z}$, that is, is principal.

The sum of ideals is defined in the same way as the sum of subspaces of a vector space (or of subgroups of an abelian group). The **support** of a family of elements $(x_\lambda)_{\lambda \in \Lambda}$ in A is the set $\{\lambda \in \Lambda : x_\lambda \neq 0\}$. A family $(x_\lambda)_{\lambda \in \Lambda}$ of elements of A is called **of finite support** if its support is a finite set, or, equivalently, if $x_\lambda = 0$ for all but at most finitely many $\lambda \in \Lambda$. We use the suggestive expression $x_\lambda = 0$ **for almost all** λ. If $(x_\lambda)_{\lambda \in \Lambda}$ is such that $x_\lambda = 0$ for almost all λ, then the seemingly infinite sum $\sum_{\lambda \in \Lambda} x_\lambda$ is actually finite and makes sense in the algebra.

I.3.8 Definition. Let $(I_\lambda)_{\lambda \in \Lambda}$ be a family of ideals of an algebra A. Their **sum** is the set of the finite sums of the $x_\lambda \in I_\lambda$, that is:

$$\sum_{\lambda \in \Lambda} I_\lambda = \left\{ \sum_\lambda x_\lambda : x_\lambda \in I_\lambda, \text{ for any } \lambda, \text{ and } x_\lambda = 0, \text{ for almost all } \lambda \right\}.$$

We leave to the reader the straightforward verification that $\sum_\lambda I_\lambda$ is an ideal whenever each I_λ is so, and, actually, is the smallest one containing all the I_λ.

We need another construction. Let I, J be ideals of an algebra A, then the set IJ of all sums of the form $\sum_\lambda x_\lambda y_\lambda$, where $x_\lambda \in I, y_\lambda \in J$ are such that $x_\lambda y_\lambda = 0$ for almost all λ, that is, of all finite sums of the form $\sum_\lambda x_\lambda y_\lambda$, is an ideal, called the **product** of I and J. Clearly, $IJ \subseteq I \cap J$. But in general, equality does not hold true, see Exercise I.3.10(d).

This allows us to define inductively the **powers** of an ideal I: we set $I^1 = I$ and $I^n = I^{n-1} I$ for any integer $n > 1$. Each of the I^n is an ideal.

One can characterise fields in terms of their ideals.

I.3.9 Proposition. *A nonzero commutative algebra is a field if and only if it has no nonzero proper ideals.*

Proof. Necessity. Let A be a field and I a nonzero ideal of A. Let $a \in I, a \neq 0$. Because A is a field, a^{-1} exists in A. Therefore, $1 = a^{-1} a \in I$, because I is an ideal. But then $I = A$.

Sufficiency. Assume that A has no nonzero proper ideal. Let $a \in A$ be nonzero. The ideal Aa generated by a is nonzero. Hence, $Aa = A$. Therefore, $1 \in Aa$ and so a is invertible. ∎

The assumption about commutativity of A is essential in the statement: see Example I.3.11(b) below.

Ideals allow to define quotients of algebras, just as normal subgroups allow to define quotients of groups. Let A be an algebra and I an ideal of A. In particular, I is an abelian subgroup of A, so the relation defined by $a \equiv b \pmod{I}$ if and only if $a - b \in I$ is an equivalence relation on A. The equivalence class of $a \in A$ is the so-called **coset** $a + I = \{a + x : x \in I\}$. The element a is called a **representative** of the coset $a + I$. Coset representatives are usually not unique: indeed, one has $a \equiv b \pmod{I}$ if and only if $a + I = b + I$. The quotient set $A/I = \{a + I : a \in A\}$ is an abelian group under the operation (unambiguously) defined, for $a, b \in A$, by

$$(a + I) + (b + I) = (a + b) + I,$$

which can be expressed by saying that the sum of two cosets is the coset of the sum of their representatives. The zero element of the group A/I is the coset $0 + I = I$.

We prove that A/I has actually a K-algebra structure.

(a) We define the multiplication of $a + I \in A/I$ with $\alpha \in K$ by

$$(a + I)\alpha = a\alpha + I,$$

that is, the product of a coset by a scalar is the coset of the product of the representative by this scalar. This multiplication is unambiguously defined: $a + I = a' + I$ implies $a - a' \in I$, hence $a\alpha - a'\alpha = (a - a')\alpha \in I$, because I is a K-submodule of A. Therefore, $a\alpha + I = a'\alpha + I$.

(b) We define the multiplication of $a + I, b + I \in A/I$ by

$$(a + I)(b + I) = ab + I,$$

that is, the product of cosets is the coset of the product of their representatives. This operation is again unambiguously defined: $a + I = a' + I$ and $b + I = b' + I$ imply

$a - a', b - b' \in I$, hence $ab - a'b' = a(b - b') + (a - a')b \in I$, because I is an ideal. Therefore, $ab + I = a'b' + I$.

I.3.10 Proposition. *Let A be an algebra and I an ideal of A. Then A/I is an algebra under the operations defined, for $a, b \in A$ and $\alpha \in K$, by*

$$(a + I) + (b + I) = (a + b) + I,$$

$$(a + I)\alpha = a\alpha + I,$$

$$(a + I)(b + I) = ab + I.$$

Proof. The proof consists in verifying the axioms one by one. We prove, for example, associativity of (internal) multiplication: let $a, b, c \in A$ then $a(bc) = (ab)c$ in A implies that:

$$(a + I)((b + I)(c + I)) = (a + I)(bc + I) = a(bc) + I = (ab)c + I$$

$$= (ab + I)(c + I) = ((a + I)(b + I))(c + I)$$

as required. The class $1 + I$ is the identity of A/I, because, for any $a \in A$,

$$(a + I)(1 + I) = (a \cdot 1) + I = a + I = (1 \cdot a) + I = (1 + I)(a + I). \qquad \square$$

The algebra A/I is called the **quotient algebra** of A by I. Its zero element is the class $0 + I = I$.

The proof of the proposition illustrates a useful fact: A/I inherits many properties of A. For example, if A is commutative, then so is A/I, because, if $ab = ba$ in A, then $(a + I)(b + I) = ab + I = ba + I = (b + I)(a + I)$. Again, 'many' does not mean all. For instance, quotients of integral domains do not have to be integral: $\mathbb{Z}$ is an integral domain, but not $\mathbb{Z}/6\mathbb{Z}$.

The quotient may have additional desirable properties that A does not possess. For instance, the fact that I is the zero in A/I implies that a class $a + I$ equals the zero class if and only if $a \in I$: that is, elements of I are equated to zero in the quotient.

I.3.3 Examples of algebras

I.3.11 Examples.
(a) Consider the ring $\mathbb{H}$ of Hamilton's quaternions, see Example I.2.8(a). Any quaternion can be written uniquely in the form

$$\begin{pmatrix} a + bi & c + di \\ -c + di & a - bi \end{pmatrix}$$

with $a, b, c, d \in \mathbb{R}$. Therefore, $\mathbb{H}$ is a four-dimensional real vector space. Because it is a ring, and matrix multiplication is compatible with multiplication by real numbers, it is a four-dimensional real algebra. An algebra which is also a division ring (such as $\mathbb{H}$) is called a **division algebra**.

(b) Let A be a K-algebra, $n > 0$ an integer and $M_n(A)$ the set of $n \times n$-matrices with coefficients in A. Under ordinary matrix operations, $M_n(A)$ is a K-algebra. A particular role is played by the matrices e_{ij} having coefficient 1 at the intersection of line i and column j, and 0 elsewhere. Indeed, if $\mathbf{a} = [a_{ij}] \in M_n(A)$, then $\mathbf{a} = \sum_{i,j=1}^{n} e_{ij}a_{ij}$, that is, any matrix can be written (uniquely!) as a linear combination of the e_{ij}.

The matrices e_{ij} multiply as follows:

$$e_{ij}e_{kl} = \begin{cases} e_{il} & \text{if } j = k, \\ 0 & \text{otherwise.} \end{cases}$$

If $A = K$ is a field (or, more generally, a division ring), then the matrices e_{ij} form a basis of the K-vector space $M_n(K)$. An important property of $M_n(K)$ is that this (noncommutative) algebra has no proper nonzero ideal (hence, Proposition I.3.9 does not extend to the noncommutative case). Let indeed $I \trianglelefteq M_n(K)$ be nonzero. Then there exists a nonzero matrix $\mathbf{a} = [a_{ij}]$ in I. Let r, s be such that $a_{rs} \neq 0$. Then a_{rs}^{-1} exists in K. But then, for any pair of indices i, j, we have:

$$e_{ij} = a_{rs}^{-1}(e_{ir}\mathbf{a}e_{sj}) \in I$$

because I is an ideal. All basis vectors of $M_n(K)$ thus lie in I, hence $I = M_n(K)$.

(c) Several subalgebras of $M_n(K)$ are of interest. For instance, let $T_n(K) = \{\mathbf{a} = [a_{ij}] \in M_n(K) : a_{ij} = 0 \text{ for } i < j\}$ be the set of lower tri-angular matrices. Applying Lemma I.3.4, one verifies easily that it is a subalgebra of $M_n(K)$. Another example, for $n = 4$, is the set of matrices

$$A = \begin{pmatrix} K & 0 & 0 & 0 \\ K & K & 0 & 0 \\ K & 0 & K & 0 \\ K & K & K & K \end{pmatrix} = \left\{ \begin{pmatrix} \alpha_1 & 0 & 0 & 0 \\ \alpha_2 & \alpha_3 & 0 & 0 \\ \alpha_4 & 0 & \alpha_5 & 0 \\ \alpha_6 & \alpha_7 & \alpha_8 & \alpha_9 \end{pmatrix} : \alpha_i \in K, \text{ for all } i \right\}$$

which is a subalgebra of $T_4(K)$, and hence of $M_4(K)$.

(d) Let $A_1, \cdots, A_n$ be K-algebras. Their cartesian product

$$\prod_{i=1}^{n} A_i - A_1 \times \cdots \times A_n = \{(a_1, \cdots, a_n) : a_l \in A_l\}$$

admits a K-algebra structure, with operations defined componentwise

$$(a_1, \cdots, a_n) + (b_1, \cdots, b_n) = (a_1 + b_1, \cdots, a_n + b_n)$$
$$(a_1, \cdots, a_n)\alpha = (a_1\alpha, \cdots, a_n\alpha)$$
$$(a_1, \cdots, a_n)(b_1, \cdots, b_n) = (a_1b_1, \cdots, a_nb_n)$$

where $a_i, b_i \in A_i$, for each i, and $\alpha \in K$. The algebra $\prod_{i=1}^{n} A_i$ is the **product** of the A_i. If $A_1 = \cdots = A_n = A$, then we denote the product by A^n.

(e) One also considers infinite products of algebras. Let K be a commutative ring. A **sequence** of elements of K is a map $a : \mathbb{N} \to K$. We write $a(i) = a_i$, for $i \in \mathbb{N}$, and denote the sequence as $a = (a_i)_{i \geq 0}$. The set $K^{\mathbb{N}}$ of sequences of elements of K becomes a K-algebra if we set

$$(a_i)_{i \geq 0} + (b_i)_{i \geq 0} = (a_i + b_i)_{i \geq 0}$$

$$(a_i)_{i \geq 0} \alpha = (a_i \alpha)_{i \geq 0}$$

$$(a_i)_{i \geq 0} (b_i)_{i \geq 0} = \left(\sum_i a_j b_{i-j} \right)_{i \geq 0}$$

with $a_i, b_i \in K$, for any i, and $\alpha \in K$. The rule for multiplication being similar to that for polynomials, we agree to denote the previous sequence as $(a_i)_{i \geq 0} = \sum_{i=0}^{\infty} a_i t^i$ with t an indeterminate. The notation of sum clearly makes no sense here, because one does not assume the a_i almost all zero, but it makes calculations easier: in order to multiply two such elements, we multiply term-by-term, then we group together the summands corresponding to the same powers of t. The resulting algebra is denoted by $K[[t]]$ and called the **algebra of formal power series** in an indeterminate t, with coefficients in K. This algebra is commutative (because so is K) and contains the polynomial algebra $K[t]$ as a subalgebra.

(f) Let E be a finite poset (that is, partially ordered set), ordered by $\leqslant$. The **incidence algebra** KE of E is the set of linear combinations of the pairs $e_{ij} = (i, j) \in E \times E$, with $j \leqslant i$, having coefficients in K. The product of the pairs e_{ij} and e_{kl} is defined to be $e_{ij} e_{kl} = e_{il}$ if $j = k$ and zero otherwise. We then extend this product to arbitrary elements by distributivity. Thus, the e_{ij} multiply exactly as the matrices $\mathbf{e}_{ij}$ of Example (b) above. Consequently, KE can be viewed as a subalgebra of $M_n(K)$, where n is the cardinality of E. Actually, KE is a subalgebra of $T_n(K)$, because $i \leqslant j$ and $i \neq j$ imply $j \not\leqslant i$.

   For instance, if E is the finite poset $1 \leqslant 2 \leqslant \cdots \leqslant n$, then KE is another realisation of the lower triangular matrix algebra $T_n(K)$. Similarly, if $E = \{1, 2, 3, 4\}$, ordered by $1 \leqslant 2 \leqslant 4, 1 \leqslant 3 \leqslant 4$ (and $2, 3$ non comparable), then KE is another realisation of the subalgebra A of $T_4(K)$ of Example (c) above.

(g) Let G be a group. The **group algebra** KG of G is the set of linear combinations of elements of G with coefficients in K. The product of $g \in G$ by $h \in G$ in KG is their product gh in the group G, and is extended to other elements of KG by distributivity. Thus, if $\sum_{g \in G} g \alpha_g, \sum_{h \in G} h \beta_h$ are elements of KG, where the elements α_g, β_h of K are almost all zero, then we have:

$$\left(\sum_{g \in G} g \alpha_g \right) \cdot \left(\sum_{h \in G} h \beta_h \right) = \sum_{k = gh} k \alpha_g \beta_h.$$

In Examples (e) and (f), we defined the algebra as the set of K-linear combinations of elements in a given generating set, we next defined multiplication on elements of this generating set, then we extended this definition by distributivity. The same technique is used in the next example.

(h) A **quiver** $Q = (Q_0, Q_1, s, t)$ is a quadruple consisting of two sets: Q_0, whose elements are called **points**, and Q_1, whose elements are called **arrows**, and two maps $s, t : Q_1 \rightarrow Q_0$ associating to an arrow $\alpha \in Q_1$ its **source** $s(\alpha)$ and its **target** $t(\alpha)$, both lying in Q_0.

An arrow α of source x and target y is denoted as $\alpha : x \rightarrow y$ or $x \xrightarrow{\alpha} y$. The quiver Q is called a **finite quiver** if Q_0, Q_1 are both finite sets. For an integer $\ell \geq 1$, a **path** $\alpha_1 \alpha_2 \cdots \alpha_\ell$ of length ℓ from x to y is a sequence of ℓ arrows such that $s(\alpha_1) = x, t(\alpha_\ell) = y$ and $t(\alpha_i) = s(\alpha_{i+1})$, for all i satisfying $1 \leqslant i < \ell$. It may be depicted as

$$x = x_0 \xrightarrow{\alpha_1} x_1 \xrightarrow{\alpha_2} \cdots \xrightarrow{\alpha_\ell} x_\ell = y$$

(so, arrows are composed from left to right). We associate to each point $x \in Q_0$ a path ϵ_x of length zero from x to itself, called the **stationary path** at x. A quiver Q is called **acyclic** if it contains no cycle, that is, no path of length at least one from a point to itself. Examples of quivers are:

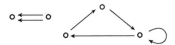

The first one is acyclic but not the second.

A **subquiver** $Q' = (Q'_0, Q'_1, s', t')$ of $Q = (Q_0, Q_1, s, t)$ is a quiver such that $Q'_0 \subseteq Q_0, Q'_1 \subseteq Q_1$ and if an arrow $x \xrightarrow{\alpha} y$ belongs to Q'_1, then $s'(\alpha) = x = s(\alpha), t'(\alpha) = y = t(\alpha)$, that is, the restrictions of s, t to Q'_1 equal s', t' respectively. We say that Q' is a **full subquiver** of Q provided $Q'_1 = \{\alpha \in Q_1 : s(\alpha), t(\alpha) \in Q'_0\}$, that is, the arrows of Q' are precisely those arrows of Q whose source and target belong to Q'. Thus, a full subquiver is completely determined by its set of points.

Let K be a commutative ring. To a finite quiver Q, we associate its **path algebra** KQ. The elements of KQ are the linear combinations of paths in Q, including the stationary ones. The product of paths is defined to be their composite if they are composable, and zero if they are not. That is, if $\alpha_1 \alpha_2 \cdots \alpha_\ell$ and $\beta_1 \beta_2 \cdots \beta_k$ are paths, then their product is

$$(\alpha_1 \alpha_2 \cdots \alpha_\ell) \cdot (\beta_1 \beta_2 \cdots \beta_k) = \begin{cases} \alpha_1 \alpha_2 \cdots \alpha_\ell \beta_1 \beta_2 \cdots \beta_k & \text{if } t(\alpha_\ell) = s(\beta_1), \\ 0 & \text{if } t(\alpha_\ell) \neq s(\beta_1). \end{cases}$$

It is then extended to all elements of KQ using distributivity. The identity is the sum $\sum_{x \in Q_0} \epsilon_x$ of all stationary paths, which exists in KQ because Q_0 is finite.

For instance, let K be a field and Q the quiver $3 \xrightarrow{\alpha} 2 \xrightarrow{\beta} 1$. A K-basis of KQ is given by the paths $\{\epsilon_1, \epsilon_2, \epsilon_3, \alpha, \beta, \alpha\beta\}$ with multiplication table given by:

	ϵ_1	ϵ_2	ϵ_3	α	β	$\alpha\beta$
ϵ_1	ϵ_1	0	0	0	0	0
ϵ_2	0	ϵ_2	0	0	β	0
ϵ_3	0	0	ϵ_3	α	0	$\alpha\beta$
α	0	α	0	0	$\alpha\beta$	0
β	β	0	0	0	0	0
$\alpha\beta$	$\alpha\beta$	0	0	0	0	0

In particular, $\dim_K KQ = 6$. Also, if Q is the quiver

then KQ is generated by its unique stationary path ϵ_x, which is therefore the identity of KQ, and by the cycles α^i, for all $i \geq 1$. Multiplication of cycles is defined by the exponent rule $\alpha^i \alpha^j = \alpha^{i+j}$. In other words, KQ is another realisation of the polynomial algebra $K[t]$. In particular, if K is a field, then KQ is infinite-dimensional.

Because KQ is generated by its paths, it has finitely many generators if and only if Q is acyclic. If Q contains cycles, but we decide to equate to zero all paths long enough, then we still have only finitely many generators. This is achieved through factoring by an ideal containing these paths.

Let K be a field and KQ^+ the ideal of KQ generated by the arrows. It contains all paths of positive length. For an integer $m \geq 2$, let KQ^{+m} be the mth power of KQ^+, see the definition before Proposition I.3.9. That is, KQ^{+m} is the ideal generated by paths of length m. It contains all paths of length greater than or equal to m. An ideal I of KQ is called **admissible** if there exists $m \geq 2$ such that

$$KQ^{+m} \subseteq I \subseteq KQ^{+2}.$$

To say that $I \subseteq KQ^{+2}$ means that I contains only linear combinations of paths of length at least two. Note that we suppose that $I \subseteq KQ^{+2}$ and not $I \subseteq KQ^+$: for, if an element of I belongs to $KQ^+ \setminus KQ^{+2}$, then at least one arrow of the quiver would be superfluous as a generator of KQ. To say that there exists $m \geq 2$ such that $KQ^{+m} \subseteq I$ means that all paths of length at least m (that is, long enough) belong to I.

If I is admissible, then the quotient algebra KQ/I is generated by the residual classes modulo I of the paths in Q, and only finitely many of them are nonzero in the quotient because $KQ^{+m} \subseteq I$. We call KQ/I a **bound quiver algebra**. The pair (Q, I) is called a **bound quiver**. We have just proved that bound quiver algebras have finite bases, that is, they are finite-dimensional.

For example, for the quiver $3 \xrightarrow{\alpha} 2 \xrightarrow{\beta} 1$ considered above, the ideal I generated by $\alpha\beta$ is admissible. The quotient KQ/I, in this case, has dimension 5, and a basis is given by the set $\{\epsilon_1 + I, \epsilon_2 + I, \epsilon_3 + I, \alpha + I, \beta + I\}$. In the one-loop quiver above, we can consider I to be the ideal $\langle \alpha^2 \rangle$ generated by α^2. So, I contains all α^i with $i \geq 2$, hence $(\alpha + I)^i = 0$ for $i \geq 2$, and the two elements $\epsilon_x + I, \alpha + I$ constitute a basis of KQ/I.

In a bound quiver algebra, one can visualise on the quiver a basis and also how to multiply its vectors. We shall see later that one can also express in terms of points and arrows many structural features of this class of algebras. This makes it particularly suitable for constructing illustrative examples.

For instance, it is easy to visualise the opposite algebra of a path algebra: given a quiver Q, construct the opposite quiver by keeping the same points and reversing the sense of arrows, then the path algebra of the opposite quiver is another realisation of the opposite algebra of the path algebra of Q, see Exercise I.4.8.

Exercises of Section I.3

1. Let A be an associative K-algebra which does not necessarily have a multiplicative identity. Consider the set $A_1 = A \times K = \{(a, \alpha) : a \in A, \alpha \in K\}$ equipped with the operations

$$(a, \alpha) + (b, \beta) = (a + b, \alpha + \beta)$$
$$(a, \alpha)\beta = (a\beta, \alpha\beta)$$
$$(a, \alpha)(b, \beta) = (ab + a\beta + b\alpha, \alpha\beta)$$

for $a, b \in A, \alpha, \beta \in K$. Verify that A_1 is an associative K-algebra having $(0, 1)$ as an identity (so, any associative algebra can be embedded in a unitary associative algebra). One says that A_1 is obtained from A by **adjoining an identity**.

2. Let K be a field. Prove that each of the following sets of matrices, under the usual matrix operations, is a subalgebra of $T_4(K)$.

(a) $A_0 = \begin{pmatrix} K & 0 & 0 & 0 \\ 0 & K & 0 & 0 \\ 0 & 0 & K & 0 \\ 0 & 0 & 0 & K \end{pmatrix} = \left\{ \begin{pmatrix} \alpha & 0 & 0 & 0 \\ 0 & \beta & 0 & 0 \\ 0 & 0 & \gamma & 0 \\ 0 & 0 & 0 & \delta \end{pmatrix} : \alpha, \beta, \gamma, \delta \in K \right\}$

(b) $A_1 = \begin{pmatrix} K & 0 & 0 & 0 \\ 0 & K & 0 & 0 \\ 0 & 0 & K & 0 \\ K & K & K & K \end{pmatrix} = \left\{ \begin{pmatrix} \alpha_1 & 0 & 0 & 0 \\ 0 & \alpha_2 & 0 & 0 \\ 0 & 0 & \alpha_3 & 0 \\ \alpha_4 & \alpha_5 & \alpha_6 & \alpha_7 \end{pmatrix} : \alpha_i \in K \text{ for all } i \right\}$

(c) $A_2 = \left\{ \begin{pmatrix} \alpha & 0 & 0 & 0 \\ \beta & \alpha & 0 & 0 \\ \gamma & \beta & \alpha & 0 \\ \delta & \gamma & \beta & \alpha \end{pmatrix} : \alpha, \beta, \gamma, \delta \in K \right\}$

(d) $A_3 = \left\{ \begin{pmatrix} \alpha & 0 & 0 & 0 \\ 0 & \alpha & 0 & 0 \\ \beta & \gamma & \alpha & 0 \\ \delta & \epsilon & \lambda & \alpha \end{pmatrix} : \alpha, \beta, \gamma, \delta, \epsilon, \lambda \in K \right\}$

3. Let A be a K-algebra. Prove that the centre $Z(A)$ of A is a subalgebra of A.

4. We show an infinite-dimensional analogue of the lower triangular matrix algebra $T_n(K)$. A matrix $\mathbf{a} = (a_{ij})_{i,j \geq 1}$, with rows and columns indexed by the positive integers, is **row-finite** (or **column-finite**) if it has at most finitely many nonzero coefficients in each row (or column, respectively). It is **lower triangular** if $a_{ij} = 0$ whenever $i < j$. Let $TF(K)$ be the set of lower triangular matrices with coefficients in K which are both row-finite and column-finite. Prove that:
   (a) $TF(K)$ is a K-algebra under the usual matrix operations.
   (b) The set of matrices $\mathbf{a} = (a_{ij})_{i,j \geq 1} \in TF(K)$ such that $a_{ii} = 0$, for any i, is an ideal in $TF(K)$.

5. Let B be at the same time a subalgebra and an ideal of an algebra A. Prove that $B = A$.

6. Prove that a nonempty subset I of a K-algebra A is an ideal if and only if, for any $x, y \in I, a, b, c, d \in A$, we have $axb + cyd \in I$.

7. Let I, J be ideals of an algebra A. Prove that $I \cup J$ is an ideal of A if and only if $I \subseteq J$ or $J \subseteq I$.

8. Let A be an algebra and $(I_\lambda)_{\lambda \in \Lambda}$ a family of ideals of A which is a **chain**, that is, for any $\lambda, \mu \in \Lambda$, we have $I_\lambda \subseteq I_\mu$ or $I_\mu \subseteq I_\lambda$. Prove that $\cup_{\lambda \in \Lambda} I_\lambda$ is an ideal of A.

9. Let A, B be algebras. Prove that any ideal of the product $A \times B$ is of the form $I \times J$, for some $I \trianglelefteq A, J \trianglelefteq B$.

10. Let I, J, L be ideals of an algebra A. Prove that:
    (a) $(IJ)L = I(JL)$.
    (b) $I(J + L) = IJ + IL$ and $(J + L)I = JI + LI$.
    (c) If $J \subseteq I$, then $I \cap (J + L) = J + (I \cap L)$. This identity is known as the **modular law**.
    (d) In general, $IJ \neq I \cap J$.

11. Let A be a commutative algebra and $X, Y \subseteq A$ subsets such that each element of one of them is a linear combination (with coefficients in A) of elements of the other. Prove that X and Y generate the same ideal in A.

12. Let A be an algebra and $I \trianglelefteq A$. Prove that A/I is commutative if and only if, for any $a, b \in A$, we have $ab - ba \in I$. Deduce that any algebra has a commutative quotient.

13. Let A be a commutative algebra. An element $a \in A$ is called **nilpotent** if there exists an integer $n > 0$ such that $a^n = 0$. Prove that the set N of nilpotent elements of A is an ideal of A and the quotient A/N contains no nonzero nilpotent elements.

14. Let K be a nonzero commutative ring. Prove that:
   (a) There exists a unique ring homomorphism $\varphi : \mathbb{Z} \to K$.
   (b) For $x \in \mathbb{Z}$, we have $\varphi(x) = 0$ if and only if x is divisible by the characteristic of K, if the latter is positive, and $x = 0$ otherwise. See Exercise I.2.5.

15. Let K be a commutative ring and $K[[t]]$ the algebra of formal power series in t, see Example I.3.11(e). The **order** $\omega(a)$ of a nonzero element $a = (a_i)_{i \geq 0} \in K[[t]]$ is the least nonnegative integer i_0 such that $a_{i_0} \neq 0$. Prove that:
   (a) If $a, b \in K[[t]]$, then either $ab = 0$, or $\omega(ab) \geq \omega(a) + \omega(b)$, with equality if K is an integral domain.
   (b) If $a, b \in K[[t]]$, then either $a + b = 0$, or $\omega(a + b) \geq \min\{\omega(a), \omega(b)\}$.
   (c) If K in an integral domain, then so is $K[[t]]$.
   (d) The element $1 - t$ is invertible in $K[[t]]$, and its inverse is $\sum_{i=0}^{\infty} t^i$.
   (e) An element $a = (a_i)_{i \geq 0} \in K[[t]]$ is invertible in $K[[t]]$ if and only if a_0 is invertible in K.
   (f) If K is a field, then any nonzero element of $K[[t]]$ can be written in the form $t^i a$, for some $i \geq 0$ and an invertible $a \in K[[t]]$.

16. Let K be a field and Q the quiver

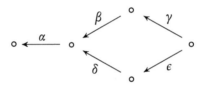

   (a) Prove that the path algebra KQ of Q is finite-dimensional and find its dimension.
   (b) Show that the ideal $I = \langle \gamma\beta - \varepsilon\delta \rangle$ of KQ is admissible and write a basis for $A = KQ/I$.
   (c) Give generators for admissible ideals I_i, with $i = 1, 2$, such that the corresponding quotient algebras $A_i = KQ/I_i$ have dimensions 10 and 13, respectively.

I.4 Algebra morphisms

I.4.1 Morphisms

A homomorphism between sets having the same algebraic structure is a map which preserves the structure, in the sense that it is compatible with the operations defining it. An (associative and unitary) algebra is defined by three operations: an addition, an external multiplication and an internal one, the latter admitting an identity. We need compatibility with respect to each.

I.4.1 Definition. Let A, B be K-algebras. A **morphism**, or **homomorphism**, of K-algebras from A to B is a map $\varphi : A \to B$ satisfying the following conditions:

(AM1) $\varphi(a_1 + a_2) = \varphi(a_1) + \varphi(a_2)$,
(AM2) $\varphi(a\alpha) = \varphi(a)\alpha$,
(AM3) $\varphi(a_1 a_2) = \varphi(a_1)\varphi(a_2)$,
(AM4) $\varphi(1) = 1$,
for any $a, a_1, a_2 \in A, \alpha \in K$.

We use the term morphism, rather than homomorphism, not only for brevity but because it fits into a general scheme which we introduce in Chapter III. The reader certainly noticed that in (AM4), we used the same symbol 1 to denote the identities of A and B.

The previous definition says that a map is an algebra morphism if and only if it is a ring homomorphism satisfying the additional axiom (AM2), expressed by saying that it is K-linear.

If $K = \mathbb{Z}$, then axiom (AM2) is a consequence of (AM1) and induction. Because rings are $\mathbb{Z}$-algebras, this implies that any ring homomorphism is a morphism of $\mathbb{Z}$-algebras.

If A is an algebra, then a morphism from A to itself is called an **endomorphism** of A.

We give examples of algebra morphisms.

I.4.2 Examples.

(a) Let A, B be K-algebras, with B a subalgebra of A. The inclusion map $\iota : B \to A$ defined by $b \mapsto b$, for $b \in B$, is a morphism, called the **inclusion morphism** or the **injection morphism**. If $B = A$, this morphism is just the identity map on A (or B) denoted by 1_A. This notation is not to be confused with the multiplicative identity of A, because, as pointed out, the latter is now simply denoted by 1.

(b) Let A be a K-algebra and I an ideal of A. The map $\pi : A \to A/I$ given by $a \mapsto a + I$, for $a \in A$, is a morphism. Indeed, let $a, a_1, a_2 \in A$ and $\alpha \in K$, then

$$\pi(a_1 + a_2) = (a_1 + a_2) + I = (a_1 + I) + (a_2 + I) = \pi(a_1) + \pi(a_2),$$

$$\pi(a_1 a_2) = (a_1 a_2) + I = (a_1 + I)(a_2 + I) = \pi(a_1)\pi(a_2),$$

$$\pi(a\alpha) = (a\alpha) + I = (a + I)\alpha = \pi(a)\alpha,$$

because of the definition of the operations on A/I, while $\pi(1) = 1 + I$ follows from the definition of π. Actually, the previous equalities say more: we defined exactly those operations on A/I which make π an algebra morphism. The map π is called the **projection** or **surjection morphism**.

(c) Let $\varphi : A \to B$ and $\psi : B \to C$ be algebra morphisms. Then so is their composite $\psi \circ \varphi : A \to C$.

(d) Let $\lambda \in K$ be fixed and define $\varphi : K[t] \to K$ as follows: for a polynomial $p = \sum_{i=0}^{d} \alpha_i t^i$, set $\varphi(p) = \sum_{i=0}^{d} \alpha_i \lambda^i$, that is, $\varphi(p)$ is the evaluation $p(\lambda)$ of the polynomial at λ. It follows from properties of evaluation that φ is a morphism: for instance, if $p, q \in K[t]$, then

$$\varphi(p + q) = (p + q)(\lambda) = p(\lambda) + q(\lambda) = \varphi(p) + \varphi(q)$$

and similarly for axioms (AM2) and (AM3). Finally, $\varphi(1) = 1$ because the 1 in $K[t]$ is the constant polynomial equal to 1.

Because algebra morphisms are ring homomorphisms which are also K-linear, they satisfy the usual properties of these maps. We recall some definitions. Let $\varphi : A \to B$ be a map. For any $A' \subseteq A$, its *image* is the subset $\varphi(A') = \{\varphi(a) \in B : a \in A'\}$, and the *preimage* of $B' \subseteq B$ is the subset $\varphi^{-1}(B') = \{a \in A : \varphi(a) \in B'\}$. The *image* Im φ of the morphism φ is $\varphi(A)$ and its *kernel* Ker φ is $\varphi^{-1}(0)$. We warn the reader that this notation does *not* mean that φ^{-1} exists as a map, it only does in case φ is bijective, see Lemma I.4.6 below. We list some properties of algebra morphisms.

I.4.3 Proposition. *Let $\varphi : A \to B$ be an algebra morphism. Then:*
  (a) $\varphi(0) = 0$.
  (b) $\varphi(-a) = -\varphi(a)$, *for any $a \in A$.*
  (c) $\varphi(na) = n\varphi(a)$, *for any $a \in A, n \in \mathbb{Z}$.*
  (d) $\varphi(a^n) = \varphi(a)^n$, *for any $a \in A$ and integer $n \geq 0$.*
  (e) Im φ *is a subalgebra of B.*
  (f) Ker φ *is an ideal of A.*
  (g) φ *is injective if and only if* Ker $\varphi = 0$.

Proof. (a), (b), (c), (g) are properties of abelian group homomorphisms and (d) a property of monoid homomorphisms. We prove (e) and (f). Because φ is a ring homomorphism, Im φ is a subring of B. It is actually a subalgebra of B, because, if $c \in$ Im φ and $\alpha \in K$, then there exists $a \in A$ such that $c = \varphi(a)$, so $c\alpha = \varphi(a)\alpha = \varphi(a\alpha) \in$ Im φ. Similarly, $a \in$ Ker φ implies $\varphi(ca) = 0 = \varphi(ac)$, for any $c \in A$, thus Ker φ is an ideal of A considered as ring. Then it is also an ideal of A viewed as algebra, see comments after Definition I.3.5. $\square$

I.4.4 Example. Let K be a field, $\lambda \in K$ and $\varphi : K[t] \to K$ the evaluation at λ, see Example I.4.2(d) above. Then φ is surjective, that is, Im $\varphi = K$, because, for any $\alpha \in K$, the constant polynomial equal to α evaluates (obviously!) to α. We show that Ker φ is the ideal $\langle t - \lambda \rangle$ generated by the polynomial $t - \lambda$. Because $K[t]$ is commutative, this ideal is the set of (polynomial) multiples of $t - \lambda$. Plainly, $(t - \lambda) \in$ Ker φ and hence $\langle t - \lambda \rangle \subseteq$ Ker φ. Conversely, let $p \in$ Ker φ. Dividing p by $t - \lambda$, we get $p = (t - \lambda)q + r$ where q is the quotient and the remainder r is zero or of degree less than one, thus it is a constant. Evaluating both sides at λ yields $0 = p(\lambda) = (\lambda - \lambda)q(\lambda) + r$, because evaluation is an algebra morphism, and so $p \in$ Ker φ. Hence, $r = 0$ and $p = (t - \lambda)q \in \langle t - \lambda \rangle$. This completes the proof.

In all areas of algebra (and other fields of mathematics), there exists a notion of iso-morphism, meant to express that two structures are 'identical except for the names of their elements'. This somewhat vague idea can be expressed rigorously.

I.4.5 Definition. A morphism of K-algebras $\varphi : A \to B$ is an *isomorphism* if there exists a morphism $\varphi' : B \to A$ such that $\varphi' \circ \varphi = 1_A$ and $\varphi \circ \varphi' = 1_B$. The algebras A and B are then said to be *isomorphic*, which we denote as $A \cong B$.

The definition of isomorphism implies that it is invertible as a map, and hence bijective. Moreover, the morphism φ' of the definition is the uniquely determined inverse map of φ. We thus write φ^{-1} instead of φ'.

Clearly, the identity morphism $1_A : A \to A$ is an isomorphism. If $\varphi : A \to B$ is an isomorphism, then so is $\varphi^{-1} : B \to A$ and we have $(\varphi^{-1})^{-1} = \varphi$. Finally, if $\varphi : A \to B$, $\psi : B \to C$ are isomorphisms, then so is $\psi\varphi : A \to C$ and we have $(\psi\varphi)^{-1} = \varphi^{-1}\psi^{-1}$. In particular, the relation $\cong$ between algebras is reflexive, symmetric and transitive.

An isomorphism from an algebra A to itself is called an *automorphism* of A. The previous considerations imply that the set $\operatorname{Aut} A$ of automorphisms of A is a group under composition of morphisms. It is called the *automorphism group* of A.

Isomorphisms are bijective morphisms. The remarkable thing is that the converse also holds true: any bijective morphism is an isomorphism.

I.4.6 Lemma. *A morphism of algebras is an isomorphism if and only if it is bijective.*

Proof. We just need to prove sufficiency. Let $\varphi : A \to B$ be a bijective morphism. We must prove that the inverse map φ^{-1} is an algebra morphism. Let $b_1, b_2 \in B$, then:

$$\varphi^{-1}(b_1 + b_2) = \varphi^{-1}(1_B(b_1) + 1_B(b_2)) = \varphi^{-1}((\varphi \circ \varphi^{-1})(b_1) + (\varphi \circ \varphi^{-1})(b_2))$$

$$= \varphi^{-1}[\varphi(\varphi^{-1}(b_1)) + \varphi(\varphi^{-1}(b_2))] = (\varphi^{-1} \circ \varphi)[\varphi^{-1}(b_1) + \varphi^{-1}(b_2)]$$

$$= 1_A[\varphi^{-1}(b_1) + \varphi^{-1}(b_2)] = \varphi^{-1}(b_1) + \varphi^{-1}(b_2)$$

where, at the fourth equality, we used that φ is a morphism. The other conditions are verified in the same way. $\qquad\square$

Isomorphic algebras have the same properties. For instance, it is easily seen that if $A \cong B$, then A is commutative if and only if so is B.

In order to see examples of isomorphisms, recall that we used in Examples I.3.11(f) and (h) the expression 'is another realisation of': by this, we meant 'is isomorphic to'. The reader is encouraged to construct explicitly the isomorphism in each case.

I.4.2 The isomorphism theorems

It may be useful, in a given problem, to replace an algebra by another one which has the same properties inasmuch as our problem is concerned but is easier to handle. Hence the usefulness of the isomorphism theorems which show how to transfer information from one algebra to another.

We need a convention: if, in the sequel, a statement asserts the existence of a map completing a diagram, then this map is represented by dotted lines on the diagram (like the morphism $\bar{\varphi}$ in the theorem below). To say that a diagram consisting of objects (here, algebras) and maps (here, morphisms) is commutative, or commutes, means that if one composes maps on a path from an object to another, then the result of the composition depends only on these two objects, and does not depend on the path followed.

I.4.7 Theorem. *Let $\varphi : A \to B$ be an algebra morphism, $\iota : \operatorname{Im} \varphi \to B$ the inclusion and $\pi : A \to A/\operatorname{Ker} \varphi$ the projection. Then there exists a unique morphism $\bar{\varphi} : A/\operatorname{Ker} \varphi \to \operatorname{Im} \varphi$ which makes the following diagram commutative*

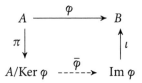

that is, such that $\varphi = \iota \circ \bar{\varphi} \circ \pi$. Moreover, $\bar{\varphi}$ is an isomorphism. In particular, $A/\operatorname{Ker} \varphi \cong \operatorname{Im} \varphi$.

Proof. We outline the strategy: we start by assuming that $\bar{\varphi}$ exists, and prove its uniqueness by finding a formula giving it explicitly. Then we use the same formula to prove its existence. Set for short $I = \operatorname{Ker} \varphi$. An arbitrary element of A/I is of the form $a + I = \pi(a)$, for some $a \in A$. One must then have

$$\bar{\varphi}(a + I) = \bar{\varphi}\pi(a) = \iota\bar{\varphi}\pi(a) = \varphi(a),$$

where the second equality holds true because $b \in B$ lies in $\operatorname{Im} \varphi$ if and only if $b = \iota(b)$. This shows uniqueness of $\bar{\varphi}$. But this formula also defines unambiguously a map, because if $a, a' \in A$ are such that $a + I = a' + I$, then $a - a' \in I = \operatorname{Ker} \varphi$ gives $\varphi(a - a') = 0$ and hence $\varphi(a) = \varphi(a')$. It is easy to verify that $\bar{\varphi}$ is an algebra morphism, and that it makes the shown diagram commutative.

Because the very definition of $\bar{\varphi}$ implies its surjectivity, we prove injectivity. Let $a \in A$ be such that $\bar{\varphi}(a+I) = 0$. Then $\varphi(a) = 0$ gives $a \in \operatorname{Ker} \varphi = I$ and so $a+I = I$, which is the zero of A/I. $\qquad\square$

Theorem I.4.7 can be thought of as saying that, given an algebra morphism $\varphi : A \to B$, 'part' of A (more precisely, a quotient algebra of A) is the same as, that is, is isomorphic to, 'part' of B (more precisely, a subalgebra of B).

A situation which occurs frequently is when $\varphi : A \to B$ is surjective. In this case, $\operatorname{Im} \varphi = B$ and $\iota = 1_B$ so $\bar{\varphi}$ defines an isomorphism $A/\operatorname{Ker} \varphi \cong B$.

In Example I.4.4, we proved that if K is a field, then evaluation at $\lambda \in K$ induces a surjective algebra morphism from $K[t]$ to K whose kernel is the ideal $\langle t - \lambda \rangle$. Hence, $K[t]/\langle t - \lambda \rangle \cong K$.

On the other hand, if $\varphi : A \to B$ is injective, then $\operatorname{Ker} \varphi = 0$. In this case, A is isomorphic to a subalgebra of B.

The next isomorphism theorem compares two quotients of the same algebra.

I.4.8 Theorem. *Let A be an algebra, I, J ideals of A such that $I \subseteq J$ and $\pi_I : A \to A/I$, $\pi_J : A \to A/J$ the respective projections. Then there exists a unique morphism*

$\varphi : A/I \to A/J$ such that $\varphi \circ \pi_I = \pi_J$. Moreover, φ is surjective and its kernel is J/I.

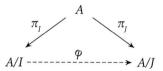

Proof. We follow the same strategy of proof as in the previous theorem. An element of A/I is of the form $a + I = \pi_I(a)$, for some $a \in A$. One must then have

$$\varphi(a + I) = \varphi\pi_I(a) = \pi_J(a) = a + J.$$

This shows uniqueness. The same formula defines unambiguously a map, because if $a, a' \in A$ are such that $a + I = a' + I$, then $a - a' \in I \subseteq J$ gives $a + J = a' + J$. Again, one checks easily that φ is an algebra morphism.

Surjectivity of φ follows from its definition. One has $a + I \in \operatorname{Ker} \varphi$ if and only if $a + J = J$, that is, $a \in J$, so that $\operatorname{Ker} \varphi = \{a + I : a \in J\} = J/I$, as required. $\qquad \square$

As a consequence, J/I is an ideal of the algebra A/I, because of Proposition I.4.3(f), and then Theorem I.4.7 gives an algebra isomorphism $(A/I)/(J/I) \cong A/J$.

The third isomorphism theorem compares quotients of subalgebras of a given algebra. Let A be an algebra, B a subalgebra and I an ideal of A. Using Lemma I.3.4, one proves easily that $B + I = \{b + x : b \in B, x \in I\}$ is a subalgebra of A, in which I is an ideal. Thus, the quotient $(B + I)/I$ makes sense (and is an algebra).

I.4.9 Theorem. *Let A be an algebra, B a subalgebra and I an ideal of A. Then there exists an algebra isomorphism*

$$\frac{B}{B \cap I} \cong \frac{B + I}{I}.$$

Proof. Our strategy here consists in constructing a surjective algebra morphism $\varphi : B \to (B + I)/I$ whose kernel is $B \cap I$, then applying Theorem I.4.7.

Set $\varphi : b \mapsto b + I$, for $b \in B$. Then φ is easily seen to be a morphism. It is surjective because an arbitrary element of $(B + I)/I$ is of the form $(b + x) + I$, for some $b \in B, x \in I$, and we have $b + x + I = b + I = \varphi(b)$, because $x \in I$. The kernel consists of all $b \in B$ such that $b + I = I$ or, equivalently, $b \in I$: this is $B \cap I$, as required. Theorem I.4.7 yields the result. $\qquad \square$

In particular, $B \cap I$ is an ideal of B.

This theorem is called the 'parallelogram law': it asserts that the notched opposite sides of the following parallelogram are isomorphic.

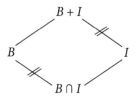

Here, $B \cap I$ is the largest ideal (in fact, the largest subset) contained in both B and I. On the other hand, $B + I$ is the smallest abelian group containing both B and I, and it is a subalgebra of A, hence it is the smallest subalgebra containing both B and I. The segment (side) from $B \cap I$ to B represents the quotient $B/(B \cap I)$. Similarly, the opposite segment represents the quotient $(B + I)/I$. The similar notches mean that they are isomorphic. The other two (unnotched) sides of the parallelogram do not correspond to algebras: they are just K-modules.

We finish this subsection with a theorem comparing the ideals in an algebra and one of its quotients. For this reason, it is called the **correspondence of ideals theorem**. Recall that if $\varphi : A \to B$ is a map and $A' \subseteq A$, then one always has $A' \subseteq \varphi^{-1}(\varphi(A'))$ and equality holds true when φ is injective. Also, if $B' \subseteq B$, then one has $\varphi(\varphi^{-1}(B')) \subseteq B'$ and equality holds true when φ is surjective.

I.4.10 Theorem. *Let A be an algebra, I an ideal and $\pi : A \to A/I$ the projection. The map $J \mapsto \pi(J)$ is an inclusion-preserving bijection from the set of ideals of A containing I to the set of ideals of A/I. The inverse bijection is $\bar{J} \mapsto \pi^{-1}(\bar{J})$.*

Proof. Let J be an ideal of A containing I. Because of Theorem I.4.8, $\pi(J) = J/I = \{x + I : x \in J\}$ is an ideal of A/I. Clearly, if $J_1 \subseteq J_2$, then $\pi(J_1) \subseteq \pi(J_2)$. Moreover, $J \subseteq \pi^{-1}\pi(J)$. We show the reverse inclusion. An element $a \in A$ lies in $\pi^{-1}\pi(J)$ if and only if $\pi(a) \in \pi(J)$, that is, there exists $x \in J$ such that $\pi(a) = \pi(x)$, or, equivalently, $a - x \in \operatorname{Ker} \pi = I \subseteq J$. So, $a \in J$ and hence $J = \pi^{-1}\pi(J)$. On the other hand, if $\bar{J}$ is an ideal in A/J, then $\pi^{-1}(\bar{J}) = \{a \in A : a + I \in \bar{J}\}$ is an ideal of A containing $I = \operatorname{Ker} \pi$, see Exercise I.4.4(d). Finally, $\pi\pi^{-1}(\bar{J}) = \bar{J}$, because π is surjective. The proof is complete. $\qquad\square$

As a corollary, we characterise maximal ideals in commutative algebras. A proper ideal I in an algebra A is called **maximal** if no ideal can be properly inserted between I and A, that is, if J is an ideal such that $I \subseteq J \subseteq A$, then $J = I$ or $J = A$.

I.4.11 Corollary. *Let A be a nonzero commutative algebra. A proper ideal I in A is maximal if and only if A/I is a field.*

Proof. Because A is commutative, so is A/I. Hence, according to Proposition I.3.9, A/I is a field if and only if it is nonzero and has no nonzero proper ideal. Because of Theorem I.4.10, this happens if and only if $A \neq I$ and no ideal is properly contained between A and I, that is, I is a maximal ideal. $\qquad\square$

For instance, if K is a field and $\lambda \in K$, then $K[t]/\langle t - \lambda \rangle \cong K$, see the example just after Theorem I.4.7. Hence, the ideal $\langle t - \lambda \rangle$ is maximal in $K[t]$, for any λ.

Exercises of Section I.4

1. Let $\varphi : A \to B, \varphi' : A' \to B'$ be algebra morphisms. Define $\Phi : A \times A' \to B \times B'$ by $(a, a') \mapsto (\varphi(a), \varphi'(a'))$, for $a \in A, a' \in A'$. Prove that:

    (a) Φ is an algebra morphism.

    (b) Ker Φ = Ker $\varphi \times$ Ker φ'.

    (c) Im Φ = Im $\varphi \times$ Im φ'.

2. Given a non necessarily unitary algebra A, denote by A_1 the algebra obtained from A by adjoining an identity, see Exercise I.3.1.

    (a) Prove that, if A is already unitary, then A_1 is isomorphic to $A \times K$. In particular, A_1 is not the smallest unitary algebra containing A.

    (b) Let $\varphi : A \to B$ be a morphism of nonunitary algebras, that is, a map such that $\varphi(a + a') = \varphi(a) + \varphi(a')$, $\varphi(a\alpha) = \varphi(a)\alpha$ and $\varphi(aa') = \varphi(a)\varphi(a')$, for $a, a' \in A$ and $\alpha \in K$. Prove that φ uniquely extends to a morphism $\varphi_1 : A_1 \to B_1$.

    (c) Prove that, conversely, any morphism $\varphi_1 : A_1 \to B_1$ restricts to a morphism $\varphi : A \to B$.

    (d) Prove that $A \cong B$ if and only if $A_1 \cong B_1$.

3. Let K be a field and A a finite-dimensional K-algebra with basis $\{e_1, \cdots, e_n\}$.

    (a) Prove that A is uniquely determined, up to isomorphism, by a family of constants $(\alpha_{ij}^k)_{1 \leqslant i,j,k \leqslant n}$ from K defined by $e_i e_j = \sum_{k=1}^{n} \alpha_{ij}^k e_k$. The α_{ij}^k are called the **structure constants** of the algebra.

    (b) Deduce that, if K is a finite field, then there are, up to isomorphism, only finitely many algebras of a given dimension n.

    (c) Prove that associativity of multiplication in A is expressed by the relations $\sum_{\ell=1}^{n} \alpha_{ij}^\ell \alpha_{\ell k}^m = \sum_{\ell=1}^{n} \alpha_{jk}^\ell \alpha_{i\ell}^m$, for all i, j, k, ℓ, m (so, structure constants cannot be chosen arbitrarily).

4. Let $\varphi : A \to B$ be an algebra morphism. Prove the following:

    (a) For any subalgebra A' of A, its image $\varphi(A')$ is a subalgebra of B.

    (b) For any subalgebra B' of B, its preimage $\varphi^{-1}(B')$ is a subalgebra of A.

    (c) For any ideal I of A, its image $\varphi(I)$ is an ideal of Im φ. Is it necessarily an ideal of B?

    (d) For any ideal J of B, its preimage $\varphi^{-1}(J)$ is an ideal of A containing Ker φ.

5. Let A, B be finite-dimensional algebras over the same field K. If $(e_i)_{1 \leqslant i \leqslant n}$ is a basis of A, and $\varphi : A \to B$ a K-linear map such that $(\varphi(e_i))_{1 \leqslant i \leqslant n}$ is a basis of B, then φ is an algebra isomorphism if and only if $\varphi(e_i)\varphi(e_j) = \varphi(e_i e_j)$, for any i, j.

6. Let A be an algebra containing a field K as a subalgebra, B an algebra and $\varphi : A \to B$ an injective algebra morphism. Prove that B contains a field isomorphic to K.

7. Prove that the matrix algebra of Exercise I.2.9 is isomorphic to the $\mathbb{R}$-algebra of the quaternions.

8. Given a quiver $Q = (Q_0, Q_1, s, t)$, its **opposite quiver** $Q^{\mathrm{op}} = (Q_0', Q_1', s', t')$ is obtained from Q by reversing the direction of arrows, that is, it is defined as follows: $Q_0' = Q_0$, $Q_1' = Q_1$ and, if $\alpha \in Q_1$, then $s'(\alpha) = t(\alpha), t'(\alpha) = s(\alpha)$. Let K be a field and Q a finite quiver. Prove that:

    (a) $(KQ)^{\mathrm{op}} \cong K(Q^{\mathrm{op}})$.

    (b) If I is an admissible ideal in KQ, then there exists an admissible ideal I^{op} in KQ^{op} such that $KQ^{\mathrm{op}}/I^{\mathrm{op}} \cong (KQ/I)^{\mathrm{op}}$.

9. Prove that the map $z \mapsto \bar{z}$ is an automorphism of $\mathbb{C}$ considered as an $\mathbb{R}$-algebra.

10. Let K be a field and $n > 0$ an integer. Prove that we have isomorphisms between:

(a) The centre of the full matrix algebra $M_n(K)$ and K.

(b) The incidence algebra of the poset $E = \{1, 2, 3\}$, ordered by $1 \leqslant 3, 2 \leqslant 3$, and the subalgebra of $T_3(K)$ given by

$$\begin{pmatrix} K & 0 & 0 \\ 0 & K & 0 \\ K & K & K \end{pmatrix} = \left\{ \begin{pmatrix} \alpha_1 & 0 & 0 \\ 0 & \alpha_2 & 0 \\ \alpha_3 & \alpha_4 & \alpha_5 \end{pmatrix} : \alpha_i \in K \right\}.$$

(c) $K[s, t]/\langle t - s \rangle$ and $K[t]$.

(d) The path algebra of the quiver

$$n \to n - 1 \to \cdots \to 2 \to 1$$

and the lower triangular matrix algebra $T_n(K)$.

(e) The path algebra of the one-loop quiver

bound by the ideal $\langle \alpha^n \rangle$ and $K[t]/\langle t^n \rangle$.

(f) The path algebra of the two-loops quiver

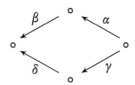

bound by the ideal $\langle \alpha\beta - \beta\alpha \rangle$ and $K[s, t]$.

(g) The path algebra of the quiver

$$\begin{array}{ccc} & \beta \quad \circ \quad \alpha & \\ \circ & & \circ \\ & \delta \quad \quad \gamma & \\ & \circ & \end{array}$$

bound by the ideal $\langle \alpha\beta - \gamma\delta \rangle$ and the subalgebra of $T_4(K)$ given by

$$\begin{pmatrix} K & 0 & 0 & 0 \\ K & K & 0 & 0 \\ K & 0 & K & 0 \\ K & K & K & K \end{pmatrix} = \left\{ \begin{pmatrix} \alpha_1 & 0 & 0 & 0 \\ \alpha_2 & \alpha_3 & 0 & 0 \\ \alpha_4 & 0 & \alpha_5 & 0 \\ \alpha_6 & \alpha_7 & \alpha_8 & \alpha_9 \end{pmatrix} : \alpha_i \in K \text{ for all } i \right\}.$$

(h) The path algebra of the quiver

$$\circ \rightleftarrows \circ$$

and the algebra of lower triangular matrices

$$\begin{pmatrix} \alpha & 0 & 0 \\ \beta & \gamma & 0 \\ \delta & 0 & \gamma \end{pmatrix}$$

with $\alpha, \beta, \gamma, \delta \in K$ and the usual matrix operations.

11. Let K be a field. Consider the set of matrices

$$A = \begin{pmatrix} K & 0 & 0 \\ K & K & 0 \\ 0 & K & K \end{pmatrix} = \left\{ \begin{pmatrix} \alpha_1 & 0 & 0 \\ \alpha_2 & \alpha_3 & 0 \\ 0 & \alpha_4 & \alpha_5 \end{pmatrix} : \alpha_i \in K \right\}.$$

(a) Prove that A becomes an algebra, when given the usual K-vector space structure of $T_3(K)$ and the multiplication defined by

$$\begin{pmatrix} \alpha_1 & 0 & 0 \\ \alpha_2 & \alpha_3 & 0 \\ 0 & \alpha_4 & \alpha_5 \end{pmatrix} \cdot \begin{pmatrix} \beta_1 & 0 & 0 \\ \beta_2 & \beta_3 & 0 \\ 0 & \beta_4 & \beta_5 \end{pmatrix} = \begin{pmatrix} \alpha_1\beta_1 & 0 & 0 \\ \alpha_2\beta_1 + \alpha_3\beta_2 & \alpha_2\beta_2 & 0 \\ 0 & \alpha_4\beta_3 + \alpha_5\beta_4 & \alpha_3\beta_3 \end{pmatrix}.$$

for $\alpha_i, \beta_i \in K$ for all i.

(b) Prove that $A \cong T_3(K)/I$, where $I = \begin{pmatrix} 0 & 0 & 0 \\ 0 & 0 & 0 \\ K & 0 & 0 \end{pmatrix}$.

12. Let A be a K-algebra. Prove that:
(a) The set $\mathrm{End}_K A$ admits a natural K-algebra structure.
(b) Given $a \in A$, the map $\rho_a : A \to A$ defined by $x \mapsto xa$, for $x \in A$, is a K-linear endomorphism of A.
(c) The map $a \mapsto \rho_a$, for $a \in A$, is an injective algebra morphism from A to $\mathrm{End}_K A$.
(d) If K is a field, then any finite-dimensional algebra is a subalgebra of a full matrix algebra.

13. Let $n > 0$ be an integer. Construct a surjective algebra morphism from $T_n(K)$ to K^n and compute its kernel.

14. Let A be an algebra, I an ideal and $\pi : A \to A/I$ the projection. Prove that, for any $J \trianglelefteq A$, we have $\pi(J) = (I + J)/I$.

15. Let A be an algebra and I an ideal. Prove that:
(a) $M_n(I)$ is an ideal in the algebra $M_n(A)$.
(b) $M_n(A)/M_n(I) \cong M_n(A/I)$.

16. Prove that the ideal $\langle t^2 + 1 \rangle$ is maximal in $\mathbb{R}[t]$.

17. Let d be a positive integer which is not a perfect square. Prove that:
(a) $\mathbb{Q}[\sqrt{d}] = \{a + b\sqrt{d} : a, b \in \mathbb{Q}\}$ is a subalgebra of $\mathbb{R}$, and is a two-dimensional $\mathbb{Q}$-algebra.
(b) The map $a + b\sqrt{d} \mapsto a - b\sqrt{d}$, for $a, b \in \mathbb{Q}$, is an automorphism of $\mathbb{Q}[\sqrt{d}]$ as a $\mathbb{Q}$-algebra.
(c) $A = \left\{ \begin{pmatrix} a & b \\ db & a \end{pmatrix} : a, b \in \mathbb{Q} \right\}$, with usual matrix operations, is a subalgebra of $M_2(\mathbb{Q})$.
(d) The algebra A of (c) above is isomorphic to $\mathbb{Q}[\sqrt{d}]$.
(e) $\mathbb{Q}[\sqrt{2}]$ and $\mathbb{Q}[\sqrt{3}]$ are not isomorphic. Can one deduce a necessary and sufficient condition on two positive integers d, d' so that $\mathbb{Q}[\sqrt{d}] \cong \mathbb{Q}[\sqrt{d'}]$?

18. Let A be a commutative algebra. An ideal P of A is called a **prime ideal** if $ab \in P$ implies $a \in P$ or $b \in P$. Prove that:
   (a) An ideal in $\mathbb{Z}$ is prime if and only if it is generated by a prime integer.
   (b) An ideal P in A is prime if and only if A/P is an integral domain.
   (c) Any maximal ideal is prime.
   (d) The ideal $\langle t^2 + 1 \rangle$ is prime but not maximal in $\mathbb{Z}[t]$.

19. This exercise outlines another proof of Theorem I.4.9. Let A be an algebra, B a subalgebra and I an ideal.
   (a) Prove that $B + I = \{b + x : b \in B, x \in I\}$ is a subalgebra of A, and in fact the smallest subalgebra containing B and I.
   (b) Construct a surjective algebra morphism $\psi : B+I \rightarrow B/(B \cap I)$ with kernel I.
   (c) Deduce Theorem I.4.9.

20. Let $\varphi : A \rightarrow B$ be a surjective algebra morphism. Prove that:
   (a) If I is an ideal of A containing Ker φ, then $A/I \cong B/\varphi(I)$.
   (b) If J is an ideal of B, then $B/J \cong A/\varphi^{-1}(J)$.

21. Let $\varphi : A \rightarrow B$ be an algebra morphism, $I \trianglelefteq A$ and $J \trianglelefteq B$. Denote by $\pi_I : A \rightarrow A/I, \pi_J : B \rightarrow B/J$ the respective projections. Prove that $\varphi(I) \subseteq J$ if and only if there exists a unique morphism $\bar{\varphi} : A/I \rightarrow B/J$ such that $\bar{\varphi} \circ \pi_I = \pi_J \circ \varphi$. If this is the case, then compute the kernel and the image of $\bar{\varphi}$.

22. Let K be a field of characteristic two, see Exercise I.2.5, and $G = \{1, g, h, gh\}$ the Klein Four-group. Define $\varphi : K[s, t] \rightarrow KG$ by $\varphi(s) = 1 + g, \varphi(t) = 1 + h$.
   (a) Prove that φ is surjective of kernel $\langle s^2, t^2 \rangle$.
   (b) Deduce that $KG \cong K[s, t]/\langle s^2, t^2 \rangle$.
   (c) Let Q be the two-loops quiver

Prove that $KG \cong KQ/\langle \alpha^2, \beta^2, \alpha\beta - \beta\alpha \rangle$.

I.5 Principal ideal domains

I.5.1 Irreducible elements

The first examples of rings one encounters are usually the ring $\mathbb{Z}$ of integers and the ring $K[t]$ of polynomials in t with coefficients in a field K. Both are integral domains which share the additional property that in each, one can perform efficiently a division. They also share another, less evident, property: in both, any ideal of the ring is principal, that is, is generated by a single element. This leads to the following definition.

I.5.1 Definition. An algebra (or a ring) A is a **principal ideal domain** if it is an integral domain in which any ideal is principal, that is, is of the form $\langle a \rangle = Aa$, for some $a \in A$.

In particular, principal ideal domains are commutative. Any field K is a principal ideal domain because it has only two ideals: $0 = \langle 0 \rangle$ and $K = \langle 1 \rangle$. We already observed after Lemma I.3.7 that $\mathbb{Z}$ is a principal ideal domain.

The next examples use the division of polynomials, that we recall here. Let A be an integral domain, then the algebra $A[t]$ of polynomials in t with coefficients in A is an integral domain, see the comments after Definition I.2.4. The division theorem for polynomials asserts that if $f, g \in A[t]$ are such that the leading coefficient of g is invertible (in particular, $g \neq 0$), then there exist unique polynomials q (the **quotient**) and r (the **remainder**) such that $f = gq + r$ and either $r = 0$ or the degree $\deg(r)$ of r is strictly less than the degree $\deg(g)$ of g. In particular, division by a nonzero polynomial is always possible if $A = K$ is a field.

I.5.2 Examples.

(a) Recall that the **Well-Ordering Principle** states that any set can be ordered so that any of its nonempty subsets has a least element. The Well-Ordering Principle is equivalent to the Axiom of Choice (hence to Zorn's lemma).

A polynomial algebra $K[t]$ with coefficients in a *field* K is a principal ideal domain. Let I be a nonzero ideal in $K[t]$. Because $I \neq 0$, the set $\{\deg(f) : f \neq 0, f \in I\}$ is a nonempty set of natural numbers. Because of the Well-Ordering Principle, there exists in I a nonzero polynomial p of lesser degree. Clearly, $\langle p \rangle \subseteq I$. Conversely, let $f \in I$. Because of the division theorem, and because $p \neq 0$, there exist $q, r \in K[t]$ such that $f = pq + r$ and $r = 0$ or $\deg(r) < \deg(p)$. Because $f, p \in I$, we have $r = f - qp \in I$. But p is of lesser degree in I. Hence, $r = 0$ and $f = qp \in \langle p \rangle$. Therefore, $I = \langle p \rangle$, as required.

(b) On the other hand, a polynomial algebra $K[t]$ with coefficients in a *ring* K (even a principal ideal domain) is usually not a principal ideal domain. For instance, $\mathbb{Z}[t]$ is an integral domain, but not a principal ideal domain. Indeed, division by a nonzero polynomial is not always possible in $\mathbb{Z}[t]$, such as, for instance, division by an integer $n > 1$. Consider the ideal $\langle 2, t \rangle$ generated by 2 and t in $\mathbb{Z}[t]$. If it is principal and generated by p, say, then p divides both 2 and t. Hence, $p \in \{-1, +1\}$. But $\langle 1 \rangle = \langle -1 \rangle = \mathbb{Z}[t] \neq \langle 2, t \rangle$, a contradiction. Therefore, $\langle 2, t \rangle$ is not a principal ideal.

The proof in example (a) above rests on the notion of degree of a polynomial. One can generalise this notion by defining a class of algebras, called **Euclidean domains**, in which division is possible, and there is an integer-valued function, called **valuation**, which behaves like the degree. Examples are the ring of Gaussian integers $\mathbb{Z}[i] = \{a + bi : a, b \in \mathbb{Z}\}$, see Exercise I.2.8, the ring $\mathbb{Z}[\sqrt{2}] = \{a + b\sqrt{2} : a, b \in \mathbb{Z}\}$ and others, a proof is outlined in Exercise I.5.8 below.

We need considerations on division in integral domains. Let A be an integral domain and $a, b \in A$. Then b is said to **divide** a, in symbols $b \mid a$, if there exists $q \in A$ such that $a = qb$. The element q is the **quotient** of a by b. If $b \neq 0$, then, because of Lemma I.2.5, q is unique.

One word of caution: when we say that b divides a, it does *not* mean that we can multiply a by the multiplicative inverse of b. Indeed, inverses usually do not exist in integral domains. For instance, one has $0 \mid 0$, which means that 0 is a multiple of 0, because, for any $q \in A$, one has $0 = q \cdot 0$. Actually, for any $a \in A$, we have $1 \mid a$ and $a \mid 0$.

I.5.3 Lemma. *Let A be an integral domain and $a, b \in A$ nonzero. Then:*

(a) *$Aa \subseteq Ab$ if and only if $b \mid a$.*

(b) *$Aa = Ab$ if and only if there exists $u \in A$ invertible such that $a = ub$.*

Proof. (a) $Aa \subseteq Ab$ happens if and only if $a \in Ab$, that is, if and only if there exists $q \in A$ such that $a = qb$. This amounts to saying that $b \mid a$.

(b) Assume that $Aa = Ab$. Because of (a), we have $a \mid b$ and $b \mid a$, so there exist q, q' such that $b = qa$ and $a = q'b$. Then $a = qq'a$ and $a \neq 0$ yield $qq' = 1$, because of Lemma I.2.5. Therefore, q and q' are invertible. Conversely, if u is invertible and such that $a = ub$, then $b \mid a$ implies $Aa \subseteq Ab$. But u is invertible. So $b = u^{-1}a$ and $Ab \subseteq Aa$ as well. □

Consequently, $u \in A$ is invertible if and only if $Au = A$, or, equivalently, if and only if $u \mid 1$. If $a, b \in A$ are such that $Aa = Ab$, then they are called **associate**. Equivalently, a, b are associate if and only if $a \mid b$ and $b \mid a$, or, if and only if there exists $u \in A$ invertible such that $a = ub$.

For example, if K is a field, an element of $K[t]$ is invertible if and only if it is a nonzero element of K. Thus, two polynomials $p, q \in K[t]$ are associate if and only if there exists $c \in K, c \neq 0$ such that $q = cp$. The consequence is that any $p \in K[t]$ is associate to a unique **normalised** polynomial that is, such that its leading coefficient equals +1.

It is well-known that any pair of integers, or of polynomials, has a greatest common divisor and a least common multiple. This property holds true in any principal ideal domain. We define these concepts.

I.5.4 Definition. Let A be a principal ideal domain and $a, b \in A$.

(a) A **greatest common divisor** of a, b, abbreviated to gcd, is an element $d \in A$ such that:

(i) $d \mid a$ and $d \mid b$, and

(ii) if $c \mid a$ and $c \mid b$, then $c \mid d$.

(b) A **least common multiple** of a, b, abbreviated to lcm, is an element $m \in A$ such that:

(i) $a \mid m$ and $b \mid m$, and

(ii) if $a \mid c$ and $b \mid c$, then $m \mid c$.

If it exists, a gcd is unique up to associates. For, if d, d' are both gcd of a, b, then, because d' is a common divisor of a, b, applying the definition of gcd to d yields $d \mid d'$. Similarly, $d' \mid d$. Hence, d and d' are associate. By abuse of notation (because d is not unique), we denote by $d = (a, b)$ a gcd of a, b.

In the same way, we can prove that an lcm m of a, b is unique up to associates, and, by abuse of notation, we write $m = [a, b]$ for an lcm of a, b,

For instance, for any $a \in A$, we have $(a, 0) = a$ and $(u, 1) = 1$. Also, $[a, 0] = 0$ and $[a, 1] = a$. We prove our existence theorem for gcd and lcm.

I.5.5 Theorem. *Any pair of elements in a principal ideal domain has a gcd and an lcm.*

Proof. Let A be a principal ideal domain and $a, b \in A$. The sum and the intersection of ideals are ideals. Hence, there exist $d, m \in A$ such that $Aa + Ab = Ad$ and $Aa \cap Ab = Am$. It is easy to verify that d and m satisfy respectively the definition of a gcd and an lcm of a, b.   □

I.5.6 Corollary. *Let A be a principal ideal domain and $a, b \in A$. Then:*
   (a) *If $(a, b) = d$, then there exist $s, t \in A$ such that $sa + tb = d$.*
   (b) *$(a, b) = 1$ if and only if there exist $s, t \in A$ such that $sa + tb = 1$.*

Proof. (a) Because of the proof of Theorem I.5.5, a gcd d of a, b belongs to the ideal $Aa + Ab$. Hence, it can be written in the form $d = sa + tb$, for some $s, t \in A$.

(b) Necessity follows from (a). Conversely, let s, t be such that $sa + tb = 1$. We certainly have $1 \mid a$ and $1 \mid b$. If $d' \mid a$ and $d' \mid b$, then $d' \mid sa + tb = 1$. So d' is invertible and $(a, b) = 1$.   □

Two elements having 1 as a gcd are called **coprime**. Corollary I.5.6(b) is a criterion allowing to verify whether two elements are coprime or not. It is known as the **Bézout–Bachet theorem**.

We arrived at the main result of this subsection.

I.5.7 Theorem. *Let A be a principal ideal domain and p a nonzero element of A. The following conditions are equivalent:*
   (a) *p is not invertible and $p \mid ab$ implies $p \mid a$ or $p \mid b$.*
   (b) *p is not invertible and $p = ab$ implies that a or b is invertible.*
   (c) *Ap is a maximal ideal in A.*

Proof. (a) implies (b). Assume that $p = ab$. Then $p \mid ab$. Consequently, $p \mid a$ or $p \mid b$. Assume the former, then there exists q such that $a = qp$. Then $p = ab = pqb$ and $p \neq 0$ yield $qb = 1$, because of Lemma I.2.5. Hence, b is invertible. Similarly, if $p \mid b$, then a is invertible.

(b) implies (c). Let $I \trianglelefteq A$ be such that $Ap \subseteq I \subseteq A$. Because A is principal, there exists $a \in A$ such that $I = Aa$. Because $Ap \subseteq Aa$, there exists $b \in A$ such that $p = ab$. The hypothesis (b) says that a or b is invertible. If a is invertible, then, because of Lemma I.5.3, we have $I = Aa = A$. If b is invertible, then, for the same reason, $I = Aa = Ap$.

(c) implies (a). Assume that Ap is maximal. Then $Ap \neq A$ shows that p is not invertible. If $p \mid ab$, then $ab \in Ap$, and so, in the quotient A/Ap, we have $(a + Ap)(b + Ap) = ab + Ap = 0$. Because Ap is maximal, it follows from

Corollary I.4.11 that A/Ap is a field. Hence, $a + Ap = 0$ or $b + Ap = 0$. Thus, $a \in Ap$ or $b \in Ap$. Equivalently, $p \mid a$ or $p \mid b$. □

Elements satisfying the equivalent conditions of the theorem are called **irreducible elements** in the principal ideal domain. Because of Lemma I.5.3, an element $p \in A$ is irreducible if and only if, for any invertible element $u \in A$, the element up is also irreducible, that is, if and only if the associates of p are irreducible. In this case, we call p and up **associate irreducible** elements. It follows from condition (a) of the theorem and induction that, if p is irreducible and $p \mid a_1 \cdots a_m$, for any integer $m > 1$, and the a_i elements of A, then there exists i such that $p \mid a_i$.

It is reasonable to ask whether an integral domain has 'enough' irreducible elements. Because of Theorem I.5.7, this is equivalent to the existence of 'enough' maximal ideals. We start by proving that maximal ideals exist in any nonzero algebra. This is known as **Krull's theorem**.

I.5.8 Theorem. Krull's Theorem. *Let I_0 be a proper ideal in a nonzero algebra A. Then there exists a maximal ideal of A containing I_0. In particular, any nonzero algebra has maximal ideals.*

Proof. Let $\mathcal{E}$ be the set of proper ideals of A containing I_0. Because $I_0 \in \mathcal{E}$, we have $\mathcal{E} \neq \emptyset$. The set $\mathcal{E}$ is partially ordered by inclusion. We show that it is inductive. Let $\mathcal{F}$ be a chain in $\mathcal{E}$ and set $I = \cup_{J \in \mathcal{F}} J$. Then I is an ideal of A, see Exercise I.3.8, and contains I_0. Assume that $I = A$, then $A = \cup_{J \in \mathcal{F}} J$ gives $1 \in J$, for some $J \in \mathcal{F}$, and hence J equals A, which contradicts $J \in \mathcal{F} \subseteq \mathcal{E}$. Hence, I is a proper ideal and $I \in \mathcal{E}$. So, $\mathcal{E}$ is an inductive poset. Applying Zorn's lemma finishes the proof. The last statement follows by taking I_0 equal to the zero ideal. □

We explain what we meant by saying that a principal ideal domain has 'enough' irreducible elements.

I.5.9 Corollary. *Let A be a principal ideal domain. For any $a \in A$ which is neither zero nor invertible, there exists an irreducible element p such that $p \mid a$.*

Proof. Because a is not invertible, Aa is a proper ideal of A. Because of Krull's Theorem I.5.8, there exists a maximal ideal I containing Aa. But A is principal. Hence, there exists $p \in A$ such that $I = Ap$. Because I is maximal, p is irreducible, due to Theorem I.5.7. Because $Aa \subseteq Ap$, we have $p \mid a$. □

For instance, the irreducible elements in $\mathbb{Z}$ are the prime integers, as follows from Example I.2.8(b) and Corollary I.4.11. In $K[t]$, with K a field, we have seen at the end of Section I.4 that any polynomial of the form $t - \lambda$, for some λ, is irreducible. In general, polynomials of this form are not the only irreducible ones. For instance, Exercise I.4.18(c) says that $t^2 + 1$ is irreducible in $\mathbb{R}[t]$. The Fundamental Theorem of Algebra says that the only irreducible polynomials in $\mathbb{C}[t]$ are of the form $t - \lambda$, for some $\lambda \in \mathbb{C}$, and their associates, that is, their multiples by nonzero complex numbers.

I.5.10 Example. We apply Example I.5.2(a) above to algebraically closed fields. A commutative ring K is called ***algebraically closed*** if, for any nonconstant polynomial $p \in K[t]$, there exists $\alpha \in K$ such that $p(\alpha) = 0$. Then, α is called a ***root*** of p. Algebraically closed rings are actually fields: indeed, for any nonzero $\alpha \in K$, the polynomial $p = \alpha t - 1$ has a root, hence α has an inverse. The Fundamental Theorem of Algebra asserts that the field $\mathbb{C}$ of complex numbers is algebraically closed. One can prove that any field has an algebraically closed field extension.

Let D be a division ring extension of a field K, see Definition I.2.9, which is finite-dimensional as K-vector space. Then, for any $a \in D$, the infinite sequence of elements $\{1, a, \cdots, a^i, \cdots\}$ cannot be linearly independent over K. Hence, there exist $d > 0$ and $\alpha_0, \cdots, \alpha_d \in K$ such that $\alpha_0 + \alpha_1 a + \cdots + \alpha_d a^d = 0$, that is, the polynomial $p = \alpha_0 + \alpha_1 t + \cdots + \alpha_d t^d$ has a as a root. Therefore, the set of polynomials in $K[t]$ having a as a root is nonzero. Because this set is easily seen to be an ideal, there exists a unique normalised polynomial $m_a \in K[t]$ which generates it. Because m_a is of lesser degree, due to Example I.5.2(a), it is irreducible. It is called the ***minimal polynomial*** of a.

If K is algebraically closed, then no proper division ring extension D is finite-dimensional as K-vector space. For, let $a \in D$ and m_a its minimal polynomial. Because K is algebraically closed, the irreducible polynomial m_a is of degree one, hence of the form $m_a = t - a$. Therefore, $a \in K$. Thus, $D = K$.

I.5.2 The Unique Factorisation Theorem

The reader is certainly familiar with the so-called Fundamental Theorem of Arithmetics which asserts that any positive integer distinct from 1 decomposes uniquely as product of positive prime (=irreducible) integers, and this decomposition is unique up to order of the factors. This statement easily extends to any nonzero integer distinct from ± 1. A similar one holds true for polynomials over a field. Integers and polynomials are our main examples of principal ideal domains. We now prove that this unique factorisation property holds true in any principal ideal domain. We need a lemma (which will take its real significance in Chapter VII).

I.5.11 Lemma. *Let $(I_i)_{i \geq 0}$ be a sequence of ideals in a principal ideal domain A such that, for any $i \geq 0$, we have $I_i \subseteq I_{i+1}$. Then there exists $j \geq 0$ such that $I_j = I_{j+1} = I_{j+2} = \cdots$.*

Proof. The set $I = \cup_{i \geq 0} I_i$ is an ideal in A, see Exercise I.3.8. Because A is a principal ideal domain, there exists $a \in A$ such that $I = Aa$. But then there exists $j \geq 0$ such that $a \in I_j$. Hence, $I = Aa \subseteq I_j$, so that $I = I_j$. Therefore, $I_j = I_{j+k}$, for any $k \geq 0$. □

We prove the Unique Factorisation Theorem.

I.5.12 Theorem. *Let A be a principal ideal domain and $a \in A$ nonzero and noninvertible. Then:*

(a) *a can be written in the form $a = p_1 \cdots p_m$, where each p_i is irreducible, and*
(b) *this factorisation is essentially unique: if $p_1 \cdots p_m = q_1 \cdots q_n$, with the p_i, q_j irreducible, then $m = n$ and there exists a permutation σ of $\{1, \cdots, m\}$ such that, for each i, p_i and $q_{\sigma(i)}$ are associate.*

Proof. (a) *Existence.* Because of Corollary I.5.9, there exists p_1 irreducible such that $p_1 \mid a$. Let a_1 be such that $a = p_1 a_1$. Then $a_1 \neq 0$. Inductively, we may assume that we have a factorisation $a = p_1 \cdots p_k a_k$, where all p_i are irreducible and $a_k \neq 0$. If a_k is invertible or irreducible, then we are done. Otherwise, because of Corollary I.5.9, there exist p_{k+1} irreducible and $a_{k+1} \in A$ such that $a_k = p_{k+1} a_{k+1}$. Because each p_i is irreducible, we have strict inclusions:

$$Aa \subsetneqq Aa_1 \subsetneqq \cdots \subsetneqq Aa_k \subsetneqq Aa_{k+1}.$$

Because of Lemma I.5.11, this chain stabilises, that is, there exists k such that a_k is invertible or irreducible. This completes the proof of existence.

(b) *Uniqueness.* Assume that $p_1 \cdots p_m = q_1 \cdots q_n$, with the p_i, q_j irreducible. Then $p_1 \mid q_1 \cdots q_n$. As seen after Theorem I.5.7, there exists q_j such that $p_1 \mid q_j$. Changing the order if necessary, we may assume that $j = 1$ so $p_1 \mid q_1$. But q_1 is irreducible. Hence, there exists u_1 invertible such that $q_1 = u_1 p_1$, that is, p_1 and q_1 are associates. Simplifying, we get $p_2 \cdots p_m = u_1 q_2 \cdots q_n$. We finish by induction. $\qquad\square$

For instance, if K is a field, any nonconstant polynomial in $K[t]$ decomposes in the form $p = cp_1 \cdots p_m$, where $c \in K, c \neq 0$, and the p_i are irreducible polynomials, unique if normalised.

The number m of irreducible factors in the unique factorisation of a is sometimes called its **length**. It follows from Theorem I.5.12 that the length of a is unambiguously defined. Several proofs on principal ideal domains are done by induction on length of their nonzero, noninvertible elements.

Exercises of Section I.5

1. Prove that each of the following is a subalgebra of $\mathbb{Q}$, then that it is a principal ideal domain.
    (a) The fractions $\frac{a}{b}$, with b an odd integer.
    (b) Given a prime integer p, the fractions $\frac{a}{p^n}$, with $a, n \in \mathbb{Z}$.

2. Let K be a field. Prove that the algebra $K[[t]]$ of formal power series in t, see Example I.3.11(e), is a principal ideal domain, of which any ideal is of the form $\langle t^i \rangle$, for some $i \geq 0$. (Hint: use the order of an element, see Exercise I.3.15).

3. Let a, b be nonzero elements in a principal ideal domain A. Let d be a gcd and m an lcm of a, b. Prove that:
    (a) If $a = da', b = db'$, then a', b' are coprime.
    (b) There exists $u \in A$ invertible such that $ab = udm$.

4. Let A be a principal ideal domain and $a, b, c \in A$. Prove that:
   (a) $(a, (b, c)) = ((a, b), c)$.
   (b) $(ca, cb) = c(a, b)$.
   (c) $(a + bc, b) = (a, b)$.
   (d) $(a, b) = d, a \mid c, b \mid c$ imply $ab \mid cd$.
   (e) $(a, b) = 1$ if and only if $Aa + Ab = A$, or if and only if $Aa \cap Ab = A(ab)$.
   (f) $(a, b) = 1$ if and only if, for any $c \in A, a \mid bc$ implies $a \mid c$.
   (g) If $(a, b) = 1$, then $(a + b, ab) = 1$.
   (h) If $(a, b) = 1$ and $(a, c) = 1$, then $(a, bc) = 1$.
   (i) If $(a, b) = 1$ and $a \mid bc$, then $a \mid c$.
   (j) If $(a, b) = 1, a \mid c$ and $b \mid c$, then $ab \mid c$.
   (k) If $(a, b) \neq 1$, then there exists $p \in A$ irreducible such that $p \mid a$ and $p \mid b$.
   (l) If m, n are coprime integers such that $a^m = b^m$ and $a^n = b^n$, then $a = b$.

5. Let A be a principal ideal domain and $a_1, \cdots, a_n \in A$.
   (a) Define what are meant by the gcd and the lcm of the set $\{a_1, \cdots, a_n\}$. Prove that each set containing $n \geq 2$ elements in A has a gcd $(a_1, \cdots, a_n)$ and an lcm $[a_1, \cdots, a_n]$ in A. Moreover, the gcd may be written as $s_1 a_1 + \cdots + s_n a_n$, for some $s_1, \cdots, s_n \in A$.
   (b) Prove that $[a_1, \cdots, a_n] = a_1 \cdots a_n$ if and only if $(a_i, a_j) = 1$, for $i \neq j$.
   (c) If $a_1, \cdots, a_n$ and $b_1, \cdots, b_n$ are elements of A such that $a_1 b_1 = \cdots = a_n b_n = c$, prove that $(a_1, \cdots, a_n)[b_1, \cdots, b_n] = c$.

6. Prove that:
   (a) A normalised polynomial in $\mathbb{C}[t]$ is irreducible if and only if it is of the form $t - \lambda$, for some $\lambda \in \mathbb{C}$.
   (b) If a normalised polynomial in $\mathbb{R}[t]$ is irreducible, then its degree is at most two, and if $p(t) = t^2 + bt + c$, then it is irreducible if and only if $b^2 - 4c < 0$.
   (c) There exist irreducible polynomials in $\mathbb{Q}[t]$ of any positive degree.

7. Let p be a polynomial in $\mathbb{R}[t]$ having a complex root α. Prove that the complex conjugate $\bar{\alpha}$ of α is also a root of p.

8. We claim that the ring $\mathbb{Z}[i] = \{a + bi : a, b \in \mathbb{Z}\}$ of Gaussian integers is a principal ideal domain.
   (a) Prove that the function $\varphi : z \mapsto z\bar{z} = |z|^2$ from $\mathbb{Z}[i]$ to the natural numbers satisfies the following properties:
      (i) $\varphi(z_1 z_2) = \varphi(z_1)\varphi(z_2)$, for any $z_1, z_2 \in \mathbb{Z}[i]$.
      (ii) $\varphi(z_1) \leqslant \varphi(z_1 z_2)$ for any $z_1, z_2 \neq 0$ in $\mathbb{Z}[i]$.
      (iii) $\varphi(z) = 1$ if and only if z is invertible in $\mathbb{Z}[i]$.
      (iv) If $\varphi(z)$ is a prime integer, then, for any decomposition $z = z'z''$, with $z', z'' \in \mathbb{Z}[i]$, we have z' or z'' invertible (that is, z is irreducible).
   (b) Let $z, z' \in \mathbb{Z}[i]$ with $z' \neq 0$. Then $\frac{z}{z'} = x + yi$, with $x, y \in \mathbb{Q}$. Let $a, b \in \mathbb{Z}$ be such that $|x - a| \leqslant \frac{1}{2}$ and $|y - b| \leqslant \frac{1}{2}$ and set $q = a + bi$. Prove that $\varphi(\frac{z}{z'} - q) < 1$ and deduce that $z = qz' + r$, with $r = 0$ or $\varphi(r) < \varphi(z')$.
   (c) Show that the quotient q and the remainder r obtained in (b) are not necessarily unique.
   (d) Prove that $\mathbb{Z}[i]$ is a principal ideal domain.

9. Prove that $\mathbb{Z}[\sqrt{2}]$ is a principal ideal domain (Hint: use the technique of the previous exercise).

10. Prove that two irreducible elements p, q in a principal ideal domain are associate if and only if $p \mid q$.

11. Prove that the ideal $< 2, t >$ is maximal in $\mathbb{Z}[t]$.

12. Prove that an ideal in a principal ideal domain is maximal if and only if it is prime, see Exercise I.4.18.

13. Let A be a principal ideal domain. Denote by $\ell(a)$ the length of a nonzero, non-invertible element $a \in A$. Assume that $a, b \in A$ are nonzero, noninvertible and such that $a \mid b$. Prove that:
    (a) $\ell(a) \leqslant \ell(b)$.
    (b) $\ell(a) = \ell(b)$ if and only if a, b are associate.

14. Prove that any algebraically closed field is infinite.

Chapter II
Modules

II.1 Introduction

As seen before, a module is an abelian group whose elements can be multiplied in a reasonable manner by elements of a ring. Modules appear among other places in the representation theory of finite groups: let K be a field and G a finite group, a K-linear representation of G is a group homomorphism from G to the automorphism group of some K-vector space (so elements of G are 'represented' by invertible matrices). One can prove that the classification of K-linear representations of G is equivalent to the classification of modules over the group algebra KG, see Example I.3.11(g). Moreover, it turns out that working with modules is easier and more elegant than working with representations. One defines similarly representations of algebraic structures other than groups, and their study always reduces to that of modules over some algebra. In the present chapter, we study elementary properties of modules and appropriate maps between them. Throughout, K is a commutative ring and our algebras are associative and unitary K-algebras.

II.2 Modules and submodules

II.2.1 Basic definitions and examples

As seen in Definition I.2.12, modules take their scalars inside rings. Because rings and algebras are equivalent concepts, one can talk about modules over an algebra. The definition is the same.

II.2.1 Definition. Let A be a K-algebra, a **right A-module** M is an additive abelian group equipped with an external multiplication, or right A-action, $M \times A \to M, (x, a) \mapsto xa$, for $x \in M, a \in A$, such that, for any $x, y \in M$ and $a, b \in A$, we have:
   (B1) $x(ab) = (xa)b$,
   (B2) $x1 = x$,
   (C1) $x(a + b) = xa + xb$,
   (C2) $(x + y)a = xa + ya$.

The 1 which appears in (B2) is the multiplicative identity of A. By analogy with vector spaces, elements of M are sometimes called **vectors** and those of A **scalars**.

One defines **left A-modules** in the same way. We write M_A or $_A M$ to indicate whether M is a right, or left, A-module, respectively. As we have seen, left A-modules can be

An Introduction to Module Theory. Ibrahim Assem and Flávio U. Coelho, Oxford University Press.
© Ibrahim Assem and Flávio U. Coelho (2024). DOI: 10.1093/9780198904939.003.0003

considered as right modules over the opposite algebra A^{op}. This remark allows to work only with right modules, except in cases where a right and a left module structures appear simultaneously.

Any A-module M has a natural K-module structure. Indeed, let $x \in M$ and $\alpha \in K$, then set

$$x\alpha = x(1 \cdot \alpha).$$

It is easily verified that M becomes a right K-module under this operation. Because K is commutative, M is also a left K-module for the same operation (so, we may write $x\alpha = \alpha x$). Thus, if K is a field, an A-module is also a K-vector space.

The arithmetic properties of modules stated in Proposition I.2.14 obviously hold true, but there is an additional one: if M is an A-module, then its multiplication by A is compatible with its multiplication by K. Given $x \in M, \alpha \in K$ and $a \in A$, we have

$$x(\alpha a) = (x\alpha)a = (\alpha x)a = \alpha(xa)$$

because of axiom (B1) in Definition II.2.1 and commutativity of K.

We repeat the definition of submodule.

II.2.2 Definition. Let A be a K-algebra and M an A-module. A subset L of M is a **submodule** if it is itself an A-module for the same operations as M.

Any module M has at least two submodules, namely the zero submodule 0, which is the one-element set $\{0\}$, and M itself. We restate the criterion of Proposition I.2.16.

II.2.3 Proposition. *Let A be an algebra, M an A-module and L a nonempty subset of M. The following conditions are equivalent:*
  (a) *L is a submodule of M.*
  (b) *L is an additive subgroup of M and $x \in L, a \in A$ imply $xa \in L$.*
  (c) *For any $x, y \in L$ and $a, b \in A$, we have $xa + yb \in L$.*

Proof. Similar to that of Proposition I.2.16 and left to the reader. □

We draw the reader's attention to nonemptiness of L: the empty set satisfies (trivially) condition (c) of the previous proposition, but is not a submodule, because it does not contain the zero element.

Examples of modules and submodules are found in I.2.13 and I.2.17. We give additional ones.

II.2.4 Examples.
  (a) Let A be an algebra. The product $(x, a) \mapsto xa$ of $x, a \in A$ endows A itself with a right A-module structure. This module is denoted by A_A. A submodule I of A_A is, according to Proposition II.2.3(b), an additive subgroup of A such that $x \in I, a \in A$ imply $xa \in I$. Such a submodule I is called a **right ideal**

of A, because its multiplication by elements of A is right stable (a property sometimes expressed by writing $IA \subseteq I$).

Similarly, A has a left A-module structure, given by the product $(a,x) \mapsto ax$ of $a,x \in A$ and denoted by $_AA$. A submodule I of $_AA$ is an additive subgroup of A such that $x \in I, a \in A$ imply $ax \in I$ (which can be written as $AI \subseteq I$). Then, I is called a **left ideal** of A.

It follows from these definitions that a subset I of A is an ideal if and only if it is at the same time a right and a left ideal of A. In this case, it is a submodule of both A_A and $_AA$. If A is commutative, then the three notions of ideal, right ideal and left ideal coincide, but this is not the case in non-commutative algebras. For instance, let $A = M_2(K)$ be the 2×2-full matrix algebra, then the set

$$I = \begin{pmatrix} 0 & 0 \\ K & K \end{pmatrix} = \left\{ \begin{pmatrix} 0 & 0 \\ \alpha & \beta \end{pmatrix} : \alpha, \beta \in K \right\}$$

is a right ideal, but not a left ideal of A because A has no nonzero proper ideals, see Example I.3.11(b). Similarly, $I' = \begin{pmatrix} K & 0 \\ K & 0 \end{pmatrix}$ is a left, but not a right ideal of A.

(b) Let A be an algebra and B a subalgebra. Then A admits a right B-module structure defined by $(a, b) \mapsto ab$, for $a \in A, b \in B$, the product being computed in A. For instance, the polynomial algebra $A[t]$ can be considered as a right, and also as a left, A-module.

(c) Let $M_1, \ldots, M_n$ be A-modules, their cartesian product

$$\prod_{i=1}^{n} M_i = M_1 \times \cdots \times M_n = \{(x_1, \cdots, x_n) : x_i \in M_i\}$$

admits an A-module structure, with operations defined componentwise:

$$(x_1, \cdots, x_n) + (y_1, \cdots, y_n) = (x_1 + y_1, \cdots, x_n + y_n)$$

$$(x_1, \cdots, x_n) \cdot a = (x_1 a, \cdots, x_n a)$$

for $x_i, y_i \in M_i, a \in A$. The module $\prod_{i=1}^{n} M_i$ is called the **product module** of the M_i.

(d) Let A be an algebra and n a positive integer. Because of Example (c) above, the product $A^n = \{(x_1, \cdots, x_n) : x_i \in A\}$ admits a right, and also a left, A-module structure as product module. But it has another right module structure, this time over the full matrix algebra $M_n(A)$: we give A^n the same additive structure as the product module, but define multiplication of $x = (x_1, \cdots, x_n) \in A^n$ by $\mathbf{a} = [a_{ij}] \in M_n(A)$ as

$$\mathbf{x} \cdot \mathbf{a} = \left(\sum_{i=1}^{n} x_i a_{i1}, \cdots, \sum_{i=1}^{n} x_i a_{in} \right).$$

That is, the product of the row vector **x** by the matrix **a** is their product as matrices. This is the structure of Exercise I.2.10. Similarly, one defines a left $M_n(A)$-module structure on the set of column vectors with n coordinates in A.

(e) Let A, B be algebras and $\varphi : A \to B$ an algebra morphism. A B-module M_B admits an A-module structure defined by

$$xa = x\varphi(a)$$

for $x \in M, a \in A$. This A-module structure is said to be induced by **change of scalars** (because we pass from scalars in B to scalars in A). There are two important particular cases. The first is when A is a subalgebra of B and φ is the inclusion, this is the situation of Example (b) above. The second is when $B = A/I$, for some ideal I of A, and $\varphi : A \to A/I$ is the projection: then, the right A-action on an A/I-module M is given by $xa = x(a + I)$, for $x \in M, a \in A$.

(f) Let K be a field, E a K-vector space and $f : E \to E$ a K-linear map. Define inductively the powers of f as follows: set $f^0 = 1_E$, the identity map, and, for $i \geq 1$, set $f^i = f \circ f^{i-1}$. This data induces a left $K[t]$-module structure on E as follows: addition is the same as in the vector space but multiplication of the polynomial $p = \alpha_0 + \alpha_1 t + \cdots + \alpha_d t^d$ by the vector $v \in E$ is given by

$$p \cdot v = (\alpha_0 1_E + \alpha_1 f + \cdots + \alpha_d f^d)(v) = \alpha_0 v + \alpha_1 f(v) + \cdots + \alpha_d f^d(v).$$

This module is denoted by $_fE$ in order to indicate dependence on f. Because $K[t]$ is commutative, E is also a right $K[t]$-module for the same multiplication. However, maps act on the left of elements, so it is more convenient to look at $_fE$ as a left module. This module structure will be much used in Chapter V.

We claim that a subset F of E is a $K[t]$-submodule of $_fE$ if and only if it is a vector subspace which is moreover **f-invariant**, that is, such that $f(F) \subseteq F$.

Indeed, suppose that F is an f-invariant subspace of E. Because addition is the same in the module as in the vector space, we need only verify stability of multiplication by an element of $K[t]$. Let $v \in F$. Then, f-invariance of F gives $f(v) \in F$ and induction yields $f^i(v) \in F$, for any integer $i > 0$. Because F is stable under K-linear combinations, then, for any $p \in K[t], v \in F$, we have $p \cdot v \in F$. Therefore, F is a $K[t]$-submodule of $_fE$. Conversely, if F is a $K[t]$-submodule, it is an additive subgroup of E and, moreover, for $v \in F, \alpha \in K$, we have $\alpha v \in F$, so that F is a vector subspace of E. Also, for $v \in F$, we have $f(v) = t \cdot v \in F$, and so F is f-invariant. This establishes our claim.

For instance, given $\lambda \in K$, the set $E_\lambda = \{v \in E : f(v) = \lambda v\}$ is an f-invariant subspace of E. It is nonzero if and only if λ is a proper value of f.

(g) Let K be a field and Q a finite quiver. A K-linear **representation** V of Q is the following data:

(i) to each point $x \in Q_0$ is attached a K vector space $V(x)$, and

(ii) to each arrow $\alpha \in Q_1$, say $x \xrightarrow{\alpha} y$, is attached a K-linear map $V(\alpha) : V(x) \to V(y)$.

For example, if Q is the quiver

$$3 \xrightarrow{\alpha} 2 \xrightarrow{\beta} 1$$

a representation consists of a triple of vector spaces $V(1), V(2), V(3)$ and a pair of K-linear maps $V(\alpha) : V(3) \to V(2), V(\beta) : V(2) \to V(1)$. It may be depicted visually as

$$V(3) \xrightarrow{V(\alpha)} V(2) \xrightarrow{V(\beta)} V(1).$$

If V is a representation, then a path $\gamma = \alpha_1 \cdots \alpha_\ell$ in Q from x to y of length $\ell \geq 1$ induces a unique K-linear map from $V(x)$ to $V(y)$, namely the composite $V(\gamma) = V(\alpha_\ell) \cdots V(\alpha_1)$, because maps compose on the opposite direction to arrows. If $\gamma \in KQ$ is of the form $\gamma = \sum_i \lambda_i \gamma_i$, where each λ_i is a scalar, and γ_i is a path from x to y, say, then it induces likewise the linear map $V(\gamma) = \sum_i \lambda_i V(\gamma_i)$ from $V(x)$ to $V(y)$. On the other hand, any element in KQ may be assumed to be of this form, that is, is a linear combination of paths having all the same source and the same target, because, for any $w \in KQ$, we have $w = \sum_{x,y \in Q_0} \varepsilon_x w \varepsilon_y$, and each term of the sum $\varepsilon_x w \varepsilon_y$ is a linear combination of paths from x to y.

We claim that a representation V of Q defines a module over the path algebra KQ, see Example I.3.11(h). Consider the K-vector space $M = \bigoplus_{x \in Q_0} V(x)$. We define a right KQ-action on M, thus making M a right KQ-module. Because elements of KQ are K-linear combinations of paths, it suffices to define multiplication by paths, then extend by distributivity. Let $v = (v_z)_{z \in Q_0} \in M$ and γ be a path. If γ is of length zero, then it is a stationary path ε_x, for some $x \in Q_0$, and we set $v \cdot \varepsilon_x = v_x$ (considered as element of M). If γ is of positive length from x to y, say, then we set $(v \cdot \gamma)_z = \delta_{yz} V(\gamma)(v_x)$, for any $z \in Q_0$, where δ_{yz} is the Kronecker delta and $V(\gamma)$ the linear map associated to the path γ as defined above. In other words, $v\gamma$ is the element of $M = \bigoplus_{x \in Q_0} V(x)$ whose only nonzero coordinate is $(v\gamma)_y = V(\gamma)(v_x)$.

For example, consider the representation

$$Ke_4 \xrightarrow{V(\alpha)} Ke_2 \oplus Ke_3 \xrightarrow{V(\beta)} Ke_1$$

of the previous quiver $3 \xrightarrow{\alpha} 2 \xrightarrow{\beta} 1$, where $e_1, \cdots, e_4$ are basis vectors, $V(\alpha) : \lambda e_4 \mapsto \lambda e_2$ and $V(\beta) : \lambda e_2 + \mu e_3 \mapsto \mu e_1$. In this case, $M = \bigoplus_{i=1}^4 Ke_i = \{\lambda_1 e_1 + \cdots + \lambda_4 e_4 : \lambda_i \in K\}$ and the right KQ-action is defined by

$$(\lambda_1 e_1 + \cdots + \lambda_4 e_4) \cdot \alpha = V(\alpha)(\lambda_1 e_1 + \cdots + \lambda_4 e_4) = \lambda_4 e_2$$

$$(\lambda_1 e_1 + \cdots + \lambda_4 e_4) \cdot \beta = V(\beta)(\lambda_1 e_1 + \cdots + \lambda_4 e_4) = \lambda_3 e_1.$$

Moreover $V(\alpha\beta) = V(\beta)V(\alpha) = 0$, so we have $v \cdot \alpha\beta = 0$, for any $v \in M$. Because any element of KQ is a linear combination of paths in Q, that is, of $\{\varepsilon_1, \varepsilon_2, \varepsilon_3, \alpha, \beta, \alpha\beta\}$, this defines the KQ-module structure of M. For instance, if $\gamma = \varepsilon_1 - \varepsilon_3 + \alpha + \beta - \alpha\beta \in KQ$ and $v = \lambda_1 e_1 + \cdots + \lambda_4 e_4 \in M$ as above, then

$$x.\gamma = V(\gamma)(x) = (V(\varepsilon_1) - V(\varepsilon_3) + V(\alpha) + V(\beta) - V(\alpha\beta))(x)$$

$$= \lambda_1 e_1 - \lambda_4 e_4 + \lambda_4 e_2 + \lambda_3 e_1 = (\lambda_1 + \lambda_3)e_1 + \lambda_4 e_2 - \lambda_4 e_4.$$

We define subrepresentations. Let $V = (V(x), V(\alpha))_{x \in Q_0, \alpha \in Q_1}$ be a representation of Q. A **subrepresentation** $V' = (V'(x), V'(\alpha))_{x \in Q_0, \alpha \in Q_1}$ is a representation of Q such that:

(i) for any $x \in Q_0$, the space $V'(x)$ is a subspace of $V(x)$, and

(ii) for any arrow $x \xrightarrow{\alpha} y$ in Q_1, the map $V'(\alpha)$ is the restriction of $V(\alpha)$ to $V'(x)$.

Because $V'(\alpha)$ maps into $V'(y)$, the second condition says that, for any arrow $\alpha : x \to y$, we have a commutative diagram

$$
\begin{array}{ccc}
V'(x) & \xrightarrow{\;\;V'(\alpha)\;\;} & V'(y) \\
\big\uparrow & & \big\uparrow \\
V(x) & \xrightarrow{\;\;V(\alpha)\;\;} & V(y)
\end{array}
$$

where the vertical maps are inclusions of the respective vector spaces.

Subrepresentations yield submodules: indeed, $L = \bigoplus_{x \in Q_0} V'(x)$ is a vector subspace of $M = \bigoplus_{x \in Q_0} V(x)$ and it is stable under right multiplication by paths, because $V'(\alpha)$ is the restriction of $V(\alpha)$, for any $\alpha \in Q_1$.

In the example above, we have a subrepresentation

$$
\begin{array}{ccccc}
0 & \xrightarrow{\;\;V'(\alpha)\;\;} & 0 \oplus Ke_3 & \xrightarrow{\;\;V'(\beta)\;\;} & Ke_1 \\
\big\uparrow & & \big\uparrow & & \big\uparrow \\
Ke_4 & \xrightarrow{\;\;V(\alpha)\;\;} & Ke_2 \oplus Ke_3 & \xrightarrow{\;\;V(\beta)\;\;} & Ke_1
\end{array}
$$

where $V'(\alpha)$ is (has to be) the zero map, while $V'(\beta)$ maps $\lambda_3 e_3 \in 0 \oplus Ke_3$ to $\lambda_3 e_1 \in Ke_1$: it is indeed the restriction of $V(\beta)$.

As seen in Example II.2.4(d), the abelian group A^n can be equipped with two distinct module structures: as an A-module (either on the right or on the left), and also as a right $M_n(A)$-module. This situation occurs frequently and calls for a definition.

II.2.5 Definition. Let A, B be K-algebras. An abelian group M which is at the same time a left A-module and a right B-module is an A-B-**bimodule** if these two structures are compatible, that is, if $a \in A, x \in M$ and $b \in B$, then $a(xb) = (ax)b$.

The notation for an A-B-bimodule M is ${}_A M_B$. This definition includes a subtle point: because an A-B-bimodule M is also a K-module, and K is commutative, we should have $\lambda \cdot x = x \cdot \lambda$ for $x \in M, \lambda \in K$. But, in this context, left multiplication by λ is actually left multiplication by $1_A.\lambda \in A$, while right multiplication is right multiplication by $1_B.\lambda \in B$. In order to consider M as a K-module, we should therefore have that the left action and the right action of K on elements of M are the same. This is expressed by saying that K operates centrally on M. Of course, if A and B are only viewed as rings (thus as $\mathbb{Z}$-algebras), then this condition is automatically satisfied.

Considering, in Example II.2.4(d), A^n as left A-module and right $M_n(A)$-module, associativity of matrix multiplication implies that A^n is an A-$M_n(A)$-bimodule. We have proved, see the remarks just before Definition II.2.2, that any right A-module is a K-A-bimodule. Also, associativity of multiplication in an algebra A shows that ideals of A are A-A-bimodules. This is notably the case for A itself. If B is a subalgebra of A, then A, or any ideal of A, can be considered as B-A-bimodule, A-B-bimodule or B-B-bimodule. A typical example is the set $\mathbb{C}$ of complex numbers, which admits two vector space structures, over itself and over the real numbers $\mathbb{R}$. Thus we have bimodules $_\mathbb{C}\mathbb{C}_\mathbb{C}$, $_\mathbb{R}\mathbb{C}_\mathbb{C}$, $_\mathbb{C}\mathbb{C}_\mathbb{R}$ and $_\mathbb{R}\mathbb{C}_\mathbb{R}$.

II.2.2 Linear combinations and quotients

Linear combinations in modules are defined as in vector spaces. Let A be an algebra, M an A-module and $X \subseteq M$ a subset. A **linear combination** of elements of X is a sum of the form $\sum_\lambda x_\lambda a_\lambda$ with $x_\lambda \in X, a_\lambda \in A$ and the a_λ almost all zero. For instance, it follows from Proposition II.2.3 and induction that a nonempty subset of a module is a submodule if and only if it is stable under linear combinations.

II.2.6 Lemma. *Let X be a nonempty subset of an A-module M, then the set XA of linear combinations of elements of X is a submodule of M.*

Proof. This is a straightforward application of Proposition II.2.3. □

As in linear algebra, XA is the smallest submodule of M containing X. In order to prove it, we need a lemma.

II.2.7 Lemma. *Let $(M_\lambda)_\lambda$ be a family of submodules of M. Then so is their intersection $\cap_\lambda M_\lambda$.*

Proof. This is another easy application of Proposition II.2.3. □

If X is a subset of a module M, then the family of all submodules of M containing X is nonempty, because it contains M itself. Because of Lemma II.2.7, its intersection $\langle X \rangle$ is a submodule containing X. Due to its construction, $\langle X \rangle$ is the smallest submodule of M containing X. For this reason, $\langle X \rangle$ is said to be **generated** by X, and the elements of X are its **generators**. For instance, if $X = \emptyset$, then $\langle X \rangle = 0$.

II.2.8 Corollary. *Let X be a nonempty subset of M, then $\langle X \rangle = XA$.*

Proof. Due to Lemma II.2.6, XA is a submodule containing X, hence $\langle X \rangle \subseteq XA$. Conversely, linear combinations of elements of X belong to $\langle X \rangle$, because of Proposition II.2.3. So $XA \subseteq \langle X \rangle$. □

If $X = \{x\}$ is a one-element set, then $XA = xA = \{xa : a \in A\}$. A (sub)module of the form xA, for some x, is called a **cyclic (sub)module** generated by x. For instance, the right ideal $I = \begin{pmatrix} 0 & 0 \\ K & K \end{pmatrix}$ of $A = M_2(K)$ of Example II.2.4(a) is cyclic and generated by the matrix $\begin{pmatrix} 0 & 0 \\ 1 & 0 \end{pmatrix}$, because, for any matrix $\begin{pmatrix} \alpha & \beta \\ \gamma & \delta \end{pmatrix} \in A$, one has $\begin{pmatrix} 0 & 0 \\ 1 & 0 \end{pmatrix} \begin{pmatrix} \alpha & \beta \\ \gamma & \delta \end{pmatrix} = \begin{pmatrix} 0 & 0 \\ \alpha & \beta \end{pmatrix}$. Of course, generators of cyclic modules need not be unique. In the former example, the reader can easily verify that the matrix $\begin{pmatrix} 0 & 0 \\ 0 & 1 \end{pmatrix}$ is another generator of I.

There is a fruitful generalisation of the notion of cyclic module: an A-module M is called a **finitely generated module** if it has a finite set of generators, that is, if there exists a finite set X such that $M = \langle X \rangle$. In particular, cyclic modules are finitely generated. If K is a field, then a K-vector space is finite-dimensional if and only if it is finitely generated as K-module.

Any finite abelian group is clearly finitely generated. The infinite group $\mathbb{Z}$ is finitely generated, because it is cyclic generated by 1 (or –1). On the other hand, the group $\mathbb{Q}$ of rationals is not finitely generated. For, if it were, there would exist a finite set of rationals $\{a_1/b_1, \ldots, a_n/b_n\}$ such that any rational is a linear combination of the a_i/b_i with *integral* coefficients. But the denominators b_i contain only finitely many prime factors, while the set of prime integers is infinite. Let p be a prime which is not a factor of the b_i. Then $1/p$ is not a linear combination of the a_i/b_i, a contradiction.

As in a vector space, a basis of a module is a linearly independent generating subset. In a module M, a subset X of M is called **linearly independent** if, for any finite subset $\{x_1, \cdots, x_n\}$ of X, a relation of the form

$$x_1 a_1 + \cdots + x_n a_n = 0$$

with the a_i in A, implies $a_i = 0$, for any i. Thus, the empty set is trivially linearly independent. If a set $X \subseteq M$ is not linearly independent, then it is called **linearly dependent**. A **basis** of a module M is a linearly independent subset which generates M.

II.2.9 Proposition. *Let M be an A-module. A subset X of M is a basis if and only if any element of M can be expressed uniquely as linear combination of elements of X.*

Proof. Same as in linear algebra and left to the reader. □

We underline the crucial fact that, contrary to vector spaces, modules in general do *not* have bases. Indeed, let M be a finite nonzero abelian group ($\mathbb{Z}$-module) having, say, m elements. Then $mx = 0$, for any $x \in M$. Therefore, no nonempty subset of M is linearly independent and M has no basis. This forces us to devise new methods for studying modules.

The following theorem implies that any vector space has a basis.

II.2.10 Theorem. *Let A be a division ring and M an A-module. If X is a linearly independent subset of M, then there exists a basis of M containing X. In particular, M has a basis.*

Proof. Let $\mathcal{E}$ be the family of linearly independent subsets of M containing X. Then $\mathcal{E} \neq \emptyset$ because $X \in \mathcal{E}$. We show that $\mathcal{E}$ is inductive. Let $(X_\lambda)_\lambda$ be a chain in $\mathcal{E}$. We claim that $X_0 = \cup_\lambda X_\lambda$ is in $\mathcal{E}$.

If not, then X_0 contains a linearly dependent finite subset $\{x_1, \cdots, x_n\}$. For any i, with $1 \leqslant i \leqslant n$, there exists λ_i such that $x_i \in X_{\lambda_i}$. However, $(X_\lambda)_\lambda$ is a chain, therefore there exists λ_0 such that $\{x_1, \cdots, x_n\} \subseteq X_{\lambda_0}$. But X_{λ_0} is linearly independent, a contradiction which establishes our claim.

Because of Zorn's lemma, there exists a maximal linearly independent subset X_1 of M containing X. We claim that X_1 is a basis of M. Because X_1 is linearly independent, we just have to prove that it generates M. If not, then there exists $y \in M$ which is not a linear combination of elements of X_1. If we could prove that $X_1 \cup \{y\}$ is linearly independent, then we would get a contradiction to maximality of X_1 and our claim would be proved.

So, assume that $x_1, \cdots, x_n \in X_1$ and $a, a_1, \cdots, a_n \in A$ are such that

$$ya + x_1 a_1 + \cdots + x_n a_n = 0.$$

Then $ya = -\sum_{i=1}^n x_i a_i$. Assume that $a \neq 0$. Because A is a division ring, a is invertible and so $y = -\sum_{i=1}^n x_i a_i a^{-1}$, contradicting the assumption that y is not a linear combination of elements of X_1. Therefore, $a = 0$. Hence, $\sum_{i=1}^n x_i a_i = 0$. Because the x_i belong to X_1, which is linearly independent, $a_i = 0$, for all i. So, $X_1 \cup \{y\}$ is linearly independent, a contradiction to maximality of X_1. Therefore, X_1 is a basis of M. This proves the first statement.

The second follows by considering the case $X = \emptyset$. $\qquad\square$

It follows from the proof of the theorem that bases coincide with maximal linearly independent sets.

Because modules over division rings have bases, we may (and shall) consider them equivalently as vector spaces over that division ring.

II.2.11 Examples. We give examples of modules having bases. Let A be an algebra.
  (a) For any positive integer n, the product module A^n has the so-called **canonical basis** $e_1 = (1, 0, \cdots, 0), e_2 = (0, 1, \cdots, 0), \cdots, e_n = (0, 0, \cdots, 1)$.
  (b) The polynomial algebra $A[t]$ has as basis the infinite set of polynomials $(t^i)_{i \geqslant 0}$.
  (c) Let $n > 0$ be an integer. Because of Example I.3.11(b), the matrix algebra $M_n(A)$, viewed as an A-module, has as basis the set of matrices e_{ij}, with i, j such that $1 \leqslant i, j \leqslant n$.

We turn to quotient modules. As in any algebraic structure, the purpose of passing to the quotient is to cancel, that is, equate to zero, unwanted elements of the module, grouped inside a submodule, while retaining the pleasant properties of the original module.

Let M be an A-module and N a submodule. In particular, N is an abelian subgroup of M, so the relation defined by $x \equiv y \pmod{N}$ if and only if $x-y \in N$ is an equivalence on M. The equivalence class of $x \in M$ is called its **coset** $x + N = \{x + y : y \in N\}$. The element x is a **representative** of the coset $x + N$. Coset representatives are generally not unique: we have $x + N = x' + N$ if and only if $x \equiv x' \pmod{N}$, that is $x - x' \in N$. The quotient set $M/N = \{x + N : x \in M\}$ is an abelian group under the operation defined, for $x, y \in M$, by

$$(x + N) + (y + N) = (x + y) + N.$$

In order to make M/N an A-module, we define a multiplication of cosets by elements of A. Set, for $x \in M, a \in A$,

$$(x + N)a = (xa) + N$$

that is, the product of a coset by an element of A is the coset of the product of the representative by this element. This product is unambiguously defined: for, let $x, x' \in M$ be such that $x + N = x' + N$, then, for any $a \in A$, we have $xa - x'a = (x - x')a \in N$, because $x - x' \in N$ and N is a submodule of M. Therefore, $xa + N = x'a + N$.

II.2.12 Proposition. *Let M be an A-module and N a submodule. Then M/N has a right A-module structure given, for $x, y \in M, a \in A$, by:*

$$(x + N) + (y + N) = (x + y) + N \quad and \quad (x + N)a = (xa) + N.$$

Proof. This is done by verifying axioms as in Proposition I.3.10. □

The resulting module M/N is called the **quotient** of M by N. The zero element of M/N is the coset $0 + N = N$ and we have $x + N = N$ if and only if $x \in N$. Thus, the elements of N are 'equated to zero' in M/N. The nonzero elements of M/N are (the classes of) those elements of M which do not belong to N. That is, the quotient is the algebraic analogue of the set-theoretical notion of difference between a set and one of its subsets.

II.2.13 Examples.

(a) Let A be an algebra, I an ideal of A and M an A-module. Define MI by

$$MI = \left\{ \sum_\lambda x_\lambda a_\lambda : x_\lambda \in M, a_\lambda \in I, x_\lambda a_\lambda \text{ almost all zero} \right\}.$$

It is easy to see that MI is a submodule of M, using Proposition II.2.3. Because of Proposition II.2.12, M/MI inherits from M an A-module structure.

An interesting property of M/MI is that it is not only an A-module, it is also an A/I-module. The A/I-right action is defined, for $x \in M, a \in A$, by

$$(x + MI)(a + I) = xa + MI.$$

We prove that this action is unambiguously defined. Assume that $x + MI = x' + MI$ and $a + I = a' + I$. Then $x - x' \in MI$ and $a - a' \in I$. Consequently, $xa - x'a' = x(a - a') + (x - x')a' \in MI$. Therefore, $xa + MI = x'a' + MI$. The rest of the proof that this definition makes M/MI an A/I-module is a simple verification.

(b) Let A be an algebra and I a right ideal of A. The construction of Proposition II.2.12 yields a quotient module A/I. This module is cyclic, and generated by $1 + I$: indeed, for any $a \in A$, we have $a + I = (1 + I)a$.

The notation A/I may be confusing: we warn the reader that A/I is not a quotient algebra in general, because the latter is only defined when I is an ideal, not just a right ideal, see Proposition I.3.10. However, if I is indeed an ideal, then Proposition II.2.12 shows that A/I has not only an algebra structure, but also an A-A-bimodule structure.

II.2.3 Lattice of submodules

Given a module, its set of submodules is partially ordered by inclusion. We describe here the structure of this poset.

II.2.14 Definition. Let E be a poset, partially ordered by $\leqslant$.
  (a) If $(x_\lambda)_\lambda$ is a family of elements of E, an element $y \in E$ is:
      (i) a **supremum**, or **least upper bound**, of the x_λ, denoted as $y = \vee_\lambda x_\lambda$, if:
          (1) $x_\lambda \leqslant y$, for any λ, and
          (2) if $x_\lambda \leqslant z$, for any λ, then $y \leqslant z$.
      (ii) an **infimum**, or **greatest lower bound**, of the x_λ, denoted as $y = \wedge_\lambda x_\lambda$, if:
          (1) $y \leqslant x_\lambda$, for any λ, and
          (2) if $z \leqslant x_\lambda$, for any λ, then $z \leqslant y$.
  (b) A poset in which any pair of elements has supremum and infimum is a **lattice**. If an arbitrary family of elements in a poset has supremum and infimum, then this poset is a **complete lattice**.
  (c) A lattice E is **modular** if, for $x, y, z \in E$, the inequality $x \leqslant y$ implies that $y \wedge (x \vee z) = x \vee (y \wedge z)$. This identity is known as the **modular law**.

Suprema and infima, if they exist, are unique. We prove it for suprema. Let y, y' be suprema of the family $(x_\lambda)_\lambda$. Because of condition (1) in the definition, $x_\lambda \leqslant y$ for any λ, then condition (2) says that $y' \leqslant y$. Exchanging the roles of y and y', we also have $y \leqslant y'$. Antisymmetry of $\leqslant$ forces $y = y'$. The proof is similar for infima.

Inductively, if any two elements of a poset have a supremum, or an infimum, then so does any finite family. Therefore, a finite lattice is complete. A complete lattice has a greatest element and a smallest one (namely, the supremum and the infimum, respectively, of all elements). Consequently, so does any finite lattice.

One can think of the modular law as a weak form of distributivity. Indeed, assume that $x \leqslant y$ and that $\wedge$ is distributive over $\vee$, then the modular law holds true, indeed we have

$$y \wedge (x \vee z) = (y \wedge x) \vee (y \wedge z) = x \vee (y \wedge z).$$

because $x \leqslant y$ implies $y \wedge x = x$. Similarly, if $\vee$ is distributive over $\wedge$, the modular law also holds true.

II.2.15 Examples.

(a) In an arbitrary poset, if $x \leqslant y$, then $x \wedge y = x$ and $x \vee y = y$. This implies in particular that any chain is a lattice.

(b) Consider the set of positive integers partially ordered by divisibility. It is a lattice, the supremum of a pair of integers is their least common multiple, and the infimum is their greatest common divisor. Actually, the proof we just gave of uniqueness of supremum and infimum is the same as the one given in Subsection I.5.1 for uniqueness of gcd and lcm, taking into account that the only positive associate of a positive integer under divisibility is itself. This lattice is not complete, because it has no greatest element. On the other hand, it has a least element, namely 1.

(c) Let X be a set. Its power set (the set of its subsets) $P(X)$, ordered by inclusion, is a complete lattice, the supremum of a family of subsets is their union, and the infimum their intersection.

Finite posets are often visualised using quivers. Let $(E, \leqslant)$ be a finite poset. Its **Hasse quiver** Q_E has as points the elements of E and there is an arrow (and only one arrow) $x \to y$ if y **covers** x, that is, $x < y$ and $x \leqslant z \leqslant y$ implies $z = x$ or $z = y$ (that is, one cannot insert any element between x and y).

As an example, let $X = \{a, b, c\}$. The Hasse quiver of $P(X)$ ordered by inclusion is

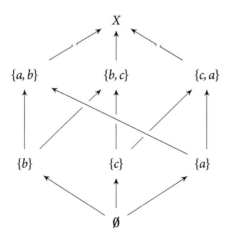

One can prove, we shall not do it here, that a lattice is modular if and only if it contains no subposet with Hasse quiver

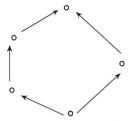

The arrows of the Hasse quiver always point upwards, and for this reason, one can equivalently replace an arrow by an edge, that is, draw the arrow without its head.

We prove that the set of submodules of a given module is a complete modular lattice. Let A be an algebra, M an A-module and $S(M)$ the set of its submodules, ordered by inclusion.

Infima are easy to construct in $S(M)$. Let $(M_\lambda)_\lambda$ be a family of submodules. Because $\cap_\lambda M_\lambda$ is the largest subset of M contained in the M_λ, and it is a submodule due to Lemma II.2.7, it is also the largest submodule of M contained in the M_λ, and therefore their infimum in $S(M)$.

We turn to suprema. The examples of vector spaces and abelian groups suggest the candidacy of the sum of submodules, which is defined exactly as the sum of vector subspaces, or of subgroups of an abelian group.

II.2.16 Definition. Let $(M_\lambda)_\lambda$ be a family of submodules of a given module M. Its **sum**, denoted by $\sum_\lambda M_\lambda$, is the set of finite sums of the $x_\lambda \in M_\lambda$, that is:

$$\sum_\lambda M_\lambda = \left\{ \sum_\lambda x_\lambda : x_\lambda \in M_\lambda, \text{ for any } \lambda, \text{ and the } x_\lambda \text{ almost all zero} \right\}.$$

The sum of a finite family $\{M_1, \cdots, M_t\}$ of submodules is denoted by $\sum_{i=1}^{t} M_i$ or $M_1 + \cdots + M_t$. We prove that the sum of a family of submodules is their supremum in $S(M)$.

II.2.17 Lemma. *Let $(M_\lambda)_\lambda$ be a family of submodules of a given module M. Then $\sum_\lambda M_\lambda$ is a submodule of M and actually the smallest submodule containing all the M_λ.*

Proof. The first statement is an application of Proposition II.2.3, and the second is proved by the same argument as in Corollary II.2.8.   □

The second statement of the lemma says that $\sum_\lambda M_\lambda$ is the submodule generated by the M_λ, that is, $\sum_\lambda M_\lambda = \langle \cup_\lambda M_\lambda \rangle$.

In particular, if M is a finitely generated module, with generating set $\{x_1, \cdots, x_n\}$, then $M = \sum_{i=1}^{n} x_i A$. Thus, any finitely generated module is a sum of cyclic submodules. We prove the announced result.

II.2.18 Proposition. *The set $S(M)$ of submodules of a given module M, ordered by inclusion, is a complete modular lattice, having 0 as smallest element, and M as greatest element.*

Proof. We know that any family $(M_\lambda)_\lambda$ of submodules of M has an infimum $\cap_\lambda M_\lambda$ and a supremum $\sum_\lambda M_\lambda$. Plainly, 0 is the smallest element and M the greatest element of $S(M)$. There remains to prove modularity.

Let M_1, M_2, M_3 be submodules of M such that $M_1 \subseteq M_2$. Then we have inclusions $M_1 \subseteq M_2 \cap (M_1 + M_3)$ and also $M_2 \cap M_3 \subseteq M_2 \cap (M_1 + M_3)$. Lemma II.2.17 implies that $M_1 + (M_2 \cap M_3) \subseteq M_2 \cap (M_1 + M_3)$. Conversely, let $x_2 = x_1 + x_3$ be an element of $M_2 \cap (M_1 + M_3)$, with $x_i \in M_i$ for $i = 1, 2, 3$. Then $x_3 = x_2 - x_1$ and $M_1 \subseteq M_2$ give $x_3 \in M_2$, so $x_3 \in M_2 \cap M_3$. Therefore, $x_2 = x_1 + x_3 \in M_1 + (M_2 \cap M_3)$. This proves that $M_1 + (M_2 \cap M_3) = M_2 \cap (M_1 + M_3)$, as required. $\square$

We make a digression about the sum of a family of submodules $(M_\lambda)_\lambda$ of M. Each element of this sum can be written as $x_{\lambda_1} + \cdots + x_{\lambda_n}$, where $x_{\lambda_i} \in M_{\lambda_i}$, for any i. One can ask whether this expression is unique. Easy examples, such as Example II.2.22 below, show that uniqueness does not hold true in general. When it happens, we say that the sum is direct.

II.2.19 Definition. Let M be a module. A family of submodules $(M_\lambda)_\lambda$ of M is in *direct sum* if any element of $\sum_\lambda M_\lambda$ can be written uniquely in the form $\sum_\lambda x_\lambda$ where $x_\lambda \in M_\lambda$, for any λ, and the x_λ are almost all zero.

In this case, we denote $\sum_\lambda M_\lambda = \oplus_\lambda M_\lambda$. For finitely many submodules $M_1, \cdots, M_t$, we also write $M_1 \oplus \cdots \oplus M_t$ or $\oplus_{i=1}^t M_i$. Each M_λ is called a *direct summand* of this sum.

As an illustration, Proposition II.2.9 can be reformulated to say that a subset X of an A-module M is a basis of M if and only if $M = \oplus_{x \in X} xA$.

We give criteria allowing to verify whether a family of submodules is in direct sum or not.

II.2.20 Theorem. *Let $(M_\lambda)_\lambda$ be a family of submodules of a module M. The following conditions are equivalent:*

(a) *The $(M_\lambda)_\lambda$ are in direct sum.*
(b) *$\sum_\lambda x_\lambda = 0$, with $x_\lambda \in M_\lambda$, for any λ, and the x_λ almost all zero, implies $x_\lambda = 0$, for any λ.*
(c) *$M_\lambda \cap (\sum_{\mu \neq \lambda} M_\mu) = 0$, for any λ.*

Proof. (a) implies (b). This is clear.

(b) implies (c). Let $x_\lambda \in M_\lambda \cap (\sum_{\mu \neq \lambda} M_\mu)$, then $x_\lambda = \sum_{\mu \neq \lambda} x_\mu$, with $x_\mu \in M_\mu$, for any μ and the x_μ almost all zero. This implies $-x_\lambda + \sum_{\mu \neq \lambda} x_\mu = 0$. Hence $x_\lambda = 0$, because of (b).

(c) implies (a). Assume that $\sum_\lambda x_\lambda = \sum_\lambda y_\lambda$ with $x_\lambda, y_\lambda \subseteq M_\lambda$ for any λ and the x_λ, y_λ almost all zero. For any λ, we have $x_\lambda - y_\lambda = \sum_{\mu \neq \lambda} (y_\mu - x_\mu)$ which belongs to $M_\lambda \cap (\sum_{\mu \neq \lambda} M_\mu)$. Because of (c), $x_\lambda - y_\lambda = 0$ and so $x_\lambda = y_\lambda$, for any λ, as required. □

The situation is especially nice when we deal with only two submodules.

II.2.21 Corollary. *Let L, N be submodules of M. Their sum is direct if and only if $L \cap N = 0$.*

Proof. This follows from the equivalence between (a) and (c) in Theorem II.2.20 above. □

II.2.22 Example. A common mistake is to think that more than two submodules are in direct sum if and only if their intersections two-by-two are zero. This is not true (and we really require condition (c) of Theorem II.2.20).

Let A be a nonzero algebra and assume that $M = A^2$, let $M_1 = \{(a, 0) : a \in A\}$, $M_2 = \{(a, a) : a \in A\}$, and $M_3 = \{(0, a) : a \in A\}$. It is easily seen that the M_i are submodules of M. We have $M_1 \cap M_2 = M_2 \cap M_3 = M_3 \cap M_1 = 0$, but the sum $M_1 + M_2 + M_3 = M$ is not direct. Indeed, let $(x, y) \in A^2$ be arbitrary, then, for any $t \in A$, we have

$$(x, y) = (x - t, 0) + (t, t) + (0, y - t)$$

with $(x - t, 0) \in M_1, (t, t) \in M_2$ and $(0, y - t) \in M_3$, a decomposition which depends on t and so is not unique. The problem here is that $M_1 + M_3 = M$, so $M_2 \cap (M_1 + M_3) = M_2 \neq 0$.

We prove an important lattice-theoretical property of finitely generated modules. Let $(E, \leqslant)$ be a poset, having a greatest element x_0. An element $x \in E$ is called **maximal** if x_0 covers x, that is, $x < x_0$ and $x \leqslant y \leqslant x_0$ implies $y = x$ or $y = x_0$. Thus, maximal elements are those which are covered by greatest elements. This definition resembles that of a maximal ideal in an algebra, seen in Subsection I.4.2. The following statement and proof are adaptations to modules of Krull's Theorem I.5.8.

II.2.23 Proposition. *Let M be a nonzero finitely generated A-module. Any proper submodule of M is contained in a maximal submodule. In particular, M has maximal submodules.*

Proof. Let $M = \langle x_1, \ldots, x_m \rangle$, N a proper submodule of M and $\mathcal{E}$ the set of proper submodules of M containing N, ordered by inclusion. Then $\mathcal{E} \neq \emptyset$ because $N \in \mathcal{E}$. We prove that $\mathcal{E}$ is inductive.

Let $(L_\lambda)_\lambda$ be a chain in $\mathcal{E}$. Then $L = \cup_\lambda L_\lambda$ is a submodule of M. Indeed, it contains the zero element, so it is nonempty, and, if $x, y \in L, a, b \in A$, then there exist μ, ν

such that $x \in L_\mu, y \in L_\nu$. Because $(L_\lambda)_\lambda$ is a chain, we have $L_\nu \subseteq L_\mu$ or $L_\mu \subseteq L_\nu$. Assume the former, then $x, y \in L_\mu$, and so $xa + yb \in L_\mu \subseteq L$, which is thus a submodule of M.

We claim that L is a proper submodule. Assume that $L = M$. Because $M = \langle x_1, \ldots, x_m \rangle$, there exists, for any i with $1 \leqslant i \leqslant m$, an index λ_i such that $x_i \in L_{\lambda_i}$. But $(L_\lambda)_\lambda$ is a chain, hence there exists λ_0 such that $L_{\lambda_i} \subseteq L_{\lambda_0}$, for any i, and so $x_i \in L_{\lambda_0}$, for any i. But then $M = \langle x_1, \ldots, x_m \rangle \subseteq L_{\lambda_0}$ and thus $M = L_{\lambda_0}$, a contradiction because L_{λ_0} is a proper submodule. This establishes our claim.

Because $N \subseteq L_\lambda$, for any λ, implies $N \subseteq L$, the set $\mathcal{E}$ is inductive. Zorn's lemma yields a maximal element in $\mathcal{E}$, thus proving the first statement. The second comes upon applying the first to the zero submodule. $\qquad\square$

The statement of Proposition II.2.23 does not hold true if the module is not finitely generated. For instance, we prove in Example VIII.2.2(c) below that $\mathbb{Q}$ is a $\mathbb{Z}$-module without maximal submodule.

We deduce a version of Krull's theorem for right ideals.

II.2.24 Corollary. *Let A be a nonzero algebra. Any proper right ideal of A is contained in a maximal right ideal. In particular, A has maximal right ideals.*

Proof. Indeed, $A_A = \langle 1 \rangle$ is finitely generated and even cyclic. $\qquad\square$

II.2.25 Examples.

  (a) From group theory, a group of order four is abelian and isomorphic either to the cyclic group $M = \mathbb{Z}/4\mathbb{Z}$ or the Klein Four-group $N = \mathbb{Z}/2\mathbb{Z} \times \mathbb{Z}/2\mathbb{Z}$. We compute the submodule lattice of each. Because of Lagrange's theorem, any nonzero proper subgroup of each is of order two, hence cyclic and generated by an element of order two. The cyclic group M has only one element of order two, namely the residual class $\bar{2} = 2 + 4\mathbb{Z}$. Thus, there is only one nonzero proper subgroup $\langle \bar{2} \rangle = \{\bar{0}, \bar{2}\}$. The Hasse quiver of the submodule lattice of M is

$$M$$
$$\uparrow$$
$$\langle \bar{2} \rangle$$
$$\uparrow$$
$$\langle \bar{0} \rangle$$

In contrast, the Four-group N has three elements of order two, namely $(\bar{1}, \bar{0})$, $(\bar{1}, \bar{1})$ and $(\bar{0}, \bar{1})$, where residual classes are taken modulo 2. The Hasse quiver of its submodule lattice is

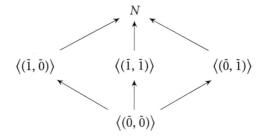

The latter is an example of a modular, but not distributive, lattice (that is, the $\vee$ and the $\wedge$ do not distribute on each other). Indeed, one sees that

$$\langle(\bar{1},\bar{0})\rangle \cap (\langle(\bar{1},\bar{1})\rangle + \langle(\bar{0},\bar{1})\rangle) = \langle(\bar{1},\bar{0})\rangle \cap N = \langle(\bar{1},\bar{0})\rangle$$

while

$$(\langle(\bar{1},\bar{0})\rangle \cap \langle(\bar{1},\bar{1})\rangle) + (\langle(\bar{1},\bar{0})\rangle \cap \langle(\bar{0},\bar{1})\rangle) = \langle(\bar{0},\bar{0})\rangle + \langle(\bar{0},\bar{0})\rangle = \langle(\bar{0},\bar{0})\rangle.$$

This example is rather typical. One can prove, we shall not do it here, that a lattice is distributive if and only if it contains no sublattice of the previous form.

The submodule lattice of an abelian group of order p^2, with p a prime integer, is computed exactly in the same manner and its Hasse quiver has a similar form.

(b) Let $M = \mathbb{Z}/12\mathbb{Z}$. Because M is a finite cyclic group of order 12, its submodules are in bijection with the divisors of 12, so we can draw its submodule lattice as:

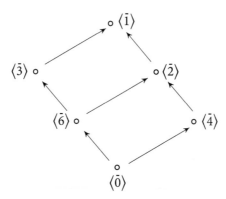

where residual classes are taken modulo 12. It is easy to prove that $M \cong \mathbb{Z}/4\mathbb{Z} \oplus \mathbb{Z}/3\mathbb{Z}$ (a more general statement will be shown later on, see Proposition V.3.14). As submodules of M, we have $\mathbb{Z}/4\mathbb{Z} \cong 3\mathbb{Z}/12\mathbb{Z} = \langle\bar{3}\rangle$ and $\mathbb{Z}/3\mathbb{Z} \cong 4\mathbb{Z}/12\mathbb{Z} = \langle\bar{4}\rangle$. They appear in the lattice as the lower sides of the rectangle which are attached to $\langle \bar{0} \rangle$, that is, whose intersection is zero (as it should be).

Because the zero submodule is the only source of the Hasse quiver of a module M, the submodules of M correspond to subquivers in which no arrow enters (but arrows may leave). These are those which are in 'the lower part' of the Hasse quiver. Dually, the quotients of M correspond to subquivers from which no arrow leaves (but arrows may enter). These lie in 'the upper part' of the Hasse quiver.

Exercises of Section II.2

1. Let A be a non necessarily unitary algebra and A_1 the algebra obtained from it by adjoining an identity, see Exercises I.3.1 and I.4.2. Let M be an A-module and define a right A_1-action on M by setting, for $x \in M, (a, \alpha) \in A_1$,

$$x(a, \alpha) = xa + x\alpha.$$

   Prove that this action and the addition of M make M an A_1-module.

2. Let A be an algebra, M a right A-module and X a subset of M. The **annihilator** of X is

$$\mathrm{Ann}X = \{a \in A : xa = 0, \text{ for all } x \in X\}.$$

   (a) Show that $\mathrm{Ann}X$ is a right ideal of A.
   (b) Assume that X is a submodule of M. Prove that $\mathrm{Ann}X$ is an ideal of A.
   (c) Prove that M has an $A/\mathrm{Ann}M$-module structure defined by $x(a + \mathrm{Ann}M) = xa$ for $x \in M, a \in A$.
   (d) An A-module N is called **faithful** if $\mathrm{Ann}N = 0$. Prove that any A-module M becomes faithful when viewed as an $A/\mathrm{Ann}M$-module.

3. Let $L = \{(a, b, 0) : a, b \in \mathbb{Z}\}$ and $M = \langle (1, 1, 0), (0, 1, 1) \rangle$ be submodules of the $\mathbb{Z}$-module $\mathbb{Z}^3$. Prove that $L + M = \mathbb{Z}^3$ and compute $L \cap M$.

4. Let p, p' be distinct prime integers. Prove that $\frac{1}{p} + \mathbb{Z} \neq \frac{1}{p'} + \mathbb{Z}$ in $\mathbb{Q}/\mathbb{Z}$. Deduce that the latter is an infinite abelian group. Then prove that the empty set is the only linearly independent subset in $\mathbb{Q}/\mathbb{Z}$.

5. Let K be a field, consider the vector space $E = K^2$ with the linear map $f : E \to E$ given in the canonical basis by the matrix $\begin{pmatrix} -1 & 1 \\ 1 & -1 \end{pmatrix}$. Find all submodules of the $K[t]$-module $_fE$.

6. Let K be a field and Q the quiver $3 \overset{\alpha}{\to} 2 \overset{\beta}{\to} 1$. Consider the representations U, V given respectively by:

$$K \overset{1}{\to} K \overset{1}{\to} K \text{ and } K \overset{1}{\to} K \overset{0}{\to} K.$$

Describe the action of the path algebra KQ on each of the modules M, N corresponding to U, V, respectively.

7. Let K be a field and Q the quiver $3 \xrightarrow{\alpha} 2 \xrightarrow{\beta} 1$.
   (a) Describe the KQ-module structures of the two following representations
       $P = (K \xrightarrow{1} K \xrightarrow{1} K)$ and $P' = (0 \to K \xrightarrow{1} K)$, where 1 denotes the identity map $1_K : K \to K$.
   (b) Prove that P' is a subrepresentation of P. Is $I = (K \xrightarrow{1} K \to 0)$ also a subrepresentation of P?

8. Let K be a field and Q the quiver

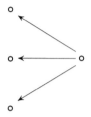

Prove that the representation

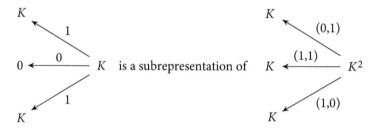

9. Let K be a field and Q the quiver

$$\underset{1}{\circ} \xleftarrow{\beta} \underset{2}{\circ} \xleftarrow{\alpha} \underset{3}{\circ} \xrightarrow{\gamma} \underset{4}{\circ} \xrightarrow{\delta} \underset{5}{\circ}$$

   (a) Construct a basis of the path algebra KQ, and the multiplication table of its elements.
   (b) Let I be the ideal $\langle \alpha\beta, \gamma\delta \rangle$ of KQ generated by the paths $\alpha\beta$ and $\gamma\delta$. Construct a basis of the bound quiver algebra KQ/I, and the multiplication table of its elements.
   (c) For each of the elements $e_2 = \epsilon_2 + I$ and $\tilde{\alpha} = \alpha + I$, compute a basis of the right ideal, the left ideal and the ideal it generates inside KQ/I.

10. Let K be a field, $A = KQ/I$ a bound quiver algebra and x a point in the quiver Q. Find a necessary and sufficient condition so that the right (or left) ideal generated by $e_x = \epsilon_x + I$ is also an ideal.

11. Let A, B be K-algebras and $_B M_A$ a B-A-bimodule.

(a) Prove that the set

$$R = \begin{pmatrix} A & 0 \\ M & B \end{pmatrix} = \left\{ \begin{pmatrix} a & 0 \\ x & b \end{pmatrix} : a \in A, b \in B \text{ and } x \in M \right\}$$

with the usual matrix operations and the multiplication induced from the bimodule structure of M, is a K-algebra.

(b) Prove that any right ideal of R is of the form $X \oplus I$, with I a right ideal of B and X an A-submodule of $(A \oplus M)_A$ containing IM, that is, if $b \in I, x \in M$, then $bx \in X$.

(c) Prove that any left ideal of R is of the form $J \oplus Y$, with J a left ideal of A and Y a B-submodule of $_B(M \oplus B)$ containing MJ.

(d) Compute all right ideals and all left ideals of the $\mathbb{Z}$-algebra

$$\begin{pmatrix} \mathbb{Q} & 0 \\ \mathbb{Q} & \mathbb{Z} \end{pmatrix} = \left\{ \begin{pmatrix} a & 0 \\ b & k \end{pmatrix} : a, b \in \mathbb{Q} \text{ and } k \in \mathbb{Z} \right\}.$$

12. Let A be an algebra, M a module and L, N submodules of M. Prove that $L \cup N$ is a submodule of M if and only if $L \subseteq N$ or $N \subseteq L$.

13. Let A be an algebra, M a module and $(M_\lambda)_\lambda$ a chain of submodules of M. Prove that $\cup_\lambda M_\lambda$ is a submodule of M.

14. Let A be an algebra, M a module and U, V, W submodules of M such that we have $U \subseteq W, U + V = W + V$ and $U \cap V = W \cap V$. Prove that $U = W$.

15. Let A be an algebra, M a module and N a submodule of M. Prove that:
    (a) If M is finitely generated, then so is M/N.
    (b) If both N and M/N are finitely generated, then so is M.
    (c) If N and M/N are respectively generated by n and m elements, then M is generated by (at least) $n + m$ elements.

16. Let A be an algebra and M a finitely generated module. Prove that any set of generators of M contains a finite subset which generates M.

17. Let A be an algebra. Given a family $(M_\lambda)_{\lambda \in \Lambda}$ of submodules of a module M, prove that $M = \oplus_{\lambda \in \Lambda} M_\lambda$ if and only if:
    (a) $M = \sum_{\lambda \in \Lambda} M_\lambda$, and
    (b) $M_{\lambda_0} \cap (M_{\lambda_1} + \cdots + M_{\lambda_n}) = 0$ for any finite subset $\{\lambda_1, \cdots, \lambda_n\} \subseteq \Lambda \setminus \{\lambda_0\}$.

18. Let A be an algebra and M an A-module. Assume that $M = L \oplus N$ and $L \subseteq L' \subseteq M$. Prove that $L' = L \oplus (L' \cap N)$, where M, L, L', N are A-modules.

19. Compute the submodule lattice, and draw the Hasse quiver of the latter, for each of the following $\mathbb{Z}$-modules: $\mathbb{Z}/13\mathbb{Z}, \mathbb{Z}/27\mathbb{Z}, \mathbb{Z}/6\mathbb{Z}$ and $\mathbb{Z}/42\mathbb{Z}$.

20. Compute the submodule lattice of the real vector space $\mathbb{R}^2$.

21. Let E be a lattice under the partial order $\leqslant$. Prove that, for any $x, y, z \in E$, the following identities hold true:
    (a) The associativity laws: $x \wedge (y \wedge z) = (x \wedge y) \wedge z$ and $x \vee (y \vee z) = (x \vee y) \vee z$.

(b) The commutativity laws: $x \wedge y = y \wedge x$ and $x \vee y = y \vee x$.

(c) The absorption laws: $x \vee (x \wedge y) = x$ and $x \wedge (x \vee y) = x$.

(d) The idempotence laws: $x \wedge x = x$ and $x \vee x = x$.

Then prove that, conversely, if E is a set having two internal operations $\wedge$ and $\vee$ satisfying (a), (b), (c), (d) above, then a partial order $\leqslant$ may be defined on E by the rule:

$$x \leqslant y \quad \text{if and only if} \quad x \wedge y = x.$$

Further, prove that, with respect to this order $\leqslant$, the set E is a lattice where the supremum and the infimum of $x, y \in E$ are respectively given by $x \vee y$ and $x \wedge y$.

22. Let A be an algebra, M an A-module, N a submodule and $x \in M$ such that $x \notin N$. Prove that M has a maximal submodule containing N and not containing x.

II.3 Module morphisms

II.3.1 Definition and examples

Together with an algebraic structure come the maps compatible with it. The structure of a module is determined by addition and scalar multiplication. Therefore, a 'good' map between modules must preserve these operations, and this amounts to requiring that it preserves linear combinations. Thus, it is a direct generalisation of the notion of linear map between vector spaces.

II.3.1 Definition. Let A be an algebra. A map $f : M \to N$ between A-modules is a **morphism** of A-modules, or **homomorphism**, or **A-linear map**, if, for $x, y \in M$ and $a \in A$, we have:

(a) $f(x + y) = f(x) + f(y)$, and

(b) $f(xa) = f(x)a$.

We use most of the time the terms A-linear maps and morphisms. The properties of the definition are equivalent to requiring that, for $x, y \in M$ and $a, b \in A$, one has:

$$f(xa + yb) = f(x)a + f(y)b.$$

Induction shows that the latter amounts to saying that, if $(x_\lambda)_\lambda$ is a family of elements of M, and $(a_\lambda)_\lambda$ a family of elements of A almost all zero, then

$$f\left(\sum_\lambda x_\lambda a_\lambda\right) = \sum_\lambda f(x_\lambda)a_\lambda.$$

In other words, a map $f : M \to N$ is A-linear if and only if it preserves A-linear combinations.

If $f : M \to N$ is A-linear, then it is also K-linear because, if $x \in M, \alpha \in K$, then:

$$f(x\alpha) = f(x(1 \cdot \alpha)) = f(x)(1 \cdot \alpha) = f(x)\alpha.$$

An A-linear map from a module to itself is called an **endomorphism** of this module. Linear maps satisfy the expected properties.

II.3.2 Proposition. *Let A be an algebra and $f : M \to N$ an A-linear map between A-modules. Then we have:*
- (a) $f(0) = 0$,
- (b) $f(-x) = -f(x)$, *for any $x \in M$,*
- (c) $f(nx) = nf(x)$, *for any $x \in M, n \in \mathbb{Z}$,*
- (d) $\mathrm{Im}\, f = \{f(x) : x \in M\}$ *is a submodule of N,*
- (e) $\mathrm{Ker}\, f = \{x \in M : f(x) = 0\}$ *is a submodule of M,*
- (f) *f is injective if and only if $\mathrm{Ker}\, f = 0$.*

Proof. Similar to that of Proposition I.4.3 and left to the reader.   □

II.3.3 Examples.
- (a) If $A = K$ is a field, then Definition II.3.1 is the usual definition of a linear map between K-vector spaces.
- (b) If $A = \mathbb{Z}$, and M, N are abelian groups, a $\mathbb{Z}$-linear map $f : M \to N$ is a group homomorphism, because of condition (a) of Definition II.3.1. Conversely, if $f : M \to N$ is a group homomorphism, then it follows from induction and the property $f(-x) = -f(x)$ for $x \in M$, that $f(xn) = f(x)n$ for $x \in M, n \in \mathbb{Z}$. Thus, $\mathbb{Z}$-linear maps coincide with group homomorphisms.
- (c) Let A be an algebra and M, N be A-modules. The map $M \to N$ defined by $x \mapsto 0$, for any $x \in M$, is linear and called the **zero map**. It is usually denoted by 0. A map whose domain or codomain is the zero module can only be the zero map. Also, the composite of the zero map with any morphism (on either side) is the zero map.
- (d) Let A be an algebra, M an A-module and L a submodule of M. The map $i : L \to M$ defined by $x \mapsto x$, for $x \in L$, is linear and called the **inclusion** or **injection**. If $L = M$, the inclusion reduces to the identity $i = 1_M$.
- (e) Let A be an algebra, M an A-module and L a submodule of M. The map $p : M \to M/L$ defined by $x \mapsto x + L$, for $x \in M$, is linear and called the **projection** or **surjection**. In fact, we defined the operations on the quotient module exactly so that projection becomes a linear map, see Example I.4.2(b).
- (f) The composite of linear maps $f : L \to M, g : M \to N$ is a linear map $gf = g \circ f : L \to N$.
- (g) Let K be a field and E, F be K-vector spaces. Because of Example II.2.4(f), two K-linear maps $f : E \to E, g : F \to F$ induce on E and F structures of $K[t]$-modules, $_f E$ and $_g F$, respectively. We claim that a morphism $u :_f E \to_g F$ of $K[t]$-modules is just a K-linear map $u : E \to F$ such that $uf = gu$, that is, the following square commutes:

$$E \xrightarrow{\ f\ } E$$

$$u \downarrow \qquad\qquad u \downarrow$$

$$F \xrightarrow{\ g\ } F$$

Indeed, let $u : E \to F$ be K-linear and such that $uf = gu$. Because addition is the same in the vector space and the corresponding $K[t]$-module, we only have to prove that u preserves multiplication by polynomials. Let $x \in E$, we have:

$$u(t \cdot x) = u(f(x)) = (uf)(x) = (gu)(x) = g(u(x)) = t \cdot u(x)$$

A straightforward induction yields $u(t^i x) = t^i u(x)$, for any integer $i \geq 0$. Therefore, for any $p \in K[t]$, we have $u(p \cdot x) = p \cdot u(x)$, as required.

Conversely, if $u :_f E \to_g F$ is $K[t]$-linear, it is in particular K-linear. Also, if $x \in E$, we have $uf(x) = u(t \cdot x) = t \cdot u(x) = gu(x)$. Therefore, $uf = gu$.

(h) Let K be a field, Q a finite quiver, and V, W representations of Q, see Example II.2.4(g). A **morphism of representations** $f : V \to W$ is a family of K-linear maps $f = (f_x : V(x) \to W(x))_{x \in Q_0}$ indexed by the points and compatible with the arrows in Q, that is, if $\alpha : x \to y$ is an arrow, then $f_y V(\alpha) = W(\alpha) f_x$. In other words, the following square commutes:

$$V(x) \xrightarrow{\ V(\alpha)\ } V(y)$$

$$f_x \downarrow \qquad\qquad f_y \downarrow$$

$$W(x) \xrightarrow{\ W(\alpha)\ } W(y)$$

If $\gamma = \sum_i \lambda_i \gamma_i$, where the λ_i belong to K and the γ_i are paths from x to y, say, then to each path γ_i corresponds a linear map $V(\gamma_i) : V(x) \to V(y)$ and to γ corresponds $V(\gamma) = \sum_i \lambda_i V(\gamma_i)$, as in Example II.2.4(g). Then the definition of morphism of representations implies that $f_y V(\gamma) = W(\gamma) f_x$.

We claim that a morphism of representations induces a KQ-linear map between the corresponding KQ-modules, see Example II.2.4(g). Let $f : V \to W$ be a morphism of representations, and M, N the KQ-modules corresponding to V, W respectively. As vector spaces, $M = \bigoplus_{x \in Q_0} V(x)$, $N = \bigoplus_{x \in Q_0} W(x)$. Consequently, f induces a K-linear map $f' : M \to N$ given by $(v_x)_{x \in Q_0} \mapsto (f_x(v_x))_{x \in Q_0}$, for $(v_x)_{x \in Q_0} \in M$. We verify that f' is compatible with the right KQ-action. For this, it suffices to check its compatibility with multiplication by paths. Due to definition of multiplication, it suffices to consider the case where we multiply the coordinate v_x by a path γ from x to y. In this case, we indeed have

$$f'(v_x \gamma) = f' V(\gamma)(v_x) = f_y V(\gamma)(v_x) = W(\gamma) f_x(v_x) = f_x(v_x)\gamma = f'(v_x)\gamma.$$

In particular, the inclusion of a subrepresentation V into a representation W, as defined in Example II.2.4(g), is a morphism from V to W.

For example, let Q be the quiver

$$\circ \rightleftarrows \circ$$

and define representations of Q as follows:

$$K \underset{(0\ 1)}{\overset{(1\ 0)}{\rightleftarrows}} K^2 \quad \text{and} \quad K^2 \underset{\left(\begin{smallmatrix} 0 & 1 \\ 0 & 0 \end{smallmatrix}\right)}{\overset{\left(\begin{smallmatrix} 1 & 0 \\ 0 & 1 \end{smallmatrix}\right)}{\rightleftarrows}} K^2$$

Then we have a morphism of representations

$$
\begin{array}{ccc}
K^2 & \underset{\left(\begin{smallmatrix} 0 & 1 \\ 0 & 0 \end{smallmatrix}\right)}{\overset{\left(\begin{smallmatrix} 1 & 0 \\ 0 & 1 \end{smallmatrix}\right)}{\rightleftarrows}} & K^2 \\
{\scriptstyle (1\ 0)}\Big\downarrow & & \Big\downarrow{\scriptstyle \left(\begin{smallmatrix} 1 & 0 \\ 0 & 1 \end{smallmatrix}\right)} \\
K & \underset{(0\ 1)}{\overset{(1\ 0)}{\rightleftarrows}} & K^2
\end{array}
$$

because of the matrix identities:

$$(1\ \ 0)\begin{pmatrix} 1 & 0 \\ 0 & 1 \end{pmatrix} = (1\ \ 0)\begin{pmatrix} 1 & 0 \\ 0 & 1 \end{pmatrix} \quad \text{and} \quad (1\ \ 0)\begin{pmatrix} 0 & 1 \\ 0 & 0 \end{pmatrix} = (0\ \ 1)\begin{pmatrix} 1 & 0 \\ 0 & 1 \end{pmatrix}.$$

(i) Let A, B be algebras and M, N both A-B-bimodules. A morphism of abelian groups $f : M \to N$ such that $f(axb) = af(x)b$ for $a \in A, x \in M, b \in B$, is at the same time A-linear and B-linear. We call it a **bimodule morphism**.

Isomorphisms are defined as for algebras.

II.3.4 Definition. Let M, N be A-modules. An A-linear map $f : M \to N$ is an *isomorphism* if there exists an A-linear map $f' : N \to M$ such that $f' \circ f = 1_M$ and $f \circ f' = 1_N$. In this case, M, N are *isomorphic*, which we denote as $M \cong N$.

In particular, an isomorphism is an invertible map and therefore bijective. The map f' of the definition is the unique inverse of f, and so we may write $f' = f^{-1}$.

For any module M, the identity map $1_M : M \to M$ is an isomorphism. If $f : M \to N$ is an isomorphism, then so is f^{-1} and we have $(f^{-1})^{-1} = f$. Finally, if $f : L \to M, g : M \to N$ are isomorphisms, then so is $g \circ f : L \to N$ and we have $(g \circ f)^{-1} = f^{-1} \circ g^{-1}$. Therefore, the relation $\cong$ between modules is reflexive, symmetric and transitive.

An isomorphism from a module M to itself is called an **automorphism** of M. The previous considerations imply that the set $\mathrm{Aut}M$ of automorphisms of M is a group under composition of maps, called the **automorphism group** of M.

As in the case of algebras, bijective morphisms are isomorphisms.

II.3.5 Lemma. *A module morphism is an isomorphism if and only if it is bijective.*

Proof. We need to prove that, if $f : M \to N$ is a bijective A-linear map between A-modules, then $f^{-1} : N \to M$ is also A-linear. Indeed, let $y \in N, a \in A$, then

$$f^{-1}(ya) = f^{-1}(1_N(y)a) = f^{-1}((ff^{-1})(y)a)$$
$$= f^{-1}(f(f^{-1}(y))a) = f^{-1}(f(f^{-1}(y)a))$$
$$= (f^{-1}f)(f^{-1}(y)a) = 1_M(f^{-1}(y)a) = f^{-1}(y)a$$

where we used A-linearity of f at the fourth equality. One shows similarly that f^{-1} preserves sums, see the proof of Lemma I.4.6. $\qquad\square$

As is the case for algebras, two isomorphic modules have the same module-theoretical properties. For instance, if $f : M \to N$ is an isomorphism and M is a cyclic module generated by x, say, then N is cyclic generated by $f(x)$. Two isomorphic modules may thus be identified for all useful purposes, which we shall do in the sequel.

II.3.6 Examples.
(a) A module may be isomorphic to one of its proper submodules. Let $n > 0$ be an integer. The map $f : \mathbb{Z} \to n\mathbb{Z}$ defined by $x \mapsto nx$, for $x \in \mathbb{Z}$, is linear because of

$$f(x + y) = n(x + y) = nx + ny = f(x) + f(y)$$

and Example II.3.3(b). Moreover, it is bijective with inverse $n\mathbb{Z} \to \mathbb{Z}$ given by $ny \mapsto y$, for $y \in \mathbb{Z}$. So, it is an isomorphism.
(b) Let p be a prime integer and $f : \mathbb{Z}/p^2\mathbb{Z} \to \mathbb{Z}/p^2\mathbb{Z}$ the map induced from multiplication by p, that is, the map $x + p^2\mathbb{Z} \mapsto px + p^2\mathbb{Z}$, for $x \in \mathbb{Z}$. Clearly, f is $\mathbb{Z}$-linear. We compute its image and kernel. Because $f(1+p^2\mathbb{Z}) = p+p^2\mathbb{Z}$, the image is the submodule $p\mathbb{Z}/p^2\mathbb{Z}$ of the codomain. On the other hand, $f(x + p^2\mathbb{Z}) = 0 + p^2\mathbb{Z}$ if and only if $px \in p^2\mathbb{Z}$, that is, if and only if p divides x. Consequently, the kernel of f is the submodule $p\mathbb{Z}/p^2\mathbb{Z}$ of the domain. We now look at the effect of the map on the submodule lattices of its domain and codomain. The submodule lattice of $\mathbb{Z}/p^2\mathbb{Z}$ is calculated as in Example II.2.25(a) and has a similar shape. The effect of the map on the lattices may be visualised as follows:

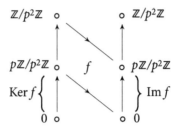

Because Ker f and Im f are submodules of the domain and the codomain, respectively, both are located in the *lower part* of the respective submodule lattices.

II.3.2 Sets of linear maps

Let A be an algebra and M, N be A-modules. We look at the algebraic structure of the set $\mathrm{Hom}_A(M, N)$ of A-linear maps from M to N. The notation 'Hom' refers to homomorphisms and these sets are usually called **Hom-sets**. We define operations on $\mathrm{Hom}_A(M, N)$. Let $f, g : M \to N$ be linear and define $f + g : M \to N$ for $x \in M$ by:

$$(f + g)(x) = f(x) + g(x).$$

Also, let $f : M \to N$ be linear and $\alpha \in K$, then define $f\alpha : M \to N$ for $x \in M$ by:

$$(f\alpha)(x) = \alpha f(x).$$

II.3.7 Lemma. *With these operations, $\mathrm{Hom}_A(M, N)$ is a K-module.*

Proof. It is easily shown that, given $f, g \in \mathrm{Hom}_A(M, N)$, and $\alpha \in K$, the maps $f + g$ and $f\alpha$ are A-linear. The axioms of K-modules are easily verified except perhaps mixed associativity. So, let $f \in \mathrm{Hom}_A(M, N)$, $\alpha, \beta \in K$ and $x \in M$, then we have

$$(f(\alpha\beta))(x) = \alpha\beta f(x) = \beta\alpha f(x) = \beta(f\alpha)(x) = ((f\alpha)\beta)(x).$$

Because x is arbitrary, this implies that $f(\alpha\beta) = (f\alpha)\beta$, as required. $\square$

One can ask whether $\mathrm{Hom}_A(M, N)$ also has an A-module structure or not. In the previous proof, we used essentially commutativity of multiplication in K, therefore one cannot repeat this proof using scalars from A instead of scalars from K, except if A is commutative. Thus, $\mathrm{Hom}_A(M, N)$ is not in general an A-module, but it is so when A is commutative.

If one deals with a bimodule over two algebras, then one gets a module structure on the Hom-set, over one of these two algebras.

II.3.8 Lemma. *Let A, B be algebras.*

(a) *If $_AM_B$ and N_B are given, then $\mathrm{Hom}_B(M, N)$ has a right A-module structure given, for $f : M \to N, a \in A, x \in M,$ by $(fa)(x) = f(ax)$.*

(b) *If $_AM_B$ and $_AN$ are given, then $\mathrm{Hom}_A(M, N)$ has a left B-module structure given, for $f : M \to N, b \in M, x \in M,$ by $(bf)(x) = f(xb)$.*

(c) *If $_AM$ and $_AN_B$ are given, then $\mathrm{Hom}_A(M, N)$ has a right B-module structure given, for $f : M \to N, b \in B, x \in M,$ by $(fb)(x) = f(x)b$.*

(d) *If M_B and $_AN_B$ are given, then $\mathrm{Hom}_B(M, N)$ has a left A-module structure given, for $f : M \to N, a \in A, x \in M,$ by $(af)(x) = af(x)$.*

Proof. Left as an exercise to the reader. □

These rules are not hard to remember: if the additional scalar occurs in the first variable of Hom, then it is multiplied the *opposite* way it occurs (in (a): $_AM$ is a *left* A-module, but $\mathrm{Hom}_B(M, N)$ becomes a *right* A-module, similarly for B in (b)), while, if it occurs in the second variable, then it is multiplied the *same* way (in (c): both N_B and $\mathrm{Hom}_A(M, N)$ are right B-modules). As to how the vector is multiplied by the scalar, in each case, we have the only multiplication that makes sense (for instance, in (a), $x \in M$ which is a *left* A-module, so only ax makes sense).

An algebra A has a natural A-A-bimodule structure, so, for any right A-module M_A, the K-module $\mathrm{Hom}_A(A, M)$ has a right A-module structure given as in (a) above:

$$(fa)(x) = f(ax),$$

for $f : A \to M, a, x \in A$. It turns out that, as modules, $\mathrm{Hom}_A(A, M)$ is isomorphic to M.

II.3.9 Theorem. *Let A be an algebra and M an A-module. We have an isomorphism of A-modules*

$$\mathrm{Hom}_A(A, M) \cong M.$$

Proof. Define $u : \mathrm{Hom}_A(A, M) \to M$ by $f \mapsto f(1)$, for $f \in \mathrm{Hom}_A(A, M)$. This is an A-linear map because, if $f, g \in \mathrm{Hom}_A(A, M)$ and $a, b \in A$, then

$$u(fa + gb) = (fa + gb)(1) = (fa)(1) + (gb)(1) = f(a) + g(b) =$$
$$= f(1 \cdot a) + g(1 \cdot b) = f(1)a + g(1)b = u(f)a + u(g)b.$$

Then, consider the map $v : M \to \mathrm{Hom}_A(A, M)$ defined by $x \mapsto (a \mapsto xa)$, for $x \in M, a \in A$. It is easy to verify that u and v are mutually inverse, □

A useful remark: when one is to construct explicitly module isomorphisms, the 'right' maps are usually the most natural ones. Consider, for instance, the isomorphism $u : \mathrm{Hom}_A(A, M) \to M$ above: it must take $f : A \to M$ to some element of M. It is reasonable to choose an element in A and map f to the image of this element. But A has a distinguished element 1. Hence the idea of setting $u : f \mapsto f(1)$. Similarly for its inverse v : it maps an $x \in M$ to a linear map $A \to M$, which sends an arbitrary $a \in A$

to some element of M. But M is a right A-module, so one can right-multiply x by a, hence the reasonable choice is the product xa. This gives the map $x \mapsto (a \mapsto xa)$. Of course, once this construction is done, one should verify that the map satisfies the required conditions, as we did in the proof of the lemma.

Composition of linear maps has some obvious properties which we state for future reference.

II.3.10 Lemma. *Let A be an algebra, L, M, N be A-modules, $f, f_1, f_2 \in \mathrm{Hom}_A(L, M)$, $g, g_1, g_2 \in \mathrm{Hom}_A(M, N)$ and $\alpha_1, \alpha_2 \in K$, then :*
   (a) $g(f_1\alpha_1 + f_2\alpha_2) = (gf_1)\alpha_1 + (gf_2)\alpha_2$.
   (b) $(g_1\alpha_1 + g_2\alpha_2)f = (g_1f)\alpha_1 + (g_2f)\alpha_2$.
   (c) $g(-f) = (-g)f = -(gf)$.   □

A first consequence is the next lemma.

II.3.11 Lemma. *Let A be a K-algebra and M be an A-module. The set $\mathrm{End}_A M$ of endomorphisms of M has a K-algebra structure with multiplication given by composition of maps.*

Proof. Because of Lemma II.3.7, $\mathrm{End}_A M$ is a K-module. Lemma II.3.10 and associativity of composition of maps show that $\mathrm{End}_A M$ is a K-algebra with identity 1_M.   □

The K-algebra $\mathrm{End}_A M$ is called the ***endomorphism algebra*** of the module M.

Thus, an A-module M can also be viewed as an $\mathrm{End}_A M$-A-bimodule, because the relation

$$(f(x))a = f(x)a = f(xa)$$

for $f \in \mathrm{End}_A M, x \in M, a \in A$, follows from linearity of f. Because of the definition of the K-module structure of $\mathrm{End}_A M$, the ring K operates centrally on M, that is, the left and right actions of K on M give the same result: a scalar $\alpha \in K$ corresponds to the linear map $1_M \cdot \alpha : M \to M$, which is just multiplication by α, and the latter operates in the same way on the right and on the left on elements of M.

Specialising to $M = A$, we get the following corollary.

II.3.12 Corollary. *Let A be an algebra. We have an isomorphism of K-algebras $\mathrm{End}_A A \cong A$.*

Proof. The endomorphism algebra $\mathrm{End}_A A$ of A_A, is a K-module, and because of Theorem II.3.9, we have an isomorphism of K-modules $u : \mathrm{End}_A A \to A$ given by $u : f \mapsto f(1)$. It suffices to prove that u preserves multiplication. Indeed, let $f, g \in \mathrm{End}_A A$, then

$$u(gf) = (gf)(1) = g(f(1)) = g(1.f(1)) = g(1)f(1) = u(g)u(f).   \qquad \square$$

II.3.13 Example. Because of Theorem II.3.9, we have $\mathrm{Hom}_{\mathbb{Z}}(\mathbb{Z}, \mathbb{Q}) \cong \mathbb{Q}$. We prove that $\mathrm{Hom}_{\mathbb{Z}}(\mathbb{Q}, \mathbb{Z}) = 0$. Let $f : \mathbb{Q} \to \mathbb{Z}$ be a $\mathbb{Z}$-linear map, that is, a homomorphism of abelian groups, and n a positive integer. Then $nf(\frac{1}{n}) - f(n \cdot \frac{1}{n}) = f(1)$, which means that $f(1)$ is divisible by any positive integer n. This is only possible if $f(1) = 0$, that is, $f = 0$.

Exercises of Section II.3

1. Let A be an algebra, M an A-module and a an element of the centre $Z(A)$ of A. Prove that the map $f_a : M \to M$ given by $x \mapsto xa$, for $x \in M$, is A-linear.

2. Let A be a K-algebra and M a K-module.
   (a) Assume that M has a *left* A-module structure and let $a \in A$. Prove that the map $\lambda_a : M \to M$ defined by $x \mapsto ax$, for $x \in M$, is a K-endomorphism of M.
   (b) Deduce that the map $a \mapsto \lambda_a$ is an algebra morphism from A to $\mathrm{End}_K M$.
   (c) Conversely, prove that any algebra morphism from A to $\mathrm{End}_K M$ defines a left A-module structure on M.

3. Let A be an algebra, M an A-module and M_1, M_2 submodules of M. Consider the map $f : M_1 \times M_2 \to M$ given by $(x_1, x_2) \mapsto x_1 + x_2$, for $x_1 \in M_1, x_2 \in M_2$. Prove that:
   (a) f is linear.
   (b) $\mathrm{Ker} f = \{(x, -x) : x \in M_1 \cap M_2\}$.
   (c) $\mathrm{Im} f = M_1 + M_2$.

4. Let A be an algebra and $f : A^3 \to A^3$ the map defined, for $x_1, x_2, x_3 \in A$, by

$$f(x_1, x_2, x_3) = (-x_1 + x_2, -x_2 + x_3, -x_3 + x_1).$$

   (a) Prove that f is linear.
   (b) Compute $\mathrm{Ker} f$ and $\mathrm{Im} f$.
   (c) Let $M = \langle (1, 1, 0), (0, 1, 1) \rangle$. Compute $f(M)$ and $f^{-1}(M)$.

5. Consider the abelian groups $\mathbb{Z}/18\mathbb{Z}$ and $\mathbb{Z}/27\mathbb{Z}$.
   (a) Prove that the rule $x + 18\mathbb{Z} \mapsto 3x + 27\mathbb{Z}$, for $x \in \mathbb{Z}$, defines unambiguously a map $f : \mathbb{Z}/18\mathbb{Z} \to \mathbb{Z}/27\mathbb{Z}$.
   (b) Prove that f is $\mathbb{Z}$-linear.
   (c) Compute $\mathrm{Ker} f$ and $\mathrm{Im} f$.
   (d) Show each of $\mathrm{Ker} f$ and $\mathrm{Im} f$ on the submodule lattices of the domain and codomain, respectively.

6. Let A be an algebra, M an A-module and X a subset of M. Prove that the following conditions are equivalent:
   (a) M is generated by X.
   (b) If $f : L \to M$ is an A-linear map such that $X \subseteq \mathrm{Im} f$, then f is surjective.
   (c) If $g : M \to N$ is an A-linear map such that $g(x) = 0$ for any $x \in X$, then $g = 0$.

7. Let A be an algebra, M an A-module, X a subset of M and $f : M \to N$ a surjective A-linear map. Prove that $M = \langle X \rangle$ if and only if $N = \langle f(X) \rangle$ and $\operatorname{Ker} f \subseteq \langle X \rangle$.

8. Let K be a field with $\operatorname{char} K \neq 2$, see Exercise I.2.5, and A a K-algebra. Prove that the map $f : A^2 \to A^2$ given by $(a_1, a_2) \mapsto (a_1 + a_2, a_1 - a_2)$, for $a_1, a_2 \in A$, is an isomorphism.

9. Let A be an algebra and $f : M \to N$ an A-linear map between A-modules. Denote by $S(M), S(N)$ the submodule lattices of M, N, respectively. Prove that:
   (a) If M' is a submodule of M, then $f(M')$ is a submodule of N.
   (b) If N' is a submodule of N, then $f^{-1}(N')$ is a submodule of M.
   (c) We have inclusion-preserving maps between the submodule lattices
   $$S(M) \xrightarrow{f} S(N) \text{ and } S(N) \xrightarrow{f^{-1}} S(M).$$
   (d) On these submodule lattices, we have $f \circ f^{-1} \circ f = f$ and $f^{-1} \circ f \circ f^{-1} = f^{-1}$.

10. Let A be an algebra, $f : M \to N$ a linear map between A-modules and M', N' submodules of M, N, respectively. Prove that:
    (a) $f(M' \cap f^{-1}(N')) = f(M') \cap N'$.
    (b) $f^{-1}(N' + f(M')) = f^{-1}(N') + M'$.
    (c) $ff^{-1}(N') = N' \cap \operatorname{Im} f$, and $ff^{-1}(N') = N'$, for any N', if and only if f is surjective.
    (d) $f^{-1}f(M') = M' + \operatorname{Ker} f$ and $f^{-1}f(M') = M'$, for any M', if and only if f is injective.

11. Let A be an algebra and $f : L \to M, g : M \to N$ linear maps between A-modules. Prove that:
    (a) $\operatorname{Ker} f = \operatorname{Ker}(gf)$ if and only if $\operatorname{Im} f \cap \operatorname{Ker} g = 0$.
    (b) $\operatorname{Im} g = \operatorname{Im}(gf)$ if and only if $\operatorname{Im} f + \operatorname{Ker} g = M$.

12. Let A be an algebra. An endomorphism f of an A-module M is called **idempotent** if $f \circ f = f$. Prove that, if f is idempotent, then $M = \operatorname{Im} f \oplus \operatorname{Ker} f$.

13. Let A be an algebra and f, g endomorphisms of an A-module M such that $g \circ f \circ g = g$ and $f \circ g \circ f = f$. Prove that $M = \operatorname{Im} f \oplus \operatorname{Ker} g$.

14. Let A be an algebra and f an endomorphism of an A-module M. Prove that $\operatorname{Im} f \cap \operatorname{Ker} f = 0$ if and only if $\operatorname{Ker} f^3 \subseteq \operatorname{Ker} f$. Here, as usual, $f^3 = f \circ f \circ f$.

15. Let A be an algebra. An endomorphism f of an A-module M is called **nilpotent** if there exists an integer $n > 0$ such that $f^n = 0$. Prove that, if f is nilpotent, then $1_M - f$ is an isomorphism.

16. Let A be a K-algebra and M an A-module. Assume that there exists a set X and a bijection $f : X \to M$. Define an A-module structure on X such that the map f becomes an isomorphism. This procedure is called **transportation of structure**.

17. Let A be an algebra. A nonzero A-module S is called a **simple module** if it has only two submodules, namely 0 and itself. Prove that:
    (a) If $n > 0$ is an integer, then $\mathbb{Z}/n\mathbb{Z}$ is a simple $\mathbb{Z}$-module if and only if n is prime.

(b) If M is a module, S a simple module, then any nonzero morphism $S \to M$ is injective.

(c) If M is a module, S a simple module, then any nonzero morphism $M \to S$ is surjective.

(d) If S, S' are simple modules, then any nonzero morphism $S \to S'$ is an isomorphism.

(e) If S is a simple module, then $\operatorname{End} S$ is a division ring.

18. Let $m, n > 0$ be integers. Compute each of the abelian groups:
   (a) $\operatorname{Hom}_{\mathbb{Z}}(\mathbb{Z}, \mathbb{Z}/n\mathbb{Z})$.
   (b) $\operatorname{Hom}_{\mathbb{Z}}(\mathbb{Z}/n\mathbb{Z}, \mathbb{Z})$.
   (c) $\operatorname{Hom}_{\mathbb{Z}}(\mathbb{Z}/m\mathbb{Z}, \mathbb{Z}/n\mathbb{Z})$.

19. Let A be a nonunitary algebra and A_1 obtained from A by adjoining an identity, see Exercise I.3.1. Recall that, given an A-module M, there is a right A_1-action on M making it an A_1-module, denoted as M_1, see Exercise II.2.1. Let M, N be A-modules.
   (a) Prove that there exists an isomorphism of K-modules $\operatorname{Hom}_A(M, N) \cong \operatorname{Hom}_{A_1}(M_1, N_1)$.
   (b) Deduce that $M \cong N$ if and only if $M_1 \cong N_1$.

20. Let K be a field, E, F vector spaces and $f : E \to E, g : F \to F$ linear maps. Show that the induced $K[t]$-modules $_fE, {_g}F$ are isomorphic if and only if there exists an isomorphism of vector spaces $u : E \to F$ such that $f = u^{-1}gu$. Deduce that the $K[t]$-module structure induced by a linear map on a vector space is independent of the changes of bases.

21. Let K be a field and Q the quiver

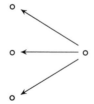

Consider the representations

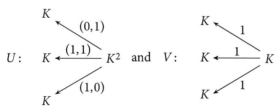

Prove that:
   (a) $\operatorname{Hom}_{KQ}(U, V) \cong K$.

    (b) $\mathrm{Hom}_{KQ}(V, U) \cong K^2$.
    (c) $\mathrm{End}\, V \cong K$.

22. Let K be a field and Q the quiver

Define representations of Q as follows:

$$U \text{ is given by } K \underset{(0\ 1)}{\overset{(1\ 0)}{\rightleftarrows}} K^2 \text{ and } V \text{ is given by } K^2 \underset{\left(\begin{smallmatrix}0 & 1 \\ 0 & 0\end{smallmatrix}\right)}{\overset{\left(\begin{smallmatrix}1 & 0 \\ 0 & 1\end{smallmatrix}\right)}{\rightleftarrows}} K^2$$

Prove each of the following statements:
    (a) $\mathrm{End}\, U \cong K$.
    (b) $\mathrm{End}\, V \cong K[t]/\langle t^2 \rangle$.
    (c) $\mathrm{Hom}_{KQ}(U, V) = 0$.
    (d) $\mathrm{Hom}_{KQ}(V, U) \cong K^2$.

23. Let K be a field, Q a quiver and $f : V \to W$ a morphism between representations of Q. Prove that f is injective, or surjective, or bijective, if and only if so is each of the component maps f_x, with $x \in Q_0$.

II.4 The isomorphism theorems

II.4.1 Monomorphisms and epimorphisms

Before proving the isomorphism theorems for modules, we need to look in a more systematic way at morphisms. Because, most of the time, we shall not be concerned with just one morphism, but with many of them, it is practical to devise methods of working 'without elements', that is, express properties of morphisms in terms of maps only, not considering elements of the sets on which these maps act. We start with two simple 'element-free' definitions which should give the flavour of the coming arguments. Throughout, A stands for a K-algebra.

II.4.1 Definition. Let M, N be A-modules and $f : M \to N$ an A-linear map.
    (a) f is a **monomorphism** if, for any linear maps $g, h : L \to M$ such that $fg = fh$, we have $g = h$.

$$L \underset{h}{\overset{g}{\rightrightarrows}} M \overset{f}{\longrightarrow} N$$

(b) f is an *epimorphism* if, for any linear maps $g, h : N \to L'$ such that $gf = hf$, we have $g = h$.

$$M \xrightarrow{\ f\ } N \underset{h}{\overset{g}{\rightrightarrows}} L'$$

Thus, monomorphisms and epimorphisms are, respectively, left cancellable and right cancellable linear maps. Setting $k = g - h$ in Definition II.4.1, we get a reformulation:

(a) f is a monomorphism if, for any k such that $fk = 0$, we have $k = 0$.

(b) f is an epimorphism if, for any k such that $kf = 0$, we have $k = 0$.

If one 'reverses the sense of arrows' in the diagram of Definition II.4.1(a), one obtains that of Definition II.4.1(b) and conversely, that is, monomorphisms become epimorphisms and vice versa. In this sense, we may (and shall) consider them as dual notions.

Although these properties may seem new, they represent well-known types of maps.

II.4.2 Lemma. *Let M, N be A-modules and $f : M \to N$ an A-linear map.*

(a) *f is a monomorphism if and only if it is injective.*

(b) *f is an epimorphism if and only if it is surjective.*

(c) *f is an isomorphism if and only if it is both a monomorphism and an epimorphism.*

Proof. (a) Let f be a monomorphism and consider the inclusion $i : \mathrm{Ker}\, f \hookrightarrow M$. Then $fi = 0$. Because f is a monomorphism, $i = 0$, that is, $\mathrm{Ker}\, f = 0$. Therefore, f is injective.

Conversely, if f is injective, assume that $fg = 0$. For any x in the domain of g, we have $f(g(x)) = (fg)(x) = 0$. Injectivity of f yields $g(x) = 0$. Because x is arbitrary, $g = 0$.

(b) Let f be an epimorphism and consider the projection $p : N \to N/\mathrm{Im}\, f$. Then $pf = 0$. Because f is an epimorphism, $p = 0$, that is, $N = \mathrm{Im}\, f$. Therefore, f is surjective.

Conversely, if f is surjective and $gf = 0$, then, for y in the domain N of g, there exists $x \in M$ such that $y = f(x)$. Then $g(y) = g(f(x)) = (gf)(x) = 0$. Because $y \in N$ is arbitrary, $g = 0$.

(c) Due to Lemma II.3.5, f is an isomorphism if and only if it is bijective, hence the statement. $\qquad\square$

We have expressed set-theoretical properties of linear maps (injectivity, surjectivity and bijectivity) in terms of intrinsic properties of maps, making no reference to elements. Observe indeed that Definition II.3.4 of isomorphism is also 'element-free'.

A word of warning: one can define monomorphisms and epimorphisms for ring, or algebra, morphisms exactly as for module morphisms. But then Lemma II.4.2 ceases to hold true, as one sees from the next example.

II.4.3 Example. The inclusion map $j : \mathbb{Z} \hookrightarrow \mathbb{Q}$ is a ring homomorphism which is injective, but certainly not surjective. We prove that j is a $\mathbb{Z}$-algebra (ring) epimorphism.

Let A be a $\mathbb{Z}$-algebra and $g, h : \mathbb{Q} \to A$ algebra morphisms such that $gj = hj$. Then, for $m \in \mathbb{Z}$, we have $g(m) = h(m)$ and, for $n \in \mathbb{Z}$ nonzero, we have

$$g(n)g\left(\frac{1}{n}\right) = g\left(n \cdot \frac{1}{n}\right) = g(1) = 1 = h(1) = h\left(n \cdot \frac{1}{n}\right) = h(n)h\left(\frac{1}{n}\right).$$

Hence, $g(1/n) = g(n)^{-1} = h(n)^{-1} = h(1/n)$. But then, for $m/n \in \mathbb{Q}$, we get

$$g\left(\frac{m}{n}\right) = g\left(m \cdot \frac{1}{n}\right) = g(m)g\left(\frac{1}{n}\right) = h(m)h\left(\frac{1}{n}\right) = h\left(m \cdot \frac{1}{n}\right) = h\left(\frac{m}{n}\right),$$

that is, $g = h$. Thus, j is an epimorphism which is not surjective.

On the other hand, j is a $\mathbb{Z}$-algebra monomorphism. If A is a $\mathbb{Z}$-algebra and $g, h : A \to \mathbb{Z}$ are algebra morphisms such that $jg = jh$, then, for $a \in A$, we have $jg(a) = jh(a)$. Injectivity of j implies $g(a) = h(a)$, so $g = h$. Thus j is a monomorphism and epimorphism but not an isomorphism (because it is not bijective).

This example shows that, while one can freely identify the terms 'monomorphism' and 'injective', and similarly 'epimorphism' and 'surjective', when dealing with linear maps between modules, one must exercise caution in general. We return to this question in Chapter IV.

The proof of part (a) of Lemma II.4.2 is based on the equivalence between injectivity of a linear map and vanishing of its kernel. This leads us to an 'element-free' definition of the kernel.

II.4.4 Theorem. *Let M, N be A-modules and $f : M \to N$ an A-linear map. The kernel of f is the unique module L, up to isomorphism, which satisfies the following properties:*
  (a) *There exists an A-linear map $k : L \to M$ such that $fk = 0$.*
  (b) *If there exists an A-linear map $k' : L' \to M$ such that $fk' = 0$, then there exists a unique A-linear map $u : L' \to L$ such that $k' = ku$.*

$$L \xrightarrow{\ k\ } M \xrightarrow{\ f\ } N$$

with $u : L' \to L$ and $k' : L' \to M$.

Proof. We first prove uniqueness of L up to isomorphism. Assume that L_1, L_2 both satisfy conditions (a) and (b), and that $k_1 : L_1 \to M, k_2 : L_2 \to M$ are the linear maps attached to L_1 and L_2, respectively, according to (a). Because $fk_1 = 0$, it follows from (b) applied to L_2 that there exists $u : L_1 \to L_2$ such that $k_1 = k_2 u$. Exchanging the roles of L_1 and L_2 yields $v : L_2 \to L_1$ such that $k_2 = k_1 v$.

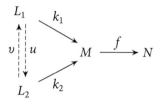

These equalities yield $k_1(vu) = (k_1 v)u = k_2 u = k_1 = k_1 1_{L_1}$. Because of uniqueness in (b), we get $vu = 1_{L_1}$. Similarly, $uv = 1_{L_2}$. Thus, u and v are inverse isomorphisms, and $L_1 \cong L_2$.

We prove existence. Because we know that there is, up to isomorphism, at most one module that satisfies conditions (a) and (b), it suffices to prove that $\operatorname{Ker} f = \{x \in M : f(x) = 0\}$ satisfies these conditions. Let k be the inclusion of the kernel as a submodule of M, then certainly $fk = 0$. Let $k' : L' \to M$ be an A-linear map such that $fk' = 0$. For any $x \in L'$, we have $fk'(x) = 0$, so $k'(x) \in \operatorname{Ker} f$, which means that $\operatorname{Im} k' \subseteq \operatorname{Ker} f$. Then the map $u = k' : L' \to \operatorname{Ker} f$ satisfies $k' = ku$. Its uniqueness follows from the fact that the inclusion k is injective, hence a monomorphism: if u_1, u_2 are such that $ku_1 = ku_2$, then $u_1 = u_2$. □

Some comments are in order. The proof of the uniqueness part is 'element-free': it rests on the uniqueness of the morphism u as asserted in (b). As seen in the proof, uniqueness of u says exactly that k is a monomorphism.

The property stated in the theorem is called the **universal property** of the kernel. The reason for this name is that the morphism k receives an arrow from any morphism which vanishes when composed with f (that is, in the universe of all such maps). As seen in the proof, the universal property implies the uniqueness of L, up to isomorphism. We show more universal properties in Chapter III.

Returning to Lemma II.4.2, the quotient module $N/\operatorname{Im} f$ plays in part (b) of the lemma a role corresponding to that of $\operatorname{Ker} f$ in part (a). For this reason, it is called the cokernel of f.

II.4.5 Definition. Let M, N be A-modules and $f : M \to N$ an A-linear map. The **cokernel** of f is the quotient module $\operatorname{Coker} f = N/\operatorname{Im} f$.

A linear map f is injective (a monomorphism) if and only if its kernel is zero. In the same way, it is surjective (that is, an epimorphism) if and only if its cokernel is zero. Indeed, $\operatorname{Coker} f = 0$ if and only if $N = \operatorname{Im} f$, that is, if and only if f is surjective.

II.4.6 Example. Let K be a field, and consider the linear map $f : K \to K^2$ given by $\alpha \mapsto (\alpha, \alpha)$, for $\alpha \in K$. Its image is the 'diagonal' subspace $D = \{(\alpha, \alpha) : \alpha \in K\}$ of K^2. Because $\dim_K D = 1$, the cokernel K^2/D is also one-dimensional, and isomorphic to any supplement of D inside K^2. For instance, $\operatorname{Coker} f = K^2/D \cong \{(\alpha, 0) : \alpha \in K\}$.

Because monomorphisms are obtained from epimorphisms by reversing the sense of arrows and vice-versa, it is logical to think that one obtains a universal property for cokernels by reversing the sense of arrows in that for kernels, stated in Theorem II.4.4.

II.4.7 Theorem. *Let M, N be A-modules and $f : M \to N$ an A-linear map. The cokernel of f is the unique module L, up to isomorphism, which satisfies the following properties:*
  (a) *There exists an A-linear map $k : N \to L$ such that $kf = 0$.*
  (b) *If there exists an A-linear map $k' : N \to L'$ such that $k'f = 0$, then there exists a unique A-linear map $u : L \to L'$ such that $k' = uk$*

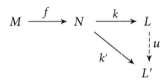

Proof. We first prove uniqueness as we did it in Theorem II.4.4. Assume that L_1, L_2 both satisfy conditions (a) and (b), and that $k_1 : N \to L_1, k_2 : N \to L_2$ are the linear maps attached to L_1, L_2 according to (a). Because $k_1 f = 0$, applying (b) to L_2 yields $u : L_1 \to L_2$ such that $k_1 = uk_2$. Exchanging the roles of L_1 and L_2, we get $v : L_2 \to L_1$ such that $k_2 = vk_1$.

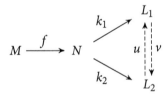

Hence, $uvk_1 = k_1 = 1_{L_2}k_1$. Uniqueness in (b) yields $uv = 1_{L_2}$. Similarly, $vu = 1_{L_1}$.

For existence, we verify that $\operatorname{Coker} f = N/\operatorname{Im} f$ satisfies conditions (a) and (b). Let $k : N \to \operatorname{Coker} f$ be the projection. For $x \in M$, we have $f(x) \in \operatorname{Im} f$ and therefore $kf(x) = 0$ in $\operatorname{Coker} f$. Thus, $kf = 0$. Let $k' : N \to L'$ be such that $k'f = 0$. We construct an A-linear map $u : \operatorname{Coker} f \to L'$ such that $uk = k'$. Indeed, if $x + \operatorname{Im} f \in \operatorname{Coker} f$, set $u(x + \operatorname{Im} f) = k'(x)$. Then u is defined unambiguously: $x_1 + \operatorname{Im} f = x_2 + \operatorname{Im} f$ implies the existence of $y \in M$ such that $x_1 = x_2 + f(y)$, but then $k'(x_1) = k'(x_2) + k'f(y) = k'(x_2)$ because $k'f = 0$. Uniqueness of u follows from the fact that the projection $k : N \to \operatorname{Coker} f$ is surjective and hence an epimorphism. □

The proof of the uniqueness part was obtained simply by reversing the sense of arrows in the proof of uniqueness in Theorem II.4.4. The property of cokernels stated in Theorem II.4.7 above is called their **universal property**: It says that the morphism k sends an arrow to any morphism which vanishes when composed with f. Uniqueness in part (b) amounts to saying that k is an epimorphism.

II.4.2 Exact sequences

As a first approximation, exact sequences can be viewed as an attempt to arrange linear maps in a manner which allows to visualise properties of interest. They are an efficient tool for constructing linear maps with predictable properties, and to compute properties of modules by means of linear maps.

II.4.8 Definition. A sequence of linear maps between modules $L \xrightarrow{f} M \xrightarrow{g} N$ is **exact at** M if $\operatorname{Im} f = \operatorname{Ker} g$. A longer sequence of linear maps

$$\cdots \to M_{i+1} \xrightarrow{f_{i+1}} M_i \xrightarrow{f_i} M_{i-1} \to \cdots$$

is **exact** if it is exact at each M_i, that is, if $\operatorname{Im} f_{i+1} = \operatorname{Ker} f_i$, for each i.

Assume that $L \xrightarrow{f} M \xrightarrow{g} N$ is a sequence of linear maps. To say that $gf = 0$ amounts to saying that $\operatorname{Im} f \subseteq \operatorname{Ker} g$, that is, the image of the first map is sent to zero by the second. Thus, the sequence is exact if and only if $gf = 0$ and also $\operatorname{Im} f \supseteq \operatorname{Ker} g$. In other words, an element is sent to zero by the second map if and only if it belongs to the image of the first map. The figure below illustrates this situation.

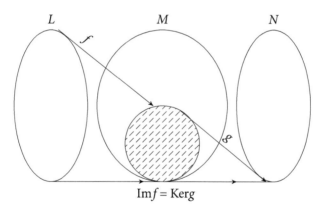

$$\operatorname{Im} f = \operatorname{Ker} g$$

We give an example of exact sequence.

II.4.9 Example. Let $A = \mathbb{Z}$, p a prime integer and $L = M = N = \mathbb{Z}/p^2\mathbb{Z}$. Let $f : \mathbb{Z}/p^2\mathbb{Z} \to \mathbb{Z}/p^2\mathbb{Z}$ be the map induced from multiplication by p, that is, $f : x + p^2\mathbb{Z} \mapsto px + p^2\mathbb{Z}$, for $x \in \mathbb{Z}$. Then f is linear. Consider the sequence

$$\mathbb{Z}/p^2\mathbb{Z} \xrightarrow{f} \mathbb{Z}/p^2\mathbb{Z} \xrightarrow{f} \mathbb{Z}/p^2\mathbb{Z}.$$

Because of Example II.3.6(b), both the kernel and the image of f equal the unique nonzero proper submodule $p\mathbb{Z}/p^2\mathbb{Z}$ of $\mathbb{Z}/p^2\mathbb{Z}$. Therefore, the sequence is exact. The following picture showing the submodule lattices helps giving an intuitive idea of the situation.

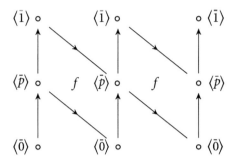

Namely, the image of the first map coincides (exactly!) with the part of the middle term which is sent to zero by the second map, that is, the kernel of the second map.

We claimed that one can visualise properties of linear maps by looking at exact sequences. The following proposition summarises some of these properties.

II.4.10 Proposition. *Let M, N be A-modules and $f : M \to N$ an A-linear map.*

(a) *f is a monomorphism if and only if $0 \to M \xrightarrow{f} N$ is exact.*

(b) *f is an epimorphism if and only if $M \xrightarrow{f} N \to 0$ is exact.*

(c) *f is an isomorphism if and only if $0 \to M \xrightarrow{f} N \to 0$ is exact.*

(d) *The sequence $0 \to \operatorname{Ker} f \xrightarrow{i} M \xrightarrow{f} N$, where i is the inclusion, is exact.*

(e) *The sequence $M \xrightarrow{f} N \xrightarrow{p} \operatorname{Coker} f \to 0$, where p is the projection, is exact.*

(f) *With i, p as in (d),(e), the sequence $0 \to \operatorname{Ker} f \xrightarrow{i} M \xrightarrow{f} N \xrightarrow{p} \operatorname{Coker} f \to 0$ is exact.*

Proof. Follows directly from the definitions. □

An exact sequence of the form $0 \to L \to M \xrightarrow{f} N$ is said to be **left exact**. Thus, the sequence of (d) above is left exact. One asks whether conversely, given a left exact sequence $0 \to L \to M \xrightarrow{f} N$, is L isomorphic to the kernel of f? The answer is positive, and we prove it after the next proposition.

II.4.11 Proposition. *Assume that we have a commutative diagram with exact rows of A-modules and A-linear maps*

$$
\begin{array}{ccccccc}
0 & \longrightarrow & L & \xrightarrow{f} & M & \xrightarrow{g} & N \\
 & & \downarrow{\scriptstyle u} & & \downarrow{\scriptstyle v} & & \downarrow{\scriptstyle w} \\
0 & \longrightarrow & L' & \xrightarrow{f'} & M' & \xrightarrow{g'} & N'
\end{array}
$$

Then there exists a unique A-linear map $u : L \to L'$ making the left square commute. Moreover, if v is a monomorphism, then so is u.

Proof. Let $x \in L$, we claim that $vf(x) \in \operatorname{Im} f'$. Indeed, $g'vf(x) = wgf(x) - 0$ because $gf = 0$, due to exactness of the upper row. Hence, $vf(x) \in \operatorname{Ker} g' = \operatorname{Im} f'$, as required. Because f' is a monomorphism, there exists a unique $x' \in L'$ such that $vf(x) = f'(x')$. Thus, the map $u : L \to L'$ defined by $u : x \mapsto x'$, for $x \in L$, satisfies $f'u = vf$. We prove its linearity: let $x_1, x_2 \in L$ and $a_1, a_2 \in A$. There exist unique $x_1', x_2' \in L'$ such that $f'(x_1') = vf(x_1)$ and $f'(x_2') = vf(x_2)$. Because f', f, v are linear, we have

$$f'(x_1'a_1 + x_2'a_2) = f'(x_1')a_1 + f'(x_2')a_2 = vf(x_1)a_1 + vf(x_2)a_2 = vf(x_1a_1 + x_2a_2).$$

Therefore, $u(x_1a_1 + x_2a_2) = x_1'a_1 + x_2'a_2 = u(x_1)a_1 + u(x_2)a_2$.

Assume that v is a monomorphism. Because so is f, then so is $f'u = vf$. Hence so is u. $\qquad\square$

At the last step of the proof, we used the well-known property of maps saying that if a composite gf is injective, or surjective, then so is f, or g, respectively. This property is used freely in the sequel. Also, in the proof, linearity of u follows from the fact that it is constructed using linear maps, this will be the case for all the maps we construct. The morphism u is said to be obtained by **passing to kernels**. We prove the announced result.

II.4.12 Corollary. *Let $0 \to L \xrightarrow{f} M \xrightarrow{g} N$ be a left exact sequence of A-modules and A-linear maps and $i : \operatorname{Ker} g \to M$ the inclusion. Then there exists a unique isomorphism $u : L \to \operatorname{Ker} g$ making the following diagram commute:*

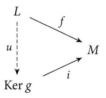

Proof. Because of Propositions II.4.10(d) and II.4.11, there exists a unique map $u : L \to \operatorname{Ker} g$ making the following diagram commute:

$$
\begin{array}{ccccc}
0 & \longrightarrow & L & \xrightarrow{f} & M & \xrightarrow{g} & N \\
& & \downarrow{u} & & \downarrow{1_M} & & \downarrow{1_N} \\
0 & \longrightarrow & \operatorname{Ker} g & \xrightarrow{i} & M & \xrightarrow{g} & N
\end{array}
$$

Moreover, u is a monomorphism, because so is 1_M. Let $y \in \operatorname{Ker} g$, then $gi(y) = 0$ yields a unique $x \in L$ such that $i(y) = f(x)$. Because of the definition of u, we have $u(x) = y$. Hence, u is surjective, and therefore an isomorphism. $\qquad\square$

Thus, in an exact sequence of the form $0 \to L \xrightarrow{f} M \xrightarrow{g} N$, one has $L \cong \operatorname{Ker} g$.

One could have used the opposite approach, that is, to prove directly Corollary II.4.14 then deduce Proposition II.4.11. Indeed, under the hypothesis of the Corollary, the universal property of the kernel and $gf = 0$ give a unique morphism $u : L \to$ Ker$g$ such that $f = iu$. Because f is a monomorphism, so is u. Let $x \in$ Ker$g$. Then $gi(x) = 0$ gives $i(x) \in$ Ker$g =$ Imf and there exists $y \in L$ such that $i(x) = f(y) = iu(y)$. Because i is injective, $x = u(y)$. Therefore u is surjective, and so an isomorphism. This completes the proof of the Corollary. Next, with the hypothesis of the Proposition II.4.11, the universal property of $L' =$ Kerg' and $g'vf = wgf = 0$ give the morphism $u : L \to L'$ such that $f'u = vf$. If v is a monomorphism, this same equality implies that so is u.

Cokernels are obtained from kernels by reversing arrows. An exact sequence of the form $L \to M \to N \to 0$ is called **right exact**. We now prove the statements corresponding to the previous proposition and corollary when one reverses arrows.

II.4.13 Proposition. *Assume that we have a commutative diagram with exact rows of A-modules and A-linear maps*

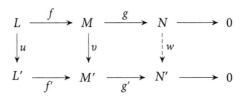

Then there exists a unique A-linear map $w : N \to N'$ making the right square commute. Moreover, if v is an epimorphism, then so is w.

Proof. Let $z \in N$. Because g is an epimorphism, there exists (at least one) $y \in M$ such that $z = g(y)$. Set $w(z) = g'v(y) \in N'$. We claim that this formula defines unambiguously a map $w : N \to N'$. Assume that $z = g(y_1) = g(y_2)$. Then we have $y_1 - y_2 \in$ Ker$g =$ Imf, so there exists $x \in L$ such that $y_1 - y_2 = f(x)$. Therefore,

$$g'v(y_1) = g'v(y_2 + f(x)) = g'v(y_2) + g'vf(x) = g'v(y_2) + g'f'u(x) = g'v(y_2)$$

because $g'f' = 0$. This establishes our claim. It is easy to prove, as in Proposition II.4.11, that w is linear. The last statement follows from $wg = g'v$. □

II.4.14 Corollary. *Let $L \xrightarrow{f} M \xrightarrow{g} N \to 0$ be a right exact sequence of A-modules and A-linear maps and $p : M \to$ Coker$f$ the projection. Then there exists a unique isomorphism $w : N \to$ Coker$f$ making the following diagram commute:*

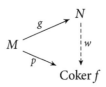

Proof. Because of Propositions II.4.10(e) and II.4.13, there exists a unique A-linear map $w : N \to \operatorname{Coker} f$ making the following diagram commute:

$$
\begin{array}{ccccccc}
L & \xrightarrow{\;f\;} & M & \xrightarrow{\;g\;} & N & \longrightarrow & 0 \\
\downarrow{\scriptstyle 1_L} & & \downarrow{\scriptstyle 1_M} & & \downarrow{\scriptstyle w} & & \\
L & \xrightarrow{\;f\;} & M & \xrightarrow{\;p\;} & \operatorname{Coker} f & \longrightarrow & 0
\end{array}
$$

Moreover, w is an epimorphism. Let $z \in N$ be such that $w(z) = 0$. Because g is an epimorphism, there exists $y \in M$ such that $z = g(y)$. Hence, $p(y) = wg(y) = w(z) = 0$, so $y \in \operatorname{Ker} p = \operatorname{Im} f$ and there exists $x \in L$ such that $y = f(x)$. But then $z = g(y) = gf(x) = 0$ because $gf = 0$. This proves that w is injective and hence an isomorphism. □

Thus, in an exact sequence of the form $L \xrightarrow{f} M \xrightarrow{g} N \to 0$, one always has $N \cong \operatorname{Coker} f$. The morphism w of Proposition II.4.13 is said to be obtained by **passing to cokernels**.

Exactly as for the kernel, one could have taken the opposite approach, and prove directly this Corollary and then deduce Proposition II.4.13. We leave the details to the reader as an exercise.

Results of this subsection and the previous must have convinced the reader that, given a result involving a diagram, the statement obtained by reversing the sense of arrows is very likely to hold true. We formalise this idea in the next chapter.

A sequence which is both left and right exact is called short exact.

II.4.15 Definition. An exact sequence of the form $0 \to L \xrightarrow{f} M \xrightarrow{g} N \to 0$ is a **short exact sequence**.

If a sequence $0 \to L \xrightarrow{f} M \xrightarrow{g} N \to 0$ is short exact, then f is a monomorphism and g an epimorphism, because of Proposition II.4.10(a)(b). Moreover, $L \cong \operatorname{Ker} g$ and $N \cong \operatorname{Coker} f$ because of Corollaries II.4.12, II.4.14, respectively. Any monomorphism $f : L \to M$ induces a short exact sequence $0 \to L \xrightarrow{f} M \to \operatorname{Coker} f \to 0$ and, dually, any epimorphism $g : M \to N$ induces a short exact sequence $0 \to \operatorname{Ker} g \to M \xrightarrow{g} N \to 0$.

In general, a sequence of the form $0 \to L \xrightarrow{f} M \xrightarrow{g} N \to 0$ is short exact if and only if:

(a) f is a monomorphism,
(b) g is an epimorphism, and
(c) $\operatorname{Im} f = \operatorname{Ker} g$.

II.4.16 Examples.

(a) Let A be an algebra and L, N be A-modules. Consider the sequence

$$
0 \to L \xrightarrow{j} L \times N \xrightarrow{p} N \to 0
$$

where $j : x \mapsto (x, 0)$ and $p : (x, y) \mapsto y$. Then j is injective and p is surjective. Moreover, $\text{Im } j = \{(x, 0) \in L \times N : x \in L\} = \text{Ker} p$. The sequence is short exact.

(b) Let A be a commutative algebra and $a \in A$. Multiplication by a induces an A-linear map $\mu_a : A_A \to A_A$ defined by $x \mapsto ax$, for $x \in A$. This map is injective if and only if a is not a zero divisor. The image of μ_a is the submodule (or ideal) aA of A, and hence its cokernel is A/aA, which is nonzero if and only if a is not invertible in A. Thus, if a is neither a zero divisor nor invertible, we have a short exact sequence

$$0 \to A \xrightarrow{\mu_a} A \to A/aA \to 0.$$

(c) Let p be a prime integer and consider the sequence of abelian groups

$$0 \to \mathbb{Z}/p\mathbb{Z} \xrightarrow{f} \mathbb{Z}/p^2\mathbb{Z} \xrightarrow{g} \mathbb{Z}/p\mathbb{Z} \to 0$$

with $f : 1 + p\mathbb{Z} \mapsto p + p^2\mathbb{Z}, g : 1 + p^2\mathbb{Z} \mapsto 1 + p\mathbb{Z}$. Then f is a monomorphism because $0 = f(x + p\mathbb{Z}) = px + p^2\mathbb{Z}$ implies $px \in p^2\mathbb{Z}$ so that p divides x and hence $x + p\mathbb{Z} = 0$. Because of Example II.3.6(b), the image of f is the set of all $px + p^2\mathbb{Z}$, with $x \in \mathbb{Z}$, that is, the unique nonzero proper submodule $p\mathbb{Z}/p^2\mathbb{Z}$ of $\mathbb{Z}/p^2\mathbb{Z}$. The map g is clearly an epimorphism and its kernel consists of the $x + p^2\mathbb{Z} \in \mathbb{Z}/p^2\mathbb{Z}$ such that $x + p\mathbb{Z} = 0$ in $\mathbb{Z}/p\mathbb{Z}$. But this means that p divides x so $\text{Ker } g = p\mathbb{Z}/p^2\mathbb{Z}$. Hence, the sequence is exact. Drawing Hasse quivers as in Examples II.3.6(b) and II.4.9, this short exact sequence may be visualised as follows:

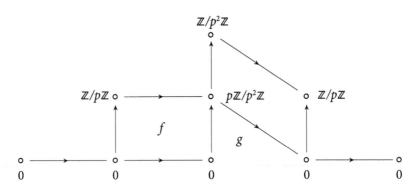

(d) Let K be a field, Q a finite quiver and V a representation of Q. For any subrepresentation V' of V, the inclusion $i : V' \to V$ is a monomorphism, see Exercise II.3.24. We compute its cokernel W, so as to obtain a short exact sequence $0 \to V' \xrightarrow{i} V \xrightarrow{p} W \to 0$. If we identify representations with the KQ-modules they represent, we have $W = V/\text{Im } i$. The image of the inclusion being V' itself, we get $W = V/V'$. We compute the latter.

For each $x \in Q_0$, the space $V'(x)$ is a subspace of $V(x)$, hence $W(x) = V(x)/V'(x)$ is isomorphic, as a vector space, to any supplement of $V'(x)$ inside $V(x)$. For any arrow $\alpha : x \to y$, the map $V'(\alpha)$ is the restriction of $V(\alpha)$ to $V'(x)$, so, due to Proposition II.4.13, there exists a unique K-linear map $W(\alpha) : W(x) \to W(y)$ induced by passing to cokernels, that is, such that the following diagram commutes:

$$
\begin{array}{ccccccccc}
0 & \longrightarrow & V'(x) & \xrightarrow{\ i_x\ } & V(x) & \xrightarrow{\ p_x\ } & W(x) & \longrightarrow & 0 \\
& & \downarrow{\scriptstyle V'(\alpha)} & & \downarrow{\scriptstyle V(\alpha)} & & \downarrow{\scriptstyle W(\alpha)} & & \\
0 & \longrightarrow & V'(y) & \xrightarrow{\ i_y\ } & V(y) & \xrightarrow{\ p_y\ } & W(y) & \longrightarrow & 0
\end{array}
$$

The representation $W = (W(x), W(\alpha))_{x \in Q_0, \alpha \in Q_1}$ so constructed is the required cokernel.

Assuming Q to be the quiver $3 \xrightarrow{\alpha} 2 \xrightarrow{\beta} 1$ considered in Example II.2.4(g) and using the representations V and V' defined there, we get W as the lower line in the diagram

$$
\begin{array}{ccccc}
0 & \xrightarrow{\ V'(\alpha)\ } & 0 \oplus Ke_3 & \xrightarrow{\ V'(\beta)\ } & Ke_1 \\
\uparrow & & \uparrow & & \uparrow \\
Ke_4 & \xrightarrow{\ V(\alpha)\ } & Ke_2 \oplus Ke_3 & \xrightarrow{\ V(\beta)\ } & Ke_1 \\
\downarrow{\scriptstyle 1} & & \downarrow{\scriptstyle (1\ 0)} & & \downarrow{\scriptstyle 0} \\
Ke_4 & \xrightarrow{\ W(\alpha)\ } & Ke_2 \oplus 0 & \xrightarrow{\ W(\beta)\ } & 0
\end{array}
$$

where $W(\beta)$ is necessarily the zero map, while $W(\alpha)$ is given by $\lambda e_4 \mapsto \lambda e_2$, for $\lambda \in K$ (if one identifies bases, this is just the identity map).

II.4.3 Diagrammatic lemmata

Diagrammatic lemmata replace calculations about morphisms by arguments of a visual nature.

II.4.17 Lemma. The Four Lemma. *Assume that we have a commutative diagram with exact rows of A-modules and A-linear maps*

$$
\begin{array}{ccccccc}
M_1 & \xrightarrow{\ u_1\ } & M_2 & \xrightarrow{\ u_2\ } & M_3 & \xrightarrow{\ u_3\ } & M_4 \\
\downarrow{\scriptstyle f_1} & & \downarrow{\scriptstyle f_2} & & \downarrow{\scriptstyle f_3} & & \downarrow{\scriptstyle f_4} \\
N_1 & \xrightarrow{\ v_1\ } & N_2 & \xrightarrow{\ v_2\ } & N_3 & \xrightarrow{\ v_3\ } & N_4
\end{array}
$$

(a) *If f_2, f_4 are monomorphisms, and f_1 is an epimorphism, then f_3 is a monomorphism.*

(b) *If f_1, f_3 are epimorphisms, and f_4 is a monomorphism, then f_2 is an epimorphism.*

Proof. (a) Assume that $x_3 \in \operatorname{Ker} f_3$, then $f_4 u_3(x_3) = v_3 f_3(x_3) = 0$. Now, f_4 is a monomorphism. Hence, $u_3(x_3) = 0$, that is, $x_3 \in \operatorname{Ker} u_3 = \operatorname{Im} u_2$, so there exists $x_2 \in M_2$ such that $x_3 = u_2(x_2)$. Moreover, $v_2 f_2(x_2) = f_3 u_2(x_2) = f_3(x_3) = 0$, hence $f_2(x_2) \in \operatorname{Ker} v_2 = \operatorname{Im} v_1$. That is, there exists $y_1 \in N_1$ such that $f_2(x_2) = v_1(y_1)$. But f_1 is an epimorphism, so there exists $x_1 \in M_1$ such that $y_1 = f_1(x_1)$. Therefore, $f_2(x_2) = v_1(y_1) = v_1 f_1(x_1) = f_2 u_1(x_1)$ implies $x_2 = u_1(x_1)$, because f_2 is a monomorphism. Consequently, $x_3 = u_2(x_2) = u_2 u_1(x_1) = 0$, because $u_2 u_1 = 0$. Therefore, f_3 is a monomorphism.

(b) Let $y_2 \in N_2$. Because f_3 is an epimorphism, there exists $x_3 \in M_3$ such that $v_2(y_2) = f_3(x_3)$. Now $f_4 u_3(x_3) = v_3 f_3(x_3) = v_3 v_2(y_2) = 0$, because $v_3 v_2 = 0$. Because f_4 is a monomorphism, $u_3(x_3) = 0$ so $x_3 \in \operatorname{Ker} u_3 = \operatorname{Im} u_2$ and there exists $x_2 \in M_2$ such that $x_3 = u_2(x_2)$. Therefore, $v_2(y_2) = f_3(x_3) = f_3 u_2(x_2) = v_2 f_2(x_2)$ or $v_2(y_2 - f_2(x_2)) = 0$ and so $y_2 - f_2(x_2) \in \operatorname{Ker} v_2 = \operatorname{Im} v_1$. There exists $y_1 \in N_1$ such that $y_2 - f_2(x_2) = v_1(y_1)$, Because f_1 is an epimorphism, there exists $x_1 \in M_1$ such that $y_1 = f_1(x_1)$. Therefore, $y_2 = f_2(x_2) + v_1(y_1) = f_2(x_2) + v_1 f_1(x_1) = f_2(x_2) + f_2 u_1(x_1) = f_2(x_2 + u_1(x_1))$. We have found a preimage for y_2. $\quad\square$

We point out that all hypotheses made about the f_i have been used in the proof.

The technique used in the proof of this lemma is known as ***diagram chasing***. It consists in following the construction of elements in the different modules appearing in the diagram. The construction itself is based either on exactness of certain sequences or on properties of the maps (injectivity, surjectivity) stated in, or following from, the hypothesis. The proofs may appear long, but each step is so easy that they offer no real challenge. Actually, we already used diagram chasing without naming it in the proofs of statements II.4.4 to II.4.14.

II.4.18 Lemma. The Five Lemma. *Assume that we have a commutative diagram with exact rows of A-modules and A-linear maps*

$$
\begin{array}{ccccccccc}
M_1 & \xrightarrow{u_1} & M_2 & \xrightarrow{u_2} & M_3 & \xrightarrow{u_3} & M_4 & \xrightarrow{u_4} & M_5 \\
\downarrow{f_1} & & \downarrow{f_2} & & \downarrow{f_3} & & \downarrow{f_4} & & \downarrow{f_5} \\
N_1 & \xrightarrow{v_1} & N_2 & \xrightarrow{v_2} & N_3 & \xrightarrow{v_3} & N_4 & \xrightarrow{v_4} & N_5
\end{array}
$$

(a) *If f_2, f_4 are monomorphisms, and f_1 is an epimorphism, then f_3 is a monomorphism.*

(b) *If f_2, f_4 are epimorphisms, and f_5 is a monomorphism, then f_3 is an epimorphism.*

(c) *If f_2, f_4 are isomorphisms, f_1 is an epimorphism and f_5 is a monomorphism, then f_3 is an isomorphism.*

Proof. (a) Apply the Four Lemma II.4.17, part (a), to the right-hand three squares.

(b) Apply the same, part (b), to the left-hand three squares.

(c) It is the conjunction of (a) and (b). $\quad\square$

In the previous lemmata, diagram chasing was used to prove properties of morphisms in an existing diagram. In the next, we construct a morphism yielding a new exact sequence.

II.4.19 Lemma. The Snake Lemma. *Assume that we have a commutative diagram with exact rows of A-modules and A-linear maps*

$$
\begin{array}{ccccccc}
L & \xrightarrow{u} & M & \xrightarrow{v} & N & \longrightarrow & 0 \\
\downarrow{\scriptstyle f} & & \downarrow{\scriptstyle g} & & \downarrow{\scriptstyle h} & & \\
0 & \longrightarrow & L' & \xrightarrow{u'} & M' & \xrightarrow{v'} & N'
\end{array}
$$

Then there exists an exact sequence

$$
\operatorname{Ker} f \xrightarrow{u_1} \operatorname{Ker} g \xrightarrow{v_1} \operatorname{Ker} h \xrightarrow{\delta} \operatorname{Coker} f \xrightarrow{u_2} \operatorname{Coker} g \xrightarrow{v_2} \operatorname{Coker} h
$$

where u_1, v_1 are deduced by passing to the kernels, u_2, v_2 by passing to the cokernels and δ is explicited below. Moreover:
  (a) *If u is a monomorphism, then so is u_1.*
  (b) *If v' is an epimorphism, then so is v_2.*

Proof. Let $i : \operatorname{Ker} f \to L, j : \operatorname{Ker} g \to M, k : \operatorname{Ker} h \to N$ be the inclusions and $p : L' \to \operatorname{Coker} f, q : M' \to \operatorname{Coker} g, r : N' \to \operatorname{Coker} h$ the projections. We wish to construct A-linear maps $u_1, v_1, \delta, u_2, v_2$ making exact the dotted line in the diagram below, where all columns and the two middle rows are exact

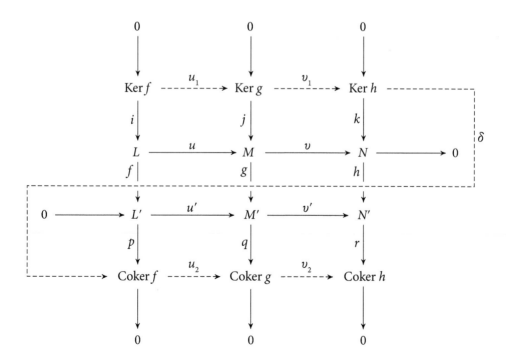

(the design of the dotted line explains the name of the lemma). Because of Propositions II.4.11 and II.4.13, there exist morphisms u_1, v_1 deduced by passing to kernels and making the two upper squares commutative, as well as morphisms u_2, v_2 deduced by passing to cokernels and making the two lower squares commutative. We divide the proof into several steps.

(1) The sequence $\operatorname{Ker} f \xrightarrow{u_1} \operatorname{Ker} g \xrightarrow{v_1} \operatorname{Ker} h$ is exact. Indeed, $kv_1 u_1 = vui = 0$ gives $v_1 u_1 = 0$, because k is a monomorphism. So, $\operatorname{Im} u_1 \subseteq \operatorname{Ker} v_1$. Let $y \in \operatorname{Ker} v_1$. Then $vj(y) = kv_1(y) = 0$, so $j(y) \in \operatorname{Ker} v = \operatorname{Im} u$. Hence, there exists $x \in L$ such that $j(y) = u(x)$. But $u' f(x) = gu(x) = gj(y) = 0$ because $gj = 0$. Now u' is a monomorphism, therefore $f(x) = 0$, so $x \in \operatorname{Ker} f$, that is, $x = i(x)$. Hence, $j(y) = u(x) = ui(x) = ju_1(x)$ gives $y = u_1(x)$ because j is a monomorphism. Consequently $y \in \operatorname{Im} u_1$ and we are done.

(2) The sequence $\operatorname{Coker} f \xrightarrow{u_2} \operatorname{Coker} g \xrightarrow{v_2} \operatorname{Coker} h$ is exact. The proof is similar to that of (1) and omitted.

(3) The construction of the A-linear map $\delta : \operatorname{Ker} h \to \operatorname{Coker} f$. Let $z \in \operatorname{Ker} h$. Because v is an epimorphism, there exists $y \in M$ such that $v(y) = k(z)$. Moreover, $v'g(y) = hv(y) = hk(z) = 0$. Therefore, $g(y) \in \operatorname{Ker} v' = \operatorname{Im} u'$. Because u' is injective, there exists a *unique* $x' \in L'$ such that $g(y) = u'(x')$. We set $\delta(z) = p(x') = x' + \operatorname{Im} f \in \operatorname{Coker} f$.

We prove that this definition is unambiguous, that is, does not depend on the choice of the preimage y of $k(z)$. Assume that $v(y_1) = v(y_2)$. Then $y_1 - y_2 \in \operatorname{Ker} v = \operatorname{Im} u$, so there exists $x \in L$ such that $y_1 - y_2 = u(x)$. Hence, $g(y_1) - g(y_2) = gu(x) = u'f(x)$. Let x'_1, x'_2 be the unique preimages of $g(y_1), g(y_2)$, respectively, under u'. We get $u'(x'_1 - x'_2) = u'(x'_1) - u'(x'_2) = g(y_1) - g(y_2) = u'f(x)$. Therefore, $x'_1 - x'_2 = f(x)$, because u' is a monomorphism. Hence, $p(x'_1) - p(x'_2) = pf(x) = 0$ because $pf = 0$.

Thus, we defined a map $\delta : \operatorname{Ker} h \to \operatorname{Coker} f$. We prove its linearity. Let $z_1, z_2 \in \operatorname{Ker} h$ and $a_1, a_2 \in A$. Take $y_1, y_2 \in M$ such that $v(y_1) = k(z_1), v(y_2) = k(z_2)$. Then $v(y_1 a_1 + y_2 a_2) = k(z_1 a_1 + z_2 a_2)$, because v, k are linear. Take $x'_1, x'_2 \in L'$ such that $u'(x'_1) = g(y_1), u'(x'_2) = g(y_2)$. Then $g(y_1 a_1 + y_2 a_2) = u'(x'_1 a_1 + x'_2 a_2)$, because g, u' are linear. Hence,

$$\delta(z_1 a_1 + z_2 a_2) = p(x'_1 a_1 + x'_2 a_2) = p(x'_1)a_1 + p(x'_2)a_2 = \delta(z_1)a_1 + \delta(z_2)a_2.$$

(4) The sequence $\operatorname{Ker} g \xrightarrow{v_1} \operatorname{Ker} h \xrightarrow{\delta} \operatorname{Coker} f$ is exact. We claim that $\operatorname{Im} v_1 \subseteq \operatorname{Ker} \delta$. Let $z \in \operatorname{Im} v_1$ and $y \in \operatorname{Ker} g$ be such that $z = v_1(y)$. Then $z = k(z) = kv_1(y) = vj(y)$. Thus $j(y)$ is a preimage of z under v. But $gj(y) = 0$ because $gj = 0$. Hence, the unique preimage of $gj(y)$ under the monomorphism u' is $x' = 0$. Therefore, $\delta(z) = p(x') = p(0) = 0$, which proves our claim.

We next prove that $\operatorname{Ker} \delta \subseteq \operatorname{Im} v_1$. Suppose that $z \in \operatorname{Ker} h$ satisfies $\delta(z) = 0$. Construct x' as before, then $p(x') = 0$ yields $x' \in \operatorname{Ker} p = \operatorname{Im} f$. So, there exists $x \in L$ such that $x' = f(x)$. Let $y \in M$ be such that $v(y) = k(z)$. Then $g(y) = u'(x') = u'f(x) = gu(x)$ so that $y - u(x) \in \operatorname{Ker} g = \operatorname{Im} j$. Let

$y_0 \in \operatorname{Ker} g$ be such that $y = j(y_0) + u(x)$. Then $k(z) = v(y) - vj(y_0) + vu(x) = vj(y_0) = kv_1(y_0)$. Because k is a monomorphism, this gives $z = v_1(y_0)$ and so $\operatorname{Ker} \delta \subseteq \operatorname{Im} v_1$.

(5) The sequence $\operatorname{Ker} h \xrightarrow{\delta} \operatorname{Coker} f \xrightarrow{u_2} \operatorname{Coker} g$ is exact. The proof is similar to that of (4) and omitted.

(6) If u is a monomorphism, then so is u_1. Indeed, because u and i are monomorphisms, then so is $ju_1 = ui$. The statement follows.

(7) If v' is an epimorphism, then so is v_2. This is due, as above, to the equality $v_2 q = rv'$. $\qquad\square$

The morphism δ is called the ***connecting morphism*** (it connects kernels with cokernels). A simple (and very informal) way to remember its construction is to write

$$\delta(z) = pu'^{-1}gv^{-1}k(z)$$

but one should not forget that while u' is a monomorphism (and so u'^{-1} is uniquely defined on the image of u'), the notation v^{-1} corresponds to the choice of an arbitrary preimage of $k(z)$, all of which giving the same final result. In the diagram, the element z travels as follows:

$$\operatorname{Ker} h \to N \leftarrow M \to M' \leftarrow L' \to \operatorname{Coker} f$$
$$z \mapsto k(z) \mapsto y \mapsto g(y) \mapsto x' \mapsto p(x').$$

The exact sequence of the lemma, namely

$$\operatorname{Ker} f \xrightarrow{u_1} \operatorname{Ker} g \xrightarrow{v_1} \operatorname{Ker} h \xrightarrow{\delta} \operatorname{Coker} f \xrightarrow{u_2} \operatorname{Coker} g \xrightarrow{v_2} \operatorname{Coker} h$$

is sometimes called the ***Ker-Coker sequence***. We shall present several applications of the Snake Lemma. The most immediate is the following.

II.4.20 Lemma. The 3×3 Lemma. *Assume that we have a commutative diagram with exact columns of A-modules and A-linear maps*

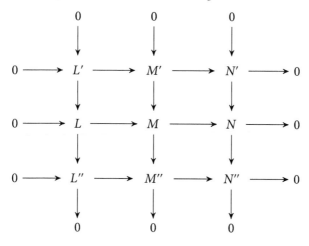

(a) *If the two upper rows are exact, then so is the lower one.*
(b) *If the two lower rows are exact, then so is the upper one.*

Proof. For (a), apply the Snake Lemma II.4.19 to the two upper rows, and, for (b), to the two lower rows. □

The symmetry of the diagram in the 3 × 3 Lemma implies that the statement remains true if we exchange the words 'rows' and 'columns'.

II.4.4 The isomorphism theorems

As for algebras (and groups, and other algebraic structures), the main isomorphism theorem asserts that any morphism between modules induces an isomorphism between 'part' of the domain (a quotient) and 'part' of the codomain (a submodule). Consequently, any morphism can be written as the composite of an epimorphism, an isomorphism and a monomorphism.

II.4.21 Theorem. *Let $f : M \to N$ be an A-linear map between A-modules, $i : \operatorname{Im} f \to N$ the inclusion and $p : M \to M/\operatorname{Ker} f$ the projection. Then there exists a unique A-linear map $\bar{f} : M/\operatorname{Ker} f \to \operatorname{Im} f$ making the following diagram commute:*

$$
\begin{array}{ccc}
M & \xrightarrow{\quad f \quad} & N \\
\downarrow{\scriptstyle p} & & \uparrow{\scriptstyle i} \\
M/\operatorname{Ker} f & \dashrightarrow{\ \bar{f}\ } & \operatorname{Im} f
\end{array}
$$

that is, such that $f = i\bar{f}p$. Moreover, $\bar{f}$ is an isomorphism.

Proof. We assume that $\bar{f}$ exists and try to find an explicit formula giving it. Set $L = \operatorname{Ker} f$ for short and $\bar{x} = x + L \in M/L$. We must have

$$\bar{f}(\bar{x}) = \bar{f}(x + L) = \bar{f}p(x) = i\bar{f}p(x) = f(x)$$

where, at the third equality, we used that $y \in \operatorname{Im} f$ if and only if $y = i(y)$. The formula obtained guarantees the uniqueness of $\bar{f}$, if it exists.

But this same formula defines unambiguously a map, for, if $x_1 + L = x_2 + L$, then $x_1 - x_2 \in L = \operatorname{Ker} f$ so that $f(x_1 - x_2) = 0$, hence $f(x_1) = f(x_2)$. Thus, $\bar{f}$ exists as a map.

We prove its linearity. Assume that $\bar{x}_1 = x_1 + L, \bar{x}_2 = x_2 + L \in M/L$ and $a_1, a_2 \in A$. Then

$$\bar{f}(\bar{x}_1 a_1 + \bar{x}_2 a_2) = f(x_1 a_1 + x_2 a_2) = f(x_1)a_1 + f(x_2)a_2 = \bar{f}(\bar{x}_1)a_1 + \bar{f}(\bar{x}_2)a_2.$$

This completes the proof that there exists a unique A-linear map $\bar{f}$ such that $f = i\bar{f}p$.

We prove that $\bar{f}$ is an isomorphism. Surjectivity follows from the fact that any element of the image of f can be written as $f(x) = \bar{f}(x + L)$. To prove injectivity, assume that $\bar{f}(x + L) = 0$, then $f(x) = 0$ gives $x \in \mathrm{Ker}\, f = L$. Therefore, $x + L = L$ and so $\mathrm{Ker}\, \bar{f} = 0$. $\qquad\square$

The isomorphism $\bar{f}$ is called the **canonical isomorphism** and the decomposition $f = i\bar{f}p$ is the **canonical decomposition** of the morphism f.

If, in particular, $f : M \to N$ is an epimorphism, then $\mathrm{Im}\, f = N, i = 1_N$ and so $\bar{f}$ is an isomorphism $M/\mathrm{Ker}\, f \cong N$, that is, N is (isomorphic to) a quotient of M.

If, on the other hand, f is a monomorphism, then $\mathrm{Ker}\, f = 0, p = 1_M$ and so $\bar{f}$ is an isomorphism $M \cong \mathrm{Im}\, f$, that is, M is (isomorphic to) a submodule of N.

The Isomorphism Theorem II.4.21 can also be interpreted as saying that any morphism $f : M \to N$ induces a short exact sequence of the form

$$0 \to \mathrm{Ker}\, f \to M \to \mathrm{Im}\, f \to 0$$

where the morphism on the left is the inclusion $\mathrm{Ker}\, f \to M$, and the morphism on the right is f. Indeed, the cokernel of the inclusion is the quotient $M/\mathrm{Ker}\, f$. Therefore, the given sequence is exact if and only if $M/\mathrm{Ker}\, f \cong \mathrm{Im}\, f$.

As an application, we prove a characterisation of cyclic modules.

II.4.22 Proposition. *An A-module M is cyclic if and only if there exists a right ideal I of A such that $M \cong A_A/I$.*

Proof. If I is a right ideal, then A_A/I is cyclic, generated by $1 + I$. Conversely, if $M = xA$ for some $x \in M$, then the map $f_x : A_A \to M_A$ given by $a \mapsto xa$, for $a \in A$, is A-linear and surjective. Moreover, $I = \mathrm{Ker}\, f_x$ is a right ideal of A. Theorem II.4.21 gives $M \cong A/I$. $\qquad\square$

II.4.23 Example. Consider the morphism $f : \mathbb{Z}/p^2\mathbb{Z} \to \mathbb{Z}/p^2\mathbb{Z}$ of Example II.4.9. We proved there that its image and its kernel are both equal to the unique nonzero proper submodule $p\mathbb{Z}/p^2\mathbb{Z}$ of $\mathbb{Z}/p^2\mathbb{Z}$, so that f can be depicted in terms of the submodule lattices as

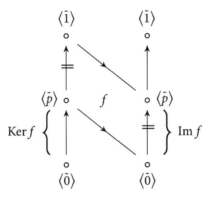

Here, the bar over an integer denotes a residual class modulo $p^2\mathbb{Z}$. Passing from M to $M/\mathrm{Ker}f$ means 'equating to zero' the elements of $\mathrm{Ker}f$, that is, 'contracting' $\mathrm{Ker}f$ to zero. This yields (in the domain) the upper notched segment. Theorem II.4.21 says that this quotient is isomorphic to $\mathrm{Im}f$, namely the lower notched segment in the codomain. The canonical factorisation is:

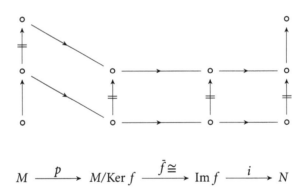

$$M \xrightarrow{\ p\ } M/\mathrm{Ker}\,f \xrightarrow{\ \bar{f} \cong\ } \mathrm{Im}\,f \xrightarrow{\ i\ } N$$

which can be interpreted as saying that the 'upper part' of M is isomorphic to the 'lower part' of N. Because the projection $p : M \to M/\mathrm{Ker}f$ is an epimorphism, the 'lower part' of the codomain $M/\mathrm{Ker}f$ equals the whole of it. Dually, because the inclusion $i : \mathrm{Im}f \to N$ is a monomorphism, the 'upper part' of the domain equals the whole of it.

These observations are of a general nature: a module morphism $f : M \to N$ determines submodules $\mathrm{Ker}f$ of M and $\mathrm{Im}f$ of N, both located in the lower parts of the respective submodules lattices

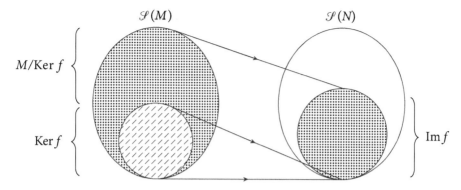

The isomorphism theorem says that the dotted parts are isomorphic, that is, the quotient $M/\mathrm{Ker}\,f$ is obtained from M by contracting the submodule $\mathrm{Ker}\,f$ to zero, and the result of contraction is (isomorphic to) the submodule $\mathrm{Im}\,f$ of N. Thus, the mere existence of a morphism from M to N, implies that part of the structure of M appears as part of the structure of N. We stress that, while these pictures may help intuition and forming mental images, they are by no means substitutes for rigorous mathematical proofs.

We prove the second isomorphism theorem.

II.4.24 Theorem. *Let M be an A-module and L, N submodules of M. Then*

$$\frac{L+N}{L} \cong \frac{N}{L \cap N}.$$

Proof. The inclusion morphisms $u : L \cap N \to L, v : N \to L + N, f : L \cap N \to N$ and $f' : L \to L + N$ induce a commutative diagram with exact rows of A-modules and A-linear maps.

$$
\begin{array}{ccccccccc}
0 & \longrightarrow & L \cap N & \overset{f}{\longrightarrow} & N & \overset{g}{\longrightarrow} & N/(L \cap N) & \longrightarrow & 0 \\
& & \downarrow{\scriptstyle u} & & \downarrow{\scriptstyle v} & & \downarrow{\scriptstyle w} & & \\
0 & \longrightarrow & L & \overset{f'}{\longrightarrow} & L + N & \overset{g'}{\longrightarrow} & (L+N)/L & \longrightarrow & 0
\end{array}
$$

where g, g' are the projections and $w : x + L \cap N \mapsto x + L$, for $x \in N$, is induced by passing to cokernels, see Proposition II.4.13. We must prove that w is an isomorphism.

Assume that $x \in N$ is such that $w(x + L \cap N) = 0$, then $x + L = 0$, that is, $x \in L$. Hence, $x \in L \cap N$ and $\mathrm{Ker}\,w = 0$. Thus, w is a monomorphism.

Let $(x+y)+L \in (L+N)/L$ with $x \in L, y \in N$, then $(x+y)+L = y+L = w(y+L\cap N)$, hence w is also an epimorphism. This finishes the proof. $\qquad \square$

Because w is an isomorphism, it also follows from the Snake Lemma II.4.19 that $(L + N)/N \cong \mathrm{Coker}\,v \cong \mathrm{Coker}\,u \cong L/(L \cap N)$, as was to be expected.

For instance, if L and N are in direct sum, then $L \cap N = 0$, so $(L \oplus N)/L \cong N$, as expected.

The previous theorem is called the *parallelogram law.* It asserts that in the diagram

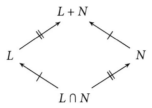

$$L + N$$
$$L \qquad\qquad N$$
$$L \cap N$$

the opposite sides, which represent the respective quotients, are isomorphic. These isomorphisms are indicated by similar notches. In contrast to the case of algebras, see Theorem I.4.9, the situation here is completely symmetric because all four quotient modules make sense. In the lattice $\mathcal{S}(M)$, the submodules $L + N$ and $L \cap N$ are respectively the supremum and the infimum of the pair $\{L, N\}$.

II.4.25 Example. Let a,b be integers. Denote by (a,b) and $[a,b]$ respectively the gcd and the lcm of a and b. Due to the proof of Theorem I.5.5, we have $a\mathbb{Z} + b\mathbb{Z} = (a,b)\mathbb{Z}$ and $a\mathbb{Z} \cap b\mathbb{Z} = [a,b]\mathbb{Z}$. The isomorphism theorem II.4.24 says that we have an isomorphism of abelian groups ($\mathbb{Z}$-modules)

$$\frac{(a,b)\mathbb{Z}}{a\mathbb{Z}} \cong \frac{b\mathbb{Z}}{[a,b]\mathbb{Z}}.$$

The third isomorphism theorem can be proved exactly as we proved the corresponding statement for algebras, see Theorem I.4.8, but we prefer to give a proof based on the Snake Lemma.

II.4.26 Theorem. *Let M be an A-module and L, N submodules of M such that $L \subseteq N$. Denote by $p_L : M \to M/L, p_N : M \to M/N$ the respective projections. Then there exists a unique A-linear map $f : M/L \to M/N$ making the following diagram commute:*

$$M$$
$$p_L \qquad\qquad p_N$$
$$M/L \dashrightarrow[f] M/N$$

that is, such that $fp_L = p_N$. Moreover, f is an epimorphism with kernel N/L and induces an isomorphism $(M/L)/(N/L) \cong M/N$.

Proof. We apply the Snake Lemma II.4.19 to the commutative diagram with exact rows of A-modules and A-linear maps

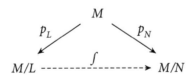

$$\begin{array}{ccccccccc}
0 & \longrightarrow & L & \longrightarrow & M & \xrightarrow{p_L} & M/L & \longrightarrow & 0 \\
& & \downarrow & & \downarrow{\scriptstyle 1_M} & & \downarrow{\scriptstyle f} & & \\
0 & \longrightarrow & N & \longrightarrow & M & \xrightarrow{p_N} & M/N & \longrightarrow & 0
\end{array}$$

where the maps $L \to M, L \to N$ and $N \to M$ are the inclusions and f results by passing to cokernels, see Proposition II.4.13. The Snake Lemma gives an exact dotted sequence in the diagram

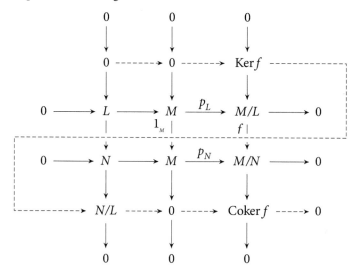

Exactness of this sequence yields $\mathrm{Ker} f \cong N/L$ and $\mathrm{Coker} f = 0$ (so, f is surjective). The last statement follows from Theorem II.4.21. □

II.4.27 Example. Let $A = \mathbb{Z}, M = \mathbb{Z}/12\mathbb{Z}, N = \mathbb{Z}/8\mathbb{Z}$ and consider the map $f : M \to N$ given by $x + 12\mathbb{Z} \mapsto 2x + 8\mathbb{Z}$, for $x \in \mathbb{Z}$. We show its effect on the submodule lattices of $\mathbb{Z}/12\mathbb{Z}$ and $\mathbb{Z}/8\mathbb{Z}$. The first lattice is computed in Example II.2.25(b), and the second is easy to compute.

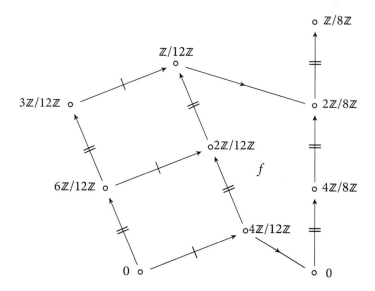

Notches are as explained after the proof of Theorem II.4.24. Here $\operatorname{Im} f = 2\mathbb{Z}/8\mathbb{Z} \cong \mathbb{Z}/4\mathbb{Z}$ while $\operatorname{Ker} f = 4\mathbb{Z}/12\mathbb{Z} \cong \mathbb{Z}/3\mathbb{Z}$. Hence,

$$M/\operatorname{Ker} f = (\mathbb{Z}/12\mathbb{Z})/(4\mathbb{Z}/12\mathbb{Z}) \cong \mathbb{Z}/4\mathbb{Z} \cong \operatorname{Im} f,$$

as expected. Moreover, the Isomorphism Theorems II.4.21 and II.4.25 yield the isomorphisms:

$$(\mathbb{Z}/12\mathbb{Z})/(3\mathbb{Z}/12\mathbb{Z}) \cong (2\mathbb{Z}/12\mathbb{Z})/(6\mathbb{Z}/12\mathbb{Z}) \cong 4\mathbb{Z}/12\mathbb{Z} \cong \mathbb{Z}/3\mathbb{Z}$$

and

$$(\mathbb{Z}/12\mathbb{Z})/(2\mathbb{Z}/12\mathbb{Z}) \cong (3\mathbb{Z}/12\mathbb{Z})/(6\mathbb{Z}/12\mathbb{Z}) \cong (2\mathbb{Z}/12\mathbb{Z})/(4\mathbb{Z}/12\mathbb{Z})$$
$$\cong 6\mathbb{Z}/12\mathbb{Z} \cong \mathbb{Z}/2\mathbb{Z}.$$

We finish with the so-called *correspondence of submodules* theorem.

II.4.28 Theorem. *Let M be an A-module, N a submodule and $p : M \to M/N$ the projection. The map $L \mapsto p(L)$ is an inclusion-preserving bijection from the set of submodules of M containing N to the set of submodules of M/N. The inverse bijection is p^{-1}.*

Proof. Let L be a submodule of M containing N. Because of Exercise II.3.10(a), $p(L) = L/N$ is a submodule of M/N. It is clear that p preserves inclusions. Because of Exercise II.3.11(d), we have $L \subseteq p^{-1}p(L)$. We show the reverse inclusion. Let $x \in p^{-1}p(L)$. Then $p(x) \in p(L)$, so there exists $y \in L$ such that $p(x) = p(y)$, that is, $x - y \in \operatorname{Ker} p = N \subseteq L$. Therefore, $x \in L$ and we indeed have $L = p^{-1}p(L)$ for any submodule L containing N.

   Conversely, let $\bar{L}$ be a submodule of M/N. Because of Exercise II.3.10(b), we have $p^{-1}(\bar{L}) = \{x \in M : x + N \in \bar{L}\}$, this is a submodule of M which certainly contains $p^{-1}(0) = N$. On the other hand, p is surjective, therefore $\bar{L} = pp^{-1}(\bar{L})$, due to Exercise II.3.11(c).   □

This result may be interpreted in terms of submodule lattices. Let $(E, \leqslant), (F, \leq)$ be posets. A map $f : E \to F$ is called a *morphism*, or *homomorphism*, of posets if $x \leqslant y$ in E implies $f(x) \leq f(y)$ in F. For examples of such morphisms, see Exercise II.3.10. A morphism $f : E \to F$ is an *isomorphism* between the posets E and F if there exists a morphism $f' : F \to E$ such that $f'f = 1_E$ and $ff' = 1_F$. Clearly, then $f' = f^{-1}$. In the situation of the theorem, consider the posets E of submodules of M containing N and F of submodules of M/N (that is, $F = S(M/N)$), both ordered by inclusion. The theorem says that p and p^{-1} are inverse isomorphisms between the posets E and F. In particular, F is a complete lattice, due to Proposition II.2.18, hence so is E.

Exercises of Section II.4

1. Let A be an algebra, $f : M \to N$ an epimorphism of A-modules, L a submodule of M and $i : L \to M$ the inclusion. Prove that:
   (a) If $L \cap \mathrm{Ker} f = 0$, then $fi : L \to N$ is a monomorphism.
   (b) If $L + \mathrm{Ker} f = M$, then $fi : L \to N$ is an epimorphism.

2. Let A be an algebra. Prove that an A-module M is finitely generated if and only if there exist $m \geq 0$ and an epimorphism $A_A^m \to M$.

3. Let A be an algebra, M a module and L, N submodules of M. Prove that there exists a short exact sequence of A-modules and A-linear maps
$$0 \to L \cap N \to L \times N \to L + N \to 0.$$

4. Let $a, b > 1$ be integers. Prove that there exists a short exact sequence of $\mathbb{Z}/(ab)\mathbb{Z}$-modules
$$0 \to a\mathbb{Z}/(ab)\mathbb{Z} \xrightarrow{f} \mathbb{Z}/(ab)\mathbb{Z} \xrightarrow{g} b\mathbb{Z}/(ab)\mathbb{Z} \to 0$$
where f is the inclusion and g the multiplication by b.

5. Let p be a prime integer. Construct an exact sequence of abelian groups
$$0 \to \mathbb{Z}/p\mathbb{Z} \to \mathbb{Z}/p^2\mathbb{Z} \to \mathbb{Z}/p^2\mathbb{Z} \to \mathbb{Z}/p\mathbb{Z} \to 0$$
and deduce an infinite exact sequence
$$\ldots \to \mathbb{Z}/p^2\mathbb{Z} \to \mathbb{Z}/p^2\mathbb{Z} \to \mathbb{Z}/p^2\mathbb{Z} \to \ldots$$

6. Let A be an algebra. Prove that:
   (a) The data of an exact sequence $L \xrightarrow{f} M \xrightarrow{g} N$ of A-modules is equivalent to the data of three short exact sequences: $0 \to L' \to L \xrightarrow{u} M' \to 0$, $0 \to M' \xrightarrow{v_1} M \xrightarrow{v_2} M'' \to 0$ and $0 \to M'' \xrightarrow{w} N \to N' \to 0$ such that $v_1 u = f$, $wv_2 = g$.
   (b) The data of an exact sequence of A-modules
$$\cdots \to M_{i+1} \xrightarrow{f_{i+1}} M_i \xrightarrow{f_i} M_{i-1} \to \cdots$$
   is equivalent to the data, for each i, of a short exact sequence $0 \to \mathrm{Ker} f_i \to M_i \to \mathrm{Ker} f_{i-1} \to 0$.

7. Let K be a field and E_1, E_2, E_3 vector spaces such that there exists an exact sequence
$$0 \to E_1 \to E_2 \to E_3 \to 0.$$
Prove that $\dim_K E_2 = \dim_K E_1 + \dim_K E_3$. Deduce that, if we have an exact sequence of K-vector spaces
$$0 \to E_1 \to E_2 \to \cdots \to E_n \to 0,$$
then $\sum_{i=1}^{n} (-1)^i \dim_K E_i = 0$.

8. Let A be an algebra. Assume that we have a commutative diagram with exact rows of A-modules and A-linear maps

$$
\begin{array}{ccccccc}
L & \longrightarrow & M & \longrightarrow & N & \longrightarrow & 0 \\
\downarrow{\scriptstyle f} & & \downarrow{\scriptstyle g} & & \downarrow{\scriptstyle h} & & \\
L' & \longrightarrow & M' & \longrightarrow & N' & \longrightarrow & 0
\end{array}
$$

Prove that, if f, g are isomorphisms, then so is h.

9. Let A be an algebra. Assume that we have a commutative diagram with exact rows of A-modules and A-linear maps

$$
\begin{array}{ccccccc}
0 & \longrightarrow & L & \longrightarrow & M & \longrightarrow & N \\
& & \downarrow{\scriptstyle f} & & \downarrow{\scriptstyle g} & & \downarrow{\scriptstyle h} \\
0 & \longrightarrow & L' & \longrightarrow & M' & \longrightarrow & N'
\end{array}
$$

Prove that, if g, h are isomorphisms, then so is f.

10. Let A be an algebra. Assume that we have a commutative diagram with exact rows of A-modules and A-linear maps

$$
\begin{array}{ccccccccc}
0 & \longrightarrow & L & \longrightarrow & M & \longrightarrow & N & \longrightarrow & 0 \\
& & \downarrow{\scriptstyle f} & & \downarrow{\scriptstyle g} & & \downarrow{\scriptstyle h} & & \\
0 & \longrightarrow & L' & \longrightarrow & M' & \longrightarrow & N' & \longrightarrow & 0
\end{array}
$$

Prove that, if f, h are isomorphisms, then so is g.

11. Let A be an algebra, L, M, N modules and $f : L \to M, g : M \to N$ linear maps. Prove that $\operatorname{Im}(gf) \cong \operatorname{Im} f/(\operatorname{Im} f \cap \operatorname{Ker} g)$.

12. Let A be an algebra. Assume that we have a commutative diagram with exact rows of A-modules and A-linear maps

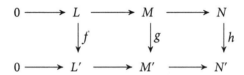

Prove that:
(a) $\operatorname{Im} v \cap \operatorname{Im} f' = v(\operatorname{Ker}(wg))$.
(b) $\operatorname{Ker} v + \operatorname{Ker} g = v^{-1}(\operatorname{Im}(f'u))$.
(c) $(\operatorname{Im} v \cap \operatorname{Im} f')/\operatorname{Im}(vf) \cong \operatorname{Ker}(wg)/(\operatorname{Ker} v + \operatorname{Ker} g)$.

13. Let A be an algebra. Assume that we have a commutative diagram with exact rows of A-modules and A-linear maps

$$0 \xrightarrow{\quad} L \xrightarrow{\ u\ } M \xrightarrow{\ v\ } N \xrightarrow{\quad} 0$$

with vertical maps $f: L \to L'$, $g: M \to M'$, $h: N \to N'$

$$0 \xrightarrow{\quad} L' \xrightarrow{\ u'\ } M' \xrightarrow{\ v'\ } N' \xrightarrow{\quad} 0$$

Assume that f, h are given. Prove that:

(a) If $g_1, g_2 : M \to M'$ both make the diagram commute, then there exists an A-linear map $k : N \to L'$ such that $g_1 - g_2 = u'kv$.

(b) Conversely, if $g : M \to M'$ makes the diagram commute and $k : N \to L'$ is arbitrary, then $g + u'kv : M \to M'$ also makes the diagram commute.

14. Let A be an algebra. Assume that we have a commutative diagram with exact rows and columns of A-modules and A-linear maps

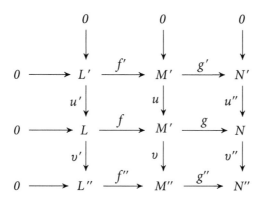

Prove that $\operatorname{Im}(uf') = \operatorname{Ker} g \cap \operatorname{Ker} v$.

15. Let A be an algebra. Assume that we have a commutative diagram with exact rows and columns of A-modules and A-linear maps

$$
\begin{array}{ccccccc}
L' & \xrightarrow{f'} & M' & \xrightarrow{g'} & N' & \longrightarrow & 0 \\
\downarrow{\scriptstyle u'} & & \downarrow{\scriptstyle u} & & \downarrow{\scriptstyle u''} & & \\
L & \xrightarrow{f} & M' & \xrightarrow{g} & N & \longrightarrow & 0 \\
\downarrow{\scriptstyle v'} & & \downarrow{\scriptstyle v} & & \downarrow{\scriptstyle v''} & & \\
L'' & \xrightarrow{f''} & M'' & \xrightarrow{g''} & N'' & \longrightarrow & 0 \\
\downarrow & & \downarrow & & \downarrow & & \\
0 & & 0 & & 0 & &
\end{array}
$$

Prove that $\operatorname{Ker}(v''g) = \operatorname{Im} u + \operatorname{Im} f$.

16. Let A be an algebra.

(a) Assume that we have a commutative diagram with exact rows and columns of A-modules and A-linear maps

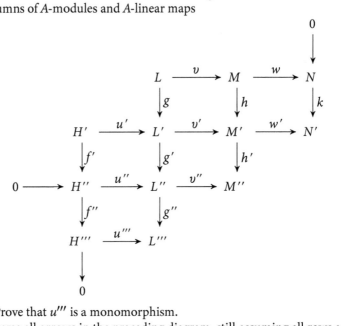

Prove that u''' is a monomorphism.

(b) Reverse all arrows in the preceding diagram, still assuming all rows and columns exact. Prove that u''' is an epimorphism.

17. Let A be an algebra. Assume that we have a commutative diagram of A-modules and A-linear maps

$$\cdots \longrightarrow N_{n+1} \xrightarrow{w_{n+1}} L_n \xrightarrow{u_n} M_n \xrightarrow{v_n} N_n \xrightarrow{w_n} L_{n-1} \xrightarrow{u_{n-1}} \cdots$$
$$\downarrow{h_{n+1}} \quad \downarrow{f_n} \quad \downarrow{g_n} \quad \downarrow{h_n} \quad \downarrow{f_{n-1}}$$
$$\cdots \longrightarrow N'_{n+1} \xrightarrow{w'_{n+1}} L'_n \xrightarrow{v'_n} M'_n \xrightarrow{v'_n} N'_n \xrightarrow{w'_n} L'_{n-1} \xrightarrow{u'_{n-1}} \cdots$$

where all the h_n are isomorphisms. Deduce the exact sequence

$$\cdots \longrightarrow L_n \xrightarrow{\binom{f_n}{g_n}} L'_n \oplus M_n \xrightarrow{(u'_n \ \ g_n)} M'_n \xrightarrow{w_n h_n^{-1} v'_n} L_{n-1} \longrightarrow \cdots$$

18. Let A be an algebra. Assume that we have a commutative diagram with exact rows of A-modules and A-linear maps

$$0 \longrightarrow L \longrightarrow M \longrightarrow N \longrightarrow 0$$
$$\downarrow{f} \quad \downarrow{g} \quad \downarrow{h}$$
$$0 \longrightarrow L' \longrightarrow M' \longrightarrow N' \longrightarrow 0$$

with g an isomorphism. Prove that:

    (a) f is a monomorphism and h an epimorphism.

    (b) f is an epimorphism if and only if h is a monomorphism.

19. Let A be an algebra. Assume that we have a commutative diagram with exact (oblique) rows of A-modules and A-linear maps

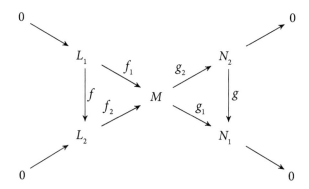

    Prove that $g_2 f_1 = 0, g_1 f_2 = 0$ and that f, g are isomorphisms.

20. Let A be an algebra. Assume that we have a commutative diagram of A-modules and A-linear maps

$$
\begin{array}{ccccccc}
L & \xrightarrow{f} & M & \xrightarrow{g} & N & \longrightarrow & 0 \\
\downarrow{\scriptstyle u} & & \downarrow{\scriptstyle v} & & & & \\
L' & \xrightarrow{f'} & M' & \xrightarrow{g'} & N' & &
\end{array}
$$

    with the upper row exact and $g'f' = 0$. Prove that there exists a unique A-linear map $w : N \to N'$ such that $wg = g'v$.

21. Let A be an algebra. Assume that we have a commutative diagram of A-modules and A-linear maps

$$
\begin{array}{ccccccc}
 & L' & \xrightarrow{f'} & M' & \xrightarrow{g'} & N' & \\
 & & & \downarrow{\scriptstyle v} & & \downarrow{\scriptstyle w} & \\
0 & \longrightarrow & L & \xrightarrow{f} & M & \xrightarrow{g} & N
\end{array}
$$

    with the lower row exact and $g'f' = 0$. Prove that there exists a unique A-linear map $u : L' \to L$ such that $fu = vf'$.

22. Let A be an algebra. Assume that we have a commutative diagram with exact rows of A-modules and A-linear maps

$$
\begin{array}{ccccc}
L & \xrightarrow{\ f\ } & M & \xrightarrow{\ g\ } & N \\
\downarrow{\scriptstyle u} & & \downarrow{\scriptstyle v} & & \downarrow{\scriptstyle w} \\
L' & \xrightarrow{\ f'\ } & M' & \xrightarrow{\ g'\ } & N'
\end{array}
$$

Prove that:
   (a) If u, w, f' are monomorphisms, then so is v.
   (b) If u, w, g are epimorphisms, then so is v.

23. Let A be an algebra and $M = xA$ a cyclic A-module generated by x. Prove that $M \cong A/\mathrm{Ann}(x)$, where $\mathrm{Ann}(x)$ denotes the annihilator of the element x, see Exercise II.2.2.

24. Let A be an algebra, M a module, L, N submodules of M and $p : M \to M/N$ the projection. Prove that $p(L) = (L + N)/N$.

25. Let A be an algebra, M a module, N, N', L, L' submodules of M such that $N \subseteq L$ and $N' \subseteq L'$. Prove that:

$$
\frac{N + (L \cap L')}{N + (L \cap N')} \cong \frac{L \cap L'}{(N \cap L') + (N' \cap L)} \cong \frac{N' + (L \cap L')}{N' + (N \cap L')}.
$$

This result is known as **Zassenhaus' Butterfly Lemma**.

26. Let A be an algebra, M, M' modules, $f : M \to M'$ an A-linear map, N, N' submodules of M, M', respectively, such that $f(N) \subseteq N'$, and $p : M \to M/N, p' : M' \to M'/N'$ the respective projections. Prove that there exists a unique A-linear map $\bar{f} : M/N \to M'/N'$ such that the following diagram commutes:

$$
\begin{array}{ccc}
M & \xrightarrow{\ f\ } & M' \\
\downarrow{\scriptstyle p} & & \downarrow{\scriptstyle p'} \\
M/N & \dashrightarrow{\ \bar{f}\ } & M'/N'
\end{array}
$$

27. Let A be an algebra, L, L' modules and M, M' submodules of L, L', respectively. Prove that:

$$
\frac{L \oplus L'}{M \oplus M'} \cong \frac{L}{M} \oplus \frac{L'}{M'}.
$$

28. Let G be an abelian group. A subgroup G' of G is called a **pure subgroup** if, for any $m \in \mathbb{Z}$, we have $G' \cap mG = mG'$. Prove that:
   (a) Any direct summand of G is pure in G.
   (b) If G' is pure in G and G'' is pure in G', then G'' is pure in G.

(c) If $G' = \cup_{i=1}^{\infty} G_i'$ with $G_i' \subseteq G_{i+1}'$ and G_i' is pure in G for any i, then G' is pure in G.

(d) For any positive integer m, the subgroup $m\mathbb{Z}/m^2\mathbb{Z}$ of $\mathbb{Z}/m^2\mathbb{Z}$ is not pure.

(e) A subgroup G' of G is pure if and only if, for any $m \in \mathbb{Z}$, a short exact sequence $0 \to G' \to G \to G'' \to 0$ induces a short exact sequence $0 \to G'/mG' \to G/mG \to G''/mG'' \to 0$.

Chapter III
Categories and functors

III.1 Introduction

We need a new language. In order to see its necessity, remember that often, according to context, the same mathematical entity assumes different roles. For example, the quotient of an algebra by an ideal may be viewed as a quotient algebra, or a module (even a bimodule) over the original algebra (and over the quotient algebra!). Hence the idea of categories: working inside a category may be thought of as fixing a mathematical context. We may for instance talk about the category of K-algebras, or the category of modules over some algebra A if we decide, at some stage, to work only with K-algebras, or with A-modules, respectively. Together with each structure come the maps compatible with this structure. Thus, the category of K-algebras contains not only algebras, but also algebra morphisms, and the module category contains also module morphisms.

If, in the course of a problem, we need to change contexts, that is, to pass from one category to another, we use a device called functor. Roughly speaking, functors are to categories what maps are to sets. For instance, a basic idea of algebraic topology is to attach to a topological space a group, then to translate a topological problem, considered as difficult, into a group theoretical one, hopefully easier to solve. This means defining a functor from the category of topological spaces to the category of groups (well-known examples are the fundamental group and the singular homology groups). This construction is blind to elements: an element of the topological space does not necessarily correspond to an element of the associated group. Hence, we should consider topological spaces and groups as objects together with appropriate maps, but without considering their elements, that is, without looking at them as sets. In this case, we cannot speak about maps as such. Hence the need to axiomatise the relevant properties of maps and work only starting from axioms. The objective of this chapter is to introduce rudiments of the categorical language and apply them in module theory.

III.2 Categories and functors

III.2.1 Categories

In this section, we introduce the basic vocabulary of category theory. A category is a class of mathematical entities, called objects, having a common structure together with the maps compatible with that structure. We wish to axiomatise basic properties of these maps: we want to be able to compose maps (when they are composable, of course) so that composition is associative, and we want to attach to each object an identity map.

An Introduction to Module Theory. Ibrahim Assem and Flávio U. Coelho, Oxford University Press.
© Ibrahim Assem and Flávio U. Coelho (2024). DOI: 10.1093/9780198904939.003.0004

III.2.1 Definition. A *category* C is defined by the following data:

    (a) A class C_0, or ObC, called the class of *objects* of C.

    (b) For each pair (X, Y) of objects of C, a set denoted by $C(X, Y)$, or $\mathrm{Hom}_C(X, Y)$, whose elements are *morphisms* from X to Y in C.

    (c) For each triple (X, Y, Z) of objects in C, a map $C(X, Y) \times C(Y, Z) \to C(X, Z)$ given by $(f, g) \mapsto g \circ f$, or, briefly, gf, called *composition* of morphisms, such that:

        (c1) If $f \in C(X, Y)$, $g \in C(Y, Z)$, $h \in C(Z, W)$, then $h(gf) = (hg)f$.

        (c2) To each object X in C is attached a morphism $1_X \in C(X, X)$, called the *identity* on X, such that for $f \in C(X, Y)$, $g \in C(Z, X)$, we have $f1_X = f$ and $1_X g = g$.

We shall not examine the logical foundations of the theory. The term 'class' used in (a) above will be taken as primary notion, just like 'set', except that we reserve the term sets to classes small enough to have a cardinal number. On the other hand, we speak of the class of all sets, of all modules etc. A category whose object class is a set is called a *small category*.

Even though object classes are generally not sets, we sometimes write $X \in C_0$ to express the fact that X is an object in the category C.

Given objects X, Y in a category C, a morphism $f \in C(X, Y)$ is said to have X as *source*, or *domain*, and Y as *target*, or *codomain*. We write as usual $f : X \to Y$ or $X \xrightarrow{f} Y$. A morphism determines uniquely its source and its target, a condition sometimes expressed (very informally!) by writing that $(X, Y) \neq (X', Y')$ implies $C(X, Y) \cap C(X', Y') = \emptyset$. In particular, a category is entirely determined by its morphisms. The composite gf of the morphisms g and f is defined if and only if the target of f coincides with the source of g. The morphisms are then called *composable*.

A morphism from an object X of C to itself is called an *endomorphism* of X. We denote by $C(X, X) = \mathrm{End}_C X$ their set. Two endomorphisms of X are always composable, and axioms (c1) and (c2) above say that $\mathrm{End}_C X$ is a monoid under composition of morphisms.

The identity morphism at an object X is uniquely determined: for, if both 1_X and $1'_X$ satisfy axiom (c2) above, then

$$1_X = 1_X 1'_X = 1'_X.$$

The existence of a one-to-one correspondence between objects of a category and its identity morphisms implies that a category may be defined in terms of its morphisms only (thus, morphisms matter more than objects).

A morphism $f : X \to Y$ in a category C is called an *isomorphism* if there exists a morphism $f' : Y \to X$ in C such that $f'f = 1_X$ and $ff' = 1_Y$. In this case, f' is uniquely determined by f. For, if $f', f'' : Y \to X$ satisfy $f'f = 1_X$ and $ff'' = 1_Y$, then associativity of composition (c1) yields

$$f'' = 1_X f'' = (f'f)f'' = f'(ff'') = f' 1_Y = f'.$$

This unique morphism f' is called the **inverse** of f and denoted by $f' = f^{-1}$. If there exists an isomorphism $f: X \to Y$, then X and Y are said to be **isomorphic**, denoted by $X \cong Y$. An isomorphism from an object X to itself is called an **automorphism** of X. For any object X, the identity 1_X is an isomorphism (even, an automorphism). If $f: X \to Y$ is an isomorphism, then so is $f^{-1}: Y \to X$ and we have $(f^{-1})^{-1} = f$. Finally, if $f: X \to Y$ and $g: Y \to Z$ are isomorphisms, then so is $gf: X \to Z$ and we have $(gf)^{-1} = f^{-1}g^{-1}$. Hence, the relation $\cong$ is reflexive, symmetric and transitive. The previous developments are identical to those of the corresponding notions for algebras or for modules.

III.2.2 Examples.

(a) The category $Sets$ of sets has as objects the sets, as morphisms the maps, and composition is the usual composition of maps. In this category, the isomorphisms are the bijections.

(b) The category Gr of groups has as objects the groups, as morphisms the group homomorphisms, and composition is the usual composition of maps. This makes sense, because the composite of group homomorphisms is a group homomorphism. Here, isomorphisms are group isomorphisms.

(c) The category Ab of abelian groups has as objects the abelian groups, as morphisms the group homomorphisms, and composition is the usual composition of maps. Again, isomorphisms are group isomorphisms.

(d) The category Top of topological spaces has as objects the topological spaces, as morphisms the continuous maps, and composition is the usual composition of maps. This makes sense, because the composite of continuous maps is continuous. In Top, isomorphisms, which are continuous bijections having continuous inverses, are usually called **homeomorphisms**.

(e) Let K be a commutative ring. The category Alg_K of K-algebras has as objects K-algebras, as morphisms algebra morphisms, and composition is the usual composition of maps. Isomorphisms are algebra isomorphisms.

(f) Let A be an algebra. The category $\operatorname{Mod}A$ of A-modules has as objects right A-modules, as morphisms A-linear maps, and composition is the usual composition of maps. We define similarly the category $\operatorname{Mod}A^{op}$ of left A-modules (left A-modules coincide with right A^{op}-modules, hence the notation). Isomorphisms are module isomorphisms.

(g) A monoid M can be viewed as a category $\mathcal{M}$ having a unique object X and $\mathcal{M}(X,X) = M$, the composition of morphisms being the operation of M. The isomorphisms are the invertible elements of M.

(h) A poset $(E, \leqslant)$ may be considered as a small category S: its objects are the elements of E and, for $x, y \in E$, the set $S(x,y)$ contains a single element ρ_{yx} if $x \leqslant y$ and is empty otherwise. Transitivity of the partial order implies that $\rho_{zy}\rho_{yx} = \rho_{zx}$ if $x \leqslant y \leqslant z$, which defines composition. The only isomorphisms are the identity morphisms.

(i) Let K be a field. The category $\mathcal{E}$ has as objects the pairs (E,f), where E is a K-vector space and $f: E \to E$ an endomorphism. A morphism of pairs $u: (E,f) \to (F,g)$ is a K-linear map $u: E \to F$ such that $uf = gu$, that is, we have a commutative square

$$E \xrightarrow{\ f\ } E$$
$$u \downarrow \qquad \downarrow u$$
$$F \xrightarrow{\ g\ } F$$

Let $u: (E, f) \to (F, g)$ and $v: (F, g) \to (G, h)$ be morphisms, then $uf = gu$ and $vg = hv$ imply $h(vu) = (hv)u = (vg)j = v(gu) = v(uf) = (vu)f$. Therefore, the usual composition of maps makes $\mathcal{E}$ a category. A morphism of pairs $u: (E, f) \to (F, g)$ is an isomorphism if and only if $u : E \to F$ is an isomorphism of K-vector spaces.

(j) Let K be a field and Q a finite quiver. The category $\mathrm{Rep}_K Q$ has as objects the representations of Q, see Example II.2.4(g), and as morphisms the morphisms of representations, see Example II.3.2(h). Let $f = (f_x)_{x \in Q_0}: U \to V$, $g = (g_x)_{x \in Q_0}: V \to W$ be morphisms. For any arrow $\alpha: x \to y$ in Q, we have $f_y U(\alpha) = V(\alpha) f_x$ and $g_y V(\alpha) = W(\alpha) g_x$. These relations imply that $(g_y f_y) U(\alpha) = W(\alpha)(g_x f_x)$, so $(g_x f_x)_{x \in Q_0}$ is a morphism from U to W. Defining composition componentwise $gf = (g_x f_x)_{x \in Q_0}$ makes thus $\mathrm{Rep}_K Q$ into a category. A morphism of representations $f = (f_x)_{x \in Q_0}$ is an isomorphism if and only if each f_x is an isomorphism of K-vector spaces, see Exercise II.3.24.

Clearly, this list of examples can be continued indefinitely. In most of them (all except (g) and (h)), the object class is not a set, that is, the shown categories are not small. In Examples (a) to (f), the object classes consist of sets equipped with a given mathematical structure, the morphisms are maps which are compatible with this structure, and composition is the usual composition of maps. Such categories are called **concrete categories**.

We now give precise meanings to the term 'dual' and the expression 'reversing arrows' used in Subsection II.4.1.

III.2.3 Definition. Let C be a category. Its **dual category** or **opposite category** C^{op} has the same objects as C but, for objects X, Y in C, we have $C^{op}(X, Y) = C(Y, X)$. The composite of $f \in C^{op}(X, Y)$ and $g \in C^{op}(Y, Z)$ is the element $fg \in C^{op}(X, Z) = C(Z, X)$.

Thus, for instance, $(C^{op})^{op} = C$. If C is constructed as in Example III.2.2(g) starting from a monoid, then C^{op} is constructed in the same way starting from the opposite monoid.

A notion or property of C^{op} is the **dual** of a notion or property of C if it is obtained from the latter by reversing the direction of morphisms. A notion or property is called **selfdual** if it coincides with its dual.

The **duality principle** asserts that if a *categorical* property holds in a category C, then the dual property holds in C^{op}. Indeed, one is obtained (and proved!) from the other by reversing arrows. A categorical property is one which follows from purely categorical notions and results and not from the particularities of the category under study.

Exactly as one has subgroups, subspaces or submodules, there is a natural notion of subcategory.

III.2.4 Definition. Let C, D be categories. Then D is a **subcategory** of C if any object in D is an object in C, any morphism in D is a morphism in C, the composition in D is the same as in C, and for any object X in D, the identity 1_X is in D.

If X, Y are objects in D, then $D(X, Y) \subseteq C(X, Y)$. If, for any pair of objects (X, Y) in D, we have equality $D(X, Y) = C(X, Y)$, then D is called a **full subcategory** of C. Thus, full subcategories are uniquely determined by their object classes.

III.2.5 Examples.
- (a) The category Ab of abelian groups is a full subcategory of the category Gr of groups.
- (b) If A is an algebra, then the category $\mathrm{mod}A$ of finitely generated A-modules and A-linear maps between them is a full subcategory of the category $\mathrm{Mod}A$ of A-modules.
- (c) Let M be a monoid, considered as a category $\mathcal{M}$, as in Example III.2.2(g), then a subcategory is simply a submonoid. Moreover, the only nonempty full subcategory of $\mathcal{M}$ is $\mathcal{M}$ itself.
- (d) Let K be a field and Q a finite quiver. The subcategory $\mathrm{rep}_K Q$ of $\mathrm{Rep}_K Q$ consisting of the finite-dimensional representations, that is, those representations V such that $\dim_K V(x) < \infty$, for any $x \in Q_0$, (or, equivalently, because Q_0 is finite, $\Sigma_{x \in Q_0} \dim_K V(x) < \infty$) is full.

A word of caution: when we say that D is a subcategory of C, we mean that distinct objects in D remain distinct when viewed as objects in C. For instance, even though any K-algebra is, in particular, an abelian group, the category Alg_K is not a subcategory of Ab: indeed, the multiplication of the algebra may be defined on a given abelian group in many different ways.

III.2.2 Functors

Functors allow to pass from a category to another. One can think of a functor as a "morphism between categories". Because categories are defined by objects, morphisms and composition, a functor should associate to each object, or morphism, of the first category, an object, or a morphism, respectively, of the second, and this association should respect composition and preserve identity morphisms. There are two ways to orient a morphism between two given objects, so we have two types of functors.

III.2.6 Definition. Let C and D be categories.

(a) A **covariant functor** $F: C \to D$ associates to each object X of C an object $F(X)$, or, briefly, FX, of D, and to each morphism $f: X \to Y$ in C a morphism $F(f): FX \to FY$, or, briefly, $Ff: FX \to FY$, in D such that:
- (F1) If g, f are composable in C, then so are Fg, Ff in D and we have $F(gf) = (Fg)(Ff)$.
- (F2) For any object X in C, we have $F(1_X) = 1_{FX}$.

(b) A **contravariant functor** $F: C \to D$ associates to each object X of C an object $F(X)$, or, briefly, FX, of D, and to each morphism $f: X \to Y$ in C a morphism $F(f): FY \to FX$, or, briefly, $Ff: FY \to FX$, in D such that:

(F1) If g, f are composable in C, then so are Ff, Fg in D and we have
$F(gf) = (Ff)(Fg)$.

(F2) For any object X in C, we have $F(1_X) = 1_{FX}$.

Thus, a functor may be viewed as a 'map' from the class of objects of C to that of D, and, for any pair (X, Y) of objects of C, of a map (a real one!)

$$C(X, Y) \to D(FX, FY) \quad \text{if } F \text{ is covariant, and}$$

$$C(X, Y) \to D(FY, FX) \quad \text{if } F \text{ is contravariant.}$$

given by $f \mapsto Ff$, for $f \in C(X, Y)$.

A functor from a category to itself is called an **endofunctor** of this category.

It follows from the definition that a contravariant functor $F: C \to D$ may be viewed as a covariant functor $F: C^{op} \to D$ (or $F: C \to D^{op}$). We may thus restrict ourselves to covariant functors, except when our discussion involves at the same time covariant and contravariant functors.

The compatibility of functors with composition of morphisms has nice consequences: for instance, applying a functor to all objects and morphisms in a commutative diagram yields another commutative diagram.

III.2.7 Examples.

(a) Let D be a subcategory of C. The **inclusion functor** $D \to C$ sends any object, or morphism, of D to itself, considered as object, or morphism, respectively, of C. If $C = D$, then the inclusion functor becomes the **identity functor** 1_C.

(b) Let A, B, C be categories, $F: A \to B$ and $G: B \to C$ functors. Their composite $G \circ F$ or $GF: A \to C$ is defined by setting, for an object X of A, $(GF)(X) = G(FX)$ and for a morphism f of A, $(GF)(f) = G(Ff)$. This definition guarantees associativity of the composition of functors.

One sees easily that the composite of two covariant, or two contravariant, functors is covariant, while the composite of a covariant and a contravariant functor (in either order) is contravariant.

(c) **Forgetful functors.** As the name indicates, a forgetful functor 'forgets' *all* or *part* of the structure of the object considered. For instance, we define a functor $|-|: Gr \to Sets$ as follows. For a group G, the set $|G|$ is its underlying set and, for a group homomorphism $f: G \to H$, the map $|f|: |G| \to |H|$ is its underlying map. One also defines forgetful functors $Top \to Sets$, $Alg_K \to Ab$, $ModA \to Ab$, with A an algebra, etc. If D is a full subcategory of C, then the inclusion functor may be viewed as a forgetful functor. This is the case, for instance, of the inclusion of Ab into Gr (one considers an abelian group simply as a group, 'forgetting' commutativity of the operation).

(d) Let C be a category and X an object in C. The functor $C(X,-):C \to Sets$ associates to an object M in C the set $C(X,M)$ and to a morphism $f: M \to N$ in C the map $C(X,f): C(X,M) \to C(X,N)$ defined for $g \in C(X,M)$ by $g \mapsto fg$.

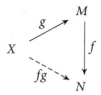

This functor is covariant. If $C = \mathrm{Mod}A$, with A a K-algebra, and K a commutative ring, then the functor $C(X,-)$ is usually denoted as $\mathrm{Hom}_A(X,-)$. It actually sends $\mathrm{Mod}A$ to $\mathrm{Mod}K$, see Lemma II.3.7.

Similarly, the functor $C(-,X):C \to Sets$ associates to an object M in C the set $C(M,X)$ and to a morphism $f: M \to N$ in C the map $C(f,X): C(N,X) \to C(M,X)$ defined for $g \in C(N,X)$ by $g \mapsto gf$.

This functor is contravariant. If $C = \mathrm{Mod}A$, with A an algebra over a commutative ring K, then the functor $C(-,X) = \mathrm{Hom}_A(-,X)$ sends $\mathrm{Mod}A$ into $\mathrm{Mod}K$, see again Lemma II.3.7.

This situation is expressed by saying that $C(-, ?)$ is a **bifunctor**, or **functor in two variables**, contravariant in the first variable, and covariant in the second. Different symbols – and ? are used for stressing that, usually, the same variable does not occur in both places.

When we deal with bimodules, contravariance in the second variable and covariance in the first is also reflected in the way scalars are multiplied. Let A, B be algebras and ${}_BM_A$ a B-A-bimodule. Because of Lemma II.3.8, if U_A is an A-module, then the left B-module structure of M induces a right B-module structure on $\mathrm{Hom}_A(M, U)$ by

$$(fb)(x) = f(bx)$$

for $f \in \mathrm{Hom}_A(M, U), b \in B, x \in M$. Assume that $u : U \to V$ is an A-linear map, then $\mathrm{Hom}_A(M, u) : f \mapsto uf$ is B-linear, because $u(fb) = (uf)b$ for $f \in \mathrm{Hom}_A(M, U), b \in B$. Thus, the bimodule ${}_BM_A$ induces a functor $\mathrm{Hom}_A(M, -) : \mathrm{Mod}A \to \mathrm{Mod}B$.

Similarly, the bimodule ${}_BM_A$ induces a functor $\mathrm{Hom}_B(M,-) : \mathrm{Mod}B^{op} \to \mathrm{Mod}A^{op}$.

(e) Let K be a field and $\mathcal{E}$ the category of pairs (E,f) with E a K-vector space, and $f: E \to E$ an endomorphism, see Example III.2.2(i). The functor $\Phi:\mathcal{E} \to \mathrm{Mod}K[t]$ sends a pair (E,f) to the $K[t]$-module ${}_fE$ constructed from it, see Example II.2.4(f), and a morphism $u:(E,f) \to (F,g)$ to itself.

Because u is a morphism in $\mathcal{E}$, it satisfies $uf = gu$, and so is a morphism of $K[t]$-modules, see Example II.3.3(g).

(f) Let K be a field, Q a finite quiver and $\text{Rep}_K Q$ the category of K-linear representations of Q, see Example III.2.2(j). Define a functor $F: \text{Rep}_K Q \to \text{Mod}KQ$ by sending a representation V to the KQ-module associated to V, see Example II.2.4(g), and sending a morphism of representations $f: V \to W$ to the associated morphism of KQ-modules, see Example II.3.3(h).

(g) Let A, B be K-algebras. A morphism $\varphi : A \to B$ induces a functor $\Phi : \text{Mod}B \to \text{Mod}A$ as follows: given a B-module M, let $\Phi(M)$ denote M considered as A-module, see Example II.2.4(e). It is easy to check that, under this definition, any B-module morphism becomes an A-module morphism. The functor Φ is called a **change of scalars functor**.

III.2.3 Functorial morphisms

In order to compare functors, we need a notion of morphism between them. We consider a pair of functors which are both covariant, or both contravariant, between the same categories.

III.2.8 Definition. Let C, D be categories.

(a) If $F, G: C \to D$ are covariant functors, a **functorial morphism**, or **natural transformation**, $\varphi: F \to G$ is the data, for any object X in C, of a morphism $\varphi_X: FX \to GX$ such that, for any morphism $f: X \to Y$ in C, we have that $\varphi_Y(Ff) = (Gf)\varphi_X$, that is, we have a commutative square

$$
\begin{array}{ccc}
FX & \xrightarrow{\varphi_X} & GX \\
{\scriptstyle Ff}\downarrow & & \downarrow{\scriptstyle Gf} \\
FY & \xrightarrow{\varphi_Y} & GY
\end{array}
$$

(b) If $F, G: C \to D$ are contravariant functors, a **functorial morphism**, or **natural transformation**, $\varphi: F \to G$ is the data, for any object X in C, of a morphism $\varphi_X: FX \to GX$ such that, for any morphism $f: X \to Y$, we have that $\varphi_X(Ff) = (Gf)\varphi_Y$, that is, we have a commutative square

$$
\begin{array}{ccc}
FX & \xrightarrow{\varphi_X} & GX \\
{\scriptstyle Ff}\uparrow & & \uparrow{\scriptstyle Gf} \\
FY & \xrightarrow{\varphi_Y} & GY
\end{array}
$$

The commutation of the squares, in both definitions, is a compatibility condition (with the morphisms of the source category). We call it **functoriality** of the morphism φ. Most commutative squares appearing in this book are produced using functoriality.

III.2.9 Examples.

(a) To a functor F is associated the **identity** functorial morphism 1_F defined by $(1_F)_X = 1_{FX}$, for any object X.

(b) Let $\varphi: F \to G$ and $\psi: G \to H$ be functorial morphisms, it is immediate that $\psi\varphi: F \to H$ defined by $(\psi\varphi)_X = \psi_X\varphi_X$, for any object X, is a functorial morphism. This defines an associative composition of functorial morphisms.

(c) One may ask whether there exists a category whose objects are functors and morphisms are functorial morphisms. One has to exercise caution because morphisms must constitute sets. Let $\mathcal{I}, \mathcal{C}$ be categories and $F, G : \mathcal{I} \to \mathcal{C}$ functors. A functorial morphism $\varphi : F \to G$ is a collection of morphisms $\varphi_X \in \mathcal{C}(FX, GX)$, indexed by the objects X of $\mathcal{I}$, and satisfying the functoriality conditions. Thus, $\varphi \in \prod_{X \in \mathcal{I}_0} \mathcal{C}(FX, GX)$ but the latter is a set if and only if so is $\mathcal{I}_0$, that is, if and only if $\mathcal{I}$ is a small category. Hence, if $\mathcal{I}$ is small, then we have a category of functors from $\mathcal{I}$ to $\mathcal{C}$, denoted by $\mathrm{Fun}(\mathcal{I}, \mathcal{C})$.

(d) Let $u: M \to N$ be a morphism in a category $\mathcal{C}$. We associate to u a functorial morphism $\mathcal{C}(u, -): \mathcal{C}(N, -) \to \mathcal{C}(M, -)$ as follows: for an object X of $\mathcal{C}$, the map $\mathcal{C}(u, X): \mathcal{C}(N, X) \to \mathcal{C}(M, X)$ is defined by $\mathcal{C}(u, X)(f) = fu$:

We have to prove functoriality, that is, to a morphism $f: X \to Y$ in $\mathcal{C}$ should correspond a commutative square

$$\begin{array}{ccc}
\mathcal{C}(N, X) & \xrightarrow{\mathcal{C}(u, X)} & \mathcal{C}(M, X) \\
{\scriptstyle \mathcal{C}(N, f)} \downarrow & & \downarrow {\scriptstyle \mathcal{C}(M, f)} \\
\mathcal{C}(N, Y) & \xrightarrow{\mathcal{C}(u, Y)} & \mathcal{C}(M, Y)
\end{array}$$

and this follows from the fact that, for $g \in \mathcal{C}(N, X)$, we have

$$\begin{aligned}
\mathcal{C}(u, Y)\mathcal{C}(N, f)(g) = \mathcal{C}(u, Y)(fg) &= fgu \\
&= \mathcal{C}(M, f)(gu) = \mathcal{C}(M, f)\mathcal{C}(u, X)(g).
\end{aligned}$$

The common composite $\mathcal{C}(u, Y)\mathcal{C}(N, f) = \mathcal{C}(M, f)\mathcal{C}(u, X)$ is a map from $\mathcal{C}(N, X)$ to $\mathcal{C}(M, Y)$, denoted by $\mathcal{C}(u, f)$. If $L \xrightarrow{u} M \xrightarrow{v} N$ and $X \xrightarrow{f} Y \xrightarrow{g} Z$ are morphisms in $\mathcal{C}$, then $\mathcal{C}(u, g)\mathcal{C}(v, f) = \mathcal{C}(vu, gf)$.

Similarly, $u: M \to N$ defines a functorial morphism between contravariant functors $\mathcal{C}(-, u): \mathcal{C}(-, M) \to \mathcal{C}(-, N)$.

Notation in this example (and others) may seem heavy because we must indicate dependence on the variables of our functors. However, as one can see, there is no conceptual difficulty.

Let $\mathcal{C}, \mathcal{D}$ be categories, $F, G: \mathcal{C} \to \mathcal{D}$ functors and $\varphi: F \to G$ a functorial morphism. We say that φ is a ***functorial isomorphism*** if φ_X is an isomorphism for any object X in $\mathcal{C}$. In this case, it is easily seen that the collection of morphisms φ_X^{-1} defines a functorial morphism $G \to F$, called the ***inverse*** of φ, and denoted by φ^{-1}. Hence, $\varphi: F \to G$ is a functorial isomorphism if and only if there exists a functorial morphism $\psi: G \to F$

such that $\varphi\psi = 1_G$ and $\psi\varphi = 1_F$. As expected, if φ is a functorial isomorphism, so is φ^{-1}, and moreover $(\varphi^{-1})^{-1} = \varphi$, and, if φ, ψ are functorial isomorphisms, so is $\varphi\psi$ and moreover $(\varphi\psi)^{-1} = \psi^{-1}\varphi^{-1}$.

III.2.10 Examples.

(a) In Example III.2.9(d), if u is an isomorphism of C, then $C(u, -)$ and $C(-, u)$ are functorial isomorphisms, with respective inverses $C(u^{-1}, -)$ and $C(-, u^{-1})$.

(b) Let A be an algebra, and consider the functors $\mathrm{Hom}_A(A, -)$ and $1_{\mathrm{Mod}A}$ from $\mathrm{Mod}A$ to itself. The functorial morphism

$$\varphi: \mathrm{Hom}_A(A, -) \to 1_{\mathrm{Mod}A} \quad \text{defined by} \quad \varphi_M: \mathrm{Hom}_A(A, M) \to M, f \mapsto f(1),$$

for $f \in \mathrm{Hom}_A(A, M)$, is a functorial isomorphism with inverse

$$\psi: 1_{\mathrm{Mod}A} \to \mathrm{Hom}_A(A, -) \text{ defined by } \psi_M: M \to \mathrm{Hom}_A(A, M), x \mapsto (a \mapsto xa)$$

for $x \in M, a \in A$. Indeed, because of Theorem II.3.9, φ_M and ψ_M are inverse isomorphisms, for any object M. We verify functoriality, that is, given $u: M \to N$, we prove that the following square commutes:

$$
\begin{array}{ccc}
\mathrm{Hom}_A(A, M) & \xrightarrow{\varphi_M} & M \\
\Big\downarrow{\scriptstyle \mathrm{Hom}_A(A, u)} & & \Big\downarrow{\scriptstyle u} \\
\mathrm{Hom}_A(A, N) & \xrightarrow{\varphi_N} & N
\end{array}
$$

For any $f \in \mathrm{Hom}_A(A, M)$, we have

$$\varphi_N\,\mathrm{Hom}_A(A, u)(f) = \varphi_N(uf) = (uf)(1) = u(f(1)) = u\varphi_M(f).$$

This proves functoriality of φ. Functoriality of ψ follows from the fact that, for any module M, we have $\psi_M = \varphi_M^{-1}$. Thus, φ and ψ are inverse functorial isomorphisms.

III.2.4 Yoneda's lemma

Suppose we wish to study an object X in a category C. In general, X is not a set, so we cannot talk about its elements or subsets. It can only be characterised by means of the morphisms starting or ending in it. This means studying the sets $C(X, Y)$ and $C(Y, X)$, for *any* object Y, or, equivalently, the functors $C(X, -)$ and $C(-, X)$. The same philosophy, applied to these functors, leads to studying the functorial morphisms between them and an arbitrary functor F.

Given functors $F, G : C \to D$, denote by $\mathrm{Fun}(F, G)$ the class of functorial morphisms from F to G. Yoneda's lemma states that, for any functor $F : C \to Sets$, there exists a bijection between $\text{Fun}(C(X,-),F)$ and the set FX. This bijection is constructed explicitly. As pointed out in Subsection II.3.2, the construction should be 'natural' in the sense that it is difficult to imagine another one. We wish to map a functorial morphism $\varphi : C(X,-) \to F$ to an element of the set FX. Because FX is the evaluation of the functor F on the set X, it seems 'natural' to consider the evaluation $\varphi_X : C(X,X) \to FX$ of the functorial morphism φ on the object X. The set $C(X,X)$ has a distinguished element, namely the identity 1_X, whose image under φ_X is an element $\varphi_X(1_X)$ of FX. This defines a map $\varphi \mapsto \varphi_X(1_X)$ from $\text{Fun}(C(X,-),F)$ to FX.

III.2.11 Theorem. Yoneda's lemma. *Let C be a category, X an object in C and $F : C \to Sets$ a functor. Then there exists a bijection*

$$\eta : \text{Fun}(C(X,-),F) \to FX \quad \text{given by } \varphi \mapsto \varphi_X(1_X).$$

Proof. We construct an inverse bijection

$$\sigma : FX \to \text{Fun}(C(X,-),F).$$

To $x \in FX$, we want to attach a functorial morphism $\sigma(x) : C(X,-) \to F$. The latter, evaluated on an object Y becomes a morphism $\sigma(x)_Y : C(X,Y) \to FY$. Take $f \in C(X,Y)$. We want to map it on an element $\sigma(x)_Y(f) \in FY$. But $x \in FX$, and Ff is a map from FX to FY. It is thus natural to set: $\sigma(x)_Y(f) = F(f)(x)$. We prove that $\sigma(x)$ is indeed a functorial morphism. Let $u : Y \to Z$ be a morphism in C. We must prove commutativity of the square:

$$
\begin{array}{ccc}
C(X,\,Y) & \xrightarrow{\;\sigma(x)_Y\;} & FY \\
{\scriptstyle C(X,u)}\Big\downarrow & & \Big\downarrow{\scriptstyle F(u)} \\
C(X,\,Z) & \xrightarrow{\;\sigma(x)_Z\;} & FZ
\end{array}
$$

Now, for $f \in C(X,Y)$, we have

$$F(u)\sigma(x)_Y(f) = F(u)F(f)(x) = F(uf)(x) = F(C(X,u)(f))(x) = \sigma(x)_Z C(X,u)(f).$$

Because f is arbitrary, we get $F(u)\sigma(x)_Y = \sigma(x)_Z C(X,u)$, as required.

We next prove that η and σ are inverse bijections. First, for $x \in FX$, the definitions give $\eta\sigma(x) = \sigma(x)_X(1_X) = F(1_X)(x) = x$. That is, $\eta\sigma(x) = x$, for any x.

Second, let $\varphi : C(X,-) \to F$ be a functorial morphism. In order to prove that $\sigma\eta(\varphi) = \varphi$, we must prove that $\sigma\eta(\varphi)_Y = \varphi_Y$, for any object Y in C. For $f \in C(X,Y)$, we have

$$\sigma\eta(\varphi)_Y(f) = F(f)\eta(\varphi) = F(f)\varphi_X(1_X) = \varphi_Y C(X,f)(1_X) = \varphi_Y(f)$$

because of commutativity of the square

$$
\begin{array}{ccc}
C\,(X,X) & \xrightarrow{\;\varphi_X\;} & FX \\[2mm]
{\scriptstyle C(X,f)}\big\downarrow & & \big\downarrow{\scriptstyle F(f)} \\[2mm]
C\,(X,Y) & \xrightarrow{\;\varphi_Y\;} & FY
\end{array}
$$

which results from functoriality of φ. This proves that $\sigma\eta(\varphi)_Y = \varphi_Y$, for any Y, and hence $\sigma\eta(\varphi) = \varphi$. Finally, functoriality of η is an exercise left to the reader. ☐

In practice, not only the statement of the Lemma is useful, but also the explicit formulae giving η and σ. For instance, if $F = C(Y,-)$, for some object Y, then we have inverse bijections:

$$
\mathrm{Fun}\left(C(X,-),C(Y,-)\right) \; \underset{\sigma}{\overset{\eta}{\rightleftarrows}} \; C\,(Y,X)
$$

III.2.12 Corollary. *Let C be a category and X, Y objects in C:*

(a) *There exists a bijection*

$$
C(Y,X) \to \mathrm{Fun}\big(C(X,-),C(Y,-)\big) \quad \textit{given by} \quad f \mapsto C(f,-).
$$

(b) *The bijection in (a) restricts to a bijection between the isomorphisms of $C(Y,X)$ and the functorial isomorphisms of $\mathrm{Fun}(C(X,-),C(Y,-))$. In particular, $X \cong Y$ if and only if $C(X,-) \cong C(Y,-)$.*

Proof. (a) Given a morphism $f: Y \to X$, the functorial morphism of Yoneda's lemma $\sigma(f): C(X,-) \to C(Y,-)$ is evaluated on an object U as $\sigma(f)_U : C(X,U) \to C(Y,U)$ which is given by the formula

$$
\sigma(f)_U(u) = C(Y,u)(f) = uf = C(f,U)(u)
$$

for $u : X \to U$. Therefore, $\sigma(f)_U = C(f,U)$, for any object U, and hence $\sigma: f \mapsto C(f,-)$.

(b) Let $f: Y \to X$ be an isomorphism with inverse g. Then $C(f,-): C(X,-) \to C(Y,)$ is a functorial isomorphism with inverse $C(g,-)$, because

$$
C(f,-)C(g,-) = C(gf,-) = C(1_Y,-) = 1_{C(N,-)}
$$

and, similarly, $C(g,-)C(f,-) = 1_{C(X,-)}$.

Conversely, let $\varphi : C(X,-) \to C(Y,-)$ be a functorial isomorphism with inverse ψ. Because of (a), there exist unique morphisms $f: Y \to X, g: X \to Y$ such that $\varphi = C(f,-), \psi = C(g,-)$. Then,

$$
C(gf,-) = C(f,-)C(g,-) = \varphi\psi = 1_{C(Y,-)} = C(1_Y,-)
$$

from which we deduce $gf = 1_Y$ because of (a). Similarly, $fg = 1_X$. ☐

The proof of Corollary III.2.12 illustrates the practical use of Yoneda's lemma: it is a powerful tool allowing to prove existence and uniqueness of morphisms. Assume that A is an algebra and M, N are A-modules. Part (a) of the corollary says that, to a functorial morphism $\varphi : \text{Hom}_A(M, -) \to \text{Hom}_A(N, -)$ corresponds a unique A-linear map $f : N \to M$ such that $\varphi = \text{Hom}_A(f, -)$, and part (b) that $f : N \to M$ is an isomorphism if and only if, for any module U, the map $\text{Hom}_A(f, U) : \text{Hom}_A(M, U) \to \text{Hom}_A(N, U)$ given by $u \mapsto uf$, for $u \in \text{Hom}_A(M, U)$, is bijective, that is, $M \cong N$ if and only if $\text{Hom}_A(M, -) \cong \text{Hom}_A(N, -)$. That is, a module is characterised up to isomorphism by the morphisms starting in it.

There is also a version of Yoneda's lemma for contravariant functors. It is stated as follows.

III.2.13 Theorem. Yoneda's lemma. *Let C be a category, X an object in C and $F : C \to$ Sets a contravariant functor. Then there exists a bijection*

$$\eta : \text{Fun}(C(-, X), F) \to FX \quad \text{defined by} \quad \varphi \mapsto \varphi_X(1_X).$$

Proof. The inverse bijection $\sigma : FX \to \text{Fun}(C(-, X), F)$ is given by the same formula as in Theorem III.2.11: given $x \in FX$ and an object Y in C, we define $\sigma(x)_Y : C(Y, X) \to FY$ by setting

$$\sigma(x)_Y(f) = F(f)(x)$$

for $f \in C(Y, X)$. The rest is an exercise which can safely be left to the reader. $\quad\square$

III.2.14 Corollary. *Let C be a category and X, Y objects in C:*
  (a) *There exists a bijection*

$$C(X, Y) \to \text{Fun}(C(-, X), C(-, Y)) \quad \text{given by} \quad f \mapsto C(-, f).$$

  (b) *The bijection in (a) restricts to a bijection between the isomorphisms of $C(X, Y)$ and the functorial isomorphisms of $\text{Fun}(C(-, X), C(-, Y))$. In particular, $X \cong Y$ if and only if $C(-, X) \cong C(-, Y)$.*

Proof. Similar to that of Corollary III.2.12. $\quad\square$

If A is an algebra, then it follows from the corollary that a module morphism $f : M \to N$ is an isomorphism if and only if, for any module U, the map $\text{Hom}_A(U, f) : \text{Hom}_A(U, M) \to \text{Hom}_A(U, N)$ given by $u \mapsto fu$, for $u \in \text{Hom}(U, M)$, is bijective. Thus, a module is characterised up to isomorphism by the morphism ending in it.

Yoneda's lemma allows to translate statements about modules to statements about functors and vice-versa. Actually, functor categories are, in many aspects, better behaved than module categories. This is because, as the reader will see, module categories are not selfdual, that is, not identical to their duals, and so in many contexts 'dual statements' and 'dual proofs' do not hold true for modules, while working with functors is smoother. Moreover, the Hom-functors which are used in Yoneda's lemma have particularly nice properties, see Lemma VI.4.10.

Exercises of Section III.2

1. Let C be a category. Define an equivalence relation $\sim$ in each set $C(X, Y)$ and assume it is compatible with compositions, that is, such that $f \sim g, f' \sim g'$ imply $ff' \sim gg'$ whenever these composites are defined.
  - (a) Show how to form a 'quotient category' $C/\sim$ whose objects are the same as those of C, and whose morphisms are equivalence classes of morphisms of C.
  - (b) Define a 'projection' functor $F: C \rightarrow C/\sim$.

2. Let E be a set. Define the **discrete category** C_E on E whose objects are the elements of E and, for $x, y \in E$, the morphism set $C_E(x, y)$ is the one-element set $\{1_x\}$ if $x = y$ and $\emptyset$ otherwise.
  - (a) Verify that C_E is a category.
  - (b) Let E, F be sets. Prove that the functors $C_E \rightarrow C_F$ coincide with the maps $E \rightarrow F$.
  - (c) Let $f, g: C_E \rightarrow C_F$ be functors. Prove that there exists a functorial morphism $f \rightarrow g$ if and only if $f = g$.

3. Let C be the category whose objects are posets and the morphisms are poset morphisms (see at the end of Subsection II.4.4). Give an example of a bijective morphism in C which is not an isomorphism.

4. Let C, D be categories. The **product category** $C \times D$ has as objects the pairs (X, U) with X an object in C and U an object in D, and as morphisms the pairs $(f, u): (X, U) \rightarrow (Y, V)$ with $f: X \rightarrow Y$ a morphism in C and $u: U \rightarrow V$ a morphism in D.
  - (a) Verify that this data defines a category.
  - (b) Let C', D' be subcategories of C, D, respectively. Prove that $C' \times D'$ is a subcategory of $C \times D$.

5. Let D be a subcategory of C. Prove that D^{op} is a subcategory of C^{op}.

6. Let C be a category and X an object in C. Prove that the set $\mathrm{Aut}_C X$ of automorphisms of X is a group under composition of morphisms.

7. Let C be a category. Its **morphism category** $\mathrm{Mor}C$ has as objects the triples (X, Y, f) where X, Y are objects in C and $f: X \rightarrow Y$ a morphism in C, a morphism in $\mathrm{Mor}C$ from (X, Y, f) to (X', Y', f') is a pair of morphisms (u, v) in C such that $u: X \rightarrow X'$, $v: Y \rightarrow Y'$ satisfy $vf = f'u$, that is, the following square commutes:

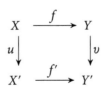

and finally, define composition componentwise $(u, v)(u', v') = (uu', vv')$. Prove that:

  (a) $\mathrm{Mor}C$ is a category.
  (b) A morphism (u, v) is an isomorphism in $\mathrm{Mor}C$ if and only if each of u, v is an isomorphism in C, and then we have $(u, v)^{-1} = (u^{-1}, v^{-1})$.
  (c) One defines functors $S, T: \mathrm{Mor}C \to C$ by setting $S(X, Y, f) = X$ and $T(X, Y, f) = Y$.

8. Let M, N be monoids and $\mathcal{M}, \mathcal{N}$ the associated categories as in Example III.2.2(g). Prove that covariant functors $\mathcal{M} \to \mathcal{N}$ coincide with monoid homomorphisms $M \to N$.

9. Show that, for any set E, there exists a functor $E \times -: Sets \to Sets$ and for any map $f: E \to E'$, there exists a functorial morphism $f \times -: E \times - \to E' \times -$.

10. Let E be a nonempty set, G an abelian group and G^E the set of maps from E to G. Prove that:
   (a) G^E becomes an abelian group if one defines, for $f, g \in G^E$, their sum $f + g$ by $x \mapsto f(x) + g(x)$, for $x \in E$.
   (b) The correspondence $(-)^E: G \mapsto G^E$ defines a functor $Ab \to Ab$.

11. Let G be an abelian group.
   (a) Prove that $T(G) = \{x \in G : x \text{ is of finite order}\}$ is a subgroup of G called its **torsion subgroup.**
   (b) Prove that the torsion subgroup $T(G)$ is a pure subgroup of G, see Exercise II.4.28.
   (c) Prove that $G \mapsto T(G)$ defines a functor $T : Ab \to Ab$.
   (d) Prove that, if $f: G \to H$ is a monomorphism, then so is $T(f)$.
   (e) Give an example of an epimorphism $f: G \to H$ such that $T(f)$ is not an epimorphism.

12. Let $p \in \mathbb{Z}$ be a prime integer.
   (a) Prove that $G \mapsto G/pG$ defines a functor $S: Ab \to Ab$.
   (b) Prove that, if $f: G \to H$ is an epimorphism, then so is $S(f)$.
   (c) Give an example of a monomorphism $f: G \to H$ such that $S(f)$ is not a monomorphism.

13. Let A be a K-algebra and $f : M \to N$ an A-linear map. Prove that:
   (a) f is an epimorphism if and only if the induced K-linear map $\mathrm{Hom}_A(f, U) : \mathrm{Hom}_A(N, U) \to \mathrm{Hom}_A(M, U)$ is injective, for any A-module U.
   (b) f is a monomorphism if and only if the induced K-linear map $\mathrm{Hom}_A(U, f) : \mathrm{Hom}_A(U, M) \to \mathrm{Hom}_A(U, N)$ is injective, for any A-module U.
   (c) The following conditions are equivalent:
      (i) f is the zero morphism.
      (ii) The induced map $\mathrm{Hom}_A(f, U) : \mathrm{Hom}_A(N, U) \to \mathrm{Hom}_A(M, U)$ is zero, for any A-module U.
      (iii) The induced map $\mathrm{Hom}_A(U, f) : \mathrm{Hom}_A(U, M) \to \mathrm{Hom}_A(U, N)$ is zero, for any A-module U.

14. Let A, B be algebras and $_A M_B$ an A-B-bimodule. Prove that:
   (a) $\mathrm{Hom}_B(-, M)$ defines a contravariant functor from $\mathrm{Mod}B$ to $\mathrm{Mod}A^{op}$.
   (b) $\mathrm{Hom}_A(-, M)$ defines a contravariant functor from $\mathrm{Mod}A^{op}$ to $\mathrm{Mod}B$.

15. Assume that we have four categories A, B, C and D, four functors $F : A \to B$, $G_1, G_2 : B \to C$ and $H : C \to D$ as well as a functorial morphism $\varphi : G_1 \to G_2$. Prove that:

    (a) There exist functorial morphisms $\varphi F : G_1 F \to G_2 F$ and $H\varphi : HG_1 \to HG_2$.

    (b) If φ is a functorial isomorphism, then so are φF and $H\varphi$.

    (c) We have $H(\varphi F) = (H\varphi)F$.

    (d) For any functors $F' : A' \to A$ and $H' : D \to D'$, we have $(\varphi F)F' = \varphi(FF')$ and $H'(H\varphi) = (H'H)\varphi$.

16. Let X, X' be objects in a category C.

    (a) If $F, F' : C \to Sets$ are functors such that we have functorial isomorphisms $\psi : C(X, -) \to F$ and $\psi' : C(X', -) \to F'$, prove that there exists a bijection between functorial morphisms $\varphi : F \to F'$ and morphisms $f : X' \to X$ such that $\varphi\psi = \psi'C(f, -)$.

    (b) Redo part (a) in case F, F' are contravariant functors.

III.3 Products and coproducts of modules

III.3.1 Products of modules

We introduced categories as a new language, that is, a terminology. In order to justify its introduction, we need to see whether it allows or not a better understanding of modules. We start with the well-known notion of product. Products exist for sets, groups, topological spaces, algebras and modules. We give a unifying definition valid in any category. Because categories involve only objects and morphisms, we can only use these concepts for our definition, that is, make the latter 'element-free', as we did in Subsection II.4.1. For instance, the product of two modules M_1, M_2 over the same algebra A, see Example II.2.4(d), is their cartesian product:

$$M_1 \times M_2 = \{(x_1, x_2) : x_1 \in M_1, x_2 \in M_2\}$$

equipped with componentwise addition and scalar multiplication. To $M_1 \times M_2$ are associated morphisms, namely the projections $p_1 : M_1 \times M_2 \to M_1$, given by $(x_1, x_2) \mapsto x_1$ and $p_2 : M_1 \times M_2 \to M_2$, given by $(x_1, x_2) \mapsto x_2$, for $(x_1, x_2) \in M_1 \times M_2$. If elements of $M_1 \times M_2$ are written as column vectors, then $p_1 = (1 \ 0)$ and $p_2 = (0 \ 1)$ in matrix notation. The definition consists in axiomatising properties of the projections considered as morphisms in the category.

III.3.1 Definition. Let $(M_\lambda)_{\lambda \in \Lambda}$ be a family of objects in a category C. A ***product*** of this family is a pair $(M, (p_\lambda)_{\lambda \in \Lambda})$, consisting of an object M and a family of morphisms $(p_\lambda : M \to M_\lambda)_{\lambda \in \Lambda}$ in C such that, for any pair $(M', (p'_\lambda)_{\lambda \in \Lambda})$ consisting of an object M' and a family of morphisms $(p'_\lambda : M' \to M_\lambda)_{\lambda \in \Lambda}$ in C, there exists a unique morphism $f : M' \to M$ in C satisfying $p_\lambda f = p'_\lambda$, for any $\lambda \in \Lambda$:

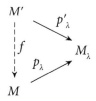

The property in the definition is called the **universal property** of the product. While the p_λ *start* from M, the morphism f *ends* in M. One can reformulate the uniqueness assertion in the definition: if f and g are morphisms in C such that $p_\lambda f = p_\lambda g$, for any $\lambda \in \Lambda$, then $f = g$.

This universal property is satisfied by the cartesian product $M_1 \times M_2$ of two modules: indeed, if M' is an A-module, and $p'_1 : M' \to M_1, p'_2 : M' \to M_2$ are A-linear maps, the unique A-linear map $f : M' \to M_1 \times M_2$ such that $p_1 f = p'_1, p_2 f = p'_2$ is given by $x \mapsto (p'_1(x), p'_2(x))$, for $x \in M'$.

A universal property implies uniqueness (up to isomorphism) of the object it defines. All uniqueness proofs for universal objects follow the same pattern, shown in the proofs of Theorems II.4.4 and II.4.7. One assumes the existence of two objects satisfying the universal property, uses this property in order to construct two morphisms, of opposite directions, between the objects, and finally applies the uniqueness assertion in the universal property to prove that these morphisms are inverse isomorphisms.

III.3.2 Lemma. *If a family of objects $(M_\lambda)_{\lambda \in \Lambda}$ admits a product in a category C, then this product is unique up to isomorphism.*

Proof. Let $(M, (p_\lambda)_{\lambda \in \Lambda})$ and $(M', (p'_\lambda)_{\lambda \in \Lambda})$ be products of the same family $(M_\lambda)_{\lambda \in \Lambda}$. The universal property applied successively to each of M and M' yields morphisms $f : M' \to M, f' : M \to M'$ such that $p_\lambda f = p'_\lambda$ and $p'_\lambda f' = p_\lambda$, for any $\lambda \in \Lambda$:

$$
\begin{array}{ccc}
M & \xrightarrow{\ p_\lambda\ } & M_\lambda \\
{\scriptstyle f'}\big\downarrow & & \big\downarrow{\scriptstyle 1_{M_\lambda}} \\
M' & \xrightarrow{\ p'_\lambda\ } & M_\lambda \\
{\scriptstyle f}\big\downarrow & & \big\downarrow{\scriptstyle 1_{M_\lambda}} \\
M & \xrightarrow{\ p'_\lambda\ } & M_\lambda
\end{array}
$$

Therefore,

$$p_\lambda f f' = p'_\lambda f' = p_\lambda = p_\lambda 1_M,$$

for any λ. Uniqueness in the universal property yields $f f' = 1_M$. Similarly, $f' f = 1_{M'}$. Thus, f, f' are inverse isomorphisms. $\qquad\square$

By abuse of language, we say that M, if it exists, is the **product** of the M_λ and we write $M = \prod_{\lambda \in \Lambda} M_\lambda$. If $\Lambda = \{1, 2, \cdots, n\}$, then we write $M = \prod_{i=1}^n M_i$, or $M = M_1 \times \cdots \times M_n$. If all the M_λ are equal to N, say, then we write $M = N^\Lambda$ (or N^n, if $\Lambda = \{1, 2, \cdots, n\}$). The morphisms p_λ are called the **projections**.

In a given category, it is not obvious that the product of an arbitrary family of objects exists, and if it does, then its construction depends on properties of the category under consideration. We construct the product of a family of modules. In the rest of this section, K denotes a commutative ring.

III.3.3 Theorem. *Let A be a K-algebra. Any family $(M_\lambda)_{\lambda \in \Lambda}$ of A-modules admits a product in ModA, unique up to isomorphism.*

Proof. Because of Lemma III.3.2, it suffices to establish existence and, for this, to find a candidate which satisfies the conditions of Definition III.3.1. Let $M = \{(x_\lambda)_{\lambda \in \Lambda} : x_\lambda \in M_\lambda\}$ be the cartesian product of the M_λ and, for $\lambda \in \Lambda$, set $p_\lambda : (x_\mu)_{\mu \in \Lambda} \mapsto x_\lambda$, for $(x_\mu)_{\mu \in \Lambda} \in M$. Define operations in M componentwise

$$(x_\lambda)_{\lambda \in \Lambda} + (y_\lambda)_{\lambda \in \Lambda} = (x_\lambda + y_\lambda)_{\lambda \in \Lambda} \quad \text{and} \quad (x_\lambda)_{\lambda \in \Lambda} \cdot a = (x_\lambda \cdot a)_{\lambda \in \Lambda}$$

for $(x_\lambda)_{\lambda \in \Lambda}, (y_\lambda)_{\lambda \in \Lambda} \in M, a \in A$. Then M becomes an A-module and the p_λ become linear maps. Let $(M', (p'_\lambda)_{\lambda \in \Lambda})$ be a pair with M' an A-module and $(p'_\lambda : M' \to M_\lambda)_{\lambda \in \Lambda}$ linear maps. Because $p_\lambda f = p'_\lambda$ means $p_\lambda f(x) = p'_\lambda(x)$, for $x \in M'$, the only morphism $f : M' \to M$ satisfying the conditions of the definition is defined by $f(x) = (p'_\lambda(x))_{\lambda \in \Lambda}$, for $x \in M'$. □

This construction generalises Example II.2.4(d). At this point, it is natural to ask why do we indulge in such generality and introduce categories to describe such classical concepts as products. Category theory, like other branches of mathematics, furnishes a continuous supply of new ideas and problems to module theorists. But the main point is that categorical techniques are useful to solve classical problems of module theory. A typical example is the elegant proof of Lemma III.3.2, which is based on intrinsic properties of the morphisms considered only, forgetting about elements of objects. Our objective is to develop in the reader the ability to think categorically, something that is perhaps best translated by saying 'think maps'.

III.3.4 Example. Let K be a field, Q a finite quiver and $(V_\lambda)_\lambda$ a family of representations of Q. The product $V = \prod_\lambda V_\lambda$ in the category of representations is constructed componentwise. For $x \in Q_0$, set $V(x) = \prod_\lambda V_\lambda(x)$ and, for $\alpha : x \to y$ in Q_1, let $V(\alpha) : V(x) \to V(y)$ be the unique K-linear map obtained by applying the universal property of the product to the square

$$
\begin{array}{ccc}
V(x) & \xrightarrow{\;p_\lambda(x)\;} & V_\lambda(x) \\
{\scriptstyle V(\alpha)}\Big\downarrow & & \Big\downarrow{\scriptstyle V_\lambda(\alpha)} \\
V(y) & \xrightarrow{\;p_\lambda(y)\;} & V_\lambda(y)
\end{array}
$$

where the horizontal maps are the vector space projections onto the λ-component of the product. This definition ensures that the maps $p_\lambda : V \to V_\lambda$ defined by $p_\lambda = (p_\lambda(x))_{x \subset Q_0}$ are morphisms of representations. Finally, because each $p_\lambda(x) : V(x) \to V_\lambda(x)$ is a projection in the sense of Definition III.3.1, so is each $p_\lambda : V \to V_\lambda$. Therefore, V is the product of the V_λ in the category of representations.

III.3.2 Coproducts

We dualise the definition of product by reversing arrows as seen in Definition III.2.3.

III.3.5 Definition. Let $(M_\lambda)_{\lambda \in \Lambda}$ be a family of objects in a category C. A **coproduct** of this family is a pair $(M, (q_\lambda)_{\lambda \in \Lambda})$, consisting of an object M and a family of morphisms $(q_\lambda : M \to M_\lambda)_{\lambda \in \Lambda}$ in C such that, for any pair $(M', (q'_\lambda)_{\lambda \in \Lambda})$ consisting of an object M' and a family of morphisms $(q'_\lambda : M' \to M_\lambda)_{\lambda \in \Lambda}$ in C, there exists a unique morphism $f : M \to M'$ in C satisfying $fq_\lambda = q'_\lambda$, for any $\lambda \in \Lambda$:

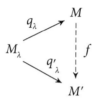

This is again a universal property. While the q_λ end in M, the morphism f *starts* there. Uniqueness in the universal property can be reformulated: if f and g are morphisms in C such that $fq_\lambda = gq_\lambda$, for any $\lambda \in \Lambda$, then $f = g$. Coproducts are also called **direct sums**.

III.3.6 Lemma. *If a family of objects $(M_\lambda)_{\lambda \in \Lambda}$ admits a coproduct in a category C, then this coproduct is unique up to isomorphism.*

Proof. Dual to (obtained by reversing arrows in) Lemma III.3.2 and left to the reader. □

By abuse of language, we say that M, if it exists, is the **coproduct** of the M_λ, or their **direct sum**, denoted as $M = \coprod_{\lambda \in \Lambda} M_\lambda$ or $M = \bigoplus_{\lambda \in \Lambda} M_\lambda$. If $\Lambda = \{1, 2, \cdots, n\}$, then we write $M = \coprod_{i=1}^{n} M_i$, or $M = \bigoplus_{i=1}^{n} M_i$, or $M = M_1 \coprod \cdots \coprod M_n$, or $M = M_1 \oplus \cdots \oplus M_n$. If all the M_λ are equal to N, say, then we write $M = N^{(\Lambda)}$ (or $N^{(n)}$, if $\Lambda = \{1, 2, \cdots, n\}$). The morphisms q_λ are called the **injections**.

We construct coproducts in module categories. While the purely categorical notions of product and coproduct are dual, the actual constructions of the corresponding objects in ModA (which depend on particular properties of this category) are not. This phenomenon will occur again and again, and is what we meant by saying that ModA is not selfdual. In order to construct the coproduct of modules, we think of the construction of direct sums as in Definition II.2.19, where we were considering families of elements of the M_λ which are almost all zero.

III.3.7 Theorem. *Let A be a K-algebra. Any family $(M_\lambda)_{\lambda \in \Lambda}$ of A-modules admits a direct sum (coproduct) in ModA, unique up to isomorphism.*

Proof. Because of Lemma III.3.6, it suffices to establish existence, and we do so by exhibiting a candidate module. Consider the set

$$M = \left\{ (x_\lambda)_{\lambda \in \Lambda} \in \prod_{\lambda \in \Lambda} M_\lambda : x_\lambda \in M_\lambda \text{ almost all zero} \right\}.$$

It is easily proved, using Proposition II.2.3, that M is a submodule of the product of the M_λ, and therefore is a module itself. Given $\lambda \in \Lambda$, define $q_\lambda : M_\lambda \to M$ by $q_\lambda : x_\lambda \mapsto (y_\mu)_{\mu \in \Lambda}$, where $y_\lambda = x_\lambda$ and $y_\mu = 0$ for $\mu \neq \lambda$. Then q_λ is A-linear.

Let $(M', (p'_\lambda)_{\lambda \in \Lambda})$ be a pair consisting of a module M' and A-linear maps $(q'_\lambda : M_\lambda \to M')_{\lambda \in \Lambda}$. We construct the unique morphism $f : M \to M'$ required in the definition as follows. If $(x_\lambda)_{\lambda \in \Lambda} \in M$ then, because the x_λ are almost all zero, $(x_\lambda)_{\lambda \in \Lambda} = \Sigma_{\lambda \in \Lambda} q_\lambda(x_\lambda)$. Hence,

$$f((x_\lambda)_{\lambda \in \Lambda}) = f\left(\sum_{\lambda \in \Lambda} q_\lambda(x_\lambda) \right) = \sum_{\lambda \in \Lambda} (fq_\lambda)(x_\lambda) = \sum_{\lambda \in \Lambda} q'_\lambda(x_\lambda)$$

defines the unique A-linear map f such that $fq_\lambda = q'_\lambda$, for any $\lambda \in \Lambda$. $\square$

The construction of the theorem shows that, if the index set Λ is finite, then product and direct sum of modules coincide. A finite direct sum has attached to it both projections p_λ and injections q_λ. For instance, if M_1, M_2 are A-modules, then $M_1 \oplus M_2 = \{ \binom{x_1}{x_2} : x_1 \in M_1, x_2 \in M_2 \}$ and the injections are $q_1 : x_1 \mapsto \binom{x_1}{0}, q_2 : x_2 \mapsto \binom{0}{x_2}$, for $x_1 \in M_1, x_2 \in M_2$. In matrix terms, $q_1 = \binom{1}{0}$ and $q_2 = \binom{0}{1}$. If we compare these injections to the projections $p_1 = (1 \ 0) : M_1 \oplus M_2 \to M_1, p_2 = (0 \ 1) : M_1 \oplus M_2 \to M_2$ we see that $p_1 q_1 = 1_{M_1}, p_1 q_2 = 0, p_2 q_1 = 0, p_2 q_2 = 1_{M_2}$ and finally $q_1 p_1 + q_2 p_2 = 1_{M_1 \oplus M_2}$. More generally, we have the following lemma.

III.3.8 Lemma. *Let A be a K-algebra and $\{M_1, \cdots, M_m\}$ a finite family of A-modules. The projections and injections associated to the direct sum $M = \bigoplus_{i=1}^m M_i$ satisfy the relations.*

$$(a) \ p_i q_j = \begin{cases} 1_{M_i} & \text{if } i = j \\ 0 & \text{if } i \neq j \end{cases} \qquad\qquad (b) \ \sum_{i=1}^m q_i p_i = 1_M$$

Proof. This is a direct calculation. $\square$

We now justify the term direct sum which is used concurrently with coproduct. Assume that $(M_\lambda)_{\lambda \in \Lambda}$ is a family of submodules of a given A-module M, and consider their sum $\Sigma_{\lambda \in \Lambda} M_\lambda$. For any index $\mu \in \Lambda$, there exists an inclusion $i_\mu : M_\mu \to \Sigma_{\lambda \in \Lambda} M_\lambda$ obtained by mapping any element of M_μ to itself, viewed as element of the sum. The universal property of coproduct yields a unique A-linear map $f : \coprod_{\lambda \in \Lambda} M_\lambda \to \Sigma_{\lambda \in \Lambda} M_\lambda$ such that $fq_\lambda = i_\lambda$, for any λ. As seen in the proof of Theorem III.3.7, this map is given by

$$f((x_\lambda)_{\lambda \in \Lambda}) = \sum_{\lambda \in \Lambda} i_\lambda(x_\lambda) = \sum_{\lambda \in \Lambda} x_\lambda$$

for $(x_\lambda)_{\lambda \in \Lambda} \in \coprod_{\lambda \in \Lambda} M_\lambda$, because each i_λ is the inclusion. Any element of $\Sigma_{\lambda \in \Lambda} M_\lambda$ is of this form, so f is surjective. It is an isomorphism if and only if its kernel vanishes, that is, if and only if $\Sigma_{\lambda \in \Lambda} x_\lambda = 0$, with the x_λ almost all zero, yields $x_\lambda = 0$ for any λ. But this says exactly that the M_λ are in direct sum, see Theorem II.2.20. Thus, the submodules M_λ are in direct sum if and only if their coproduct is isomorphic to their sum. This explains the terminology.

III.3.9 Example. Let K be a field, Q a finite quiver and $(V_\lambda)_\lambda$ a family of representations of Q. Their coproduct, or direct sum, $V = \oplus_\lambda V_\lambda$ in the category of representations is constructed componentwise. For $x \in Q_0$, set $V(x) = \oplus_\lambda V_\lambda(x)$ and, for $\alpha : x \to y$ in Q_1, let $V(\alpha) : V(x) \to V(y)$ be the unique K-linear map obtained by applying the universal property of the coproduct (direct sum) to the commutative square

$$
\begin{array}{ccc}
V_\lambda(x) & \xrightarrow{\;q_\lambda(x)\;} & V(x) \\
{\scriptstyle V_\lambda(\alpha)}\big\downarrow & & \big\downarrow{\scriptstyle V(\alpha)} \\
V_\lambda(y) & \xrightarrow{\;q_\lambda(y)\;} & V(y)
\end{array}
$$

where the horizontal maps are the vector space injections from the λ-component into the direct sum. The maps $q_\lambda : V_\lambda \to V$ defined by $q_\lambda = (q_\lambda(x))_{x \in Q_0}$ are morphisms of representations and they satisfy the conditions of Definition III.3.5. Therefore, V is the direct sum of the V_λ. This is proved as in Example III.3.4.

Let, for instance, Q be the quiver $1 \longrightarrow 2 \longrightarrow 3$ and consider the representations $V = (K \xrightarrow{1} K \xrightarrow{1} K)$ and $V' = (0 \to K \to 0)$. Their direct sum is given by

$$V \oplus V' = (K \xrightarrow{\binom{1}{0}} K^2 \xrightarrow{(1\;0)} K).$$

This sum being finite, it is also the product of V and V'.

III.3.3 Functorial properties

We study the behaviour of product and coproduct with respect to the Hom-functor.

III.3.10 Theorem. *Let C be a category, $(M_\lambda)_{\lambda \in \Lambda}$, $(N_\sigma)_{\sigma \in \Sigma}$ families of objects having, respectively, a coproduct $(M, (q_\lambda)_{\lambda \in \Lambda})$ and a product $(N, (p_\sigma)_{\sigma \in \Sigma})$ in C. Then the map*

$$\varphi : C(M, N) \to \prod_{(\lambda,\sigma) \in \Lambda \times \Sigma} C(M_\lambda, N_\sigma)$$

given by $f \mapsto (p_\sigma f q_\lambda)_{(\lambda,\sigma) \in \Lambda \times \Sigma}$, for $f \in C(M, N)$, is bijective.

Proof. Indeed, the diagram

$$
\begin{array}{ccc}
M & \xrightarrow{\ f\ } & N \\[4pt]
{\scriptstyle q_\lambda}\big\uparrow & & \big\downarrow{\scriptstyle p_\sigma} \\[4pt]
M_\lambda & \xdashrightarrow{\ p_\sigma f q_\lambda\ } & N_\sigma
\end{array}
$$

shows that φ is a map. We prove bijectivity.

(1) φ is surjective: assume that

$$
(f_{\sigma\lambda})_{(\lambda,\sigma)\in\Lambda\times\Sigma} \in \prod_{(\lambda,\sigma)\in\Lambda\times\Sigma} C(M_\lambda, N_\sigma).
$$

For a given $\lambda \in \Lambda$, the universal property of the product N gives $f_\lambda\colon M_\lambda \to N$ such that $p_\sigma f_\lambda = f_{\sigma\lambda}$, for any $\sigma \in \Sigma$. Then, the universal property of the coproduct M yields $f\colon M \to N$ such that $f q_\lambda = f_\lambda$, for any λ. Hence, $p_\sigma f q_\lambda = p_\sigma f_\lambda = f_{\sigma\lambda}$, for any pair (λ, σ).

(2) φ is injective: if $p_\sigma f q_\lambda = p_\sigma g q_\lambda$, for any (λ, σ), then uniqueness in the definition of the product N implies $f q_\lambda = g q_\lambda$, for any λ. Next, uniqueness in the definition of the coproduct M gives $f = g$, as required. □

If, in the statement, $C = \mathrm{Mod}A$, with A a K-algebra, so we are dealing with modules, then the bijection of the theorem

$$
\varphi\colon \mathrm{Hom}_A(M, N) \to \prod_{(\lambda,\sigma)\in\Lambda\times\Sigma} \mathrm{Hom}_A(M_\lambda, N_\sigma)
$$

is an isomorphism of K-modules. For, let $f, g \in \mathrm{Hom}_A(M, N)$ and $\alpha, \beta \in K$, then, using Lemma II.3.10, we have

$$
\begin{aligned}
\varphi(f\alpha + g\beta) &= (p_\sigma(f\alpha + g\beta)q_\lambda)_{(\lambda,\sigma)} = ((p_\sigma f\alpha + p_\sigma g\beta)q_\lambda)_{(\lambda,\sigma)} = \\
&= ((p_\sigma f q_\lambda)\alpha + (p_\sigma g q_\lambda)\beta)_{(\lambda,\sigma)} = (p_\sigma f q_\lambda)_{(\lambda,\sigma)}\alpha + (p_\sigma g q_\lambda)_{(\lambda,\sigma)}\beta = \\
&= \varphi(f)\alpha + \varphi(g)\beta.
\end{aligned}
$$

Assuming that Λ or Σ is a one-element set, we get the following corollaries.

III.3.11 Corollary. *Let A be a K-algebra, M, $(N_\sigma)_{\sigma\in\Sigma}$ be A-modules, then the map*

$$
\varphi\colon \mathrm{Hom}_A\left(M, \prod_{\sigma\in\Sigma} N_\sigma\right) \to \prod_{\sigma\in\Sigma} \mathrm{Hom}_A(M, N_\sigma),
$$

given by $f \mapsto (p_\sigma f)_{\sigma\in\Sigma}$, for $f \in \mathrm{Hom}_A(M, \prod_{\sigma\in\Sigma} N_\sigma)$, is an isomorphism of K-modules. □

III.3.12 Corollary. *Let A be a K-algebra, $(M_\lambda)_{\lambda \in \Lambda}$, N be Λ-modules, then the map*

$$\varphi: \operatorname{Hom}_A\left(\bigoplus_{\lambda \in \Lambda} M_\lambda, N\right) \to \prod_{\lambda \in \Lambda} \operatorname{Hom}_A(M_\lambda, N),$$

given by $f \mapsto (fq_\lambda)_{\sigma \in \Sigma}$, for $f \in \operatorname{Hom}_A(\bigoplus_{\lambda \in \Lambda} M_\lambda, N)$, is an isomorphism of K-modules. □

Thus, the covariant functor $\operatorname{Hom}_A(M, -)$ commutes with products, while the contravariant functor $\operatorname{Hom}_A(-, N)$ transforms direct sums into products (the dual notion, because of contravariance). Because finite direct sums of modules coincide with finite products, both the covariant and contravariant Hom functor commute with *finite* direct sums. We give, in Example IV.2.8(b) below, a sufficient condition so that the covariant Hom functor commutes with *arbitrary* direct sums.

Consider the case where $\Lambda = \{1, 2, \cdots, m\}$ and $\Sigma = \{1, 2, \cdots, n\}$ are both finite. Then, products and direct sums coincide. On the other hand, the image of the bijection of Theorem III.3.10 can be written in matrix form. Indeed, consider the set of matrices

$$H = \begin{pmatrix} \operatorname{Hom}_A(M_1, N_1) & \operatorname{Hom}_A(M_2, N_1) & \cdots & \operatorname{Hom}_A(M_m, N_1) \\ \operatorname{Hom}_A(M_1, N_2) & \operatorname{Hom}_A(M_2, N_2) & \cdots & \operatorname{Hom}_A(M_m, N_2) \\ \vdots & \vdots & & \vdots \\ \operatorname{Hom}_A(M_1, N_n) & \operatorname{Hom}_A(M_2, N_n) & \cdots & \operatorname{Hom}_A(M_m, N_n) \end{pmatrix}$$

$$= \{[f_{ij}]_{ij} : f_{ij} \in \operatorname{Hom}_A(M_j, N_i) \text{ for } 1 \leqslant i \leqslant n, 1 \leqslant j \leqslant m\}.$$

Then H has a natural K-module structure given, for $[f_{ij}]_{i,j}, [g_{ij}]_{i,j} \in H$ and $\alpha, \beta \in K$ by

$$[f_{ij}]_{i,j}\alpha + [g_{ij}]_{i,j}\beta = [f_{ij}\alpha + g_{ij}\beta]_{i,j}.$$

III.3.13 Corollary. *With this notation, we have an isomorphism of K-modules*

$$\operatorname{Hom}_A\left(\bigoplus_{j=1}^{m} M_j, \bigoplus_{i=1}^{n} N_i\right) \cong H.$$

Proof. Because of Theorem III.3.10, we have

$$\operatorname{Hom}_A\left(\bigoplus_{j=1}^{m} M_j, \bigoplus_{i=1}^{n} N_i\right) \cong \prod_{(i,j)} \operatorname{Hom}_A(M_j, N_i)$$

as K-modules. But it is clear that the latter is isomorphic to H, as K-modules. □

This corollary allows us to write an A-linear map as a matrix of maps between smaller modules.

Assume finally that the M_i coincide with the N_j. In this case, Corollary III.3.13 says that we have an isomorphism of K-modules

$$H \cong \operatorname{Hom}_A\left(\bigoplus_{i=1}^{m} M_i, \bigoplus_{i=1}^{m} M_i\right) = \operatorname{End}_A\left(\bigoplus_{i=1}^{m} M_i\right).$$

But the latter is a K-algebra. We ask whether this K-module isomorphism is also an algebra isomorphism. In our case, H is a set of square matrices, which has a product given by the usual matrix rule:

$$[g_{ij}]_{i,j}[f_{ij}]_{i,j} = \left[\sum_k g_{ik}f_{kj}\right]_{i,j}.$$

Indeed, $f_{kj} \colon M_j \to M_k$ and $g_{ik} \colon M_k \to M_i$, so the composite $g_{ik}f_{kj} \colon M_j \to M_i$ makes sense.

III.3.14 Proposition. *With this notation, H is a K-algebra, isomorphic to* $\operatorname{End}_A\left(\bigoplus_{i=1}^{m} M_i\right)$.

Proof. It is easily verified that the product just defined makes H a K-algebra. We prove that the K-module isomorphism $\varphi \colon \operatorname{End}_A\left(\bigoplus_{i=1}^{m} M_i\right) \to H$, given by $f \mapsto [p_i f q_i]_{i,j}$, for $f \in \operatorname{End}_A(\bigoplus_i M_i)$, is an algebra isomorphism. Using the relations of Lemma III.3.8, we have

$$\varphi(gf) = [p_j(gf)q_i]_{i,j} = \left[p_j g\left(\sum_{k=1}^{m} q_k p_k\right) f q_i\right]_{i,j}$$

$$= \left[\sum_{k=1}^{m}(p_j g q_k)(p_k f q_i)\right]_{i,j} = [p_j g q_i]_{i,j}[p_j f q_i]_{i,j}$$

$$= \varphi(g)\varphi(f).$$

Because $\varphi(1_{\bigoplus_{i=1}^{m} M_i}) = [p_j q_i]_{i,j}$ is the identity matrix, this completes the proof. $\square$

Thus, if M is an A-module and m a positive integer, then $\operatorname{End}_A(M^{(m)}) \cong M_m(\operatorname{End}_A M)$.

Exercises of Section III.3

1. Prove that, in the category Alg_K, any finite family of objects admits a product.

2. Prove that, in the category $Sets$, the disjoint union of a family of sets is their coproduct.

3. Prove that, in the category Gr, of (not necessarily abelian) groups, the free product of two groups is their coproduct.

4. Let A be a K-algebra, $M, M_1, \cdots, M_m$ be A-modules and $p_i' : M \to M_i$, $q_i' : M_i \to M$ be A-linear maps such that: (a) $p_i' q_i' = 1_{M_i}$ and $p_i' q_j' = 0$ if $i \neq j$; and (b) $q_1' p_1' + \cdots q_m' p_m' = 1_M$. Prove that $M \cong \oplus_{i=1}^m M_i$.

5. Let A be an algebra, L, M_1, M_2, N be A-modules and $f_1 : L \to M_1, f_2 : L \to M_2$, $g_1 : M_1 \to N, g_2 : M_2 \to N$ be A-linear maps. Prove that:
   (a) The maps $\binom{f_1}{f_2} : L \to M_1 \oplus M_2$ given by $x \mapsto \binom{f_1(x)}{f_2(x)}$, for $x \in L$,, and $(g_1 \ g_2) : M_1 \oplus M_2 \to N$, given by $\binom{x_1}{x_2} \mapsto g_1(x_1) + g_2(x_2)$, for $\binom{x_1}{x_2} \in M_1 \oplus M_2$, are A-linear.
   (b) If f_1 or f_2 is injective, then so is $\binom{f_1}{f_2}$ and, if g_1 or g_2 is surjective, then so is $(g_1 \ g_2)$

6. (a) Let A be a K-algebra, $(M_\lambda)_{\lambda \in \Lambda}$ and $(M_\lambda')_{\lambda \in \Lambda}$ families of A-modules with respective products $(M, (p_\lambda)_{\lambda \in \Lambda})$ and $(M', (p_\lambda')_{\lambda \in \Lambda})$ and assume that, for any λ, there exists an A-linear map $f_\lambda : M_\lambda \to M_\lambda'$. Prove that there exists a unique A-linear map $f : M \to M'$ such that $p_\lambda' f = f_\lambda p_\lambda$, then prove that f is functorial. The morphism f is said to be obtained by **passing to products**.
   (b) State and prove the dual of (a). The resulting morphism is said to be obtained by **passing to coproducts**.

7. Let $(L_\lambda)_{\lambda \in \Lambda}, (M_\lambda)_{\lambda \in \Lambda}, (N_\lambda)_{\lambda \in \Lambda}$ be families of A-modules such that, for any λ, we have an exact sequence $L_\lambda \xrightarrow{f_\lambda} M_\lambda \xrightarrow{g_\lambda} N_\lambda$.
   (a) Prove that the sequences

$$\prod_\lambda L_\lambda \xrightarrow{f} \prod_\lambda M_\lambda \xrightarrow{g} \prod_\lambda N_\lambda \quad \text{and} \quad \bigoplus_\lambda L_\lambda \xrightarrow{f'} \bigoplus_\lambda M_\lambda \xrightarrow{g'} \bigoplus_\lambda N_\lambda$$

(where f, g and f', g' are obtained respectively by passing to products and coproducts, see Exercise III.3.4) are exact.
   (b) Prove that f and f' are monomorphisms if and only if each f_λ is a monomorphism, and, dually, g and g' are epimorphisms if and only if each g_λ is an epimorphism.

8. Let A be a K-algebra, I an ideal of A and $(M_\lambda)_\lambda$ a family of A-modules. Prove that:
   (a) $(\oplus_\lambda M_\lambda) I \cong \oplus_\lambda (M_\lambda I)$.
   (b) $(\oplus_\lambda M_\lambda)/(\oplus_\lambda M_\lambda) I \cong \oplus_\lambda (M_\lambda / M_\lambda I)$.

9. Let K be a field and Q the quiver $1 \leftarrow 3 \rightarrow 2$. Consider the representations

$$I = (K \xleftarrow{1} K \rightarrow 0), I' = (0 \leftarrow K \xrightarrow{1} K), S = (0 \leftarrow K \rightarrow 0) \text{ and } P = (K \xleftarrow{1} K \xrightarrow{1} K).$$

Compute each of the direct sums $I \oplus I'$ and $P \oplus S$.

10. Let A be a K-algebra and L, N submodules of an A-module M. Prove that:

   (a) We have a short exact sequence

$$0 \rightarrow \frac{M}{L \cap N} \xrightarrow{f} \frac{M}{L} \oplus \frac{M}{N} \xrightarrow{g} \frac{M}{L+N} \rightarrow 0$$

   where $f(x + L \cap N) = (x + L, x + N)$, for $x \in M$, and $g(x + L, y + N) = (x - y) + (L + N)$, for $x, y \in M$.

   (b) There exists an isomorphism

$$\frac{L+N}{L \cap N} \cong \frac{L+N}{L} \oplus \frac{L+N}{N}.$$

11. Let A be a K-algebra and L, N submodules of an A-module M. Prove that, if $L+N$ and $L \cap N$ are finitely generated, then so are L and N.

12. Let A, B be K-algebras.

   (a) Let $_BM_A$ be a B-A-bimodule and $(N_\sigma)_{\sigma \in \Sigma}$ be A-modules. Prove that the isomorphism of Corollary III.3.11 is an isomorphism of B-modules.

   (b) Let $_BN_A$ be a B-A-bimodule and $(M_\lambda)_{\lambda \in \Lambda}$ be A-modules. Prove that the isomorphism of Corollary III.3.12 is an isomorphism of B^{op}-modules.

13. Let A be an algebra. Assume that we have a commutative diagram with exact rows of A-modules and A-linear maps

$$
\begin{array}{ccccccccc}
0 & \longrightarrow & L & \xrightarrow{f} & M & \xrightarrow{g} & N & \longrightarrow & 0 \\
& & \downarrow{u} & & \downarrow{v} & & \downarrow{w} & & \\
0 & \longrightarrow & L' & \xrightarrow{f'} & M' & \xrightarrow{g'} & N' & \longrightarrow & 0
\end{array}
$$

Prove that there exists a short exact sequence $0 \rightarrow M \times L' \xrightarrow{\phi} M \times M' \xrightarrow{\eta} N' \rightarrow 0$ where $\phi(x, x') = (x, v(x)+f'(x'))$, for $(x, x') \in M \times M'$, and $\eta(y, y') = wg(y)-g'(y')$, for $(y, y') \in M \times M'$.

14. (a) Let A be a K-algebra and M an A-module. Assume that M has submodules M_1, M_2, M_3 such that $M = M_1+M_2+M_3$, $M_1 \cap (M_2+M_3) = 0$ and $M_2 \cap M_3 = 0$. Prove that $M = M_1 \oplus M_2 \oplus M_3$.

   (b) Generalise (a) to the case of $n > 3$ submodules.

15. Let A be a K-algebra and M an A-module. Denote by $p_1: M^2 \rightarrow M$, $p_2: M^2 \rightarrow M$, respectively, the projections on the first and the second coordinate. Prove that there exists a unique morphism $f\colon M^2 \to M^2$ such that the following diagram commutes

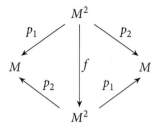

III.4 Free modules

III.4.1 Definition and basic properties

One reason why vector spaces are so pleasant to handle is the existence of bases: it implies that vectors can be expressed in terms of coordinates and linear maps in terms of matrices. In general, modules do not have bases, see Subsection II.2.2. But some modules do, as in Examples II.2.11. They are called free modules and this section is devoted to their study. Throughout, K denotes a commutative ring and A a K-algebra. We define free modules via a universal property.

III.4.1 Definition. Let X be a set. A *free module* on X is a pair $(L(X), i_X)$, where $L(X)$ is an A-module, and $i_X\colon X \to L(X)$ is a map such that, if (M, f) is a pair with M an A-module and $f\colon X \to M$ a map, then there exists a unique A-linear map $\bar{f}\colon L(X) \to M$ such that $\bar{f}i_X = f$. In other words, the following diagram commutes

We stress that the module $L(X)$ is associated to the *set* X, that is, X is considered without algebraic structure. For this reason, $i_X\colon X \to L(X)$ and $f\colon X \to M$ are just maps, whereas $\bar{f}\colon L(X) \to M$ is a morphism (that is, an A-linear map).

Uniqueness in the definition means that, if $\bar{f}, \bar{g}$ are A-linear maps such that $\bar{f}i_X = \bar{g}i_X$, then $\bar{f} = \bar{g}$. As in Section III.3, universality guarantees uniqueness up to isomorphism.

III.4.2 Lemma. *Let X be a set. A free A-module on X, if it exists, is unique up to isomorphism.*

Proof. Let $(L, i), (L', i')$ be free modules on the same set X. The universal property provides linear maps $f\colon L \to L', g\colon L' \to L$ such that $fi = i'$ and $gi' = i$, that is, the following diagram commutes

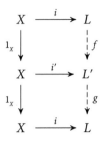

But then, $gfi = i = 1_L i$. Uniqueness in the universal property gives $gf = 1_L$. Similarly, $fg = 1_{L'}$. Hence, f, g are inverse isomorphisms. $\square$

III.4.3 Theorem. *For any set X, there exists a free A-module on X, unique up to isomorphism.*

Proof. It suffices to establish existence. Let $L(X) = A_A^{(X)}$, that is, $L(X)$ is the direct sum of copies of A_A, indexed by X. Define $i_X \colon X \to L(X)$ by $\lambda \mapsto e^\lambda = (e^\lambda_\mu)_{\mu \in X}$, where $e^\lambda_\lambda = 1$ and $e^\lambda_\mu = 0$ for $\mu \neq \lambda$ (thus, e^λ_μ is the Kronecker delta). Let $a \in L(X)$. Then $a = (a_\lambda)_{\lambda \in X}$, with the a_λ almost all zero, so we can write $a = \sum_{\lambda \in X} e^\lambda a_\lambda$. If f maps X to an A-module M, then the only A-linear map $\bar{f} \colon L(X) \to M$ such that $\bar{f} i_X = f$ is given by the formula

$$\bar{f}(a) = \bar{f}\left(\sum_{\lambda \in X} e^\lambda a_\lambda\right) = \sum_{\lambda \in X} \bar{f}(e^\lambda) a_\lambda = \sum_{\lambda \in X} \bar{f} i_X(\lambda) a_\lambda = \sum_{\lambda \in X} f(\lambda) a_\lambda.$$

Because this formula actually defines an A-linear map, the proof is complete. $\square$

By abuse of language, we call $L(X)$ the ***free module*** on X. An A-module L is called ***free*** if there exists a set X such that $L \cong L(X)$.

III.4.4 Examples. Let A be an algebra.
(a) If $X = \{1, 2, \cdots, n\}$ is finite, then

$$L(X) = A^{(n)} = A^n = \{(a_1, \cdots, a_n) \colon a_i \in A\}.$$

   The elements e^λ are here $e^1 = (1, 0, \cdots, 0), e^2 = (0, 1, \cdots, 0), \cdots,$ $e^n = (0, 0, \cdots, 1)$: this is the canonical basis of $A^{(n)}$, see Example II.2.11(a).
(b) Let $X = \mathbb{N}$. An element of $L(\mathbb{N}) = A^{(\mathbb{N})}$ is a sequence $(a_n)_{n \geq 0}$, with the a_n almost all zero or, equivalently, such that there exists $d \geq 0$ with $a_n = 0$ for $n > d$. Because operations on $A^{(\mathbb{N})}$ are defined componentwise, we have $A^{(\mathbb{N})} \cong A[t]$, as A-modules. Under this isomorphism, the elements e^n correspond to the polynomials t^n, with $n \in \mathbb{N}$, see Example II.2.11(b).
(c) Let $n > 0$, the matrix algebra $M_n(A)$ is a free A-module, having as basis the matrices e_{ij}, with $1 \leq i, j \leq n$. Because its basis contains n^2 elements, it is isomorphic, as an A-module, to $A^{(n^2)}$. See Example II.2.11(c).

We now show that free modules are exactly those having bases.

III.4.5 Theorem. *An A-module L is free if and only if it has a basis.*

Proof. *Necessity.* Assume that $L \cong L(X) = A^{(X)}$, for some set X. We claim that the set $(e^{\lambda})_{\lambda \in X}$ constructed in the proof of Theorem III.4.3 is a basis of L. Because any element of $A^{(X)}$ is a linear combination of the e^{λ}, it is a generating set. We prove linear independence. Let $\sum_{\lambda \in X} e^{\lambda} a_{\lambda} = 0$, for some a_{λ}, almost all zero. This means that $(a_{\lambda})_{\lambda \in X} = 0$ in $A^{(X)}$. Theorem II.2.20 gives $a_{\lambda} = 0$, for any λ, proving our claim and hence necessity.

Sufficiency. Assume that L has a basis $(e^{\lambda})_{\lambda \in X}$ indexed by a set X. We claim that $L \cong L(X)$. Define $i: X \to L$ by $\lambda \mapsto e^{\lambda}$, for $\lambda \in X$. Any $a \in L$ can be written as linear combination of the basis elements $a = \sum_{\lambda \in X} e^{\lambda} a_{\lambda}$, with the a_{λ} almost all zero. Let $f: X \to M$ be a map with M an A-module. The unique A-linear map $\bar{f}: L \to X$ such that $\bar{f} i = f$ is given by

$$\bar{f}(a) = \bar{f}\left(\sum_{\lambda \in X} e^{\lambda} a_{\lambda}\right) = \sum_{\lambda \in X} \bar{f}(e^{\lambda}) a_{\lambda} = \sum_{\lambda \in X} \bar{f} i(\lambda) a_{\lambda} = \sum_{\lambda \in X} f(\lambda) a_{\lambda}.$$

Because this formula indeed defines an A-linear map such that $\bar{f} i = f$, the module L satisfies the universal property of a free module on X. Hence, $L \cong L(X)$. □

Thus, an A-module L is free with basis $(e_{\lambda})_{\lambda}$ if and only if $L = \oplus_{\lambda} e_{\lambda} A$. For instance, an abelian group ($\mathbb{Z}$-module) is free if and only if it is a direct sum of copies of $\mathbb{Z}$.

The proof of the theorem shows that the set X may be identified to a basis of $L(X)$. The universal property may thus be understood as saying that an A-linear map starting at a free module is uniquely determined by the values it takes on a basis, a well-known statement of linear algebra, usually referred to as the **linear extension of maps theorem**.

III.4.6 Corollary. *Let A be a division ring. Any A-module is free.*

Proof. This follows from Theorems III.4.5 and II.2.10 □

A well-known result of linear algebra asserts that two bases of a vector space have the same cardinality. The same statement holds true for free modules over commutative algebras.

III.4.7 Theorem. *Any two bases of a free module over a commutative algebra have the same cardinality.*

Proof. Let A be commutative. Because of Krull's Theorem I.5.8, it has a maximal ideal I. Because of Corollary I.4.11, A/I is a field. Let L be a free A-module with basis $(e_{\lambda})_{\lambda \in \Lambda}$, that is, $L = \oplus_{\lambda \in \Lambda} e_{\lambda} A$. As seen in Example II.2.13(a), the quotient module L/LI can be considered as an A/I-module, that is, an A/I-vector space. Because

$LI = \bigoplus_{\lambda \in \Lambda} e_\lambda I$, we get $L/LI \cong \bigoplus_{\lambda \in \Lambda}(e_\lambda A/e_\lambda I)$, see Exercise III.3.5. Now, for any λ, we have $e_\lambda A/e_\lambda I \cong A/I$. Thus, the dimension of $L/LI \cong \bigoplus_{\lambda \in \Lambda} e_\lambda(A/I)$ as an A/I-vector space equals the cardinality of Λ. If L has another basis of cardinality equal to that of, say, the set Σ, then the same reasoning gives that L/LI is an A/I-vector space of dimension the cardinality of Σ. Hence, Λ and Σ have the same cardinality. □

Therefore, two free modules over a commutative algebra are isomorphic if and only if they have bases with same cardinality, see Exercise III.4.2 below. This cardinality is a complete invariant for free modules over commutative algebras, exactly as dimension is a complete invariant for vector spaces over fields. The case when this cardinality is finite is considered in detail in Chapter V.

III.4.2 Free modules inside ModA

In this subsection, we study how free modules behave with respect to morphisms of the module category. We start by generalising Exercise II.4.2.

III.4.8 Proposition. *Any module is a quotient of a free module. If, moreover, the given module is finitely generated, then one can choose the free module to be finitely generated as well.*

Proof. Let A be an algebra, M an A-module and $X \subseteq M$ a generating set for M (one can take, for instance, $X = M$). The inclusion $i\colon X \to M$ induces a morphism $f\colon L(X) \to M$ making the following triangle commute

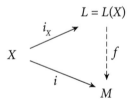

$$L = L(X)$$

Because any $x \in M$ is a linear combination of elements of X, and these are all contained in the image of f, then f is surjective. Hence, M is a quotient of L.

If M is finitely generated, say $M = \langle x_1, \cdots, x_m \rangle$, we apply the same reasoning to $L = A^{(m)}$ with a basis $\{e^1, \cdots, e^m\}$ in bijection with the x_i. The morphism f is then uniquely determined by the relations $f(e^i) = x_i$, for any i such that $1 \leqslant i \leqslant m$. Because the x_i generate M, the morphism f is surjective. □

For instance, if $M = xA$ is cyclic, then $L = A_A$ and the morphism $f : A \to M$ is given by $x \mapsto xa$, for $a \in A$

III.4.9 Corollary. *Let K be a field and A a finite-dimensional K-algebra. An A-module is finitely generated if and only if it is finite-dimensional as a K-vector space.*

Proof. An A-module M has a natural K-module structure, hence, in our case, is a K-vector space. If M is finite-dimensional, then it is finitely generated because one can take the vectors of a basis as generators. Conversely, if M is finitely generated, say $M = \langle x_1, \cdots, x_m \rangle$, then there exists an epimorphism from $A^{(m)}$ to M sending each vector e^i of the canonical basis of $A^{(m)}$ to x_i. Viewing both as vector spaces, the existence of this epimorphism implies that $\dim_K M \leqslant \dim_K A^{(m)} = m \cdot \dim_K A$. Because $\dim_K A$ is finite, we are done. □

III.4.10 Corollary. *Let M be an A-module. Then there exist free modules L_i, with $i \geq 0$, and an exact sequence*

$$\cdots \to L_i \xrightarrow{f_i} L_{i-1} \to \cdots \to L_1 \xrightarrow{f_1} L_0 \xrightarrow{f_0} M \to 0.$$

Proof. By induction on i. The existence of f_0 follows from Proposition III.4.8. For $i > 0$, we apply Proposition III.4.8 to $\mathrm{Ker}\, f_{i-1}$. □

An exact sequence as in Corollary III.4.10 is called a **free resolution** of M. It may happen that there exists i such that $L_j = 0$ for all $j \geq i$, thus, this exact sequence may actually be finite. If we consider just the two-terms exact sequence $L_1 \xrightarrow{f_1} L_0 \xrightarrow{f_0} M \to 0$, then the latter is called a **free presentation** of M. Free resolutions and presentations may be viewed as approximations of a module by free modules. They are useful when a calculation is done more easily on a free module than on an arbitrary one (which occurs frequently, because free modules are direct sums of copies of the algebra).

If an A-module M has a free presentation as above with the free modules L_0, L_1 finitely generated, then M is said to be **finitely presented**. Thus, a finitely presented module is finitely generated, but the converse is not always true: if M is finitely generated, then one can choose L_0 finitely generated, see Proposition III.4.8, but the kernel of the surjection $L_0 \to M$ is not necessarily finitely generated, so it is not always possible to choose L_1 finitely generated. We return to this point in Chapter VII.

The next property of free modules is called **projectivity**, and we study it in detail in Chapter VI. Given an algebra A, an A-module P is called a **projective module** if, for any epimorphism $f : M \to N$ and morphism $u : P \to N$, there exists a morphism $u' : P \to M$ such that $u = fu'$, that is, the following diagram commutes

$$
\begin{array}{ccc}
 & & P \\
 & \overset{u'}{\swarrow} & \downarrow u \\
M & \underset{f}{\longrightarrow} N & \longrightarrow 0
\end{array}
$$

III.4.11 Lemma. *Any free module is projective.*

Proof. Let $f : M \to N$ be an epimorphism of A-modules and $u : L \to N$ an A-linear map with L free. We claim that there exists an A-linear map $u' : L \to M$ such that $u = fu'$, that is, we have a commutative diagram

$$L$$
$$u' \overset{\nearrow}{\underset{\nwarrow f}{\dashrightarrow}} \quad \downarrow u$$
$$M \overset{f}{\longrightarrow} N \longrightarrow 0$$

Let $(e_\lambda)_\lambda$ be a basis of L. Because f is an epimorphism, there exists, for any λ, an element $x_\lambda \in M$ such that $f(x_\lambda) = u(e_\lambda)$. The universal property grants the existence of an A-linear map $u': L \to M$ such that $u'(e_\lambda) = x_\lambda$. But then $fu'(e_\lambda) = u(e_\lambda)$, for any λ. Uniqueness in the universal property gives $fu' = u$, as required.   □

The map u' constructed in the lemma is far from being unique: it depends first on the choice of a basis for L, then on the choice of a preimage in M for each basis element.

We use the universal property of free modules in order to construct a functor $L : Sets \to \mathrm{Mod}A$. This procedure is called **functorisation**, and L is called the **free module functor**.

Let X, Y be sets and $f: X \to Y$ a map. The universal property of free modules gives a unique A-linear map $L(f): L(X) \to L(Y)$ making the following square commute

$$
\begin{array}{ccc}
X & \overset{i_X}{\longrightarrow} & L(X) \\
\downarrow{\scriptstyle f} & & \vdots{\scriptstyle L(f)} \\
Y & \overset{i_Y}{\longrightarrow} & L(Y)
\end{array}
$$

where i_X, i_Y are as in Definition III.4.1. It is obvious that $L(1_X) = 1_{L(X)}$. Let $f: X \to Y$, $g: Y \to Z$ be maps, we have a commutative diagram

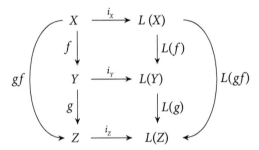

Then $L(g)L(f)i_X = i_Z gf = L(gf)i_X$ and uniqueness gives $L(gf) = L(g)L(f)$. Thus, we have defined a covariant functor $L: Sets \to \mathrm{Mod}A$.

Exercises of Section III.4

1. Let A be a K-algebra, L an A-module and $(L_\lambda)_\lambda$ a family of submodules such that $L = \oplus_\lambda L_\lambda$. Assume that each L_λ is free with basis X_λ. Prove that L is free with basis $\cup_\lambda X_\lambda$.

2. Let A be a K-algebra, L, M free A-modules with respective bases $(e_\lambda)_{\lambda \in \Lambda}$ and $(f_\sigma)_{\sigma \in \Sigma}$ such that there exists a bijection between Λ and Σ. Prove that $L \cong M$.

3. Prove that an algebra is a division ring if and only if each of its modules is free.

4. Let X be a subset of an A-module M and $f : L(X) \to M$ the unique A-linear map which extends the inclusion $X \hookrightarrow M$. Prove that:
   (a) $\operatorname{Im} f = \langle X \rangle$.
   (b) f is surjective if and only if X generates M.
   (c) f is injective if and only if X is linearly independent.
   (d) f is an isomorphism if and only if X is a basis of M.

5. Let A be a K-algebra, $X = \{1, \cdots, n\}$ a finite set and $L(X)$ a free A-module on X. Prove that $\operatorname{Hom}_A(L(X), A)$ is a free A-module on X.

6. Let a, b be positive integers and $A = \mathbb{Z}/(ab)\mathbb{Z}$. Prove that the submodule $b\mathbb{Z}/(ab)\mathbb{Z}$ ($\cong \mathbb{Z}/a\mathbb{Z}$) of A is not free (thus, in general, a submodule of a free module is not free).

7. Prove that $\mathbb{Q}$ is not a free $\mathbb{Z}$-module (abelian group).

8. Let A be a K-algebra. Assume that we have a diagram with exact rows of A-modules and A-linear maps with L free.

$$
\begin{array}{ccccccccc}
0 & \longrightarrow & L' & \overset{f}{\longrightarrow} & L & \overset{g}{\longrightarrow} & L'' & \longrightarrow & 0 \\
& & & & & & \downarrow{w} & & \\
0 & \longrightarrow & M' & \overset{h}{\longrightarrow} & M & \overset{k}{\longrightarrow} & M'' & \longrightarrow & 0
\end{array}
$$

   (a) Prove that there exist $u: L' \to M', v: L \to M$ linear making the diagram commutative.
   (b) Prove that, if $v_1: L \to M, v_2: L \to M$ both make the right square commute, then there exists $v': L \to M'$ such that $v_1 - v_2 = hv'$.

9. Let A be a K-algebra, L a free A-module with basis $(e_\lambda)_{\lambda \in \Lambda}$ and $E = \operatorname{End} L$. For any $f \in E$ and any $\lambda \in \Lambda$, there exists a set $(f_{\sigma\lambda})_{\sigma \in \Lambda}$, with the $f_{\sigma\lambda}$ almost all zero, such that

$$f(e_\lambda) = \sum_\sigma f_{\sigma\lambda} e_\sigma$$

   Show that the map $f \mapsto [f_{\sigma\lambda}]_{(\sigma\lambda) \in \Lambda \times \Lambda}$ defines an isomorphism between E and the set of column-finite matrices over A of Exercise I.3.4.

10. Let A, B be algebras and $\varphi : A \to B$ an algebra morphism. For a B-module M_B, let $M_{[A]}$ denote the induced A-module under the change of scalars functor of

Examples II.2.4(e) and III.2.7(g). Assume that $(b_\lambda)_{\lambda \in \Lambda}$ is a set of generators (or a linearly independent subset, or a basis) of $B_{[A]}$ and $(x_\sigma)_{\sigma \in \Sigma}$ is a set of generators (or a linearly independent subset, or a basis, respectively) of the B-module M. Prove that $(x_\sigma b_\lambda)_{(\lambda, \sigma) \in \Lambda \times \Sigma}$ is a set of generators (or a linearly independent subset, or a basis, respectively) of $M_{[A]}$.

11. Let K be a field and X a set. A **free algebra** over X is a pair $(K\langle X \rangle, j)$, where $K\langle X \rangle$ is a K-algebra, and $j : X \to K\langle X \rangle$ a map, such that, if A is a K-algebra and $f : X \to A$ is a map, then there exists a unique morphism of K-algebras $\tilde{f} : K\langle X \rangle \to A$ such that $\tilde{f} j = f$.

(a) Prove that, for any set X, there exists a free K-algebra $K\langle X \rangle$ on this set (whose elements can be viewed as K-linear combinations of the monomials on the elements of X, considered as noncommuting indeterminates).

(b) If X has only one element, prove that $K\langle X \rangle \cong K[t]$.

(c) If $X = \{s, t\}$, prove that $K\langle s, t \rangle$ is isomorphic to the path algebra of the two-loops quiver

$$\alpha \overset{\circ}{\circlearrowright} \circlearrowleft \beta$$

(d) Let $n > 0$ be an integer. Consider the subalgebra B of $K\langle s, t \rangle$ generated by the elements $u_i = st^i$, with i such that $0 \leqslant i \leqslant n$. Prove that B is a free algebra in $n + 1$ generators.

(e) Construct a subalgebra of $K\langle s, t \rangle$ which is free in countably many generators.

12. Let K be a field, E an infinite-dimensional vector space and $A = \mathrm{End}_K E$. Prove that $E \cong E \oplus E$, as vector spaces. Next, evaluating the functor $\mathrm{Hom}_K(E, -)$ on this isomorphism, deduce that, as A-modules, $A \cong A \oplus A$ (therefore, contrary to the situation of Theorem III.4.7, if A is arbitrary, then two bases of a free A-module do not necessarily have the same cardinality).

Chapter IV
Abelian categories

IV.1 Introduction

The notion of a category is a very general one. In a sense, it is too general to describe adequately what happens in a module category. Hence the idea of imposing axioms on categories, equipping them with structures which describe best what happens for modules. Endowing a category with a structure is similar to endowing a set with an algebraic or an analytic structure: it opens technical possibilities to work while sticking to a concrete situation which one wants to study at the 'right' level of generality. It is therefore not surprising that the study of a category with a structure leads to more specialised results, and gets closer to what happens in our model, namely a module category. Historically, the simplest and best-studied structures are those of abelian groups and vector spaces. Now, the Hom-sets in a module category over a K-algebra are abelian groups and even K-modules (vector spaces if K is a field). For instance, a basic feature of module categories is the existence of zeros, the zero module and the zero morphism. Hence the idea of imposing a linear structure on the Hom-sets in a category.

IV.2 Linear and abelian categories

IV.2.1 K-categories

Let C be a category and K a commutative ring. We wish to equip each Hom-set of C with a K-module structure (that is, addition and scalar multiplication). But there exists another operation between morphisms, namely composition. It is reasonable to require compatibility of composition with the module operations.

IV.2.1 Definition. Let K be a commutative ring. A category C is a **K-category** or a **category enriched in ModK** if:

(L1) For any pair of objects (X, Y) in C, the set $C(X, Y)$ has a K-module structure.

(L2) Composition of morphisms is compatible with the K-module structures of the Hom-sets, that is, it is K-bilinear: if $f, f_1, f_2 : X \to Y, g, g_1, g_2 : Y \to Z$ are morphisms and $\alpha_1, \alpha_2, \beta_1, \beta_2 \in K$, then:

(i) $g(f_1\alpha_1 + f_2\alpha_2) = (gf_1)\alpha_1 + (gf_2)\alpha_2$, and

(ii) $(g_1\beta_1 + g_2\beta_2)f = (g_1 f)\beta_1 + (g_2 f)\beta_2$.

The notion of K-category is selfdual, that is, if C is a K-category, then so is C^{op}. A $\mathbb{Z}$-category is sometimes called an **Ab-category**. Clearly, for any K, a K-category is a $\mathbb{Z}$-category.

An Introduction to Module Theory. Ibrahim Assem and Flávio U. Coelho, Oxford University Press.
© Ibrahim Assem and Flávio U. Coelho (2024). DOI: 10.1093/9780198904939.003.0005

If C is a K-category and X an object in C, it follows from the axioms that the set End $_C X = C(X,X)$ has a K-module structure. Because composition of morphisms endows it with an additional monoid structure, compatible with the former, it becomes a K-algebra, called the **K-algebra of endomorphisms** of X.

Modules have zero elements, hence, for any objects X, Y in a K-category C, the set $C(X,Y)$ contains a zero morphism $0: X \to Y$. As expected, zero morphisms are absorbing elements: if $f: Y \to Z$, then $f0 = 0$ (apply the usual argument: $f0 = f(0+0) = f0 + f0$, then subtract $f0$ from both sides). Similarly, if $g: W \to X$, then $0g = 0$.

IV.2.2 Examples.

(a) Let A be a K-algebra. Because of Lemmata II.3.7 and II.3.10, the module category Mod$A$ is a K-category. In particular, Ab is a $\mathbb{Z}$-category. Also, the full subcategory mod$A$ of Mod$A$ consisting of finitely generated A-modules is a K-category.

(b) The categories $Sets, Gr, Top$ are not K-categories, because their morphism sets are not equipped with module structures.

(c) A K-algebra A may be considered as a K-category $\mathcal{A}$ having a unique object x, which can be identified with the multiplicative identity of A, the elements of A are endomorphisms of x, addition is addition in A and composition is multiplication in A. Conversely, one may think of a K-category as being an algebra having 'many identities'.

(d) An abelian group G is called a **topological abelian group** if it has a topology compatible with group operations, that is, if the maps $G \times G \to G$, $(x,y) \mapsto x+y$ and $G \to G, x \mapsto -x$, for $x, y \in G$, are continuous in this topology. The category Tag has as objects the topological abelian groups and as morphisms the continuous group homomorphisms. The sum of continuous group homomorphisms is a continuous group homomorphism and their composition is $\mathbb{Z}$-bilinear. Therefore, Tag is a $\mathbb{Z}$-category.

(e) An abelian group G is called **divisible** if, for any integer $n > 0$, we have $nG = G$, or, equivalently, if, for any $g \in G$ and integer $n > 0$, there exists $g' \in G$ such that $g = ng'$. For instance, $\mathbb{Q}$ is divisible, but neither $\mathbb{Z}$ nor any of its subgroups $k\mathbb{Z}$, with $k > 0$ an integer. The full subcategory Div of Ab consisting of the divisible abelian groups is, clearly, a $\mathbb{Z}$-category.

(f) Let K be a field and $\mathcal{E}$ the category whose objects are the pairs (E,f), with E a K-vector space and f an endomorphism of E. A morphism $(E,f) \to (F,g)$ is a K-linear map $u: E \to F$ such that $uf = gu$, and composition is the usual composition of maps, see Example III.2.2(i). We claim that $\mathcal{E}$ is a K-category. Let $u, v: (E,f) \to (F,g)$ be morphisms in $\mathcal{E}$ and $\alpha, \beta \in K$. Then $uf = gu$ and $vf = gv$ imply $(u\alpha + v\beta)f = (uf)\alpha + (vf)\beta = (gu)\alpha + (gv)\beta = g(u\alpha+v\beta)$, so Hom-sets are K-modules. Because composition of morphisms is ordinary composition of maps, it is K-bilinear. This establishes our claim.

(g) Let K be a field, Q a finite quiver and Rep $_K Q$ the category of representations of Q, see Example III.2.2(j). One proves exactly as in Example (f) above that Rep $_K Q$ is a K-category.

(h) Let $\mathcal{I}$ be a small category and C a K-category. The category Fun$(\mathcal{I},C)$ of functors from $\mathcal{I}$ to C, see Example III.2.9(c), is a K-category. We define a

K-module structure on morphism sets. Let $\varphi, \psi : F \to G$ be functorial morphisms and $\alpha, \beta \in K$. Set:

$$(\alpha\varphi + \beta\psi)_X = \alpha\varphi_X + \beta\psi_X$$

for any $X \in \mathcal{I}_0$. With this definition, the set of functorial morphisms from F to G becomes a K-module, and composition is bilinear. Hence, $\mathrm{Fun}(\mathcal{I}, C)$ is a K-category.

This is the first instance of a well-known fact: if the target category C has some nice structure (in our case, that of K-category), then $\mathrm{Fun}(\mathcal{I}, C)$ inherits this nice structure of C.

When we define an algebraic structure, we define, together with it, the maps which preserve this structure. Similarly, when we define a categorical structure, we need appropriate functors.

IV.2.3 Definition. Let K be a commutative ring and C, D be K-categories. A functor $F: C \to D$ is **K-linear**, or simply **linear**, if, for any morphisms $f, g: X \to Y$ in C and scalars $\alpha, \beta \in K$, we have

$$F(f\alpha + g\beta) = F(f)\alpha + F(g)\beta.$$

In other words, a covariant, or contravariant, functor $F: C \to D$ is K-linear if and only if the map $f \mapsto Ff$ it induces on the morphism sets is a K-linear map. This definition implies that $F(0) = 0$. A $\mathbb{Z}$-linear functor is called **additive**. A functor $F: C \to D$ is additive if and only if, for any morphisms $f, g: X \to Y$ in C, we have $F(f + g) = Ff + Fg$. Thus, for any K, a K-linear functor is additive.

IV.2.4 Examples.
    (a) Let A be a K-algebra, the forgetful functor $|-|: \mathrm{Mod}A \to \mathrm{Mod}K$ is K-linear.
    (b) If C is a K-category and M an object in C, the two Hom-functors

$$C(M, -): X \mapsto C(M, X), f \mapsto C(M, f) \quad \text{and}$$
$$C(-, M): X \mapsto C(X, M), f \mapsto C(f, M)$$

    are K-linear.

Unless otherwise specified, all functors between K-categories we shall work with in the sequel are K-linear.

We repeat in our context the definitions of monomorphisms and epimorphisms, see Definition II.4.1.

IV.2.5 Definition. Let C be a K-category.
    (a) A morphism f in C is a **monomorphism** if $fg = fh$ implies $g = h$, or equivalently, if $fk = 0$ implies $k = 0$.
    (b) A morphism f in C is an **epimorphism** if $gf = hf$ implies $g = h$, or equivalently, if $kf = 0$ implies $k = 0$.

Clearly the dual of a monomorphism is an epimorphism, and conversely. We give examples.

IV.2.6 Examples.

(a) Because of Lemma II.4.2, monomorphisms in Mod$A$ coincide with injective morphisms, and epimorphisms with surjective morphisms.

(b) Let C be a concrete category, see the comments following Examples III.2.2. Elementary properties of maps imply that injective morphisms are always monomorphisms and surjective morphisms are always epimorphisms. The converse is generally not true as shown in Examples (e) and (f) below. It is however useful to know that, in the category $\mathcal{G}r$ of groups, which is not a K-category, monomorphisms coincide with injective morphisms, and epimorphisms with surjective morphisms (the proof of the second statement is not trivial).

(c) Let K be a field and $\mathcal{E}$ the category of pairs of Examples III.2.2(i) and IV.2.2(f). A morphism $u: (E, f) \to (F, g)$ in $\mathcal{E}$ is in particular a morphism of K-vector spaces $u: E \to F$. Consequently, it is a monomorphism (or an epimorphism, or an isomorphism) of K-vector spaces if and only if it is injective (or surjective, or bijective, respectively). This proves that u is an isomorphism in $\mathcal{E}$ if and only if it is both a monomorphism and an epimorphism.

(d) Let K be a field, Q a finite quiver and $f: V \to W$ a morphism in the category Rep $_K Q$ of representations of Q, see Examples III.2.2(j) and IV.2.2(g). Because composition of morphisms in Rep $_K Q$ is done componentwise, $f = (f_x)_{x \in Q_0}$ is a monomorphism (or epimorphism, or isomorphism) if and only if so is each f_x, with $x \in Q_0$. Because the f_x are K-linear maps, this is the case if and only if each f_x is injective (or surjective, or bijective). As in Exercise II.3.24, this amounts to saying that f is injective (or surjective, or bijective, respectively). Consequently, f is an isomorphism in Rep $_K Q$ if and only if it is both a monomorphism and an epimorphism.

(e) Consider the $\mathbb{Z}$-category Div of Example IV.2.2(e). The image of a divisible abelian group is divisible: for, let $f : G \to H$ be a surjective group homomorphism, with G divisible. Then, for any integer $n > 0$, we have $G = nG$. Hence, $H = f(G) = f(nG) = nf(G) = nH$. For instance, $\mathbb{Q}$ is divisible, hence so is $\mathbb{Q}/\mathbb{Z}$. The projection $p: \mathbb{Q} \to \mathbb{Q}/\mathbb{Z}$ is surjective, hence it is an epimorphism because of (b) above. We prove that it is also a monomorphism (though it is clearly not injective).

Let $f: G \to \mathbb{Q}$ be a nonzero morphism in Div. Because Im f is divisible, and no nonzero subgroup of $\mathbb{Z}$ is divisible, see Example IV.2.2(e), Im $f \not\subseteq \mathbb{Z}$. So there exists $x \in G$ such that $f(x) = \frac{r}{s} \neq 0$, with r, s coprime and $s \neq \pm 1$. Let $y \in G$ be such that $ry = x$. Then $f(y) = \frac{1}{s}$ and so $pf(y) \neq 0$. Hence, $f \neq 0$ implies $pf \neq 0$, that is, p is a monomorphism. Because Div is a full subcategory of $\mathcal{A}b$, any monomorphism in $\mathcal{A}b$ (injective morphism) remains a monomorphism in Div. But the latter contains additional monomorphisms which are not monomorphisms of $\mathcal{A}b$, as in the case for p.

This furnishes an example of a monomorphism and epimorphism in a K-category which is not an isomorphism.

(f) Let, as in Example IV.2.2(d), $\mathcal{T}ag$ be the category of topological abelian groups. A topological abelian group G is called **separated** (or a **Hausdorff** *space*) if the one-element set {0} is closed in G. The full subcategory *Stag* of *T ag* consisting of the separated topological abelian groups is clearly a $\mathbb{Z}$-category.

We claim that a morphism $f: X \to Y$ in *Stag* is an epimorphism if and only if its image Im f is dense in Y, that is, the closure Y' of Im f equals Y. Indeed, assume that $Y' \neq Y$. A result of topology implies that Y/Y' is an object in *Stag*. Let $p: Y \to Y/Y'$ be the projection. Because $Y' \neq Y$, we have $p \neq 0$. But $pf = 0$, so f is not an epimorphism. Conversely, assume that $g: Y \to Z$ is a morphism in *Stag* such that $gf = 0 = 0f$. The continuous maps g and 0 coincide on Im f, which is dense in Y. Hence, $g = 0$. This establishes our claim.

Now, give $\mathbb{R}$ its Euclidean topology and $\mathbb{Q}$ the subspace topology. Because $\mathbb{Q}$ is dense in $\mathbb{R}$, the inclusion $\mathbb{Q} \hookrightarrow \mathbb{R}$ is an epimorphism (though it is not surjective). Because it is injective, it is also a monomorphism. On the other hand, it is not bijective, hence it is not an isomorphism (that is, a homeomorphism). This is another example of a monomorphism and epimorphism which is not an isomorphism.

(g) Let A be an algebra and $i_X : X \to L(X)$ the embedding of a set X as a basis of the free A-module on X, see Definition III.4.1. Uniqueness in the universal property of free modules says that, whenever M is an A-module and $f, g : L(X) \to M$ are linear maps such that $f i_X = g i_X$, then $f = g$. This does *not* mean that i_X is an epimorphism. Indeed, these maps are taken from different categories: i_X is a morphism in *Sets* while f, g are morphisms in Mod$A$.

Particular types of monomorphisms and epimorphisms play an important role in the sequel. Monomorphisms and epimorphisms are respectively left and right cancellable morphisms. In an algebraic structure, left and right invertible elements are left and right cancellable, respectively. This remark suggests the following definition.

IV.2.7 Definition. Let C be a K-category.

(a) A morphism $s: X \to Y$ in C is a *section* if there exists a morphism $s': Y \to X$ in C such that $s's = 1_X$.

(b) A morphism $r: X \to Y$ in C is a *retraction* if there exists a morphism $r': Y \to X$ in C such that $rr' = 1_Y$.

The notions of section and retraction are plainly dual one of the other. They appear together, for, if $s : Y \to X$ is a section, and s' is such that $s's = 1_X$, then s' is a retraction (associated to s). Similarly, if $r : X \to Y$ is a retraction and r' is such that $rr' = 1_Y$, then r' is a section (associated to r).

IV.2.8 Examples.

(a) Let $(X_\lambda)_{\lambda \in \Lambda}$ be a family of objects in a K-category such that the product $X = \prod_{\lambda \in \Lambda} X_\lambda$ exists in this category. We claim that, for any λ, the projection $p_\lambda : X \to X_\lambda$ is a retraction.

Indeed, fix an index λ and, for $\mu \in \Lambda$, let $\delta_{\lambda\mu} : X_\mu \to X_\lambda$ be the Kronecker delta defined by $\delta_{\lambda\lambda} = 1_{X_\lambda}$ and $\delta_{\lambda\mu} = 0$ for $\mu \neq \lambda$. The universal property of the product yields a unique morphism $q_\mu : X_\mu \to X$ such that $p_\lambda q_\mu = \delta_{\lambda\mu}$, for any μ.

But then $p_\lambda q_\lambda = 1_{X_\lambda}$ (and also $p_\lambda q_\mu = 0$ for $\mu \neq \lambda$). Hence, p_λ is a retraction, and our claim is proved.

Dually, if $(X_\lambda)_{\lambda \in \Lambda}$ is a family of objects in a K-category such that the coproduct $X = \coprod_{\lambda \in \Lambda} X_\lambda$ exists in this category, then, for any λ, the injection $q_\lambda : X_\lambda \to X$ is a section.

(b) The construction in (a) above allows to say more about the commutation of the Hom functor with arbitrary direct sums, see Subsection III.3.3. Let A be a K-algebra, M an A-module and $(N_\lambda)_{\lambda \in \Lambda}$ an arbitrary family of A-modules. Denote by $q_\mu : N_\mu \to \oplus_\lambda N_\lambda$ and $q'_\mu : \mathrm{Hom}_A(M, N_\mu) \to \oplus_\lambda \mathrm{Hom}_A(M, N_\lambda)$, respectively, the injections associated to the direct sums. Set for brevity $q^*_\mu = \mathrm{Hom}_A(M, q_\mu)$. Because of the universal property, there exists a unique morphism

$$f : \oplus_{\lambda \in \Lambda} \mathrm{Hom}_A(M, N_\lambda) \to \mathrm{Hom}_A(M, \oplus_{\lambda \in \Lambda} N_\lambda)$$

such that $fq'_\mu = q^*_\mu$, for any $\mu \in \Lambda$.

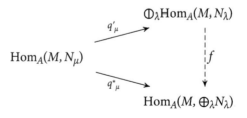

This morphism is expressed as follows: an element of its domain $\oplus_\lambda \mathrm{Hom}_A(M, N_\lambda)$ is a family $(u_\lambda)_{\lambda \in \Lambda}$ of A-linear maps $u_\lambda : M \to N_\lambda$ which are almost all zero. Then

$$f((u_\lambda)_\lambda) = f\left(\sum_\lambda q'_\lambda u_\lambda\right) = \sum_\lambda f(u_\lambda) = \sum_\lambda q^*_\lambda(u_\lambda)$$
$$= \sum_\lambda \mathrm{Hom}_A(M, q_\lambda)(u_\lambda) = \sum_\lambda q_\lambda u_\lambda$$

where the sums make sense because the u_λ are almost all zero.

The morphism f is injective: indeed, suppose that $\sum_\lambda q_\lambda u_\lambda = 0$, then, for any μ,

$$0 = p_\mu\left(\sum_\lambda q_\lambda u_\lambda\right) = \sum_\lambda p_\mu q_\lambda u_\lambda = \sum_\lambda \delta_{\mu\lambda} u_\lambda = u_\mu.$$

Thus, the family $(u_\lambda)_\lambda$ is zero. On the other hand, a morphism $v \in \mathrm{Hom}_A(M, \oplus_\lambda N_\lambda)$ belongs to the image of f if and only if $v(M)$ is contained in a finite sum of the N_λ: indeed, if v satisfies this condition and $v(M) \subseteq \oplus_{i=1}^m N_{\lambda_i}$, then $v = 1_{\oplus_{i=1}^m N_{\lambda_i}} \circ v = \sum_{i=1}^m q_{\lambda_i}(p_{\lambda_i} v)$, where we have used Lemma III.3.8, that is, v lies in the image of f. The converse is evident.

Assume that M is finitely generated, say, by $\langle x_1, \cdots, x_n \rangle$. Then, for any i with $1 \leqslant i \leqslant n$, there exists a finite set Λ_i such that $v(x_i) \in \oplus_{\lambda \in \Lambda_i} N_\lambda$. Hence, $\Lambda_0 = \cup_{i=1}^n \Lambda_i$ is finite, and such that $v(x_i) \in \oplus_{\lambda \in \Lambda_0} N_\lambda$, for any i. Therefore, $v(M) \subseteq \oplus_{\lambda \in \Lambda_0} N_\lambda$, and f is surjective. This proves that, if M is finitely generated, then f is an isomorphism of K-modules

$$\oplus_{\lambda \in \Lambda} \mathrm{Hom}_A(M, N_\lambda) \cong \mathrm{Hom}_A(M, \oplus_{\lambda \in \Lambda} N_\lambda).$$

(c) Any epimorphism with target a projective module is a retraction. Recall that a module P is projective if, for any epimorphism $f : M \to N$ and any morphism $u : P \to N$, there exists a morphism $f' : P \to M$ such that $u = ff'$, see the discussion before Lemma III.4.11. Taking $P = N$ and $u = 1_P$, we get $f' : P \to M$ such that $ff' = 1_P$.

$$\begin{array}{ccc}
& & P \\
& {\scriptstyle f'} \swarrow & \downarrow {\scriptstyle 1_P} \\
M & \xrightarrow{\ f\ } P & \longrightarrow 0
\end{array}$$

Because free modules are projective, due to Lemma III.4.11, this implies that an epimorphism with target a free module is a retraction. In particular, an epimorphism between vector spaces over a field is a retraction.

Monomorphisms, epimorphisms, sections and retractions satisfy the expected properties.

IV.2.9 Proposition. *Let C be a K-category and g, f morphisms such that gf exists.*
  (a) *If f, g are monomorphisms, then so is gf.*
  (a') *If f, g are epimorphisms, then so is gf.*
  (b) *If gf is a monomorphism, then so is f.*
  (b') *If gf is an epimorphism, then so is g.*

> (c) *If f is a section, then it is a monomorphism.*
> (c') *If f is a retraction, then it is an epimorphism.*
> (d) *f is an isomorphism if and only if it is both a section and a retraction.*
> (e) *If gf is an isomorphism, then f is a section and g is a retraction.*

Proof. (a) Assume that $(gf)h = 0$. Because g is a monomorphism, $g(fh) = (gf)h = 0$ yields $fh = 0$. But f is also a monomorphism, hence $h = 0$.

(b) Assume that $fh = 0$. Then $(gf)h = g(fh) = g0 = 0$. Because gf is a monomorphism, we get $h = 0$.

(c) If f is a section, then there exists f' such that $f'f = 1$. Assume that $fh = 0$. Then $h = 1 \cdot h = (f'f)h = f'(fh) = f'0 = 0$.

(a'), (b'), (c') are the respective duals of (a), (b), (c).

(d) Let f be an isomorphism, then $f^{-1}f = 1$ shows that f is a section. Similarly, it is a retraction. Conversely, if f is both a section and a retraction, then there exist f', f'' such that $f'f = 1$ and $ff'' = 1$. In order to prove that f is an isomorphism, it suffices to prove that $f' = f''$. But this follows from $f' = f' \cdot 1 = f'(ff'') = (f'f)f'' = 1 \cdot f'' = f''$.

(e) Assume that gf is an isomorphism, and let $h = (gf)^{-1}$. Then $hgf = 1$ gives that f is a section and $gfh = 1$ gives that g is a retraction.   □

It follows from (d) that an isomorphism is at the same time a monomorphism and an epimorphism. But the converse is generally not true, see Examples IV.2.6(e),(f).

Another comment: a section is a monomorphism and, in a module category, this means an injective morphism. Conversely, assume that $s: M \to N$ is an injective linear map between A-modules. Elementary set theory tells us that there always exists a *map* $s': N \to M$ such that $s's = 1_M$. The linear map s is a section if it is possible to find a *linear map* s' satisfying this condition and this is not always the case. A similar comment applies to retractions.

IV.2.10 Example. Consider the $\mathbb{Z}$-linear map $f: \mathbb{Z} \to \mathbb{Z}$ defined by $x \mapsto 2x$, for $x \in \mathbb{Z}$. It is injective, so there exists a map $g: \mathbb{Z} \to \mathbb{Z}$ such that $gf = 1_\mathbb{Z}$ (for instance, send any even number to its half, and any odd number to zero). Assume that there exists a $\mathbb{Z}$-linear map $g : \mathbb{Z} \to \mathbb{Z}$ such that $gf = 1_\mathbb{Z}$, then $2g(1) = g(2 \cdot 1) = gf(1) = 1$, an absurdity because the equation $2x = 1$ has no integral solution. Therefore, f is not a section.

IV.2.2 Linear categories

One of the nicest properties of K-categories is that, in such a category, a finite family of objects admits a product if and only if it admits a coproduct, and in this case, they are isomorphic, exactly as in the case of a module category, see Subsection III.3.2. In order to prove this statement, we introduce an *ad hoc* notion, inspired from Lemma III.3.8.

IV.2.11 Definition. Let $\{X_1, \cdots, X_m\}$ be a finite family of objects in a K-category C. A **biproduct** of this family is a triple $(X, (p_i)_{1 \leqslant i \leqslant m}, (q_i)_{1 \leqslant i \leqslant m})$, with X an object in C and $(p_i: X \to X_i)_{1 \leqslant i \leqslant m}, (q_i: X_i \to X)_{1 \leqslant i \leqslant m}$ families of morphisms in C such that

(a) $\sum_{i=1}^{m} q_i p_i = 1_X$.

(b) $p_i q_j = \delta_{ij}$, where δ_{ij} is the Kronecker delta, equal to the identity 1_{X_i} if $i = j$, and 0 otherwise.

The notion of biproduct is selfdual: the dual of a biproduct is a biproduct. It also has the advantage of being intrinsic, that is, the biproduct is defined only in terms of the morphisms p_i, q_j between X and the X_i, whereas the definitions of product and coproduct involve morphisms between all objects of the category and the X_i.

It follows from Definition IV.2.11 that each q_i is a section while each p_i is a retraction.

Lemma III.3.8 shows that the coproduct of a finite family of modules, which equals their product, is a biproduct. We prove that, in a K-category, the product of a finite family exists if and only if its biproduct exists, and then they are isomorphic. The proof is inspired from Example IV.2.8(a).

IV.2.12 Theorem. *Let $\{X_1, \cdots, X_m\}$ be a finite family of objects in a K-category C. Then $(X, (p_i)_{1 \leqslant i \leqslant m})$ is a product of the X_i if and only if there exist morphisms $(q_i \colon X_i \to X)_{1 \leqslant i \leqslant m}$ making $(X, (p_i)_{1 \leqslant i \leqslant m}, (q_i)_{1 \leqslant i \leqslant m})$ a biproduct of the X_i.*

Proof. Necessity. We define the q_i as in Example IV.2.8(a). For a pair (i,j) with $1 \leqslant i, j \leqslant m$, consider the morphism $\delta_{ji} \colon X_i \to X_j$, where δ_{ji} is the Kronecker delta, see Definition IV.2.11. The definition of product yields a unique morphism $q_i \colon X_i \to X$ such that $p_j q_i = \delta_{ji}$.

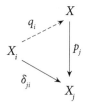

In order to prove that $(X, (p_i)_{1 \leqslant i \leqslant m}, (q_i)_{1 \leqslant i \leqslant m})$ is a biproduct, it suffices to prove that $\sum_{i=1}^{m} q_i p_i = 1_X$. Now, for any j, we have

$$p_j \left(\sum_{i=1}^{m} q_i p_i \right) = \sum_{i=1}^{m} (p_j q_i) p_i = \sum_{i=1}^{m} \delta_{ji} p_i = p_j = p_j 1_X.$$

Applying the universal property of the product completes the proof of necessity.

Sufficiency. It suffices to verify that $(X, (p_i)_{1 \leqslant i \leqslant m})$ satisfies the universal property of the product. Let $(f_i \colon Y \to X_i)_{1 \leqslant i \leqslant m}$ be morphisms in C. If $f \colon Y \to X$ satisfies $p_i f = f_i$, for any i, then necessarily

$$f = 1_X f = \left(\sum_{i=1}^{m} q_i p_i \right) f = \sum_{i=1}^{m} q_i (p_i f) = \sum_{i=1}^{m} q_i f_i.$$

This proves uniqueness of f, if it exists, and gives at the same time a formula defining it. Conversely, $f = \sum_{i=1}^{m} q_i f_i$ satisfies

$$p_j f = p_j \left(\sum_{i=1}^{m} q_i f_i \right) = \sum_{i=1}^{m} (p_j q_i) f_i = \sum_{i=1}^{m} \delta_{ji} f_i = f_j$$

for any j. This completes the proof. $\qquad\qquad\qquad\qquad\qquad\qquad\square$

IV.2.13 Theorem. *Let $\{X_1, \cdots, X_m\}$ be a finite family of objects in a K-category C. Then $(X, (q_i)_{1 \leqslant i \leqslant m})$ is a coproduct of the X_i if and only if there exist morphisms $(p_i \colon X_i \to X)_{1 \leqslant i \leqslant m}$ making $(X, (p_i)_{1 \leqslant i \leqslant m}, (q_i)_{1 \leqslant i \leqslant m})$ a biproduct of the X_i.*

Proof. Dual to the proof of Theorem IV.2.12 and left as an exercise. $\qquad\square$

We have proved our statement.

IV.2.14 Corollary. *The product of a finite family of objects in a K-category exists if and only if its coproduct exists and, in this case, they are isomorphic.*

Because the product and the coproduct of a finite family of objects are isomorphic, one usually calls them the ***direct sum*** of this family. This leads to the definition.

IV.2.15 Definition. A K-category C is a ***K-linear category*** or, briefly, ***linear category***, if any finite family of objects in C has a product (or, equivalently, a coproduct) in this category.

The notion of linear category is selfdual. A $\mathbb{Z}$-linear category is called an ***additive category***. Hence, for any K, a K-linear category is additive.

Before giving examples, we see a nice consequence. The empty family of objects is certainly finite. The empty product is an object F, unique up to isomorphism (because of universality of product), such that, for any object X, there exists a unique morphism $X \to F$ (in a K-category, this is necessarily the zero morphism): F is called a ***final object*** of the category. Dually, the coproduct of the empty family is an object I, unique up to isomorphism, such that, for any object Y, there exists a unique morphism $I \to Y$ (also zero if we are inside a K-category). Then, I is called an ***initial object***.

In Mod$A$, with A an algebra, the zero module is at the same time an initial and a final object. By analogy, if, in a category, there exists an object which is at the same time initial and final, we call it a ***zero object*** and denote it by 0. It follows from Corollary IV.2.14 that initial and final objects in a linear category coincide. Hence, any linear category contains a zero object, unique up to isomorphism (because so are initial and final objects).

Many examples of K-categories considered in Subsection IV.2.1 are actually K-linear.

A nice consequence is the following. Let X, Y be objects in a linear category C, then the zero of the K-module $C(X, Y)$ equals the composite of the unique morphism $X \to 0$

with the unique morphism $0 \to Y$: indeed, this composite is zero in $C(X, Y)$ and the latter has a unique zero element.

IV.2.16 Examples.

   (a) Assume that A is a K-algebra. Because of Theorems III.3.3 and III.3.7, the module category $\mathrm{Mod}A$ is K-linear. In particular, $\mathcal{A}b$ is an additive category. Also, the full subcategory $\mathrm{mod}A$ of $\mathrm{Mod}A$ consisting of finitely generated A-modules is K-linear, because in Definition IV.2.15, only finite products and coproducts are required.

   (b) Let A be a K-algebra considered as a K-category having a unique object x, as in Example IV.2.2(c). This K-category is not K-linear, because finite direct sums of x do not exist in it.

   (c) The category $\mathcal{T}ag$ is an additive category. Indeed, the cartesian product $G \times H$ of the topological abelian groups G, H, equipped with the product topology is both their product and their coproduct in $\mathcal{T}ag$.

   (d) The category $\mathcal{D}iv$ is additive. In order to prove it, we verify that two divisible abelian groups have a coproduct (product) which is divisible. Let G, H be divisible abelian groups. Their coproduct (and product) in $\mathcal{A}b$ is the direct sum $G \oplus H$. We prove that $G \oplus H$ is divisible. Let $n > 0$ be an integer. Then $nG = G$ and $nH = H$. Hence, for any $(g, h) \in G \oplus H$, there exist $g' \in G, h' \in H$, such that $g = ng', h = nh'$. Therefore, $(g, h) = n(g', h') \in n(G \oplus H)$ and so $G \oplus H = n(G \oplus H)$.

   (e) Let K be a field, and $\mathcal{E}$ the category of pairs (E, f) with E a K-vector space and f an endomorphism of E, see Examples III.2.2(i) and IV.2.2(f). Then $\mathcal{E}$ is K-linear. Given objects $(E, f), (F, g)$ in $\mathcal{E}$, the pair $(E \oplus F, f \oplus g)$ where $f \oplus g : E \oplus F \to E \oplus F$ is the unique endomorphism induced by passing to coproducts (direct sums), see Exercise III.3.4, is at the same time the product and the coproduct of the pairs $(E, f), (F, g)$ in $\mathcal{E}$.

   (f) Let K be a field, Q a finite quiver and $\mathrm{Rep}_K Q$ the category of representations of Q, see Example III.2.2(j). It follows from Examples III.3.4 and III.3.9 that $\mathrm{Rep}_K Q$ is K-linear.

   (g) Let $\mathcal{I}$ be a small category and C a K-linear category, then the functor category $\mathrm{Fun}(\mathcal{I}, C)$ of Examples III.2.9(c) and IV.2.2 is K-linear. We define the direct sum of the functors F and G as being the functor given by $(F \oplus G)(X) = FX \oplus GX$, for an object X in $\mathcal{I}$, and, for a morphism $f : X \to Y$ in $\mathcal{I}$, we let $(F \oplus G)(f)$ be the unique morphism from $FX \oplus GX$ to $FY \oplus GY$ obtained by passing to coproducts. This establishes our claim. The zero functor is the one which sends any object of $\mathcal{I}$ on the zero object of C, which exists, because the latter is a linear category.

IV.2.3 Abelian categories

As the name indicates, abelian categories are built on the model of the category of abelian groups or, more generally, of module categories. Among other pleasant properties of abelian categories, any morphism has a kernel and a cokernel. We thus start by defining these notions. Because we look for categorical notions (that is, 'element-free'), it is natural to think of their universal properties, introduced in Theorems II.4.4 and II.4.7.

IV.2.17 Definition. Let $f: X \to Y$ be a morphism in a K-category C. A **kernel** of f is a pair (U, u), with U an object and $u: U \to X$ a morphism in C, such that:

(a) $fu = 0$.
(b) If $u': U' \to X$ is such that $fu' = 0$, then there exists a unique $v: U' \to U$ such that $uv = u'$

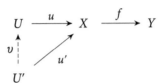

It is not obvious that a morphism in a K-category has a kernel, but if it does, then the latter is unique up to isomorphism (see the proof of uniqueness in Theorem II.4.4). We denote $U = \mathrm{Ker}\, f$ (with capital K) and $u = \mathrm{ker}\, f$ (with small k) in order to distinguish them. In Mod$A$, this notion coincides with the usual notion of kernel because of Theorem II.4.4. As pointed out there, uniqueness in (b) is equivalent to saying that u is a monomorphism. The morphism u is called the **injection** or **inclusion**.

Cokernels being dual to kernels, their definition is obtained from that of kernels by reversing arrows.

IV.2.18 Definition. Let $f: X \to Y$ be a morphism in a K-category C. A **cokernel** of f is a pair (U, u), with U an object and $u: Y \to U$ a morphism in C, such that:

(a) $uf = 0$.
(b) If $u': Y \to U'$ is such that $u'f = 0$, then there exists a unique $v: U \to U'$ such that $vu = u'$

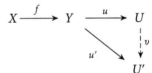

It is not obvious that a morphism in a K-category has a cokernel, but if it does, then the latter is unique up to isomorphism (see the proof of uniqueness in Theorem II.4.7). We write $U = \mathrm{Coker}\, f$ (with capital C) and $u = \mathrm{coker}\, f$ (with small c). Because of Theorem II.4.7, this notion coincides in Mod$A$ with the usual notion of cokernel. Uniqueness in (b) amounts to saying that u is an epimorphism. The morphism u is called the **projection** or **surjection**.

IV.2.19 Lemma. *Let $f: X \to Y$ be a morphism in a K-category C in which any morphism has a kernel and a cokernel. Then:*
(a) *f is a monomorphism if and only if $\mathrm{Ker}\, f = 0$.*
(b) *f is an epimorphism if and only if $\mathrm{Coker}\, f = 0$.*

Proof. This follows directly from the definitions.   □

The definitions allow to compute easily the kernel and the cokernel of a zero morphism.

IV.2.20 Lemma. *Let C be a K-category in which any morphism has a kernel and a cokernel, and X an object in C. Then:*

- (a) *The kernel of the zero morphism $X \to Y$ is 1_X.*
- (b) *The cokernel of the zero morphism $W \to X$ is 1_X.*

Proof. We prove (a), because (b) is dual. For any $f : W \to X$, we have $0 \cdot f = 0$ and $1_X \cdot f = f$. Hence, the kernel of the zero morphism $X \to Y$ is 1_X.   □

Exact sequences are one of our main working tools in module categories. Their definition involves the notion of image of a morphism. It turns out that, in a categorical setting, two definitions correspond to that of image.

IV.2.21 Definition. Let $f: X \to Y$ be a morphism in a K-category C. Then:

- (a) The ***image*** of f is the kernel of the cokernel of f.
- (b) The ***coimage*** of f is the cokernel of the kernel of f.

Image and coimage may or may not exist, but if they do, then they are unique up to isomorphism (because so are kernels and cokernels). The objects image and coimage are denoted by $\mathrm{Im}\, f = \mathrm{Ker}(\mathrm{coker}\, f)$ and $\mathrm{Coim}\, f = \mathrm{Coker}(\mathrm{ker}\, f)$, respectively, and the associated morphisms by $\mathrm{im}\, f = \mathrm{ker}(\mathrm{coker}\, f)$ and $\mathrm{coim}\, f = \mathrm{coker}(\mathrm{ker}\, f)$.

As an example, we compute the coimage of a monomorphism and the image of an epimorphism.

IV.2.22 Corollary. *Let $f : X \to Y$ be a morphism in a K-category C in which any morphism has a kernel and a cokernel. Then:*

- (a) *If f is a monomorphism, then $\mathrm{coim}\, f = 1_X$.*
- (b) *If f is an epimorphism, then $\mathrm{im}\, f = 1_Y$.*

Proof. This follows directly from Lemmata IV.2.19 and IV.2.20.   □

Before the next proposition, consider the situation in a module category Mod$A$, with A a K-algebra. Let $f: M \to N$ be a linear map between A-modules. Because of Theorems II.4.4 and II.4.7, the kernel morphism is the inclusion of the submodule $\mathrm{Ker}\, f = \{x \in M: f(x) = 0\}$ into M, while the cokernel morphism is the projection of N onto the quotient $N/f(M)$, with $f(M) = \{f(x) \in N: x \in M\}$. Therefore, $\mathrm{Im}\, f = \mathrm{Ker}(\mathrm{coker}\, f) = f(M)$ (that is, we get the classical notion of image), while $\mathrm{Coim}\, f = \mathrm{Coker}(\mathrm{ker}\, f) = M/\mathrm{Ker}\, f$.

Now, the Isomorphism Theorem II.4.21 in Mod$A$ asserts the existence of a unique A-linear map $\bar{f}: \mathrm{Coim}\, f = M/\mathrm{Ker}\, f \to f(M) = \mathrm{Im}\, f$ making the square below commute

$$M \xrightarrow{\quad f \quad} N$$
$$\downarrow p \qquad\qquad \uparrow i$$
$$M/\mathrm{Ker}\, f \dashrightarrow^{\bar f} f(M)$$

(with i the inclusion and p the projection) and moreover $\bar f$ is an isomorphism.

The next proposition says that, if kernel and cokernel exist in a K-category, then such a morphism $\bar f$ always exists.

IV.2.23 Proposition. *Let $f: X \to Y$ be a morphism in a K-category C in which any morphism admits a kernel and a cokernel. Then there exists a unique morphism $\bar f$ such that the following diagram commutes*

$$\mathrm{Ker} f \xrightarrow{\ k\ } X \xrightarrow{\ f\ } Y \xrightarrow{\ c\ } \mathrm{Coker} f$$
$$\downarrow p \qquad\qquad \uparrow i$$
$$\mathrm{Coim} f \dashrightarrow^{\bar f} \mathrm{Im} f$$

where k, i are the injections and p, c the projections.

Proof. Existence. Because $cf = 0$, and $i = \ker c$, there exists $h: X \to \mathrm{Im}\, f$ such that $f = ih$. Then $fk = 0$ gives $ihk = 0$. Because i is a monomorphism, this implies $hk = 0$. Then $p = \mathrm{coker}\, k$ yields $\bar f: \mathrm{Coim}\, f \to \mathrm{Im}\, f$ such that $h = \bar f p$. Hence, $i\bar f p = ih = f$, as required.

Uniqueness. The morphisms i, p are respectively a monomorphism and an epimorphism, hence $i\bar f p = i\bar f' p$ implies $\bar f p = \bar f' p$, then $\bar f = \bar f'$. □

The factorisation of the proposition is called the **canonical factorisation** of f, and $\bar f$ the **canonical morphism** induced by f.

The isomorphism theorem in ModA states that $\bar f$ is an isomorphism. Abelian categories can be thought of as being those linear categories in which the isomorphism theorem holds true.

IV.2.24 Definition. A K-linear category C is **K-abelian**, or simply **abelian**, if:
   (Ab1) Any morphism in C admits a kernel and a cokernel.
   (Ab2) For any morphism f, the canonical morphism $\bar f: \mathrm{Coim}\, f \to \mathrm{Im}\, f$ is an isomorphism.

The notion of abelian category is selfdual. This has an enormous advantage: if we prove a categorical statement in an abelian category, then, due to the duality principle, its dual holds true automatically.

IV.2.25 Lemma. *A morphism in an abelian category is an isomorphism if and only if it is both a monomorphism and an epimorphism.*

Proof. Assume that $f: X \to Y$ is both a monomorphism and an epimorphism. Because f is a monomorphism, $\operatorname{Ker} f = 0$. Hence, $1_X: X \to X$ is the projection $X \to \operatorname{Coim} f$, due to Corollary IV.2.22. Similarly, $1_Y: Y \to Y$ is the injection $\operatorname{Im} f \to Y$. Uniqueness of $\bar{f}$ and $f = 1_Y f 1_X$ yield $f = \bar{f}$. So f is an isomorphism. The converse follows from Proposition IV.2.9(d). □

IV.2.26 Examples.

(a) An obvious example of abelian category is the module category $\operatorname{Mod} A$, with A an algebra. This includes $\mathcal{A}b$ (= $\operatorname{Mod}\mathbb{Z}$). A better example is the category $\operatorname{mod} A$ of finitely generated A-modules. This is because the axioms of linear categories require only existence of finite products and coproducts. In this sense, $\operatorname{Mod} A$ and $\mathcal{A}b$ are particular: they are abelian categories having arbitrary products and coproducts. This explains why $\operatorname{mod} A$ should be considered as the real prototype of an abelian category.

(b) Due to Lemma IV.2.25, the categories *Div* and *Stag* of Examples IV.2.6(e),(f) are not abelian.

(c) Let K be a field and $\mathcal{E}$ the category of pairs (E, f) of Example IV.2.6(c). A morphism $u: (E, f) \to (F, g)$ in C being also a morphism $u: E \to F$ of K-vector spaces, we consider the kernel (U, i) of u viewed as a K-linear map. By passing to kernels, there exists an endomorphism $h: U \to U$ making the following diagram commute

$$
\begin{array}{ccccccc}
0 & \longrightarrow & U & \overset{i}{\longrightarrow} & E & \overset{u}{\longrightarrow} & F \\
 & & {\scriptstyle h}\downarrow & & {\scriptstyle f}\downarrow & & {\scriptstyle g}\downarrow \\
0 & \longrightarrow & U & \overset{i}{\longrightarrow} & E & \overset{u}{\longrightarrow} & F
\end{array}
$$

Hence, the pair (U, h) is the kernel of u in C. One constructs similarly cokernel, image and coimage of u in $\mathcal{E}$. Because the isomorphism theorem holds true in the category of K-vector spaces, it is also valid in $\mathcal{E}$, which is therefore abelian.

(d) Let K be a field and Q a finite quiver. The category $\operatorname{Rep}_K Q$ of representations of Q, see Example IV.2.6(d), is abelian. Let indeed $f = (f_x)_{x \in Q_0}: V \to W$ be a morphism of representations. Because morphisms in $\operatorname{Rep}_K Q$ compose componentwise, the kernel of f is the representation $U = (U(x), U(\alpha))_{x \in Q_0, \alpha \in Q_1}$, where, if $x \in Q_0$, then $U(x) = \operatorname{Ker} f_x$ and, if $\alpha: x \to y$ in Q, then $U(\alpha)$ is obtained by passing to kernels in the diagram

$$
\begin{array}{ccccccc}
0 & \longrightarrow & U(x) & \overset{i_x}{\longrightarrow} & V(x) & \overset{f_x}{\longrightarrow} & W(x) \\
 & & {\scriptstyle U(\alpha)}\downarrow & & {\scriptstyle V(\alpha)}\downarrow & & {\scriptstyle W(\alpha)}\downarrow \\
0 & \longrightarrow & U(y) & \overset{i_y}{\longrightarrow} & V(y) & \overset{f_y}{\longrightarrow} & W(y)
\end{array}
$$

with i_x, i_y the injections. The construction is similar for cokernel, image and coimage. Because f_x is a morphism of K-vector spaces, the canonical morphism $\bar{f}_x$ is an isomorphism, for any x. Exercise II.3.24 implies that so is $\bar{f}$. Thus, $\operatorname{Rep}_K Q$ is an abelian category.

(e) Let $\mathcal{I}$ be a small category and C an abelian category. Then the functor category $\mathrm{Fun}(\mathcal{I}, C)$ is abelian. We first prove that it contains kernels and cokernels. Let $F, G : \mathcal{I} \to C$ be functors and $\varphi : F \to G$ a functorial morphism. We define the kernel functor $\mathrm{Ker}\,\varphi$ as follows. For an object $X \in \mathcal{I}_0$, set $(\mathrm{Ker}\,\varphi)(X) = \mathrm{Ker}\,\varphi_X$ and for a morphism $f : X \to Y$ in $\mathcal{I}$, let $(\mathrm{Ker}\,\varphi)(f)$ be the unique morphism from $(\mathrm{Ker}\,\varphi)(X)$ to $(\mathrm{Ker}\,\varphi)(Y)$ induced by passing to kernels. Cokernels are defined dually.

Also, for an object $X \in \mathcal{I}_0$, the canonical morphism $\overline{\varphi}_X$ is an isomorphism because C is abelian. Therefore, $\overline{\varphi} = (\overline{\varphi}_X)_{X \in \mathcal{I}_0}$ is a functorial isomorphism. This completes the proof.

Now that we have a categorical notion of image, we may define exact sequences as we did in a module category: a sequence of objects and morphisms:

$$\cdots \longrightarrow X_{i-1} \xrightarrow{f_{i-1}} X_i \xrightarrow{f_i} X_{i+1} \longrightarrow \cdots$$

in an abelian category is **exact at** X_i if $\mathrm{Im}\,f_{i-1} = \mathrm{Ker}\,f_i$. It is **exact** if it is exact at any X_i. A **short exact sequence** is an exact sequence of the form

$$0 \to X \xrightarrow{f} Y \xrightarrow{g} Z \to 0.$$

Proposition II.4.10 holds *verbatim* in an abelian category. Thus, a short sequence like the one above is exact if and only if f is a monomorphism, g is an epimorphism and $\mathrm{Im}\,f = \mathrm{Ker}\,g$.

Before returning to short exact sequences, let f be a morphism in an abelian category. Due to Proposition IV.2.23, it can be written as $f = \bar{i}\bar{f}p$ with i a monomorphism and p an epimorphism. Because our category is abelian, $\bar{f}$ is an isomorphism. Hence, f is the composite of a monomorphism and an epimorphism. We prove that this decomposition is essentially unique.

IV.2.27 Proposition. *Let $f : X \to Y$ be a morphism in an abelian category, and $i = \mathrm{im}\,f, q = \mathrm{coim}\,f$. Assume that $f = jp$, where $p : X \to Z$ is an epimorphism and $j : Z \to Y$ a monomorphism. Then there exists an isomorphism $k : Z \to \mathrm{Im}\,f$ such that $j = ik$ and $p = k^{-1}q$.*

Proof. Let (U, u) and (V, v) be respectively the kernel and the cokernel of f. Because we are working up to isomorphism, we may assume that $\bar{f} = 1_{\mathrm{Im}\,f}$ so we have a commutative diagram

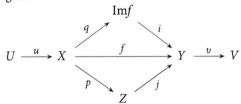

where $i = \ker v$ and $q = \mathrm{coker}\,u$. We have $vjp = vf = 0$. Because p is an epimorphism, $vj = 0$. But $i = \ker v$. Hence, there exists $k : Z \to \mathrm{Im}\,f$ such that $j = ik$. Dually, there exists $\ell\colon \mathrm{Im}\, f \to Z$ such that $p = \ell q$. It suffices to prove that k, ℓ are mutually inverse. Indeed, $iq = f = jp = ik\ell q$, hence $k\ell = 1_{\mathrm{Im}\, f}$, because i is a monomorphism and q an epimorphism. Similarly, $j\ell k = ik\ell k = ik = j$ yields $\ell k = 1_Z$, because j is a monomorphism. □

IV.2.28 Corollary. *In an abelian category:*

(a) *A sequence* $0 \longrightarrow X \overset{f}{\longrightarrow} Y \overset{g}{\longrightarrow} Z$ *is exact if and only if* $f = \ker g$.

(b) *A sequence* $X \overset{f}{\longrightarrow} Y \overset{g}{\longrightarrow} Z \longrightarrow 0$ *is exact if and only if* $g = \mathrm{coker}\, f$.

(c) *For a sequence* $0 \longrightarrow X \overset{f}{\longrightarrow} Y \overset{g}{\longrightarrow} Z \longrightarrow 0$, *the following are equivalent:*

(i) *The sequence is exact.*

(ii) f *is a monomorphism and* $g = \mathrm{coker}\, f$.

(iii) g *is an epimorphism and* $f = \ker g$.

Proof. Because (b) is dual to (a), and (c) follows from (a) and (b), we need only prove (a). If the sequence is exact, then f is a monomorphism. Proposition IV.2.27 gives its canonical factorisation as $f = \mathrm{im}\, f$. Then $\mathrm{im}\, f = \ker g$ implies $f = \ker g$ as required. Conversely, if $f = \ker g$, then f is a monomorphism, so $f = \mathrm{im}\, f$. Hence, $\mathrm{im}\, f = \ker g$ and the sequence is exact. □

In order to show how to handle these concepts, we offer a categorical (and very simplified) version of the Five Lemma II.4.18.

IV.2.29 Lemma. *In an abelian category, assume that we have a commutative diagram with exact rows*

$$
\begin{array}{ccccccccc}
0 & \longrightarrow & X & \overset{u}{\longrightarrow} & Y & \overset{v}{\longrightarrow} & Z & \longrightarrow & 0 \\
 & & \downarrow{\scriptstyle f} & & \downarrow{\scriptstyle g} & & \downarrow{\scriptstyle h} & & \\
0 & \longrightarrow & X' & \overset{u'}{\longrightarrow} & Y' & \overset{v'}{\longrightarrow} & Z' & \longrightarrow & 0
\end{array}
$$

If f, h *are monomorphisms, or epimorphisms, or isomorphisms, then so is* g.

Proof. Assume that f and h are monomorphisms, and that $gw = 0$. Then $hvw = v'gw = 0$. Because h is a monomorphism, $vw = 0$. Because $u = \ker v$, there exists w' such that $w = uw'$. Hence, $u'fw' = guw' = gw = 0$. But both u', f are monomorphisms, hence $w' = 0$ and so $w = 0$. Therefore, g is a monomorphism. The statement about epimorphisms is dual and the one about isomorphisms is their conjunction because of Lemma IV.2.25. □

IV.2.30 Example. We redo Example II.4.16(a) in an abelian category. Let X, Z be objects and $Y = X \oplus Z$ their direct sum. Denote by $p_1\colon Y \to X, p_2\colon Y \to Z$ and

$q_1: X \to Y, q_2: Z \to Y$ the projections and injections, respectively. We claim that the sequence

$$0 \longrightarrow X \xrightarrow{q_1} Y \xrightarrow{p_2} Z \longrightarrow 0$$

is short exact. Indeed, p_2 is an epimorphism (even a retraction) because $p_2 q_2 = 1_Z$, and q_1 a monomorphism (even a section) because $p_1 q_1 = 1_X$. Because Im $q_1 = X$, there remains to prove that $X \cong \text{Ker } p_2$. We have $p_2 q_1 = 0$. So let $f: Y \to X$ be such that $p_2 f = 0$. Then $f = 1 \cdot f = (q_1 p_1 + q_2 p_2)f = q_1(p_1 f)$ factors through q_1 as required. This factorisation is unique because q_1 is a monomorphism. This proves our claim.

This example leads to the following definition.

IV.2.31 Definition. A short exact sequence

$$0 \longrightarrow X \xrightarrow{f} Y \xrightarrow{g} Z \longrightarrow 0$$

in an abelian category *splits*, or is *split*, if there exists an isomorphism $h: Y \to X \oplus Z$ such that the following diagram commutes

$$
\begin{array}{ccccccccc}
0 & \longrightarrow & X & \xrightarrow{\ f\ } & Y & \xrightarrow{\ g\ } & Z & \longrightarrow & 0 \\
& & \downarrow{\scriptstyle 1_X} & & \downarrow{\scriptstyle h} & & \downarrow{\scriptstyle 1_Z} & & \\
0 & \longrightarrow & X & \xrightarrow{\ q_1\ } & X \oplus Z & \xrightarrow{\ p_2\ } & Z & \longrightarrow & 0
\end{array}
$$

where $q_1: X \to X \oplus Z, p_2: X \oplus Z \to Z$ are respectively the injection and the projection.

Because of Lemma IV.2.29, any morphism h which makes the diagram of the definition commute is necessarily an isomorphism. We have the following characterisation of split exact sequences.

IV.2.32 Theorem. *The following conditions are equivalent for a short exact sequence* $0 \to X \xrightarrow{f} Y \xrightarrow{g} Z \to 0$ *in an abelian category:*

(a) *The sequence splits.*
(b) *f is a section.*
(c) *g is a retraction.*

Proof. We prove the equivalence of (a) and (b), that of (a) and (c) being dual.

(a) implies (b). If the sequence splits, then we have a commutative diagram as in Definition IV.2.31. Let $p_1 : X \oplus Z \to X$ be the projection, so $p_1 q_1 = 1_X$. Set $f' = p_1 h$. Then $f'f = p_1 h f = p_1 q_1 = 1_X$, and f is a section.

(b) implies (a). Assume that $f' : Y \to X$ is such that $f'f = 1_X$. Then $1_Y - ff' : Y \to Y$ satisfies $(1_Y - ff')f = f - ff'f = f - f1_X = 0$. Because $g = \operatorname{coker} f$, there exists $g' : Z \to Y$ such that $1_Y - ff' = g'g$. In order to prove (a), we prove that (Y, f', g, f, g') is a biproduct of X and Z. We know that $f'f = 1_X, ff' + g'g = 1_Y$ and $gf = 0$. Also, $g = g \cdot 1_Y = g(ff' + g'g) = gg'g$ yields $gg' = 1_Z$, because g is an epimorphism. Finally, $g' = 1_Y g' = (ff' + g'g)g' = ff'g' + g'$ yields $ff'g' = 0$. Hence, $f'g' = 0$, because f is a monomorphism. □

Thus, if a short exact sequence $0 \to X \xrightarrow{f} Y \xrightarrow{g} Z \to 0$ splits, we necessarily have $Y \cong X \oplus Z$. But the converse is not always true: as Theorem IV.2.32 says, splitting of a sequence is really a property of the morphisms. See Example IV.2.33(c) below.

IV.2.33 Examples.

    (a) Example IV.2.8(c) says that a short exact sequence of modules $0 \to X \to Y \to L \to 0$ with L free splits. Because a vector space over a field K is a free K-module, see Corollary III.4.6, any short exact sequence of vector spaces splits.

    (b) We prove that the short exact sequence

$$0 \to \frac{\mathbb{Z}}{\mathbb{Z}a} \xrightarrow{f} \frac{\mathbb{Z}}{\mathbb{Z}(ab)} \xrightarrow{g} \frac{\mathbb{Z}}{\mathbb{Z}b} \to 0$$

with $f: 1 + \mathbb{Z}a \mapsto b + \mathbb{Z}(ab), g: 1 + \mathbb{Z}(ab) \mapsto 1 + \mathbb{Z}b$ splits if and only if the integers a, b are coprime. Assume first that a, b are coprime. Because of the Bézout–Bachet Theorem, there exist $r, s \in \mathbb{Z}$ such that $ra + sb = 1$. Define

$$h: \frac{\mathbb{Z}}{\mathbb{Z}(ab)} \to \frac{\mathbb{Z}}{\mathbb{Z}a} \oplus \frac{\mathbb{Z}}{\mathbb{Z}b} \quad \text{by} \quad h(1 + \mathbb{Z}(ab)) = \begin{pmatrix} s + \mathbb{Z}a \\ 1 + \mathbb{Z}b \end{pmatrix}.$$

Then the diagram

$$
\begin{array}{ccccccccc}
0 & \longrightarrow & \mathbb{Z}/\mathbb{Z}a & \xrightarrow{\;f\;} & \mathbb{Z}/\mathbb{Z}(ab) & \xrightarrow{\;g\;} & \mathbb{Z}/\mathbb{Z}b & \longrightarrow & 0 \\
& & \downarrow{\scriptstyle 1} & {\scriptstyle \binom{1}{0}} & \downarrow{\scriptstyle h} & {\scriptstyle (0\ 1)} & \downarrow{\scriptstyle 1} & & \\
0 & \longrightarrow & \mathbb{Z}/\mathbb{Z}a & \longrightarrow & \mathbb{Z}/\mathbb{Z}a \oplus \mathbb{Z}/\mathbb{Z}b & \longrightarrow & \mathbb{Z}/\mathbb{Z}b & \longrightarrow & 0
\end{array}
$$

commutes because

$$hf(1 + \mathbb{Z}a) = h(b + \mathbb{Z}(ab)) = \begin{pmatrix} sb + \mathbb{Z}a \\ b + \mathbb{Z}b \end{pmatrix} = \begin{pmatrix} 1 - ra + \mathbb{Z}a \\ b + \mathbb{Z}b \end{pmatrix} = \begin{pmatrix} 1 + \mathbb{Z}a \\ 0 \end{pmatrix}$$

$$= \begin{pmatrix} 1 \\ 0 \end{pmatrix}(1 + \mathbb{Z}a)$$

and

$$(0 \quad 1)\begin{pmatrix} s + \mathbb{Z}a \\ 1 + \mathbb{Z}b \end{pmatrix} = 1 + \mathbb{Z}b = g(1 + \mathbb{Z}(ab)).$$

Lemma IV.2.29 gives that h is an isomorphism. Hence, the sequence splits.

On the other hand, if a, b are not coprime, they have a common divisor distinct from ± 1. Because of Corollary I.5.9, they also have a common prime divisor p, say. Let $s, t > 0$ be integers such that $p^s \mid a, p^t \mid b$ but $p^{s+1} \nmid a, p^{t+1} \nmid b$. Then $\mathbb{Z}/\mathbb{Z}(ab)$ contains an element of order p^{s+t} whereas $\mathbb{Z}/\mathbb{Z}a \oplus \mathbb{Z}/\mathbb{Z}b$ does not. Thus, they cannot be isomorphic, and the sequence is not split.

(c) We give an example of a nonsplit short exact sequence of abelian groups whose middle term is the direct sum of the endterms. Let $(\mathbb{Z}/2\mathbb{Z})^{\mathbb{N}}$ denote the product of $\mathbb{N}$ copies of $\mathbb{Z}/2\mathbb{Z}$, that is, the additive group of sequences $(y_0, y_1, \cdots)$, with the $y_i \in \mathbb{Z}/2\mathbb{Z}$. Consider the sequence

$$0 \to \mathbb{Z} \xrightarrow{f} \mathbb{Z} \oplus (\mathbb{Z}/2\mathbb{Z})^{\mathbb{N}} \xrightarrow{g} (\mathbb{Z}/2\mathbb{Z})^{\mathbb{N}} \to 0$$

where $f : x \mapsto (2x, 0)$ and $g : (x, (y_0, y_1, \cdots)) \mapsto (x + 2\mathbb{Z}, y_0, y_1, \cdots)$, for $x \in \mathbb{Z}$, and $y_i \in \mathbb{Z}/2\mathbb{Z}$, for all i. It is easily seen that f is injective, g is surjective and $\operatorname{Ker} g = 2\mathbb{Z} \oplus 0 = \operatorname{Im} f$. So the sequence is exact. If it splits, then there exists a morphism $g' : (\mathbb{Z}/2\mathbb{Z})^{\mathbb{N}} \to \mathbb{Z} \oplus (\mathbb{Z}/2\mathbb{Z})^{\mathbb{N}}$ such that $gg' = 1$. Now the element $x = (1 + 2\mathbb{Z}, 0, 0 \cdots) \in (\mathbb{Z}/2\mathbb{Z})^{\mathbb{N}}$ is of order 2. Hence, $g'(x)$ is also of order two in $g^{-1}((\mathbb{Z}/2\mathbb{Z})^{\mathbb{N}}) = (1, 0) + (2\mathbb{Z} \oplus 0)$. But the latter contains no element of order two, a contradiction.

The situation is better when one deals with finitely generated modules over finite-dimensional algebras, see Example VI.3.9(c).

Exercises of Section IV.2

1. Let f be a morphism in a K-category C. Prove that the following conditions are equivalent:
   (a) f is an isomorphism.
   (b) f is a retraction and a monomorphism.
   (c) f is a section and an epimorphism.

2. Let C be a K-category and f, g morphisms in C such that gf exists. Prove that:

(a) If g, f are sections, then so is gf.

(b) If g, f are retractions, then so is gf.

(c) If gf is a section, then so is f.

(d) If gf is a retraction, then so is g.

(e) If f is a section, then it is a kernel.

(f) If g is a retraction, then it is a cokernel.

3. Let $f : X \to Y$ be a morphism in a K-category. Prove that:

(a) f is a section if and only if any morphism of source X factors though f.

(b) f is a retraction if and only if any morphism of target Y factors through f.

4. Let $F : C \to D$ be a linear functor between K-categories. Prove that, if $f \in C(X, Y)$ is a section, or a retraction, or an isomorphism, then so is Ff.

5. This exercise gives another proof of the fact that, if M is finitely generated, then $\mathrm{Hom}_A(M, -)$ commutes with arbitrary coproducts, see Example IV.2.8(b). Let A be an algebra.

(a) Let $p : L \to M$ be an epimorphism of A-modules. Prove that, if $\mathrm{Hom}_A(L, -)$ commutes with arbitrary coproducts, then so does $\mathrm{Hom}_A(M, -)$.

(b) Prove that, for any integer $m > 0$, the functor $\mathrm{Hom}_A(A^{(m)}, -)$ commutes with arbitrary coproducts.

(c) Deduce that, if M is finitely generated, then $\mathrm{Hom}_A(M, -)$ commutes with arbitrary coproducts.

6. Let C be a K-category. Prove that the inverse bijections of Yoneda's Lemma III.2.11 are isomorphisms of K-modules.

7. Let C be a category.

(a) Assume that C contains a final object F. Prove that F is unique up to isomorphism.

(b) Same question if C contains an initial object I.

8. Prove that in the category whose objects are the rings, and morphisms are the ring homomorphisms, with the usual composition of maps, $\mathbb{Z}$ is an initial object and the zero ring a final object. Thus, this category has no zero object and is not abelian.

9. Let A be an algebra. Prove that any epimorphism $M_A \to A_A$, with M an A-module, is a retraction, but there exist monomorphisms $A_A \to M_A$ which are not sections.

10. Prove that a divisible subgroup of an abelian group is pure in it, in the sense of Exercise II.4.28.

11. Let $F : C \to D$ be a linear functor between linear categories. Prove that, for any objects X, Y in C, we have $F(X \oplus Y) \cong FX \oplus FY$. What are the inverse isomorphisms?

12. Let I, J be right ideals of an algebra A such that $I + J = A$. Prove that $I \oplus J \cong A \oplus (I \cap J)$.

13. Let A be an algebra, M an A-module and N a submodule such that $M/N \cong A_A$. Prove that there exists $x \in M$ such that $M = N \oplus xA$.

14. (a) Let A be an algebra. Assume that we have a commutative diagram with exact rows in Mod$A$

$$
\begin{array}{ccccccccc}
0 & \longrightarrow & L & \longrightarrow & M & \longrightarrow & N & \longrightarrow & 0 \\
& & \downarrow{\scriptstyle f} & & \downarrow{\scriptstyle g} & & \downarrow{\scriptstyle h} & & \\
0 & \longrightarrow & L' & \longrightarrow & M' & \longrightarrow & N' & \longrightarrow & 0
\end{array}
$$

Prove that, if f is a retraction and g is a section, then h is a section.
(b) State and prove the dual of (a).

15. Let C be a linear category, X, Y, X', Y' objects in C and a morphism $u: X \oplus Y \to X' \oplus Y'$ of the form

$$
u = \begin{pmatrix} f & g \\ 0 & h \end{pmatrix}
$$

with $f: X \to X', g: Y \to X', h: Y \to Y'$. Prove that:
(a) If u is an isomorphism, then f is a section and h is a retraction.
(b) If u is an isomorphism and f or h is an isomorphism, then so is the other.
(c) If f, h are isomorphisms, then so is u.

16. Let $f: X \to Y$ be a morphism in a K-category C. Prove that:
(a) (U, u) is a kernel of f if and only if, for any object V in C, we have an exact sequence of K-modules

$$
0 \to C(V, U) \xrightarrow{C(V,u)} C(V, X) \xrightarrow{C(V,f)} C(V, Y).
$$

(b) (U, u) is a cokernel of f if and only if, for any object V in C, we have an exact sequence of K-modules

$$
0 \to C(U, V) \xrightarrow{C(u,V)} C(Y, V) \xrightarrow{C(f,V)} C(X, V).
$$

17. Let C be a K-category.
(a) Assume that any morphism in C has a kernel. Let $g : Y \to Z$ and $g' : Y' \to Z'$ be morphisms with respective kernels (X, f) and (X', f'). Assume that there exist morphisms $v : Y \to Y', w : Z \to Z'$ such that $wg = g'v$. Prove that there exists a unique morphism $u : X \to X'$ such that $f'u = vf$, then prove that u is functorial.
(b) State and prove the dual of (a).

18. Let C be a K-category in which any morphism has a kernel and a cokernel, and $f: X \to Y$ a morphism in C. Prove that:
(a) If g is a monomorphism and gf exists, then Ker (gf) = Ker f, and Im (gf) = Im f.
(b) If h is an epimorphism and fh exists, then Coker (fh) = Coker(f) and Coim (fh) = Coim f.

19. We claim that the morphism $\bar{f}$ arising from the canonical factorisation is functorial. Let C be a K-category in which any morphism has a kernel and a cokernel, $f: X \to Y, f': X' \to Y'$ morphisms in C, and $u : X \to X', v : Y \to Y'$ morphisms such that $vf = f'u$. Prove that there exists morphisms $u': \operatorname{Coim} f \to \operatorname{Coim} f'$ and $v': \operatorname{Im} f \to \operatorname{Im} f'$ such that $\bar{f}'u' = v'\bar{f}$.

20. Let g, f be morphisms in an abelian category such that gf exists.
    (a) Assume that f is an epimorphism having same kernel as gf. Prove that g is a monomorphism.
    (b) State and prove the dual of (a).

21. Let C be the morphism category of $\operatorname{Mod}A$, see Exercise III.2.7. Prove that:
    (a) C is an abelian category.
    (b) Ker and Coker define functors $C \to \operatorname{Mod}A$.

22. Let $X \xrightarrow{f} Y \xrightarrow{g} Z$ be a sequence in an abelian category. Prove that $gf = 0$ if and only if there exists a monomorphism $h: \operatorname{Im} f \to \operatorname{Ker} g$ such that $kh = j$, where $j: \operatorname{Im} f \to Y$ and $k: \operatorname{Ker} g \to Y$ are the injections.

23. Let $0 \to L \xrightarrow{f} M \xrightarrow{g} N \to 0$ be a short exact sequence in $\operatorname{Mod}A$. Prove that:
    (a) $f' : M \to L$ satisfies $f'f = 1_L$ if and only if $M = \operatorname{Ker} f' \oplus \operatorname{Im} f$.
    (b) $g' : N \to M$ satisfies $gg' = 1_N$ if and only if $M = \operatorname{Im} g' \oplus \operatorname{Ker} g$.

24. Let A be an algebra. Prove that a short exact sequence $0 \to L \xrightarrow{f} M \xrightarrow{g} N \to 0$ in $\operatorname{Mod}A$ splits if and only if there exist $g': N \to M$ and $f': M \to L$ such that $ff' + g'g = 1_M$.

25. Prove that the short exact sequence of representations of Example II.4.16(d) splits.

26. With the notation of Exercise III.3.7, construct a monomorphism $j : P \hookrightarrow I \oplus I'$ and prove that no such monomorphism is a section. What is Coker j?

27. Prove that the full subcategory of $\mathcal{A}b$ consisting of the finite abelian groups is abelian.

28. Let p be a prime integer. An abelian group G is a **p-group** if, for any $x \in G$, there exists an integer $n > 0$ such that $p^n x = 0$. Prove that the full subcategory of $\mathcal{A}b$ consisting of the p-groups is abelian.

29. Let K be a field, Q a finite quiver, U, V, W representations of Q and $f : U \to V$, $g : V \to W$ morphisms.
    (a) Prove that the sequence $0 \to U \xrightarrow{f} V \xrightarrow{g} W \to 0$ is short exact if and only if, for any $x \in Q_0$, the component sequence $0 \to U(x) \xrightarrow{f_x} V(x) \xrightarrow{g_x} W(x) \to 0$ is short exact.
    (b) Give an example showing that even though each of the component sequences $0 \to U(x) \xrightarrow{f_x} V(x) \xrightarrow{g_x} W(x) \to 0$ splits (because it is a sequence of vector spaces), the sequence $0 \to U \xrightarrow{f} V \xrightarrow{g} W \to 0$ does not necessarily split.

IV.3 Fibered products and amalgamated sums

IV.3.1 Fibered products

So far, we used universal properties to give alternative descriptions of known objects, like products, modules with bases or kernels. We now use them to define new objects, fibered products and amalgamated sums. The concepts and names come from algebraic topology. In module theory, these constructions can be used to transform an exact sequence into another. In this section, K denotes a commutative ring and C a K-linear category.

IV.3.1 Definition. Let $f_1: X_1 \to X, f_2: X_2 \to X$ be morphisms in C with the same target X. A ***fibered product*** of f_1 and f_2 is a triple (P, p_1, p_2), where P is an object in C, and $p_1: P \to X_1, p_2: P \to X_2$ morphisms in C, such that:

(a) $f_1 p_1 = f_2 p_2$.

(b) For any triple (Y, g_1, g_2), where Y is an object and $g_1: Y \to X_1, g_2: Y \to X_2$ morphisms in C such that $f_1 g_1 = f_2 g_2$, there exists a unique morphism $u: Y \to P$ such that $p_1 u = g_1$ and $p_2 u = g_2$.

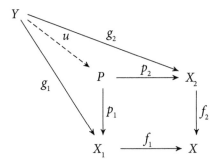

The commutative square formed by p_1, p_2, f_1, f_2 is called ***cartesian square***. Uniqueness of u means that, if $p_1 u = p_1 u'$ and $p_2 u = p_2 u'$, then $u = u'$. Universality implies that, if a fibered product of f_1 and f_2 exists, then it is unique up to isomorphism. A fibered product of f_1, f_2 is also called ***pullback*** of this pair. We prefer to use the term 'fibered product' because it evokes the topological origin of the concept, and also, because the universal property it satisfies resembles that of the product, one indeed thinks of p_1, p_2 as 'projections' on coordinate spaces. Finally, as we shall see, the fibered product is a subobject of the product, hence this term hints at its actual construction.

IV.3.2 Theorem. *Assume that C is a linear category in which any morphism has a kernel. Any pair of morphisms with same target has a fibered product, unique up to isomorphism.*

Proof. It suffices to prove existence. Let $f_1: X_1 \to X, f_2: X_2 \to X$ be morphisms in C. Denote by $\pi_1: X_1 \times X_2 \to X_1, \pi_2: X_1 \times X_2 \to X_2$ the projections associated to the product. In the matrix notation of Subsection III.3.2, we have $\pi_1 = (1\ 0)$ and

$\pi_2 = (0\ 1)$. We claim that the kernel (P, p) of $f_1\pi_1 - f_2\pi_2 = (f_1\ \ -f_2): X_1 \times X_2 \to X$ is the required fibered product.

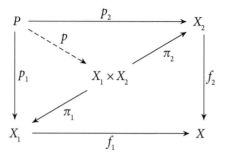

Set $p_1 = \pi_1 p$ and $p_2 = \pi_2 p$. Then $(f_1\pi_1 - f_2\pi_2)p = 0$ yields $f_1 p_1 = f_1\pi_1 p = f_2\pi_2 p = f_2 p_2$. Suppose (Y, g_1, g_2) is a triple as in the definition such that $f_1 g_1 = f_2 g_2$. The definition of product yields $v: Y \to X_1 \times X_2$ such that $\pi_1 v = g_1, \pi_2 v = g_2$. Hence, $f_1 g_1 = f_2 g_2$ implies $(f_1\pi_1 - f_2\pi_2)v = 0$. The definition of kernel yields $u: Y \to P$ such that $pu = v$. Therefore, $p_1 u = \pi_1 pu = \pi_1 v = g_1$ and similarly $p_2 u = g_2$. This establishes our claim.

We prove uniqueness of u: if $p_1 u = p_1 u', p_2 u = p_2 u'$, then $p_1(u - u') = 0$ and $p_2(u - u') = 0$ give $\pi_1 p(u - u') = 0 = \pi_2 p(u - u')$. The universal property of product gives $p(u - u') = 0$. Because p is a monomorphism, we get $u - u' = 0$ and so $u = u'$. □

More generally, fibered products exist in a category in which any finite family of objects has a product, and any morphism has a kernel, as for instance, the category of groups, or that of K-algebras (which are not linear). The proof is easy to adapt.

The previous proof is constructive: it shows not only that fibered products exist in C, but also how to construct them. It says that, in an abelian category, the square

$$
\begin{array}{ccc}
P & \xrightarrow{p_2} & X_2 \\
{\scriptstyle p_1}\downarrow & & \downarrow{\scriptstyle f_2} \\
X_1 & \xrightarrow{f_1} & X
\end{array}
$$

is cartesian if and only if we have an exact sequence

$$
0 \longrightarrow P \xrightarrow{\binom{p_1}{p_2}} X_1 \times X_2 \xrightarrow{(f_1\ -f_2)} X.
$$

In particular, any pair of morphisms in an abelian category having the same target admits a fibered product.

For example, if $C = \mathrm{Mod}A$, then the previous construction gives

$$
P = \{(x_1, x_2) \in X_1 \times X_2 : f_1(x_1) = f_2(x_2)\}
$$

and p_1, p_2 are the restrictions to P of the projections of $X_1 \times X_2$ to X_1, X_2, respectively.

IV.3.3 Lemma. *In an abelian category, let*

$$
\begin{array}{ccc}
P & \xrightarrow{\ p_2\ } & X_2 \\
{\scriptstyle p_1}\downarrow & & \downarrow{\scriptstyle f_2} \\
X_1 & \xrightarrow{\ f_1\ } & X
\end{array}
$$

be a cartesian square. Then:
(a) f_1 *is a monomorphism if and only if so is* p_2.
(b) *If* f_1 *is an epimorphism, then so is* p_2.

Proof. (a) Let f_1 be a monomorphism and g satisfy $p_2 g = 0$. Then $f_1 p_1 g = f_2 p_2 g = 0$. Because f_1 is a monomorphism, $p_1 g = 0$. Uniqueness in the universal property yields $g = 0$. Therefore, p_2 is a monomorphism.

Conversely, if p_2 is a monomorphism and g such that $f_1 g = 0$, the universal property yields a unique u satisfying $g = p_1 u$ and $p_2 u = 0$. But p_2 is a monomorphism. Hence, $u = 0$, so $g = 0$ and f is a monomorphism.

(b) Because $f_1 = (f_1 \ {-}f_2)\begin{pmatrix} 1 \\ 0 \end{pmatrix}$ is an epimorphism, so is $(f_1 \ {-}f_2)$. We thus have a short exact sequence

$$
0 \to P \xrightarrow{\begin{pmatrix} p_1 \\ p_2 \end{pmatrix}} X_1 \times X_2 \xrightarrow{(f_1 \ {-}f_2)} X \to 0.
$$

Let $w\colon X_2 \to W$ be such that $w p_2 = 0$. Then $(0 \ w) : X_1 \times X_2 \to W$ satisfies $(0 \ w)\begin{pmatrix} p_1 \\ p_2 \end{pmatrix} = 0$. Now X is the cokernel of $\begin{pmatrix} p_1 \\ p_2 \end{pmatrix}$. Hence, there exists $v\colon X \to W$ such that $v(f_1 \ {-}f_2) = (0 \ w)$. Therefore, $v f_1 = 0$ implies $v = 0$ because f_1 is an epimorphism. Consequently, $w = -v f_2 = 0$. $\qquad\square$

IV.3.4 Theorem. *In an abelian category, a square*

$$
\begin{array}{ccc}
P & \xrightarrow{\ p_2\ } & X_2 \\
{\scriptstyle p_1}\downarrow & & \downarrow{\scriptstyle f_2} \\
X_1 & \xrightarrow{\ f_1\ } & X
\end{array}
$$

is cartesian with f_1 *an epimorphism if and only if there exists a commutative diagram with exact rows*

$$
\begin{array}{ccccccccc}
0 & \longrightarrow & U & \xrightarrow{\ p_0\ } & P & \xrightarrow{\ p_2\ } & X_2 & \longrightarrow & 0 \\
& & {\scriptstyle 1_U}\downarrow & & {\scriptstyle p_1}\downarrow & & \downarrow{\scriptstyle f_2} & & \\
0 & \longrightarrow & U & \xrightarrow{\ f_0\ } & X_1 & \xrightarrow{\ f_1\ } & X & \longrightarrow & 0
\end{array}
$$

Proof. *Necessity.* Assume that the given square is cartesian with f_1 an epimorphism of kernel (U, f_0). The universal property of fibered product applied to $f_0 \colon U \to X_1$ and $p_2 \colon U \to X_2$ yields $p_0 \colon U \to P$ such that $p_2 p_0 = 0$ and $p_1 p_0 = f_0$. Now, f_0 is a monomorphism (because it is a kernel). Hence so is p_0. We claim that $p_0 = \ker p_2$. We know that $p_2 p_0 = 0$, so let g be such that $p_2 g = 0$. Then $f_1 p_1 g = f_2 p_2 g = 0$. Because $f_0 = \ker f_1$, there exists h such that $p_1 g = f_0 h$. But then $p_1 p_0 h = f_0 h = p_1 g$ while $p_2 p_0 h = 0 = p_2 g$, and uniqueness in the universal property yield $g = p_0 h$. Therefore, g factors through p_0. This factorisation is unique because p_0 is a monomorphism. Due to Lemma IV.3.3, p_2 is an epimorphism, hence the upper sequence is exact.

Sufficiency. We start with the commutative diagram with exact rows and assume that (P', p_1', p_2') is the fibered product of f_1, f_2. We claim that (P', p_1', p_2') is isomorphic to (P, p_1, p_2). Because of the necessity part above, there exists a commutative diagram with exact rows

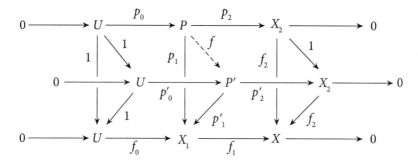

The universal property of the fibered product P' gives a unique $f \colon P \to P'$ such that $p_2' f = p_2$ and $p_1' f = p_1$. Then $p_1' f p_0 = p_1 p_0 = f_0 = p_1' p_0'$ and $p_2' f p_0 = p_2 p_0 = 0 = p_2' p_0'$. Uniqueness in the universal property yields $f p_0 = p_0'$. Finally, Lemma IV.2.29 yields that f is an isomorphism, as required.   □

Thus, given a morphism from X_2 to X, any short exact sequence $0 \to U \to X_1 \to X \to 0$, lifts to another short exact sequence having the same kernel.

IV.3.5 Examples. Let C be an abelian category.

(a) Let X_1, X_2 be objects in C and $\pi_1 \colon X_1 \times X_2 \to X_1, \pi_2 \colon X_1 \times X_2 \to X_2$ the projections associated to the product. The commutative square

$$\begin{array}{ccc} X_1 \times X_2 & \xrightarrow{\ \pi_2\ } & X_2 \\ \downarrow{\scriptstyle \pi_1} & & \downarrow \\ X_1 & \longrightarrow & 0 \end{array}$$

shows that $(X_1 \times X_2, \pi_1, \pi_2)$ is the fibered product of the zero morphisms $X_1 \to 0$ and $X_2 \to 0$, so finite products occur as fibered products.

(b) For any morphism $f \colon X \to Y$ in C, the square

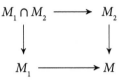

is cartesian if and only if (P, p) is a kernel of f, so kernels occur as fibered products.

(c) Let $C = \mathrm{Mod}A$ and M_1, M_2 submodules of a module M. Then the square

$$
\begin{array}{ccc}
M_1 \cap M_2 & \longrightarrow & M_2 \\
\downarrow & & \downarrow \\
M_1 & \longrightarrow & M
\end{array}
$$

where morphisms are inclusions, is cartesian, so intersections occur as fibered products.

IV.3.2 Amalgamated sums

We dualise our work of Subsection IV.3.1. Again C denotes a K-linear category.

IV.3.6 Definition. Let $f_1: X \to X_1, f_2: X \to X_2$ be morphisms in C having the same source. An *amalgamated sum* of f_1 and f_2 is a triple (Q, q_1, q_2), where Q is an object in C, and $q_1: X_1 \to Q, p_2: X_2 \to Q$ morphisms in C such that:

(a) $q_1 f_1 = q_2 f_2$.

(b) For any triple (Y, g_1, g_2), where Y is an object and $g_1: X_1 \to Y, g_2: X_2 \to Y$ morphisms in C such that $g_1 f_1 = g_2 f_2$, there exists a unique morphism $u: Q \to Y$ such that $u q_1 = g_1$ and $u q_2 = g_2$.

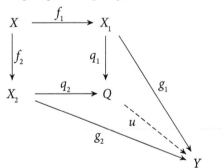

The commutative square formed by q_1, q_2, f_1, f_2 is called a *cocartesian square*. Uniqueness of u means that, if $u q_1 = u' q_1$ and $u q_2 = u' q_2$, then $u = u'$. Universality implies that, if an amalgamated sum of f_1 and f_2 exists, then it is unique up to isomorphism. The amalgamated sum of f_1, f_2 is also called their *pushout*. Again, we prefer to use 'amalgamated sum' because of the topological origin of the concept, and because its universal property resembles that of the coproduct. One thinks of q_1, q_2 as 'injections' from the coordinate spaces. Also, amalgamated sum is a quotient of the direct sum, so this term hints at its construction.

IV.3.7 Theorem. *Assume that C is a linear category in which any morphism has a cokernel. Any pair of morphisms with same source has an amalgamated sum, unique up to isomorphism.*

Proof. It suffices to prove existence. Let $f_1: X \to X_1, f_2: X \to X_2$ be morphisms in C. Denote by $\iota_1 = \begin{pmatrix} 1 \\ 0 \end{pmatrix}: X_1 \to X_1 \oplus X_2, \iota_2 = \begin{pmatrix} 0 \\ 1 \end{pmatrix}: X_2 \to X_1 \oplus X_2$ the injections associated to the direct sum. The proof dual to that of Theorem IV.3.2 shows that Q is the cokernel of $\iota_1 f_1 - \iota_2 f_2 = \begin{pmatrix} f_1 \\ -f_2 \end{pmatrix}: X \to X_1 \oplus X_2$. $\qquad\square$

Amalgamated sums exist in any category having finite coproducts and cokernels, such as, for instance, the category of groups, or that of K-algebras, in which the coproduct is usually called free product). The proof is also easy to adapt.

This proof shows that in an abelian category the square

$$
\begin{array}{ccc}
X & \xrightarrow{\ f_1\ } & X_1 \\
{\scriptstyle f_2}\big\downarrow & & \big\downarrow{\scriptstyle q_1} \\
X_2 & \xrightarrow{\ q_2\ } & Q
\end{array}
$$

is cocartesian if and only if the sequence

$$
X \xrightarrow{\begin{pmatrix} f_1 \\ -f_2 \end{pmatrix}} X_1 \oplus X_2 \xrightarrow{(q_1\ q_2)} Q \longrightarrow 0
$$

is exact. In particular, any pair of morphisms in an abelain category having the same source admits an amalgamated sum. In Mod$A$, the construction of the theorem gives

$$
Q = \frac{X_1 \oplus X_2}{X'} \quad \text{where } X' = \left\{ \begin{pmatrix} f_1(x) \\ -f_2(x) \end{pmatrix} \in X_1 \oplus X_2 : x \in X \right\}.
$$

Denoting by $p : X_1 \oplus X_2 \to \frac{X_1 \oplus X_2}{X'}$ the projection, we have $q_1 = p\iota_1$ and $q_2 = p\iota_2$, where ι_1 and ι_2 are as in the proof.

IV.3.8 Lemma. *Let, in an abelian category,*

$$
\begin{array}{ccc}
X & \xrightarrow{\ f_1\ } & X_1 \\
{\scriptstyle f_2}\big\downarrow & & \big\downarrow{\scriptstyle q_1} \\
X_2 & \xrightarrow{\ q_2\ } & Q
\end{array}
$$

be a cocartesian square. Then
(a) f_1 is an epimorphism if and only if so is q_2.
(b) If f_1 is a monomorphism, then so is q_2.

Proof. Dual to that of Lemma IV.3.3 and left to the reader. $\qquad\square$

IV.3.9 Theorem. *In an abelian category, a square*

$$
\begin{array}{ccc}
X & \xrightarrow{\ f_1\ } & X_1 \\
{\scriptstyle f_2}\big\downarrow & & \big\downarrow{\scriptstyle q_1} \\
X_2 & \xrightarrow{\ q_2\ } & Q
\end{array}
$$

is cocartesian with f_1 a monomorphism if and only if there exists a commutative diagram with exact rows

$$
\begin{array}{ccccccccc}
0 & \longrightarrow & X & \xrightarrow{\ f_1\ } & X_1 & \xrightarrow{\ f_0\ } & V & \longrightarrow & 0 \\
 & & {\scriptstyle f_2}\big\downarrow & & \big\downarrow{\scriptstyle q_1} & & \big\downarrow{\scriptstyle 1_V} & & \\
0 & \longrightarrow & X_2 & \xrightarrow{\ q_2\ } & Q & \xrightarrow{\ q_0\ } & V & \longrightarrow & 0
\end{array}
$$

Proof. Dual to that of Theorem IV.3.4 and left to the reader. □

So, given a morphism from X to X_2, any short exact sequence $0 \to X \to X_1 \to V \to 0$ can be lowered to a new short exact sequence with the same cokernel.

By now, the reader must be convinced of the extraordinary plasticity of categorical concepts. We add to the pile by observing that a square may be both cartesian and cocartesian. It is then called **bicartesian**. Bicartesian squares combine both universal properties, that of fibered product and that of amalgamated sum.

IV.3.10 Lemma. *In an abelian category, a square*

$$
\begin{array}{ccc}
X & \xrightarrow{\ f_1\ } & X_1 \\
{\scriptstyle f_2}\big\downarrow & & \big\downarrow{\scriptstyle g_1} \\
X_2 & \xrightarrow{\ g_2\ } & Y
\end{array}
$$

is bicartesian if and only if we have a short exact sequence

$$
0 \to X \xrightarrow{\ \binom{f_1}{f_2}\ } X_1 \oplus X_2 \xrightarrow{\ (g_1\, -g_2)\ } Y \to 0
$$

Proof. The square is cartesian if and only if the sequence

$$
0 \to X \xrightarrow{\ \binom{f_1}{f_2}\ } X_1 \oplus X_2 \xrightarrow{\ (g_1\, -g_2)\ } Y
$$

is exact, and cocartesian if and only if the sequence

$$X \xrightarrow{\binom{f_1}{f_2}} X_1 \oplus X_2 \xrightarrow{(g_1 \ -g_2)} Y \to 0$$

is exact. The result follows.   ☐

The apparent asymmetry in the exact sequence disappears when one realises that the minus sign can be put in front of any of the morphisms $\{f_1, f_2, g_1, g_2\}$. It is only needed in order to ensure commutativity of the square shown in the statement.

IV.3.11 Corollary. *In an abelian category, assume that we have a square*

$$\begin{array}{ccc} X & \xrightarrow{f_1} & X_1 \\ {\scriptstyle f_2}\downarrow & & \downarrow{\scriptstyle g_1} \\ X_2 & \xrightarrow{g_2} & Y \end{array}$$

(a) *If the square is cartesian, and g_2 is an epimorphism, then the square is bicartesian.*

(b) *If the square is cocartesian, and f_1 is a monomorphism, then the square is bicartesian.*

Proof. We prove (a), because (b) is dual. The hypothesis says that we have an exact sequence

$$0 \to X \xrightarrow{\binom{f_1}{-f_2}} X_1 \oplus X_2 \xrightarrow{(g_1 \ g_2)} Y$$

and that $g_2 = (g_1 \ g_2)\binom{0}{1}$ is an epimorphism. But then $(g_1 \ g_2)$ is also an epimorphism. Applying Lemma IV.3.10 completes the proof.   ☐

For instance, in Theorem IV.3.4, the right-hand square of the commutative diagram with exact rows, and dually, in Theorem IV.3.9, the left-hand square of the commutative diagram with exact rows are bicartesian.

IV.3.12 Examples.

(a) Let X_1, X_2 be objects in C and $q_1 : X_1 \to X_1 \oplus X_2, q_2 : X_2 \to X_1 \oplus X_2$ the injections associated to the direct sum. The commutative square

$$\begin{array}{ccc} 0 & \xrightarrow{} & X_1 \\ \downarrow & & \downarrow{\scriptstyle q_1} \\ X_2 & \xrightarrow{q_2} & X_1 \oplus X_2 \end{array}$$

shows that $(X_1 \oplus X_2, q_1, q_2)$ is the amalgamated sum of the zero morphisms $0 \to X_1$ and $0 \to X_2$, so finite direct sums occur as amalgamated sums.

(b) For any morphism $f: X \to Y$ in C, the square

$$
\begin{array}{ccc}
X & \xrightarrow{\ f\ } & Y \\
\downarrow & & \downarrow{\scriptstyle q} \\
0 & \longrightarrow & Q
\end{array}
$$

is cocartesian if and only if (Q, q) is a cokernel of f, so cokernels occur as amalgamated sums.

(c) Let $C = \mathrm{Mod}A$ and M_1, M_2 submodules of a module M, then the square

$$
\begin{array}{ccc}
M_1 \cap M_2 & \longrightarrow & M_2 \\
\downarrow & & \downarrow \\
M_1 & \longrightarrow & M_1 + M_2
\end{array}
$$

where morphisms are inclusions, is bicartesian, so sums occur as amalgamated sums, and intersections as fibered products.

Exercises of Section IV.3

1. (a) Let $f, g: X \to Y$ be morphisms in a linear category. Prove that the square

$$
\begin{array}{ccc}
P & \xrightarrow{\ h\ } & X \\
\downarrow{\scriptstyle h} & & \downarrow{\scriptstyle 1 \times f} \\
X & \xrightarrow{\ 1 \times g\ } & X \times Y
\end{array}
$$

is cartesian if and only if (P, h) is a kernel of $f - g$.

(b) State and prove the dual of (a).

2. (a) Let $f: X \to Y$ be a morphism in a linear category. Prove that the square

$$
\begin{array}{ccc}
X & \xrightarrow{\ 1\ } & X \\
\downarrow{\scriptstyle 1} & & \downarrow{\scriptstyle f} \\
X & \xrightarrow{\ f\ } & Y
\end{array}
$$

is cartesian if and only if f is a monomorphism.

(b) State and prove the dual of (a).

3. (a) Let A be an algebra. Assume that we have a cartesian square in $\mathrm{Mod}A$

$$
\begin{array}{ccc}
P & \xrightarrow{\ p_1\ } & M_1 \\
\downarrow{\scriptstyle p_2} & & \downarrow{\scriptstyle f_1} \\
M_2 & \xrightarrow{\ f_2\ } & M
\end{array}
$$

with f_1, f_2 monomorphisms. Prove that we may identify P to $M_1 \cap M_2$ and that there exists an amalgamated sum

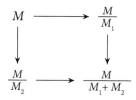

where all morphisms are projections.
(b) State and prove the dual of (a).

4. Let A be an algebra.
   (a) Assume that we have a commutative diagram in Mod$A$.

$$
\begin{array}{ccccc}
L & \xrightarrow{f} & M & \xrightarrow{g} & N \\
\downarrow{u} & & \downarrow{v} & & \downarrow{w} \\
L' & \xrightarrow{f'} & M' & \xrightarrow{g'} & N'
\end{array}
$$

   Prove that:
   (i) If the right-hand square and the left-hand square are cartesian, then the big square is cartesian.
   (ii) If the big square is cartesian and g is a monomorphism, then the left-hand square is cartesian.
   (b) State and prove the dual of (a).

5. Consider the commutative diagram in $\mathcal{A}b$

$$
\begin{array}{ccc}
\mathbb{Z} & \xrightarrow{\mu_n} & \mathbb{Z} \\
\downarrow{\mu_m} & & \downarrow{\mu_m} \\
\mathbb{Z} & \xrightarrow{\mu_n} & \mathbb{Z}
\end{array}
$$

where μ_m, μ_n denote multiplication by the positive integers m, n, respectively. Find a necessary and sufficient condition in order that this square be:
   (a) cartesian.
   (b) cocartesian.

6. Let p, q be distinct prime integers. Find in $\mathcal{A}b$ the amalgamated sum of the projections $\mathbb{Z} \to \mathbb{Z}/p\mathbb{Z}$ and $\mathbb{Z} \to \mathbb{Z}/q\mathbb{Z}$.

7. (a) Prove that a K-category admits fibered products if and only if it admits finite products and kernels.
   (b) State and prove the dual of (a).

8. Let A be an algebra and

$$P \xrightarrow{\ p_2\ } M_2$$

$$p_1 \downarrow \qquad\qquad \downarrow f_2$$

$$M_1 \xrightarrow{\ f_1\ } M$$

a cartesian square in ModA. If f_1 is the kernel of a morphism $f\colon M \to N$, prove that p_2 is the kernel of ff_2.

(b) State and prove the dual of (a).

9. (a) Let A be an algebra. Consider the following commutative diagram with exact rows in ModA

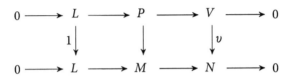

Prove that v factors through M if and only if the upper sequence is split.

(b) State and prove the dual of (a).

IV.4 Equivalences and dualities of categories

IV.4.1 Equivalences

The notion of equivalence of categories corresponds to the familiar notion of isomorphism between objects. There exists an obvious generalisation of the latter: the categories C, D are called **isomorphic** if there exist functors $F\colon C \to D, G\colon D \to C$ such that $GF = 1_C$ and $FG = 1_D$. In this case, G is uniquely determined by F and called its **inverse**. However, this notion is much too restrictive for our needs. Indeed, inside a category, two isomorphic objects should be identified for all practical purposes, while, in isomorphic categories, distinct objects of the first category map to distinct objects of the second, even if they are isomorphic, see Exercise IV.4.2. It is indeed asking too much, for two mathematical constructions, performed after one another (in either order) to give exactly the same object we started from. An example will explain better this situation: a **skeleton** of a category C is a full subcategory S of C whose object class contains one and only one object from each isomorphism class of objects of C. For instance, let K be a field and $C = \mathrm{mod}K$ the category of finite-dimensional K-vector spaces, a skeleton of C is given by the spaces of the form K^d, with $d \geq 0$ an integer. If we agree to work up to isomorphism, then a category C and one of its skeletons should have the same properties. But they are not isomorphic as categories, because an isomorphism would induce a bijection between the object classes of S and C, and this bijection does not exist in general (for instance, in $\mathrm{mod}K$, both K^d and the subspace of $K[t]$ consisting of polynomials of degree less than d correspond to a single object in the skeleton). This leads us to relax the definition of isomorphism.

IV.4.1 Definition. The categories C, D are *equivalent* if there exist functors $F: C \rightarrow D, G: D \rightarrow C$ and functorial isomorphisms $GF \cong 1_C$ and $FG \cong 1_D$. The functors F, G are then *quasi-inverse equivalences*, and we write $C \cong D$.

Isomorphic categories are equivalent, but the converse is generally not true. If the functor F, or the functor G, of Definition IV.4.1 is given, then the other is only unique up to a functorial isomorphism. Indeed, if $GF \cong 1_C$ and $FG' \cong 1_D$, then

$$G = G1_D \cong G(FG') = (GF)G' \cong 1_C G' = G'.$$

For this reason, G is called a quasi-inverse (and not an inverse) of F. It is immediate that the relation $\cong$ between categories is reflexive, symmetric and transitive.

Our objective is to give a criterion allowing to verify whether a given functor is an equivalence or not. We start by observing that equivalences should preserve morphisms.

IV.4.2 Definition. Let $F: C \rightarrow D$ be a functor.
(a) F is *faithful* if, for any objects X, Y in C, the map $C(X, Y) \rightarrow D(FX, FY)$ given by $f \mapsto Ff$, for $f \in C(X, Y)$, is injective.
(b) F is *full* if, for any objects X, Y in C, the map $C(X, Y) \rightarrow D(FX, FY)$ given by $f \mapsto Ff$, for $f \in C(X, Y)$, is surjective.
(c) F is *full and faithful* if, for any objects X, Y in C, the map $C(X, Y) \rightarrow D(FX, FY)$ given by $f \mapsto Ff$, for $f \in C(X, Y)$, is bijective.

Clearly, the composite of two faithful, or full, functors is faithful, or full, respectively. A functor that is full and faithful is sometimes called *fully faithful*, or an *embedding*.

IV.4.3 Examples.
(a) If C is a subcategory of D, then the inclusion functor $C \rightarrow D$ is always faithful. It is full if and only if C is a full subcategory of D. The identity functor on C is full and faithful.
(b) Let C be a small category, so that $\text{Fun}(C, Sets)$ is a category. Corollary III.2.12(a) can be reformulated to say that the functor $C^{op} \rightarrow \text{Fun}(C, Sets)$ given by $X \mapsto C(X, -)$, for an object X and $f \mapsto C(f, -)$, for a morphism f, is full and faithful.

Similarly, the functor $C \rightarrow \text{Fun}(C^{op}, Sets)$ given by $X \mapsto C(-, X)$, for an object X and $f \mapsto C(-, f)$, for a morphism f, is full and faithful.

These two functors are often called the *Yoneda embeddings*.
(c) A category C is concrete if and only if there exists a faithful functor from C to $Sets$. This is a forgetful functor.

We want to be able to include in the image of an equivalence, together with any object, all objects which are isomorphic to it. We start by characterising the latter.

IV.4.4 Definition. Let $F: C \rightarrow D$ be a functor. Its *essential image* is the full subcategory of D having as objects those U in D such that there exists an object X in C

with $U \cong FX$. A functor $F: C \to D$ is **essentially surjective**, or **dense**, if its essential image equals D.

We are now able to state and prove the wanted criterion.

IV.4.5 Theorem. *Let C, D be categories. A functor $F: C \to D$ is an equivalence if and only if it is full, faithful and essentially surjective.*

Proof. Necessity. Let G be a quasi-inverse for F. For any object U in D, we have $U \cong FGU$, with GU an object in C. Therefore, F is essentially surjective.

Let $f, g: X \to Y$ be morphisms in C such that $Ff = Fg$. The functorial isomorphism $\varphi: GF \to 1_C$ gives a commutative square

$$\begin{array}{ccc} GFX & \xrightarrow{\ \varphi_X \cong\ } & X \\ {\scriptstyle GF f}\downarrow & & \downarrow{\scriptstyle f} \\ GFY & \xrightarrow{\ \varphi_Y \cong\ } & Y \end{array}$$

so $f = \varphi_Y(G Ff)\varphi_X^{-1}$. Similarly, $g = \varphi_Y(G Fg)\varphi_X^{-1}$. Hence $f = g$ and so, F is faithful. Similarly, G is also faithful. Let $u: FX \to FY$ be a morphism in D and set $f = \varphi_Y(Gu)\varphi_X^{-1}$. Then $GFf = \varphi_Y^{-1}f\varphi_X = Gu$. But G is faithful, so $u = Ff$, that is, F is full. This completes the proof of necessity.

Sufficiency. In order to prove that F is an equivalence, we construct a quasi-inverse $G: D \to C$. We define G on objects. Let U be an object in D. Because F is essentially surjective, there exists an object X in C, not necessarily unique, such that $U \cong FX$. We fix one such X and an isomorphism $\psi_U: U \to FX$. We set $GU = X$. We next define G on morphisms in such a way that the family $\psi = (\psi_U)_{U \in D_0}$ becomes a functorial morphism. Let $u: U \to V$ be a morphism in D. Assume that $X = GU, Y = GV$. Then we should have a commutative diagram

$$\begin{array}{ccc} U & \xrightarrow{\ \psi_U \cong\ } & FX = FGU \\ {\scriptstyle u}\downarrow & & \downarrow{\scriptstyle F Gu} \\ V & \xrightarrow{\ \psi_V \cong\ } & FY = FGV \end{array}$$

that is, $FGu = \psi_V u\psi_U^{-1}$. Because F is full and faithful, there exists a unique $f: X \to Y$ such that $Ff = \psi_V u\psi_X^{-1}$. We set $Gu = f$. This completes the definition of G.

We must prove that G is a functor. The previous construction shows that, for any object U in D, we have $G(1_U) = 1_{GU}$. So it suffices to prove that, for any pair of morphisms $u: U \to V, v: V \to W$ in D, we have $G(vu) = GvGu$. Because F is faithful, this is equivalent to proving that $FG(vu) = F(Gv \cdot Gu) = FGv \cdot FGu$, and the latter equality follows from the commutative diagram

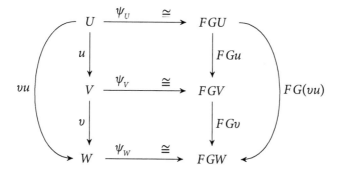

To any object X in C corresponds an isomorphism $\psi_{FX}: FX \to FGFX$. Hence, $\psi_{FX}^{-1} \in D(FGFX, FX) \cong C(GFX, X)$, where the last bijection expresses that F is full and faithful. Therefore, there exists a unique morphism $\varphi_X: GFX \to X$ such that $F\varphi_X = \psi_X^{-1}$. Moreover, a morphism $f: X \to Y$ in C induces a commutative diagram

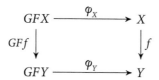

Now, F is faithful. Therefore, we have a commutative diagram

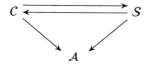

so that $(\varphi_X)_X$ is a functorial morphism $\varphi: GF \to 1_C$. Because $F\varphi = \psi^{-1}$, it is a functorial isomorphism, as required. $\qquad\square$

A consequence is that a full and faithful functor $F: C \to D$ induces an equivalence between C and its essential image, which is a full subcategory of D. Another is that, as expected, any category is equivalent to each of its skeletons. The latter remark implies an interesting fact. If a category C is not small itself, but one (and therefore any) of its skeletons S is small, then, for any category A, the functor category $\mathrm{Fun}(S, A)$ is defined. But then the equivalence between S and C implies that this functor category also describes functors from C to A.

$$C \rightleftarrows S$$
$$\searrow \quad \swarrow$$
$$A$$

A category equivalent to a small category is called an ***essentially small category***. This is the case, for instance, for the category of finite-dimensional vector spaces over a field K (a skeleton is given by the set $(K^d)_{d \geq 0}$). This is also the case for the category of finitely generated modules over a finite-dimensional algebra, see Exercise IV.4.4.

IV.4.6 Examples.

(a) Consider the change of scalars, introduced in Example II.2.4(e). Let K be a commutative ring, A a K-algebra, I an ideal of A and define a functor $\mathrm{Mod}(A/I) \to \mathrm{Mod}A$ as follows. Given an A/I-module M, we define an A-module structure on M by setting

$$xa = x(a + I) \quad \text{for } x \in M, a \in A.$$

With this definition, any A/I-linear map becomes A-linear, so we have indeed a functor. This is actually a change of scalars functor, see Example III.2.7(g). It induces, for any A/I-modules M, N, an isomorphism of K-modules

$$\mathrm{Hom}_{A/I}(M, N) \cong \mathrm{Hom}_A(M, N).$$

That is, it is full and faithful.

The essential image of this functor consists of those A-modules M on which one can define unambiguously a multiplication $M \times A/I \to M$ by setting

$$x(a + I) = xa \quad \text{for } x \in M, a \in A.$$

This is the case if and only if $xc = 0$, for any $x \in M, c \in I$, which we write as $MI = 0$. Thus, the essential image of the change of scalars functor is the full subcategory $\mathcal{A}(I)$ of $\mathrm{Mod}A$ consisting of those A-modules which are ***annihilated*** by I, that is, of the modules M such that $MI = 0$. Theorem IV.4.5 implies that we have a category equivalence

$$\mathrm{Mod}(A/I) \cong \mathcal{A}(I).$$

The subcategory $\mathcal{A}(I)$ coincides with $\mathrm{Mod}A$ if and only if $I = 0$. For, if $\mathcal{A}(I) = \mathrm{Mod}A$, then A_A would belong to $\mathcal{A}(I)$ and so $I = AI = 0$. The converse is obvious.

(b) Let K be a field. Consider the category $\mathcal{E}$ of pairs (E, f), with E a K-vector space and $f \colon E \to E$ an endomorphism, see Example III.2.2(i). We defined in Example III.2.7(e) a functor $\Phi \colon \mathcal{E} \to \mathrm{Mod}K[t]$ by setting, for a pair (E, f) in $\mathcal{E}$, $\Phi(E, f) =_f E$.

We claim that Φ is an equivalence. In order to prove it, we construct a quasi-inverse $\Psi \colon \mathrm{Mod}K[t] \to \mathcal{E}$. A $K[t]$-module M is, in particular, a K-vector space. Define $f \colon M \to M$ by $f(m) = t \cdot m$, for $m \in M$. Then f is a K-linear map so that (M, f) is an object in $\mathcal{E}$. Let $u \colon M \to N$ be a morphism of $K[t]$-modules and set $\Psi(M) = (M, f)$, $\Psi(N) = (N, g)$. For $m \in M$, we have $u(t \cdot m) = t \cdot u(m)$. Hence, $uf = gu$ and so u is a morphism in $\mathcal{E}$. We leave to the reader the (trivial) proof that Φ and Ψ are mutually quasi-inverse functors, so that $\mathcal{E} \cong \mathrm{Mod}K[t]$.

Because both functors restrict to the identity on the level of the vector spaces, the equivalence $\mathcal{E} \cong \mathrm{Mod}K[t]$ restricts to an equivalence between the full subcategory of $\mathcal{E}$ consisting of the pairs (E, f), with E finite-dimensional, and that of $\mathrm{Mod}K[t]$ consisting of the $K[t]$-modules which are finite-dimensional as K-vector spaces.

(c) Let K be a field, Q a finite quiver and I an admissible ideal of the path algebra KQ, see Example I.3.11(g). As seen there, the bound quiver algebra $A = KQ/I$ is finite-dimensional. We claim that $\mathrm{Mod}A$ is equivalent to a category $\mathrm{Rep}_K(Q, I)$ which we now define. A representation V of the quiver Q is called a **bound representation** (by the ideal I) if it vanishes when evaluated on I, that is, for any $\gamma \in I$, $\gamma = \sum_i \lambda_i w_i$, with the λ_i nonzero scalars and the w_i paths in Q, we have

$$V(\gamma) = \sum_i \lambda_i V(w_i) = 0.$$

If $\mathrm{Rep}_K Q$ denotes as usual the category of representations of Q, we let $\mathrm{Rep}_K(Q, I)$ be its full subcategory consisting the representations of Q bound by I. For instance, the representation V shown in Example II.2.4(g) is a representation of the quiver $Q : 3 \xrightarrow{\alpha} 2 \xrightarrow{\beta} 1$ bound by the ideal $\langle \alpha\beta \rangle$ generated by $\alpha\beta$, because it is proved there that $V(\alpha\beta) = 0$.

We construct quasi-inverse equivalences between $\mathrm{Mod}A$ and $\mathrm{Rep}_K(Q, I)$.

(i) A functor $\Phi : \mathrm{Rep}_K(Q, I) \to \mathrm{Mod}A$ is easy to construct. Indeed, there exists an inclusion functor $J : \mathrm{Rep}_K(Q, I) \to \mathrm{Rep}_K Q$ and a functor $F : \mathrm{Rep}_K Q \to \mathrm{Mod}KQ$, defined in Example III.2.7(f) (it is our familiar functor 'viewing a representation as a module'). We claim that the image of the composite $F \circ J : \mathrm{Rep}_K(Q, I) \to \mathrm{Mod}KQ$ lies in $\mathrm{Mod}A$.

Because of Example (a) above, $\mathrm{Mod}A$ is equivalent to the full subcategory of $\mathrm{Mod}KQ$ consisting of the KQ-modules annihilated by I. Let V be an object in $\mathrm{Rep}_K(Q, I)$ and $M = (F \circ J)(V) = FV$. Let $m \in M$ and $\gamma \in I$, $\gamma = \sum_i \lambda_i w_i$, with the λ_i nonzero scalars and the w_i paths in Q, then we have

$$m \cdot \gamma = m\left(\sum_i \lambda_i w_i\right) = \sum_i \lambda_i(mw_i) = \sum_i \lambda_i V(w_i)(m) = V(\gamma)(m) = 0$$

because V is bound by I. This establishes our claim. We may thus set $\Phi = F \circ J$.

(ii) We next construct a functor $\Psi : \mathrm{Mod}A \to \mathrm{Rep}_K(Q, I)$. To an A-module M, we associate a representation $V = \Psi(M)$ as follows. For $x \in Q_0$, let $e_x = \varepsilon_x + I$ be the residual class of the stationary path ε_x at x, and set $V(x) = Me_x$. Then $V(x)$ is a K-vector space. Let $\alpha : x \to y$ be an arrow and $\bar{\alpha} = \alpha + I$ its residual class. Define $V(\alpha) : V(x) \to V(y)$ by $V(\alpha)(m) = m\bar{\alpha}$, for $m \in V(x)$. Because M is an A-module, $V(\alpha)$ is a K-linear map. We prove that V is bound by I (hence lies in $\text{Rep}_K(Q, I)$) : for, let $\gamma \in I, \gamma = \sum_i \lambda_i w_i$, with the λ_i nonzero scalars and the w_i paths in Q, then $\bar{\gamma} = 0$, so

$$V(\gamma)(m) = \sum_i \lambda_i V(w_i)(m) = \sum_i \lambda_i (m\bar{w}_i) = m \left(\sum_i \lambda_i \bar{w}_i \right) = m\bar{\gamma} = m \cdot 0 = 0.$$

for $m \in V(x)$. This defines our functor on objects. Let $f : M \to N$ be an A-linear map between A-modules. We want to define a morphism of representations $\Psi f : \Psi M \to \Psi N$. Set $V = \Psi M, W = \Psi N$, let $x \in Q_0$ and $m = me_x \in Me_x = V(x)$. Then

$$f(m) = f(me_x) = f(m)e_x \in Ne_x = W(x)$$

so the restriction f_x of f to $V(x)$ is a linear map, mapping into $W(x)$. We set $\Psi(f) = (f_x)_{x \in Q_0}$. In order to prove that $f = (f_x)_{x \in Q_0}$ is a morphism of representations, we have to prove that, for any arrow $\alpha : x \to y$, there exists a commutative square

$$
\begin{array}{ccc}
V(x) & \xrightarrow{V(\alpha)} & V(y) \\
\downarrow{\scriptstyle f_x} & & \downarrow{\scriptstyle f_y} \\
W(x) & \xrightarrow{W(\alpha)} & W(y)
\end{array}
$$

Indeed, for $m \in V(x)$, we have

$$f_y V(\alpha)(m) = f_y(m\alpha) = f(m\alpha) = f(m)\alpha = f_x(m)\alpha = W(\alpha)f_x(m).$$

This completes the definition of the functor Ψ. Again we leave to the reader the verification that Φ and Ψ are mutually quasi-inverse.

This equivalence expresses the abstract structure of an A-module in terms which are reasonably concrete: vector spaces connected by linear maps. Here is an immediate application. We have computed in Examples III.3.4 and III.3.9 the product and the coproduct of a family of representations $(V_\lambda)_\lambda$. It follows from the theorem above that $\Phi(\prod_\lambda V_\lambda) \cong \prod_\lambda \Phi(V_\lambda)$ and $\Phi(\oplus_\lambda V_\lambda) \cong \oplus_\lambda \Phi(V_\lambda)$, that is, the module corresponding to a product of representations is isomorphic to the product of the modules corresponding to the representations, and similarly for the coproduct.

Let Q be a finite quiver. A representation $V = (V(x), V(\alpha))_{x \in Q_0, \alpha \in Q_1}$ is called a **finite-dimensional representation** if each $V(x)$ is a finite-dimensional K-vector space. If V is a finite-dimensional representation of Q bound by I, and $A = KQ/I$, then the A-module $\Phi V = \oplus_{x \in Q_0} V(x)$ is also finite-dimensional as K-vector space, because Q is finite. Conversely, if an A-module M is finite-dimensional as K-vector space, so is each Me_x (because the latter is a subspace of M). Hence, ΨM is a finite-dimensional representation. Therefore, the equivalence $\text{Rep}_K(Q, I) \cong \text{Mod}A$ restricts to an equivalence between the full subcategory $\text{rep}_K(Q, I)$ of $\text{Rep}_K(Q, I)$ of the finite-dimensional representations and the full subcategory of ModA consisting of the modules which are finite-dimensional as K-vector spaces. Because of Corollary III.4.9, the latter coincide with modA. That is, the equivalence $\text{Rep}_K(Q, I) \cong \text{Mod}A$ restricts to an equivalence $\text{rep}_K(Q, I) \cong$ modA. In pictures, we have a commutative square of categories and functors

$$
\begin{array}{ccc}
\text{Rep}_K(Q, I) & \xrightarrow{\;\cong\, \Phi\;} & \text{Mod}A \\
\uparrow & & \uparrow \\
\text{rep}_K(Q, I) & \xrightarrow{\;\cong\, \Phi_1\;} & \text{mod}A
\end{array}
$$

where the vertical functors are the inclusions and Φ_1 is the restriction of Φ.

For instance, let Q be the quiver

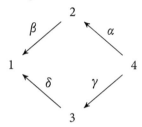

and $I = < \alpha\beta - \gamma\delta >$, that is, the elements of I are multiples of $\alpha\beta - \gamma\delta \in KQ$. The ideal I is admissible, because it is generated by a linear combination of paths of length at least two, and the quiver contains no path of length larger than two. The category of KQ/I-modules is equivalent to the category of those representations V of Q such that

$$
V(\alpha\beta - \gamma\delta) \;=\; V(\beta)V(\alpha) - V(\delta)V(\gamma) \;=\; 0,
$$

or equivalently, $V(\beta)V(\alpha) = V(\delta)V(\gamma)$. An example of such a representation is

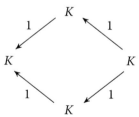

while

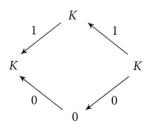

does not belong to this subcategory. Because of Exercise I.4.9(g), the bound
quiver algebra KQ/I is isomorphic to a subalgebra A of the triangular matrix
algebra. Hence, the category $\text{rep}_K(Q, I)$ of finite-dimensional bound representations is equivalent to the category $\text{mod}A$ of finitely generated modules
over this algebra.

A word on the practical use of equivalences. Suppose we want to study a category C. If
we construct an equivalence between C and another category D, then we can transfer
our problem to D. The mere fact of transferring it obviously does not suffice to solve
it, because the problem in D is equivalent to that in C. But working in a new category
gives access to new tools allowing to study our problem from different points of view.

IV.4.2 Dualities

We end this section (and chapter) by considering the contravariant case.

IV.4.7 Definition. The categories C and D are **dual categories** if there exist
contravariant functors $F: C \rightarrow D, G: D \rightarrow C$ and functorial isomorphisms
$GF \cong 1_C, FG \cong 1_C$. The functors F, G are **quasi-inverse dualities**.

Any category C is dual to its opposite category C^{op}: a duality functor $C \rightarrow C^{op}$ sends
any object of C to itself, viewed as object in C^{op}, and any morphism in $C(X, Y)$, with
X, Y objects in C, to itself, viewed as morphism in $C^{op}(Y, X)$.
    More generally, D is dual to C if and only if it is equivalent to C^{op}.

IV.4.8 Examples.
(a) Let K be a field and $\text{Mod}K$ the category of K-vector spaces. The contravari-
    ant functor $(-)^* = \text{Hom}_K(-, K)$ sends a vector space E to its dual space E^*,
    consisting of the K-linear maps from E to K (the so-called linear forms), and
    a K-linear map $u: E \rightarrow F$ to $u^*: F^* \rightarrow E^*$ defined by $f \mapsto fu$, for $f \in F^*$. If E is
    finite-dimensional, then so is E^* and in fact $\dim {}_KE^* = \dim {}_KE$. Thus, $(-)^*$
    is also an endofunctor of the category $\text{mod}K$ of finite-dimensional K-vector
    spaces. We claim that $(-)^*$ is a duality functor on this category (hence its
    name!).
        It suffices to prove that the composite $(-)^{**}$ of $(-)^*$ with itself is functo-
    rially isomorphic to the identity on $\text{mod}K$. We define a map $ev_E: E \rightarrow E^{**}$,
    called **evaluation**, as follows: if $x \in E$, then $ev_E(x): E^* \rightarrow K$ is the element
    of E^* defined by

$$ev_E(x)(f) = f(x)$$

    for $f \in E^*$. Indeed, $ev_E(x)$ is a K-linear map from E^* to K for, if $f, g \in E^*$,
    $\alpha, \beta \in K$,

$$ev_E(x)(\alpha f + \beta g) = (\alpha f + \beta g)(x) = \alpha f(x) + \beta g(x) = \alpha ev_E(x)(f) + \beta ev_E(x)(g).$$

Moreover, $ev_E \colon E \to E^*$ is itself K linear, because if $x, y \in E$, $\alpha, \beta \in K$,

$$ev_E(\alpha x + \beta y)(f) = f(\alpha x + \beta y) = \alpha f(x) + \beta f(y) = \alpha ev_E(x)(f) + \beta ev_E(y)(f)$$

for any $f \in E^*$. Hence, $ev_E(\alpha x + \beta y) = \alpha ev_E(x) + \beta ev_E(y)$ as required.

We claim that the ev_E defines a functorial morphism $ev \colon 1 \to (-)^{**}$. Let indeed $u \colon E \to F$ be a linear map between K-vector spaces. We prove commutativity of the square

$$
\begin{array}{ccc}
E & \xrightarrow{ev_E} & E^{**} \\
\downarrow{\scriptstyle u} & & \downarrow{\scriptstyle u^{**}} \\
F & \xrightarrow{ev_F} & F^{**}
\end{array}
$$

For $x \in E$, we have

$$u^{**} ev_E(x)(g) = ev_E(x)u^*(g) = ev_E(x)(gu) = gu(x) = g(u(x))$$
$$= ev_F(u(x))(g)$$

for $g \in F^*$. Hence, $u^{**} ev_E(x) = ev_F u(x)$, which establishes our claim.

There remains to prove that, if E is finite-dimensional, then $ev_E \colon E \to E^{**}$ is an isomorphism of vector spaces. Indeed, let $x \in E$ be such that $ev_E(x) = 0$. Then, for any $f \in E^*$, we have $f(x) = ev_E(x)(f) = 0$. If $x \neq 0$, then there exists a basis of E containing x. The dual basis contains a vector x^* such that $x^*(x) = 1$, a contradiction to our assumption that $f(x) = 0$ for any f. Therefore, $x = 0$ and so ev_E is injective.

Because E is finite-dimensional, we have $\dim_K E^* = \dim_K E$. Hence, $\dim_K E^{**} = \dim_K E^* = \dim_K E$, therefore the monomorphism ev_E is an isomorphism. This completes the proof.

(b) Let K be a field, A a finite-dimensional K-algebra and mod$A$ the category of finitely generated A-modules. Because of Corollary III.4.9, it coincides with the category of modules which are finite-dimensional over K. The functor $D = \mathrm{Hom}\,_K(-, K)$ sends a right A-module L_A to a left A-module DL, where the left A-module structure is defined by $(af)(x) = f(xa)$, for $a \in A$, $f \in DL, x \in L$, see Lemma II.3.8. With this definition, if $u \colon L \to M$ is a morphism of right A-modules, then $Du \colon DM \to DL$ is a morphism of left A-modules, so we have a contravariant functor $D = \mathrm{Hom}\,_K(-, K) \colon$ mod$A$ $\to$ modA^{op}.

Similarly, $D = \mathrm{Hom}\,_K(-, K)$ sends a left A-module $_AU$ to a right A-module DU with multiplication given by $(fa)(x) = f(ax)$, for $a \in A$, $f \in DU, x \in U$. So we also have a contravariant functor $D = \mathrm{Hom}\,_K(-, K) \colon$ mod $A^{op} \to$ mod$A$.

We claim that $D \colon$ mod$A \to$ modA^{op} and $D \colon$ mod$A^{op} \to$ mod$A$ are quasi-inverse dualities. It suffices, by symmetry, to construct an isomorphism of functors $1_{\text{mod}A} \to D^2 = D \circ D$. For any A-module M, we prove that evaluation $ev_L \colon L \to L^{**}$ is a morphism of A-modules. Indeed, if $x \in L, a \in A$,

$f \in DL$, we have

$$ev_L(xa)(f) = f(xa) = (af)(x) = ev_L(x)(af) = (ev_L(x)a)(f).$$

Hence, $ev_L(xa) = ev_L(x)a$. Because ev_L is a functorial isomorphism of K-vector spaces, see Example (a) above, we have finished.

Exercises of Section IV.4

1. Let $F : C \to D$ be a linear functor between linear categories. Prove that, if F is faithful and $Ff \in C(X, Y)$ is a section, or a retraction, or an isomorphism, then so is f. This is a partial converse to Exercise IV.2.4.

2. Prove that a functor $F : C \to D$ is an isomorphism of categories if and only if:
   (a) F is full and faithful, and
   (b) for any object X in D, there exists a unique object M in C such that $FM = X$.

3. Let $F : C \to D$ be an equivalence. Prove that a morphism f in C is a monomorphism, or an epimorphism, or a section, or a retraction, or an isomorphism, if and only if so is Ff.

4. Let K be a field. Using Exercise I.4.3, or otherwise, prove that:
   (a) The category of finite-dimensional K-algebras is essentially small.
   (b) Given a finite-dimensional K-algebra A, the category $\mathrm{mod}A$ of finitely generated A-modules is essentially small.

5. Let A be a K-algebra, M, N be K-modules and $\varphi : A \to \mathrm{End}_K M, \psi : A \to \mathrm{End}_K N$ algebra morphisms, see Exercise II.3.2. A morphism from the pair (M, φ) to the pair (N, ψ) is a K-linear map $f : M \to N$ such that $f\varphi(a) = \psi(a)f$, for any $a \in A$. The composition of morphisms is the ordinary composition of K-linear maps. Prove that the pairs (M, φ) with these morphisms form a category equivalent to the category $\mathrm{Mod}A^{op}$ of left A-modules.

6. Let C, D be categories, $F, F' : C \to D$ and $G, G' : D \to C$ functors such that $F \cong F', G \cong G'$. Prove that, if F, G are mutually quasi-inverse equivalences, then so are F', G'.

7. Let $F : C \to D, G : D \to C$ be mutually quasi-inverse equivalences, A a category and $H : A \to C, K : D \to A$ functors. Prove that $KF \cong H$ if and only if $HG \cong K$.

8. Let $F, G : C \to D$ be functors such that $F \cong G$. Prove that F is faithful, or full, or essentially surjective, if and only if so is G.

9. Let $\varphi : A \to B$ be an algebra morphism, where A, B are K-algebras. Consider the associated change of scalars functor $\mathrm{Mod}B \to \mathrm{Mod}A$. Prove that:
   (a) Any B-submodule of a B-module M_B is an A-submodule of the associated A-module M_A.
   (b) If φ is surjective, the submodules lattices of M_B and M_A are isomorphic.
   (c) Any morphism $f : L_B \to M_B$ in $\mathrm{Mod}B$ is A-linear, and $\mathrm{Hom}_B(L, M)$ is a K-submodule of $\mathrm{Hom}_A(L, M)$.

(d) If φ is surjective, then $\operatorname{Hom}_B(L, M) = \operatorname{Hom}_A(L, M)$.

(e) $\operatorname{End}_B M$ is a subalgebra of $\operatorname{End}_A M$, for any B-module M.

(f) If φ is surjective, then $\operatorname{End}_B M = \operatorname{End}_A M$, for any B-module M.

10. Consider the category $\mathcal{E}$ of pairs (E, f) as in Example IV.4.6(b), and the $K[t]$-modules $M_1 = K[t]/ < t >$ and $M_2 = K[t]/< t^2 >$. Construct the corresponding pairs under the equivalence $\mathcal{E} \cong \operatorname{Mod} K[t]$.

11. Let A be the 3×3-matrix algebra over a field K:

$$A = \begin{pmatrix} K & 0 & 0 \\ K & K & 0 \\ K & K & K \end{pmatrix} = \left\{ \begin{pmatrix} \alpha_1 & 0 & 0 \\ \alpha_2 & \alpha_3 & 0 \\ \alpha_4 & \alpha_5 & \alpha_6 \end{pmatrix} : \alpha_i \in K \right\}.$$

(a) Prove that $A \cong KQ$, where Q is the quiver $3 \xrightarrow{\alpha} 2 \xrightarrow{\beta} 1$.

(b) Prove that the rows of A, namely $P_1 = (K\ 0\ 0)$, $P_2 = (K\ K\ 0)$, $P_3 = (K\ K\ K)$, can be considered as A-modules under ordinary matrix multiplication by A on the right (Hint: see Exercise I.2.10).

(c) Construct the representation corresponding to each of P_1, P_2, P_3 under the equivalence $\operatorname{Mod} A \cong \operatorname{Rep}_K Q$.

(d) Construct monomorphisms $P_1 \rightarrow P_2$ and $P_2 \rightarrow P_3$. Compute the cokernel of each.

(e) Prove that $\operatorname{Hom}_A(P_1, P_3) \cong K$, and any morphism from P_1 to P_3 factors through P_2.

(f) Let $I = \langle \alpha\beta \rangle$ be the ideal of A generated by the path $\alpha\beta$. Prove that A/I is isomorphic to the algebra of Exercise I.4.11.

(g) Prove that each of $P_1' = (K\,0\,0)$, $P_2' = (K\,K\,0)$, $P_3' = (0\,K\,K)$ can be considered an A/I-module, and construct the corresponding representation of the bound quiver (Q, I).

(h) Construct nonzero morphisms from P_1' to P_2' and from P_2' to P_3'. Compute the kernel and the cokernel of each.

12. Let K be a field. Because of Exercise I.4.10(g), we know that the path algebra of the quiver

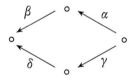

bound by the ideal $\langle \alpha\beta - \gamma\delta \rangle$ is isomorphic to the lower triangular matrix algebra

$$A = \begin{pmatrix} K & 0 & 0 & 0 \\ K & K & 0 & 0 \\ K & 0 & K & 0 \\ K & K & K & K \end{pmatrix} = \left\{ \begin{pmatrix} \alpha_1 & 0 & 0 & 0 \\ \alpha_2 & \alpha_3 & 0 & 0 \\ \alpha_4 & 0 & \alpha_5 & 0 \\ \alpha_6 & \alpha_7 & \alpha_8 & \alpha_9 \end{pmatrix} : \alpha_i \in K \text{ for all } i \right\}.$$

(a) Prove that the rows of the matrix algebra $P_1 = (K\ 0\ 0\ 0)$, $P_2 = (K\ K\ 0\ 0)$, $P_3 = (K\ 0\ K\ 0)$, $P_4 = (K\ K\ K\ K)$ can be considered as A-modules under ordinary matrix multiplication on the right.

(b) Construct the representation corresponding to each of P_1, P_2, P_3, P_4 under the equivalence between modules and bound quiver representations.

(c) Prove that there exist monomorphisms from each of P_2 and P_3 to P_4. Compute the cokernel of each.

(d) Let $f : P_2 \oplus P_3 \to P_4$ be the morphism obtained from the monomorphisms of (c) by passing to the direct sum. Compute $\operatorname{Ker} f$ and $\operatorname{Coker} f$.

13. Let K be a field and A the 2×2-matrix algebra

$$A = \begin{pmatrix} K & 0 \\ K^2 & K \end{pmatrix}$$

with the usual matrix addition and the multiplication given by

$$\begin{pmatrix} a & 0 \\ (b,c) & d \end{pmatrix}\begin{pmatrix} a' & 0 \\ (b',c') & d' \end{pmatrix} = \begin{pmatrix} aa' & 0 \\ (ba' + db', ca' + dc') & dd' \end{pmatrix}.$$

(a) Prove that $A \cong KQ$, where Q is the so-called **Kronecker quiver**.

$$\circ \leftleftarrows \circ$$

(b) Define A-module structures on the rows $P_1 = (K\ 0)$, $P_2 = (K^2\ K)$ and construct the corresponding representations under the equivalence $\operatorname{Mod} A \cong \operatorname{Rep}_K Q$.

(c) Prove that $\operatorname{Hom}_A(P_1, P_2)$ is two-dimensional. Construct two monomorphisms from P_1 to P_2 with distinct images. Are the cokernels isomorphic?

Chapter V
Modules over principal ideal domains

V.1 Introduction

We pause a bit from theory in order to apply it to a classification problem. Classifying mathematical objects means arranging them into classes in some unique way, usually up to isomorphism. A typical and well-known example of a classification theorem is that of cyclic groups: if a cyclic group G is infinite, then $G \cong \mathbb{Z}$, while, if it is finite with m elements, then $G \cong \mathbb{Z}/m\mathbb{Z}$.

Because our object of study is modules, we wish to classify a class of modules. Modules are not as easy to handle as, say, vector spaces. However, it is reasonable to expect that if the algebra is 'close' to a field, then modules should behave in a similar way to vector spaces. A class of algebras close to fields is that of principal ideal domains, see Section I.5. In this chapter, we give a complete classification of finitely generated modules over a principal ideal domain. Typical examples of principal ideal domains include the ring of integers $\mathbb{Z}$ and the algebra of polynomials $K[t]$ in one indeterminate t over a field K. Thus, our result includes the classification of finitely generated abelian groups as well as of finitely generated $K[t]$-modules.

V.2 Free modules and torsion

V.2.1 Ranks of finitely generated free modules

Because we wish our treatment to retain the flavour of linear algebra, we look for analogues, in our context, of finite-dimensional vector spaces. Reasonable candidates are finitely generated free modules (because they have bases). Assume for the moment that A is an integral domain (in particular, it is commutative). The following is a well-known property of vector spaces.

V.2.1 Lemma. *Let A be an integral domain and L a finitely generated A-module. If $\{x_1, \cdots, x_m\}$ is a linearly independent subset of L, and $\{y_1, \cdots, y_n\}$ a generating set of L, then $m \leqslant n$.*

Proof. Assume that $m > n$, and suppose that $\sum_{i=1}^{m} x_i c_i = 0$, with the $c_i \in A$. Because the y_j form a generating set for L, we have, for any i,

$$x_i = \sum_{j=1}^{n} y_j a_{ji}$$

An Introduction to Module Theory. Ibrahim Assem and Flávio U. Coelho, Oxford University Press.
© Ibrahim Assem and Flávio U. Coelho (2024). DOI: 10.1093/9780198904939.003.0006

where $a_{ji} \in A$ for any i, j. Because the x_i are linearly independent, hence nonzero, then, for any i, at least one of the a_{ji} is nonzero. Substituting x_i into the first equality yields

$$0 = \sum_{i=1}^{m} x_i c_i = \sum_{i=1}^{m} \left(\sum_{j=1}^{n} y_j a_{ji} \right) c_i = \sum_{j=1}^{n} y_j \left(\sum_{i=1}^{m} a_{ji} c_i \right).$$

Consider the homogeneous system of linear equations in $c_1, \cdots, c_m$

$$\begin{cases} a_{11} c_1 & + & \cdots & + & a_{1m} c_m & = & 0 \\ \vdots & & & & \vdots & & \\ a_{n1} c_1 & + & \cdots & + & a_{nm} c_m & = & 0 \end{cases}$$

This is a system in A, but also in its field of fractions $Q(A)$. It is nontrivial because, for each i, at least one of the a_{ji} is nonzero. Because $m > n$, it has a nonzero solution in $Q(A)$. Proposition I.2.11 says that it has also a nonzero solution in A, that is, there exists i such that $c_i \neq 0$. Therefore, the x_i are linearly dependent, a contradiction. The proof is complete.   □

The proof we gave rests on the existence of the field of fractions, hence the need for assuming that A is an integral domain.

V.2.2 Theorem. *Let A be an integral domain and L a finitely generated free A-module. Any basis of L is finite, and any two bases have the same cardinality.*

Proof. Because L is finitely generated, it has a finite generating set having, say, m elements. Now any basis of L is a linearly independent set. Due to Lemma V.2.1, it has at most m elements. In particular, it is finite. The second statement follows from Theorem III.4.7.   □

The previous theorem generalises the well-known result of linear algebra saying that a finitely generated vector space has a finite basis, and any two bases have the same cardinality. Moreover, the proof is almost the same. We define an analogue of the notion of dimension.

V.2.3 Definition. Let A be an integral domain and L a finitely generated free A-module. The **rank** of L is the cardinality of any basis of L.

For instance, the rank of $A_A^{(n)}$ is n, because of Example III.4.4(a). If $A = K$ is a field, then an A-module is a K-vector space, hence is free, and its rank coincides with its dimension.

We look at linear maps between finitely generated free modules. Let M, N be free A-modules of respective ranks m and n. The existence of a basis in each implies that a linear map from M to N can be expressed in the form of a matrix (relative to these bases) exactly as is the case for linear maps between vector spaces. For later reference, we formalise this relation here.

Let $u: M \longrightarrow N$ be a linear map and $\mathcal{E} = \{e_1, \cdots, e_m\}$ and $\mathcal{F} = \{f_1, \cdots, f_n\}$ bases of M and N, respectively. For any i with $1 \leqslant i \leqslant m$, we can write

$$u(e_i) = \sum_{j=1}^{n} f_j a_{ji}.$$

The $m \times n$-matrix $[a_{ij}]_{1 \leqslant i \leqslant m, 1 \leqslant j \leqslant n}$ is called the **matrix** of u with respect to the bases $\mathcal{E}$ and $\mathcal{F}$. In order to underline this dependence on the bases, we denote it as $[u]_{\mathcal{F}\mathcal{E}}$. Most relevant results of linear algebra remain valid in this context with similar proofs. We suggest to the reader to write down the proofs as an easy exercise.

(a) Denote by

$$[x]_{\mathcal{E}} = \begin{pmatrix} x_1 \\ \vdots \\ x_m \end{pmatrix}$$

the coordinate vector of $x = \sum_{i=1}^{m} e_i x_i \in M$ in the basis $\mathcal{E}$, and similarly, by

$$[y]_{\mathcal{F}} = \begin{pmatrix} y_1 \\ \vdots \\ y_n \end{pmatrix}$$

the coordinate vector of $y = \sum_{j=1}^{n} f_j y_j \in N$ in the basis $\mathcal{F}$. Then we have

$$[u(x)]_{\mathcal{F}} = [u]_{\mathcal{F}\mathcal{E}}[x]_{\mathcal{E}}.$$

(b) Let $u, v: M \longrightarrow N$ be linear maps, $\mathcal{E}, \mathcal{F}$ bases as above and $a, b \in A$. Then

$$[ua + vb]_{\mathcal{F}\mathcal{E}} = [u]_{\mathcal{F}\mathcal{E}} a + [v]_{\mathcal{F}\mathcal{E}} b.$$

Further, if $u: L \longrightarrow M, v: M \longrightarrow N$ are linear maps and $\mathcal{G}, \mathcal{E}, \mathcal{F}$ bases of L, M, N, respectively, then

$$[vu]_{\mathcal{F}\mathcal{G}} = [v]_{\mathcal{F}\mathcal{E}}[u]_{\mathcal{E}\mathcal{G}}$$

that is, the matrix of the composite of linear maps is the product of the matrices of these maps.

(c) Let $\mathcal{E}, \mathcal{E}'$ be bases of a module M. The matrices of the identity map $P = [1_M]_{\mathcal{E}'\mathcal{E}}$ and $P' = [1_M]_{\mathcal{E}\mathcal{E}'}$ are called the **transition matrices** between $\mathcal{E}$ and $\mathcal{E}'$. It follows from (b) that P, P' are invertible and $P' = P^{-1}$. Moreover, (a) gives $[x]_{\mathcal{E}'} = P[x]_{\mathcal{E}}$, for $x \in M$.

(d) Let $u: M \longrightarrow N$ be a linear map, $\mathcal{E}, \mathcal{E}'$ bases of M and $\mathcal{F}, \mathcal{F}'$ bases of N. Set $J = [u]_{\mathcal{F}\mathcal{E}}, J' = [u]_{\mathcal{F}'\mathcal{E}'}, P = [1_M]_{\mathcal{E}'\mathcal{E}}$ and $Q = [1_N]_{\mathcal{F}'\mathcal{F}}$. Then $J' = QJP^{-1}$. This follows from

$$J'P = [u]_{\mathcal{F}'\mathcal{E}'}[1_M]_{\mathcal{E}'\mathcal{E}} = [u]_{\mathcal{F}'\mathcal{E}} = [1_N]_{\mathcal{F}'\mathcal{F}}[u]_{\mathcal{F}\mathcal{E}} = QJ$$

and invertibility of P.

V.2.2 Free modules over principal ideal domains

One of the most remarkable properties of principal ideal domains is that submodules of free modules are free (for instance, subgroups of free abelian groups are free). This is not true for other rings, see Exercise III.4.5. This property follows from the next theorem, due to Kaplansky, which we prove for nonnecessarily commutative algebras satisfying a very general assumption, because we want to apply it to different situations later on. Recall that free modules are projective, see Lemma III.4.11.

V.2.4 Theorem. *Let A be an algebra such that any right ideal of A is projective. Then any submodule N of a free A-module L is isomorphic to the direct sum of right ideals in A.*

Proof. Let $(e_\lambda)_{\lambda \in \Lambda}$ be a basis of L. Without loss of generality, assume that the index set Λ is well-ordered, see Example I.5.2. For any $\lambda \in \Lambda$, set $L_\lambda = \oplus_{\mu < \lambda} e_\mu A$. Then $L_{\lambda+1} = L_\lambda \oplus (e_\lambda A)$, so an element x of $N \cap L_{\lambda+1}$ can be written uniquely in the form $x = y + e_\lambda a$, with $y \in L_\lambda, a \in A$. Because a is unique, there exists an A-linear map $f_\lambda : N \cap L_{\lambda+1} \to A$ defined by $x \mapsto a$, for $x \in N \cap L_{\lambda+1}$. Then $\operatorname{Ker} f_\lambda = N \cap L_\lambda$ and we have an exact sequence

$$0 \to N \cap L_\lambda \to N \cap L_{\lambda+1} \to \operatorname{Im} f_\lambda \to 0.$$

Now, $\operatorname{Im} f_\lambda$ is a right ideal of A. Because of the hypothesis, it is projective. Due to Example IV.2.8(c), the sequence splits. That is, there exists a projective submodule $I_\lambda \cong \operatorname{Im} f_\lambda$ of $N \cap L_{\lambda+1}$ such that

$$N \cap L_{\lambda+1} = (N \cap L_\lambda) \oplus I_\lambda.$$

To complete the proof, it suffices to show that $N = \oplus_\lambda I_\lambda$. This is done in two steps:
   (i) N equals $I = \sum_\lambda I_\lambda$. Indeed, L equals the union of the increasing sequence of the L_λ. Consequently, to any $y \in L$ corresponds a least index λ_y such that $y \in L_{\lambda_y+1}$.
   Because $I_\lambda \subseteq N$, for any λ, we have $I \subseteq N$. If inclusion is strict, then there exist elements of $N = N \cap L = N \cap (\cup_\lambda L_\lambda) = \cup_\lambda (N \cap L_\lambda)$ not lying in I. Because Λ is well-ordered, there exists a least index $\mu = \lambda_x$ such that $x \in N \cap L_{\mu+1}, x \notin I$. Because $x \in N \cap L_{\mu+1}$, we have $x = y + z$, with $y \in N \cap L_\mu$ and $z \in I_\mu$. Moreover, $y \notin I$ (for, otherwise, $x \in I$, a contradiction), and, clearly, $y \in N$. So, $y \in L_\mu$ implies $\lambda_y < \mu$. This contradicts minimality of μ and completes the proof of (i).
   (ii) The sum $N = \sum_\lambda I_\lambda$ is direct. For, suppose that $x_1 + \cdots + x_m = 0$, with $x_i \in I_{\lambda_i}$ where we assume, without loss of generality, that $\lambda_1 < \cdots < \lambda_m$. Then $x_1 + \cdots + x_{m-1} = -x_m \in (N \cap L_{\lambda_{m-1}}) \cap I_{\lambda_m} = 0$ gives $x_m = 0$. Descending induction yields $x_i = 0$, for any i. □

In order to apply the theorem to our situation, we need a lemma.

V.2.5 Lemma. *Let A be an integral domain. Any principal ideal of A is free with rank at most 1.*

Proof. Let $I = aA$ be a principal ideal. If $a = 0$, then $I = 0$ is trivially free. If $a \neq 0$, then the map $f : A \to aA$ given by $x \mapsto ax$, for $x \in A$, is A-linear, because A is commutative. It is surjective due to its definition, and injective because A is an integral domain. So it is an isomorphism and $I = aA \cong A_A$. $\qquad\square$

From now on, and for the rest of this subsection, we assume that A is a principal ideal domain. Because of the previous Lemma V.2.5, any ideal of A is free of rank at most one.

V.2.6 Corollary. *Any submodule L' of a free A-module L is free. Moreover, if L is finitely generated, then so is L' and its rank equals at most the rank of L.*

Proof. Because of Lemma V.2.5, any ideal of A is free. Because of Lemma III.4.11, it is projective. Theorem V.2.4 yields that L' is the direct sum of ideals of A. But any such ideal is free, hence L' itself is free.

Suppose that L is finitely generated free of rank ℓ. If the rank of L' is larger than ℓ, then L' contains a family of more than ℓ linearly independent elements, which remain linearly independent in L, and this contradicts Lemma V.2.1. Therefore, the rank of L' is at most ℓ. $\qquad\square$

A consequence is that there exist free resolutions as in Corollary III.4.10 which are short exact sequences of finitely generated A-modules.

V.2.7 Corollary. *Let M be a finitely generated A-module. Then there exists a short exact sequence $0 \to L' \to L \to M \to 0$, where L, L' are finitely generated free, and the rank of L' is less than or equal to the rank of L.*

Proof. Because of Proposition III.4.8, there exist a finitely generated free module L of rank ℓ, say, and an epimorphism from L to M. Because of Corollary V.2.6 above, the kernel of this epimorphism is a free module L' of rank $n \leqslant \ell$. $\qquad\square$

V.2.8 Corollary. *A submodule of a finitely generated A-module is finitely generated.*

Proof. Let M' be a submodule of a finitely generated A-module M. Because of Corollary V.2.7, there exist integers ℓ and $n \leqslant \ell$ and a short exact sequence $0 \to A^{(n)} \to A^{(\ell)} \to M \to 0$. Therefore, $M \cong A^{(\ell)}/A^{(n)}$. Because of Theorem II.4.27, there exists a submodule L' of $A^{(\ell)}$ such that $M' \cong L'/A^{(n)}$. Because of Corollary V.2.6, L' is free of rank at most ℓ. So M', which is a quotient of L', is generated by at most ℓ elements. $\qquad\square$

V.2.3 Torsion elements and modules

Torsion elements in a module generalise elements of finite order in abelian groups. Let G be an abelian group. An element $a \in G$ is said to be of *finite order* if there exists a nonzero integer m such that $ma = 0$. For instance, in a finite abelian group, all elements are of finite order (dividing the cardinality of the group).

V.2.9 Definition. Let A be a commutative algebra and M an A-module.
 (a) An element $x \in M$ is *torsion* if there exists $a \in A, a \neq 0$, such that $xa = 0$. It is *torsion-free* if, for any $a \in A, a \neq 0$, we have $xa \neq 0$.
 (b) The module M is *torsion* if all its elements are torsion, and *torsion-free* if all its nonzero elements are torsion-free.

We reformulate the definition of torsion-free element: let M be an A-module, then $x \in M$ is torsion-free if and only if $xa = 0$ implies $a = 0$, that is, if and only if its annihilator $\mathrm{Ann}\{x\}$, see Exercise II.2.2, is zero.

V.2.10 Examples.
 (a) In any module, the zero element is torsion.
 (b) In an abelian group, an element is torsion if and only if it has finite order. Hence, any finite abelian group is a torsion $\mathbb{Z}$-module.
 (c) Let K be a field, $K[t]$ the algebra of polynomials in t and p a nonconstant polynomial. Any element $\bar{q}$ in the $K[t]$-module $K[t]/\langle p \rangle$ is torsion because $\bar{q} \cdot p = 0$. Thus, $K[t]/\langle p \rangle$ is a torsion module.
 (d) A free module L over an integral domain A is torsion-free. Let $(e_\lambda)_\lambda$ be a basis for L and $x \in L$ a nonzero torsion element. Then there exists $a \in A, a \neq 0$, such that $xa = 0$. On the other hand, there exist $(b_\lambda)_\lambda$, almost all zero, such that $x = \sum_\lambda e_\lambda b_\lambda$. But then

$$0 = xa = \left(\sum_\lambda e_\lambda b_\lambda \right) a = \sum_\lambda e_\lambda (b_\lambda a).$$

Because $x \neq 0$, there exists an index λ_0 such that $b_{\lambda_0} \neq 0$, and this together with $a \neq 0$ implies $b_{\lambda_0} a \neq 0$, a contradiction to linear independence of the family $(e_\lambda)_\lambda$.
 If $A = K$ is a field, we get that any K-vector space is torsion-free.
 (e) Let A be a commutative ring and $f : M \to N$ a morphism between A-modules. If $x \in M$ is torsion, there exists $a \in A, a \neq 0$, such that $xa = 0$. But then $f(x)a = f(xa) = 0$, and so $f(x)$ is torsion. Hence, if N is a torsion-free module, then $f(x) = 0$. In particular, if M is a torsion module and N is a torsion-free module, then $\mathrm{Hom}_A(M, N) = 0$. Therefore, the only module which is at the same time torsion and torsion-free is the zero module. The same considerations imply that a quotient of a torsion module is torsion, and a submodule of a torsion-free module is torsion-free, see Exercise V.2.11.

Given an A-module M, denote by $T(M)$ the subset consisting of the torsion elements of M. Over an integral domain, $T(M)$ is a submodule such that the quotient is torsion-free.

V.2.11 Proposition. *Let A be an integral domain and M an A-module.*

(a) *There exists a short exact sequence* $0 \to T(M) \to M \to M/T(M) \to 0$ *with* $T(M)$ *torsion and* $M/T(M)$ *torsion-free.*

(b) *The short exact sequence of (a) is unique: if there exists a short exact sequence* $0 \to L \overset{j}{\to} M \to N \to 0$ *with L torsion and N torsion-free, then* $L \cong T(M)$ *and* $N \cong M/T(M)$.

Proof. (a) We first show that $T(M)$ is a submodule of M. Because the zero element is torsion, $T(M)$ is nonempty. Let $x, x' \in T(M)$ and $b, b' \in A$. There exist nonzero elements $a, a' \in A$ such that $xa = 0, x'a' = 0$. Because A is integral, $aa' \neq 0$. Using commutativity of A, we get

$$(xb + x'b')(aa') = (xa)(ba') + (x'a')(b'a) = 0.$$

Hence, $xb + x'b' \in T(M)$ and the latter is a submodule due to Proposition II.2.3. We get a short exact sequence

$$0 \to T(M) \overset{i}{\to} M \to M/T(M) \to 0$$

where i is the inclusion. There remains to prove that $M/T(M)$ is torsion-free.

Let $\bar{x} = x + T(M) \in M/T(M)$ be nonzero, that is, $x \notin T(M)$. If $\bar{x}$ is torsion, there exists $a \in A, a \neq 0$, such that $\bar{x}a = 0$, that is, $xa \in T(M)$. Then there exists $b \in A, b \neq 0$, such that $xab = 0$. Now, $a, b \neq 0$ and A integral imply $ab \neq 0$, so $x \in T(M)$, a contradiction.

(b) Because L is torsion, so is its image $j(L) \subseteq M$, due to Example V.2.10(e). Hence, $j(L) \subseteq T(M)$ and we have a commutative diagram with exact rows

$$
\begin{array}{ccccccccc}
0 & \longrightarrow & L & \overset{j}{\longrightarrow} & M & \longrightarrow & N & \longrightarrow & 0 \\
 & & \downarrow{\scriptstyle j'} & & \downarrow{\scriptstyle 1_M} & & \downarrow{\scriptstyle f} & & \\
0 & \longrightarrow & T(M) & \overset{i}{\longrightarrow} & M & \longrightarrow & M/T(M) & \longrightarrow & 0
\end{array}
$$

where j' is the inclusion and f is induced by passing to cokernels. Applying the Snake Lemma II.4.19, we get that $\operatorname{Coker} f = 0$ (thus f is surjective), while $\operatorname{Ker} f \cong \operatorname{Coker} j'$ is a quotient of a torsion module $T(M)$ (hence it is torsion), and also a submodule of a torsion-free module N (hence it is torsion-free). But then it is zero. Therefore, f is an isomorphism and so is j'. □

We now prove that, if A is a principal ideal domain, and M a finitely generated A-module, then the previous short exact sequence splits, that is, $M = T(M) \oplus M/T(M)$, and, moreover, this decomposition is essentially unique. In order to prove that the sequence splits, it suffices, because of Example IV.2.8(c), to prove that the torsion-free module $M/T(M)$ is actually free.

V.2.12 Lemma. *Let A be a principal ideal domain. A finitely generated A-module is torsion-free if and only if it is free.*

Proof. Sufficiency follows from Example V.2.10(d), so we prove necessity. Let M be a torsion-free module, with finite generating set $\{y_1, \cdots, y_n\}$. Because of Lemma V.2.1, a linearly independent subset of M has at most n elements. Let $\mathcal{X} = \{x_1, \cdots, x_m\}$ be a maximal linearly independent subset in M. The submodule L of M generated by $\mathcal{X}$ is free, because it has a basis.

We claim that, for any i with $1 \leqslant i \leqslant n$, there exists $a_i \in A$ nonzero and such that $y_i a_i \in L$. Indeed, because of maximality of $\mathcal{X}$, the set $\{x_1, \cdots, x_m, y_i\}$ is linearly dependent, hence there exist $b_1, \cdots, b_m, a_i$ in A, not all zero, such that

$$x_1 b_1 + \cdots + x_m b_m + y_i a_i = 0.$$

Because $\mathcal{X}$ is a linearly independent set, we have $a_i \neq 0$ and also

$$y_i a_i = -x_1 b_1 - \cdots - x_m b_m \in L,$$

establishing our claim.

Let $a = a_1 \cdots a_m$. Because A is integral and $a_i \neq 0$ for any i, we have $a \neq 0$ and $y_i a \in L$ for any i. Now, the y_i generate M, hence $Ma \subseteq L$. But L is free. Because of Corollary V.2.6, so is Ma. Consider the map $M \to Ma$ defined by $x \mapsto xa$, for $x \in M$. It is A-linear, surjective and also injective, because M is torsion-free. Hence, it is an isomorphism, and M is free. $\square$

V.2.13 Examples.
  (a) According to Lemma V.2.5, any finitely generated torsion-free A-module is isomorphic to a module of the form $A^{(n)}$, for some n. Finite generation of the module is essential for the validity of this result. For instance, the abelian group ($\mathbb{Z}$-module) $\mathbb{Q}$ is obviously torsion-free, but is not free, see Exercise III.4.6. Because of Lemma V.2.12, this implies that $\mathbb{Q}$ is not finitely generated.
  (b) Because Lemma V.2.12 characterises torsion-free finitely generated abelian groups, we may ask what are the torsion ones. We prove that a finitely generated abelian group is torsion if and only if it is finite.
  Sufficiency being clear, we prove necessity. Suppose that G is a torsion group with finite generating set $\{x_1, \cdots, x_n\}$. Then $G = \sum_{i=1}^{n} x_i \mathbb{Z}$. Each x_i is a torsion element, hence of finite order. Then each of the subgroups $x_i \mathbb{Z}$ is finite. Hence, G itself is a finite group.

We are able to prove the announced result.

V.2.14 Theorem. *Let A be a principal ideal domain and M a finitely generated A-module. Then $M \cong T(M) \oplus M/T(M)$. Moreover, if $M \cong T' \oplus L'$, with T' torsion and L' torsion-free, then $T' = T(M)$ and $L' \cong M/T(M)$.*

Proof. Because of Proposition V.2.11(a), we have a short exact sequence of A-modules

$$0 \to T(M) \to M \to M/T(M) \to 0.$$

Now, $M/T(M)$ is torsion-free, hence it is free, due to Lemma V.2.12. Because of Example IV.2.8(c), the sequence splits. Hence, $M \cong T(M) \oplus M/T(M)$, as required. Uniqueness of this decomposition follows from Proposition V.2.11(b). $\qquad\square$

For instance, the theorem, Lemma V.2.12 and Example V.2.13(b) imply that any finitely generated abelian group can be written uniquely in the form $G \oplus L$, with G a finite abelian group and L a finitely generated free abelian group (that is, $L \cong \mathbb{Z}^{(n)}$, for some $n \geq 0$).

Exercises of Section V.2

1. Prove that an integral domain A is a principal ideal domain if and only if any submodule of a finitely generated free A-module is free.

2. Assume that A is a principal ideal domain, and L a free A-module with generating set $G = \{x_1, \cdots, x_n\}$. Let $G' = \{x'_1, \cdots, x'_n\}$ be obtained from G by applying one or a sequence of the following so-called **elementary operations**.
   (a) Exchange of the order of generators.
   (b) Multiplication of a generator by an invertible element of A.
   (c) Addition to a generator of a multiple of another generator.
   Prove that L is generated by G'.

3. Apply Exercise V.2.2 to find a basis of the subgroup of $\mathbb{Z}^{(4)}$ generated by each of the sets
   (a) $(12, 6, 0, 0)$, $(18, 9, 0, 0)$.
   (b) $(3, 4, 2, 3)$, $(1, 3, 2, 2,)$, $(2, 2, 2, 3)$, $(3, 3, 0, 1)$.
   (c) $(1, 2, 6, 1)$, $(3, 5, 17, 2)$, $(3, 3, 7, 4)$, $(2, 4, 10, 3)$, $(0, 8, 22, 1)$.

4. Apply Exercise V.2.2 to find a basis of the submodule of $\mathbb{Q}[x]^{(3)}$ generated by the set $\{(x^2 + 3, x, 2x - 1), (x^2, x, x), (2x^2 - 4, 2x, x + 1)\}$.

5. Find a basis of the subgroup of $\mathbb{Z}^{(3)}$ defined by:
$$G = \{(x_1, x_2, x_3): 3x_1 + 2x_2 + x_3 = 0, 9x_1 + 4x_2 + x_3 = 0\}.$$

6. Prove that two distinct prime integers p, q always generate the $\mathbb{Z}$-module $\mathbb{Z}_{\mathbb{Z}}$, but the set $\{p, q\}$ contains no basis of $\mathbb{Z}$. So, contrary to what happens with vector spaces, a generating set for a finitely generated free module does not necessarily contain a basis.

7. Let A be a commutative ring. Prove that the A-module A_A is torsion-free if and only if A is an integral domain.

8. Let A be an integral domain. Prove that a finitely generated A-module is torsion if and only if its annihilator is nonzero.

9. Prove that the torsion submodule of the $\mathbb{Z}$-module $\mathbb{R}/\mathbb{Z}$ is $\mathbb{Q}/\mathbb{Z}$.

10. Compute the torsion submodule of the $\mathbb{Z}$-module $\mathbb{Z}^{(2)}/\langle(6, 9)\rangle$.

11. Let A be an integral domain. Show that:

(a) A submodule of a torsion-free module is torsion-free, but this does not hold true in general for quotients.

(b) Submodules and quotients of torsion modules are torsion.

(c) Finite direct sums of torsion-free modules are torsion-free modules.

(d) Finite direct sums of torsion modules are torsion.

12. Let A be an integral domain and N a submodule of an A-module M. Show that $T(N) = T(M) \cap N$.

13. Let A be an integral domain, M an A-module and N a submodule of M such that M/N is torsion-free. Show that $T(M) \subseteq N$.

14. Let A be a commutative ring and M a nonzero A-module.

(a) Assume that $M = xA$ is a cyclic module. Prove that $\operatorname{Ann}M = \operatorname{Ann}\{x\}$.

(b) If M is cyclic, prove that M is torsion-free if and only if A is an integral domain and $\operatorname{Ann}M = \{0\}$.

(c) Show that A is a field if and only if any A-module is torsion-free.

15. Let A be an algebra. A pair of full subcategories $(\mathcal{T}, \mathcal{F})$ in Mod$A$ is called a *torsion pair* if $\operatorname{Hom}_A(M, N) = 0$, for any $M \in \mathcal{T}_0, N \in \mathcal{F}_0$, and these classes are maximal for this property, that is, if $\operatorname{Hom}_A(X, N) = 0$ for any $N \in \mathcal{F}_0$, then $X \in \mathcal{T}_0$, and, dually, if $\operatorname{Hom}_A(M, Y) = 0$ for all $M \in \mathcal{T}_0$, then $Y \in \mathcal{F}_0$. The class $\mathcal{T}$ is the *torsion class* and its objects are *torsion objects* while $\mathcal{F}$ is the *torsion-free class* and its objects are *torsion-free objects*. Prove that:

(a) Let A be an integral domain. The pair of full subcategories defined by $\mathcal{T} = \{M : M$ is a torsion module$\}$ and $\mathcal{F} = \{N : N$ is a torsion-free module$\}$ is a torsion pair.

(b) Given any class $\mathcal{M}$ of modules, one gets a torsion pair by setting $\mathcal{F} = \{N : \operatorname{Hom}_A(M, N) = 0$, for any M in $\mathcal{M}\}$ and $\mathcal{T} = \{M : \operatorname{Hom}_A(M, N) = 0$, for any $N \in \mathcal{F}_0\}$. Furthermore, $\mathcal{T}$ is the smallest torsion class containing $\mathcal{M}$.

(c) Dually, starting from any class $\mathcal{N}$, there exists a torsion pair $(\mathcal{T}, \mathcal{F})$ such that $\mathcal{F}$ is the smallest torsion-free class containing $\mathcal{N}$.

(d) Assume that $(\mathcal{T}, \mathcal{F})$ is a torsion pair in Mod$A$. Then, for any A-module M, there exists a short exact sequence $0 \to tM \to M \to M/tM \to 0$, with $tM \in \mathcal{T}_0, M/tM \in \mathcal{F}_0$, unique in the sense that any short exact sequence $0 \to L \to M \to N \to 0$, with $L \in \mathcal{T}_0, N \in \mathcal{F}_0$ is isomorphic to it.

(e) A full subcategory $\mathcal{T}$ of Mod$A$ is the torsion class for some torsion pair if and only if it is stable under taking quotient objects, coproducts, and if $0 \to L \to M \to N \to 0$ is a short exact sequence, with $L, N \in \mathcal{T}_0$, then $M \in \mathcal{T}_0$.

(f) A full subcategory $\mathcal{F}$ of Mod$A$ is a torsion-free class for some torsion pair if and only if it is stable under taking subobjects, products, and if $0 \to L \to M \to N \to 0$ is a short exact sequence, with $L, N \in \mathcal{F}_0$, then $M \in \mathcal{F}_0$.

V.3 The structure theorems

V.3.1 The Smith normal form for matrices

The structure theorems for finitely generated modules over a principal ideal domain aim at describing an arbitrary finitely generated module in terms of ones which are well-understood. In this section, A stands for a principal ideal domain.

The strategy of proof of our first structure theorem is fairly simple. Let M be a finitely generated A-module. Due to Corollary V.2.7, there exists a short exact sequence

$$0 \to L' \xrightarrow{j} L \xrightarrow{f} M \to 0$$

with L free of finite rank ℓ, say, and L' free of rank $\ell' \leqslant \ell$. As seen in Subsection V.2.1, given bases of L' and L, the inclusion $j : L' \to L$ is expressed by an $\ell \times \ell'$-matrix with coefficients in A. We show that bases can be chosen so that the resulting matrix is of a particular diagonal form, and then it is essentially unique. This is the so-called Smith normal form of the matrix of j.

Because we deal in general with nonsquare matrices, we agree to say that an $\ell \times \ell'$-matrix $[a_{ij}]$ is **diagonal** whenever $a_{ij} = 0$ for $i \neq j$, that is, elements off the diagonal starting at the upper-left corner of the matrix vanish. The diagonal matrix is then denoted as $\mathrm{diag}(a_{11}, \cdots, a_{mm})$, where $m = \min(\ell, \ell')$.

We recall from the end of Subsection V.2.1 that, if $j : L' \to L$ is an A-linear map between finitely generated free modules, J, J' the matrices of j with respect to two pairs of bases (one for L', the other for L), then there exist invertible (in particular, square) matrices P, Q such that $J' = QJP^{-1}$.

V.3.1 Definition. Two matrices J, J' of same size with coefficients in A are **equivalent matrices** if there exist invertible matrices P, Q such that $J' = QJP^{-1}$.

This is plainly an equivalence relation on rectangular matrices of a fixed size. We prove that each equivalence class contains a diagonal matrix of a form we define now, and the latter is unique up to association of its diagonal elements. Recall that, in a principal ideal domain, the notation $a|b$ means that a divides b, and that a, b are associate if $a|b$ and $b|a$, see Subsection I.5.1. Also, any pair of elements admits a gcd, see Theorem I.5.5.

V.3.2 Definition. A diagonal matrix $\mathrm{diag}(d_1, \cdots, d_m)$, with coefficients in A, is a **Smith normal matrix** if $d_1|\cdots|d_m$.

Here is an example of a Smith normal matrix over the integers

$$\begin{pmatrix} 2 & 0 \\ 0 & 14 \\ 0 & 0 \end{pmatrix}$$

because $2|14$.

We start by proving that any 2×1- or 1×2-matrix is equivalent to a Smith normal matrix.

V.3.3 Lemma. *Let $a, b \in A$, and d be a gcd of a and b. Then:*

(a) *There exists an invertible matrix Q such that $Q \begin{pmatrix} a \\ b \end{pmatrix} = \begin{pmatrix} d \\ 0 \end{pmatrix}$.*

(b) *There exists an invertible matrix P such that $\begin{pmatrix} a & b \end{pmatrix} P = \begin{pmatrix} d & 0 \end{pmatrix}$.*

Proof. We prove (a) because (b) is obtained by transposing matrices. Because of Theorem I.5.5, there exist $s, t, a', b' \in A$ such that $sa + tb = d, a = a'd$ and $b = b'd$. If a, b are nonzero, which we may assume without loss of generality, then $d \neq 0$. So, $(sa' + tb')d = d$ gives $sa' + tb' = 1$. Consequently, the matrix

$$Q = \begin{pmatrix} s & t \\ -b' & a' \end{pmatrix}$$

is invertible over A and moreover $Q \begin{pmatrix} a \\ b \end{pmatrix} = \begin{pmatrix} d \\ 0 \end{pmatrix}$, as required. $\square$

For matrices of larger size, we use the elementary operations of linear algebra on rows and columns of the given matrix. It is known from linear algebra that the result of a row elementary operation is obtained by premultiplying the given matrix by the invertible matrix obtained from the identity matrix by performing on the latter the same row operation. Transposing, we get that the result of a column operation is obtained by postmultiplying the matrix by the invertible matrix obtained from the identity by performing on the latter the same column operation. These operations are:

(R1) Permuting the rows i and j.
(R2) Multiplying row i by an invertible element in A.
(R3) Adding to row i the row $j(\neq i)$ multiplied by an arbitrary element of A.
We add to these the following operation described in Lemma V.3.3, which also corresponds to premultiplication by an invertible matrix.
(R4) Altering two rows such that their first coefficients are replaced by their gcd and 0.

The corresponding column operations are denoted by $(C1), \cdots, (C4)$, respectively.
We wish to prove that an arbitrary matrix with coefficients in a principal ideal domain is equivalent to a unique (up to associates) Smith normal matrix. For ease of exposition, we split this existence and uniqueness theorem into two parts. We start with existence.

V.3.4 Theorem. *Let $\ell, \ell' > 0$ be integers. Any $\ell \times \ell'$-matrix J with coefficients in a principal ideal domain is equivalent to a Smith normal matrix.*

Proof. Without loss of generality, we assume that $J \neq 0$. We use row and column operations to bring J to a Smith normal matrix. Because these operations correspond to pre- and postmultiplication by invertible matrices, this procedure replaces J by an equivalent matrix.

First, we permute rows and columns so as to obtain a nonzero coefficient a_{11} in the (1-1)-position (the upper-left corner). Then we apply systematically (R4) to a_{11} and a_{i1}, with i such that $2 \leqslant i \leqslant \ell$, until the first column consists entirely of zeros below the coefficient a_{11} in the (1-1)-position. Of course, if $a_{11}|a_{i1}$, for some i, then we may use (R3) instead of (R4). In each operation, the length of a_{11}, namely the number of irreducible factors in the unique decomposition of a_{11}, see Theorem I.5.12, decreases.

Second, we proceed in the same way with the first column, applying (C4) systematically so that the first row consists entirely of zeros on the right of a_{11}. One of these operations may introduce a nonzero coefficient in the first column below a_{11}. However, if it does, then it also decreases the length of a_{11}. Then we apply (R4) again until all coefficients in the first column below a_{11} vanish. Because the length of a_{11} is finite, we obtain after finitely many steps a matrix of the form

$$
Q_0 J P_0^{-1} = \begin{pmatrix} a_{11} & 0 & \cdots & 0 \\ 0 & & & \\ \vdots & & J_0 & \\ 0 & & & \end{pmatrix}
$$

where P_0, Q_0 are invertible and J_0 is an $(\ell - 1) \times (\ell' - 1)$-matrix. Inductively, J_0 is equivalent to a Smith normal matrix. Therefore, there exist invertible matrices P_1, Q_1 such that

$$
Q_1 J P_1^{-1} = \begin{pmatrix} a_{11} & 0 & \cdots & 0 \\ 0 & a_{22} & \cdots & 0 \\ \vdots & \vdots & & \vdots \\ 0 & 0 & \cdots & a_{mm} \end{pmatrix}
$$

where $m = \min(\ell, \ell')$ and $a_{22} | \cdots | a_{mm}$. There remains to arrange that the first diagonal element a_{11} divides the second a_{22}. Now

$$
\begin{pmatrix} a_{11} & 0 \\ 0 & a_{22} \end{pmatrix} \begin{pmatrix} 1 & 0 \\ 1 & 1 \end{pmatrix} = \begin{pmatrix} a_{11} & 0 \\ a_{22} & a_{22} \end{pmatrix}
$$

so we apply (R4) once more to replace a_{11}, a_{22} by their gcd and zero in the first column. Because this gcd divides a_{22}, the proof is complete. □

We did not assume that $\ell' \leqslant \ell$. The theorem and its proof are valid for any rectangular matrix with coefficients in a principal ideal domain.

If $A = \mathbb{Z}$, namely, in the case of abelian groups, operations (R4) and (C4) are not needed. Indeed, assume, up to a permutation of rows and columns, that $a_{11} \neq 0$. If $a_{11}|a_{i1}$, for some i, then we use (R3) as mentioned in the proof. If not, then dividing a_{i1} by a_{11}, we get

$$
a_{i1} = q a_{11} + r,
$$

with $0 < r < |a_{11}|$. Taking row i minus q times row 1 replaces a_{i1} by r. Permuting rows 1 and i brings r into the (1-1)-position. The result follows inductively. The same remark holds true for $K[t]$, with K a field, because, in the previous proof, we just needed the division theorem.

V.3.5 Example. Consider the integral matrix

$$J = \begin{pmatrix} 2 & 10 \\ -4 & -6 \\ -2 & 4 \end{pmatrix}.$$

Applying successively the operations: row $2 + 2\times$ row 1, and row 3 + row 1 yields the matrix

$$\begin{pmatrix} 2 & 10 \\ 0 & 14 \\ 0 & 14 \end{pmatrix}.$$

Taking column 2 – 5× column 1 yields

$$\begin{pmatrix} 2 & 0 \\ 0 & 14 \\ 0 & 14 \end{pmatrix}.$$

Finally, row 3– row 2 yields

$$\begin{pmatrix} 2 & 0 \\ 0 & 14 \\ 0 & 0 \end{pmatrix}.$$

This is a Smith normal matrix. In terms of matrix multiplication, we premultiplied successively by

$$\begin{pmatrix} 1 & 0 & 0 \\ 2 & 1 & 0 \\ 0 & 0 & 1 \end{pmatrix}, \begin{pmatrix} 1 & 0 & 0 \\ 0 & 1 & 0 \\ 1 & 0 & 1 \end{pmatrix}, \text{ then by } \begin{pmatrix} 1 & 0 & 0 \\ 0 & 1 & 0 \\ 0 & -1 & 1 \end{pmatrix}.$$

Therefore,

$$Q = \begin{pmatrix} 1 & 0 & 0 \\ 0 & 1 & 0 \\ 0 & -1 & 1 \end{pmatrix} \begin{pmatrix} 1 & 0 & 0 \\ 0 & 1 & 0 \\ 1 & 0 & 1 \end{pmatrix} \begin{pmatrix} 1 & 0 & 0 \\ 2 & 1 & 0 \\ 0 & 0 & 1 \end{pmatrix} = \begin{pmatrix} 1 & 0 & 0 \\ 2 & 1 & 0 \\ -1 & -1 & 1 \end{pmatrix}.$$

It is postmultiplied by

$$P^{-1} = \begin{pmatrix} 1 & -5 \\ 0 & 1 \end{pmatrix},$$

and it is easily verified that

$$QJP^{-1} = \begin{pmatrix} 2 & 0 \\ 0 & 14 \\ 0 & 0 \end{pmatrix}.$$

One has to be careful when multiplying together the matrices of premultiplication and postmultiplication, because they do not multiply in the same order.

We next prove uniqueness up to associates. A **minor** of order k in an $\ell \times \ell'$-matrix J is the determinant of the $k \times k$ matrix extracted from J by deleting $\ell - k$ rows and $\ell' - k$ columns.

V.3.6 Theorem. *Let $D = \mathrm{diag}(d_1, \cdots, d_m)$ and $D' = \mathrm{diag}(d'_1, \cdots, d'_m)$ be equivalent Smith normal matrices with coefficients in A. Then d_i and d'_i are associate, for any i.*

Proof. Let $J = [a_{ij}]$ be an $\ell \times \ell'$-matrix and $Q = [b_{st}]$ an $\ell \times \ell$-matrix. The (i,j)-coefficient of QJ is $\sum_{h=1}^{\ell} b_{ih} a_{hj}$. Therefore, the rows of QJ are linear combinations of the rows of J. So, a minor of QJ of order k, say, is a linear combination of minors of J of order k. Similarly, if P is an invertible $\ell' \times \ell'$-matrix, the minors of JP^{-1} of order k are linear combinations of minors of J of the same order k.

Assume that $D = QJP^{-1}$, as in Theorem V.3.4. Because $J = Q^{-1}DP$, the minors of J of order k are linear combinations of minors of D of order k, and conversely. Therefore, these two collections of minors generate the same ideal in A, see Exercise I.3.11. Because A is principal, this ideal is of the form $c_k A$, with c_k a gcd of the minors of order k, see the proof of Theorem I.5.5. For instance, $d_1 = c_1$ is a gcd of the minors of order 1, that is, of the coefficients of J. Similarly, $c_2 = d_1 d_2$ is a gcd of the minors of order 2. Moreover, d_2 is the quotient of c_2 by c_1, which we denote, abusing notation, by $d_2 = c_2 c_1^{-1}$. Inductively, $d_k = c_k c_{k-1}^{-1}$, for $k \geq 2$. The gcd of minors of a fixed order being unique up to associates, so are the d_i. $\qquad\square$

We have proved that there exists a, unique up to associates, Smith normal matrix which belongs to the equivalence class of a given matrix J. It is called its **Smith normal form**. The diagonal elements $d_1, \cdots, d_m$ are called the **invariant factors** of J.

As seen in the proof of Theorem V.3.6, if c_k is the gcd of the minors of order k in J, then the Smith normal form of J is $D = \mathrm{diag}(d_1, \cdots, d_m)$ where $d_1 = c_1$, $d_2 = c_2 c_1^{-1}, \cdots d_m = c_m c_{m-1}^{-1}$. This allows to compute the Smith normal form (at least for matrices of reasonably small size).

In Example V.3.5 above, the gcd of the coefficients of the matrix J is $d_1 = c_1 = 2$. There are three minors of order two, namely

$$\begin{vmatrix} 2 & 10 \\ -4 & -6 \end{vmatrix} = 28, \begin{vmatrix} 2 & 10 \\ -2 & 4 \end{vmatrix} = 28 \text{ and } \begin{vmatrix} -4 & -6 \\ -2 & 4 \end{vmatrix} = -28.$$

Therefore, $c_2 = 28$ and $d_2 = c_2 c_1^{-1} = \frac{28}{2} = 14$. The Smith normal form is then

$$D = \mathrm{diag}(2, 14) = \begin{pmatrix} 2 & 0 \\ 0 & 14 \\ 0 & 0 \end{pmatrix}.$$

Applying Theorem V.3.4 to an inclusion morphism between finitely generated free modules, we obtain the following corollary.

V.3.7 Corollary. *Let L be a free module of finite rank ℓ and L' a submodule of L. Then there exists a basis $\{e_1, \cdots, e_\ell\}$ of L and elements $d_1, \cdots, d_\ell$ of A such that $d_1 | \cdots | d_\ell$ and the nonzero elements of $\{e_1 d_1, \cdots, e_\ell d_\ell\}$ constitute a basis of L'.*

Proof. Because of Corollary V.2.6, L' is finitely generated of rank $\ell' \leqslant \ell$. If $L' = 0$, taking any basis of L and setting all d_i equal to zero will do. Otherwise, let $\mathcal{E}' = \{e'_1, \cdots, e'_\ell\}$ and $\mathcal{F}' = \{f'_1, \cdots, f'_{\ell'}\}$ be bases of L, L', respectively, such that J is the matrix of the inclusion of L' into L with respect to these bases. Theorem V.3.4 asserts the existence of invertible matrices P, Q such that $QJP^{-1} = D = \mathrm{diag}(d_1, \cdots, d_{\ell'})$ is the Smith normal form of J. Because P, Q are invertible, they transform $\mathcal{E}', \mathcal{F}'$ into new bases $\mathcal{E} = \{e_1, \cdots, e_\ell\}$ and $\mathcal{F} = \{f_1, \cdots, f_{\ell'}\}$, respectively. With respect to $\mathcal{E}, \mathcal{F}$, the matrix of inclusion is $D = \mathrm{diag}(d_1, \cdots, d_{\ell'})$. Therefore, $f_1 = e_1 d_1, \cdots, f_{\ell'} = e_{\ell'} d_{\ell'}$. We finish the proof by setting $d_{\ell'+1} = \cdots = d_\ell = 0$. □

V.3.8 Examples.

(a) Let $A = K$ be a field, L a vector space and L' a subspace of L. We assume that L and L' have dimensions ℓ and ℓ', respectively. The Incomplete Basis Theorem of linear algebra asserts that any basis $\{e_1, \cdots, e_{\ell'}\}$ of L' can be completed to a basis $\{e_1, \cdots, e_{\ell'}, e_{\ell'+1}, \cdots, e_\ell\}$ of L. In the Corollary, this corresponds to setting $d_1 = \cdots = d_{\ell'} = 1$ and $d_{\ell'+1} = \cdots = d_\ell = 0$. The Smith normal matrix is $\mathrm{diag}(1, \cdots, 1, 0, \cdots, 0)$. In this example, we have $d_i = 1$, for $i \leqslant \ell'$, and $d_i = 0$, for $i > \ell'$. One indeed has $1 | \cdots | 1 | 0 | \cdots | 0$, see the remarks just preceding Lemma I.5.3.

(b) The proof of Corollary V.3.7 shows how to compute the new bases $\mathcal{E}, \mathcal{F}$ starting from the old ones $\mathcal{E}', \mathcal{F}'$. If, in the notation of the Corollary V.3.7, we have $Q^{-1} = [a_{ij}], P^{-1} = [b_{ij}]$, then $e_i = \sum_{j=1}^{\ell} e'_j a_{ji}$, for any i, and $f_h = \sum_{k=1}^{n} f'_k b_{kh}$, for any h.

For instance, in Example V.3.5, we have $A = \mathbb{Z}, L = \mathbb{Z}^{(3)}$ with the canonical basis $\{e'_1, e'_2, e'_3\}$ and L' the submodule of L generated by the vectors

$$\begin{cases} f'_1 = 2e'_1 - 4e'_2 - 2e'_3 \\ f'_2 = 10e'_1 - 6e'_2 + 4e'_3. \end{cases}$$

Because

$$Q = \begin{pmatrix} 1 & 0 & 0 \\ 2 & 1 & 0 \\ -1 & -1 & 1 \end{pmatrix}, P^{-1} = \begin{pmatrix} 1 & -5 \\ 0 & 1 \end{pmatrix}$$

we have

$$Q^{-1} = \begin{pmatrix} 1 & 0 & 0 \\ -2 & 1 & 0 \\ -1 & 1 & 1 \end{pmatrix}.$$

Therefore,

$$\begin{cases} e_1 = e_1' - 2e_2' - e_3' \\ e_2 = e_2' + e_3' \\ e_3 = e_3' \end{cases}$$

and

$$\begin{cases} f_1 = f_1' = 2e_1' - 4e_2' - 2e_3' \\ f_2 = -5f_1' + f_2' = 14e_2' + 14e_3'. \end{cases}$$

We have indeed $f_1 = 2e_1$ and $f_2 = 14e_2$, as expected.

V.3.2   The Invariant Factors Theorem

The first classification theorem for finitely generated modules over a principal ideal domain follows from the existence and uniqueness of the Smith normal form. It says that such a module can be written as direct sum of cyclic modules determined by the invariant factors of the matrix of the inclusion morphism in a free resolution of the module. Moreover, this direct sum decomposition is unique up to isomorphism. Again, for clarity, we separate the existence and uniqueness parts of the theorem. Throughout, we let A be a principal ideal domain.

V.3.9 Theorem. The Invariant Factors Theorem.   *Let A be a principal ideal domain and M a finitely generated A-module. Then there exist $d_1, \cdots, d_\ell \in A$ such that $d_1 | \cdots | d_\ell$ and*

$$M \cong \bigoplus_{i=1}^{\ell} (A/d_i A).$$

Proof.   Because M is finitely generated, there exists a short exact sequence $0 \to L' \to L \to M \to 0$ with L, L' free of respective ranks ℓ, ℓ' such that $\ell' \leqslant \ell$, see Corollary V.2.7. Because of Corollary V.3.7, there exists a basis

$\{e_1, \cdots, e_\ell\}$ of L and elements $d_1, \cdots, d_\ell \in A$ such that $d_1 | \cdots | d_\ell$ and the nonzero elements of $\{e_1 d_1, \cdots, e_\ell d_\ell\}$ form a basis of L'. Then $L = \oplus_{i=1}^{\ell} e_i A$ and $L' = \oplus_{i=1}^{\ell} e_i d_i A$. Because $e_i d_i A \subseteq e_i A$, for any i, Exercise II.4.27 implies that

$$M \cong L/L' \cong \left(\bigoplus_{i=1}^{\ell} e_i A \right) \Big/ \left(\bigoplus_{i=1}^{\ell} (e_i d_i) A \right) \cong \bigoplus_{i=1}^{\ell} (e_i A/(e_i d_i) A).$$

Each e_i is a basis element, and, in the isomorphism $A \to e_i A$ given by $x \mapsto e_i x$, for $x \in A$, the image of the ideal $d_i A$ is $(e_i d_i) A$. Therefore, $A/d_i A \cong e_i A/(e_i d_i) A$. The statement follows. □

If n is the largest positive integer such that $d_n \neq 0$, then the formula in the theorem becomes

$$M \cong \bigoplus_{i=1}^{n} (A/d_i A) \bigoplus A^{(\ell-n)}.$$

One can assume that, in the previous decomposition, no d_i is invertible, because otherwise $A/d_i A = 0$. We make this assumption in the sequel.

For instance, if G is a finitely generated abelian group, then there exists a direct sum decomposition of the form

$$G \cong (\mathbb{Z}/d_1\mathbb{Z}) \oplus \cdots \oplus (\mathbb{Z}/d_n\mathbb{Z}) \oplus \mathbb{Z}^{(\ell-n)}$$

where $d_1 | \cdots | d_n \neq 0$, see the remarks after Theorem V.2.14.

We deduce the torsion and torsion-free parts of a finitely generated module. On the other hand, if A is a field, then a finitely generated A-module M is a finite-dimensional vector space, hence a free module, and we have $M \cong A^{(\ell)}$ (that is, the torsion part is zero).

V.3.10 Corollary. *In the notation of the theorem, if $n \leqslant \ell$ is the least positive integer such that $d_{n+1} = 0$, then*

$$T(M) \cong \bigoplus_{i=1}^{n} (A/d_i A) \quad and \quad M/T(M) \cong A^{(\ell-n)}.$$

Proof. Indeed, $A^{(\ell-n)}$ is free, hence torsion-free, because of Lemma V.2.12. On the other hand, any $A/d_i A$, with $i \leqslant n$, is torsion (because it is annihilated by d_i and $d_i \neq 0$, due to the definition of n). Therefore, $\oplus_{i=1}^{n} (A/d_i A)$ is torsion, because of Exercise V.2.11(d). The assertion then follows from uniqueness in Theorem V.2.14. □

We now prove that the direct sum decomposition of the Invariant Factors Theorem V.3.9 is unique up to isomorphism. Let M be a finitely generated A-module and $a \in A$. The map $f_a : M \to M$ defined by $x \mapsto xa$, for $x \in M$, is A-linear, because A is commutative. Its image $Ma = \{xa : x \in M\}$ and its kernel $M_a = \{x \in M : xa = 0\}$ are submodules of M and we have a short exact sequence $0 \to M_a \to M \to Ma \to 0$.

Because M is finitely generated, so is its image Ma, see Exercise II.2.12. Because A is principal, M_a is also finitely generated due to Corollary V.2.8. It is thus an exact sequence of finitely generated A-modules.

V.3.11 Lemma. *Let $a \in A$ be nonzero.*

(a) *If M, N are finitely generated A-modules, then $(M \oplus N)a = Ma \oplus Na$ and $(M \oplus N)_a = M_a \oplus N_a$.*

(b) *If $M = A/dA$, with d nonzero and coprime with a, then $Ma = M$ and $M_a = 0$.*

(c) *If $M = A/dA$, with d nonzero and $d = ab$, then $Ma \cong A/bA$ and $M_a \cong A/aA$.*

Proof. (a) The first statement is obvious and the second comes from the fact that $(x + y)a = 0$ with $x \in M, y \in N$, happens if and only if both $xa = 0$ and $ya = 0$, because the sum is direct.

(b) Because a, d are coprime, there exist $s, t \in A$ such that $sa + td = 1$. Therefore, for $x \in M$, we have $x = x(sa + td) = xsa \in Ma$, so $M = Ma$. The second statement follows from the short exact sequence $0 \to M_a \to M \to Ma \to 0$.

(c) The morphism f_a maps A/dA onto aA/dA. Define $g : A \to aA$ by $x \mapsto ax$, for $x \in A$. Because $a \neq 0$ and A is a principal ideal domain, it is injective. Because it is obviously surjective, it is an isomorphism. It maps the ideal bA onto $baA = dA$. Passing to cokernels, we get $A/bA \cong aA/dA$, hence the first statement. Let $x + dA \in M_a$. Then $(x + dA)a = 0$, hence $xa \in dA$ which gives $y \in A$ such that $ax = dy = aby$, that is, $x = by \in bA$, because $a \neq 0$. Therefore, $M_a = bA/dA \cong A/aA$. $\square$

We prove uniqueness of the invariant factors decomposition.

V.3.12 Theorem. *Assume that*

$$\bigoplus_{i=1}^{n}(A/d_iA) \oplus A^{(\ell-n)} \cong \bigoplus_{j=1}^{s}(A/d_j'A) \oplus A^{(t-s)}$$

where $d_1|\cdots|d_n \neq 0$ and $d_1'|\cdots|d_s' \neq 0$ then $\ell = t, n = s$, and d_i, d_i' are associate and non invertible for any i.

Proof. It follows from uniqueness in Theorem V.2.14 that

$$\bigoplus_{i=1}^{n}(A/d_iA) \cong \bigoplus_{i=1}^{s}(A/d_i'A) \text{ and } A^{(\ell-n)} \cong A^{(t-s)}.$$

Because the rank of a finitely generated free module is unique, see Theorem V.2.2, $\ell - n = t - s$. We may thus assume, without loss of generality, that M is torsion, that is,

$$M = \bigoplus_{i=1}^{n}(A/d_iA) \cong \bigoplus_{i=1}^{s}(A/d_i'A).$$

We claim that $n = s$. If not, then we may assume, again without loss of generality, that $n > s$. Because $d_1 \neq 0$, (for, $n > 1$!) there exists an irreducible element $p \in A$

which divides d_1. Then p divides d_i, for any i. Due to Lemma V.3.11(c), we have $(A/d_iA)_p \cong A/pA$, for any i. Hence,

$$M_p = \left(\bigoplus_{i=1}^{n}(A/d_iA)\right)_p \cong \bigoplus_{i=1}^{n}(A/d_iA)_p \cong (A/pA)^{(n)}.$$

Because p is irreducible, A/pA is a field. So, M_p is an n-dimensional A/pA-vector space.

We apply the same argument to the second direct sum decomposition. The irreducible element p does not necessarily divide d_1', so there exists $r \leqslant s$ such that p does not divide $d_1', \cdots, d_r'$ but does divide $d_{r+1}', \cdots, d_s'$. Because p is irreducible, it is coprime with each of $d_1', \cdots, d_r'$. Lemma V.3.11 implies that $(A/d_j'A)_p = 0$, if $j \leqslant r$, and $(A/d'jA)_p \cong A/pA$, if $j > r$. Then

$$M_p \cong \bigoplus_{j=1}^{s}(A/d_j'A)_p \cong \bigoplus_{j>r}(A/d_j'A)_p \cong (A/pA)^{(s-r)}.$$

Therefore, M_p is an A/pA-vector space of dimension $s - r \leqslant s < n$, a contradiction. This establishes our claim that $n = s$.

There remains to prove that d_i and d_i' are associate, for any i. We use induction on the length of d_n, that is, on the number of irreducible factors in the decomposition of Theorem I.5.12. If the length of d_n is one, then d_n is irreducible. Because $pM = d_nM = 0$, we have $p(A/d_iA) = 0$, for any i, hence $pA \subseteq d_iA$. But p is irreducible, hence pA is maximal, see Theorem I.5.7. So $pA = d_iA$, for any i. Also, $pM = 0$ and the same reasoning applied to the second decomposition give $pA = d_i'A$, for any i. Hence, $d_iA = d_i'A$, for any i, so d_i and d_i' are associate, see Lemma I.5.3.

In general, let p be an irreducible divisor of d_n. There exists m such that $1 \leqslant m \leqslant n$, the element p divides $d_{m+1}, \cdots, d_n$ but divides none of $d_1, \cdots, d_m$. If $i \leqslant m$, then p, d_i are coprime, hence $p(A/d_iA) \cong A/d_iA$, because of Lemma V.2.12. If $i > m$, then the same lemma yields $p(A/d_iA) \cong A/c_iA$, where c_i is such that $d_i = pc_i$. Hence,

$$pM = \bigoplus_{i=1}^{m}(A/d_iA) \oplus \bigoplus_{j=m+1}^{n}(A/c_jA)$$

and we have $d_1|\cdots|d_m|c_{m+1}|\cdots|c_n$.

The same procedure applied to the second decomposition yields

$$pM = \bigoplus_{i=1}^{m'}(A/d_i'A) \oplus \bigoplus_{j=m'+1}^{n}(A/c_j'A)$$

where $1 \leqslant m' \leqslant n$, $d_j' = pc_j'$, for $j > m'$ and $d_1'|\cdots|d_{m'}'|c_{m'+1}'|\cdots|c_n'$. The length of c_n equals that of d_n minus one. The induction hypothesis applied to pM gives $m = m'$, the elements d_i, d_i' are associate for $i \leqslant m$, and c_i, c_i' associate for $i > m$. Because

$d_i = pc_i$ and $d'_i = pc'_i$ for $i > m$, this implies that d_i, d'_i are also associate. The proof is now complete. $\square$

V.3.13 Examples.

(a) In Example V.3.8(b), we considered the free $\mathbb{Z}$-module $L = \mathbb{Z}^{(3)}$ and its submodule $L' = \mathbb{Z}^{(2)}$ such that the inclusion of L' into L is given in the canonical bases by the matrix

$$\begin{pmatrix} 2 & 10 \\ -4 & -6 \\ -2 & 4 \end{pmatrix}.$$

Its Smith normal form, obtained in Example V.3.5, is $D = \text{diag}(2, 14)$, that is, has invariant factors $d_1 = 2, d_2 = 14$. Because of the Invariant Factors Theorem, $M = L/L'$ is the abelian group $M \cong \mathbb{Z}/2\mathbb{Z} \oplus \mathbb{Z}/14\mathbb{Z}$. In this example, $M = T(M)$ and so $M/T(M) = 0$.

(b) Assume we wish to describe the abelian group

$$G = <a, b, c : 3a + 6b = 0, 3b + 3c = 0, 6a + 3b + 15c>,$$

that is, the group having three generators $\{a, b, c\}$ satifying the relations $3a + 6b = 0, 3b + 3c = 0, 6a + 3b + 15c = 0$. Then G is the image of the free abelian group $\mathbb{Z}^{(3)}$ with basis $\{a, b, c\}$ having as kernel the subgroup G' generated by $3a + 6b, 3b + 3c, 6a + 3b + 15c$. Let $g : \mathbb{Z}^{(3)} \to G$ be the projection. Take another copy of $\mathbb{Z}^{(3)}$ with basis $\{a', b', c'\}$ and let $f : \mathbb{Z}^{(3)} \to \mathbb{Z}^{(3)}$ be the composite of the map $\mathbb{Z}^{(3)} \to G'$ given by

$$a' \mapsto 3a + 6b, \ b' \mapsto 3b + 3c, \ c' \mapsto 6a + 3b + 15c$$

with the inclusion of G' into $\mathbb{Z}^{(3)}$. We have an exact sequence

$$\mathbb{Z}^{(3)} \xrightarrow{f} \mathbb{Z}^{(3)} \xrightarrow{g} G \to 0$$

The matrix of f in the canonical bases is

$$\begin{pmatrix} 3 & 0 & 6 \\ 0 & -3 & 3 \\ 6 & -3 & 15 \end{pmatrix}.$$

Its Smith normal form is easily found to equal $D = \text{diag}(3, 3, 0)$, with invariant factors $d_1 = d_2 = 3, d_3 = 0$. The theorem gives $G \cong \mathbb{Z}/3\mathbb{Z} \oplus \mathbb{Z}/3\mathbb{Z} \oplus \mathbb{Z}$. Here, $T(G) \cong \mathbb{Z}/3\mathbb{Z} \oplus \mathbb{Z}/3\mathbb{Z}$ and $G/T(G) \cong \mathbb{Z}$.

V.3.3 The primary decomposition

Although the idea of the proof of the Invariant Factors Theorem is simple, the proof itself is relatively long and combines several tools, coming from different areas (linear algebra, module theory, arithmetic of principal ideal domains). This is the case for many classification theorems. However, the Invariant Factors Theorem still does not answer all questions. It says that a finitely generated module over a principal ideal domain decomposes as direct sum of cyclic modules. But cyclic modules themselves may decompose as direct sums of smaller cyclic modules. For instance, $\mathbb{Z}/6\mathbb{Z} \cong \mathbb{Z}/2\mathbb{Z} \oplus \mathbb{Z}/3\mathbb{Z}$. It is natural to ask whether or not a finitely generated module over a principal ideal domain decomposes uniquely as direct sum of indecomposable summands, that is, summands which cannot be further decomposed nontrivially into a direct sum, and to characterise the latter. Answering these questions is the purpose of the present subsection. Our first proposition should be compared to Example IV.2.33(b). As usual, A stands for a principal ideal domain. We first consider the case of cyclic modules.

V.3.14 Proposition. *Let $a, b \in A$ be nonzero. Then there exists an A-linear map $f : A/(ab)A \to A/aA \oplus A/bA$ defined by $x + (ab)A \mapsto (x + aA, x + bA)$, for $x \in A$. Moreover, the following conditions are equivalent:*
  (a) *a, b are coprime.*
  (b) *f is injective.*
  (c) *f is surjective.*
  (d) *f is an isomorphism.*

Proof. Let $p_a : A \to A/aA, p_b : A \to A/bA$ be the projections. The universal property (of the product, which equals the coproduct in mod$A$) yields the morphism $p : A \to A/aA \oplus A/bA$ given by $x \mapsto (x + aA, x + bA)$, for $x \in A$.

The kernel of p consists of the $x \in A$ such that $x \in aA$ and also $x \in bA$. That is, Ker $p = aA \cap bA$. Now $(ab)A \subseteq aA \cap bA$. Therefore, p induces, by passing to quotients, an A-linear map $f : A/(ab)A \to A/aA \oplus A/bA$ given by the formula $x + (ab)A \mapsto (x + aA, x + bA)$, for $x \in A$, whose kernel is $(aA \cap bA)/(ab)A$, as seen from the commutative diagram with exact rows and columns.

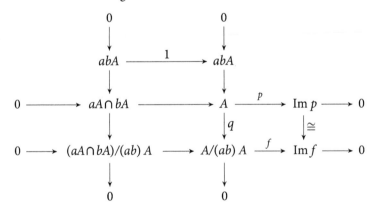

where $q : A \to A/(ab)A$ is the projection, so that we have $fq - p$ and, consequently, $\operatorname{Im} p \cong \operatorname{Im} f$.

Hence, f is injective if and only if $aA \cap bA = (ab)A$, that is, if and only if the lcm of a, b equals their product ab. But this amounts to saying that a, b are coprime, see Exercise I.5.3. This proves the equivalence of (a) and (b).

Assume that a, b are coprime. Then we have $aA + bA = A$. Let us assume that $(y+aA, z+bA) \in A/aA \oplus A/bA$. Because $y - z \in A$, there exist $s, t \in A$ such that $y - z = as + bt$ and then $x = y - as = z + bt$ satisfies $(y + aA, z + bA) = (x + aA, x + bA) = f(x + (ab)A)$. So $f : A/(ab)A \to A/aA \oplus A/bA$ is surjective, and (a) implies (c).

Conversely, if f is surjective, then there exists $x \in A$ such that $f(x + (ab)A) = (1 + aA, 0 + bA)$ from which we get $s, t \in A$ such that $x - 1 = -as$ and $x - 0 = tb$. But then $1 = x + as = as + bt$. Therefore, a, b are coprime, that is, (c) implies (a).

Finally, (d) is the conjunction of (b) and (c). $\qquad\square$

If $A = \mathbb{Z}$, then the groups $\mathbb{Z}/(ab)\mathbb{Z}$ and $\mathbb{Z}/a\mathbb{Z} \oplus \mathbb{Z}/b\mathbb{Z}$ are finite sets with the same cardinality, so it is not surprising that f is injective if and only if it is surjective. The point here is that the statement holds true over an arbitrary principal ideal domain.

V.3.15 Corollary. *Let $a \in A$ be nonzero and $a = up_1^{t_1} \cdots p_m^{t_m}$ the decomposition of a into pairwise nonassociate irreducible elements p_i, with u invertible, then*

$$A/aA \cong \bigoplus_{i=1}^{m} (A/p_i^{t_i}A).$$

Proof. This follows from Proposition V.3.14 above and induction on m. The induction step uses that $p_m^{t_m}$ is coprime with $up_1^{t_1} \cdots p_{m-1}^{t_{m-1}}$. $\qquad\square$

This shows how to decompose into direct sum a cyclic module. Because of Corollary V.3.10, finitely generated torsion modules are direct sums of cyclic modules of the form $A/d_1A \oplus \cdots \oplus A/d_nA$, with $d_1 | \cdots | d_n \neq 0$. In order to apply Corollary V.3.15 above to each summand, we use the fact that if an irreducible element of A divides d_i, then it divides d_j, for any j such that $j \geq i$. The next lemma shows how to handle these modules.

V.3.16 Lemma. *Let M be an A-module and $a \in A, a \neq 0$, be such that $Ma = 0$. If $a = up_1^{t_1} \cdots p_m^{t_m}$ is the decomposition of a into pairwise nonassociate irreducible elements p_i, with u invertible, then*

$$M = M_1 \oplus \cdots \oplus M_m$$

where $M_i p_i^{t_i} = 0$, for any i. Moreover, the M_i are uniquely determined by these conditions.

Proof. For any i, let $a_i = u(\prod_{j \neq i} p_j^{t_j}) \in A$, then $a = p_i^{t_i} a_i$. We first prove that if

$$M = M_1 \oplus \cdots \oplus M_m$$

where $M_i p_i^{t_i} = 0$, for any i, then $M_i = Ma_i$ (in particular, the M_i are uniquely determined). Indeed, if $j \neq i$, then $p_j^{t_j} | a_i$ so $M_j a_i = 0$. Using Lemma V.3.14, we have $Ma_i = M_1 a_i \oplus \cdots \oplus M_m a_i = M_i a_i \subseteq M_i$. On the other hand, a_i and $p_i^{t_i}$ are coprime, so there exist $s, t \in A$ such that $1 = a_i s + p_i^{t_i} t$. Then, for any $x \in M_i$, we have $x p_i^{t_i} = 0$, hence $x = x a_i s = (xs)a_i \in Ma_i$. Therefore, $M_i \subseteq Ma_i$. Our claim is established.

We next prove the existence of a direct sum decomposition

$$M = Ma_1 \oplus \cdots \oplus Ma_m.$$

Indeed, $(Ma_i)p_i^{t_i} = Ma = 0$, so such a decomposition satisfies our requirements. Because 1 is a gcd of $a_1, \cdots, a_m$, there exist $s_1, \cdots s_m \in A$ such that $1 = a_1 s_1 + \cdots + a_m s_m$, see Exercise I.5.5. Therefore, any $x \in M$ can be written as

$$x = (xs_1)a_1 + \cdots + (xs_m)a_m \in \sum_{i=1}^{m} Ma_i,$$

that is, $M = \sum_{i=1}^{m} Ma_i$. We prove that the sum is direct. Take $y \in Ma_i \cap (\sum_{j \neq i} Ma_j)$. Because $y \in Ma_i$, we have $yp_i^{t_i} = 0$. But also $y \in \sum_{j \neq i} Ma_j$, hence $ya_i = 0$. But $a_i, p_i^{t_i}$ are coprime, so there exist $s, t \in A$ as above such that $1 = a_i s + p_i^{t_i} t$. But then $y = ya_i s + yp_i^{t_i} t = 0$, and the proof is complete. $\square$

For instance, let G be a torsion finitely generated abelian group. Then it is finite, see Example V.3.17(b). Let n be the cardinality of G. Decomposing n into prime numbers as $n = p_1^{t_1} \cdots p_m^{t_m}$, we get a direct sum decomposition $G = \oplus_{i=1}^{m} G_i$, where $G_i p_i^{t_i} = 0$, for any i.

Let p be an irreducible element in A. An A-module M is called *p-primary* if any of its elements is annihilated by a power of p, that is, if, for any $x \in M$, there exists an integer $t > 0$ such that $xp^t = 0$. In the decomposition of Lemma V.3.16 above, each M_i is p_i-primary. This decomposition is called the *primary decomposition* of M and the M_i its *primary components*.

V.3.17 Example. Consider the abelian group

$$M = \mathbb{Z}/12\mathbb{Z} \oplus \mathbb{Z}/18\mathbb{Z} \oplus \mathbb{Z}/20\mathbb{Z}.$$

Applying Corollary V.3.15, we get

$$\mathbb{Z}/12\mathbb{Z} \cong \mathbb{Z}/3\mathbb{Z} \oplus \mathbb{Z}/4\mathbb{Z}, \quad \mathbb{Z}/18\mathbb{Z} \cong \mathbb{Z}/2\mathbb{Z} \oplus \mathbb{Z}/9\mathbb{Z}$$

and

$$\mathbb{Z}/20\mathbb{Z} \cong \mathbb{Z}/4\mathbb{Z} \oplus \mathbb{Z}/5\mathbb{Z}.$$

The primary decomposition of M is obtained by grouping together the p-primary summands for any fixed p:

$$M \cong (\mathbb{Z}/2\mathbb{Z} \oplus \mathbb{Z}/4\mathbb{Z} \oplus \mathbb{Z}/4\mathbb{Z}) \oplus (\mathbb{Z}/3\mathbb{Z} \oplus \mathbb{Z}/9\mathbb{Z}) \oplus \mathbb{Z}/5\mathbb{Z}.$$

The 2-primary component of M is $\mathbb{Z}/2\mathbb{Z} \oplus \mathbb{Z}/4\mathbb{Z} \oplus \mathbb{Z}/4\mathbb{Z}$, its 3-primary component is $\mathbb{Z}/3\mathbb{Z} \oplus \mathbb{Z}/9\mathbb{Z}$ and its 5-primary component is $\mathbb{Z}/5\mathbb{Z}$. The decomposition of the Invariant Factors Theorem V.3.9 is obtained by rearranging the summands, first grouping together the largest power of each irreducible element, then continuing by descending induction:

$$M \cong \mathbb{Z}/2\mathbb{Z} \oplus (\mathbb{Z}/4\mathbb{Z} \oplus \mathbb{Z}/3\mathbb{Z}) \oplus (\mathbb{Z}/4\mathbb{Z} \oplus \mathbb{Z}/9\mathbb{Z} \oplus \mathbb{Z}/5\mathbb{Z}).$$

At the next step, we apply Corollary V.3.15 to each of the bracketed terms, obtaining

$$M \cong \mathbb{Z}/2\mathbb{Z} \oplus \mathbb{Z}/12\mathbb{Z} \oplus \mathbb{Z}/180\mathbb{Z}.$$

The invariant factors are thus $2|12|180$.

We summarise our findings in the following theorem.

V.3.18 Theorem. *Let A be a principal ideal domain and M a finitely generated A-module. Then there exist unique integers ℓ, m, n, irreducible elements $p_1, \cdots, p_m$ in A, unique up to associates, and integers $t_{ij} \geq 0$, with $1 \leqslant i \leqslant n, 1 \leqslant j \leqslant m$, such that*

$$M \cong \bigoplus_{i=1}^{n} \left(\bigoplus_{j=1}^{m} A/p_j^{t_{ij}}A \right) \oplus A^{(\ell)}.$$

Proof. Because of Theorem V.2.14, $M \cong T(M) \oplus M/T(M)$. Because of Lemma V.2.12, $M/T(M)$ is free. It is finitely generated, because it is a quotient of M, so there exists a unique $\ell \geq 0$ such that $M/T(M) \cong A^{(\ell)}$. The Invariant Factors Theorem V.3.9 and the Uniqueness Theorem V.3.12 assert the existence of a unique n and elements $d_1, \cdots, d_n \in A$, unique up to associates, such that $d_1|\cdots|d_n \neq 0$ and $T(M) \cong \bigoplus_{i=1}^{n}(A/d_iA)$. Let $\{p_1, \cdots, p_m\}$ be the irreducible factors of d_n (hence of all d_i). Then, for any i, j with $1 \leqslant i \leqslant n, 1 \leqslant j \leqslant m$, there exists $t_{ij} \geq 0$ such that

$$A/d_iA \cong \bigoplus_{j=1}^{m} A/p_j^{t_{ij}}A.$$

We get the decomposition in the statement. The p_j are unique up to associates, because so are the d_i, and m is unique because it is the length of d_n. $\square$

V.3.4 Indecomposable finitely generated modules

In the introduction to the previous subsection, we asked whether there exists a unique direct sum decomposition of a finitely generated module into modules which cannot be further decomposed in a nontrivial way. Our aim is to prove that this is the case with the decomposition of Theorem V.3.18 above. We first define indecomposability.

V.3.19 Definition. Let A be an algebra. A nonzero module M is **indecomposable** if, whenever $M \cong M_1 \oplus M_2$, we have $M_1 = 0$ or $M_2 = 0$. Otherwise, M is **decomposable**.

In particular, the zero module is not indecomposable.

V.3.20 Example.
  (a) Assume that $A = K$ is a field. A K-vector space is indecomposable if and only if it is one-dimensional.
  (b) Let A be any algebra. Recall from Exercise II.3.18 that an A-module S is called simple if it has non nonzero proper submodule. Plainly, simple modules are indecomposable.
      In particular, let A be a principal ideal domain and $p \in A$ an irreducible element. Because of Theorem I.5.7, A/pA is a field. So, it has no nonzero proper ideals (submodules). Therefore, it is simple and hence indecomposable.
  (c) Let A be an integral domain. The module A_A is indecomposable. Indeed, if $A \cong M_1 \oplus M_2$, with M_1, M_2 nonzero submodules (ideals) of A, then there exist nonzero elements $a_1 \in M_1, a_2 \in M_2$. But then $a_1 a_2 \neq 0$ and $a_1 a_2 \in M_1 \cap M_2$, a contradiction.

V.3.21 Lemma. *Let A be a principal ideal domain and $p \in A$ irreducible. The only submodules of $A/p^t A$ are the $p^i M \cong A/p^{t-i}A$, for $0 \leqslant i \leqslant t$.*

Proof. Because of Theorem II.4.27, the submodules of M are in bijection with the submodules of A containing $p^t A$. Any submodule (ideal) of A is of the form aA, for some a. The inclusion $p^t A \subseteq aA$ amounts to saying that $a|p^t$, that is, $a = up^i$, for some i such that $0 \leqslant i \leqslant t$ and u invertible. The submodules of M are thus of the form $p^i M \cong p^i A/p^t A \cong A/p^{t-i}A$, where we applied Lemma V.3.14. Clearly, these submodules are pairwise distinct. $\qquad\square$

Thus, the sequence of submodules of M can be written as

$$\{0\} = p^t M \subsetneq p^{t-1} M \subsetneq \cdots \subsetneq pM \subsetneq M.$$

In this sequence, the quotient of two consecutive submodules is always equal to A/pA, indeed

$$p^i M/p^{i+1} M \cong (p^i A/p^t A)/(p^{i+1} A/p^t A) \cong p^i A/p^{i+1} A \cong A/pA$$

using again Proposition V.3.14. More generally, because the quotient of A/p^tA by A/p^iA is isomorphic to $A/p^{t-i}A$, the quotients of M are isomorphic to its submodules. From another point of view, A/pA is a field, because p is irreducible, hence A/p^tA is an A/pA-vector space and Lemma V.3.21 implies that $\dim_{A/pA}(A/p^tA) = t$.

The Hasse quiver of the submodule lattice of M is of the form

where similar notches on the arrows mean that the corresponding quotients are isomorphic.

We prove that the direct summands in the decomposition of Theorem V.3.18 are indecomposable.

V.3.22 Corollary. *Let A be a principal ideal domain. The indecomposable objects in the category $\mathrm{mod}A$ of finitely generated A-modules are, up to isomorphism, A_A and the modules of the form A/p^tA, with p irreducible and $t > 0$ an integer.*

Proof. Because of Example V.3.20(c), A_A is indecomposable. We now prove that $M = A/p^tA$, with p irreducible and $t > 0$ an integer, is indecomposable. If $M = M_1 \oplus M_2$ with $M_1, M_2 \neq 0$, then, because of Lemma V.3.21, the submodules M_1 and M_2 of M both contain the minimal nonzero submodule $p^{t-1}A/p^tA \cong A/pA$ of A/p^tA. Therefore, $M_1 \cap M_2 \neq 0$, a contradiction.

Conversely, assume that M is an indecomposable A-module. Because of Theorem V.3.18, M is isomorphic either to A_A or to some A/p^tA, with p irreducible and $t > 0$. $\qquad\square$

For instance, let K be an algebraically closed field. A polynomial in $K[t]$ is irreducible if and only if it is associated to a polynomial of the form $(t - \alpha)$ with $\alpha \in K$. Hence, the indecomposable finitely generated $K[t]$-modules are $K[t]$ and the $K[t] / \langle (t - \alpha)^n \rangle$ for some scalar $\alpha \in K$ and integer $n > 0$. Recall from Example I.3.11(h) that $K[t]$ is isomorphic to the path algebra A of the quiver

$$\circ \overset{\curvearrowleft}{} \alpha$$

Understanding a category means understanding its objects and its morphisms. Due to Theorem V.3.18, any object in mod$A$ decomposes uniquely as direct sum of indecomposables, and these are listed in Corollary V.3.22 . We now describe morphisms between indecomposable objects. Because of Corollary III.3.13, this gives a good description of morphism in mod$A$.

V.3.23 Lemma. *Let A be a principal ideal domain, p, q irreducible elements in A and $t, s > 0$ integers. Then we have:*

(a) $\text{Hom}_A(A, A) \cong A$.

(b) $\text{Hom}_A(A, A/p^t A) \cong A/p^t A$.

(c) $\text{Hom}_A(A/p^t A, A) = 0$.

(d) $\text{Hom}_A(A/p^t A, A/q^s A) \neq 0$ *if and only if p, q are associate.*

Proof. (a) and (b) follow from Theorem II.3.9. We prove (c). Because $A/p^t A$ is torsion, so are its images. But A is free, and, in particular, torsion-free. Hence, so are its submodules. Thus, the image of a morphism from $A/p^t A$ is both torsion and torsion-free: it must be zero.

There remains to prove (d). Assume that p, q are associate. Then, for any integer $n > 0$, p^n and q^n are associate (hence generate the same ideal). Let $m \leqslant \min(s, t)$ be positive, then $A/p^m A$ is at the same time a quotient of $A/p^t A$ and a submodule of $A/q^s A = A/p^s A$. Therefore, there exists a nonzero projection $p_m : A/p^t A \to A/p^m A$ and a nonzero injection $j_m : A/p^m A \to A/p^s A$. The composite $f_m = j_m p_m$ is a nonzero morphism from $A/p^t A$ to $A/p^s A$, because j_m is a nonzero monomorphism and p_m a nonzero epimorphism.

Conversely, assume that there exists a nonzero morphism from $A/p^t A$ to $A/q^s A$. Because of the Isomorphism Theorem II.4.21, its image is at the same time a quotient of $A/p^t A$ and a submodule of $A/q^s A$. That is, there exists a positive integer $m \leqslant \min(s, t)$ such that $A/p^m A \cong A/q^m A$. Because these modules are isomorphic, so are their minimal nonzero submodules A/pA and A/qA. Considering an isomorphism $A/pA \to A/qA$, the fact that $p(A/pA) = 0$ implies that $p(A/qA) = 0$ as well, so $pA \subseteq qA$. Considering the inverse isomorphism gives $qA \subseteq pA$. Hence, $pA = qA$ and p, q are associate due to Lemma I.5.3. $\qquad\square$

One can construct explicitly these morphisms. Indeed, the proof of Theorem II.3.9 gives an explicit expression for the isomorphisms in cases (a) and (b). As for (d), because of the Isomorphism Theorem II.4.21, any morphism from $A/p^t A$ to $A/p^s A$ can be written in the form f_m as in the proof of sufficiency above. The figure below explains the construction of f_m.

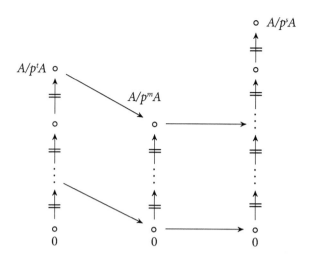

While the preceding results give a precise description of the category mod$A$, with A a principal ideal domain, they do not say anything about infinitely generated A-modules (so they give only partial information on the category Mod$A$).

For instance, it follows from Corollary V.3.22 that the indecomposable objects in mod$\mathbb{Z}$ are $\mathbb{Z}$ and the $\mathbb{Z}/p^t\mathbb{Z}$, with p prime and $t > 0$. But there are several important infinitely generated abelian groups, such as $\mathbb{Q}$, for instance, see Example V.2.13(a). The previous results do not describe their properties.

Exercises of Section V.3

1. Reduce each of the following integral matrices to its Smith normal form, giving the transformation matrices in each case:

(a) $\begin{pmatrix} 2 & 8 & 6 \\ 4 & -2 & 10 \end{pmatrix}$ (b) $\begin{pmatrix} 1 & 2 & 3 \\ 4 & 5 & 6 \\ 7 & 8 & 9 \end{pmatrix}$ (c) $\begin{pmatrix} 2 & 12 & 4 \\ 6 & 14 & -4 \\ -8 & 6 & 8 \end{pmatrix}$

(d) $\begin{pmatrix} 2 & -4 & -2 \\ 10 & -6 & 4 \\ 6 & -12 & -6 \end{pmatrix}$ (e) $\begin{pmatrix} 0 & 6 & -9 & 3 \\ 12 & 24 & 9 & 9 \\ 66 & 78 & 81 & 63 \\ 30 & 42 & 45 & 27 \end{pmatrix}$

2. Repeat the previous exercise with the following matrices in $\mathbb{Q}[t]$:

(a) $\begin{pmatrix} 1-t & t & 1+t \\ 1+t & 1-t & 2t \\ t & 1 & 1 \end{pmatrix}$ (b) $\begin{pmatrix} t & 0 & 0 \\ 0 & 2-t & 0 \\ 0 & 0 & 4-t^2 \end{pmatrix}$ (c) $\begin{pmatrix} 1+t & 1 & 1 \\ 2 & t & 1 \\ 6 & 3 & 4-t \end{pmatrix}$

3. Let K be a field. Prove that any rectangular matrix over K is equivalent to a matrix of the form $\mathrm{diag}(1, 1, \cdots, 1, 0, 0, \cdots, 0)$. How many equivalence classes of matrices do we have?

4. Let A be a principal ideal domain. Prove that a square matrix over A is invertible if and only if it is equivalent to the identity matrix.

5. Let A be a principal ideal domain. Find the Smith normal form of an $n \times n$-matrix J if:
   (a) The determinant of J is not divisible by the square of an irreducible.
   (b) The determinant of J is of the form $p^2 q$, with p, q nonassociate irreducibles.
   (c) The determinant of J is of the form $p^2 q^2$, with p, q nonassociate irreducibles.

6. Let A be a principal ideal domain, $a \in A$ and $M = A/aA$
   (a) Let N be a submodule of M. Prove that there exists $b \in A$ such that $N \cong A/bA$.
   (b) Assume that $b|a$. Prove that M has exactly one submodule isomorphic to A/bA.

7. Let A be a principal ideal domain and $a, b \in A$. Prove that the invariant factors of $A/aA \oplus A/bA$ are the gcd and the lcm of a and b.

8. Let A be a principal ideal domain and $a \in A$. Prove that, in the notation of Lemma V.3.11:
   (a) For any A-linear map between modules $f : L \to M$, we have $f(La) \subseteq Ma$ and $f(L_a) \subseteq M_a$,
   (b) The assignments $M \mapsto Ma$ and $M \mapsto M_a$ induce $\mathbb{Z}$-linear functors from $\mathrm{Mod}A$ to itself.

9. Let A be a principal ideal domain and L a finitely generated free module.
   (a) Let N be a submodule of L such that L/N is torsion. Prove that L and N have same rank.
   (b) Let $f : L \to L'$ be an A-linear map with L' free of same rank as L. Prove that f is injective if and only if $L'/\mathrm{Im}f$ is torsion.

10. Let $a, b > 1$ be coprime integers.
   (a) Prove that the association $x + a\mathbb{Z} \mapsto bx + (ab)\mathbb{Z}$, for $x \in \mathbb{Z}$, defines a morphism of abelian groups $f : \mathbb{Z}/a\mathbb{Z} \to \mathbb{Z}/(ab)\mathbb{Z}$.
   (b) Prove that the abelian group $\mathrm{Hom}_{\mathbb{Z}}(\mathbb{Z}/a\mathbb{Z}, \mathbb{Z}/(ab)\mathbb{Z})$ is generated by f.
   (c) Compute the annihilator of f inside this abelian group.
   (d) Deduce that $\mathrm{Hom}_{\mathbb{Z}}(\mathbb{Z}/a\mathbb{Z}, \mathbb{Z}/(ab)\mathbb{Z}) \cong \mathbb{Z}/a\mathbb{Z}$.

11. Let L, M, N be finitely generated modules over a principal ideal domain such that $L \oplus N \cong M \oplus N$. Prove that $L \cong M$.

12. Let A be a commutative algebra and $a \in A$. Prove that a map $f : A/aA \to A/aA$ is A-linear if and only if there exists $c \in A$ such that f is induced from the multiplication by c.

13. Let K be a field, we endow $E = K^n$ with a $K[t]$-module structure using the linear map $f : (x_1, x_2, \cdots, x_n) \mapsto (x_2, \cdots, x_n, 0)$, for $x_i \in K$.

    (a) For $i, j \in \{1, \cdots, n\}$, compute the products $t^i \cdot e_j$, where e_j is the jth vector of the canonical basis of E.

    (b) Prove that $_f E$ is a cyclic $K[t]$-module, then compute its annihilator $\text{Ann}_{K[t]}(_f E)$, see Exercise II.2.2.

14. For each of the following abelian groups, give the primary decomposition and compute the invariant factors.

    (a) $\mathbb{Z}/12\mathbb{Z} \oplus \mathbb{Z}/42\mathbb{Z} \oplus \mathbb{Z}/72\mathbb{Z}$.

    (b) $\mathbb{Z}/36\mathbb{Z} \oplus \mathbb{Z}/56\mathbb{Z} \oplus \mathbb{Z}/84\mathbb{Z}$.

    (c) $\mathbb{Z}/1200\mathbb{Z}$.

15. Same exercise for the $K[t]$-module

$$M = K[t]/\langle fg \rangle \oplus K[t]/\langle gh \rangle \oplus K[t]/\langle hf \rangle$$

where $f = t^2, g = (t^2 + 1)^2, h = (t^2 - 3)^2$, in each of the following cases: $K = \mathbb{Q}, \mathbb{R}, \mathbb{C}$.

16. Let A be a principal ideal domain. Prove that a finitely generated module M is simple, see Exercise II.3.18, if and only if there exists an irreducible element $p \in A$ such that $M \cong A/pA$.

17. Let A be a principal ideal domain and $p \in A$ an irreducible element. Prove that the direct sum of finitely many p-primary modules is also p-primary.

18. Let A be a principal ideal domain. Prove that a torsion A-module M is cyclic if and only if there exist irreducible elements $p_1, \cdots, p_m$ pairwise nonassociate and positive integers $t_1, \cdots, t_m$ such that $M \cong \oplus_{i=1}^{m}(A/p_i^{t_i}A)$.

19. Let G be an abelian group and H a subgroup. Assume that G has $d_1 | \cdots | d_\ell$ as invariant factors while H has $c_1 | \cdots | c_n$. Prove that $n \leq \ell$ and $c_i | d_i$, for any i.

20. Prove that an abelian group of order n has a subgroup of order m if and only if $m | n$.

21. In each case, establish the list of all pairwise nonisomorphic abelian groups having the shown order: (a) 42; (b) 27; (c) 63; (d) 720.

22. Same exercise as above if p, q are distinct prime integers and the order of the group is: (a) $p^2 q$; (b) $p^3 q$; (c) $p^4 q$; (d) $p^3 q^2$.

23. Describe the structure of each of the following abelian groups given by generators and relations.

    (a) $< a, b, c : 2a + 8b + 18c = 0, 16b + 32c = 0, 6a + 4b + 14c = 0 >$.

    (b) $< a, b, c : a - 2b = 0, b - 3a = 0, a + b + c = 0 >$.

V.4 An application: the Jordan form of a matrix

V.4.1 The Jordan form

The reader has certainly heard about the Jordan form of a complex matrix. Our present objective is to show that the theory of modules over principal ideal domains allows for an easy proof, and a concrete calculation, of the Jordan form.

We work in a (slightly) more general context than complex numbers, namely, we assume that K is an algebraically closed field, see Example I.5.10. So, a polynomial in $K[t]$ is irreducible if and only if it is of degree one. If, moreover, it is normalised, that is, its leading coefficient is one, then it is of the form $t - \alpha$, for some $\alpha \in K$.

The Fundamental Theorem of Algebra asserts that the field $\mathbb{C}$ of complex numbers is algebraically closed. One can prove that any field admits an algebraically closed extension.

We wish to prove that, if K is algebraically closed, E a finite-dimensional K-vector space and $f : E \to E$ a linear map, then there exists a basis of E such that the matrix of f in this basis is a so-called Jordan matrix. Moreover, this matrix is unique up to order of its blocks.

We start with the relevant definitions. Let K be a field. A **block diagonal matrix** B over K is a square matrix of the form

$$B = \begin{pmatrix} B_1 & 0 & \cdots & 0 \\ 0 & B_2 & \cdots & 0 \\ \vdots & \vdots & & \vdots \\ 0 & 0 & \cdots & B_t \end{pmatrix}$$

where, for any i, B_i is a square matrix of order n_i having coefficients in K, such that the sum of the n_i equals the order of B. The B_i are the **blocks** of B.

Given an integer $k > 0$ and $\alpha \in K$, a **Jordan block** of order k is a $k \times k$ matrix of the form

$$J_k(\alpha) = \begin{pmatrix} \alpha & 0 & 0 & \cdots & 0 \\ 1 & \alpha & 0 & \cdots & 0 \\ 0 & 1 & \alpha & \cdots & 0 \\ \vdots & \vdots & \vdots & & \vdots \\ 0 & 0 & 0 & \cdots & \alpha \end{pmatrix}.$$

Its coefficients are α on the main diagonal, 1 just below it, and 0 elsewhere. A **Jordan matrix** is a block-diagonal matrix in which each block is a Jordan block.

For instance, a scalar is a Jordan block of order one (thus, a diagonal matrix is a Jordan matrix). Also, the following is an example of a Jordan block of order three:

$$J_3(-1) = \begin{pmatrix} -1 & 0 & 0 \\ 1 & -1 & 0 \\ 0 & 1 & -1 \end{pmatrix}.$$

We proved in Example IV.4.6(b) that the category $\mathcal{E}$ of pairs (E, f), with E a vector space and $f : E \to E$ a linear map is equivalent to the category $\mathrm{Mod}K[t]$, the product of a polynomial by a vector being as in Example II.2.4(f). Recall that, in this equivalence, a $K[t]$-submodule of E corresponds to a subspace F of E which is f-invariant, that is, such that $f(F) \subseteq F$. For the rest of this section, E is a finite dimensional vector space over an algebraically closed field K, and $f : E \to E$ a nonzero K-linear map.

V.4.1 Lemma. *E is a finitely generated torsion $K[t]$-module.*

Proof. Because any K-basis of E is a set of generators of E as a $K[t]$-module, finite dimensionality of E implies finite generation over $K[t]$. In order to prove it is torsion, we must prove that, for any vector $v \in E$, there exists a nonzero polynomial p such that $p.v = 0$. We may assume that $v \neq 0$. If $d = \dim_K E$, the $d + 1$ vectors $\{v, f(v), \cdots, f^d(v)\}$ are linearly dependent: there exist scalars $a_0, a_1, \cdots, a_d \in K$, not all zero, such that

$$a_0 v + a_1 f(v) + \cdots + a_d f^d(v) = 0.$$

Then, the nonzero polynomial $p = a_0 + a_1 t + \cdots + a_d t^d$ satisfies $p.v = 0$. □

The structure theorems for modules over principal ideal domains yield the following corollary.

V.4.2 Corollary. *The $K[t]$-module E can be decomposed as a direct sum of cyclic submodules in two ways:*
(a) *$E = \oplus_{i=1}^s K[t]/K[t]p_i$, where the p_i are unique nonzero nonconstant normalised polynomials such that $p_1 | \cdots | p_s$.*
(b) *$E = \oplus_{i=1}^s \oplus_{j=1}^{s_i} K[t]/K[t]q_{ij}^{s_{ij}}$, where the $s_{ij} \geq 0$ are integers and the q_{ij} are irreducible, normalised polynomials and unique up to order.*

Proof. (a) This is the decomposition of the Invariant Factors Theorem V.3.9. Because E is torsion, $p_s \neq 0$, hence $p_i \neq 0$, for any i. Also, $E_i = K[t]/K[t]p_i$ is nonzero if and only if p_i is not constant. The invariant factors p_i are unique up to associates, hence unique if normalised.

(b) The decomposition of Theorem V.3.18 for each E_i yields $E_i = \oplus_{j=1}^{s_i} E_{ij}$ where $E_{ij} = K[t]/K[t]q_{ij}^{s_{ij}}$, with the q_{ij} irreducible normalised polynomials, unique up to order, and the s_{ij} nonnegative integers. □

Because we decomposed E as direct sum of the $K[t]$-submodules $E_i = K[t]/K[t]p_i$ and also of $K[t]$-submodules $E_{ij} = K[t]/K[t]q_{ij}^{s_{ij}}$, each of these is an f-invariant vector subspace of E, that is, the restriction of f to this subspace is an endomorphism of the latter.

V.4.3 Corollary. *The matrix of f in a basis of E that is a union of bases of the E_{ij} is block-diagonal.*

Proof. For any pair (i, j), let $\{e_{ij}^k : 1 \leqslant k \leqslant d_{ij}\}$ be a basis of E_{ij}. The f-invariance of E_{ij} says that $f(e_{ij}^k) = \sum_{\ell=1}^{d_{ij}} a_{ij}^\ell e_{ij}^\ell$, for any k. On the other hand, because $E_i = \bigoplus_{j=1}^{s_i} E_{ij}$, the union $\{e_{ij}^k : 1 \leqslant k \leqslant d_{ij}, 1 \leqslant j \leqslant s_i\}$ is a basis of E_i. In this basis, the matrix of the restriction of f to E_i takes the form

$$
B_i = \begin{pmatrix} B_{i1} & 0 & \cdots & 0 \\ 0 & B_{i2} & \cdots & 0 \\ \vdots & \vdots & & \vdots \\ 0 & 0 & \cdots & B_{is_i} \end{pmatrix}
$$

where B_{ij} is the matrix of the restriction of f to E_{ij}. On the other hand, $E = \bigoplus_{i=1}^s E_i$ so the union $\{e_{ij}^k : i, j, k\}$ is a basis of E in which the matrix of f is of the form

$$
\begin{pmatrix} B_1 & 0 & \cdots & 0 \\ 0 & B_2 & \cdots & 0 \\ \vdots & \vdots & & \vdots \\ 0 & 0 & \cdots & B_s \end{pmatrix}.
$$

$\square$

Therefore, if we want the matrix of f to have a pleasant form, there remains to choose an appropriate basis of each $E_{ij} = K[t]/K[t]q_{ij}^{s_{ij}}$, viewed as K-vector space (not as module). Because q_{ij} is an irreducible normalised polynomial and K is algebraically closed, we have $q_{ij} = t - \alpha_{ij}$, for some $\alpha_{ij} \in K$. Thus, the subspace E_{ij} is of the form considered in the next proposition.

V.4.4 Proposition. *Let E be a K-vector space and $f : E \to E$ a linear map such that, as $K[t]$-module, $E = K[t]/\langle (t - \alpha)^k \rangle$, for some integer $k > 0$ and $\alpha \in K$. Then:*
  (a) *There exists $v \in E$ such that $\{v, (f - \alpha \cdot 1_E)(v), \cdots, (f - \alpha \cdot 1_E)^{k-1}(v)\}$ is a basis of E as a K-vector space.*
  (b) *The matrix of f in the basis of (a) is the Jordan block $J_k(\alpha)$.*

Proof. (a) As a $K[t]$-module, E is cyclic. Assume it is generated by v, say. For any $u \in E$, there exists $p \in K[t]$ such that $u = p \cdot v$.

We claim that we may assume the degree of p to be smaller than k. Indeed, dividing p by $(t - \alpha)^k$ yields $q, r \in K[t]$ such that $p = q(t - \alpha)^k + r$, where the remainder r is zero or of degree less than k. Hence, $p \cdot v = (q(t - \alpha)^k) \cdot v + r \cdot v = r \cdot v$, establishing our claim.

Because any polynomial of degree strictly less than k is a linear combination of the polynomials $1, t - \alpha, \cdots, (t - \alpha)^{k-1}$, the given set of vectors generates E as a K-vector space.

There remains to prove linear independence. Assume that $a_0, a_1, \cdots, a_{k-1} \in K$ are such that

$$
a_0 v + a_1 (f - \alpha \cdot 1_E)(v) + \cdots + a_{k-1}(f - \alpha \cdot 1_E)^{k-1}(v) = 0.
$$

The polynomial $p = a_0 + a_1(t - \alpha) + \cdots + a_{k-1}(t - \alpha)^{k-1}$ annihilates v, hence annihilates the $K[t]$-module E. Because its degree is at most $k - 1$, it must be zero. That is, $a_0 = \cdots = a_{k-1} = 0$. This completes the proof of (a).

(b) In order to compute the matrix of f in this basis, we describe the action of f on each basis vector. For any i such that $1 \leqslant i \leqslant k - 1$, we have

$$f(f - \alpha \cdot 1_E)^i(v) = \alpha(f - \alpha \cdot 1_E)^i(v) + (f - \alpha \cdot 1_E)^i(v) + (f - \alpha \cdot 1_E)^{i+1}(v)$$

and

$$f(f - \alpha \cdot 1_E)^{k-1}(v) = \alpha(f - \alpha \cdot 1_E)^{k-1}(v) + (f - \alpha \cdot 1_E)^k(v) + \alpha(f - \alpha \cdot 1_E)^{k-1}(v)$$
$$= \alpha(f - \alpha \cdot 1_E)^{k-1}(v).$$

The statement follows. □

V.4.5 Theorem. *Let K be an algebraically closed field, E a finite-dimensional K-vector space and $f : E \to E$ a nonzero linear map. Then there exists a basis of E in which the matrix of f is a Jordan matrix. Moreover, the latter is unique up to the order of blocks.*

Proof. Because of Proposition V.4.4, there exists a basis of E_{ij} in which the matrix of the restriction of f to E_{ij} is a Jordan block. Applying Corollary V.4.3, the union of these bases is a basis of E in which the matrix of f is block-diagonal, with each block a Jordan block. This proves existence. Uniqueness up to order of blocks follows from the fact that the $q_{ij}^{s_{ij}} = (t - \alpha_{ij})^{s_{ij}}$ are unique up to order. □

V.4.2 Computing the Jordan form

The explicit computation of the Jordan form of f is transparent from the proof of Theorem V.4.5. Indeed, it is block-diagonal and its blocks are the B_{ij}. To find them, we start by finding the blocks B_i. These are the matrices of the restriction of f to each of the submodules $E_i = K[t]/\langle p_i \rangle$. Hence, we must compute the invariant factors $p_1, \cdots, p_s$ of the $K[t]$-module E.

In practice, the map $f : E \to E$ is generally given in the form of its matrix $A = [a_{ij}]$ in some basis $\{e_1, \cdots, e_n\}$ of E as K-vector space. In order to compute the invariant factors, we constructed in Theorem V.3.9 a short exact sequence $0 \to L' \overset{j}{\to} L \overset{q}{\to} E \to 0$ with L and (hence) L' free. The invariant factors are those of the matrix of the inclusion morphism $j : L' \to L$ with respect to bases of L' and L. In the sequel, we denote by I the identity matrix.

V.4.6 Proposition. *With the previous notation, there exist bases of L' and L such that $p_1, \cdots, p_s$ are the invariant factors of the matrix $t \cdot I - A$.*

Proof. Because $\dim_K E = n$, we take $L = K[t]^n$ with basis $\{v_1, \cdots, v_n\}$. There exists an epimorphism of $K[t]$-modules $q : L \to E$ given by $v_i \mapsto e_i$, for i such that $1 \leqslant i \leqslant n$. Then $L' = \text{Ker } q$. For any i, we set

$$u_i = t \cdot v_i - \sum_{i=1}^{n} a_{ji} v_j.$$

We claim that $\{u_1, \cdots, u_n\}$ is a basis of the free $K[t]$-module L'. This is done in three steps.

(i) Each u_i belongs to L'. Indeed, any $v \in L$ can be written as $v = \sum_{i=1}^{n} g_i v_i$ for some $g_i \in K[t]$. Hence,

$$q(v) = \sum_{i=1}^{n} g_i e_i = \sum_{i=1}^{n} g_i(f) e_i.$$

In particular, for any i,

$$q(u_i) = f(e_i) - \sum_{j=1}^{n} a_{ji} e_j = 0$$

because of the definition of the matrix A. Hence, $u_i \in \text{Ker } q = L'$, for any i.

(ii) The u_i generate L' as a $K[t]$-module. Let $u \in L'$. Because L' is a submodule of L, there exist $g_i \in K[t]$ such that $u = \sum_{i=1}^{n} g_i v_i$. Also, $g_i = \alpha_i + t h_i$, for some $\alpha_i \in K, h_i \in K[t]$. Then

$$u = \sum_{i=1}^{n} \alpha_i v_i + \sum_{i=1}^{n} h_i(t v_i).$$

Substituting $t v_i = u_i + \sum_{j=1}^{n} a_{ji} v_j$ yields $u = \sum_{i=1}^{n} h_i u_i + \sum_{i=1}^{n} \beta_i v_i z$, for some $\beta_i \in K$. Now u, and the u_i, belong to the kernel of q. Hence,

$$0 = q(u) = \sum_{i=1}^{n} h_i q(u_i) + \sum_{i=1}^{n} \beta_i q(v_i) = \sum_{i=1}^{n} \beta_i e_i.$$

Because $\{e_1, \cdots, e_n\}$ is a basis of E, we have $\beta_i = 0$, for any i. So $u = \sum_{i=1}^{n} h_i u_i$, as required.

(iii) The u_i are linearly independent over $K[t]$. Assume that $\sum_{i=1}^{n} h_i u_i = 0$, for some $h_i \in K[t]$. Replacing u_i by its value yields $\sum_i h_i t v_i = \sum_{i,j} a_{ji} h_i v_j$. Now $\{v_1, \cdots, v_n\}$ is a basis of L. Hence, $h_i t = \sum_{j=1}^{n} a_{ij} h_j$, for any i. Assume that there exists i such that $h_i \neq 0$. We may choose i so that h_i is of maximal degree d. Then $h_i t$ is of degree $d + 1$, while the sum on the right is of degree at most d. This contradiction implies that $h_i = 0$, for any i. We have established our claim.

For any i, we have

$$u_i = -a_{1i} v_1 - a_{2i} v_2 - \cdots + (t - a_{ii}) v_i - \cdots - a_{ni} v_n$$

from which we deduce the matrix of the inclusion

$$
\begin{pmatrix}
t - a_{11} & -a_{12} & \cdots & -a_{1n} \\
-a_{21} & t - a_{22} & \cdots & -a_{2n} \\
\vdots & \vdots & & \vdots \\
-a_{n1} & -a_{n2} & \cdots & t - a_{nn}
\end{pmatrix} = t \cdot I - A.
$$

The statement follows.   □

V.4.7 Example. Let $E = \mathbb{C}^3$, equipped with the endomorphism whose matrix in the canonical basis is

$$
A = \begin{pmatrix} 3 & 3 & -2 \\ 0 & -1 & 0 \\ 8 & 6 & -5 \end{pmatrix} \quad \text{then} \quad tI - A = \begin{pmatrix} t - 3 & -3 & 2 \\ 0 & t + 1 & 0 \\ -8 & -6 & t + 5 \end{pmatrix}.
$$

The invariant factors are computed by reducing the latter matrix to its Smith normal form using the following sequence of elementary operations:

$$
tI - A \;\rightarrow\; \begin{pmatrix} t - 3 & -1 & 2 \\ 0 & t + 1 & 0 \\ -8 & t - 1 & t + 5 \end{pmatrix} \rightarrow \begin{pmatrix} -1 & t - 3 & 2 \\ t + 1 & 0 & 0 \\ t - 1 & -8 & t + 5 \end{pmatrix} \rightarrow
$$

$$
\rightarrow \begin{pmatrix} -1 & 0 & 0 \\ 0 & t^2 - 2t - 3 & 2t + 2 \\ 0 & t^2 - 4t - 5 & 3t + 3 \end{pmatrix} \rightarrow \begin{pmatrix} 1 & 0 & 0 \\ 0 & t^2 - 2t - 3 & 2t + 2 \\ 0 & -2t - 2 & t + 1 \end{pmatrix} \rightarrow
$$

$$
\rightarrow \begin{pmatrix} 1 & 0 & 0 \\ 0 & 2t + 2 & t^2 - 2t - 3 \\ 0 & t + 1 & -2t - 2 \end{pmatrix} \rightarrow \begin{pmatrix} 1 & 0 & 0 \\ 0 & t + 1 & -2t - 2 \\ 0 & 2t + 2 & t^2 - 2t + 3 \end{pmatrix} \rightarrow
$$

$$
\rightarrow \begin{pmatrix} 1 & 0 & 0 \\ 0 & t + 1 & 0 \\ 0 & 0 & (t + 1)^2 \end{pmatrix}
$$

Alternatively, we could have computed the gcd's of minors, as in Theorem V.3.6. We have obtained $p_1 = 1, p_2 = t + 1$ and $p_3 = (t + 1)^2$. We deduce the Jordan form J of A

$$
J = \begin{pmatrix} -1 & 0 & 0 \\ 0 & -1 & 0 \\ 0 & 1 & -1 \end{pmatrix}.
$$

Because J is the matrix of the same linear map as A in another basis, these matrices are similar, that is, there exists an invertible matrix P such that $J = P^{-1}AP$. Therefore, J and A have the same proper values. We recall the definitions. Let E be a finite-dimensional vector space. A scalar $\alpha \in K$ is a **proper value** of a linear map $f : E \to E$ if there exists a nonzero vector $v \in E$, called **proper vector**, such that $f(v) = \alpha v$. Because this relation reads as $(\alpha.1_E - f)(v) = 0$, a scalar α is a proper value of f if and only if it is a root of the determinant $\chi_f = \det(t.1_E - f)$, called the **characteristic polynomial** of f.

Let $A(f)$ be the set of those polynomials $p \in K[t]$ that annihilate f, that is, such that $p(f) = 0$ on E (in other words, $p(f)(v) = 0$, for any $v \in E$). Because E is finite-dimensional, the Cayley–Hamilton theorem of linear algebra asserts that $\chi_f \in A(f)$.

V.4.8 Lemma. *There exists a unique nonzero normalised polynomial m_f such that $A(f) = K[t]m_f$.*

Proof. It is easy to verify that $A(f)$ is an ideal in $K[t]$. It is nonzero because $\chi_f \in A(f)$. Because $K[t]$ is a principal ideal domain, $A(f)$ consists of the multiples of a nonzero polynomial m_f of lesser degree, unique if normalised. ☐

The polynomial m_f is called the **minimal polynomial** of f. It follows from its definition that m_f is the unique normalised polynomial such that:
  (a) $m_f(f) = 0$, and
  (b) if $p(f) = 0$, then $m_f | p$. In particular, $m_f | \chi_f$.

V.4.9 Corollary. *The roots of m_f are the proper values of f.*

Proof. Because $m_f | \chi_f$, any root of m_f is a root of χ_f, hence a proper value. Conversely, let $\alpha \in K$ be a proper value. There exists a nonzero vector v such that $f(v) = \alpha v$. Induction gives $f^k(v) = \alpha^k v$, for any integer $k \geq 0$. Assume that $m_f = a_0 + a_1 t + \cdots + a_d t^d$, then $m_f(f) = 0$ implies

$$m_f(\alpha)v = (a_0 + a_1\alpha + \cdots + a_d\alpha^d)v = (a_0 1_E + a_1 f + \cdots + a_d f^d)(v) = m_f(f)(v) = 0.$$

Because $v \neq 0$, we get $m_f(\alpha) = 0$, that is, α is a root of m_f. ☐

V.4.10 Examples.
  (a) A **projector** f is a linear map such that $f^2 = f$. The polynomial $t^2 - t$ annihilates f. Hence, $m_f | (t^2 - t)$. If $m_f = t$, then $f = 0$. If $m_f = t - 1$, then $f = 1_E$. So the minimal polynomial of a projector which is neither zero nor the identity equals $t^2 - t$.
  (b) A linear map f is **nilpotent** of **nilpotency index** n if $f^n = 0$, but $f^{n-1} \neq 0$. Its minimal polynomial is $m_f = t^n$.
  (c) Corollary V.4.9 implies that m_f is the normalised divisor of χ_f, having exactly the same roots, which annihilates f and is of least degree for this property. In Example V.4.7, one sees that $\chi_A(t) = \det(tI - a) = (t + 1)^3$. Hence, we have $m_A \in \{t + 1, (t + 1)^2, (t + 1)^3\}$. If $m_A = t + 1$, then $A + I = 0$ and this is not the case. On the other hand,

$$(A + I)^2 = \begin{pmatrix} 4 & 3 & -2 \\ 0 & 0 & 0 \\ 8 & 6 & -4 \end{pmatrix}^2 = 0.$$

Hence, $m_A = (t + 1)^2$, which is the same as the invariant factor p_3 computed in Example V.4.7. This is not a coincidence.

V.4.11 Corollary. *With the previous notation, $m_f = p_s$.*

Proof. Let, as usual, $E_i = K[t]/\langle p_i \rangle$, where $1 \leqslant i \leqslant s$. If $v_i \in E_i$, then $p_i | p_s$ implies $p_s \cdot v_i = 0$. Hence, $p_s \cdot v = 0$, for any $v \in E = \oplus_i E_i$, that is, $p_s \in \mathcal{A}(f)$. Therefore, $m_f | p_s$. On the other hand, $m_f(f) = 0$ implies $m_f \cdot v_s = 0$. Hence, $p_s | m_f$. So p_s and m_f are associate. Because both are normalised, they are equal. $\qquad\square$

There is, of course, much more to say on the fascinating subject of canonical forms of matrices. But that would take us too far from module theory.

Exercises of Section V.4

1. Let E be a vector space, $f : E \to E$ an invertible linear map and F an f-invariant subspace. Prove that F is f^{-1}-invariant.

2. Let E be a vector space, $f : E \to E$ a linear map and $\alpha \in K$. Let E_α be the set of all $v \in E$ such that $f(v) = \alpha v$. Prove that:
   (a) E_α is an f-invariant subspace of E.
   (b) The Jordan form of f is a diagonal matrix if and only if E is the direct sum of all the E_α.

3. Let E be a vector space, $f : E \to E$ a linear map such that, as $K[t]$-module, $E = K[t]/\langle p \rangle$ where $p = a_0 + a_1 t + \cdots + a_{d-1} t^{d-1} + t^d$.
   (a) Prove that, if v generates the cyclic $K[t]$-module E, then $\{v, f(v), \cdots, f^{d-1}(v)\}$ is a basis of E.
   (b) Find a basis of E in which f has the matrix

$$\begin{pmatrix} 0 & 0 & \cdots & 0 & -a_0 \\ 1 & 0 & \cdots & 0 & -a_1 \\ 0 & 1 & \cdots & 0 & -a_2 \\ \vdots & \vdots & & \vdots & \vdots \\ 0 & 0 & \cdots & 1 & -a_{d-1} \end{pmatrix}.$$

   This matrix is called the ***companion matrix*** of p.
   (c) Prove that $m_f = \chi_f = p$.

4. Prove that a complex matrix and its transpose have the same Jordan form.

5. Using the notation of this section, prove that, for any finite-dimensional vector space E and linear map $f : E \to E$, one has $\chi_f = p_1 \cdots p_s$.

6. For a Jordan block $J_k(\alpha)$, prove that $m_{J_k(\alpha)} = \chi_{J_k(\alpha)} = (t - \alpha)^k$. Deduce the invariant factors of this block.

7. Find all possible Jordan forms of matrices having the following characteristic and minimal polynomials:
   (a) $\chi = (t - 3)^5, m = (t - 3)^2$.
   (b) $\chi = (t + 1)^4 (t - 2)^2, m = (t + 1)^2 (t - 2)$.

(c) $\chi = (t-1)^3(t-3)^7, m = (t-1)^2(t-3)^3$.

8. Find the Jordan form of a matrix having the following invariant factors:
   (a) $p_1 = p_2 = t-1, p_3 = p_4 = (t-1)(t+1)$.
   (b) $p_1 = t+1, p_2 = (t+1)^2, p_3 = (t+1)^2(t-2)$.

9. Discuss, according to the values of the parameters a, b, the Jordan form of the matrix

$$\begin{pmatrix} -1 & a & b \\ 0 & 3 & 2 \\ 0 & 2 & 0 \end{pmatrix}.$$

10. Compute the Jordan form J of each of the following matrices A, then find an invertible matrix P such that $J = P^{-1}AP$.

(a) $\begin{pmatrix} 3 & 0 \\ 2 & 3 \end{pmatrix}$ (b) $\begin{pmatrix} 12 & 14 \\ -7 & -9 \end{pmatrix}$ (c) $\begin{pmatrix} 1 & 0 & -2 \\ 2 & 2 & -2 \\ 0 & 0 & -1 \end{pmatrix}$ (d) $\begin{pmatrix} 4 & -3 & -3 \\ 6 & -5 & -6 \\ 0 & 0 & 1 \end{pmatrix}$

(e) $\begin{pmatrix} 8 & -6 & -6 \\ 30 & -19 & -23 \\ -14 & 9 & 11 \end{pmatrix}$ (f) $\begin{pmatrix} 4 & -2 & -1 \\ 5 & -2 & -1 \\ -2 & 1 & 1 \end{pmatrix}$ (g) $\begin{pmatrix} 1 & -2 & -1 \\ -3 & -6 & -4 \\ 3 & 13 & 18 \end{pmatrix}$

(h) $\begin{pmatrix} 7 & 10 & 12 \\ -12 & -19 & -24 \\ 6 & 10 & 13 \end{pmatrix}$ (i) $\begin{pmatrix} 1 & 2 & -1 & -1 \\ -4 & 0 & 1 & 4 \\ -1 & 5 & -2 & -1 \\ -4 & -4 & 3 & 6 \end{pmatrix}$ (j) $\begin{pmatrix} -1 & 4 & 0 & -2 \\ 6 & -6 & 0 & 8 \\ 1 & -1 & -1 & 1 \\ 6 & -9 & 0 & 10 \end{pmatrix}$

Chapter VI
Functors between modules

VI.1 Introduction

A consequence of Yoneda's Lemma III.2.13 is that a module M over a K-algebra A is uniquely determined up to isomorphism by the covariant Hom functor $\mathrm{Hom}_A(M, -)$, and also by the contravariant one $\mathrm{Hom}_A(-, M)$, from $\mathrm{Mod}\,A$ to $\mathrm{Mod}\,K$. This explains the importance of these functors. In this chapter, taking our inspiration from multilinear algebra, we introduce another bifunctor between module categories, called the tensor functor, and study its relation to the Hom functor. Then we return to exact sequences. It is legitimate to ask whether a given functor maps an exact sequence onto another exact sequence or not. If it does, then it is said to *preserve exactness*. Unfortunately, neither the Hom nor the tensor functors preserve exactness in general. In order to correct this situation, we introduce special classes of modules, the so-called projective, injective and flat modules which have the property that the corresponding Hom or tensor functor preserves exactness. These modules play an important role in understanding a module category. In fact, in some cases, their knowledge suffices for obtaining a good description of it. Throughout this chapter, K denotes a commutative ring, and algebras are K-algebras.

VI.2 The tensor product of modules

VI.2.1 Definition and functorisation

Several algebraic problems involve the use of modules L, M, N and a bilinear map $L \times M \to N$. In order to apply the techniques developed for the study of linear maps, we need to replace the pair (L, M) by a single module T, and the given bilinear map by a linear map $T \to N$. That is, we need to linearise bilinear maps, and more generally multilinear maps, between modules. Loosely speaking, we are looking for the 'best' linear map that can replace a bilinear, or multilinear, one. This formulation leads immediately to a universal property.

Let A be a K-algebra. Given a right A-module L_A and a left A-module $_AM$, a map g from the product K-module $L \times M$ to a K-module X is called *A-bilinear* or *A-balanced* if

$$g(x_1\alpha_1 + x_2\alpha_2, y) = g(x_1, y)\alpha_1 + g(x_2, y)\alpha_2,$$
$$g(x, y_1\beta_1 + y_2\beta_2) = g(x, y_1)\beta_1 + g(x, y_2)\beta_2,$$
$$g(xa, y) = g(x, ay).$$

for $x, x_1, x_2 \in L, y, y_1, y_2 \in M, \alpha_1, \alpha_2, \beta_1, \beta_2 \in K$ and $a \in A$.

An Introduction to Module Theory. Ibrahim Assem and Flávio U. Coelho, Oxford University Press.
© Ibrahim Assem and Flávio U. Coelho (2024). DOI: 10.1093/9780198904939.003.0007

VI.2.1 Definition. Let A be a K-algebra. A ***tensor product*** of a right A-module L_A and a left A-module $_AM$ is a pair (T, t), with T a K-module and $t : L \times M \to T$ an A-bilinear map, such that, for any pair (X, g), with X a K-module and $g : L \times M \to X$ an A-bilinear map, there exists a unique K-linear map $\bar{g} : T \to X$ satisfying $\bar{g}t = g$

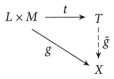

Let $\mathrm{Bil}(L \times M, X)$ denote the set of bilinear maps from $L \times M$ to X, the definition says that the K-linear map $t : L{\times}M \to T$ establishes a bijection $\mathrm{Hom}_K(T, X) \to \mathrm{Bil}(L{\times}M, X)$ given by $\bar{g} \mapsto \bar{g}t$, for a K-linear map $\bar{g} : T \to X$.

We prove existence and essential uniqueness of the tensor product of modules.

VI.2.2 Theorem. *Let A be a K-algebra, L a right A-module and M a left A-module. Then there exists a tensor product of L and M in $\mathrm{Mod}\, K$, unique up to isomorphism.*

Proof. Because the tensor product is defined via a universal property, it is unique up to isomorphism, provided it exists. One may, without difficulty, write a proof on the model of those in Chapter III. We prove existence by constructing the tensor product of L and M.

Let $K^{(L \times M)}$ be the free K-module with basis $L \times M$, that is, $K^{(L \times M)}$ is the set of linear combinations $\sum_\lambda (x_\lambda, y_\lambda)\alpha_\lambda$, with $x_\lambda \in L, y_\lambda \in M$ and the $\alpha_\lambda \in K$ almost all zero. Define R to be the submodule of $K^{(L \times M)}$ generated by all elements of the forms

$$(x_1\alpha_1 + x_2\alpha_2, y) - (x_1, y)\alpha_1 - (x_2, y)\alpha_2,$$
$$(x, y_1\beta_1 + y_2\beta_2) - (x, y_1)\beta_1 - (x, y_2)\beta_2,$$
$$(xa, y) - (x, ay)$$

with $x, x_1, x_2 \in L, y, y_1, y_2 \in M, \alpha_1, \alpha_2, \beta_1, \beta_2 \in K$ and $a \in A$. Let $T = K^{(L \times M)}/R$ and $t : L \times M \to X$ the composite of the inclusion j of $L \times M$ as a basis of $K^{(L \times M)}$ with the projection $p : K^{(L \times M)} \to K^{(L \times M)}/R$. It follows from the definition of R that t is an A-bilinear map.

Let (X, g) be a pair as in the definition. Because $K^{(L \times M)}$ is the free module with basis $L \times M$, there exists a unique K-linear map $g' : K^{(L \times M)} \to X$ such that $g = g'j$. Because g is A-bilinear, g' vanishes on the generators of R, so that $g'(R) = 0$. Hence, g' factors through p : there exists a unique K-linear map $\bar{g} : K^{(L \times M)}/R \to X$ such that $\bar{g}p = g'$. Then $\bar{g}t = \bar{g}pj = g'j = g$, as required.

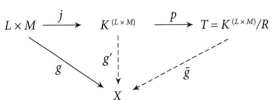

Finally, $\bar{g}$ is uniquely determined by g, for, if $\bar{g}_1, \bar{g}_2$ are such that $\bar{g}_1 t - \bar{g}_2 t$, then $\bar{g}_1 pj = \bar{g}_2 pj$. Because two maps coinciding on a basis of a free module are equal, we get $\bar{g}_1 p = \bar{g}_2 p$. But p is an epimorphism, hence $\bar{g}_1 = \bar{g}_2$. $\qquad\square$

By abuse of language, T is called the **tensor product** of the modules L_A and $_A M$ over A, which we denote as $T = L \otimes_A M$. Because of its construction, the K-module $L \otimes_A M$ is generated by the elements $t(x, y) = x \otimes y = (x, y) + R$, with $x \in L, y \in M$, called **tensors**. An arbitrary element of $L \otimes_A M$ is of the form $\sum_\lambda (x_\lambda \otimes y_\lambda) \alpha_\lambda$, with $x_\lambda \in L, y_\lambda \in M$ and the $\alpha_\lambda \in K$ almost all zero, or else $\sum_\lambda (x_\lambda \otimes y_\lambda)$, because M is a K-module. It is *not* true that any element of $L \otimes_A M$ can be written in the form $x \otimes y$, with $x \in L, y \in M$, see Exercise VI.2.5.

VI.2.3 Lemma. *Let A be a K-algebra, L a right A-module and M a left A-module. The following relations hold true inside the tensor product $L \otimes_A M$:*
   (a) $(x_1 \alpha_1 + x_2 \alpha_2) \otimes y = (x_1 \otimes y)\alpha_1 + (x_2 \otimes y)\alpha_2,$
   (b) $x \otimes (y_1 \beta_1 + y_2 \beta_2) = (x \otimes y_1)\beta_1 + (x \otimes y_2)\beta_2,$
   (c) $xa \otimes y = x \otimes ay,$
   (d) $x \otimes 0 = 0 = 0 \otimes x,$
   (e) $x \otimes ny = n(x \otimes y) = nx \otimes y,$
for $x, x_1, x_2 \in L, y, y_1, y_2 \in M, \alpha_1, \alpha_2, \beta_1, \beta_2 \in K, a \in A$ and $n \in \mathbb{Z}$.

Proof. (a), (b) and (c) follow from the construction of the tensor product while (d) follows from (a), (b) by setting the scalars equal to zero. Finally, (e) follows from (d) and induction if $n \geq 0$, and is extended to $n \in \mathbb{Z}$ using the formula $(-x) \otimes y = -(x \otimes y) = x \otimes (-y)$, which itself follows from (a), (b). $\qquad\square$

VI.2.4 Example. More results are needed in order to compute explicitly tensor products. At present, we content ourselves with an easy example showing the *a priori* surprising fact that the tensor product of nonzero modules may be zero. Let $a, b > 1$ be coprime integers. Because of the Bézout–Bachet theorem, there exist $s, t \in \mathbb{Z}$ such that $sa + tb = 1$. Consider the tensor product of the abelian groups $\mathbb{Z}/a\mathbb{Z}$ and $\mathbb{Z}/b\mathbb{Z}$. For any $x \in \mathbb{Z}/a\mathbb{Z}$ and $y \in \mathbb{Z}/b\mathbb{Z}$, we have:

$$x \otimes y = 1(x \otimes y) = (sa + tb)(x \otimes y) = s(ax \otimes y) + t(x \otimes by) = s \cdot 0 + t \cdot 0 = 0.$$

Therefore, $\mathbb{Z}/a\mathbb{Z} \otimes_\mathbb{Z} \mathbb{Z}/b\mathbb{Z} = 0$.

Any universal property leads to a functorisation procedure. Let L, L' be right A-modules, M, M' left A-modules and $f : L \to L', g : M \to M'$ linear maps. The K-linear map $f \times g : L \times M \to L' \times M'$ induced from f, g by passing to products, namely the map $f \times g : (x, y) \mapsto (f(x), g(y))$, for $x \in L, y \in M$, becomes A-bilinear when composed with the map $t' : L' \times M' \to L' \otimes_A M'$ of Definition VI.2.1. Therefore, there exists a unique K-linear map $f \otimes g : L \otimes_A M \to L' \otimes_A M'$ (or, more precisely, $f \otimes_A g$) such that the following square commutes

$$L \times M \xrightarrow{\ t\ } L \otimes_A M$$

$$f \times g \downarrow \qquad\qquad \vdots\, f \otimes g$$

$$L' \times M' \xrightarrow{\ t'\ } L' \otimes_A M'$$

The map $f \otimes g$ is thus defined on a generator $x \otimes y$ of $L \otimes_A M$ by the formula $(f \otimes g)(x \otimes y) = (f \otimes g)t(x, y) = t'(f \times g)(x, y) = t'(f(x), g(y)) = f(x) \otimes g(y)$. Clearly, if $L = L', M = M', f = 1_L$ and $g = 1_M$, we get $1_L \otimes 1_M = 1_{L \otimes M}$.

Let $f' : L'_A \to L''_A, g' : {}_A M' \to {}_A M''$ be another pair of linear maps. We have as before a commutative diagram

$$L \times M \xrightarrow{\ t\ } L \otimes_A M$$

$$f \times g \downarrow \qquad\qquad \vdots\, f \otimes g$$

$$L' \times M' \xrightarrow{\ t'\ } L' \otimes_A M'$$

$$f' \times g' \downarrow \qquad\qquad \vdots\, f' \otimes g'$$

$$L'' \times M'' \xrightarrow{\ t''\ } L'' \otimes_A M''$$

Uniqueness in the universal property implies

$$(f' \otimes g')(f \otimes g) = (f'f) \otimes (g'g).$$

Setting $M = M' = M''$ and $g = g' = 1_M$, we get $(f'f) \otimes 1_M = (f' \otimes 1_M)(f \otimes 1_M)$. Hence, the correspondence $- \otimes_A M : L \mapsto L \otimes_A M, f \mapsto f \otimes 1_M$ defines a covariant functor $- \otimes_A M : \operatorname{Mod} A \to \operatorname{Mod} K$. Following general functorial notation, the morphism $f \otimes 1_M$ is sometimes denoted as $f \otimes M$. Similarly, the correspondence $L \otimes_A - : M \mapsto L \otimes_A M, g \mapsto 1_L \otimes g$, defines a covariant functor $L \otimes_A - : \operatorname{Mod} A^{op} \to \operatorname{Mod} K$. Thus, the tensor product gives rise to a bifunctor $- \otimes_A ? : \operatorname{Mod} A \times \operatorname{Mod} A^{op} \to \operatorname{Mod} K$, covariant in both variables.

If at least one of L or M is a bimodule, then we can say more.

VI.2.5 Lemma. *Let A, B be K-algebras and ${}_A M_B$ an A-B-bimodule. Then:*
  (a) *The correspondence $L \mapsto L \otimes_A M_B, f \mapsto f \otimes 1_M$ defines a covariant functor from $\operatorname{Mod} A$ to $\operatorname{Mod} B$.*
  (b) *The correspondence $N \mapsto {}_A M \otimes_B N, g \mapsto 1_M \otimes g$ defines a covariant functor from $\operatorname{Mod} B^{op}$ to $\operatorname{Mod} A^{op}$.*

Proof. (a) We give a B-module structure to the tensor product $L \otimes_A M$. Let $x \in L$, $y \in M$ and $b \in B$. We define a right B-action by setting

$$(x \otimes y)b = x \otimes (yb).$$

This action is easily seen to satisfy module axioms. Moreover, if $f : L \to L'$ is A-linear, then $f \otimes 1_M$ becomes B-linear.

(b) We confer a left A-module structure to $M \otimes_B N$. Let $y \in M, z \in N$ and $a \in A$. Set

$$a(y \otimes z) = (ay) \otimes z.$$

The rest of the proof proceeds as in (a).   □

Hence, if A, B, C are K-algebras, $_ALB$ an A-B-bimodule and $_BM_C$ a B-C-bimodule, then $L \otimes_B M$ has a natural A-C-bimodule structure given, for $x \in L, y \in M$, $a \in A, c \in C$, by

$$a(x \otimes y)c = (ax) \otimes (yc).$$

In particular, if A is commutative and M an A-bimodule, then both $- \otimes_A M$ and $M \otimes_A -$ are covariant functors from Mod$A$ to itself.

VI.2.6 Examples.

(a) Let A, B be algebras, $_BT_A$ a B-A-bimodule and M_A an A-module. Due to Lemma II.3.8, the Hom-set $\mathrm{Hom}_A(T, M)$ has a natural right B-module structure, given as follows: if $u \in \mathrm{Hom}_A(M, A), b \in B$, then $ub \in \mathrm{Hom}_A(T, M)$ is defined by $(ub)(x) = u(bx)$, for $x \in T$. This immediately implies that the map $g : \mathrm{Hom}_A(T, M) \times T \to M$ given by $(u, x) \mapsto u(x)$, for u, x as before, is B-bilinear. The resulting morphism $\bar{g} : \mathrm{Hom}_A(T, M) \otimes_B T \to M$, given by $u \otimes x \mapsto u(x)$, is called an **evaluation morphism**. Because of Lemma VI.2.5, the tensor product $\mathrm{Hom}_A(T, M) \otimes_B T$ has a right A-module structure. It is then easy to verify that the K-linear morphism $\bar{g}$ becomes A-linear.

(b) Multiplication inside an algebra may be interpreted in terms of tensor product. Let A be a K-module. Due to definition of tensor product, the data of a K-bilinear map $A \times A \to A$ amounts to that of a K-linear map $\mu : A \otimes_K A \to A$. But such a map defines a multiplication on A if we set $ab = \mu(a \otimes b)$, for $a, b \in A$. Because of Lemma VI.2.3, this multiplication is right and left distributive over addition. It is associative if and only if the following square commutes:

$$
\begin{array}{ccc}
A \otimes_K A \otimes_K A & \xrightarrow{\mu \otimes 1_A} & A \otimes_K A \\
{\scriptstyle 1_A \otimes \mu} \downarrow & & \downarrow {\scriptstyle \mu} \\
A \otimes_K A & \xrightarrow{\quad \mu \quad} & A
\end{array}
$$

Existence of a multiplicative identity and eventual commutativity can also be expressed by means of the tensor product, but this is left to the Exercise VI.2.9 below.

If A is commutative, then the notation $f \otimes g$ or $f \otimes_A g$ for the tensor product of linear maps $f' : L_A \to L'_A, g : {}_AM \to_A M'$ may be ambiguous. As defined, it represents an element of $\mathrm{Hom}_A(L \otimes_A M, L' \otimes_A M')$. But, because of general conventions, it also represents an element of $\mathrm{Hom}_A(L, L') \otimes_A \mathrm{Hom}_A(M, M')$ (for, the latter two are A-modules because A is commutative). In practice, we shall not need the latter product.

VI.2.2 Elementary properties of the tensor product

VI.2.7 Proposition.

    (a) *Let A be a commutative K-algebra and L_A, M_A modules. We have an isomorphism of K-modules*

$$L \otimes_A M \cong M \otimes_A L$$

    functorial in each variable.

    (b) *Let A, B be K-algebras and $L_{A,A}$ $M_{B,B}$ N modules. We have an isomorphism of K-modules*

$$L \otimes_A (M \otimes_B N) \cong (L \otimes_A M) \otimes_B N$$

    functorial in each variable.

    (c) *Let A be a K-algebra, $(L_\lambda)_\lambda$ a family of right A-modules and M a left A-module. We have an isomorphism of K-modules*

$$\left(\bigoplus_\lambda L_\lambda \right) \otimes_A M \cong \bigoplus_\lambda (L_\lambda \otimes_A M)$$

    functorial in each variable.

    (d) *Let A be a K-algebra, L a right A-module and $(M_\sigma)_\sigma$ a family of left A-modules. We have an isomorphism of K-modules*

$$L \otimes_A \left(\bigoplus_\sigma M_\sigma \right) \cong \bigoplus_\sigma (L \otimes_A M_\sigma)$$

    functorial in each variable.

Proof. The proofs all follow the same pattern: we verify that one term of the stated isomorphism satisfies the universal property applied to the other term.

    (a) Because A is commutative, the map $L \times M \to M \otimes_A L$ defined by $(x, y) \mapsto y \otimes x$, for $x \in L, y \in M$, is A-bilinear, hence induces an A-linear map $f : L \otimes_A M \to M \otimes_A L$ such that $f : x \otimes y \mapsto y \otimes x$. Similarly, we construct $g : M \otimes_A L \to L \otimes_A M$ such that $g : y \otimes x \mapsto x \otimes y$. Because the composites gf and fg equal the identity on any tensor (that is, generator of the tensor product), f and g are inverse isomorphisms. Functoriality in each variable is easily verified.

    (b) Let $z \in N$ be given. Define $f_z : M \to M \otimes_B N$ by $f_z : y \mapsto y \otimes z$, for $y \in M$. This morphism of left A-modules induces one of K-modules $g_z = 1_L \otimes f_z : L \otimes_A M \to L \otimes_A (M \otimes_B N)$ such that $g_z : x \otimes y \mapsto x \otimes (y \otimes z)$, for $x \in L, y \in M$. The map $\varphi : (L \otimes_A M) \times N \to L \otimes_A (M \otimes_B N)$ defined by $\varphi(x \otimes y, z) = g_z(x \otimes y)$ is B-bilinear. Hence, it induces a morphism $\Phi : (L \otimes_A M) \otimes_B N \to L \otimes_A (M \otimes_B N)$ such that $\Phi : (x \otimes y) \otimes z \mapsto x \otimes (y \otimes z)$, for $x \in L, y \in M, z \in N$.

    Similarly, we construct a morphism $\Psi : L \otimes_A (M \otimes_B N) \to (L \otimes_A M) \otimes_B N$ of K-modules such that $\Psi : x \otimes (y \otimes z) \mapsto (x \otimes y) \otimes z$. The same argument as in (a) shows that Φ, Ψ are inverse isomorphisms, functorial in each variable.

(c) The map $(\oplus_\lambda L_\lambda) \times M \to \oplus_\lambda(L_\lambda \otimes_A M)$ defined by $((x_\lambda)_\lambda, y) \mapsto (x_\lambda \otimes y)_\lambda$ for $x_\lambda \in L_\lambda, y \in M$, is A-bilinear, so induces a map $f : (\oplus_\lambda L_\lambda) \otimes_A M \to \oplus_\lambda(L_\lambda \otimes_A M)$ such that $f : (x_\lambda)_\lambda \otimes y \mapsto (x_\lambda \otimes y)_\lambda$.

Conversely, let $q_\lambda : L_\lambda \to \oplus_\mu L_\mu$ and $q'_\lambda : L_\lambda \otimes_A M \to \oplus_\mu(L_\mu \otimes_A M)$ be the respective inclusions. The universal property of direct sum (coproduct) yields a unique morphism $g : \oplus_\lambda (L_\lambda \otimes_A M) \to (\oplus_\lambda L_\lambda) \otimes_A M$ such that $gq'_\lambda = q_\lambda \otimes 1_M$. Then

$$g((x_\lambda \otimes y)_\lambda) = g\left(\sum_\lambda q'_\lambda(x_\lambda \otimes y)\right) = \sum_\lambda gq'_\lambda(x_\lambda \otimes y) = \sum_\lambda q_\lambda(x_\lambda) \otimes y = (x_\lambda)_\lambda \otimes y.$$

Therefore, $gf = 1$. Similarly, $fg = 1$ and f and g are inverse isomorphisms. The proof of functoriality is left to the reader.

(d) Similar to (c). $\square$

As a consequence of (c), (d), if $(L_\lambda)_\lambda$ is a family of right A-modules, and $(M_\sigma)_\sigma$ a family of left A-modules, then we have a functorial isomorphism of K-modules

$$\left(\bigoplus_\lambda L_\lambda\right) \otimes_A \left(\bigoplus_\sigma M_\sigma\right) \cong \bigoplus_{\lambda,\sigma}(L_\lambda \otimes_A M_\sigma).$$

Proposition VI.2.7(b) says that the tensor product is associative. Thus, one can treat multilinear maps in the same way as bilinear.

Because of Theorem II.3.9, for any module M_A, we have an isomorphism of A-modules $\text{Hom}_A(A, M) \cong M$. The corresponding statement for tensor product is stronger, because it works for both variables.

VI.2.8 Theorem. *Let A be a K-algebra, L a right A-module and M a left A-module. We have functorial isomorphisms of A-modules $L \otimes_A A_A \cong L_A$ and $_A A \otimes_A M \cong_A M$.*

Proof. The maps $L \otimes_A A_A \to L_A$ and $L_A \to L \otimes_A A_A$ defined respectively by $x \otimes a \mapsto xa$ and $x \mapsto x \otimes 1$, for $x \in L, a \in A$, are A-linear, functorial and mutually inverse. This proves the first isomorphism. The second is established in the same way. $\square$

For instance, for any integer $m > 1$, we have $\mathbb{Z}/m\mathbb{Z} \otimes_{\mathbb{Z}} \mathbb{Z} \cong \mathbb{Z}/m\mathbb{Z} \cong \mathbb{Z} \otimes_{\mathbb{Z}} \mathbb{Z}/m\mathbb{Z}$.

VI.2.9 Corollary. *Let A be a K-algebra, L a right A-module and M a left A-module. Then, for any set Λ, we have functorial isomorphisms of A-modules $L \otimes_A A_A^{(\Lambda)} \cong L_A^{(\Lambda)}$ and $_A A_A^{(\Lambda)} \otimes_A M \cong_A M^{(\Lambda)}$.*

Proof. This follows from Theorem VI.2.8 and Proposition VI.2.7(c)(d). $\square$

If, in Corollary VI.2.9, Λ is finite and has n elements, then the statement becomes $L \otimes_A A^{(n)} \cong L^{(n)}$. Setting, in particular, $L = A^{(m)}$, we get $A^{(m)} \otimes_A A^{(n)} \cong A^{(mn)}$.

VI.2.10 Corollary. *Let K be a field, and E, F vector spaces with the respective bases $\{e_1, \cdots, e_m\}$ and $\{f_1, \cdots, f_n\}$. Then $\{e_i \otimes f_j : 1 \leqslant i \leqslant m, 1 \leqslant j \leqslant n\}$ is a basis of $E \otimes_K F$. In particular, $\dim_K (E \otimes_K F) = \dim_K E \cdot \dim_K F$.*

Proof. If K, E, F are as in the statement, then $E \cong K^{(m)}$ and $F \cong K^{(n)}$ so that $E \otimes_K F \cong K^{(m)} \otimes_K K^{(n)} \cong K^{(mn)}$. That is, $\dim_K (E \otimes_K F) = mn$. On the other hand, the set of vectors $e_i \otimes f_j$ generates the K-vector space $E \otimes_K F$. Because its cardinality equals the dimension, this generating set constitutes a basis. $\qquad\square$

Thus, the tensor product solves the old problem of linear algebra of constructing a vector space whose dimension is the product of the dimensions of two given ones. The next corollary is a special case of Proposition VI.4.39 below, but our proof illustrates the previous Corollary VI.2.10.

VI.2.11 Corollary. *Let K be a field and E, F, G finite-dimensional vector spaces. If $f : E \to F$ is a monomorphism, then so is $f \otimes 1_G : E \otimes_K G \to F \otimes_K G$.*

Proof. Let $\{e_1, \cdots, e_m\}$ be a basis of E. Because $f : E \to F$ is injective, the vectors $\{f(e_1), \cdots, f(e_m)\}$ are linearly independent in F, so there exists a basis of F of the form $\{f(e_1), \cdots, f(e_m), f_{m+1}, \cdots, f_n\}$ where $f_i \in F$ for any $i > m$. Let $\{g_1, \cdots, g_p\}$ be a basis of G. Then $\{e_i \otimes g_j : 1 \leqslant i \leqslant m, 1 \leqslant j \leqslant p\}$ is a basis of $E \otimes_K G$. It is sent by $f \otimes 1_G$ to the set $\{f(e_i) \otimes g_j : 1 \leqslant i \leqslant m, 1 \leqslant j \leqslant p\}$ which is part of a basis of $F \otimes_K G$. In particular, it is linearly independent. Thus, $f \otimes 1_G$ is injective. $\qquad\square$

The previous corollary does *not* hold true for modules in general, as is shown by the following example (which also shows that the tensor product of nonzero morphisms may be zero).

VI.2.12 Example. Let p be a prime integer and $j : \mathbb{Z}/p\mathbb{Z} \to \mathbb{Z}/p^2\mathbb{Z}$ the monomorphism given by $x + p\mathbb{Z} \mapsto px + p^2\mathbb{Z}$, for $x \in \mathbb{Z}$. Then $j \otimes 1_{\mathbb{Z}/p\mathbb{Z}} : \mathbb{Z}/p\mathbb{Z} \otimes_\mathbb{Z} \mathbb{Z}/p\mathbb{Z} \to \mathbb{Z}/p^2\mathbb{Z} \otimes_\mathbb{Z} \mathbb{Z}/p\mathbb{Z}$ is the zero map (in particular, it is not injective). Indeed, let $\bar{x}, \bar{y} \in \mathbb{Z}/p\mathbb{Z}$, then

$$(j \otimes 1)(\bar{x} \otimes \bar{y}) = j(\bar{x}) \otimes \bar{y} = p\bar{x} \otimes \bar{y} = \bar{x} \otimes p\bar{y} = \bar{x} \otimes 0 = 0.$$

VI.2.3 The adjunction isomorphism

Given an algebra A and modules $L_A, {}_A M$, the definition of the tensor product $L \otimes_A M$ says that, for any K-module X, there exists a bijection between $\text{Bil}(L \times M, X)$ and $\text{Hom}_K(L \otimes_A M, X)$, see after Definition VI.2.1. A bilinear map is in particular a map in two variables, and it is intuitively clear that a map in two variables may be evaluated by first substituting one variable, then the other. Formalising the latter statement yields the following example.

VI.2.13 Example. Let E, F, G be sets, that is, objects in the category *Sets*, we have a map

$$\psi_{E,G} : Sets(E \times F, G) \to Sets(E, Sets(F, G))$$

defined, for $f : E \times F \to G, x \in E, y \in F$ by $f \mapsto (x \mapsto (y \mapsto f(x, y)))$. This map indeed evaluates f first on x, then on y. It is not hard to see that ψ is a bijection with inverse given by $g \mapsto (((x, y) \mapsto g(x)(y))$, where $g \colon F. \to Sets(F, G)$ and $(x, y) \in E \times F$. Moreover, ψ is functorial in each of the three variables E, F and G. Therefore, there exists a functorial bijection $Sets(E \times F, G) \cong Sets(E, Sets(F, G))$.

VI.2.14 Definition. Let C, D be categories, $F \colon C \to D, G \colon D \to C$ functors. We say that F is **left adjoint** to G, or that G is **right adjoint** to F, or that (F, G) is an **adjoint pair** if, for any objects M in C and X in D, there exists a bijection

$$\varphi_{M,X} \colon C(M, GX) \overset{\cong}{\to} D(FM, X),$$

called the **adjunction**, functorial in each variable.

Adjunction is thus a tool to transform a morphism in C into one in D, and vice-versa. It is used to transfer information from one category to the other. A very obvious example of adjunction is furnished by an equivalence and its quasi-inverse. One can think roughly of an adjoint pair as 'almost quasi-inverse equivalences'.

An adjoint pair (F, G) is sometimes denoted as $F \dashv G$. We shall not use this notation.

Functoriality in each variable means that, if $f \colon M \to N$ is a morphism in C, then we have a commutative square

$$
\begin{array}{ccc}
C(M, GX) & \overset{\varphi_{M, X} \cong}{\longrightarrow} & D(FM, X) \\
\uparrow{\scriptstyle C(f, GX)} & & \uparrow{\scriptstyle D(Ff, X)} \\
C(N, GX) & \overset{\varphi_{N, X} \cong}{\longrightarrow} & D(FN, X)
\end{array}
$$

and, if $u \colon X \to Y$ is a morphism in D, then we have a commutative square

$$
\begin{array}{ccc}
C(M, GX) & \overset{\varphi_{M, X} \cong}{\longrightarrow} & D(FM, X) \\
\downarrow{\scriptstyle C(M, Gu)} & & \downarrow{\scriptstyle D(FM, u)} \\
C(M, GY) & \overset{\varphi_{M, Y} \cong}{\longrightarrow} & D(FM, Y)
\end{array}
$$

For instance, Example VI.2.13 above says that the functor $- \times F \colon Sets \to Sets$ is left adjoint to $Sets(F, -) \colon Sets \to Sets$. We give another example.

VI.2.15 Example. The free module functor $L \colon Sets \to \operatorname{Mod} A$, see at the end of Subsection III.4.2, is left adjoint to the forgetful functor $|-| \colon \operatorname{Mod} A \to Sets$.

Indeed, the universal property of free modules says that, given a set X, and an A-module M, viewed as a set (that is, as the set $|M|$, where $|-|$ is the forgetful functor from A-modules to sets), then there exists a bijection between maps $f \colon X \to |M|$ in $Sets$ and A-linear maps $\bar{f} \colon L(X) \to M$ such that $\bar{f} i_X = f$, where $i_X \colon X \to L(X)$ is the morphism of Definition III.4.1. This can be written as

$$Sets(X, |M|) \cong \mathrm{Hom}_A(L(X), M).$$

Functoriality in both variables is an easy exercise.

It is remarkable that, in an adjoint pair, each functor determines essentially the other.

VI.2.16 Lemma. *Let (F, G) be an adjoint pair. Then each of F and G determines the other up to isomorphism.*

Proof. Suppose that $F: C \to D$ and $G, G': D \to C$ are functors such that $(F, G), (F, G')$ are adjoint pairs, then, for any objects X in D and M in C, we have functorial isomorphisms:

$$C(M, GX) \cong D(FM, X) \cong C(M, G'X).$$

Because of Corollary III.2.14, there exists an isomorphism $\varphi_X : GX \to G'X$. Functoriality of the previous isomorphisms implies that of φ_X. Therefore, $G \cong G'$. □

Returning to the comments at the beginning of this subsection, Example VI.2.13 suggests to compare $\mathrm{Hom}_K(L \otimes_A M, N)$ with $\mathrm{Hom}_K(L, \mathrm{Hom}_K(M, N))$. In case M is an A–B-bimodule, and N a right B-module, then $\mathrm{Hom}_B(M, N)$ becomes a right A-module and $L \otimes_A M$ a right B-module. We get the following theorem.

VI.2.17 Theorem. The adjunction isomorphism. *Let A, B be K-algebras and L_A, $_AM_B, N_B$ modules. Then we have an isomorphism of K-modules, functorial in each variable*

$$\mathrm{Hom}_A(L, \mathrm{Hom}_B(M, N)) \cong \mathrm{Hom}_B(L \otimes_A M, N).$$

That is, $- \otimes_A M$ is left adjoint to $\mathrm{Hom}_B(M, -)$.

Proof. We construct explicitly the inverse isomorphisms. As explained after Theorem II.3.9, the 'right' maps should be the most natural ones. We first construct the map

$$\varphi : \mathrm{Hom}_A(L, \mathrm{Hom}_B(M, N)) \to \mathrm{Hom}_B(L \otimes_A M, N).$$

It should send an A-linear map $f : L \to \mathrm{Hom}_B(M, N)$ to a B-linear map $\varphi(f) : L \otimes_A M \to N$. To define the latter, it suffices to specify the image of a generator $x \otimes y \in L \otimes_A M$. Now, $x \in L$ implies $f(x) \in \mathrm{Hom}_B(M, N)$, so $f(x)$ is a morphism which sends $y \in M$ to some $f(x)(y) \in N$. We may set

$$\varphi(f)(x \otimes y) = f(x)(y)$$

for $f: L \to \mathrm{Hom}_B(M, N), x \in L, y \in M$.

Conversely, the map

$$\psi : \operatorname{Hom}_B(L \otimes_A M, N) \to \operatorname{Hom}_A(L, \operatorname{Hom}_B(M, N))$$

should send a B-linear map $u : L \otimes_A M \to N$ to an A-linear one $\psi(u) : L \to \operatorname{Hom}_B(M, N)$. The latter is defined once we define the image of an arbitrary $x \in L$. But $\psi(u)(x)$ should be itself a B-linear map from M to N, thus should map $y \in M$ to $\psi(u)(x)(y) \in N$. With this notation, we have $x \otimes y \in L \otimes_A M$ and $u(x \otimes y) \in N$. We may thus define

$$\psi(u)(x)(y) = u(x \otimes y)$$

for $u : L \otimes_A M \to N, x \in L, y \in M$.

One verifies easily that φ, ψ are K-linear, mutually inverse and functorial.   □

Applications of adjunction will appear in the next sections.

If A, B, C, D are K-algebras and $_C L_{A}, _A M_{B,D} N_B$ are bimodules, then it is easy to check that the adjunction isomorphism is an isomorphism of D-C-bimodules.

We have, of course, the corresponding statement for left modules, which we state because of its importance: if A, B are K-algebras and $_A L, _B M_A, _B N$ are modules, then there exists an isomorphism of K-modules, functorial in each variable

$$\operatorname{Hom}_A(L, \operatorname{Hom}_B(M, N)) \cong \operatorname{Hom}_B(M \otimes_A L, N).$$

Thus, $M \otimes_A -$ is left adjoint to $\operatorname{Hom}_B(M, -)$. The proof is similar to that of Theorem VI.2.17.

VI.2.18 Example. In one special case, the adjunction isomorphism is already familiar to us. Indeed, assume that $A = B$ and $M =_A A_A$, then we have

$$\operatorname{Hom}_A(L, \operatorname{Hom}_A(A, N)) \cong \operatorname{Hom}_A(L, N) \cong \operatorname{Hom}_A(L \otimes_A A, N),$$

due to Theorems II.3.9 and VI.2.8.

Exercises of Section VI.2

1. Let $\{e_1, e_2\}$ and $\{f_1, f_2, f_3\}$ be bases of $\mathbb{C}^2$ and $\mathbb{C}^3$, respectively. Find the coordinates of $x \otimes y$, where $x = \begin{pmatrix} 2 \\ -1 \end{pmatrix}, y = \begin{pmatrix} 1 \\ -1 \\ 3 \end{pmatrix}$, in the basis $e_i \otimes f_j$ of $\mathbb{C}^2 \otimes_\mathbb{C} \mathbb{C}^3$.

2. Prove each of the following isomorphisms:
   (a) $\mathbb{Q} \otimes_\mathbb{Z} \mathbb{Q} \cong \mathbb{Q}$.
   (b) $\mathbb{R} \otimes_\mathbb{Z} \mathbb{Z}[i] \cong \mathbb{C}$.
   (c) $\mathbb{Z}/m\mathbb{Z} \otimes_\mathbb{Z} \mathbb{Z}/n\mathbb{Z} \cong \mathbb{Z}/d\mathbb{Z}$, where d is the gcd of m, n.
   (d) $G \otimes_\mathbb{Z} \mathbb{Q} = 0$, for any torsion additive abelian group G.

3. Let $\varphi : A \to B$ be an algebra morphism and L a free A-module with basis $(e_\lambda)_{\lambda \in \Lambda}$. Prove that $L \otimes_A B$ is a free B-module with basis $(e_\lambda \otimes 1)_{\lambda \in \Lambda}$.

4. **Complexification of a real vector space.** Let E be a real vector space and consider $\mathbb{C}$ as real vector space. Compare a basis of E to a basis of the complex vector space $C(E) = E \otimes_{\mathbb{R}} \mathbb{C}$. Prove that, if $C(E)$ is considered as real vector space, then E is functorially isomorphic to a subspace of $C(E)$.

5. Give an example of an element in a tensor product $L \otimes_A M$ which cannot be written in the form $x \otimes y$, for any $x \in L, y \in M$ (Hint: take $A = K, L = M = K^{(2)}$).

6. Let K be a field and E, F vector spaces. Prove that, for any $x \in E, y \in F$, one has $x \otimes y = 0$ if and only if $x = 0$ or $y = 0$.

7. Let M be a K-module. Prove that the elements of $M \otimes_K K[t]$ can be written as polynomials in t with coefficients in M.

8. Let A be a K-algebra. Prove that:
   (a) If L is a right A-module, and L_1, L_2 submodules of M, then $L \cong L_1 \oplus L_2$ if and only if $L \otimes_A - \cong (L_1 \otimes_A -) \oplus (L_2 \otimes_A -)$.
   (b) If M is a left A-module, and M_1, M_2 submodules of M, then $M \cong M_1 \oplus M_2$ if and only if $- \otimes_A M \cong (- \otimes_A M_1) \oplus (- \otimes_A M_2)$.

9. Let K be a commutative ring and A a K-module. According to Example VI.2.6(b), a K-linear map $\mu : A \otimes_K A \to A$ defines a multiplication over A by setting $ab = \mu(a \otimes b)$, for $a, b \in A$. This multiplication is right and left distributive over addition, and is associative if and only if the diagram in Example VI.2.6(b) is commutative.
   (a) Let $\varphi : K \to A$ be a K-linear map. Prove that $e = \varphi(1)$ is a multiplicative identity for A if and only if the composites $\mu(\varphi \otimes 1_A) : K \otimes_K A \to A$ and $\mu(1_A \otimes \varphi) : A \otimes_K K \to A$ are respectively equal to the two isomorphisms of Theorem VI.2.8.
   (b) Let $\sigma : A \otimes_K A \to A \otimes_K A$ be the isomorphism defined by $a \otimes b \mapsto b \otimes a$, for $a, b \in A$. Prove that the multiplication of A is commutative if and only if $\mu\sigma = \mu$.

10. Let A be a K-algebra, L a right A-module and $(M_\lambda)_{\lambda \in \Lambda}$ a family of left A-modules. Prove that there exists a functorial morphism

$$L \otimes_A \left(\prod_{\lambda \in \Lambda} M_\lambda \right) \to \prod_{\lambda \in \Lambda} (L \otimes_A M_\lambda)$$

which is generally not an isomorphism (Hint: take $\Lambda = \mathbb{N}, M_\lambda = \mathbb{Z}/2^\lambda \mathbb{Z}$ for $\lambda \in \Lambda$ and $L = \mathbb{Q}$.)

11. Let A be a principal ideal domain. Prove that:
   (a) If M is a finitely generated A-module such that $M \otimes_A M = 0$, then $M = 0$.
   (b) If L, M are finitely generated A-modules, then $x \otimes y = 0$ in $L \otimes_A M$, for $x \in L, y \in M$, if and only if x or y is torsion.

12. Let A be a K-algebra, L a right A-module and M a left A-module. Prove that there exists an isomorphism $L \otimes_A M \cong M \otimes_{A^{op}} L$, functorial in each variable.

13. Let L, M, N be K-modules. Prove that:
    (a) The set $\mathrm{Bil}(L \times M, N)$ of bilinear maps from $L \times M$ to N becomes a K-module if, for $u, v \in \mathrm{Bil}(L \times M, N)$, $\alpha, \beta \in K$, one sets, for $x \in L, y \in M$,

    $$(\alpha u + \beta v)(x, y) = \alpha u(x, y) + \beta v(x, y)$$

    (b) There exist isomorphisms of K-modules, functorial in each variable

    $$\mathrm{Bil}(L \times M, N) \cong \mathrm{Hom}_K(L \otimes_K M, N)$$

    $$\cong \mathrm{Hom}_K(L, \mathrm{Hom}_K(M, N))$$

    $$\cong \mathrm{Hom}_K(M, \mathrm{Hom}_K(L, N)).$$

14. Let K be a field, E, E', F, F' vector spaces with respective bases $\{e_1, \cdots, e_m\}$, $\{e'_1, \cdots, e'_p\}, \{f_1, \cdots, f_n\}, \{f'_1, \cdots, f'_q\}$, and $u : E \to E', v : F \to F'$ linear maps having as matrices with respect to these bases $\mathbf{a} = [\alpha_{ij}], \mathbf{b} = [\beta_{ij}]$, respectively. Compute the matrix of $u \otimes v : E \otimes_K F \to E' \otimes_K F'$ with respect to the bases $\{e_i \otimes f_j\}$ and $\{e'_k \otimes f'_\ell\}$. It is called the ***tensor product of the matrices*** $\mathbf{a}$ and $\mathbf{b}$.

15. Let R be the triangular matrix algebra of Exercise II.2.11 and C the category whose objects are triples (U, V, f), with U an A-module, V a B-module and $f : V \otimes_B M \to U$ an A-linear map. A morphism in C is a pair $(u, v) : (U, V, f) \to (U', V', f')$ consisting of an A-linear map $u : U \to U'$ and a B-linear map $v : V \to V'$ such that $uf = f'(v \otimes M)$. Composition in C is induced from usual composition of maps. Prove that C is equivalent to the category Mod R.

16. Let G, H be abelian groups, and $x_i \in G, y_i \in H$ elements such that $\sum(x_i \otimes y_i) = 0$ in $G \otimes_{\mathbb{Z}} H$. Prove that there exist finitely generated subgroups G', H' of G, H respectively, such that $x_i \in G', y_i \in H'$ and $\sum(x_i \otimes y_i) = 0$ in $G' \otimes_{\mathbb{Z}} H'$.

17. **The field of fractions of an integral domain.** Let A be an integral domain.
    (a) Prove that a field $Q(A)$ is uniquely determined up to isomorphism by the following universal property:
       (i) There exists an injective ring morphism $\iota : A \to Q(A)$.
       (ii) For any injective ring morphism $\varphi : A \to F$, with F a field, there exists a unique ring morphism $\bar{\varphi} : Q(A) \to F$ such that $\varphi = \bar{\varphi}\iota$.

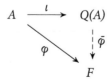

    (b) Prove that the field of fractions $Q(A)$ of A, defined in Subsection I.2.1, satisfies the universal property of (a).
    (c) Deduce that if A is itself a field, then $A \cong Q(A)$.

(d) Let A, A' be integral domains and $\eta : A \to A'$ an injective ring morphism, prove that there exists a unique injective morphism $\bar\eta : Q(A) \to Q(A')$ such that $\bar\eta(\alpha/1) = \eta(\alpha)/1$, for $\alpha \in A$.

(e) Let $\mathcal{I}d$ be the category of integral domains with injective ring morphisms and $\mathcal{F}ld$ the category of fields with ring morphisms. Construct a functor $Q : \mathcal{I}d \to \mathcal{F}ld$ which is left adjoint to the inclusion $\mathcal{F}ld \hookrightarrow \mathcal{I}d$.

18. **The abelianisation of a group.** Let G be a multiplicative, not necessarily abelian, group and G' its derived subgroup, that is, the subgroup of G whose elements are finite products of commutators $[x, y] = x^{-1}y^{-1}xy$, with $x, y \in G$.

(a) Prove that G' is a normal subgroup of G and that G/G' is abelian. Let $p_G : G \to G/G'$ be the projection.

(b) Prove that, for any group homomorphism $f : G \to A$, with A abelian, there exists a unique group homomorphism $\bar f : G/G' \to A$ such that $f = \bar f p_G$.

(c) Construct a functor $Ab : \mathcal{G}r \to \mathcal{A}b$ such that $Ab(G) = G/G'$. This functor is called the **abelianisation** functor.

(d) Prove that Ab is left adjoint to the inclusion $\mathcal{A}b \hookrightarrow \mathcal{G}r$.

19. Let A be a K-algebra and I the ideal generated by elements of the form $ab - ba$, with $a, b \in A$. Denote by $\pi : A \to A/I$ the projection.

(a) Prove that A/I is commutative.

(b) Prove that, for any algebra morphism $\varphi : A \to B$, with B commutative, there exists a unique algebra morphism $\bar\varphi : A/I \to B$ such that $\varphi = \bar\varphi\pi$.

(c) Construct a functor F from the category $\mathcal{A}lg_K$ of K-algebras to the category $\mathcal{C}alg_K$ of commutative K-algebras such that $F(A) = A/I$.

(d) Prove that F is left adjoint to the inclusion $\mathcal{C}alg_K \hookrightarrow \mathcal{A}lg_K$.

20. Let A, B be algebras and $\varphi : A \to B$ an algebra morphism. We compute right and left adjoints to the associated change of scalars functor $F : \mathrm{Mod}\, B \to \mathrm{Mod}\, A$. Prove that:

(a) For any B-module M, the K-linear isomorphism $M \to \mathrm{Hom}_B({}_A B_B, M)$ of Theorem II.3.9 is A-linear. Hence, $F \cong \mathrm{Hom}_B({}_A B_B, -)$.

(b) We also have $F \cong - \otimes_B B_A$.

(c) $- \otimes_A B_B : \mathrm{Mod}\, A \to \mathrm{Mod}\, B$ is left adjoint to F.

(d) $\mathrm{Hom}_A({}_B B_A, -) : \mathrm{Mod}\, A \to \mathrm{Mod}\, B$ is right adjoint to F.

VI.3 Exact functors

VI.3.1 General considerations

It is natural to study those functors which preserve exactness, that is, which send an exact sequence onto another. Unfortunately, the most common functors, namely the Hom and tensor functors, do not preserve it. This leads us to define weaker notions of left and right exactness, inspired from the left and right exact sequences of Subsection II.4.2.

VI.3.1 Definition. Let C, D be K-abelian categories and $F : C \to D$ a linear functor.
(a) If F is covariant:
   (i) It is **exact** if $X \xrightarrow{f} Y \xrightarrow{g} Z$ exact implies $FX \xrightarrow{Ff} FY \xrightarrow{Fg} FZ$ exact.
   (ii) It is **left exact** if $0 \to X \xrightarrow{f} Y \xrightarrow{g} Z$ exact implies $0 \to FX \xrightarrow{Ff} FY \xrightarrow{Fg} FZ$ exact.
   (iii) It is **right exact** if $X \xrightarrow{f} Y \xrightarrow{g} Z \to 0$ exact implies $FX \xrightarrow{Ff} FY \xrightarrow{Fg} FZ \to 0$ exact.
(b) If F is contravariant:
   (i) It is **exact** if $X \xrightarrow{f} Y \xrightarrow{g} Z$ exact implies $FZ \xrightarrow{Fg} FY \xrightarrow{Ff} FX$ exact.
   (ii) It is **left exact** if $X \xrightarrow{f} Y \xrightarrow{g} Z \to 0$ exact implies $0 \to FZ \xrightarrow{Fg} FY \xrightarrow{Ff} FX$ exact.
   (iii) It is **right exact** if $0 \to X \xrightarrow{f} Y \xrightarrow{g} Z$ exact implies $FZ \xrightarrow{Fg} FY \xrightarrow{Ff} FX \to 0$ exact.

In other words, whether the functor is covariant or contravariant, it is exact, left exact or right exact whenever the *induced* sequence is exact, left exact or right exact, respectively.

If $X \xrightarrow{f} Y \xrightarrow{g} Z$ is such that $gf = 0$ and F is a linear covariant functor, then $Fg \cdot Ff = F(gf) = F(0) = 0$, so that Im $Ff \subseteq$ Ker Fg. The situation is similar if F is contravariant.

The definitions of left and right exactness may be reformulated.

VI.3.2 Lemma. *Let C, D be K-abelian categories and $F : C \to D$ a linear functor.*
(a) *If F is covariant:*
   (i) *F is left exact if and only if, for any morphism f, we have $F(\ker f) = \ker Ff$.*
   (ii) *F is right exact if and only if, for any morphism f, we have $F(\operatorname{coker} f) = \operatorname{coker} Ff$.*
(b) *If F is contravariant:*
   (i) *F is left exact if and only if, for any morphism f, we have $F(\operatorname{coker} f) = \ker Ff$.*
   (ii) *F is right exact if and only if, for any morphism f, we have $F(\ker f) = \operatorname{coker} Ff$.*   □

Thus, if a covariant functor is left exact, it preserves monomorphisms. If it is right exact, it preserves epimorphisms. If it is contravariant and left exact, it transforms epimorphisms into monomorphisms, and if it is right exact, it transforms monomorphisms into epimorphisms.

Moreover, if a covariant functor F is left and right exact, then it follows from Definition IV.2.21 and the previous lemma that $F(\operatorname{im} f) = \operatorname{im} Ff$ and $F(\operatorname{coim} f) = \operatorname{coim} Ff$. If it is contravariant and left and right exact, then it interchanges images and coimages.

One expects that a functor is exact if and only if it is both left and right exact. Moreover, as we now see, one can define exact functors as being those which preserve short exact sequences.

VI.3.3 Proposition. *Let C, D be K-abelian categories and $F : C \to D$ a linear functor. The following are equivalent.*
  (a) *F is exact.*
  (b) *F preserves short exact sequences.*
  (c) *F is left and right exact*

Proof. We only prove the statement in the covariant case. Clearly, (a) implies (b) which implies (c). We prove that (c) implies (a). Let $X \xrightarrow{f} Y \xrightarrow{g} Z$ be exact. Because F is left and right exact, $\operatorname{im}(Ff) = F(\operatorname{im} f) = F(\ker g) = \ker(Fg)$, so the induced sequence is exact. $\qquad\qquad\square$

Therefore, a covariant functor is exact if and only if it preserves kernels and cokernels (and hence it preserves images and coimages), and a contravariant functor is exact if and only if it interchanges kernels and cokernels (and hence, it interchanges images and coimages).

VI.3.4 Examples.
  (a) Exact functors are scarce. The most obvious ones are the forgetful functors. Let A be a K-algebra, the forgetful functor $|-| : \operatorname{Mod} A \to \operatorname{Mod} K$ which associates to an A-module its underlying K-module, and to an A-linear map itself considered as K-linear map, is exact. Indeed, let f be an A-linear map, then $f = |f|$ as maps. Consequently, if $L \xrightarrow{f} M \xrightarrow{g} N$ is an exact sequence of A-modules, then $\operatorname{Im} |f| = \operatorname{Im} f = \operatorname{Ker} g = \operatorname{Ker} |g|$. Therefore, $|L| \xrightarrow{|f|} |M| \xrightarrow{|g|} |N|$ is exact in $\operatorname{Mod} K$.
  (b) Any equivalence between abelian categories preserves kernels and cokernels hence is an exact functor. Similarly, a duality is an exact contravariant functor.
  (c) An exact functor between abelian categories $F : C \to D$ is faithful if and only if, for any object $X \neq 0$, we have $FX \neq 0$.
       Indeed, if F is faithful and $FX = 0$, then $F(1_X) = 1_{FX} = 0 = F(0)$ gives $1_X = 0$. Hence, $X = 0$. Conversely, if $FX = 0$ implies $X = 0$, let f be a nonzero morphism. Then $\operatorname{Im} f \neq 0$, which implies $F(\operatorname{Im} f) \neq 0$, because of the hypothesis on F. Exactness gives $F(\operatorname{Im} f) = \operatorname{Im}(Ff)$. Therefore, $Ff \neq 0$ and F is faithful.
  (d) Let A be a K-algebra. The functors $\operatorname{Hom}_A(A, -)$ and $- \otimes_A A$ from $\operatorname{Mod} A$ to itself are exact, because both are isomorphic to the identity functor, and hence are equivalences between $\operatorname{Mod} A$ and itself, see Theorems II.3.9 and VI.2.8.

VI.3.2 The Hom and tensor functors

Typical examples of left exact functors between module categories are the Hom functors.

VI.3.5 Theorem. *Let A be a K-algebra. A sequence of A-modules* $0 \to L \xrightarrow{f} M \xrightarrow{g} N$ *is exact if and only if, for any A-module U, the induced sequence of K-modules*

$$0 \to \mathrm{Hom}_A(U, L) \xrightarrow{\mathrm{Hom}_A(U, f)} \mathrm{Hom}_A(U, M) \xrightarrow{\mathrm{Hom}_A(U, g)} \mathrm{Hom}_A(U, N)$$

is exact.

Proof. Necessity. Assume that $0 \to L \xrightarrow{f} M \xrightarrow{g} N$ is exact in $\mathrm{Mod}\, A$. We first prove that $\mathrm{Hom}_A(U, f)$ is a monomorphism in $\mathrm{Mod}\, K$. Take $u \in \mathrm{Hom}_A(U, L)$ such that $\mathrm{Hom}_A(U, f)(u) = 0$. Then $fu = 0$. But f is a monomorphism, hence $u = 0$, proving our assertion.

On the other hand, $gf = 0$ implies $\mathrm{Hom}_A(U, g)\mathrm{Hom}_A(U, f) = 0$, see the remarks after Definition VI.3.1. Hence, $\mathrm{Im}\, \mathrm{Hom}_A(U, f) \subseteq \mathrm{Ker}\, \mathrm{Hom}_A(U, g)$. Conversely, if $v \in \mathrm{Ker}\, \mathrm{Hom}_A(U, g)$, then $gv = \mathrm{Hom}_A(U, g)(v) = 0$. The universal property of $L = \mathrm{Ker}\, g$ gives $u \in \mathrm{Hom}_A(U, L)$ such that $v = fu = \mathrm{Hom}_A(U, f)(u)$, that is, $v \in \mathrm{Im}\, \mathrm{Hom}_A(U, f)$.

Sufficiency. We prove that f is a monomorphim. Set $U = \mathrm{Ker}\, f$ and $u : \mathrm{Ker}\, f \to L$ the injection. Then $fu = 0$, so $u \in \mathrm{Ker}\, \mathrm{Hom}_A(\mathrm{Ker}\, f, f)$. Because $\mathrm{Hom}_A(\mathrm{Ker}\, f, f)$ is a monomorphism, we have $u = 0$. Hence, $\mathrm{Ker}\, f = 0$.

Set $U = L$. Then $0 = \mathrm{Hom}_A(L, g)\mathrm{Hom}_A(L, f)(1_L) = gf$. Let $u : U \to M$ be such that $gu = 0$. Then $u \in \mathrm{Im}\, \mathrm{Hom}_A(U, f) = \mathrm{Ker}\, \mathrm{Hom}_A(U, g)$, so there exists $v : U \to L$ such that $u = fv$. But f is a monomorphism, hence v is unique with this property. This proves that $f = \mathrm{ker}\, g$ and establishes our statement. $\square$

In particular, the covariant Hom-functor $\mathrm{Hom}_A(U, -)$ is left exact, for any module U_A.

The proof given above is element-free, therefore is valid in any abelian category. We actually proved that, if C is a K-abelian category, a sequence $0 \to L \to M \to N$ in C is exact if and only if, for any object U in C, the induced sequence of K-modules

$$0 \to C(U, L) \to C(U, M) \to C(U, N)$$

is exact, see Exercise IV.2.16.

There is another proof of sufficiency in Theorem VI.3.5. Set $U = A_A$ and use the isomorphism $\mathrm{Hom}_A(A, -) \cong 1_{\mathrm{Mod}\, A}$ of functors of Theorem II.3.9 and Example III.2.10(b). Functoriality yields a commutative diagram in $\mathrm{Mod}\, A$, where the vertical arrows are isomorphisms.

$$
\begin{array}{ccccccc}
0 & \longrightarrow & \mathrm{Hom}_A(A, L) & \longrightarrow & \mathrm{Hom}_A(A, M) & \longrightarrow & \mathrm{Hom}_A(A, N) \\
& & \downarrow{\scriptstyle\cong} & & \downarrow{\scriptstyle\cong} & & \downarrow{\scriptstyle\cong} \\
0 & \longrightarrow & L & \longrightarrow & M & \longrightarrow & N
\end{array}
$$

The hypothesis implies that the upper sequence is exact. Hence, so is the lower.

This second proof, simple and conceptual, has the disadvantage that it can only be dualised (that is, applied to the contravariant Hom-functor) once we define and prove the existence of injective cogenerators, later on in Proposition VI.4.29.

VI.3.6 Corollary. *Let A be a K-algebra. A sequence of A-modules* $0 \to L \xrightarrow{f} M \xrightarrow{g} N \to 0$ *is split exact if and only if, for any A-module U, the induced sequence of K-modules*

$$0 \to \mathrm{Hom}_A(U, L) \xrightarrow{\mathrm{Hom}_A(U, f)} \mathrm{Hom}_A(U, M) \xrightarrow{\mathrm{Hom}_A(U, g)} \mathrm{Hom}_A(U, N) \to 0$$

is split exact.

Proof. Necessity. It suffices to prove that $\mathrm{Hom}_A(U, g)$ is a retraction. Because g is a retraction, there exists g' such that $gg' = 1_N$. But then, $\mathrm{Hom}_A(U, g)\mathrm{Hom}_A(U, g') = \mathrm{Hom}_A(U, gg') = \mathrm{Hom}_A(U, 1_N) = 1_{\mathrm{Hom}_A(U,N)}$.

Sufficiency. Because of Theorem VI.3.5 above, the sequence $0 \to L \xrightarrow{f} M \xrightarrow{g} N$ is exact. Setting $U = N$, there exists $g' : N \to M$ such that $1_N = \mathrm{Hom}_A(U, g)(g') = gg'$, so g is an epimorphism and even a retraction. $\qquad\qquad\square$

The proof is again element-free: the statement holds true in any abelian category.

The corresponding statements hold true for contravariant Hom-functors, with similar proofs.

VI.3.7 Theorem. *Let A be a K-algebra. A sequence of A-modules* $L \xrightarrow{f} M \xrightarrow{g} N \to 0$ *is exact if and only if, for any A-module U, the induced sequence of K-modules*

$$0 \to \mathrm{Hom}_A(N, U) \xrightarrow{\mathrm{Hom}_A(g, U)} \mathrm{Hom}_A(M, U) \xrightarrow{\mathrm{Hom}_A(f, U)} \mathrm{Hom}_A(L, U)$$

is exact.

Proof. Necessity. We first prove that $\mathrm{Hom}_A(g, U)$ is a monomorphism. Let $u : N \to U$ be such that $\mathrm{Hom}_A(g, U)(u) = 0$. Hence, $ug = 0$. But g is an epimorphism, so $u = 0$.

On the other hand, $gf = 0$ implies $\mathrm{Hom}_A(f, U)\mathrm{Hom}_A(g, U) = \mathrm{Hom}_A(gf, U) = 0$. Hence, $\mathrm{Im}\,\mathrm{Hom}_A(g, U) \subseteq \mathrm{Ker}\,\mathrm{Hom}_A(f, U)$. Finally, if $v : M \to U$ is such that $vf = \mathrm{Hom}_A(f, U)(v) = 0$, then, because $N = \mathrm{Coker}\,f$, there exists $u : N \to U$ such that $v = ug = \mathrm{Hom}_A(g, U)(u)$, so that $v \in \mathrm{Im}\,\mathrm{Hom}_A(g, U)$.

Sufficiency. In order to prove that g is an epimorphism, let $U = \mathrm{Coker}\,g$ and $p : N \to \mathrm{Coker}\,g$ the projection. Then $0 = pg = \mathrm{Hom}_A(g, U)(p)$ and injectivity of $\mathrm{Hom}_A(g, U)$ yield $p = 0$. Hence, $\mathrm{Coker}\,g = 0$ and g is an epimorphism.

Assume now that $U = N$. Then $gf = \mathrm{Hom}_A(f, U)\,\mathrm{Hom}_A(g, U)(1_N) = 0$. In order to prove that $g = \mathrm{Coker}\,f$, let $v : M \to U$ be such that $vf = 0$. Then $v \in \mathrm{Ker}\,\mathrm{Hom}_A(f, U) = \mathrm{Im}\,\mathrm{Hom}_A(g, U)$, so there exists $u : N \to U$ such that $v = ug$. Because g is an epimorphism, u is unique. Therefore, $g = \mathrm{coker}\,f$ and we are done. $\qquad\qquad\square$

In particular, the contravariant functor $\mathrm{Hom}_A(-, U)$ is left exact, for any module U_A.

This element-free proof also says that, in an abelian category $\mathcal{C}$, a sequence $L \to M \to N \to 0$ is exact if and only if, for any object U, the sequence

$$0 \to \mathcal{C}(N, U) \to \mathcal{C}(M, U) \to \mathcal{C}(L, U)$$

is exact, see again Exercise IV.2.16.

VI.3.8 Corollary. *Let A be a K-algebra. A sequence of A-modules $0 \to L \xrightarrow{f} M \xrightarrow{g} N \to 0$ is split exact if and only if, for any A-module U, the induced sequence of K-modules*

$$0 \to \operatorname{Hom}_A(N, U) \xrightarrow{\operatorname{Hom}_A(g, U)} \operatorname{Hom}_A(M, U) \xrightarrow{\operatorname{Hom}_A(f, U)} \operatorname{Hom}_A(L, U) \to 0$$

is split exact.

Proof. Similar to that of Corollary VI.3.6 and left to the reader. □

One may ask whether the Hom-functors are also right exact (so they would be exact). This may happen: for instance, if A is a K-algebra, then $\operatorname{Hom}_A(A, -)$ is exact, see Example VI.3.4(c). But this statement is generally not true: it depends on the module and on the sequence.

VI.3.9 Examples.
 (a) Take $A = \mathbb{Z}$ and $n > 1$ an integer. Consider the short exact sequence

$$0 \to \mathbb{Z} \xrightarrow{\mu_n} \mathbb{Z} \xrightarrow{p} \mathbb{Z}/n\mathbb{Z} \to 0$$

 where μ_n is multiplication by n, that is, the map $x \mapsto nx$, for $x \in \mathbb{Z}$, and p is the projection. We claim that the functor $\operatorname{Hom}_{\mathbb{Z}}(\mathbb{Z}/n\mathbb{Z}, -)$ is not exact. Indeed, $\operatorname{Hom}_{\mathbb{Z}}(\mathbb{Z}/n\mathbb{Z}, \mathbb{Z}) = 0$ because all elements of $\mathbb{Z}/n\mathbb{Z}$ are of finite order, while all nonzero elements of $\mathbb{Z}$ are of infinite order. On the other hand, $\operatorname{Hom}_{\mathbb{Z}}(\mathbb{Z}/n\mathbb{Z}, \mathbb{Z}/n\mathbb{Z}) \neq 0$: it contains at least the identity map. Therefore, $\operatorname{Hom}_{\mathbb{Z}}(\mathbb{Z}/n\mathbb{Z}, p) : \operatorname{Hom}_{\mathbb{Z}}(\mathbb{Z}/n\mathbb{Z}, \mathbb{Z}) \to \operatorname{Hom}_{\mathbb{Z}}(\mathbb{Z}/n\mathbb{Z}, \mathbb{Z}/n\mathbb{Z})$ cannot be surjective.
 (b) Take again $A = \mathbb{Z}$ and consider the short exact sequence

$$0 \to \mathbb{Z} \xrightarrow{j} \mathbb{Q} \xrightarrow{p} \mathbb{Q}/\mathbb{Z} \to 0$$

 where j is the injection and p the projection. We claim that the functor $\operatorname{Hom}_{\mathbb{Z}}(-, \mathbb{Z})$ is not exact. Indeed, $\operatorname{Hom}_{\mathbb{Z}}(\mathbb{Q}, \mathbb{Z}) = 0$, see Example II.3.13. Because, obviously, $\operatorname{Hom}_{\mathbb{Z}}(\mathbb{Z}, \mathbb{Z}) \neq 0$, the morphism $\operatorname{Hom}_{\mathbb{Z}}(j, \mathbb{Z}) : \operatorname{Hom}_{\mathbb{Z}}(\mathbb{Q}, \mathbb{Z}) \to \operatorname{Hom}_{\mathbb{Z}}(\mathbb{Z}, \mathbb{Z})$ cannot be surjective.
 (c) Let A be a K-algebra. The proof of Corollary VI.3.6 implies that a short exact sequence $0 \to L \xrightarrow{f} M \xrightarrow{g} N \to 0$ of A-modules splits if and only if the left exact sequence induced in $\operatorname{Mod} K$ upon evaluation by the covariant functor $\operatorname{Hom}_A(N, -)$:

$$0 \to \operatorname{Hom}_A(N, L) \xrightarrow{\operatorname{Hom}_A(N, f)} \operatorname{Hom}_A(N, M) \xrightarrow{\operatorname{Hom}_A(N, g)} \operatorname{Hom}_A(N, N)$$

 is also right exact. Indeed, the latter is right exact if and only if $\operatorname{Hom}_A(N, g)$ is surjective, which occurs if and only if there exists $g' \in \operatorname{Hom}_A(N, M)$ such that $1_N = \operatorname{Hom}_A(N, g)(g') = gg'$, that is, if and only if g is a retraction.

Similarly, the sequence $0 \to L \to M \to N \to 0$ splits if and only if the left exact sequence induced upon evaluation by the contravariant functor $\text{Hom}_A(-, L)$ is right exact.

This observation has a nice consequence, namely the situation of Example IV.2.33(c) cannot occur when dealing with finitely generated modules over a finite-dimensional algebra. That is, let K be a field and A a finite-dimensional K-algebra. We claim that a short exact sequence $0 \to L \to M \to N \to 0$ of finitely generated A-modules splits if and only if $M \cong L \oplus N$. We just have to prove sufficiency. Because of Corollary III.4.9, L, M, N are finite-dimensional K-vector spaces, hence so are the Hom-sets between them, and right exactness of the (above) induced left exact sequence follows from

$$\dim_K \text{Hom}_A(N, M) = \dim_K \text{Hom}_A(N, L \oplus N)$$
$$= \dim_K \text{Hom}_A(N, L) + \dim_K \text{Hom}_A(N, N).$$

Before looking at the tensor product, we give a nice application of adjunction to exactness, which actually follows from left exactness of the Hom functors.

VI.3.10 Theorem. *Let* A, B *be* K-*algebras and* $F : \text{Mod} A \to \text{Mod} B$, $G : \text{Mod} B \to \text{Mod} A$ *a pair of functors such that* (F, G) *is an adjoint pair. Then* F *is right exact and* G *is left exact.*

Proof. We prove that F is right exact, the proof that G is left exact is similar. Because (F, G) is an adjoint pair, there exists, for modules M_A and U_B, a functorial isomorphism of K-modules $\varphi_{M,U} : \text{Hom}_B(FM, U) \cong \text{Hom}_A(M, GU)$. Let $L \xrightarrow{f} M \xrightarrow{g} N \to 0$ be exact in Mod A. Functoriality of φ implies that, for any B-module U, we have a commutative diagram of K-modules and K-linear maps

$$
\begin{array}{ccccccc}
0 & \longrightarrow & \text{Hom}_B(FN, U) & \longrightarrow & \text{Hom}_B(FM, U) & \longrightarrow & \text{Hom}_B(FL, U) \\
& & \downarrow{\cong} & & \downarrow{\cong} & & \downarrow{\cong} \\
0 & \longrightarrow & \text{Hom}_A(N, GU) & \longrightarrow & \text{Hom}_A(M, GU) & \longrightarrow & \text{Hom}_A(L, GU)
\end{array}
$$

where the vertical maps are the isomorphisms φ. Because of Theorem VI.3.5, the functor $\text{Hom}_A(-, GU)$ is left exact, so the lower sequence is exact. Hence, so is the upper sequence. Because U is arbitrary, Theorem VI.3.5 yields an exact sequence $FL \xrightarrow{Ff} FM \xrightarrow{Fg} FN \to 0$. That is, F is right exact. $\qquad\square$

VI.3.11 Example. Let A, B be algebras and $\varphi : A \to B$ an algebra morphism. Because of Exercise VI.2.20, the associated change of scalars functor Mod $B \to$ Mod A has both a right and a left adjoints. Theorem VI.3.10 above implies that it is an exact functor.

We deduce right exactness of the tensor functors.

VI.3.12 Theorem. *Let A be a K algebra. A sequence of right A-modules $L \xrightarrow{f} M \xrightarrow{g} N \to 0$ is exact if and only if, for any left A-module U, the induced sequence of K-modules*

$$L \otimes_A U \xrightarrow{f \otimes 1} M \otimes_A U \xrightarrow{g \otimes 1} N \otimes_A U \to 0$$

is exact.

Proof. Necessity. Because of Theorem VI.2.17, the functor $- \otimes_A U$ is left adjoint to the functor $\mathrm{Hom}_A(U, -)$. Because of Theorem VI.3.10, it is right exact.

Sufficiency. Take $U = A$ and apply Theorem VI.2.8. $\qquad\square$

Of course, it is possible to give an *ad hoc* proof for this theorem, in the spirit of the proof of Theorem VI.3.5. But we believe that the one we gave is more elegant.

VI.3.13 Corollary. *Let A be a K-algebra. A sequence of right A-modules $0 \to L \xrightarrow{f} M \xrightarrow{g} N \to 0$ is split exact if and only if, for any left A-module U, the induced sequence of K-modules*

$$0 \to L \otimes_A U \xrightarrow{f \otimes 1} M \otimes_A U \xrightarrow{g \otimes 1} N \otimes_A U \to 0$$

is split exact.

Proof. Necessity. If f is a section and f' such that $f'f = 1$, then $(f' \otimes 1)(f \otimes 1) = (f'f) \otimes 1 = 1$. Hence $f \otimes 1$ is a section, and in particular a monomorphism, so the induced sequence is exact and splits.

Sufficiency. Set $U = A$ and apply Theorem VI.2.8. $\qquad\square$

We state the corresponding results on the other side. The proofs are easy and left to the reader.

VI.3.14 Theorem. *Let A be a K-algebra. A sequence of left A-modules $U \xrightarrow{u} V \xrightarrow{v} W \to 0$ is exact if and only if, for any right A-module L, the induced sequence of K-modules*

$$L \otimes_A U \xrightarrow{1 \otimes u} L \otimes_A V \xrightarrow{1 \otimes v} L \otimes_A W \to 0$$

is exact. $\qquad\square$

VI.3.15 Corollary. *Let A be a K-algebra. A sequence of left A-modules $0 \to U \xrightarrow{u} V \xrightarrow{v} W \to 0$ is split exact if and only if, for any right A-module L, the induced sequence of K-modules*

$$0 \to L \otimes_A U \xrightarrow{1 \otimes u} L \otimes_A V \xrightarrow{1 \otimes v} L \otimes_A W \to 0$$

is split exact.   □

The difference between the tensor and the Hom-functors is that the former is covariant in both variables, and thus behaves in the same way from both sides, while the latter is covariant with respect to the second variable and contravariant with respect to the first.

The next examples show that in general the tensor product functor is not exact (though, if A is a K-algebra, then the functors $A \otimes_A -$ and $- \otimes_A A$ are exact, because of Example VI.3.4(d)).

VI.3.16 Examples.

(a) Let $n > 1$ be an integer and consider the short exact sequence of abelian groups

$$0 \to \mathbb{Z} \xrightarrow{\mu_n} \mathbb{Z} \to \mathbb{Z}/n\mathbb{Z} \to 0$$

where μ_n is multiplication by n. Evaluating the functor $- \otimes_{\mathbb{Z}} \mathbb{Z}/n\mathbb{Z}$ yields an exact sequence

$$\mathbb{Z} \otimes_{\mathbb{Z}} \mathbb{Z}/n\mathbb{Z} \xrightarrow{\mu_n \otimes 1} \mathbb{Z} \otimes_{\mathbb{Z}} \mathbb{Z}/n\mathbb{Z} \to \mathbb{Z}/n\mathbb{Z} \otimes_{\mathbb{Z}} \mathbb{Z}/n\mathbb{Z} \to 0.$$

We prove that $\mu_n \otimes 1$ is not a monomorphism. Let $\bar{a} \in \mathbb{Z}/n\mathbb{Z}$ and $x \in \mathbb{Z}$, then

$$(\mu_n \otimes 1)(x \otimes \bar{a}) = \mu_n(x) \otimes \bar{a} = nx \otimes \bar{a} = x \otimes n\bar{a} = x \otimes 0 = 0,$$

that is, $\mu_n \otimes 1 = 0$. On the other hand, $\mathbb{Z} \otimes_{\mathbb{Z}} \mathbb{Z}/n\mathbb{Z} \cong \mathbb{Z}/n\mathbb{Z} \neq 0$, because of Theorem VI.2.8. Thus, $\mu_n \otimes 1$ is not a monomorphism.

(b) The functor $L \otimes_A -$ is not exact in general either. Indeed, when A is commutative and L, M are A-modules, we have $L \otimes_A M \cong M \otimes_A L$ functorially, due to Proposition VI.2.7(a). Now, $\mathbb{Z}$ is commutative. So we may take Example (a) above and change sides.

The following corollary is useful for computations involving quotient algebras.

VI.3.17 Corollary. *Let A be a K-algebra and I an ideal in A.*

(a) *For any right A-module L, there is a functorial isomorphism $L \otimes_A A/I \cong L/LI$ given by $x \otimes (a + I) \mapsto xa + LI$, for $x \in L, a \in A$.*

(b) *For any left A-module U, there is a functorial isomorphism $A/I \otimes_A U \cong U/IU$ given by $(a + I) \otimes x \mapsto ax + IU$, for $x \in U, a \in A$.*

Proof. We prove (a), because the proof of (b) is similar. Consider the short exact sequence of left A-modules

$$0 \to I \xrightarrow{j} A \xrightarrow{p} A/I \to 0$$

where j is the inclusion and p the projection. Evaluating the functor $L \otimes_A -$ on this sequence yields a commutative diagram with exact rows

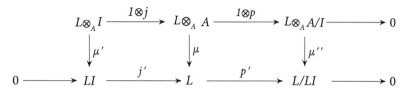

where j' is the inclusion, p' the projection, $\mu : L \otimes_A A \to L$ the multiplication map $x \otimes a \mapsto xa$, for $x \in L, a \in A$, which, according to Theorem VI.2.8, is an isomorphism, μ' the restriction of μ to $L \otimes_A I$ and μ'' the morphism obtained by passing to cokernels. In particular, μ'' is given by $x \otimes (a + I) \mapsto xa + LI$, for $x \in L, a \in A$. Because p' and μ are epimorphisms, so is $\mu''(1 \otimes p) = p'\mu$, hence so is μ''. Because LI is generated by the xa, with $x \in L, a \in I$, the map μ' is an epimorphism. Applying the Snake Lemma II.4.19 yields that μ'' is a monomorphism. Therefore, it is an isomorphism. □

It follows that, if A is an algebra, I an ideal of A, and $L \to M \to N \to 0$ a right exact sequence in Mod$A$, then the induced sequence $L/LI \to M/MI \to N/NI \to 0$ is also right exact in ModA/I.

Exercises of Section VI.3

1. Let A, B be algebras and $F : \text{Mod}\,A \to \text{Mod}\,B$ an exact covariant functor, M an A-module, and L, N submodules of M. Prove that:
   (a) FL and FN are isomorphic to submodules of FM.
   (b) F preserves finite direct sums, that is, if L, N are in direct sum, then $F(L \oplus N) \cong FL \oplus FN$.
   (c) In general, $F(L + N) \cong FL + FN$ if and only if $F(L \cap N) \cong FL \cap FN$.

2. Let A be an algebra, L, M modules and $f : L \to M$ an A-linear map. Prove that:
   (a) f is a monomorphism (or an isomorphism, or zero) if and only if, for any A-module U, the map $\text{Hom}_A(U, f)$ is a monomorphism (or an isomorphism, or zero, respectively.)
   (b) f is an epimorphism (or an isomorphism, or zero) if and only if, for any A-module U, the map $\text{Hom}_A(f, U)$ is a monomorphism (or an isomorphism, or zero, respectively.)

3. Let A be an algebra and L a free finitely generated A-module. Prove that the functors $\text{Hom}_A(L, -)$ and $L \otimes_A -$ are exact. What happens with the functor $\text{Hom}_A(-, L)$?

4. Prove that a short exact sequence of finitely generated abelian groups $0 \to G' \to G \to G'' \to 0$ splits if and only if $G \cong G' \oplus G''$.

5. Let A be a commutative algebra and $a \in A$.

    (a) Prove that, for any A-module M, the set $Ma = \{xa : x \in M\}$ is a sub-module of M, and that any morphism $f : L \to M$ restricts to a morphism $f' : La \to Ma$.

    (b) Define a functor $F : \mathrm{Mod}\,A \to \mathrm{Mod}\,A$ as follows: for a module M, set $FM = M/Ma$, and for a morphism $f : L \to M$, the morphism $Ff : L/La \to M/Ma$ is obtained by passing to the quotients. Prove that F is right exact.

    (c) Prove that F is not left exact in general (Hint: consider the short exact sequence of Example VI.3.9(a)).

6. Consider the short exact sequence of abelian groups

$$0 \to \mathbb{Z} \to \mathbb{Q} \to \mathbb{Q}/\mathbb{Z} \to 0$$

with the obvious morphisms. Let $n > 1$ be an integer. Prove that $\mathrm{Hom}_{\mathbb{Z}}(\mathbb{Z}/n\mathbb{Z}, \mathbb{Q}) = 0$. Deduce that the functor $\mathrm{Hom}_{\mathbb{Z}}(\mathbb{Z}/n\mathbb{Z}, -)$ is not exact.

7. Let A be a K-algebra and

$$0 \to L_A \xrightarrow{f} M_A \xrightarrow{g} N_A \to 0$$

$$0 \to {}_A U \xrightarrow{u} {}_A V \xrightarrow{v} {}_A W \to 0$$

short exact sequences of right and left A-modules, respectively. Prove that:

    (a) $g \otimes v$ is an epimorphism.

    (b) $\mathrm{Ker}\,(g \otimes v) = \mathrm{Im}(f \otimes 1_V) + \mathrm{Im}(1_M \otimes u)$.

    (c) $(M/N) \otimes_A (V/W) \cong (M \otimes_A V)/[(\mathrm{Im}(f \otimes 1_V) + \mathrm{Im}(1_M \otimes u)]$.

8. Let A be a K-algebra, I a right ideal and J a left ideal of A. Establish the isomorphism of K-modules $(A/I) \otimes_A (A/J) \cong A/(I + J)$.

9. Let G be an additive abelian group and $n > 1$ an integer.

    (a) Prove that $nG = \{ng : g \in G\}$ is a subgroup of G.

    (b) Establish the isomorphisms $G \otimes_{\mathbb{Z}} \mathbb{Z}/n\mathbb{Z} \cong G/nG \cong \mathbb{Z}/n\mathbb{Z} \otimes_{\mathbb{Z}} G$.

10. Let G' be a subgroup of an abelian group G and consider the exact sequence $0 \to G' \to G \to G/G' \to 0$ with the obvious morphisms. Prove that the following are equivalent:

    (a) G' is pure in G (in the sense of Exercise II.4.28).

    (b) For any positive integer m, the induced sequence

$$0 \to \mathrm{Hom}_{\mathbb{Z}}(\mathbb{Z}/m\mathbb{Z}, G') \to \mathrm{Hom}_{\mathbb{Z}}(\mathbb{Z}/m\mathbb{Z}, G) \to \mathrm{Hom}_{\mathbb{Z}}(\mathbb{Z}/m\mathbb{Z}, G/G') \to 0$$

is exact.

    (c) For any positive integer m, the induced sequence

$$0 \to \mathbb{Z}/m\mathbb{Z} \otimes_{\mathbb{Z}} G' \to \mathbb{Z}/m\mathbb{Z} \otimes_{\mathbb{Z}} G \to \mathbb{Z}/m\mathbb{Z} \otimes_{\mathbb{Z}} G/G' \to 0$$

is exact.

VI.4 Projectives, injectives and flats

VI.4.1 Projective modules

In this subsection, A denotes a K-algebra. In general, given an A-module P, the functor $\mathrm{Hom}_A(P, -)$ is only left exact. We characterise those modules P having the property that it is actually exact. We repeat the definition of projective modules, stated just before Lemma III.4.11.

VI.4.1 Definition. An A-module P is **projective** if, for any epimorphism $f : M \to N$ and morphism $u : P \to N$, there exists a morphism $v : P \to M$ such that $u = fv$.

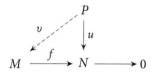

As stated in Lemma III.4.11, free modules have this property, therefore projective modules exist (and free modules are projective). The following reformulation of the definition is useful.

VI.4.2 Lemma. *An A-module P is projective if and only if the functor $\mathrm{Hom}_A(P, -)$ is exact.*

Proof. Because $\mathrm{Hom}_A(P, -)$ is always left exact, it is exact if and only if it preserves epimorphisms, that is, for any epimorphism $f : M \to N$, the induced morphism $\mathrm{Hom}_A(P, f) : \mathrm{Hom}_A(P, M) \to \mathrm{Hom}_A(P, N)$ is an epimorphism. This is the condition in the definition. $\square$

The morphism v of the definition, which is said to **lift** u, or to be a **lifting** of u, is generally not unique, see the comments after the proof of Lemma III.4.11. In particular, the lifting property of projectives is *not* a universal property.

A property of projective modules which comes directly from the definition is the following: let P be projective, then the functor $\mathrm{Hom}_A(P, -)$ is faithful if and only if, for any module $M \neq 0$, we have $\mathrm{Hom}_A(P, M) \neq 0$. This follows indeed from Example VI.3.4(c).

Recall that $P = A_A$ is projective and a module is free if and only if it is a direct sum of copies of A. This leads us to the following proposition.

VI.4.3 Proposition. *Let $(P_\lambda)_\lambda$ be a family of modules. The direct sum $\bigoplus_\lambda P_\lambda$ is projective if and only if so is each P_λ.*

Proof. *Necessity.* Assume that $\bigoplus_\lambda P_\lambda$ is projective. For any λ, let $q_\lambda : P_\lambda \to \bigoplus_\mu P_\mu$ be the injection and $p_\lambda : \bigoplus_\mu P_\mu \to P_\lambda$ the unique morphism defined by the universal property of the direct sum and the condition that $p_\lambda q_\mu$ equals 1_{P_λ} if $\lambda = \mu$ and

0 otherwise (that is, $p_\lambda q_\mu = \delta_{\lambda\mu}$), see Example IV.2.8(a). Let $f : M \to N$ be an epimorphism, and $u : P_\lambda \to N$ a morphism.

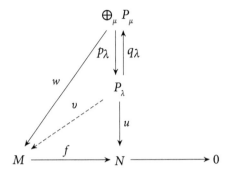

Because $\bigoplus_\mu P_\mu$ is projective, there exists $w : \bigoplus_\mu P_\mu \to M$ such that $fw = up_\lambda$. But then $f(wq_\lambda) = (fw)q_\lambda = up_\lambda q_\lambda = u$, that is, $v = wq_\lambda$ lifts u.

Sufficiency. Assume that each P_λ is projective. Denote, for any λ, by $q_\lambda : P_\lambda \to \bigoplus_\mu P_\mu$ the injection, $f : M \to N$ an epimorphism, and $u : \bigoplus_\mu P_\mu \to N$ a morphism. Then, for each λ, projectivity of P_λ yields a morphism $v_\lambda : P_\lambda \to M$ such that $fv_\lambda = uq_\lambda$.

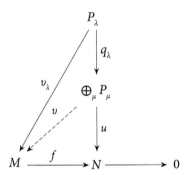

The universal property of direct sum implies the existence of $v : \bigoplus_\mu P_\mu \to M$ such that $vq_\lambda = v_\lambda$, for any λ. Then $fvq_\lambda = fv_\lambda = uq_\lambda$, for any λ, yields $fv = u$ because of uniqueness in the universal property of direct sum. □

Because finite direct sums of modules coincide with finite products, the product of a *finite* family of modules is projective if and only if each of these modules is projective.

We give a complete characterisation of projective modules.

VI.4.4 Theorem. *The following are equivalent for a module P.*
   (a) *P is projective.*
   (b) *Any short exact sequence of the form* $0 \to M \xrightarrow{f} N \xrightarrow{g} P \to 0$ *splits.*
   (c) *P is a direct summand of a free module.*

Proof. (a) implies (b). Because of Example IV.2.8(c), the morphism g is a retraction.

    (b) implies (c). Because of Proposition III.4.8, P is a quotient of a free module L. Thus, there exists a short exact sequence

$$0 \to \operatorname{Ker} f \to L \xrightarrow{f} P \to 0$$

with L free. The hypothesis says that it splits.

    (c) implies (a). Because of Lemma III.4.11, free modules are projective, and because of Proposition VI.4.3, a direct summand of a projective module is projective. $\qquad\square$

If P is finitely generated, then, because of Proposition III.4.8, it is a quotient of a finitely generated free module. Therefore, a finitely generated projective module is a direct summand of a finitely generated free module. We show that, in particular, cyclic modules generated by idempotents are projective. An element e in an algebra A is called an **idempotent** if $e^2 = e$.

VI.4.5 Corollary. *Let A be an algebra and $e \in A$ an idempotent. Then eA is a projective A-module.*

Proof. In view of Theorem VI.4.4, it suffices to prove that $A_A = eA \oplus (1-e)A$. Let $a \in A$, then $a = ea + (1 - e)a$ shows that $A = eA + (1 - e)A$. Assume that $a \in eA \cap (1 - e)A$, then there exist $b, c \in A$ such that $a = eb = (1 - e)c$. Now, e is an idempotent, so $a = eb = e^2 b = e(1 - e)c = (e - e^2)c = 0$. Hence, $eA \cap (1 - e)A = 0$ and our claim is established. $\qquad\square$

If e is an idempotent in A, then so is $1 - e$, because $(1 - e)^2 = 1 - 2e + e^2 = 1 - e$. Hence, the module $(1 - e)A$ is also projective.

If one is interested in projective *left* modules, then the cyclic left (projective) module corresponding to the idempotent e is Ae.

Usually, modules of the form eA (or Ae) are not free, see Example VI.4.6(c) below, so the previous lemma furnishes examples of projective modules which are not free.

VI.4.6 Examples.

    (a) Let A be a division ring. Because of Corollary III.4.6, any A-module is free, hence projective. If, in particular, $A = K$ is a field, then any K-vector space is a projective K-module.

    (b) We have proved, in Corollary V.2.6, that, if A is a principal ideal domain, then any submodule (in particular, any direct summand) of a free A-module is itself free. Therefore, any projective module over a principal ideal domain is free. Thus, over a principal ideal domain, a module is free if and only if it is projective. For instance, any projective abelian group is free.

(c) Let K be a field and

$$A = \begin{pmatrix} K & 0 \\ K & K \end{pmatrix} = \left\{ \begin{pmatrix} \alpha & 0 \\ \beta & \gamma \end{pmatrix} : \alpha, \beta, \gamma \in K \right\}$$

the algebra of lower triangular 2×2-matrices over K. Then $\mathbf{e}_{11} = \begin{pmatrix} 1 & 0 \\ 0 & 0 \end{pmatrix}$ is

an idempotent so $A = \mathbf{e}_{11}A \oplus (1 - \mathbf{e}_{11})A$ and $\mathbf{e}_{11}A = \begin{pmatrix} K & 0 \\ 0 & 0 \end{pmatrix}$ is projective.

But $\dim_K(\mathbf{e}_{11}A) = 1$, so it cannot be free, because a nonzero free module which is finite-dimensional is of the form $A_A^{(m)}$ for some integer $m > 0$, hence its dimension is a multiple of $\dim_K A = 3$. The same argument shows that $\mathbf{e}_{22}A = (1 - \mathbf{e}_{11})A$ is projective but not free.

(d) Because of Exercise III.4.5, if $a, b > 1$ are integers, then $b\mathbb{Z}/(ab)\mathbb{Z}$ is an example of a $\mathbb{Z}/(ab)\mathbb{Z}$-module which is projective but not free.

(e) Corollary VI.4.5 shows how to compute certain projective modules over bound quiver algebras. Let K be a field and $A \cong KQ/I$ a bound quiver algebra. As seen in Example I.3.11(g), A is a finite-dimensional K-algebra. Any point $x \in Q_0$ corresponds to a stationary path ε_x. Because $\varepsilon_x^2 = \varepsilon_x$. then $e_x = \varepsilon_x + I$ is an idempotent in A. Therefore, to the point x is associated a projective A-module $P_x = e_x A$. Now, $P_x = e_x(KQ/I) = \varepsilon_x(KQ)/\varepsilon_x I$ is spanned, as a K-vector space, by the residual classes modulo I of all paths in Q starting with x. This allows for an easy computation of the P_x, illustrated in the example below.

Consider the quiver

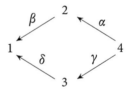

bound by the ideal $\langle \alpha\beta - \gamma\delta \rangle$, see Example IV.4.6(c). Denote as usual by $\bar{w} = w + I$ the residual class of a path w in Q. As K-vector spaces, $P_1 = e_1 A = \langle \bar{e}_1 \rangle, P_2 = e_2 A = \langle \bar{e}_2, \bar{\beta} \rangle, P_3 = e_3 A = \langle \bar{e}_3, \bar{\delta} \rangle$ and $P_4 = e_4 A = \langle \bar{e}_4, \bar{\alpha}, \bar{\gamma}, \overline{\alpha\beta} \rangle$ (because $\overline{\alpha\beta} = \overline{\gamma\delta}$ in KQ/I). One can describe the corresponding representations using the functor of Example IV.4.6(c). Indeed P_1 is a one-dimensional K-vector space, which is only nonzero at the point 1. Thus,

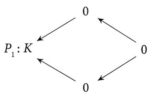

where maps between the component vector spaces are (necessarily) zero. Similarly, P_2 is two-dimensional, having one copy of K at the point 2 and another at 1 (that is, the target of the arrow β), and the morphism $P_2(2) \to P_2(1)$ is right multiplication by the class $\bar{\beta}$ of the arrow β. The remaining maps between the component spaces are zero. It is then

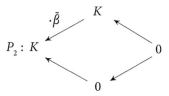

Because K is a field, a change of basis shows that this representation is isomorphic to

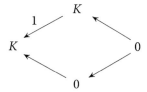

Symmetry of the quiver yields

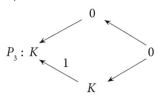

Finally, P_4 corresponds to the representation

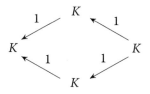

The next corollary is expressed by saying that the module category has enough projectives.

VI.4.7 Corollary. *Any module is a quotient of a projective module.*

Proof. This follows from Proposition III.4.8 and Theorem VI.4.4. □

If, in particular, a module is finitely generated, then, because of Proposition III.4.8, it is a quotient of a finitely generated free module, hence of a finitely generated projective module.

VI.4.8 Corollary. *For any A-module M, there exist projective modules P_i, with i a nonnegative integer, and an exact sequence*

$$\cdots \to P_i \xrightarrow{f_i} P_{i-1} \to \cdots \to P_1 \xrightarrow{f_1} P_0 \xrightarrow{f_0} M \to 0.$$

Proof. By induction on n. The existence of f_0 follows from Corollary VI.4.7. For $i > 0$, we apply Corollary VI.4.7 to $\operatorname{Ker} f_{i-1}$. □

A sequence as in Corollary VI.4.8 is called a ***projective resolution*** of M. For instance, any free resolution is a projective resolution, see Corollary III.4.10. A projective resolution with just two projective terms, that is, an exact sequence of the form $P_1 \to P_0 \to M \to 0$, with P_1, P_0 projective, is called a ***projective presentation*** of the module M. As is the case for free resolutions, if M is finitely generated, then one may choose P_0 finitely generated, but not necessarily P_1. Actually, there exists an exact sequence $P_1 \to P_0 \to M \to 0$, with P_1, P_0 both finitely generated projectives if and only if M is finitely presented, see Exercise VI.4.12.

VI.4.9 Examples.
(a) Let A be a commutative ring, $a \in A$ a nondivisor of zero and $\mu_a : x \mapsto ax$, for $x \in A$, the multiplication by a, then μ_a is injective and the short exact sequence

$$0 \to A \xrightarrow{\mu_a} A \to A/aA \to 0$$

is a projective resolution of A/aA.
(b) Let K be a field and Q the quiver

$$\circ \xleftarrow[\beta]{\alpha} \circ$$
$$1 \qquad 2$$

The projective modules $P_1 = e_1(KQ), P_2 = e_2(KQ)$ constructed as in Example VI.4.6(e) are given respectively by the representations P_1

$$K \leftleftarrows 0$$

and P_2

$$K^2 \xleftarrow[\binom{0}{1}]{\binom{1}{0}} K$$

It is readily verified that the pair $j = (j(1) = \binom{0}{1}, j(2) = 0)$ defines an injective morphism from P_1 to P_2 whose cokernel is the representation H_α

$$K \xleftarrow[0]{1} K$$

so that we have a short exact sequence

$$0 \to P_1 \xrightarrow{j} P_2 \xrightarrow{p} H_\alpha \to 0.$$

Here, the cokernel morphism is $p = (p(1) = (1 \ 0), p(2) = 1)$. This short exact sequence is a projective resolution of H_α. Dropping the zero, we get a projective presentation. We leave to the reader the verification that the module H_β given by the representation

$$K \underset{1}{\overset{0}{\rightleftarrows}} K$$

has a projective resolution of the form

$$0 \to P_1 \xrightarrow{j'} P_2 \xrightarrow{p'} H_\beta \to 0$$

(only the morphisms are different!).

Projective resolutions and presentations can be viewed as approximations of a module by projective modules (the kernel of each morphism on the resolution represents the 'error' at each step). They are useful when a computation is easier to perform on a projective module than on an arbitrary one, and this computation is compatible with passing to cokernels.

An interesting consequence of Yoneda's lemma is that the Hom-functors are projective as functors from Mod A to Mod K. A projective functor is a functor which is projective when viewed as object in a functor category. The definition is the same as for modules: consider functors from a category C to an abelian category D, a functor F is a **projective functor** if, for any functorial epimorphism $\varphi : G \to H$, that is, such that φ_X is an epimorphism for any object X in C, and functorial morphism $\zeta : F \to H$, there exists a functorial morphism $\psi : F \to G$ satisfying $\varphi\psi = \zeta$ that is, the following diagram commutes.

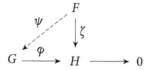

VI.4.10 Lemma. *Let A be a K-algebra and M an A-module.*
- (a) *The functor $\mathrm{Hom}_A(M, -)$ is a projective covariant functor from $\mathrm{Mod}\,A$ to $\mathrm{Mod}\,K$.*
- (b) *The functor $\mathrm{Hom}_A(-, M)$ is a projective contravariant functor from $\mathrm{Mod}\,A$ to $\mathrm{Mod}\,K$.*

Proof. We prove (a), because the proof of (b) is similar. Let $\varphi : F \to G$ be a functorial epimorphism from $\mathrm{Mod}\,A$ to $\mathrm{Mod}\,K$ and $\zeta : \mathrm{Hom}_A(M, -) \to G$ a morphism. Denote by $\eta_F : \mathrm{Fun}(\mathrm{Hom}_A(M, -), F) \to FX$ and $\eta_G : \mathrm{Fun}(\mathrm{Hom}_A(M, -), G) \to GX$ the bijections of Yoneda's Lemma III.2.11. Their functoriality yields a commutative square

$$
\begin{array}{ccc}
\mathrm{Fun}(\mathrm{Hom}_A(M,-),F) & \xrightarrow{\mathrm{Fun}(\mathrm{Hom}_A(M,-),\varphi)} & \mathrm{Fun}(\mathrm{Hom}_A(M,-),G) \\
\downarrow{\eta_F} & & \downarrow{\eta_G} \\
FM & \xrightarrow{\varphi_M} & GM
\end{array}
$$

Now η_F, η_G are bijective, and φ_M is surjective, hence

$$\mathrm{Fun}(\mathrm{Hom}_A(M, -), \varphi) : \mathrm{Fun}(\mathrm{Hom}_A(M, -), F) \to \mathrm{Fun}(\mathrm{Hom}_A(M, -), G)$$

is surjective. So, there exists ψ such that $\zeta = \mathrm{Fun}(\mathrm{Hom}_A(M, -), \varphi)(\psi) = \varphi\psi$. $\square$

This lemma explains the statement at the end of Subsection III.2.4: when passing from modules to functors, one transforms a module M (without any special property) into a *projective* functor $\mathrm{Hom}_A(M, -)$, or $\mathrm{Hom}_A(-, M)$, according to the case, which enjoys the same nice properties as projective modules.

We finish this subsection with two functorial isomorphisms that are needed later on. They furnish further relations between the Hom-functor and the tensor product. They use a nice reduction procedure: if one has to prove a functorial isomorphism involving linear functors and (among others) a finitely generated projective module P_A, then it suffices to prove it in case $P_A = A_A$. This procedure will also be used in the sequel and therefore deserves to be explained. It rests on the following lemma.

VI.4.11 Lemma. *Let A, B be K-algebras, $F, G : \mathrm{Mod}\,A \to \mathrm{Mod}\,B$ linear functors and $\varphi : F \to G$ a functorial morphism. If $L \cong M \oplus N$ in $\mathrm{Mod}\,A$, then we have a commutative square in* $\mathrm{Mod}\,B$:

$$
\begin{array}{ccc}
FM \oplus FN & \xrightarrow{\left(\begin{smallmatrix} \varphi_M & 0 \\ 0 & \varphi_N \end{smallmatrix}\right)} & GM \oplus GN \\
\Big\downarrow{\scriptstyle\cong} & & \Big\downarrow{\scriptstyle\cong} \\
FL & \xrightarrow{\varphi_L} & GL
\end{array}
$$

Furthermore, φ_L is injective, or surjective, or an isomorphism if and only if so are φ_M and φ_N.

Proof. Because of Exercise IV.2.11, one has indeed vertical isomorphisms induced by the inclusions $M \hookrightarrow L$ and $N \hookrightarrow L$. Because of Exercise III.3.4, the morphism $\left(\begin{smallmatrix} \varphi_M & 0 \\ 0 & \varphi_N \end{smallmatrix}\right) : FM \oplus FN \to GM \oplus GN$ obtained by passing to the direct sum, makes the diagram commute. Moreover, it is injective, or surjective, or an isomorphism if and only if both of its component maps φ_M and φ_N are. Hence the statement. $\square$

VI.4.12 Proposition. *Let A, B be K-algebras, P_A a finitely generated projective A-module, $_BM_A$ a bimodule and $_BN$ a left B-module. Then there exists a functorial isomorphism of K-modules*

$$\varphi : P \otimes_A \mathrm{Hom}_B(M, N) \xrightarrow{\cong} \mathrm{Hom}_B(\mathrm{Hom}_A(P, M), N)$$

defined by $x \otimes f \mapsto (g \mapsto f(g(x)))$, for $x \in P$, $f \in \mathrm{Hom}_B(M, N)$ and $g \in \mathrm{Hom}_A(P, M)$.

Proof. It is easily seen that φ is a functorial morphism. Setting $P = A_A$, a direct calculation shows that the map φ equals the composite of the well-known isomorphisms of K-modules

$$A \otimes_A \operatorname{Hom}_B(M, N) \cong \operatorname{Hom}_B(M, N) \cong \operatorname{Hom}_B(\operatorname{Hom}_A(A, M), N)$$

see Theorems II.3.9 and VI.2.8. Therefore, it remains an isomorphism if $P = A_A^{(n)}$, because the functors under consideration are K-linear. Because P is finitely generated projective, it is a direct summand of some $A_A^{(n)}$. That is, $A_A^{(n)} \cong P \oplus P'$ for some module P'. Applying the previous lemma, with $F = - \otimes_A \operatorname{Hom}_B(M, N)$, $G = \operatorname{Hom}_B(\operatorname{Hom}_A(-, M), N)$ and φ the given functorial morphism, we get that φ_P is an isomorphism because so is $\varphi_{A^{(n)}}$. $\qquad\square$

VI.4.13 Proposition. *Let A, B be K-algebras, P_A a finitely generated projective A-module, $_BM_A$ a bimodule and N_B a right B-module. Then there exists a functorial isomorphism of K-modules*

$$N \otimes_B \operatorname{Hom}_A(P, M) \xrightarrow{\;\cong\;} \operatorname{Hom}_A(P, N \otimes_B M)$$

defined by $y \otimes f \mapsto (x \mapsto y \otimes f(x))$, for $x \in P, y \in N$ and $f \in \operatorname{Hom}_A(P, M)$.

Proof. Similar to that of Proposition VI.4.12 above and left to the reader. $\qquad\square$

VI.4.2 Injective modules

Injectivity is the notion dual to projectivity. In this subsection, A denotes a K-algebra.

VI.4.14 Definition. An A-module I is ***injective*** if, for any monomorphism $f : L \to M$ and morphism $u : L \to I$, there exists a morphism $v : M \to I$ such that $u = vf$.

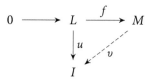

The morphism $v : M \to I$, which is generally not unique, is said to ***extend*** u, or to be an ***extension*** of u. The definition may be reformulated.

VI.4.15 Lemma. *An A-module I is injective if and only if the functor $\operatorname{Hom}_A(-, I)$ is exact.*

Proof. The functor $\operatorname{Hom}_A(-, I)$ is exact if and only if it is right exact. The definition of injectivity expresses the surjectivity of the map $\operatorname{Hom}_A(f, I) : \operatorname{Hom}_A(M, I) \to \operatorname{Hom}_A(L, I)$. $\qquad\square$

VI.4.16 Examples.

(a) Let K be a field. Then any K-vector space E is an injective K-module. For, let $f : U \to V$ be a monomorphism of vector spaces, and $u : U \to E$ a morphism. Because f is an injective linear map, there exists a basis of V of the form $(f(e_\lambda))_\lambda \cup (e'_\sigma)_\sigma$ where $(e_\lambda)_\lambda$ is a basis of U. One defines an extension $v : V \to E$ of u to V by setting $vf(e_\lambda) = u(e_\lambda)$, for any λ, and $v(e'_\sigma) = 0$, for any σ. Thus E is an injective module. We have already seen in Example VI.4.6(a) that E is projective.

(b) Let K be a field and $A = KQ/I$ a bound quiver algebra. We have seen in Example VI.4.6(e) how to attach to each point $x \in Q_0$ a projective module $P_x = e_x A$. We now see how to attach to x an injective module.

In Example IV.4.8(b), we defined a duality $D : \text{mod } A^{op} \to \text{mod } A$. It maps finitely generated projective left A-modules to finitely generated injective right A-modules (because injective is dual to projective). Because of Example VI.4.6(e), the projective left A-module corresponding to the point x is the cyclic module Ae_x having as K-basis the residual classes modulo I of the paths in Q ending at x. Therefore the corresponding injective right A-module is $I_x = D(Ae_x)$.

Consider the bound quiver of Example VI.4.6(e). Then $I_1 = D(Ae_1)$ is the dual of the vector space spanned by the classes of the paths ending at 1, namely it is the dual of the space having as K-basis $\langle e_1, \bar{\beta}, \bar{\gamma}, \overline{\alpha\beta} \rangle$ (because $\overline{\alpha\beta} = \overline{\gamma\delta}$ in KQ/I). Similarly, I_2 is the dual of the vector space $\langle e_2, \bar{\alpha} \rangle$, I_3 the dual of $\langle e_3, \bar{\gamma} \rangle$, and I_4 the dual of $\langle e_4 \rangle$. We equip each vector space with the dual basis.

The injective modules can be written as representations exactly as the projectives. Thus, I_4 is a one-dimensional representation whose only nonzero component stands at the point 4. All component morphisms are zero.

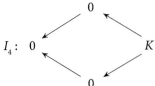

I_2 is two-dimensional, with one copy of the field K at each of the points 2 and 4, and the morphism $I_2(4) \to I_2(2)$ is the dual $D(\bar{\alpha}\cdot)$ of left multiplication by the class $\bar{\alpha}$ of the arrow α. The remaining component morphisms are zero.

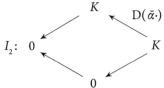

Because K is a field, a base change reduces it to

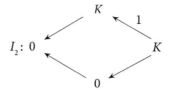

Similarly, I_3 and I_4 are respectively given by the representations

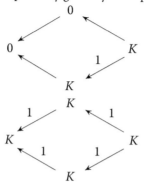

In this Example, $I_1 \cong P_4$, so it is at the same time injective and projective (it is then called a **projective-injective module**).

VI.4.17 Proposition. *Let $(I_\lambda)_\lambda$ be a family of injective modules. The product $\prod_\lambda I_\lambda$ is injective if and only if each I_λ is injective.*

Proof. The proof is dual to the proof of Proposition VI.4.3 and left to the reader. □

Because finite direct products of modules coincide with finite direct sums, the direct sum of a *finite* family of modules is injective if and only if each of these modules is injective, see however, Theorem VII.2.22 below.

There does not exist for injective modules (over arbitrary algebras) a structure theorem dual to Theorem VI.4.4. However, a module is injective if and only if any monomorphism starting at this module is a section.

VI.4.18 Theorem. *An A-module I is injective if and only if any short exact sequence of the form $0 \to I \to M \to N \to 0$ splits.*

Proof. *Necessity.* The statement follows from existence of a morphism $M \to I$ extending the identity.

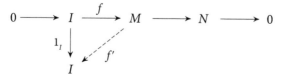

Sufficiency. Let $f : L \to M$ be a monomorphism, $u : L \to I$ a morphism and $N = \operatorname{Coker} f$. Let (Q, u', h) be the amalgamated sum of f and u. Because of Theorem IV.3.9, we have a commutative diagram with exact rows

$$
\begin{array}{ccccccccc}
0 & \longrightarrow & L & \xrightarrow{\ f\ } & M & \longrightarrow & N & \longrightarrow & 0 \\
 & & \downarrow{\scriptstyle u} & & \downarrow{\scriptstyle u'} & & \downarrow{\scriptstyle 1_N} & & \\
0 & \longrightarrow & I & \underset{h'}{\overset{h}{\rightleftarrows}} & Q & \longrightarrow & N & \longrightarrow & 0
\end{array}
$$

The hypothesis yields $h' : Q \to I$ such that $h'h = 1_I$. But then $v = h'u' : M \to I$ satisfies $vf = h'u'f = h'hu = u$. Thus, I is injective.   □

The following result, known as **Baer's criterion**, says that, in order to prove that a module is injective, instead of checking the extension property with respect to all monomorphisms of the category, it suffices to check it with respect to inclusions of right ideals in the algebra.

VI.4.19 Theorem. *An A-module I is injective if and only if, for any right ideal J_A of A and morphism $v : J \to I$, there exists a morphism $w : A \to I$ whose restriction to J equals v.*

Proof. Necessity being obvious, we prove sufficiency. Let I satisfy the condition in the statement, $f : L \to M$ a monomorphism and $u : L \to I$ an arbitrary morphism. Because f is a monomorphism, we may (and shall) view L as a submodule of M.

Let $\mathcal{E}$ be the set of pairs (L', u') where L' is a submodule of M containing L, and $u' : L' \to I$ a morphism extending u (that is, whose restriction to L equals u). Then $\mathcal{E} \neq \emptyset$ because $(L, u) \in \mathcal{E}$. We partially order $\mathcal{E}$ in the obvious way: we set $(L', u') \leqslant (L'', u'')$ if and only if $L' \subseteq L''$ and u'' extends u'.

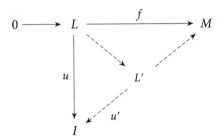

We prove that $\mathcal{E}$ is inductive for this order. Let $(L_\lambda, u_\lambda)_\lambda$ be a chain in $\mathcal{E}$. Then $\bar{L} = \cup_\lambda L_\lambda$ is a submodule of M containing all L_λ, see Exercise II.2.13. Define

$\bar{u} : \bar{L} \to I$ by setting $\bar{u}(x) = u_\lambda(x)$, if $x \in L_\lambda$. This definition is unambiguous, because, if $x \in L_\lambda \cap L_\mu$, then $L_\lambda \subseteq L_\mu$ or $L_\mu \subseteq L_\lambda$ and so $u_\lambda(x) = u_\mu(x)$. Linearity of $\bar{u}$ also follows from the fact that $(L_\lambda, u_\lambda)_\lambda$ is a chain.

Zorn's lemma yields a maximal pair (L_0, u_0) in $\mathcal{E}$. If $L_0 = M$, then we are done. So we suppose that $L_0 \neq M$ and try to reach a contradiction. Let $x \in M$ be such that $x \notin L_0$. Then $J = \{a \in A : xa \in L_0\}$ is a right ideal in A. Define $v : J \to I$ by $v(a) = u_0(xa)$, for $a \in J$. The hypothesis gives $w : A \to I$ extending v.

Consider the submodule $L_1 = L_0 + xA$ of M. Then $L_0 \subsetneqq L_1$ because $x \notin L_0$. Define $u_1 : L_1 \to I$ by $u_1(x_0 + xa) = u_0(x_0) + w(a)$, for $x_0 \in L_0, a \in A$. We prove that this definition is not ambiguous. Assume that $x_0 + xa = x_0' + xa'$, with $x_0, x_0' \in L_0, a, a' \in A$. Then $x(a'-a) = x_0-x_0' \in L_0$ gives $a'-a \in J$. Hence, $u_0(x_0-x_0')$ and $v(a'-a)$ are both defined and $u_0(x_0 - x_0') = u_0(x(a'-a)) = v(a'-a) = w(a'-a)$, because w extends v. Then $u_0(x_0) + w(a) = u_0(x_0') + w(a')$ and, therefore, u_1 is unambiguously defined. Now u_1 extends u_0, hence $(L_1, u_1) \in \mathcal{E}$. Because $L_0 \subsetneqq L_1$, this contradicts maximality of (L_0, u_0). The proof is complete. $\qquad\square$

We apply Baer's criterion to characterise injective abelian groups. Recall from Example IV.2.2(e) that an abelian group G is divisible if, for any integer $n > 0$, we have $G = nG$.

VI.4.20 Theorem. *An abelian group is injective if and only if it is divisible.*

Proof. Necessity. Let I be an injective abelian group, $x \in I$ and $n > 0$ an integer. Define $u : n\mathbb{Z} \to I$ by $u(na) = xa$, for $a \in \mathbb{Z}$. Because I is injective, there exists $v : \mathbb{Z} \to I$ which extends u.

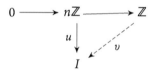

Then we have $x = u(n) = v(n) = nv(1)$. Hence, I is divisible.

Sufficiency. Let D be a divisible group. To apply Baer's criterion, take a nonzero ideal $n\mathbb{Z}$ of $\mathbb{Z}$, with $n > 0$, and a morphism $u : n\mathbb{Z} \to D$. Because D is divisible, there exists $d \in D$ such that $u(n) = nd$. Define $v : \mathbb{Z} \to D$ by $v(a) = ad$, for $a \in \mathbb{Z}$. Then v extends u, because $v(nb) = nbd = u(n)b = u(nb)$, for any $b \in \mathbb{Z}$. Baer's criterion implies that D is injective. $\qquad\square$

For instance, $\mathbb{Q}$ and $\mathbb{Q}/\mathbb{Z}$ are divisible, hence injective, abelian groups. We give another application of Baer's criterion.

VI.4.21 Example. Let A be a principal ideal domain and I a nonzero ideal of A. We claim that the quotient algebra A/I is an injective A/I-module.

Because A is a principal ideal domain, there exists a nonzero element $a \in A$ such that $I = aA$. In order to apply Baer's criterion, let $\bar{J}$ be an ideal of A/aA and $f : \bar{J} \to A/aA$ an A/aA-linear map. Because of Theorem I.4.10, there exists $b \in A$

such that $aA \subseteq bA$ (equivalently, b is a divisor of a) and $\bar{J} = bA/aA$. Set $a = bc$ and $\bar{x} = x + aA$, for any $x \in A$. Consider the diagram

$$
\begin{array}{ccc}
0 \longrightarrow & bA/aA & \longrightarrow A/aA \\
& f \downarrow \quad \swarrow \,\, f' & \\
& A/aA &
\end{array}
$$

There exists $s \in A$ such that $f(\bar{b}) = \bar{s}$. Then we have

$$\bar{s}\bar{c} = f(\bar{b})\bar{c} = f(\bar{b}\bar{c}) = f(\overline{bc}) = f(\bar{a}) = 0$$

in A/aA. Therefore, a divides sc, that is, there exists $t \in A$ such that $sc = ta = tbc$. But $c \neq 0$ (because $a \neq 0$) hence $s = tb$. Then the morphism $f' : A/aA \to A/aA$ defined by $f'(\bar{1}) = \bar{t}$ extends f, and our claim is established. Because A/I is also projective (even free!) as an A/I-module, this implies that it is a projective-injective A/I-module.

Consequently, any of the quotient algebras $\mathbb{Z}/n\mathbb{Z}$, with $n > 1$ an integer, and $K[t]/\langle p \rangle$, with p a nonzero polynomial, is projective-injective when viewed as a module over itself.

VI.4.3 Injective cogenerators

Because of Theorem VI.4.4, any module is a quotient of a projective module. Our present objective is to prove the dual statement, namely that any module is a sub-module of an injective module. Because we do not have for injectives the equivalent of Theorem VI.4.4, the proof is longer. We use the following strategy: we define a particular type of injective modules, called injective cogenerators, prove their existence (which needs a new adjunction relation), and the statement follows from this existence.

In order to motivate the definition, recall that Theorem VI.4.4 rests on existence of free modules, and, if $L = A^{(\Lambda)}$ is a nonzero free module, then, for any nonzero module M, we have $\mathrm{Hom}_A(L, M) \neq 0$, because $\mathrm{Hom}_A(A, M) \cong M \neq 0$. Dualising this property of free modules leads to the next definition.

VI.4.22 Definition. Let A be a K-algebra. A nonzero injective A-module Q is an **injective cogenerator** of $\mathrm{Mod}\,A$ if, for any nonzero module M, we have $\mathrm{Hom}_A(M, Q) \neq 0$.

Equivalently, an injective module Q is an injective cogenerator if and only if the functor $\mathrm{Hom}_A(-, Q)$ is faithful, due to Example VI.3.4(c).

The comments preceding the definition suggest that we could equally well have called Q a 'cofree' module, and actually some authors do. The origin of the term 'cogenerator' is explained in Exercise VI.4.23.

VI.4.23 Examples.
 (a) Let K be a field, then K is an injective cogenerator in the category Mod K of K-vector spaces. Indeed, it is injective because of Example VI.4.16(a) and, if E is a nonzero vector space, then its dual space $E^* = \mathrm{Hom}_K(E, K)$ is nonzero.
 (b) $\mathbb{Q}/\mathbb{Z}$ is an injective cogenerator of Mod $\mathbb{Z}$. Indeed, it is injective because of Theorem VI.4.20. Let G be a nonzero abelian group and $x \in G, x \neq 0$. The cyclic subgroup $x\mathbb{Z}$ of G is nonzero. Define $u : x\mathbb{Z} \to \mathbb{Q}/\mathbb{Z}$ as follows: if x is of infinite order in G, map x to any nonzero element of $\mathbb{Q}/\mathbb{Z}$, for instance $x \mapsto \frac{1}{2} + \mathbb{Z}$, while, if x is of finite order n, set $x \mapsto \frac{1}{n} + \mathbb{Z}$. Because $\mathbb{Q}/\mathbb{Z}$ is injective, the nonzero morphism u extends to a (necessarily) nonzero morphism $G \to \mathbb{Q}/\mathbb{Z}$. Hence, $\mathrm{Hom}_{\mathbb{Z}}(G, \mathbb{Q}/\mathbb{Z}) \neq 0$.

We need another adjunction formula.

VI.4.24 Lemma. *Let A be a K-algebra. The functor $\mathrm{Hom}_{\mathbb{Z}}({}_A A_{\mathbb{Z}}, -) : \mathrm{Mod}\,\mathbb{Z} \to \mathrm{Mod}\,A$ is right adjoint to the forgetful functor $|-| : \mathrm{Mod}\,A \to \mathrm{Mod}\,\mathbb{Z}$.*

Proof. We construct, for any A-module M and abelian group G, an isomorphism

$$\varphi : \mathrm{Hom}_{\mathbb{Z}}(|M|, G) \to \mathrm{Hom}_A(M, \mathrm{Hom}_{\mathbb{Z}}(A, G))$$

functorial in M and G. Let $f : |M| \to G$ be $\mathbb{Z}$-linear. Then the A-linear map $\varphi(f) : M \to \mathrm{Hom}_{\mathbb{Z}}(A, G)$ should map an $x \in M$ to a $\mathbb{Z}$-linear map taking $a \in A$ to $\varphi(f)(x)(a) \in G$. But $x \in M, a \in A$, imply $f(xa) \in G$. It is thus natural to set $\varphi(f)(x)(a) = f(xa)$, that is,

$$\varphi : f \mapsto (x \mapsto (a \mapsto f(xa)))$$

for $f : |M| \to G, x \in M$ and $a \in A$. Functoriality is easy to prove, so we construct the inverse

$$\psi : \mathrm{Hom}_A(M, \mathrm{Hom}_{\mathbb{Z}}(A, G)) \to \mathrm{Hom}_{\mathbb{Z}}(|M|, G).$$

Let $u : M \to \mathrm{Hom}_{\mathbb{Z}}(A, G)$ be A-linear. Given $x \in M$, the $\mathbb{Z}$-linear map $u(x)$ should take A to G. But A has a distinguished element, namely its identity. So we set

$$\psi : u \mapsto (x \mapsto u(x)(1)).$$

We prove that these maps are mutually inverse. Let $u : M \to \mathrm{Hom}_{\mathbb{Z}}(A, G)$. Then

$$\varphi(\psi(u))(x)(a) = \psi(u)(xa) = u(xa)(1) = u(x)(a).$$

for $x \in M, a \in A$. So $\varphi(\psi(u))(x) = u(x)$, for any x. Hence, $\varphi \circ \psi = 1$. Similarly, $\psi \circ \varphi = 1$. $\square$

We exhibit an injective cogenerator for any module category.

VI.4.25 Corollary. *Let A be a K-algebra, then $\mathrm{Hom}_{\mathbb{Z}}({}_A A_{\mathbb{Z}}, \mathbb{Q}/\mathbb{Z})$ is an injective cogenerator of* Mod A.

Proof. Lemma VI.4.24 yields an isomorphism of functors

$$\mathrm{Hom}_A(-, \mathrm{Hom}_{\mathbb{Z}}(A, \mathbb{Q}/\mathbb{Z})) \cong \mathrm{Hom}_{\mathbb{Z}}(|-|, \mathbb{Q}/\mathbb{Z}).$$

Because $\mathbb{Q}/\mathbb{Z}$ is an injective abelian group, the functor on the right is exact. Hence, so is the one on the left. Therefore, $\mathrm{Hom}_{\mathbb{Z}}(A, \mathbb{Q}/\mathbb{Z})$ is an injective A-module. Let M be a nonzero A-module. Because $\mathbb{Q}/\mathbb{Z}$ is an injective cogenerator of Mod$\mathbb{Z}$, see Example VI.4.23(b), we have $\mathrm{Hom}_{\mathbb{Z}}(|M|, \mathbb{Q}/\mathbb{Z}) \neq 0$. Therefore, $\mathrm{Hom}_A(M, \mathrm{Hom}_{\mathbb{Z}}(A, \mathbb{Q}/\mathbb{Z})) \neq 0$. Thus, $\mathrm{Hom}_{\mathbb{Z}}(A, \mathbb{Q}/\mathbb{Z})$ is an injective cogenerator of Mod$A$. □

For example, if $A = \mathbb{Z}$, then $\mathrm{Hom}_{\mathbb{Z}}({}_A A_{\mathbb{Z}}, \mathbb{Q}/\mathbb{Z}) \cong \mathbb{Q}/\mathbb{Z}$ and we recover the injective cogenerator of Mod $\mathbb{Z}$ seen in Example VI.4.23(b)

VI.4.26 Theorem. *Any module is a submodule of an injective module.*

Proof. Because of Corollary VI.4.25, Mod$A$ has an injective cogenerator Q. We claim that, for any A-module M, there exist a set Λ and a monomorphism $M \hookrightarrow Q^{\Lambda}$. Because products of injectives are injective, due to Proposition VI.4.17, this implies the required statement.

Let $\Lambda = \mathrm{Hom}_A(M, Q)$ and $f : M \to Q^{\Lambda}$ be given by $x \mapsto (u(x))_{u \in \Lambda}$, for $x \in M$. Then f is A-linear. We prove that f is a monomorphism. Let $x \in M$ be nonzero. Because Q is a cogenerator, there exists a nonzero morphism $u' : xA \to Q$. In particular, $u'(x) \neq 0$. But Q is injective, so there exists $u \in \Lambda$ extending u'. Then $u(x) \neq 0$, hence $f(x) \neq 0$, which proves the claim and hence the theorem.

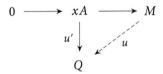

□

The previous theorem is expressed by saying that the module category has **enough injectives**.

The proof yields a reformulation of the definition of injective cogenerators. An injective module Q is an injective cogenerator if and only if, for any module M and nonzero element $x \in M$, there exists a morphism $u : M \to Q$ such that $u(x) \neq 0$. This was used in Example VI.4.23(b) above, and will be used in the proof of Proposition VI.4.29 below.

VI.4.27 Corollary. *For any A-module M, there exist injective modules I^i, with i nonnegative integers, and an exact sequence of the form*

$$0 \to M \to I^0 \to I^1 \to \cdots \to I^i \to I^{i+1} \to \cdots$$

Proof. Dual to that of Corollary VI.4.8 and left to the reader.   □

An exact sequence as in Corollary VI.4.27 above is called an **injective coresolution** of M. An injective coresolution with just two injective terms $0 \to M \to I^0 \to I^1$ is called an **injective copresentation** of M.

VI.4.28 Examples.
(a) The short exact sequence $0 \to \mathbb{Z} \to \mathbb{Q} \to \mathbb{Q}/\mathbb{Z} \to 0$, where the first morphism is the inclusion and the second the projection, is an injective coresolution of the $\mathbb{Z}$-module $\mathbb{Z}$. Deleting the final zero yields an injective copresentation of this module.
(b) Let K be a field and Q the quiver

$$\underset{1}{\circ} \overset{\alpha}{\underset{\beta}{\rightleftarrows}} \underset{2}{\circ}$$

Consider the KQ-module H_α given by the representation

$$K \overset{1}{\underset{0}{\rightleftarrows}} K$$

An injective coresolution is the exact sequence $0 \to H_\alpha \to I_1 \to I_2 \to 0$, where I_1, I_2 are the injectives attached to the points $1, 2$, respectively, as in Example VI.4.16(b).
(c) Let K be a field and A a finite-dimensional K-algebra. Let $D : \text{mod}A \to \text{mod}A^{op}$ be the duality of Example IV.4.8(b). If

$$\cdots \to P_n \to \cdots \to P_1 \to P_0 \to M \to 0$$

is a projective resolution of the right A-module M_A, with the P_i finitely generated (equivalently, finite-dimensional over K), then

$$0 \to DM \to DP_0 \to DP_1 \to \cdots DP_n \to \cdots$$

is an injective coresolution of the left A-module $_ADM$, with the DP_i finitely generated. Dually, if one applies the functor D to an injective coresolution in $\text{mod}A$, one gets a projective resolution in $\text{mod}A^{op}$.

We motivated the introduction of the injective cogenerator Q of $\text{Mod}A$ by saying that the functor $\text{Hom}_A(-, Q)$ plays a role dual to that of $\text{Hom}_A(L, -)$, with L free. Now, a sequence in $\text{Mod}A$ is exact if and only if the induced sequence under the action of $\text{Hom}_A(L, -)$ is exact (because L is projective). The corresponding statement for the injective cogenerator is the following proposition, which will be used in the next subsection and also in Chapter VII.

VI.4.29 Proposition. *Let Q be an injective cogenerator of ModA. A sequence of A-modules $L \xrightarrow{f} M \xrightarrow{g} N$ is exact if and only if the induced sequence of K-modules*

$$\text{Hom}_A(N, Q) \xrightarrow{\text{Hom}_A(g, Q)} \text{Hom}_A(M, Q) \xrightarrow{\text{Hom}_A(f, Q)} \text{Hom}_A(L, Q)$$

is exact.

Proof. Necessity follows from injectivity of Q, so we prove sufficiency. Assume that $x \in L$ is such that $gf(x) \neq 0$. As pointed out after Theorem VI.4.26, this implies the existence of $u : N \to Q$ such that $u(gf(x)) \neq 0$. But then $\text{Hom}_A(f, Q)\text{Hom}_A(g, Q)(u) = ugf \neq 0$, a contradiction. Therefore, $gf = 0$ and so $\text{Im} f \subseteq \text{Ker} g$.

Assume that $x \in \text{Ker} g$, but $x \notin \text{Im} f$. Let $p : M \to \text{Coker} f$ be the projection. Then $p(x) = x + \text{Im} f$ is nonzero in $\text{Coker} f$. Because Q is an injective cogenerator, there exists $u : \text{Coker} f \to Q$ such that $up(x) \neq 0$. Because $pf = 0$, we have $upf = 0$, so $up \in \text{Ker} \text{Hom}_A(f, Q) = \text{Im} \text{Hom}_A(g, Q)$. Hence, there exists $v : N \to Q$ such that $up = \text{Hom}_A(g, Q)(v) = vg$.

$$
\begin{array}{ccccc}
L & \xrightarrow{f} & M & \xrightarrow{g} & N \\
& & \downarrow{p} & & \downarrow{v} \\
& & \text{Coker} f & \xrightarrow{u} & Q
\end{array}
$$

Because $x \in \text{Ker} g$, we get the contradiction $0 \neq up(x) = vg(x) = v(0) = 0$. Therefore, $\text{Im} f = \text{Ker} g$. □

As an application, we present a short proof of the sufficiency part in Theorem VI.3.7. The hypothesis of this theorem implies in particular the existence of an exact sequence

$$0 \to \text{Hom}_A(N, Q) \xrightarrow{\text{Hom}_A(g, Q)} \text{Hom}_A(M, Q) \xrightarrow{\text{Hom}_A(f, Q)} \text{Hom}_A(L, Q),$$

where Q denotes an injective cogenerator of ModA. Then Proposition VI.4.29 implies that the sequence of A-modules $L \xrightarrow{f} M \xrightarrow{g} N \to 0$ is exact.

VI.4.4 Injective envelopes

While the categorical notions of projectivity and injectivity are dual to each other, this is not always the case for actual properties of projective and injective modules. This is because the module category is not selfdual. We present a property of injective modules which projective modules do *not* possess in general.

Theorem VI.4.26 asserts that any module can be embedded into an injective module. We now prove that it can be embedded into a *smallest* one.

Because our approach rests on understanding morphisms, we try to formulate this idea in terms of maps. Let A be a K-algebra, M an A-module and assume that we have two embeddings (that is, monomorphisms) $j : M \hookrightarrow I$ and $j' : M \hookrightarrow I'$, with I, I' injective. Because I is injective, there exists $f : I' \to I$ such that $fj' = j$.

To say that I' is 'smaller' than I means that I' can be embedded inside I, which is the case if f is a monomorphism. Therefore, we are looking for those monomorphisms j' such that, if $fj' = j$ is a monomorphism, then so is f.

VI.4.30 Definition. Let L, M be A-modules. A monomorphism $f : L \to M$ is **essential** if, whenever $g : M \to N$ is such that gf is a monomorphism, then g itself is a monomorphism.

As a (very trivial) example, the identity morphism is an essential monomorphism.

VI.4.31 Lemma. *Let $f : L \to M, g : M \to N$ be monomorphisms between A-modules.*
  (a) *If g, f are essential, then so is gf.*
  (b) *If gf is essential, then so is g.*

Proof. (a) If g, f are monomorphisms, so is gf. If $h(gf) = (hg)f$ is a monomorphism, then first hg is a monomorphism because f is essential, and next h is a monomorphism because g is essential.

(b) If hg is a monomorphism, then so is $(hg)f = h(gf)$. Because gf is essential, h is a monomorphism. $\square$

We give a set-theoretical interpretation of essentiality. Returning to the situation described before Definition VI.4.30, if $f : I' \to I$ is a monomorphism, then, because of Theorem VI.4.18, it is a section so there exists a submodule I'' of I such that $I = I' \oplus I''$. Because $M \subseteq I'$, we have $M \cap I'' = 0$. This leads to the following lemma.

VI.4.32 Lemma. *Let M be a module and L a submodule. The embedding $j : L \to M$ is an essential monomorphism if and only if, for any nonzero submodule N of M, we have $N \cap L \neq 0$.*

Proof. Necessity. Assume that N is a submodule of M such that $N \cap L = 0$. The composite of j with the projection $p : M \to M/N$ is a monomorphism because $pj(x) = 0$

implies $x = j(x) \in \operatorname{Ker} p = N$, so $x \in N \cap L = 0$. Because j is essential, p is a monomorphism, hence it is an isomorphism. Therefore, $N = 0$.

Sufficiency. Let $g : M \to M'$ be such that gj is a monomorphism. Assume that $N = \operatorname{Ker} g \neq 0$. Because of the hypothesis, $N \cap L \neq 0$. Let $x \in N \cap L$ be nonzero. Then $gj(x) \neq 0$, a contradiction. Therefore, $N = 0$ and g is a monomorphism. $\square$

A submodule L satisfying the equivalent conditions of Lemma VI.4.32 is called an **essential submodule** of M, or, equivalently, M is called an **essential extension** of L. We define what is meant by 'smallest' injective containing a given module.

VI.4.33 Definition. Let M be an A-module. An **injective envelope** or **injective hull** of M is a pair (E, e), where E is an injective module and $e : M \to E$ an essential monomorphism.

For brevity, we say that E is an injective envelope of M.

VI.4.34 Examples.

(a) Let A be a principal ideal domain, $p \in A$ an irreducible element and $B = A/p^t A$, with $t > 1$ an integer. Because of Example VI.4.21, B is an injective B-module. On the other hand, because of Lemma V.3.21, the submodule $M = p^{t-1} B = p^{t-1} A / p^t A \cong A/pA$ satisfies the condition of Lemma VI.4.32. It is thus an essential submodule. Therefore, the inclusion $j : M \hookrightarrow B$ is an essential monomorphism, and the pair (B, j) is an injective envelope of M. For instance, $K[t]/\langle t^2 \rangle$ is an injective envelope of $\langle t \rangle / \langle t^2 \rangle$.

(b) Lemma VI.4.32 above implies that the inclusion of abelian groups $e : \mathbb{Z} \hookrightarrow \mathbb{Q}$, is essential. Indeed, assume N to be a nonzero submodule of $\mathbb{Q}$, then there exists a rational number $\frac{r}{s} \in N$ with $r, s \in \mathbb{Z}$ and both nonzero. Because N is a submodule, we have $s \cdot \frac{r}{s} = r \in N$, and therefore $N \cap \mathbb{Z} \neq 0$. Because $\mathbb{Q}$ is divisible, it is injective. Thus, e is an essential monomorphism. Hence, the pair $(\mathbb{Q}, e)$ is an injective envelope of the abelian group $\mathbb{Z}$.

VI.4.35 Theorem. *Any module has an injective envelope, unique up to isomorphism.*

Proof. Let M be an A-module. Because of Theorem VI.4.26, there exists an injective A-module Q containing M. Let $\mathcal{E}$ be the set of those submodules L of Q, containing M and such that the inclusion $M \hookrightarrow L$ is essential. Then $\mathcal{E} \neq \emptyset$ because $M \in \mathcal{E}$. It is easy to prove that $\mathcal{E}$ is inductive. Zorn's lemma yields a maximal element E in $\mathcal{E}$. We claim that E, together with the inclusion $e : M \hookrightarrow E$, is an injective envelope of M. This is shown in two steps.

(i) First, we prove that E is maximal among the modules L containing M such that the inclusion $M \hookrightarrow L$ is essential (whether L belongs to $\mathcal{E}$ or not). If not, there exists a module $E' \supsetneq E$ such that the inclusion $M \hookrightarrow E'$ is essential. Denote by $i : E \hookrightarrow E', j : E \hookrightarrow Q$ the respective inclusions. Because Q is injective, there exists $f : E' \to Q$ such that $fi = j$.

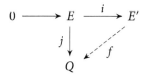

The composite $ie : M \hookrightarrow E \hookrightarrow E'$ is essential, therefore $i : E \hookrightarrow E'$ is also essential, due to Lemma VI.4.31(b). Because fi is a monomorphism, also f is a monomorphism. Maximality of E in $\mathcal{E}$ yields $E = E'$, a contradiction. This proves our statement.

(ii) Second, we prove that E is injective. Using Zorn's lemma, we find a maximal submodule L of Q with the property that $L \cap E = 0$. Let $p : Q \to Q/L$ be the projection. Then $pj : E \to Q \to Q/L$ is a monomorphism because $L \cap E = 0$. We claim that pj is essential.

Let N/L be a nonzero submodule of Q/L. Then $L \subsetneq N$ and the maximality of L yield $N \cap E \neq 0$. Therefore, $(N/L) \cap E \neq 0$. Applying Lemma VI.4.32, we establish our claim.

Now $e : M \hookrightarrow E$ essential and $pj : E \hookrightarrow Q/L$ essential yield $pje : M \hookrightarrow Q/L$ essential, because of Lemma VI.4.31(a). Hence, maximality in step (i) above implies that pj is an isomorphism. Due to Proposition IV.2.9(e), j is a section, that is, E is a direct summand of the injective module Q. Because of Proposition VI.4.17, E is injective itself.

This completes the proof that (E, e) is an injective envelope of M. We prove its uniqueness up to isomorphism: assume that (E', e') is another injective envelope of M. Then there exists a morphism $f : E' \to E$ such that $fe' = e$.

Because e' is essential and e is a monomorphism, f is a monomorphism. Due to Theorem VI.4.18, there exists E'' such that $E = E' \oplus E''$. But then $M \subseteq E'$ implies $M \cap E'' = 0$. Because e is essential, Lemma VI.4.32 yields $E'' = 0$. So $E = E'$, as required. $\square$

The considerations leading to the definition of essentiality yield the following characterisation of injective envelopes.

VI.4.36 Theorem. *Let M be an A-module. Then (E, e) is an injective envelope of M if and only if:*
  (a) *E is an injective A-module and $e : M \to E$ is a monomorphism.*
  (b) *If (I, i) is a pair with I an injective A-module and $i : M \to I$ a monomorphism, then there exists a monomorphism $f : E \to I$ such that $fe = i$.*

Proof. Necessity. Because I is injective, there exists $f : E \to I$ such that $fe = i$. Because e is essential and i a monomorphism, then f is a monomorphism.

Sufficiency. Assume that the pair (E, e) satisfies (a) and (b). Because of Theorem VI.4.35, there exists an injective envelope (E', e') of M. Applying (b) to (E', e') yields a monomorphism $f : E \to E'$. Because E is injective (due to (a)), f splits, so there exists E'' such that $E' = E \oplus E''$. Then, $M \subseteq E$ implies $M \cap E'' = 0$, a contradiction to essentiality of e'. $\square$

Of course, saying that $f : E \to I$ is a monomorphism, with E injective, amounts to saying that it embeds E as a direct summand in I. Thus, from this point of view as well, E is the 'smallest' injective containing M.

About the projective case. The categorical notions of essential monomorphisms and injective envelopes are dualised easily (to what are called superfluous epimorphisms and projective covers). But the statement corresponding to Theorem VI.4.35 is false in general: over an arbitrary algebra, it is not true that any module has a projective cover, see Example IX.2.11(a) below. We need to impose hypotheses on the algebra to ensure that this happens, see Subsection IX.2.2.

VI.4.5 Flat modules

After considering modules for which the Hom-functor is exact, we look at those for which the tensor functor is exact. In this subsection too, A denotes a K-algebra.

VI.4.37 Definition. An A-module M_A is *flat* if the functor $M \otimes_A -$ is exact.

Flat left modules are defined in the same way. Because the tensor bifunctor is covariant in both variables, the treatment is the same for right and left flat modules.

Due to Theorem VI.2.8, the module A_A is flat. So, flat modules exist. Because the functor $M \otimes_A -$ is right exact, M is flat if and only if, for any exact sequence of left modules $0 \to_A U \to_A V$, the induced sequence $0 \to M \otimes_A U \to M \otimes_A V$ is exact, that is, if and only if $M \otimes_A -$ preserves monomorphisms. Example VI.3.9(a) shows that the abelian group $\mathbb{Z}/n\mathbb{Z}$, with $n > 1$ an integer, is not flat.

VI.4.38 Lemma. *Let $(M_\lambda)_\lambda$ be a family of A-modules. Then $\oplus_\lambda M_\lambda$ is flat if and only if each M_λ is flat.*

Proof. Let $f : {}_A U \to {}_A V$ be a monomorphism of left A-modules. Proposition VI.2.7(c) yields a commutative square

$$(\oplus_\lambda M_\lambda)\otimes_A U \xrightarrow{\;1_{(\oplus_\lambda M_\lambda)}\otimes f\;} (\oplus_\lambda M_\lambda)\otimes_A V$$

$$\Big\downarrow \cong \qquad\qquad\qquad \Big\downarrow \cong$$

$$\oplus_\lambda(M_\lambda\otimes_A U) \xrightarrow{\qquad \bar{f}\qquad} \oplus_\lambda(M_\lambda\otimes_A V)$$

where the vertical arrows are isomorphisms and $\bar{f}$ is induced from the family of morphisms $(1_{M_\lambda}\otimes f)_\lambda$ by passing to the direct sums. Therefore, $1_{\oplus_\lambda M_\lambda}\otimes f$ is a monomorphism if and only if so is $\bar{f}$. Because of Exercise III.3.4, this happens if and only if so is each $1_{M_\lambda}\otimes f$, that is, if and only if each M_λ is flat. $\qquad\square$

VI.4.39 Proposition. *Any projective module is flat.*

Proof. For any algebra A, the module A_A is flat. Lemma VI.4.38 implies that any free A-module is flat. Then, Theorem VI.4.4 and another application of Lemma VI.4.38 imply than any projective A-module is flat. $\qquad\square$

As a consequence, for any A-module M, there exists an exact sequence

$$\cdots \to F_i \to F_{i-1} \to \cdots \to F_1 \to F_0 \to M \to 0$$

where the F_i are flat. Indeed, any projective resolution of M, see Corollary VI.4.8, satisfies this condition. Such a sequence is called a ***flat resolution*** of M.

The rest of this subsection is devoted to characterising flat abelian groups. We start with an *a priori* surprising relation between flats and injectives. Given a right A-module M, which we consider as $\mathbb{Z}$-A-bimodule, the left A-module $M^+ = \mathrm{Hom}_\mathbb{Z}(_\mathbb{Z}M_A, \mathbb{Q}/\mathbb{Z})$ is called the ***character module***, or ***Pontrjagin dual*** of M. Thus, we have proved in Corollary VI.4.25 that the character module A^+ of the right A-module A_A is an injective cogenerator for the category of left A-modules.

VI.4.40 Theorem. *A module is flat if and only if its character module is injective.*

Proof. Let M be an A-module. The Adjunction Isomorphism Theorem VI.2.17 gives

$$\mathrm{Hom}_A(-, M^+) = \mathrm{Hom}_A(-, \mathrm{Hom}_\mathbb{Z}(M, \mathbb{Q}/\mathbb{Z})) \cong \mathrm{Hom}_\mathbb{Z}(M \otimes_A -, \mathbb{Q}/\mathbb{Z}).$$

If M is flat, then $M \otimes_A -$ is exact. Now, the abelian group $\mathbb{Q}/\mathbb{Z}$ is divisible, hence injective, therefore the functor $\mathrm{Hom}_\mathbb{Z}(M \otimes_A -, \mathbb{Q}/\mathbb{Z})$ is also exact. Hence, so is $\mathrm{Hom}_A(-, M^+)$. That is, M^+ is an injective A-module.

Conversely, if M^+ is injective, then $\mathrm{Hom}_A(-, M^+)$ is exact. Evaluating the contravariant exact functor $\mathrm{Hom}_\mathbb{Z}(M \otimes_A -, \mathbb{Q}/\mathbb{Z})$ on an exact sequence of left A-modules $_A U \to_A V \to_A W$ yields an exact sequence

$$\mathrm{Hom}_\mathbb{Z}(M \otimes_A W, \mathbb{Q}/\mathbb{Z}) \to \mathrm{Hom}_\mathbb{Z}(M \otimes_A V, \mathbb{Q}/\mathbb{Z}) \to \mathrm{Hom}_\mathbb{Z}(M \otimes_A U, \mathbb{Q}/\mathbb{Z}).$$

But $\mathbb{Q}/\mathbb{Z}$ is an injective cogenerator of $\mathrm{Mod}\mathbb{Z}$, due to Example VI.4.23(b), so Proposition VI.4.29 yields an exact sequence $M \otimes_A U \to M \otimes_A V \to M \otimes_A W$. Therefore, M_A is flat. $\qquad\square$

We deduce a flat version of Baer's criterion.

VI.4.41 Corollary. *The following conditions are equivalent for a module M_A:*
   (a) *M is flat.*
   (b) *For any left ideal J of A, the inclusion of J into A induces a monomorphism $M \otimes_A J \to M \otimes_A A$.*
   (c) *For any left ideal J of A, the multiplication $\mu_J : M \otimes_A J \to MJ$ given by $x \otimes a \mapsto xa$, for $x \in M, a \in A$, is an isomorphism.*

Proof. (a) is equivalent to (b). Clearly, (a) implies (b). Assume that (b) holds. Because $\mathrm{Hom}_{\mathbb{Z}}(-, \mathbb{Q}/\mathbb{Z})$ is exact, there exists an epimorphism

$$\mathrm{Hom}_{\mathbb{Z}}(M \otimes_A A, \mathbb{Q}/\mathbb{Z}) \to \mathrm{Hom}_{\mathbb{Z}}(M \otimes_A J, \mathbb{Q}/\mathbb{Z}).$$

The adjunction relation in Theorem VI.4.40 translates it into an epimorphism $\mathrm{Hom}_A(A, M^+) \to \mathrm{Hom}_A(J, M^+)$. Baer's criterion VI.4.19 then says that M^+ is injective. Applying Theorem VI.4.40 completes the proof.

(b) is equivalent to (c). Let $j : J \to A$ and $j' : MJ \to M$ be the respective inclusions. We have a commutative square

$$
\begin{array}{ccc}
M \otimes_A J & \xrightarrow{\;1 \otimes j\;} & M \otimes_A A \\
\downarrow{\scriptstyle \mu_J} & & \cong \downarrow{\scriptstyle \mu_A} \\
MJ & \xrightarrow{\;\;j'\;\;} & MA
\end{array}
$$

where multiplication $\mu_A : M \otimes_A A \to M$ is an isomorphism due to Theorem VI.2.8, while the definition of MJ implies that multiplication μ_J is surjective. Assume that (b) holds true, then $1 \otimes j$ is a monomorphism, hence so is $\mu_A(1 \otimes j) = j'\mu_J$. Therefore, μ_J is a monomorphism. Consequently, it is an isomorphism. Conversely, assume that (c) holds true, then $j'\mu_J$ is injective, hence so is $1 \otimes j$. □

The proof (of the equivalence between (a) and (c)) actually shows that M is flat if and only if, for any left ideal J of A, the multiplication $\mu_J : M \otimes_A J \to MJ$ given by $x \otimes a \mapsto xa$, for $x \in A, a \in A$, is a monomorphism.

VI.4.42 Proposition. *Let L_A be a free module and $f : L \to M$ an epimorphism. Then M is flat if and only if, for any left ideal J of A, we have*

$$(LJ) \cap \mathrm{Ker}\, f = (\mathrm{Ker}\, f)J.$$

Proof. Consider the short exact sequence

$$0 \to \mathrm{Ker}\, f \to L \xrightarrow{f} M \to 0.$$

Because L is free, it is flat due to Proposition VI.4.39. For any left ideal J of A, we have a commutative diagram with exact rows

$$\begin{array}{ccccccc}
\operatorname{Ker} f \otimes_A J & \longrightarrow & I. \otimes_A J & \longrightarrow & M \otimes_A J & \longrightarrow & 0 \\
\downarrow \mu' & & \downarrow \mu & & \vdots \downarrow \mu'' & & \\
0 \longrightarrow (\operatorname{Ker} f) J & \longrightarrow & LJ & \longrightarrow & LJ/(\operatorname{Ker} f) J & \longrightarrow & 0
\end{array}$$

where μ is the multiplication, μ' its restriction to $\operatorname{Ker} f \otimes_A J$, whereas μ'' is induced by passing to cokernels. Flatness of L implies that μ is an isomorphism, due to Corollary VI.4.41. Also, μ' is surjective. The Snake Lemma II.4.19 yields that μ'' is at the same time a monomorphism and an epimorphism, hence an isomorphism. Therefore, $M \otimes_A J \cong (LJ)/(\operatorname{Ker} J)$.

Because of Corollary VI.4.41, M is flat if and only if $MJ \cong M \otimes_A J \cong (LJ)/(\operatorname{Ker} f)J$. Now, $MJ = f(L)J = f(LJ) \cong (LJ)/(LJ \cap \operatorname{Ker} f)$, because $LJ \cap \operatorname{Ker} f$ is the kernel of the restriction of f to LJ. Thus, M is flat if and only if $(LJ)/(\operatorname{Ker} f)J \cong (LJ)/(LJ \cap \operatorname{Ker} f)$. Now, $(\operatorname{Ker} f)J \subseteq LJ \cap \operatorname{Ker} f$. Hence, the Isomorphism Theorem II.4.25 gives a short exact sequence

$$0 \to (LJ \cap \operatorname{Ker} f)/(\operatorname{Ker} f)J \to LJ/(\operatorname{Ker} f)J \to LJ/(LJ \cap \operatorname{Ker} f) \to 0.$$

Therefore, M is flat if and only if $(LJ) \cap \operatorname{Ker} f = (\operatorname{Ker} f)J$, as required. $\square$

Because we always have $(\operatorname{Ker} f)J \subseteq LJ \cap \operatorname{Ker} f$, the Proposition really says that M is flat if and only if $LJ \cap \operatorname{Ker} f \subseteq (\operatorname{Ker} f)J$. The proof of the Proposition is reminiscent of that of Corollary VI.3.17, except that we used flatness of L to conclude that μ is an isomorphism.

We are now able to prove our characterisation of flat abelian groups.

VI.4.43 Theorem. *An abelian group is flat if and only if it is torsion-free.*

Proof. Let G be an abelian group. Because of Proposition III.4.8, there exist a free abelian group L and an epimorphism $f : L \to G$. Because of Proposition VI.4.42 and the fact that any ideal of $\mathbb{Z}$ is of the form $n\mathbb{Z}$, for some integer n, we get that G is flat if and only if, for any $n \in \mathbb{Z}$, we have $L(n\mathbb{Z}) \cap \operatorname{Ker} f = (\operatorname{Ker} f)(n\mathbb{Z})$, or, better said, $L(n\mathbb{Z}) \cap \operatorname{Ker} f \subseteq (\operatorname{Ker} f)(n\mathbb{Z})$. In terms of elements, this is the case if and only if, for any $n \in \mathbb{Z}, x \in L$, the assumption that $nx \in \operatorname{Ker} f$ implies $x \in \operatorname{Ker} f$. Applying the epimorphism f, that is, passing to the quotient $G \cong L/\operatorname{Ker} f$, this becomes: for any $n \in \mathbb{Z}, g \in G$, the assumption that $ng = 0$ implies that $g = 0$. This is exactly the definition of a torsion-free group. $\square$

VI.4.44 Example. We end with an example of a nonprojective flat module (so the converse of Proposition VI.4.39 does not hold true in general). The abelian group $\mathbb{Q}$ is torsion-free. Because of Theorem VI.4.43, it is flat. But it is not projective. Because if it were, then it would be free, due to Example VI.4.6(b), and this is not the case, as seen in Exercise III.4.6.

Exercises of Section VI.4

1. Let A be a K-algebra.
   (a) Assume that we have a diagram in $\mathrm{Mod}\,A$

with P projective. Prove that there exists $h : P \to M$ such that $fh = g$ if and only if $\mathrm{Im}\,g \subseteq \mathrm{Im}\,f$.
   (b) State and prove the dual of (a).

2. Let A be a K-algebra.
   (a) Assume that we have a diagram with exact row in $\mathrm{Mod}\,A$

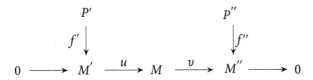

with P', P'' projective. Construct a projective module P and morphisms u', v', f such that we have a commutative diagram with exact rows

$$
\begin{array}{ccccccccc}
0 & \longrightarrow & P' & \overset{u'}{\longrightarrow} & P & \overset{v'}{\longrightarrow} & P'' & \longrightarrow & 0 \\
  & & {\scriptstyle f'}\downarrow & & {\scriptstyle f}\downarrow & & {\scriptstyle f''}\downarrow & & \\
0 & \longrightarrow & M' & \overset{u}{\longrightarrow} & M & \overset{v}{\longrightarrow} & M'' & \longrightarrow & 0
\end{array}
$$

Next, prove that if f', f'' are epimorphisms, then so is f.
   (b) State and prove the dual of (a).

3. Let A be a K-algebra.
   (a) Assume that we have a commutative diagram in $\mathrm{Mod}\,A$

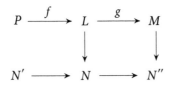

with P projective, $gf = 0$ and the lower row exact. Construct a morphism $P \to N'$ which makes the left hand square commute.
   (b) State and prove the dual of (a).

4. Let A be a K-algebra.

   (a) Prove **Schanuel's lemma**: if M is an A-module and

$$0 \to K_1 \to P_1 \to M \to 0 \text{ and } 0 \to K_2 \to P_2 \to M \to 0$$

are short exact sequences, with P_1, P_2 projective, then $K_1 \oplus P_2 \cong K_2 \oplus P_1$.

   (b) State and prove the dual of (a).

   (c) A similar result does not hold true for flat modules. Consider the two flat resolutions of the abelian group $\mathbb{Q}/\mathbb{Z}$:

$$0 \to \mathbb{Z} \to \mathbb{Q} \to \mathbb{Q}/\mathbb{Z} \to 0 \text{ and } 0 \to L' \to L \to \mathbb{Q}/\mathbb{Z} \to 0$$

where L, and hence L', are free groups. Prove that $\mathbb{Z} \oplus L \not\cong \mathbb{Q} \oplus L'$.

5. The following generalises Schanuel's lemma, Exercise VI.4.4. Let A be a K-algebra.

   (a) Let M be an A-module. Assume that we have exact sequences

$$0 \to K_n \to P_n \to \cdots \to P_1 \to M \to 0$$
$$0 \to K'_n \to P'_n \to \cdots \to P'_1 \to M \to 0$$

with the P_i, P'_i projective, then

$$K_n \oplus P'_n \oplus P_{n-1} \oplus \cdots \cong K'_n \oplus P_n \oplus P'_{n-1} \oplus \cdots$$

   (b) State and prove the dual of (a).

6. Let A be a K-algebra.

   (a) Prove that an A-module P is projective if and only if, for any epimorphism $f : I \to I'$ with I injective, and any morphism $u : P \to I'$, there exists $v : P \to I$ such that $fv = u$.

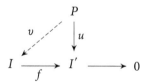

   (b) State and prove the dual of (a).

7. Let K be a field, and $A = KQ/I$ a bound quiver algebra. For each point x of each of the following bound quivers, compute as representations the corresponding projective A-module P_x and the injective A-module I_x, defined respectively in Examples VI.4.6(e) and VI.4.16(b).

   (a) $\circ \longrightarrow \circ \longrightarrow \circ$ bound by $I = 0$.

   (b) $\circ \xrightarrow{\ \alpha\ } \circ \xrightarrow{\ \beta\ } \circ$ bound by $I = \langle \alpha\beta \rangle$.

(c) $\circ \xrightarrow{\ \alpha\ } \circ \xrightarrow{\ \beta\ } \circ \xrightarrow{\ \gamma\ } \circ$ bound by $I = \langle \alpha\beta\gamma \rangle$.

(d) $\circ \xrightarrow{\ \alpha\ } \circ \xrightarrow{\ \beta\ } \circ \xrightarrow{\ \gamma\ } \circ$ bound by $I = \langle \alpha\beta, \ \beta\gamma \rangle$.

(e) $\circ \longrightarrow \circ \longrightarrow \circ$ bound by $I = 0$.

(f) $\circ \xrightarrow{\ \alpha\ } \circ \xrightarrow{\ \gamma\ } \circ$ bound by $I = \langle \alpha\gamma \rangle$.

(g) $\circ \xrightarrow{\ \alpha\ } \circ \xrightarrow{\ \gamma\ } \circ$ bound by $I = \langle \alpha\gamma, \ \beta\gamma \rangle$.

(h) $\underset{1}{\circ} \xrightarrow{\ \alpha\ } \underset{2}{\circ} \xrightarrow{\ \beta\ } \underset{3}{\circ}$ with γ bound by $I = \langle \alpha\beta \rangle$.

(i) $\underset{1}{\circ} \xrightarrow{\ \alpha\ } \underset{2}{\circ} \xrightarrow{\ \beta\ } \underset{3}{\circ}$ with γ bound by $I = 0$.

(j) $\underset{1}{\circ} \overset{\beta}{\underset{\beta'}{\rightleftarrows}} \underset{}{\circ} \overset{2 \quad \alpha}{\underset{\alpha' \quad 3}{\rightleftarrows}} \circ$ bound by $I = 0$.

8. Let $n > 1$ be an integer. Prove that the abelian group $\mathbb{Z}/n\mathbb{Z}$ is not projective by proving that the short exact sequence

$$0 \to \mathbb{Z} \xrightarrow{\ \mu_n\ } \mathbb{Z} \to \mathbb{Z}/n\mathbb{Z} \to 0$$

where μ_n is multiplication by n, is not split.

9. Prove that 0 is the only abelian group which is at the same time projective and injective.

10. Prove the **Projective Basis Theorem**: Given a K-algebra A, an A-module P is projective if and only if there exist families $(a_\lambda)_\lambda$ of elements of P, and $(f_\lambda)_\lambda$ of elements of $\mathrm{Hom}_A(P, A)$ such that, for any $x \in P$, we have:
   (a) The $(f_\lambda(x))_\lambda$ are almost all zero, and
   (b) $x = \sum a_\lambda f_\lambda(x)$.
   (Hint: If P is projective, we can take $(a_\lambda)_\lambda$ to be an arbitrary set of generators of P).

11. Let A be a K-algebra, P_1, P_2 projective A-modules, and P_1', P_2' such that $P_1 \oplus P_1', P_2 \oplus P_2'$ are free. Let M be the direct sum of countably many copies of $(P_1 \oplus P_1') \oplus (P_2 \oplus P_2')$. Prove that:
   (a) $P_1 \oplus M \cong P_2 \oplus M$ is free.
   (b) Given projective A-modules P_1, P_2, there exists a free module L such that $P_1 \oplus L \cong P_2 \oplus L$.

12. Let A be a K-algebra and M an A-module. We recall that M is called finitely presented if there exists a free presentation $L_1 \to L_0 \to M \to 0$ with L_1, L_0 finitely generated.
   (a) Prove that M is finitely presented if and only if there exists a short exact sequence $0 \to N \to P \to M \to 0$ with P finitely generated projective and N finitely generated.
   (b) Deduce that M is finitely presented if and only if there exists an exact sequence $P_1 \to P_0 \to M \to 0$ with P_1, P_0 finitely generated projective modules.
   (c) Assume that M is finitely presented. Prove that for any short exact sequence $0 \to U \to V \to M \to 0$ with V finitely generated, U is also finitely generated.
   (d) Assume that we have short exact sequences $0 \to N \to P \to M \to 0$ and $0 \to N' \to P' \to M \to 0$ with P finitely generated projective, P' projective and N' finitely generated, then P' and N are finitely generated.

13. Let A be a K-algebra, P a projective A-module and I an ideal of A. Prove that P/PI is a projective A/I-module.

14. Let A be a commutative K-algebra and P, P' projective A-modules. Prove that $P \otimes_A P'$ is a projective A-module.

15. Let A, B be K-algebras such that B is also a left A-module. Assume that P is a projective A-module. Prove that $P \otimes_A B$ is a projective B-module.

16. Let A, B be K-algebras and I an injective A-module. Assume that B is also a right A-module. Prove that $\mathrm{Hom}_A(B, I)$ is an injective B-module.

17. Let $n > 1$ be an integer and $f : \mathbb{Z} \to \mathbb{Z}/n\mathbb{Z}, g : \mathbb{Z}/n^2\mathbb{Z} \to \mathbb{Z}/n\mathbb{Z}$ the projections. Prove that the fibered product of f and g is isomorphic to $\mathbb{Z} \oplus (\mathbb{Z}/n\mathbb{Z})$.

18. Let $f : \mathbb{Z} \to \mathbb{Q}$ be the injection and $g : \mathbb{Z} \to \mathbb{Z}$ the multiplication by an integer $n > 1$. Prove that the amalgamated sum of f and g is isomorphic to $\mathbb{Q} \oplus (\mathbb{Z}/n\mathbb{Z})$.

19. Let A be a K-algebra.
   (a) Prove that, if $L, N, L \cap N$ are injective submodules of a module M, then $L + N$ is also injective.
   (b) State and prove the dual of (a).

20. Let $0 \to H \to L \to G \to 0$ be a short exact sequence of abelian groups, with L free. Prove that:
   (a) There exists an embedding of L into a direct sum of copies of $\mathbb{Q}$.
   (b) This embedding induces an embedding of G into a divisible abelian group.

21. Let $n > 1$ be an integer. Prove that the following diagram

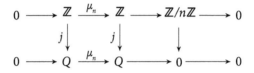

where j is the inclusion, and μ_n multiplication by n, is commutative with exact rows. Deduce an injective coresolution of the abelian group $\mathbb{Z}/n\mathbb{Z}$.

22. Let U be an A-module. A module M is said to be **generated** by U if there exist a set Λ and an epimorphism $U^{(\Lambda)} \to M$.
   (a) Prove that $U = A_A$ generates any A-module.
   (b) Prove that U generates any module if and only if the functor $\mathrm{Hom}_A(U, -)$ is faithful.
   (c) Assume that U generates any module. Prove that, for any module M, there exist sets Λ_1, Λ_2 and an exact sequence $U^{(\Lambda_1)} \to U^{(\Lambda_2)} \to M \to 0$.
   (d) Prove that, if $A = \mathbb{Z}$, an abelian group is torsion if and only if it is generated by the module $U = \oplus_{n>1}(\mathbb{Z}/n\mathbb{Z})$.
   (e) Prove that U generates a module M if and only if there exists a subset Γ of $\mathrm{Hom}_A(U, M)$ such that $M = \sum_{f \in \Gamma} \mathrm{Im} f$.
   (f) Let $\mathrm{Gen}\, U$ denote the class of A-modules generated by U. Prove that $\mathrm{Gen}\, U$ is stable under epimorphic images and direct sums.

23. This exercise is a partial dual to the preceding one. Let U be an A-module. A module M is said to be **cogenerated** by U if there exist a set Λ and a monomorphism $M \to U^\Lambda$.
   (a) Prove that an injective cogenerator cogenerates any A-module.
   (b) Prove that U cogenerates any module if and only if the functor $\mathrm{Hom}_A(-, U)$ is faithful.
   (c) Assume that U cogenerates any module. Prove that, for any module M, there exist sets Λ_1, Λ_2 and an exact sequence $0 \to M \to U^{\Lambda_1} \to U^{\Lambda_2}$.
   (d) Prove that, if $A = \mathbb{Z}$, an abelian group is torsion-free of and only if it is cogenerated by the module $U = \mathbb{Q}$.
   (e) Prove that U cogenerates a module M if and only if there exists a subset Γ of $\mathrm{Hom}_A(M, U)$ such that $0 = \cap_{f \in \Gamma} \mathrm{Ker} f$.
   (f) Let $\mathrm{Cogen}\, U$ denote the class of A-modules cogenerated by U. Prove that $\mathrm{Cogen}\, U$ is stable under submodules and products.

24. Let A be a principal ideal domain. An A-module M is called **divisible** if, for any $x \in M, a \in A$ with $a \neq 0$, there exists $y \in M$ such that $x = ya$. Prove the following statements:

    (a) An A-module is injective if and only if it is divisible.

    (b) The field of fractions Q of A is an injective A-module.

    (c) Q_A is the injective envelope of A_A.

    (d) An A-module is flat if and only if it is torsion-free.

25. Prove that $\mathbb{Q}$ is a direct summand in a direct sum of copies of $\mathbb{Q}/\mathbb{Z}$.

26. Let M be an A-module and E its injective envelope. Prove that:

    (a) M is injective if and only if $M \cong E$.

    (b) If M is not injective, then there exists no injective module I such that $M \subseteq I \subsetneq E$.

27. Let A be a K-algebra, M, M' modules with injective envelopes E, E', respectively. Prove that $E \oplus E'$ is an injective envelope of $M \oplus M'$.

28. Let A be a K-algebra. Prove that an A-module M is an essential extension of L if and only if, for any nonzero $x \in M$, there exists $a \in A$ such that $xa \in L$ and $xa \neq 0$.

29. Consider the change of scalars functor induced by an algebra morphism $\varphi : A \to B$. Prove that:

    (a) If P is a projective B-module and B is a projective A-module, then P is a projective A-module.

    (b) If P is a flat B-module and B is a flat A-module, then P is a flat A-module.

30. Let G be an abelian group and G' a subgroup of G. Prove that G' is pure in G if and only if, for any abelian group H, the inclusion $G' \hookrightarrow G$ induces a monomorphism $G' \otimes_{\mathbb{Z}} H \to G \otimes_{\mathbb{Z}} H$. (Hint: use Exercise VI.2.16)

31. Let A, B be K-algebras, L a flat A-module and $_A M_B$ an $A - B$-bimodule which is flat as a B-module. Prove that $L \otimes_A M$ is a flat B-module.

32. Let A be a K-algebra and F a flat A-module. Prove that the functor $F \otimes_A -$ is faithful if and only if, for any nonzero left A-module U, one has $F \otimes_A U \neq 0$.

33. Let A be a K-algebra. For an A-module M, let M^+ be its character module. Prove that:

    (a) If M is free, then M^+ is a product of copies of the character module A^+ of A.

    (b) A sequence of A-modules $0 \to L \to M \to N \to 0$ is exact if and only if the induced sequence of character modules $0 \to N^+ \to M^+ \to L^+ \to 0$ is exact.

    (c) Let G be a finite abelian group. Prove that $G^+ \cong G$.

    (d) Deduce from (a), (b) that, if G is finite, there exists a short exact sequence $0 \to G' \to G \to G'' \to 0$ if and only if there exists a short exact sequence $0 \to G'' \to G \to G' \to 0$.

    (e) For any A-module M, there exists an embedding $j : M \hookrightarrow M^{++}$ defined as follows: for $x \in M$, we set $j(x)(\varphi) = \varphi(x)$, where $\varphi \in M^+$.

    (f) Deduce that, for any A-module M, there exists a free left module $_A L$ and a monomorphism $M \hookrightarrow L^+$.

Chapter VII
The chain conditions

VII.1 Introduction

Given an algebra, we aim to describe efficiently its modules and the morphisms between them. Stated in this generality, this is a very ambitious goal, so we should impose conditions allowing us to work in a simple, yet nontrivial, situation. This was done in Chapter V, where we restricted ourselves to finitely generated modules over principal ideal domains. Here, we take another point of view inspired from linear algebra. Vector spaces are best behaved when they are finite-dimensional. While the notion of dimension does not exist for modules in general, chains in the submodule lattice of a module may satisfy finiteness conditions. We consider the situations where any ascending, or descending, chain of submodules is finite. This leads us to two types of modules, called noetherian and artinian, respectively. It turns out that, if A is an algebra and the module A_A satisfies one of these properties, then so does any finitely generated A-module. Moreover, the conjunction of these properties gives our first structure theorem, the Jordan–Hölder theorem. The latter allows, for modules satisfying both conditions, to define a notion of length, which behaves very much like dimension for vector spaces. Keeping a lattice-theoretical approach, we consider the situation where the submodule lattice has just two elements: such modules are called simple, and their sums are called semisimple. We obtain a complete characterisation of algebras such that any module is semisimple. These are the Wedderburn–Artin theorems, which, among others, describe group algebras in characteristic zero.

VII.2 Artinian and noetherian modules and algebras

VII.2.1 Chains in partially ordered sets

A vector space E is finite-dimensional if and only if, for any chain of subspaces of the form $F_0 \subseteq F_1 \subseteq F_2 \subseteq \cdots$, there exists an integer $m \geq 0$ such that $F_i = F_m$, for $i \geq m$, or, equivalently, if and only if, for any chain of subspaces of the form $F_0 \supseteq F_1 \supseteq F_2 \supseteq \cdots$, there exists an integer $m \geq 0$ such that $F_i = F_m$, for $i \geq m$ (that is, these chains are finite). These finiteness conditions can be translated in terms of partially ordered sets (but then they are generally not equivalent).

VII.2.1 Definition. Let $(E, \leqslant)$ be a nonempty poset.
  (a) A chain in E of the form $x_0 \leqslant x_1 \leqslant \cdots x_n \leqslant x_{n+1} \leqslant \cdots$ is an ***ascending chain***. Such a chain ***stabilises*** or ***becomes stationary*** if there exists an integer $m \geq 0$ such that $x_i = x_m$, for $i \geq m$. The poset E satisfies the ***ascending chain condition*** if any ascending chain stabilises.

An Introduction to Module Theory. Ibrahim Assem and Flávio U. Coelho, Oxford University Press.
© Ibrahim Assem and Flávio U. Coelho (2024). DOI: 10.1093/9780198904939.003.0008

(b) A chain in E of the form $x_0 \geq x_1 \geq \cdots x_n \geq x_{n+1} \geq \cdots$ is a **descending chain**. It **stabilises** or **becomes stationary** if there exists an integer $m \geq 0$ such that $x_i = x_m$, for $i \geq m$. The poset E satisfies the **descending chain condition** if any descending chain stabilises.

It follows from this definition that an ascending, or descending, chain stabilises if and only if it is finite, that is, it has only finitely many distinct elements.

VII.2.2 Examples.
(a) A finite poset satisfies both chain conditions.
(b) Let K be a field and E a finite-dimensional K-vector space. The set of its subspaces, partially ordered by inclusion, satisfies both chain conditions. Indeed, any chain of submodules, ascending or descending, yields a chain of inequalities between their dimensions, and the latter are bounded from above by $\dim_K E$ and from below by 0.

   Now, a finitely generated module over a finite-dimensional K-algebra is finite-dimensional when viewed as K-vector space, due to Corollary III.4.9, and submodules are then subspaces. Thus, the submodule lattice of a finitely generated module satisfies both chain conditions.
(c) Let K be a field, E an arbitrary K-vector space and $\mathcal{F}(E)$ the set of its finite-dimensional subspaces, ordered by inclusion. This is a modular lattice, because the sum and intersection of two finite-dimensional submodules are themselves finite-dimensional. Then $\mathcal{F}(E)$ satisfies the descending chain condition, but it satisfies the ascending chain condition if and only if E itself is finite-dimensional.
(d) Let A be a principal ideal domain. Because of Corollary V.3.22, an indecomposable finitely generated A-module is isomorphic either to $A/p^t A$, for an irreducible element $p \in A$ and an integer $t \geq 0$, or to A itself. In the first case, its submodule lattice is finite, see Lemma V.3.21, so it satisfies both chain conditions. In the second case, any submodule of A is of the form aA, for some $a \in A$, and moreover, $aA \subseteq bA$ if and only if $b|a$. Now $a \in A$, nonzero and noninvertible, has finitely many divisors, due to the Unique Factorisation Theorem I.5.12, hence any ascending chain of submodules stabilises. On the other hand, a nonzero and noninvertible has infinitely many multiples, so there exist descending chains that do not stabilise, such as, for instance, $A \supsetneq aA \supsetneq a^2 A \supsetneq a^3 A \supsetneq \cdots$.

Examples (c) and (d) above show that, in general, the ascending and descending chain conditions are not equivalent.

VII.2.3 Lemma. *Let $(E, \leq)$ be a nonempty poset.*
(a) E satisfies the ascending chain condition if and only if any nonempty subset of E has a maximal element.
(b) E satisfies the descending chain condition if and only if any nonempty subset of E has a minimal element.

Proof. We prove (a), because the proof of (b) is similar. Let E be a poset satisfying the ascending chain condition and X a nonempty subset of E without maximal element. Choose an arbitrary element $x_0 \in X$. Because x_0 is not maximal, there exists $x_1 \in X$ such that $x_0 < x_1$. Inductively, we construct an infinite sequence of pairwise distinct elements of the form $x_0 < x_1 < x_2 < \cdots$, which contradicts the hypothesis.

Conversely, assume that any nonempty subset of E has a maximal element. Let $x_0 \leqslant x_1 \leqslant x_2 \leqslant \cdots$ be an ascending chain in E. Then the nonempty subset $\{x_i : i \geq 0\}$ of E has a maximal element x_m, say. Therefore, $x_i = x_m$, for $i \geq m$. □

We need a couple of definitions. A **refinement** of a chain in E is a chain which includes the elements of the given one. Equivalently, it is obtained from the first chain by inserting further elements in it. A chain is called **maximal** if it has no proper refinement.

VII.2.4 Proposition. *Let $(E, \leqslant)$ be a poset satisfying both chain conditions. Any chain in E is finite and can be refined to a maximal chain.*

Proof. Let $\mathcal{X}$ be a chain in E. We first prove that $\mathcal{X}$ is finite by writing its elements as a finite descending chain. Because $\mathcal{X}$ satisfies the ascending chain condition, it follows from Lemma VII.2.3(a) that it has a maximal element x_0, say. Inductively, for an integer $i \geq 1$, let x_i be a maximal element in the subset $\mathcal{X} \setminus \{x_0, \cdots, x_{i-1}\}$. This yields a descending chain $x_0 > x_1 > x_2 > \cdots$ of pairwise distinct elements. Because E satisfies the descending chain condition, this chain stabilises and hence is finite, say $x_0 > x_1 > \cdots > x_m$. Because the procedure described exhausts all elements of $\mathcal{X}$, we get $\mathcal{X} = \{x_0, \cdots, x_m\}$. Thus, $\mathcal{X}$ is finite.

We next construct a maximal refinement of the chain $\mathcal{X}$. Because the ascending chain condition holds true, there exists in E a maximal element y_1 in the half-open interval $\{z \in E : x_0 > z \geq x_1\}$. Inductively, we get a descending chain

$$x_0 > y_1 > y_2 > \cdots \geq x_1.$$

It stabilises because the descending chain condition is satisfied. So there exists an integer m_1 such that

$$x_0 > y_1 > y_2 > \cdots > y_{m_1} \geq x_1$$

and this is a maximal refinement of the chain $x_0 > x_1$. Repeating this procedure for each strict inequality $x_i > x_{i+1}$, with $2 \leqslant i < m$, we end up with a maximal refinement of the chain $x_0 > x_1 > \cdots > x_m$. □

VII.2.5 Example. In general, two maximal chains between the same endpoints do not have the same number of elements, as is shown in the poset with Hasse quiver

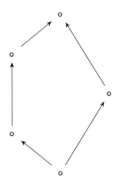

As seen in Subsection II.2.3, this is a lattice which is not modular. In particular, it cannot occur as submodule lattice of a module.

VII.2.2 Noetherian and artinian modules

From now on, let K be a commutative ring and A a K-algebra. We apply the considerations of the previous Subsection VI.2.1 to the lattice $S(M)$ of submodules of an A-module M, ordered by inclusion. Because of Proposition II.2.18, $S(M)$ is a complete modular lattice.

VII.2.6 Definition. Let A be a K-algebra and M a right A-module.
 (a) M is **right noetherian** provided $S(M)$ satisfies the ascending chain condition.
 (b) M is **right artinian** provided $S(M)$ satisfies the descending chain condition.

One defines in the same way **left noetherian** and **left artinian**, using left modules instead of right modules. Because we agreed to work with right modules throughout, we simply say noetherian and artinian instead of right noetherian and right artinian, respectively, unless we need to distinguish between a right and a left structure.
  Because of Lemma VII.2.3, a module is noetherian (or artinian) if and only if any nonempty set of submodules admits a maximal element (or a minimal element, respectively).

VII.2.7 Lemma. *Let M be an A-module.*
 (a) *If M is noetherian, then M admits maximal submodules.*
 (b) *If M is artinian, then M admits minimal submodules.*

Proof. For (a), we apply the preceding observation to the set of proper submodules of M, and for (b), to the set of its nonzero submodules. □

VII.2.8 Examples.
   (a) Let K be a field and E a finite-dimensional K-vector space. Then E, as a K-module, is both noetherian and artinian, see Example VII.2.2(b). In particular, any finitely generated module over a finite-dimensional algebra is both noetherian and artinian.

(b) The chain conditions are finiteness conditions, hence infinitely generated modules are neither noetherian nor artinian. Indeed, let A be a K-algebra and M a module having an infinite generating set $(x_\lambda)_{\lambda \in \Lambda}$, but no finite generating set. Due to the Well-Ordering Principle, see Example I.5.2(a), we may assume the index set Λ to be well-ordered. Denote by 1 the least element of Λ and i the least element of $\Lambda \setminus \{1, \cdots, i-1\}$. For $i \geq 1$, let M_i be the submodule of M generated by $\{x_1, \cdots, x_i\}$ and M_i' the submodule generated by $\{x_j : j \geq i\}$. Then we have an infinite ascending chain $M_1 \subsetneq M_2 \subsetneq \cdots$ and an infinite descending chain $M_1' \supsetneq M_2' \supsetneq \cdots$, neither of which stabilises.

A typical example of this situation is when $A = K[(t_i)_{i \in \mathbb{N}}]$, the polynomial algebra in countably many variables, and $M = A_A$.

(c) As seen in Examples VII.2.2(a)(c), any finite abelian group is both noetherian and artinian. On the other hand, the group $\mathbb{Z}$ is a noetherian but not artinian abelian group.

(d) We exhibit an abelian group which is artinian but not noetherian. Let p be a prime integer, and consider the subgroup of $\mathbb{Q}$ defined by

$$\mathbb{Z}[1/p] = \left\{ \frac{a}{p^m} : a \in \mathbb{Z}, m \in \mathbb{Z} \right\}.$$

Thus, $\mathbb{Z}$ is a subgroup of $\mathbb{Z}[1/p]$. The quotient $\mathbb{Z}(p^\infty) = \mathbb{Z}[1/p]/\mathbb{Z}$ is called the **Prüfer p-group**.

We compute its proper subgroups. Let N be a subgroup of $\mathbb{Z}[1/p]$ containing $\mathbb{Z}$, and assume that $\frac{a}{p^m} \in N$. Without loss of generality, we may assume that a and p are coprime. Then there exist $s, t \in \mathbb{Z}$ such that $sa + tp^m = 1$. Hence,

$$\frac{1}{p^m} = \frac{sa}{p^m} + t \in N.$$

That is, N contains at least one element of the form $\frac{1}{p^k}$. If all elements of this form belong to N, then $\mathbb{Z}[1/p] = N$, contrary to the hypothesis that N is a proper subgroup. On the other hand, if N contains $\frac{1}{p^k}$, for some k, then it also contains $\frac{1}{p^{k-1}} = \frac{p}{p^k}$. Hence, there exists a (unique) maximal k such that $\frac{1}{p^k} \in N$. Let $\frac{b}{p^n}$ be an arbitrary element of N, with b, p coprime. The previous argument shows that $n \leq k$. Moreover,

$$\frac{b}{p^n} = bp^{k-n} \frac{1}{p^k}.$$

Therefore, if $M = N/\mathbb{Z}$ is a proper subgroup of $\mathbb{Z}(p^\infty)$, then it is cyclic and generated by $\frac{1}{p^k} + \mathbb{Z}$. In particular, it is finite and of order p^k. Denoting this subgroup by M_k, we infer that the complete list of subgroups of the Prüfer p-group $\mathbb{Z}(p^\infty)$ is:

$$0 \subsetneq M_1 \subsetneq M_2 \subsetneq \cdots \subsetneq \mathbb{Z}(p^\infty).$$

that is, they form a chain under inclusion. Therefore, $\mathbb{Z}(p^\infty)$ is artinian. But we have also exhibited an ascending chain which does not stabilise. Thus, $\mathbb{Z}(p^\infty)$ is not noetherian.

One proves in abelian group theory that the Prüfer p-groups are the only infinite groups whose proper subgroups form a chain under inclusion, and also the only infinite groups all of whose proper subgroups are finite.

Besides the chain conditions, another finiteness condition is finite generation of modules. The relation between the two is as follows.

VII.2.9 Theorem. *A module is noetherian if and only if any of its submodules is finitely generated.*

Proof. Necessity. Let A be an algebra, M_A a noetherian module and N a submodule of M. The set $\mathcal{E}$ of finitely generated submodules of N is nonempty, because it contains the zero submodule. Hence, $\mathcal{E}$ has a maximal element N', say. We claim that $N = N'$. If not, then there exists an element $x \in N$ which does not belong to N'. Now $N'' = N' + xA$ is finitely generated, because it is the sum of two finitely generated submodules. Hence, $N'' \in \mathcal{E}$. But then $N'' \supsetneq N'$ contradicts maximality of N'. Therefore, $N' = N$ and so N is finitely generated.

Sufficiency. Assume that any submodule of M is finitely generated. Let $N_1 \subseteq N_2 \subseteq \cdots$ be an ascending chain in M. Due to Exercise II.2.13, the union $N = \cup_{i \geq 1} N_i$ is a submodule of M. The hypothesis says that N is finitely generated, say $N = \langle x_1, \cdots, x_m \rangle$. For each j, with $1 \leq j \leq m$, there exists $i_j \geq 1$ such that $x_j \in N_{i_j}$. Setting $i_0 = \max\{i_j : 1 \leq j \leq m\}$, we get $x_1, \cdots, x_m \in N_{i_0}$. Therefore, $N = N_{i_0}$ and the chain stabilises. So M is noetherian. $\qquad\square$

An obvious, though important, consequence of this theorem is that noetherian modules are themselves finitely generated. This generally does not hold true for artinian modules, for instance, the Prüfer p-groups of Example VII.2.8(d) are not finitely generated (because any finite set of elements is contained in some $M_i \subsetneq \mathbb{Z}(p^\infty)$).

One can ask whether the chain conditions are inherited by submodules and quotients. We need a lemma.

VII.2.10 Lemma. *Let L and $M' \subseteq M''$ be submodules of an A-module M. Then there exists a short exact sequence*

$$0 \to \frac{M'' \cap L}{M' \cap L} \to \frac{M''}{M'} \to \frac{M'' + L}{M' + L} \to 0$$

Proof. The Isomorphism Theorem II.4.24 yields the exact sequences

$$0 \to M' \cap L \xrightarrow{j'} M' \to (M' + L)/L \to 0$$

and

$$0 \to M'' \cap L \xrightarrow{j''} M'' \to (M'' + L)/L \to 0$$

where j', j'' are the inclusions. On the other hand, the inclusion $M' \hookrightarrow M''$ restricts to an inclusion $M' \cap L \hookrightarrow M'' \cap L$. Passing to cokernels, we get a morphism from $(M' + L)/L$ to $(M'' + L)/L$, defined by $(x' + y) + L \mapsto (x' + y) + L$, for $x' \in M', y \in L$. This morphism is also induced from the inclusion $M' \hookrightarrow M''$, hence is injective. The conclusion follows from a straightforward application of the 3×3-Lemma II.4.20 to the following diagram where all columns and the two upper rows are exact.

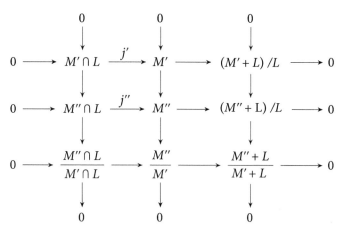

VII.2.11 Theorem. *Let $0 \to L \to M \to N \to 0$ be a short exact sequence of A-modules. Then M is noetherian, or artinian, if and only if so are L and N.*

Proof. We prove the noetherian case, the artinian one is obtained by reversing inclusions.

Necessity. Assume that M is noetherian. An ascending chain of submodules of L is also an ascending chain in M, hence stabilises. Let $N_1 \subseteq N_2 \subseteq \cdots$ be an ascending chain of submodules of $N = M/L$. Because of Theorem II.4.27, there exist submodules M_i of M, containing L, such that $N_i = M_i/L$, for $i \geq 1$, and moreover $M_1 \subseteq M_2 \subseteq \cdots$. Because M is noetherian, the latter chain stabilises. Hence, so does the former.

Sufficiency. Assume that both L, N are noetherian and that $M_1 \subsetneq M_2 \subsetneq \cdots$ is an infinite ascending chain of submodules of M. This chain induces an ascending chain of submodules of L

$$M_1 \cap L \subseteq M_2 \cap L \subseteq \cdots$$

and an ascending chain of submodules of $N = M/L$

$$(M_1 + L)/L \subseteq (M_2 + L)/L \subseteq \cdots$$

Because L, N are noetherian, both of these induced chains stabilise. But then Lemma VII.2.10 gives a contradiction.   □

VII.2.12 Corollary. *Let $\{M_1, \cdots, M_m\}$ be a finite family of A-modules. Then $\oplus_{i=1}^{m} M_i$ is noetherian, or artinian, if and only if so is each M_i.*

Proof. Necessity follows from induction and the short exact sequences

$$0 \to \bigoplus_{i<j} M_i \to \overset{j}{\underset{i=1}{\bigoplus}} M_i \to M_j \to 0$$

while sufficiency follows from induction and the short exact sequences

$$0 \to \bigoplus_{i \neq j} M_i \to \overset{m}{\underset{i=1}{\bigoplus}} M_i \to M_j \to 0.$$

□

VII.2.3 Noetherian and artinian algebras

Given an algebra A, we consider chain conditions on the right A-module A_A: in this case, these are chain conditions on right ideals.

VII.2.13 Definition. A K-algebra is *right noetherian*, or *right artinian*, in case the right A-module A_A is right noetherian, or right artinian, respectively.

One defines *left noetherian* and *left artinian* algebras in the same manner, starting from the left A-module $_A A$. If A is commutative, then left and right noetherian, or left and right artinian, coincide.

As we did for modules, we simply say noetherian and artinian instead of right noetherian and right artinian, respectively, unless we need to distinguish between a right and a left structure.

VII.2.14 Examples.
(a) A division ring D has only two right (or left) ideals, namely 0 and itself. Hence, D is right and left noetherian and artinian.
(b) Any principal ideal domain A is noetherian but not artinian, see Example VII.2.2(d). An interesting case is that of the algebra of formal power series $K[[t]]$, where K is a field: it follows from Exercise I.5.2 that this algebra is a principal ideal domain whose ideals form the infinite decreasing sequence $K[[t]] \supsetneq \langle t \rangle \supsetneq \langle t^2 \rangle \supsetneq \cdots \supsetneq \langle t^i \rangle \supsetneq \cdots \supsetneq 0$.
(c) Any finite-dimensional K-algebra is both right and left noetherian and artinian. Indeed, in this case, right and left ideals are finite-dimensional

vector spaces, hence have a dimension, so both descending and ascending chain conditions are satisfied.

As seen in Example I.3.11(h), bound quiver algebras over a field are finite-dimensional. Therefore, they are right and left noetherian and artinian.

(d) If an algebra is not commutative, then right noetherian (or artinian) is generally different from left noetherian (or artinian, respectively). Counterexamples are found among triangular matrix algebras, because these behave differently on each side. We give here an example, but must defer the proof to Examples VII.3.15 below. The $\mathbb{Z}$-algebra

$$A = \begin{pmatrix} \mathbb{Q} & 0 \\ \mathbb{Q} & \mathbb{Z} \end{pmatrix} = \left\{ \begin{pmatrix} a & 0 \\ b & n \end{pmatrix} : a, b \in \mathbb{Q}, n \in \mathbb{Z} \right\}$$

with the ordinary matrix operations, is right, but not left noetherian. We refer the reader to Exercise VII.2.11(c) for an example of a right, but not left, artinian algebra.

(e) Adding to the pile, we give another example, due to Dieudonné, of an algebra right, but not left, noetherian. Let K be a field and $A = KQ/I$, where Q is the two-loops quiver

$$\alpha \; \overset{\curvearrowright}{\underset{}{\bigcirc}} \; \circ \; \overset{}{\underset{\curvearrowleft}{\bigcirc}} \; \beta$$

bound by the ideal $I = \langle \beta^2, \alpha\beta \rangle$. We also defer the proof to Examples VII.3.15 .

(f) An algebra A is noetherian, or artinian if and only if so is $M_n(A)$, for any integer $n \geq 1$. We prove the noetherian case, the artinian one being similar. As an A-module, $M_n(A) \cong A^{(n^2)}$. Therefore, if A is noetherian, then $M_n(A)$ is noetherian as an A-module, hence, all the more so, as an $M_n(A)$-module. Conversely, if $(I_i)_{i \geq 0}$ is a strictly increasing chain of ideals in A, then $(M_n(I_i))_{i \geq 0}$, see Exercise I.4.15, is a strictly increasing chain of ideals in $M_n(A)$.

We did not give an example of an artinian algebra which is not noetherian. This is for a good reason: we prove later an important theorem, known as the **Hopkins–Levitski's Theorem** VIII.3.22, which says that artinian algebras are also noetherian.

Theorem VII.2.9 asserts that a module is noetherian if and only if its submodules are finitely generated. This implies the following proposition.

VII.2.15 Proposition. *Let A be an algebra*
(a) *A is noetherian if and only if any right ideal in A is finitely generated.*
(b) *If A is noetherian, then any finitely generated module is finitely presented.*

Proof. (a) This follows immediately from Theorem VII.2.9.
(b) Assume that M is finitely generated, say, by $x_1, \cdots, x_m$. Because of Exercise II.4.2, there exists an epimorphism $p : A^{(m)} \to M$, mapping the i-th vector of

the canonical basis of $A^{(m)}$ to x_i. Because of Corollary VII.2.12, $A^{(m)}$ is noetherian. Theorem VII.2.9 implies that Ker p, which is a submodule of $A^{(m)}$, is finitely generated, say, by n elements. Hence, there is an epimorphism $A^{(n)} \to$ Ker p. Composing with the inclusion Ker $p \hookrightarrow A^{(m)}$ yields the finite presentation

$$A^{(n)} \to A^{(m)} \xrightarrow{p} M \to 0. \qquad \square$$

The following corollary underlines the interest of working with finitely generated modules.

VII.2.16 Corollary. *Let A be a noetherian algebra. Any finitely generated A-module M has a projective resolution $\cdots \to P_i \to P_{i-1} \to \cdots \to P_1 \to P_0 \to M \to 0$ with all the P_i finitely generated projective.*

Proof. Indeed, M is the image of a finitely generated (free hence) projective module P_0, say. The kernel of the epimorphism $f : P_0 \to M$ is a submodule of P_0, which is finitely generated, hence noetherian. Therefore, Ker f is finitely generated. The statement follows from an immediate induction. $\qquad \square$

Another beautiful property of noetherian algebras is that, if a free module over a noetherian algebra has a finite basis, then all bases of the module are finite and have the same number of elements. This is Exercise VIII.4.10 below.

VII.2.17 Example. If the algebra is not noetherian, finitely generated modules are usually not finitely presented. The polynomial algebra in countably many indeterminates $A = K[(t_i)_{i \in \mathbb{N}}]$ is not noetherian, see Example VII.2.8(b). In fact, the ideal $I = \langle (t_i)_{i \in \mathbb{N}} \rangle$ is not finitely generated: indeed, if $I = \langle p_1, \cdots, p_m \rangle$, with the p_j polynomials in t_i, let i_0 be the maximal index such that t_{i_0} appears in one of the p_j, with $1 \leqslant j \leqslant m$, then $t_{i_0+1} \notin I$, a contradiction.

Now, the A-module A/I is cyclic, because it is a quotient of the cyclic module A_A (actually, $A/I \cong K$). The short exact sequence $0 \to I \to A \to A/I \to 0$ shows that A/I is finitely generated, but not finitely presented (because the kernel I is a submodule of A_A which is not finitely generated).

The following theorem allows to construct examples of noetherian algebras. Note that we do not assume commutativity of A.

VII.2.18 Theorem. The Hilbert Basis Theorem. *Let A be a noetherian algebra. Then the polynomial algebra A[t] is noetherian.*

Proof. Suppose that $A[t]$ is not noetherian. Because of Proposition VII.2.15(a), there exists a right ideal I of $A[t]$ which is not finitely generated. The Well-Ordering Principle yields a polynomial $p_1(t)$ in I of least degree. For each integer $n \geq 2$, choose $p_n(t) \in I \setminus < p_1(t), \cdots, p_{n-1}(t) >$ of least degree. This is possible, because I is not finitely generated. For $i \geq 1$, let $a_i t^{d_i}$ be the leading term of $p_i(t)$ (that is, a_i is the leading coefficient, and d_i the degree). The elements $a_1, a_2, \cdots$ of A induce an ascending chain of ideals

$$< a_1 > \subseteq < a_1, a_2 > \subseteq \cdots$$

which stabilises because A is noetherian. So, there exists at least an index n such that $a_{n+1} \in < a_1, \cdots, a_n >$, that is, $a_{n+1} = \sum_{i=1}^{n} a_i b_i$, where $b_i \in A$. Consider now

$$q(t) = p_{n+1}(t) - \sum_{i=1}^{n} t^{d_{n+1}-d_i} p_i(t) b_i.$$

Because $p_{n+1}(t)$ does not belong to $< p_1(t), \cdots, p_n(t) >$, neither does $q(t)$. A direct calculation shows that the degree of $q(t)$ is smaller than the degree of $p_{n+1}(t)$, a contradiction to the choice of this polynomial. This proves the result. $\qquad \square$

VII.2.19 Corollary. *Let A be a noetherian algebra. Then the polynomial algebra in n indeterminates $A[t_1, \cdots, t_n]$ is also noetherian.*

Proof. For $n \geq 2$, we apply induction using $A[t_1, \cdots, t_n] = (A[t_1, \cdots, t_{n-1}])[t_n]$. $\qquad \square$

So, if K is a field, the polynomial algebra in a finite number of indeterminates $K[t_1, \cdots, t_n]$ is noetherian, while the polynomial algebra in countably many indeterminates $K[(t_i)_{i \geq 0}]$ is not.

Because of Theorem VII.2.11, submodules and quotient modules of artinian, or noetherian, modules satisfy the same property. The corresponding statement does not hold true for subalgebras: for instance, $\mathbb{Q}$ is an artinian algebra (because it is a field), but its subalgebra $\mathbb{Z}$ is not. However, the statement holds true for quotient algebras.

VII.2.20 Proposition. *Let A be an algebra and I an ideal of A. If A is noetherian, or artinian, then so is its quotient A/I.*

Proof. This is because the A/I-submodules of A/I are also A-submodules of A/I, considered as an A-module, see Example IV.4.6(a). $\qquad \square$

We now turn to module-theoretical properties of noetherian and artinian algebras.

VII.2.21 Theorem. *Let A be a noetherian, or artinian, K-algebra. Then any finitely generated A-module is noetherian, or artinian, respectively.*

Proof. We give the proof in the noetherian case, the artinian case is similar.

We use induction on the number of generators of a finitely generated A-module M. If M is cyclic, then, because of Proposition II.4.22, there exists a right ideal I of A such that $M \cong A/I$, as A-modules. Because of Theorem VII.2.11, M is noetherian.

Assume that $M = \langle x_1, \cdots, x_m \rangle$. Let $L = \langle x_1, \cdots, x_{m-1} \rangle$, then we have a short exact sequence

$$0 \to L \to M \to M/L \to 0.$$

Now, L is noetherian, due to the induction hypothesis, and so is the cyclic module $M/L = \langle x_m + L \rangle$. The result follows from another application of Theorem VII.2.11. □

Noetherian algebras behave nicely with respect to injectivity. Proposition VI.4.17 says that, over any algebra, a product of injective modules is injective. Surprisingly, noetherian algebras are characterised by the property that arbitrary direct sums of injective modules are injective.

VII.2.22 Theorem. *A K-algebra is noetherian if and only if the direct sum of any family of injective modules is injective.*

Proof. Necessity. Assume that A is noetherian and let $(I_\lambda)_{\lambda \in \Lambda}$ be a family of injective modules. We apply Baer's criterion VI.4.19 to prove that $\oplus_{\lambda \in \Lambda} I_\lambda$ is injective. Let J_A be a right ideal and $f : J \to \oplus_{\lambda \in \Lambda} I_\lambda$ an A-linear map. Consider the diagram

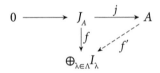

where j is the inclusion. Because of Proposition VII.2.15(a), J_A is finitely generated, say $J = \langle x_1, \cdots, x_m \rangle$. For any integer i with $1 \leqslant i \leqslant m$, there exists a finite subset $\Lambda_i \subseteq \Lambda$ such that $f(x_i) \in \oplus_{\lambda \in \Lambda_i} I_\lambda$. The set $\cup_{i=1}^m \Lambda_i$ is finite, say, equal to $\{\lambda_1, \cdots, \lambda_n\}$ and we have $f(J) \subseteq \oplus_{i=1}^n I_{\lambda_i}$. Because $\oplus_{i=1}^n I_{\lambda_i} = \prod_{i=1}^n I_{\lambda_i}$, which is injective due to Proposition VI.4.17, there exists a morphism $f' : A \to \oplus_{\lambda \in \Lambda_i} I_\lambda$ such that $f'j = f$.

Sufficiency. Assume that A is not noetherian. We construct a right ideal J and an A-linear map from J to a direct sum of injectives, which cannot be extended to A.

Because A is not noetherian, there exists an infinite ascending chain of right ideals $J_1 \subsetneqq J_2 \subsetneqq \cdots$. Due to Exercise II.2.13, $J = \cup_{i \geq 1} J_i$ is a right ideal of A. Furthermore, $J/J_i \neq 0$, for any i. Because of Theorem VI.4.26, there exists, for any i, an injective module I_i containing J/J_i. Let $p_i : J \to J/J_i$ be the projection, then we have an A-linear map $f : J \to \prod_{i \geq 1} I_i$ given by $x \mapsto (p_i(x))_{i \geq 1}$, for $x \in J$.

We claim that the image of f actually lies in $\oplus_{i \geq 1} I_i$. Indeed, for any $x \in J$, there exists a least index j, depending on x, such that $x \in J_j$. But then $p_i(x) = 0$, for any $i \geq j$, that is, $p_i(x)$ vanishes for i large enough. Therefore, f maps J into $\oplus_{i \geq 1} I_i$, and our claim is established.

If $\oplus_{i \geq 1} I_i$ is injective, then f extends to $f' : A \to \oplus_{i \geq 1} I_i$. Set $f'(1) = (y_i)_{i \geq 1}$. We claim that $y_i \neq 0$, for any i, which contradicts the fact that $f'(1)$ belongs to the direct sum and completes the proof of the theorem. For, let i be arbitrary and $x \in J$ be such that $x \notin J_i$. Then, $p_i(x) = x + J_i \neq 0$ in $J/J_i \subseteq I_i$ and the i-th coordinate

$p_i(x)$ of $f'(x) = f(x)$ is nonzero. But $f'(x) = f'(1)x = (y_i)_{i \geq 1}x = (y_i x)_{i \geq 1}$. Thus, $p_i(x) = y_i x \neq 0$ and hence $y_i \neq 0$. We are done. $\qquad\square$

Exercises of Section VII.2

1. Prove that the lattice of Example VII.2.5 is not modular.

2. Let A be a K-algebra and $\{M_1, \cdots, M_m\}$ submodules of an A-module M. Prove that all the M_i are noetherian, or artinian, if and only if so is their sum $\sum_{i=1}^{m} M_i$.

3. This exercise gives another example showing that submodules of finitely generated modules are not necessarily finitely generated. Prove that:
   (a) The subset $A = \{\Sigma_i a_i t^i \in \mathbb{Q}[t] : a_0 \in \mathbb{Z}\}$ of $\mathbb{Q}[t]$ consisting of the polynomials whose constant term is an integer is actually a subring.
   (b) $M = \{\Sigma_i a_i t^i \in A : a_0 = 0\}$ is an ideal of A, which is not finitely generated as an A-module (whereas A_A clearly is).

4. Prove that the statement of the Hilbert Basis Theorem no longer holds true if artinian is substituted for noetherian.

5. Let A be a K-algebra. Generators of $\mathrm{Mod}A$ were defined in Exercise VI.4.22. Prove the following statements:
   (a) Let U be a generator of $\mathrm{Mod}A$ and M a finitely generated A-module. Then there exist a *finite* set Λ and an epimorphism $U^{(\Lambda)} \to M$.
   (b) The following conditions are equivalent;
      (i) A is noetherian.
      (ii) $\mathrm{Mod}A$ has a generator that is a noetherian module.
      (iii) Any finitely generated A-module is noetherian.
   (c) Statement (b) above holds true if 'noetherian' is replaced with 'artinian'.

6. Let p be a prime integer. Prove the following properties of the Prüfer p-group $\mathbb{Z}(p^\infty)$.
   (a) $\mathbb{Z}(p^\infty)$ is isomorphic to the group consisting of all p^n-th roots of unity as n ranges over the nonnegative integers, that is, the group $\{z \in \mathbb{C} : z^{p^n} = 1, \text{ for some } n \geq 0\}$.
   (b) $\mathbb{Z}(p^\infty)$ is divisible (hence injective).

7. Let $n \geq 1$ be an integer. Prove that each of the following algebras is noetherian and artinian:
   (a) $M_n(K)$, with K a field.
   (b) $\mathbb{Z}/n\mathbb{Z}$.

8. Let K be a field and Q a finite quiver. Prove that the following conditions are equivalent:
   (a) The path algebra KQ is artinian.
   (b) The path algebra KQ is finite-dimensional over K.
   (c) Q is acyclic.

9. Let A be a commutative artinian algebra. Prove that, if A is an integral domain, then it is a field. Deduce that any proper prime ideal in a commutative artinian algebra is maximal.

10. Let A be a commutative algebra. Prove that, if $A[t]$ is noetherian, then so is A.

11. Consider the triangular matrix algebra of Exercise II.2.11

$$R = \begin{pmatrix} A & 0 \\ M & B \end{pmatrix} = \left\{ \begin{pmatrix} a & 0 \\ x & b \end{pmatrix} : a \in A, b \in B \text{ and } x \in M \right\}$$

(a) Prove that R is right (or left) noetherian if and only if both A and B are right (or left) noetherian and M is noetherian as right A-module (or as left B-module, respectively). (Hint: let e_A be the idempotent in R such that $A = e_A R e_A$, then, for any right ideal I in R, $I e_A$ is a right ideal in A, similarly for $B = e_B R e_B$, finally $I = I e_A \oplus I e_B$).
(b) State and prove the corresponding statement for artinian.
(c) Assume that K, K' are fields such that K' is infinite-dimensional as a K-vector space. Prove that

$$R = \begin{pmatrix} K' & 0 \\ K' & K \end{pmatrix}$$

is right artinian and right noetherian but neither left artinian nor left noetherian.

12. Let $0 \to L \to M \to N \to 0$ be a short exact sequence of modules over a noetherian algebra. Prove that M is finitely generated if and only if so are L and N.

VII.3 Decompositions of algebras

VII.3.1 Idempotent elements and decompositions as modules

Let A be a K-algebra and e an idempotent in A. Because of Corollary VI.4.5, the A-module eA is projective and the functor $\mathrm{Hom}_A(eA, -) : \mathrm{Mod}\, A \to \mathrm{Mod}\, K$ is exact. The next proposition, which generalises Theorem II.3.9, shows that this functor is right multiplication by e.

VII.3.1 Proposition. *Let $e \in A$ be an idempotent and M an A-module. Then there exists a functorial isomorphism of K-modules*

$$\mathrm{Hom}_A(eA, M) \cong Me.$$

Proof. We adapt the proof of Theorem II.3.9. For a morphism $f : eA \to M$, we have $f(e) = f(e^2) = f(e)e \in Me$. Hence, there exists a map $\mathrm{Hom}_A(eA, M) \to Me$ defined by $f \mapsto f(e)$, for $f \in \mathrm{Hom}_A(eA, M)$. It is easily seen to be K-linear. We construct its inverse: for $x \in M$, set $xe \mapsto (ea \mapsto xea)$, for $a \in A$. We leave to the reader the easy verification that these maps are mutually inverse and functorial in M. □

Thus, applying the functor $\text{Hom}_A(eA, -)$ to a short exact sequence of A-modules $0 \to L \to M \to N \to 0$ yields a short exact sequence of K-modules $0 \to Le \to Me \to Ne \to 0$.

Moreover, if M is actually a B-A-bimodule, so that $\text{Hom}_A(eA, M)$ has a natural left B-module structure, see Lemma II.3.8, then the isomorphism of the Proposition is compatible with the left action of B, and hence is an isomorphism of left B-modules.

Also, if e, f are idempotents of A, then we have an isomorphism of K-modules

$$\text{Hom}_A(eA, fA) \cong fAe.$$

In general, given an idempotent e, the module eA does not inherit from A an algebra structure, because its multiplication has no identity. However, the multiplication of A makes the K-module eAe an algebra, with identity e. The first part of next result generalises Corollary II.3.12.

VII.3.2 Corollary. *Let $e \in A$ be an idempotent. Then there exists an isomorphism of algebras $\text{End}(eA) \cong eAe$. Moreover, if A is artinian, then so is eAe.*

Proof. For the first statement, we need to check that the morphism of Proposition VII.3.1 maps the composition of endomorphisms to the multiplication of eAe, and 1_{eA} to e, and this is immediate. For the second, we must show that eAe is artinian as an eAe-module. Let I be a right ideal of eAe. Then IA is a submodule of A_A. On the other hand, $I \subseteq eAe$ implies $Ie = I$ (for, if $x \in I$, then there exists $a \in A$ such that $x = eae$, and $xe = eae^2 = eae = x$.) Therefore, the map $I \mapsto IA$ defines an injection from the submodule lattice of eAe_{eAe} into that of A_A. The second module being artinian, so is the first. □

If e is an idempotent in A then so is $1 - e$, see the remarks after Corollary VI.4.5. Moreover, $e(1 - e) = 0 = (1 - e)e$ and $1 = e + (1 - e)$. This motivates the following definitions.

VII.3.3 Definition. Let A be a K-algebra.
   (a) Two idempotents e_1, e_2 of A are **orthogonal** if $e_1e_2 = 0 = e_2e_1$.
   (b) A set of idempotents $\{e_1, \cdots, e_n\}$ in A is **complete** if $1 = e_1 + \cdots + e_n$.
   (c) A nonzero idempotent e in A is **primitive** if $e = e_1 + e_2$ with e_1, e_2 orthogonal idempotents implies $e_1 = 0$ or $e_2 = 0$.

The definition of primitive idempotent resembles those of irreducible polynomial or indecomposable module. They may be thought of as the 'smallest possible idempotents' (with respect to decompositions), see Exercise VII.3.4.

VII.3.4 Lemma. *An idempotent e in an algebra A is primitive if and only if the only idempotents of eAe are 0 and e.*

Proof. Suppose that $e = e_1 + e_2$ with e_1, e_2 orthogonal idempotents. Then $e_1 = (e_1 + e_2)e_1(e_1 + e_2) \in eAe$ and similarly $e_2 \in eAe$. Hence, $e = e_1 + e_2$ is a decomposition of the identity e of eAe into a sum of orthogonal idempotents in eAe. The statement follows. $\qquad\square$

Thus, $1 \in A$ is a primitive idempotent if and only if the only idempotents of A are 0 and 1.

Any complete set of pairwise orthogonal idempotents determines a direct sum decomposition of A as an A-module. This generalises the direct sum decomposition of Corollary VI.4.5.

VII.3.5 Lemma. *Let $\{e_1, \cdots, e_n\}$ be a complete set of pairwise orthogonal idempotents. Then*

$$A_A = e_1 A \oplus \cdots \oplus e_n A.$$

Proof. If $a \in A$, then $1 = e_1 + \cdots + e_n$ yields $a = 1 \cdot a = e_1 a + \cdots + e_n a \in \sum_{i=1}^n e_i A$. Therefore, $A = \sum_{i=1}^n e_i A$. In order to prove that the sum is direct, take $a \in e_i A \cap (\sum_{j \neq i} e_j A)$. There exist $b \in A$, and $c_j \in A$, for $j \neq i$, such that $a = e_i b = \sum_{j \neq i} e_j c_j$. Then $a = e_i b = e_i^2 b = e_i a = e_i(\sum_{j \neq i} e_j c_j) = \sum_{j \neq i} e_i e_j c_j = 0$, due to orthogonality of the e_i. $\qquad\square$

Each of the right ideals $e_i A$ is projective as an A-module. Thus, the decomposition of the Lemma is a decomposition of A_A into a direct sum of projective modules.

There is a converse to this statement: a direct sum decomposition of the A-module A into right ideals induces a decomposition of the identity 1 as sum of a complete set of pairwise orthogonal idempotents, see Proposition IX.2.1 below. We give examples of idempotents.

VII.3.6 Examples.
   (a) Any algebra has at least two idempotents, namely 0 and 1.
   (b) If A is an integral domain, then 0 and 1 are its only idempotents, because $e \neq 0, e^2 = e$ imply $e = 1$ in an integral domain. That is, 1 is a primitive idempotent.
   (c) Let $\varphi : A \to B$ be an algebra morphism. If $e \in A$ is an idempotent of A, then $\varphi(e)$ is an idempotent of B, because $\varphi(e)^2 = \varphi(e^2) = \varphi(e)$. If e_1, e_2 are orthogonal in A, then so are $\varphi(e_1), \varphi(e_2)$ in B, because $\varphi(e_1)\varphi(e_2) = \varphi(e_1 e_2) = \varphi(0) = 0$. If $\{e_1, \cdots, e_n\}$ is a complete set of pairwise orthogonal idempotents in A, then so is $\{\varphi(e_1), \cdots, \varphi(e_n)\}$ in B, because $\varphi(1) = 1$. Thus, a direct sum decomposition of A as in Lemma VII.3.5 induces one of B. A special case of interest is when $B = A/I$, for some ideal I, and φ is the projection. Then, if e is an idempotent in A, the corresponding idempotent in A/I is $e + I$.
   (d) Consider $A = \mathbb{Z}/20\mathbb{Z}$. A straightforward calculation shows that $\bar{0}, \bar{1}, \bar{5}, \bar{16}$ are the only idempotents in A. Now, $\bar{1} = \bar{5} + \bar{16}, \bar{5} \cdot \bar{16} = \bar{0}$ and both $\bar{5}, \bar{16}$ are primitive so they form a complete set of pairwise orthogonal primitive idempotents. Because of Lemma VII.3.5, we have a direct sum decomposition $\mathbb{Z}/20\mathbb{Z} \cong 5\mathbb{Z}/20\mathbb{Z} \oplus 16\mathbb{Z}/20\mathbb{Z}$ as $\mathbb{Z}/20\mathbb{Z}$-module. Recall

that, because of Proposition V.3.14, it also decomposes in this way as a $\mathbb{Z}$-module.

(e) Let K be a field. The full matrix algebra $A = M_2(K)$ contains many idempotents (infinitely many if K is an infinite field). For instance, if $\lambda \in K$, then each of the matrices:

$$\begin{pmatrix} \lambda & \lambda \\ 1-\lambda & 1-\lambda \end{pmatrix}, \begin{pmatrix} 1 & \lambda \\ 0 & 0 \end{pmatrix}, \begin{pmatrix} 0 & 0 \\ \lambda & 1 \end{pmatrix}, \begin{pmatrix} 1 & 0 \\ \lambda & 0 \end{pmatrix}, \begin{pmatrix} 0 & \lambda \\ 0 & 1 \end{pmatrix}$$

is an idempotent. It is not difficult to construct complete sets of pairwise orthogonal primitive idempotents, such as, for any $\lambda \in K$,

$$\left\{ \begin{pmatrix} 1 & \lambda \\ 0 & 0 \end{pmatrix}, \begin{pmatrix} 0 & -\lambda \\ 0 & 1 \end{pmatrix} \right\} \text{ or } \left\{ \begin{pmatrix} 1 & 0 \\ \lambda & 0 \end{pmatrix}, \begin{pmatrix} 0 & 0 \\ -\lambda & 1 \end{pmatrix} \right\}$$

It is plain to see that each of the previous is a complete set of orthogonal idempotents. Primitivity can be proved starting from the definition by a tedious calculation. A simple and conceptual proof is given in Example VIII.4.8 later on.

(f) Let K be a field and $A = KQ/I$ a bound quiver algebra. The stationary paths ε_i attached to the points obviously form a set of pairwise orthogonal idempotents of the path algebra KQ and their sum equals the identity of KQ, see Example I.3.11(h). Hence, it is a complete set of pairwise orthogonal idempotents of KQ. Because of Example (c) above, the $e_i = \varepsilon_i + I$ form a complete set of pairwise orthogonal idempotents of $A = KQ/I$. We now prove that each e_i is primitive. Because of Lemma VII.3.4, we just have to prove that the only idempotents of the algebra $e_i A e_i$ are 0 and e_i.

Let $e \in e_i A e_i$ be idempotent. There exists $a \in A$ such that $e = e_i a e_i$. Hence $e = e_i^2 a e_i^2 = e_i e e_i$. This means that the paths representing e start and end at the point i. Thus, e can be written as $e = \lambda \varepsilon_i + w + I$, where $\lambda \in K$ and w is a linear combination of paths from i to itself of length at least one (that is, cycles). Then $e^2 = e$ yields $(\lambda^2 - \lambda)\varepsilon_i + (2\lambda - 1)w + w^2 \in I$. Because $I \subseteq KQ^{+2}$, see the definition of admissible ideal in Example I.3.11(h), $\lambda^2 - \lambda = 0$. Thus, $\lambda = 0$ or $\lambda = 1$. Assume that $\lambda = 0$. Then $-w + w^2 \in I$, and induction give $w^k - w \in I$, for any integer $k \geq 1$. But there exists $m \geq 2$ such that $KQ^{+m} \subseteq I$. Therefore, $w^m \in I$. Hence, $w^m - w \in I$ gives $w \in I$. Substituting, we get $e = 0 + I$. Similarly, $\lambda = 1$ gives $e = \varepsilon_i + I = e_i$. The proof is complete.

Therefore, in a bound quiver algebra $A = KQ/I$, the set $\{e_i = \varepsilon_i + I : i \in Q_0\}$ is a complete set of primitive pairwise orthogonal idempotents of A. This, however, is not the only such set in A: let, for instance, Q be the quiver $1 \xrightarrow{\alpha} 2$, then the set $\{\varepsilon_1 + \alpha, \varepsilon_2 - \alpha\}$ is another complete set of pairwise orthogonal primitive idempotents of $A = KQ$, see Exercise VII.3.14.

VII.3.2 Decomposition of an algebra into product

Assume that we want to decompose an algebra A in the form $A = A_1 \times A_2$, where A_1, A_2 are algebras. The module decomposition $A = eA \oplus (1-e)A$, for an idempotent

$e \in A$, does not work because eA has no natural algebra structure. But eAe does. One way to make them coincide is to assume that e is central, that is, $e \in Z(A)$, for then we have $eAe = e^2A = eA$. Now, if e is central, then so is $1 - e$. Indeed, if $a \in A$, then $a(1 - e) = a - ae = a - ea = (1 - e)a$. Therefore, $(1 - e)A = (1 - e)A(1 - e)$ also inherits an algebra structure (with identity $1 - e$) and the module direct sum decomposition $A = eA \oplus (1 - e)A$ becomes an algebra decomposition into product $A = eAe \times (1 - e)A(1 - e)$. This leads us to consider central idempotents.

VII.3.7 Definition. Let A be a K-algebra.
- (a) An idempotent c in A is **central** if it belongs to the centre of A.
- (b) An idempotent c in A is **centrally primitive** if $c = c_1 + c_2$, with c_1, c_2 central orthogonal idempotents, implies $c_1 = 0$ or $c_2 = 0$.

Any algebra has 0 and 1 as central idempotents. In a commutative algebra, any idempotent is central. Plainly, a central idempotent which is primitive is centrally primitive, but the converse is not true, see Example VII.3.11(e) below. These comments lead us to the following theorem.

VII.3.8 Theorem. *Let A be a K-algebra. Then $A = \prod_{i=1}^{t} A_i$ if and only if there exists a complete set of central pairwise orthogonal idempotents $\{c_1, \cdots, c_t\}$ such that $A_i = c_iA$, for each i.*

Proof. Necessity. Assume that $A = \prod_{i=1}^{t} A_i$. Each A_i can be identified to an ideal in A. The identity of A decomposes as $1 = \sum_{i=1}^{t} c_i$, with $c_i \in A_i$. Let $a_i \in A_i$, then

$$a_i = a_i \cdot 1 = a_i \left(\sum_{j=1}^{t} c_j \right) = \sum_{i=1}^{t} (a_i c_j).$$

Because $a_i c_j \in A_i A_j \subseteq A_i \cap A_j$, we get $a_i c_j = 0$, for $i \neq j$, and $a_i c_i = a_i$. Similarly, $c_i a_i = a_i$. Hence, $c_i c_j = 0$, for $i \neq j$, and $c_i^2 = c_i$, that is, the c_i are a complete set of pairwise orthogonal idempotents.
In order to prove centrality, let $a = \sum_{j=1}^{t} a_j$, with $a_j \in A_j$. Then

$$ac_i = \left(\sum_{j=1}^{t} a_j \right) c_i = \sum_{j=1}^{t} (a_j c_i) = a_i c_i = a_i = c_i a_i = c_i \left(\sum_{j=1}^{t} a_j \right) = c_i a.$$

This completes the proof of necessity.
Sufficiency. Let $\{c_1, \cdots, c_t\}$ be a complete set of central pairwise orthogonal idempotents. Set $A_i = c_iA$. Because the c_i are central, the comments preceding Definition VII.3.7 say that each A_i is an algebra with identity c_i, which is an ideal when viewed as subset of A. Then the algebra morphism $A \to \prod_{i=1}^{t} A_i$ defined by $a \mapsto (c_i a)_{i=1}^{t}$, for $a \in A$, is an isomorphism with inverse $\prod_{i=1}^{t} A_i \to A$ given by $(a_i)_{i=1}^{t} \mapsto \sum_{i=1}^{t} a_i$, for $(a_i)_{i=1}^{t} \in \prod_{i=1}^{t} A_i$. This proves sufficiency. $\square$

Assume that $A = \prod_{i=1}^{t} A_i$. The algebras A_i are *not* subalgebras of A, because they do not have the same identity. In fact, viewed as subsets of A, they are ideals, as observed in the proof.

A decomposition such as the one of Theorem VII.3.8 is nicest when the factors A_i cannot be decomposed further.

VII.3.9 Definition. A nonzero algebra A is **connected** or **indecomposable as an algebra** if $A = A_1 \times A_2$ implies $A_1 = 0$ or $A_2 = 0$.

The following lemma, saying that an algebra is connected if and only if 1 is a centrally primitive idempotent, corresponds to Lemma VII.3.4.

VII.3.10 Lemma. *An algebra is connected if and only if its only central idempotents are 0 and 1.*

Proof. Because of Theorem VII.3.8, an algebra decomposes nontrivially into product if and only if the identity decomposes nontrivially as sum of central pairwise orthogonal idempotents. So, it is connected if and only if the identity is centrally primitive, that is, if and only if the only central idempotents are 0 and 1.   □

Consequently, the decomposition of Theorem VII.3.8 is a decomposition into connected factors A_i if and only if each of the idempotents c_i is centrally primitive. In this case, each of the connected algebras A_i is called a **block** of A, and the decomposition a **block decomposition**.

VII.3.11 Examples.
   (a) Because the only idempotents in an integral domain are 0 and 1, see Example VII.3.6(b), any integral domain is connected.
   (b) Let $\varphi : A \to B$ be a *surjective* algebra morphism and c a central element in A, then $\varphi(c)$ is central in B. For, given any $b \in B$, there exists $a \in A$ such that $b = \varphi(a)$, then $b\varphi(c) = \varphi(a)\varphi(c) = \varphi(ac) = \varphi(ca) = \varphi(c)\varphi(a) = \varphi(c)b$.
      In particular, if I is an ideal in A and $c \in A$ is a central idempotent, then $c + I$ is a central idempotent in A/I. Because of Example VII.3.6(c), if there exists a complete set of central pairwise orthogonal idempotents $\{c_1, \cdots, c_t\}$ in A, then the nonzero elements in the set $\{c_1 + I, \cdots, c_t + I\}$ form a complete set of central pairwise orthogonal idempotents in A/I. Hence, a product decomposition of A into algebras induces one of A/I.
   (c) Because the algebra $A = \mathbb{Z}/20\mathbb{Z}$ is commutative, all its idempotents are central. Hence, the complete set of primitive orthogonal idempotents $\{\bar{5}, \bar{16}\}$, see Example VII.3.6(d), is a complete set of centrally primitive idempotents. Theorem VII.3.8 gives the decomposition $\mathbb{Z}/20\mathbb{Z} = 5\mathbb{Z}/20\mathbb{Z} \times 16\mathbb{Z}/20\mathbb{Z} \cong \mathbb{Z}/4\mathbb{Z} \times \mathbb{Z}/5\mathbb{Z}$, as algebras.
   (d) Let K be a field and $M_n(K)$ the full matrix algebra, for some integer $n \geq 2$. It is known from linear algebra that the centre of $M_n(K)$ consists of the scalar matrices λI, for $\lambda \in K$ and I the identity matrix. Now, a scalar matrix λI is idempotent if and only if $\lambda^2 = \lambda$ in K, that is, $\lambda = 0$ or $\lambda = 1$. Hence, the

only central idempotents are the zero matrix 0 and the identity matrix I. Therefore, $M_n(K)$ is a connected algebra.

(e) Let K be a field and Q a finite quiver. As seen in Example VII.3.6(f), the stationary paths ϵ_i attached to the points form a complete set of pairwise orthogonal idempotents of the path algebra KQ. In general, they are not central. Indeed, consider the quiver $1 \xrightarrow{\alpha} 2$, we have $\epsilon_1 \cdot \alpha = \alpha$, while $\alpha \cdot \epsilon_1 = 0$, so that ϵ_1 is not central. A larger example will make things clearer, but we need a definition. A quiver is called connected if so is its underlying graph (the one obtained forgetting orientations of arrows). Formally, a quiver Q is **connected** provided, given two points $i, j \in Q_0$, there exist arrows $\alpha_1, \cdots, \alpha_n$ such that i is the source or the target of α_1, j is the source or the target of α_n, and for each ℓ such that $1 \leqslant \ell \leqslant n - 1$, the set $\{s(\alpha_\ell), t(\alpha_\ell)\} \cap \{s(\alpha_{\ell+1}), t(\alpha_{\ell+1})\}$ is nonempty. A quiver which is not connected is **disconnected**. Consider the disconnected quiver

$$1 \xrightarrow{\alpha} 2 \qquad 3 \xrightarrow{\beta} 4 \xrightarrow{\gamma} 5$$

Let Q', Q'' be respectively the full subquivers of Q defined by $Q'_0 = \{1, 2\}$ and $Q''_0 = \{3, 4, 5\}$, see Example I.3.11(h). Set $e' = \epsilon_1 + \epsilon_2$, $e'' = \epsilon_3 + \epsilon_4 + \epsilon_5$. Direct calculation shows that $\{e', e''\}$ is a complete set of central orthogonal idempotents. It follows from Theorem VII.3.8 that $KQ = KQ' \times KQ''$, with $KQ' = e'KQ, KQ'' = e''KQ$.

We claim that e' and e'' are centrally primitive. Now e' is the identity of the algebra KQ'. Because of Exercise I.4.10(d), we have an algebra isomorphism $KQ' \cong \begin{pmatrix} K & 0 \\ K & K \end{pmatrix}$, in which e' corresponds to the identity matrix. Direct computation shows that the only idempotents different from zero and the identity in the lower triangular matrix algebra are of one of the forms $\epsilon_\alpha = \begin{pmatrix} 1 & 0 \\ \alpha & 0 \end{pmatrix}$ and $\epsilon_\beta = \begin{pmatrix} 0 & 0 \\ \beta & 1 \end{pmatrix}$, for $\alpha, \beta \in K$.

Indeed, assume that $\begin{pmatrix} a & 0 \\ b & c \end{pmatrix}$ is an idempotent in KQ', then $a^2 = a, c^2 = c, b(a + c) = b$. This gives two cases. If $b \neq 0$, then $a, c \in \{0, 1\}$ and $a + c = 1$, which gives matrices of the required form. On the other hand, if $b = 0$, then we get one of the following four matrices:

$$\begin{pmatrix} 1 & 0 \\ 0 & 1 \end{pmatrix}, \begin{pmatrix} 1 & 0 \\ 0 & 0 \end{pmatrix}, \begin{pmatrix} 0 & 0 \\ 0 & 1 \end{pmatrix}, \begin{pmatrix} 0 & 0 \\ 0 & 0 \end{pmatrix},$$

which are also (and trivially) of the required form.

Therefore, the only possible decomposition of e' as sum of orthogonal idempotents is $e' = \epsilon_\alpha + \epsilon_\beta$, where $\beta = -\alpha$. But, for any α, neither ϵ_α not $\epsilon_{-\alpha}$ is central, hence e' is centrally primitive. Similarly, e'' is centrally primitive.

This reasoning shows that if Q is not a connected quiver, then KQ is not a connected algebra. Consequently, if I is an admissible ideal of KQ, then $A = KQ/I$ is not connected either: indeed, if Q is the disjoint union of two subquivers Q', Q'', then the ideal I, being generated by linear combinations of paths of length at least two, can be written as $I = I' + I''$, with I' an admissible ideal in KQ' and I'' an admissible ideal in KQ''. So, $KQ/I \cong KQ'/I' \times KQ''/I''$.

Conversely, assume Q to be a connected quiver. We claim that the bound quiver algebra $A = KQ/I$ is connected. In order to prove it, it suffices, because of Lemma VII.3.10, to show that its only central idempotents are 0 and 1. Let $d \neq 0$ be a central idempotent of A. In particular, setting $e_i = \epsilon_i + I$, we get $e_i d = d e_i$, for each $i \in Q_0$. Therefore, d is a linear combination of (residual classes of) cycles or stationary paths, that is, $d = \sum_{i \in Q_0} d_i$ with $d_i \in e_i A e_i$. We claim that each d_i is an idempotent. Indeed, because the e_i are pairwise orthogonal, we have $d_i d_j = 0$, whenever $i \neq j$. Therefore, $d = d^2 = \sum_{i \in Q_0} d_i^2$. Because d lies in the direct sum $\oplus_{i \in Q_0} e_i A e_i$, we get $d_i^2 = d_i$, for each i, as claimed. Because of Example VII.3.6(f), the only idempotents of $e_i A e_i$ are 0 and e_i itself, so d_i is either zero or equal to e_i. If $d \neq 0, 1$, then there exist i such that $d_i = 0$ and j such that $d_j = e_j$. Because Q is connected, we may assume that i and j are linked by an arrow, say, $\alpha: i \to j$ (the other direction is similar). Hence, $d\alpha = 0$ and $\alpha d \neq 0$, a contradiction to centrality of d.

Therefore, a bound quiver algebra $A = KQ/I$ is connected if and only if the quiver Q is connected (hence the terminology).

In the case of a noetherian or artinian algebra (for instance, a finite-dimensional algebra over a field), product decomposition occurs into a finite number of uniquely determined blocks.

VII.3.12 Theorem. *A noetherian or an artinian algebra can be decomposed into a finite product of connected algebras. Moreover, this decomposition is unique up to isomorphism.*

Proof. Existence. Let A be a noetherian algebra. If A is connected, there is nothing to prove. If not, then there exists a nonzero proper ideal I in A such that there exists an ideal I' satisfying $A = I \times I'$. Because A is noetherian, there exists a maximal such ideal A_1'. We set $A = A_1 \times A_1'$. Maximality of A_1' implies that A_1 is connected. If A_1' is also connected, then we have finished. If not, then we can write $A_1' = A_2 \times A_2'$, with A_2 connected and A_2' maximal among the ideals which are (product) factors of A_1'. Inductively, we reach a decomposition

$$A = A_1 \times \cdots \times A_t \times A_t'.$$

Because A is noetherian, the ascending chain of ideals $(A_1 \times \cdots \times A_i)_{i \geq 1}$ stabilises. That is, there exists $t \geq 1$ such that $A_t' = 0$. Because each A_i is connected, we have the required decomposition.

The artinian case is done similarly choosing at each step A_i minimal nonzero. Then the A_i' form a decreasing chain which eventually stabilises.

Uniqueness. Suppose that

$$A = A_1 \times \cdots \times A_n = A_1' \times \cdots \times A_m'$$

with the A_i, A_j' (with $1 \leqslant i \leqslant n, 1 \leqslant j \leqslant m$) connected. Viewing A_n as an ideal in A, we get

$$A_n = A_n A = A_n(A_1' \times \cdots \times A_m') = A_n A_1' \times \cdots \times A_n A_m'.$$

Because A_n is connected, there exists ℓ such that $1 \leqslant \ell \leqslant m$, $A_n = A_n A_\ell'$ and $A_n A_j' = 0$, for $j \neq \ell$. Similarly,

$$A_\ell' = A A_\ell' = (A_1 \times \cdots \times A_n) A_\ell' = A_1 A_\ell' \times \cdots \times A_n A_\ell',$$

and, because A_ℓ' is connected, we get that

$$A_\ell' = A_n A_\ell' = A_n.$$

Now,

$$\prod_{i \neq n} A_i \cong \frac{A}{A_n} = \frac{A}{A_\ell'} \cong \prod_{j \neq \ell} A_j'.$$

The result then follows from an easy induction. $\qquad\square$

Because we are mainly interested in modules, we show how block decomposition affects module categories. Products of categories were introduced in Exercise III.2.4.

VII.3.13 Proposition. *Let* $A = A_1 \times A_2$ *be a decomposition of* A. *There exists an equivalence of categories* $\operatorname{Mod} A \cong \operatorname{Mod} A_1 \times \operatorname{Mod} A_2$.

Proof. To the decomposition of A into product corresponds a complete set of orthogonal central idempotents $\{c_1, c_2\}$. Let M be an A-module, then $Mc_1 = \{xc_1 : x \in M\}$ has a natural A_1-module structure. Similarly $Mc_2 = \{xc_2 : x \in M\}$ is an A_2-module. We thus have a functor $\operatorname{Mod} A \to \operatorname{Mod} A_1 \times \operatorname{Mod} A_2$ defined on objects by $M \mapsto (Mc_1, Mc_2)$. Indeed, if $f : M \to N$ is a morphism in $\operatorname{Mod} A$, then $f(xc_i) = f(x)c_i$, for $x \in M, i = 1, 2$, so that the restriction of f to Mc_i is a morphism in $\operatorname{Mod} A_i$.

Conversely, if (M_1, M_2) is an object in $\operatorname{Mod} A_1 \times \operatorname{Mod} A_2$, then $M_1 \times M_2$ becomes an A-module if we define the product componentwise

$$(x_1, x_2)(a_1, a_2) = (x_1 a_1, x_2 a_2)$$

where $x_i \in M_i, a_i \in A_i, i = 1, 2$. A morphism $(f_1, f_2) : (M_1, M_2) \to (N_1, N_2)$ in $\operatorname{Mod} A_1 \times \operatorname{Mod} A_2$, induces a morphism $M_1 \times M_2 \to N_1 \times N_2$ by setting $(x_1, x_2) \mapsto (f_1(x_1), f_2(x_2))$, where $x_i \in M_i, i = 1, 2$. This defines a functor

$\operatorname{Mod} A_1 \times \operatorname{Mod} A_2 \to \operatorname{Mod} A$ which is easily verified to be a quasi-inverse of the one defined above. $\qquad\square$

In other words, not only does block decomposition decompose a given algebra into smaller ones, it also reduces the study of a module category into that of smaller module categories. If A is noetherian or artinian, then the study of its module category reduces to the study of the module categories of its blocks, which are connected algebras. An easy corollary is the following.

VII.3.14 Corollary. *Let A_1, A_2 be noetherian, or artinian, algebras. Then so is their product $A_1 \times A_2$.*

Proof. We prove the noetherian case, the artinian one is similar. We must prove that any ascending chain of right ideals in the product stabilises. Proposition VII.3.13 gives that this chain is equivalent to a chain consisting of pairs (I_1, I_2), with I_i a right ideal in A_i and $i = 1, 2$. The hypothesis implies that each component chain stabilises. Hence, so does the original chain. $\qquad\square$

VII.3.15 Examples. We continue here Examples VII.2.14(d) and (e).
(d) Consider the $\mathbb{Z}$-algebra

$$A = \begin{pmatrix} \mathbb{Q} & 0 \\ \mathbb{Q} & \mathbb{Z} \end{pmatrix} = \left\{ \begin{pmatrix} a & 0 \\ b & n \end{pmatrix} : a, b \in \mathbb{Q}, n \in \mathbb{Z} \right\}$$

with the ordinary matrix operations. We prove it is right noetherian by proving that any right ideal I of A is finitely generated. We consider two cases for I, see Exercise II.2.11.

Assume first that I is contained inside $L = \begin{pmatrix} \mathbb{Q} & 0 \\ \mathbb{Q} & 0 \end{pmatrix}$. The latter is itself a right

ideal of A which, as a $\mathbb{Q}$-vector space, is isomorphic to $\mathbb{Q}^{(2)} = \left\{ \begin{pmatrix} c_1 \\ c_2 \end{pmatrix} : c_1, c_2 \in \mathbb{Q} \right\}$.

Under this isomorphism, I maps onto a subspace of $\mathbb{Q}^{(2)}$. Because $\mathbb{Q}^{(2)}$ is two-dimensional, I is generated by at most two vectors.

Next, if I is not contained in L, then there exists a matrix $\begin{pmatrix} a & 0 \\ b & k \end{pmatrix}$ in I, with $k \neq 0$.
But then, for any $c \in \mathbb{Q}$, we have

$$\begin{pmatrix} a & 0 \\ b & k \end{pmatrix} \cdot \begin{pmatrix} 0 & 0 \\ c/k & 0 \end{pmatrix} = \begin{pmatrix} 0 & 0 \\ c & 0 \end{pmatrix} = \begin{pmatrix} 0 & 0 \\ 1 & 0 \end{pmatrix} c.$$

Therefore, the right ideal $J = \begin{pmatrix} 0 & 0 \\ 1 & 0 \end{pmatrix} A$ is contained in I. On the other hand,
$A/J \cong \mathbb{Q} \times \mathbb{Z}$ as algebras. Hence A/J is noetherian, due to Corollary VII.3.14. So, the right ideal I/J of A/J is finitely generated. The short exact sequence of A-modules $0 \to J \to I \to I/J \to 0$ shows that I itself is finitely generated. This proves that A is right noetherian.

In order to prove that it is not left noetherian, we exhibit an infinite ascending chain of left ideals. Let p be a prime, and, for any integer $k \geq 1$, consider the left ideal

$$J_k = \begin{pmatrix} 0 & 0 \\ \frac{\mathbb{Z}}{p^k} & 0 \end{pmatrix} = \left\{ \begin{pmatrix} 0 & 0 \\ \frac{n}{p^k} & 0 \end{pmatrix} : n \in \mathbb{Z} \right\}$$

then we have an infinite chain $J_1 \subsetneq J_2 \subsetneq J_3 \subsetneq \cdots$.

Actually, this example also follows easily from Exercise VII.2.11(a). We wanted to give an elementary proof, because it is a classical example, which we shall use again.

(e) Let K be a field, Q the two-loops quiver

and $A = KQ/I$, with $I = \langle \beta^2, \alpha\beta \rangle$. Because $I \subseteq KQ^{+2}$, we may unambiguously identify α, β with their residual classes inside A, and so consider A as generated by α, β bound by the relations $\beta^2 = 0$ and $\alpha\beta = 0$. As abelian groups, we have a decomposition $A = K[\alpha] \oplus \beta K[\alpha]$. The first summand $K[\alpha]$ is easily seen to be a sub-algebra of A, while the second $\beta K[\alpha]$ is an ideal. Now, the polynomial algebra $K[\alpha]$ in α is noetherian, and A is finitely generated (by 1 and β) as right $K[\alpha]$-module. Because of Theorem VII.2.21, A is noetherian as right $K[\alpha]$-module, hence as right A-module. Therefore, A is a right noetherian algebra.

In order to prove that A is not left noetherian, it suffices to prove that $\beta K[\alpha]$ is not finitely generated as left A-module (because of Proposition VII.2.15). Both α and β act trivially on the left of A, due to the definition of I, hence $\beta K[\alpha]$ is finitely generated as a left A-module if and only it is finite-dimensional as a K-vector space. But this is not the case, because it has the infinite basis $(\beta \alpha^i)_{i \geq 0}$.

Incidentally, because A is an infinite-dimensional vector space, the ideal I is *not* an admissible ideal, in the sense of Example I.3.11(h).

One can add that A is neither right, nor left, artinian. For, if it were, then, due to Proposition VII.2.20, the quotient $A/\beta K[\alpha] \cong K[\alpha]$, would be artinian, and this is not the case.

Finally, the algebra given by the same quiver, but bound by the ideal $I^{op} = \langle \beta^2, \beta\alpha \rangle$, is left, and not right, noetherian, see Exercise I.4.8.

VII.3.3 Product of quotients of an algebra

A classical way to construct new algebras out of a given one is to factorise through ideals. Thus, given an algebra A, and a family of ideals $\{I_1, \cdots, I_t\}$, we wish to describe the product of the A/I_i. We need a definition.

VII.3.16 Definition. Two ideals I, J in an algebra A are ***comaximal*** if $I + J = A$.

Equivalently, I, J are comaximal if and only if the identity of A decomposes as $1 - x+y$, with $x \in I, y \in J$. This decomposition, of course, is not necessarily unique.

VII.3.17 Examples.

    (a) If I, J are distinct maximal ideals in an algebra A, then they are comaximal (hence the terminology). More generally, if I is maximal and J any ideal not contained in I, then I and J are comaximal.

    (b) If A is a principal ideal domain, and $a, b \in A$ then, because of the Bézout–Bachet Theorem, Corollary I.5.6, a, b are coprime if and only if $Aa + Ab = A$, hence if and only if the ideals Aa and Ab are comaximal.

Let $\{I_1, \cdots, I_t\}$ be ideals in an algebra A. For any i, the projection $A \to A/I_i$ is an algebra morphism. The universal property of the product applied to the projections yields an algebra morphism $f : A \to \prod_{i=1}^{t}(A/I_i)$ given by $a \mapsto (a + I_i)_{1 \leqslant i \leqslant t}$, for $a \in A$.

VII.3.18 Theorem. *Let A be an algebra, $\{I_1, \cdots, I_t\}$ ideals in A and $f : A \to \prod_{i=1}^{t}(A/I_i)$ the morphism given by $a \mapsto (a + I_i)_{1 \leqslant i \leqslant t}$, for $a \in A$. Then:*

    (a) $\operatorname{Ker} f = \cap_{i=1}^{t} I_i$.

    (b) *f is surjective if and only if $i \neq j$ implies that I_i, I_j are comaximal.*

Proof. Because (a) is obvious, we only prove (b). Assume that f is surjective and $i \neq j$. Surjectivity yields $a_i \in A$ such that $a_i + I_i = 0 + I_i$ and $a_i + I_j = 1 + I_j$, that is, $a_i \in I_i$ and $1 - a_i \in I_j$. Then $1 = a_i + (1 - a_i) \in I_i + I_j$, hence I_i and I_j are comaximal.

    Conversely, assume that I_i, I_j are comaximal when $i \neq j$. For any i, set $B_i = \cap_{j \neq i} I_j$. Then, each B_i is an ideal. For $i \geq 2$, the ideals I_1 and I_i are comaximal. Hence there exist $x_i \in I_1, y_i \in I_i$ such that $x_i + y_i = 1$. Set $b_1 = y_2 \cdots y_t$, then

$$b_1 = y_2 \cdots y_t = (1 - x_2) \cdots (1 - x_t) = 1 - c_1$$

with $c_1 \in I_1$. Now $b_1 = y_2 \cdots y_t \in B_1$. Hence, $1 = c_1 + b_1 \in I_1 + B_1$, and so $A = I_1 + B_1$, that is, I_1 and B_1 are comaximal. Moreover,

$$f(b_1) = (1 - c_1 + I_1, b_1 + I_2, \cdots, b_1 + I_t) = (1 + I_1, I_2, \cdots, I_t)$$

because $B_1 \subseteq I_i$ for any $i > 1$. Similarly, for any $i > 1$, there exists $b_i \in B_i$ such that

$$f(b_i) = (I_1, \cdots, 1 + I_i, \cdots, I_t)$$

so, for any set $\{x_1, \cdots, x_t\}$ of elements of A, we have

$$f\left(\sum_{i=1}^{t} b_i x_i\right) = (x_1 + I_1, \cdots x_t + I_t)$$

and f is indeed surjective. $\qquad\square$

The theorem says that, if the I_i are pairwise comaximal, then f induces an algebra isomorphism $A/(\cap_{i=1}^t I_i) \cong \prod_{i=1}^t (A/I_i)$. If the I_i are pairwise comaximal and moreover their intersection is zero, then $A \cong \prod_{i=1}^t (A/I_i)$, as algebras. In the commutative case, the intersection can be expressed otherwise. Recall that, if I, J are ideals in an algebra A, then one always has $IJ \subseteq I \cap J$, but equality does not hold true in general, see Subsection I.3.2.

VII.3.19 Corollary. *Let A be commutative and $\{I_1, \cdots, I_t\}$ a set of pairwise comaximal ideals. Then $\cap_{i=1}^t I_i = \prod_{i=1}^t I_i$.*

Proof. We use induction on $t \geq 2$. If $t = 2$, then $1 = x_1 + x_2$, with $x_1 \in I_1, x_2 \in I_2$. Let $y \in I_1 \cap I_2$. Then $yx_1 = x_1 y \in I_1 I_2$, because $y \in I_2$. Similarly, $yx_2 \in I_1 I_2$. Hence, $y = y \cdot 1 = yx_1 + yx_2 \in I_1 I_2$. So $I_1 \cap I_2 \subseteq I_1 I_2$. The reverse inclusion is trivial, hence $I_1 \cap I_2 = I_1 I_2$.

Assume that the statement holds true for some $t \geq 2$ and let $I_1, \cdots, I_t, I_{t+1}$ be pairwise comaximal. As seen in the proof of Theorem VII.3.18, $\cap_{i=1}^t I_i$ and I_{t+1} are comaximal. Hence,

$$\left(\bigcap_{i=1}^t I_i\right) \cap I_{t+1} = \left(\bigcap_{i=1}^t I_i\right) \cdot I_{t+1} = \left(\prod_{i=1}^t I_i\right) \cdot I_{t+1} = \prod_{i=1}^{t+1} I_i. \qquad \square$$

Therefore, if A is commutative and $\{I_1, \cdots, I_t\}$ are pairwise comaximal ideals, then $A/(\prod_{i=1}^t I_i) \cong \prod_{i=1}^t (A/I_i)$, as algebras.

VII.3.20 Corollary. *Let A be a principal ideal domain, $b_1, \cdots, b_t$ pairwise coprime elements of A and $b = b_1 \cdots b_t$. Then the map $f : A/Ab \to \prod_{i=1}^t (A/Ab_i)$ given by $a + Ab \mapsto (a + Ab_i)_{1 \leq i \leq t}$ is an algebra isomorphism.*

Proof. Because the b_i are pairwise coprime, the Ab_i are pairwise comaximal, see Example VII.3.17(b). Corollary VII.3.19 yields $\cap_{i=1}^t (Ab_i) = \prod_{i=1}^t (Ab_i) = Ab$. The statement follows from Theorem VII.3.18. $\qquad \square$

In Proposition V.3.14, we proved that the map of Corollary VII.3.20 is an isomorphism of A-modules. Corollary VII.3.20 above says that it is also an *algebra isomorphism*. A consequence is the classical **Chinese Remainder Theorem**.

VII.3.21 Corollary. *Let $n_1, \cdots, n_t$ be pairwise coprime integers, then the system of congruences*

$$x \equiv a_1, \mod n_1, \cdots, x \equiv a_t, \mod n_t$$

admits a unique solution modulo $n = n_1 \cdots n_t$.

Proof. We claim that there is a unique element in $\mathbb{Z}/n\mathbb{Z}$ corresponding to the element $(a_1 + n_1\mathbb{Z}, \cdots, a_t + n_t\mathbb{Z}) \in \prod_{i=1}^t (\mathbb{Z}/n_i\mathbb{Z})$. But this is exactly the statement of Corollary VII.3.20. $\qquad \square$

Exercises of Section VII.3

1. Let $\{e_1, \cdots, e_n\}$ be a set of pairwise orthogonal idempotents in an algebra A. Prove that $e = e_1 + \cdots + e_n$ is also an idempotent.

2. Let A be an algebra and $e, f \in A$ idempotents. Prove that:
  (a) If A is commutative, then $e + f - ef$ is an idempotent and the ideal generated by e and f is principal and generated by $e + f - ef$.
  (b) If A is not commutative, then $e + f - ef$ is not necessarily an idempotent. (Hint: look at $M_2(\mathbb{Z})$).

3. Let $\{e_1, \cdots, e_n\}$ be a complete set of pairwise orthogonal idempotents in an algebra A, and M an A-module. Prove that $M \cong Me_1 \oplus \cdots \oplus Me_n$, as K-modules.

4. Let e, f be idempotents in an algebra A. Set $e \leqslant f$ if $ef = fe = e$. Prove that:
  (a) The set of idempotents in A is partially ordered by $\leqslant$, it admits 1 as largest element, and 0 as smallest element.
  (b) The idempotent e is primitive if and only if it is an **atom** in this poset (that is, it covers the zero: $0 < e$ and there exists no idempotent e' such that $0 < e' < e$).

5. Let A be an algebra and $C(A)$ the set of its central idempotents ordered by the relation $\leqslant$ of Exercise VII.3.4 above. Prove that:
  (a) If $e, f \in C(A)$, then $e \leqslant f$ if and only if $eA \subseteq fA$.
  (b) For $e, f \in C(A)$, set $e \vee f = e + f - ef$ and $e \wedge f = ef$, then $C(A)$ becomes a lattice.
  (c) If $e, f \in C(A)$, then $(e \vee f)A = eA + fA$ and $(e \wedge f)A = eA \cap fA$.

6. Let e be a nonzero idempotent in an algebra A.
  (a) Prove that $f = e + ex(1 - e)$, with $x \in A$, is also a nonzero idempotent and that $ef = f, fe = e$.
  (b) Deduce from (a) that e is central if and only if it commutes with any idempotent of A.
  (c) If in (a), we set $y = -ex(1 - e)$, prove that $e = f + fy(1 - f)$.
  (d) Deduce that, if A has no nonzero nilpotent element, then any idempotent in A is central.

7. Let A be an algebra, I an ideal and e an idempotent of A. Prove that $eIe = I \cap eAe$.

8. Let e, f be idempotents in an algebra A. Prove that $eA + fA = eA \oplus (f - ef)A$.

9. Let I be a proper ideal in an algebra A. Prove that:
  (a) If e is an idempotent in A, then $(e + I)(A/I) \cong eA/eI$, both as A-modules and as A/I-modules.
  (b) If $\{e_1, \cdots, e_n\}$ is a complete set of pairwise orthogonal idempotents, then

$$A/I = (e_1A/e_1I) \oplus \cdots \oplus (e_nA/e_nI),$$

both as A-modules and as A/I-modules.

10. Let A be an algebra and e_1, e_2 idempotents such that $e_1 A \cong e_2 A$ and e_1 is central. Prove that $e_2 = e_1 e_2 = e_2 e_1$. If e_2 is also central, prove that $e_1 = e_2$.

11. Let $n = m(m + 1)$, for some integer $m \geq 2$. Prove that the residual classes $\overline{m}^2$ and $\overline{m+1}$ are orthogonal idempotents in $\mathbb{Z}/n\mathbb{Z}$. This generalises Example VII.3.6(d).

12. Let K be a field. Prove that a matrix in $M_2(K)$, distinct from the identity or zero, is idempotent if and only if it has zero determinant and is of the form $\begin{pmatrix} a & b \\ c & 1-a \end{pmatrix}$, with $a, b, c \in K$.

13. Let A be an algebra and $c \in A$ a central idempotent. Prove that the algebra cA is connected if and only if c is centrally primitive.

14. Let K be a field and $A = KQ/I$ a bound quiver algebra. Let γ be a path from i to j with $i \neq j$. Show that $(\varepsilon_i + \gamma) + I$ and $(\varepsilon_j - \gamma) + I$ are orthogonal primitive idempotents of A.

15. In a noetherian or artinian algebra, prove that any decomposition of the identity into a sum of pairwise orthogonal central idempotents having a maximal number of terms, is unique.

16. Let A be an algebra and $\{I_1, \cdots, I_m\}, \{J_1, \cdots, J_n\}$ two sets of ideals of A such that, for any pair (i, j), the ideals I_i and J_j are comaximal. Prove that the products $\prod_{i=1}^{m} I_i$ and $\prod_{j=1}^{n} J_j$ are comaximal.

17. Let A be an algebra and I, J ideals of A such that $I \cap J = 0$. Prove that A/I and A/J are noetherian, or artinian, if and only if so is A.

18. Let m, n be positive integers. Prove that the system of congruences

$$\begin{cases} x \equiv a \mod m \\ x \equiv b \mod n \end{cases}$$

admits an integral solution if and only if $a \equiv b$ modulo the gcd of m and n.

VII.4 Composition series

VII.4.1 Simple modules

In Section VII.2, we approached module structure via properties of the submodule lattice. The nonzero modules having the simplest possible submodule lattices are certainly those having no nonzero proper submodules. These are the simple modules defined in Exercise II.3.18. Throughout this section, A denotes a K-algebra.

VII.4.1 Definition. A nonzero A-module is **simple** provided it has only two submodules, namely, 0 and itself.

The Hasse quiver of the submodule lattice of a simple module is:

From the lattice-theoretical point of view, simple modules are **atoms**, that is, posets with only two elements $\{0, 1\}$ such that $0 < 1$. Simple modules are artinian and noetherian because they have a finite submodule lattice. The Hasse quiver suggests that a simple module is cyclic. In fact, it is cyclic in a very strong sense: it is generated by any of its nonzero elements.

VII.4.2 Lemma. *An A-module S is simple if and only if, for any nonzero $x \in S$, we have $S = xA$.*

Proof. Assume that S is simple. For a nonzero $x \in S$, the submodule xA of S is nonzero, hence $xA = S$. Conversely, let S be generated by any of its nonzero elements. A nonzero submodule M of S contains a nonzero element x, say. But then $S = xA \subseteq M$ gives $S = M$. $\square$

VII.4.3 Examples.

(a) If D is a division ring, then a D-module (vector space) is simple if and only if it has no proper nonzero subspace, that is, it is one-dimensional. Because all one-dimensional D-vector spaces are isomorphic to each other and also to D, there exists, up to isomorphism, a unique simple D-module.

Conversely, if an algebra A is simple as an A-module, then A is a division ring. For, let $a \in A$ be nonzero. Then $aA \neq 0$. Hence $aA = A$. So any nonzero element has a right inverse. Because of Exercise I.2.6, A is a division ring, as claimed. We have proved that A is a division ring if and only if the module A_A is simple.

(b) Let K be a field and A a finite-dimensional K-algebra. Any A-module which is one-dimensional when considered as K-vector space is a simple A-module. The converse does not hold true in general, see Example VII.5.8 below.

(c) Let M be an A-module. A minimal submodule of M, if it exists, is a simple A-module. For instance, in an algebra, minimal right ideals are simple submodules.

(d) Because of Lemma VII.2.3, an artinian module has simple submodules. For instance, if A is a principal ideal domain, then an indecomposable module of the form A/p^tA, with p irreducible and $t \geq 1$ an integer, contains the simple module A/pA. Actually, any simple A-module is of this form, see Exercise V.3.16.

This is not true in general for noetherian modules: a principal ideal domain A has no minimal (right) ideal, and therefore no simple submodule. Indeed, any proper nonzero submodule aA, that is, a submodule with a nonzero and noninvertible, contains properly the submodule a^2A.

(e) Let K be a field and $A = KQ/I$ a bound quiver algebra. To a point $x \in Q_0$, one associates a representation $S_x = (S_x(y), S_x(\alpha))$ as follows:

$$S_x(y) = \begin{cases} K & \text{if } y = x \\ 0 & \text{if } y \neq x \end{cases} \quad \text{and } S_x(\alpha) = 0 \text{ for any arrow } \alpha \in Q_1.$$

Each S_x is a simple representation, that is, its only subrepresentations are 0 and itself. In fact, the A-module corresponding to S_x is one-dimensional as K-vector space, and hence is a simple A-module, see Example (b) above. For instance, if Q is the quiver

bound by the ideal $I = 0$, then the simple representations we constructed above are

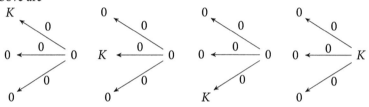

Morphisms having as source or target a simple module behave particularly well. This is the content of Exercise II.3.18 which we restate for future reference.

VII.4.4 Lemma. *Let A be an algebra and $f : M \to N$ a nonzero morphism between modules. Then:*

(a) *If M is simple, then f is a monomorphism.*
(b) *If N is simple, then f is an epimorphism.*
(c) *If M, N are both simple, then f is an isomorphism.*   □

Thus, if S is simple, then its endomorphism algebra $\text{End} S$ is a division ring, a statement known as **Schur's lemma**.

A module is simple if and only if it admits 0 as maximal submodule. On the other hand, let A be a principal ideal domain and $a \in A$ a nonzero element. Because of Example VII.4.3(d) above, the A-module A/aA is simple if and only if a is irreducible, that is, if and only if aA is maximal in A, due to Theorem I.5.7. These remarks suggest the next lemma.

VII.4.5 Lemma. *A submodule N of an A-module M is maximal if and only if M/N is a simple module.*

Proof. Suppose that N is a maximal submodule of M. If M/N is not simple, then it has a proper nonzero submodule L. Because of Theorem II.4.27, there exists a submodule $M' \subseteq M$ containing N such that $L \cong M'/N$. Because $L \neq 0$, we have $M' \supsetneq N$ and, because $L \subsetneq M/N$, also $M' \subsetneq M$. This contradicts maximality of N in M. Conversely, if M/N is simple but N is not maximal in M, there exists a submodule M' of M such that $N \subsetneq M' \subsetneq M$. Hence, M'/N is a nonzero proper submodule of M/N, a contradiction.   □

Because of Proposition II.2.23, any finitely generated module admits maximal submodules, hence, due to Lemma VII.4.5, has simple quotients. Thus, any algebra admits simple modules.

VII.4.2 Composition series

We now use simple modules as building blocks for constructing larger modules. As we have seen, a simple module, from the lattice-theoretical point of view, is an atom. Finite lattices may be constructed by arranging together finitely many atoms in some way. Our main tool is (again!) short exact sequences. A short exact sequence

$$0 \to L \to M \to N \to 0$$

is also called an **extension** of L by N: one 'extends' the module L by another module N in order to make up the larger module M. We warn the reader that this terminology is not universal: many authors call 'extension of N by L' what we call 'extension of L by N'. Let L, N be simple modules, then the Hasse quiver of an extension of L and N has one of the following two forms

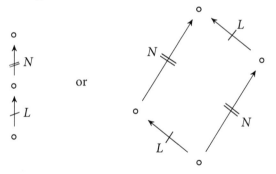

As seen in Example II.2.25, the second Hasse quiver takes different forms depending on whether the simple modules L, N are isomorphic or not. The short exact sequences corresponding to these two types of extensions are respectively

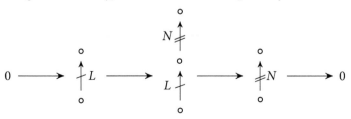

and

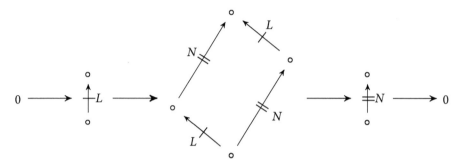

If L and N are not necessarily simple, then a reasonable way to construct a finite submodule lattice starting from simples is to proceed inductively. Because of Proposition VII.2.4, we are led to consider modules which are both noetherian and artinian.

Let M be a nonzero noetherian and artinian A-module. Because of Proposition VII.2.4, M admits maximal chains of submodules. Such chains are finite, because M is noetherian and artinian. Maximality implies that such a chain runs from the smallest submodule 0 to the largest one M, so it is of the form

$$0 = M_0 \subsetneq M_1 \subsetneq \cdots \subsetneq M_{\ell-1} \subsetneq M_\ell = M$$

and one cannot insert any submodule between M_i and M_{i+1}. Equivalently, M_i is maximal in M_{i+1}, for any i. Because of Lemma VII.4.5, this amounts to saying that M_{i+1}/M_i is simple.

VII.4.6 Definition. Let M be an A-module. A finite sequence of submodules of the form

$$0 = M_0 \subsetneq M_1 \subsetneq \cdots \subsetneq M_{\ell-1} \subsetneq M_\ell = M$$

is a ***composition series*** for M if, for any i with $0 \leqslant i < \ell$, the quotient M_{i+1}/M_i is a simple module, called ***composition factor*** of M. The integer ℓ is the ***length*** of the series.

Composition series are iterated extensions of simple modules: indeed, let

$$0 = M_0 \subsetneq M_1 \subsetneq M_2 \subsetneq \cdots \subsetneq M_{\ell-1} \subsetneq M_\ell = M$$

be a composition series for M. Then M_1 and M_2/M_1 are simple, and the short exact sequence $0 \to M_1 \to M_2 \to M_2/M_1 \to 0$ realises M_2 as an extension of M_1 by M_2/M_1, which are both simple. Inductively, each M_i, with $i > 0$, is obtained by extending M_{i-1} by the simple module M_i/M_{i-1}.

VII.4.7 Lemma. *A module is noetherian and artinian if and only if it has a composition series.*

Proof. We already proved necessity. Sufficiency is proved by induction on the least length ℓ of all composition series for an A-module M. If $\ell = 1$, then M is simple and there is nothing to prove. If $\ell > 1$, let

$$0 = M_0 \subsetneq M_1 \subsetneq \cdots \subsetneq M_{\ell-1} \subsetneq M_\ell = M$$

be a composition series of least length for M. Then $M_{\ell-1}$ has a composition series of length $\ell - 1$. The induction hypothesis implies that $M_{\ell-1}$ is noetherian and artinian. On the other hand, $M/M_{\ell-1}$ is simple, hence noetherian and artinian. The statement follows from applying Theorem VII.2.11 to the short exact sequence $0 \to M_{\ell-1} \to M \to M/M_{\ell-1} \to 0$. $\qquad\square$

A consequence of this result and Theorem VII.2.11 is that, if $0 \to L \to M \to N \to 0$ is a short exact sequence, then M admits a composition series if and only if both L and N do.

The previous considerations imply that composition series are easy to read on the Hasse quiver of the submodule lattice of M, they correspond exactly to all paths from the smallest submodule 0 to the largest one M. Each arrow on the Hasse quiver represents a simple composition factor.

VII.4.8 Examples.

(a) Let K be a field and E an n-dimensional K-vector space. The simple K-modules are the one-dimensional vector spaces, all isomorphic to K. Let $\{e_1, \cdots, e_n\}$ be a basis of E, the following is a composition series of length n :

$$0 \subsetneq Ke_1 \subsetneq Ke_1 \oplus Ke_2 \subsetneq \cdots \subsetneq \bigoplus_{i=1}^{n-1} Ke_i \subsetneq \bigoplus_{i=1}^{n} Ke_i = E.$$

Conversely, any composition series for an n-dimensional vector space is of this form. Different bases induce different composition series, but all of them have the same length n (which is the dimension of E) and K as unique simple composition factor.

(b) Let S be a simple A-module, then it has a unique composition series $0 \subsetneq S$, of length one and a unique composition factor, namely S itself.

(c) Let A be a principal ideal domain, $p \in A$ an irreducible element, $t > 0$ an integer and $M = A/p^t A$. Because of Lemma V.3.21, its submodules are:

$$\{0\} = p^t M \subsetneq p^{t-1} M \subsetneq \cdots \subsetneq pM \subsetneq M.$$

In particular, it has a unique composition series. All its composition factors $p^i M/p^{i+1} M$, are isomorphic to the simple module A/pA, and the Hasse quiver of its submodule lattice is of the form:

where the double notch shows the isomorphic composition factors.

(d) Let $M = \mathbb{Z}/2\mathbb{Z} \oplus \mathbb{Z}/2\mathbb{Z}$ be the Klein Four-group, see Example II.2.25(a). The Hasse quiver of its submodule lattice is

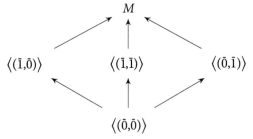

It has three distinct composition series,

$$\langle(\bar{0},\bar{0})\rangle \subsetneq \langle(\bar{0},\bar{1})\rangle \subsetneq M$$

$$\langle(\bar{0},\bar{0})\rangle \subsetneq \langle(\bar{1},\bar{0})\rangle \subsetneq M$$

$$\langle(\bar{0},\bar{0})\rangle \subsetneq \langle(\bar{1},\bar{1})\rangle \subsetneq M$$

each has length 2, and each composition factor is isomorphic to $\mathbb{Z}/2\mathbb{Z}$.

(e) The submodule lattice of the abelian group $M = \mathbb{Z}/12\mathbb{Z}$, computed in Example II.2.25(b), has the Hasse quiver

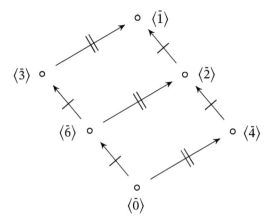

One has exactly three distinct composition series, namely

$$\langle \bar{0} \rangle \subsetneq \langle \bar{6} \rangle \subsetneq \langle \bar{3} \rangle \subsetneq \langle \bar{1} \rangle = M$$

$$\langle \bar{0} \rangle \subsetneq \langle \bar{6} \rangle \subsetneq \langle \bar{2} \rangle \subsetneq \langle \bar{1} \rangle = M$$

$$\langle \bar{0} \rangle \subsetneq \langle \bar{4} \rangle \subsetneq \langle \bar{2} \rangle \subsetneq \langle \bar{1} \rangle = M$$

each has length three, two composition factors isomorphic to $\mathbb{Z}/2\mathbb{Z}$, represented by a single notch, and one isomorphic to $\mathbb{Z}/3\mathbb{Z}$, represented by a double notch.

(f) Let K be a field and $A = KQ/I$ the K-algebra given by the quiver

$$1 \xleftarrow{\ \alpha\ } 2 \xleftarrow{\ \gamma\ } 4$$
$$\beta \uparrow$$
$$3$$

bound by the ideal $I = \langle \beta\alpha \rangle$. Consider the module M given by the representation

$$K \xleftarrow{(1\ 0)} K^2 \xleftarrow{\binom{1}{0}} K$$
$$\binom{0}{1} \uparrow$$
$$K$$

As seen in Example VII.4.3(c), to each point i in Q is associated a simple A-module, denoted by S_i, with $1 \leqslant i \leqslant 4$. Considering the submodules

$$M_1: \quad K \xleftarrow{(1\ 0)} K^2 \xleftarrow{0} 0 \qquad M_2: \quad K \xleftarrow{(1\ 0)} K^2 \xleftarrow{0} 0$$

with vertical maps $\binom{0}{1} \uparrow$ from K and $0 \uparrow$ from 0.

$$M_3: \quad K \xleftarrow{1} K \xleftarrow{0} 0 \qquad M_4 = S_1: \quad K \xleftarrow{0} 0 \xleftarrow{0} 0$$

with vertical maps $0 \uparrow$ from 0.

of M, a simple calculation shows that we have a composition series of length five

$$0 \subsetneq M_4 \subsetneq M_3 \subsetneq M_2 \subsetneq M_1 \subsetneq M$$

with composition factors S_1, S_2, S_2, S_3 and S_4 (by order of appearance). In this example, M has other composition series, all of which having length five. We leave to the reader their construction as an exercise.

Modules having a unique composition series are called **uniserial**. Simple modules are uniserial, see Example VII.4.8(b) above, but there exist uniserial modules which are not simple, see Example VII.4.8(c). In fact, to say that a module M is uniserial amounts to saying that M has at most one maximal submodule which, in its turn, has at most one maximal submodule, etc. (that is, any submodule de M has at most one maximal submodule). One can represent pictorially the Hasse quiver of a uniserial module by just showing its composition factors. For instance, the uniserial module $M = \mathbb{Z}/p^4\mathbb{Z}$, with p a prime, has a unique composition factor $S_1 = \mathbb{Z}/p\mathbb{Z}$, up to isomorphism. One can write it as repeated extension four times of S_1 as

$$\begin{array}{c} S_1 \\ S_1 \\ S_1 \\ S_1 \end{array}$$

or one can identify the simple module S_1 to its index 1 and write:

$$\begin{array}{c} 1 \\ 1 \\ 1 \\ 1 \end{array}$$

In this notation, the (unique) composition series of M is:

$$0 \subsetneq 1 \subsetneq \begin{array}{c} 1 \\ 1 \end{array} \subsetneq \begin{array}{c} 1 \\ 1 \\ 1 \end{array} \subsetneq \begin{array}{c} 1 \\ 1 \\ 1 \\ 1 \end{array}$$

Both maximal submodules and simple composition factors are plain to see. Another advantage is that, in the unique composition series of a uniserial module, *all* submodules must appear. This allows for an easy computation of images and kernels of morphisms. In order to see it, we suggest to the reader to redo the picture following Lemma V.3.23 using this new notation. Of course, in a uniserial module, simple composition factors are generally nonisomorphic. In Example VII.4.8(f), M_3 is uniserial and of the form

$$\begin{matrix} S_2 \\ S_1 \end{matrix} \qquad \text{or else} \qquad \begin{matrix} 2 \\ 1 \end{matrix}$$

Because submodule lattices are modular, they have no sublattice of the form of Example VII.2.5, so one expects all composition series to have the same length. Moreover, it turns out that a module determines uniquely its composition factors, up to order.

VII.4.9 Theorem. Jordan–Hölder Theorem. *If an A-module M admits two composition series*

$$0 = M_0 \subsetneq M_1 \subsetneq \cdots \subsetneq M_m = M$$
$$0 = N_0 \subsetneq N_1 \subsetneq \cdots \subsetneq N_n = M$$

then $m = n$ and there exists a permutation σ of $\{0, 1, \cdots, m - 1\}$ such that $N_{i+1}/N_i \cong M_{\sigma(i)+1}/M_{\sigma(i)}$, for any i such that $0 \leqslant i < m$.

Proof. We use induction on m. If $m = 0$, then $M = 0$ and there is nothing to prove. If $m = 1$, then M is simple and has a unique composition series $0 \subsetneq M$. Assume that $m > 1$ and consider the 'composite' sequence of submodules:

$$0 = N_0 \cap M_{m-1} \subseteq N_1 \cap M_{m-1} \subseteq \cdots \subseteq N_n \cap M_{m-1} = M_{m-1} = N_0 + M_{m-1}$$
$$\subseteq N_1 + M_{m-1} \subseteq \cdots \subseteq N_n + M_{m-1} = N_n = M.$$

Because M/M_{m-1} is simple, there exists a unique i, with $0 \leqslant i < n$, such that

$$M_{m-1} = N_0 + M_{m-1} = \cdots = N_i + M_{m-1} \subsetneq N_{i+1} + M_{m-1} = \cdots = N_n + M_{m-1} = M.$$

Because of Lemma VII.2.10, we have, for any j such that $0 \leqslant j < n$, a short exact sequence

$$0 \to \frac{N_{j+1} \cap M_{m-1}}{N_j \cap M_{m-1}} \to \frac{N_{j+1}}{N_j} \to \frac{N_{j+1} + M_{m-1}}{N_j + M_{m-1}} \to 0.$$

The middle term of this sequence is simple. Therefore, one of the extremal terms is isomorphic to this simple, while the other vanishes. That is, there exists an index

i such that, if $j = i$, we have

$$\frac{M}{M_{m-1}} \cong \frac{N_{i+1} + M_{m-1}}{N_i + M_{m-1}} \cong \frac{N_{i+1}}{N_i}$$

(and hence $N_{i+1} \cap M_{m-1} = N_i \cap M_{m-1}$) while, if $j \neq i$,

$$\frac{N_{j+1} \cap M_{m-1}}{N_j \cap M_{m-1}} \cong \frac{N_{j+1}}{N_j}$$

which is simple. Therefore the sequence

$$0 = N_0 \cap M_{m-1} \subsetneq N_1 \cap M_{m-1} \subsetneq \cdots \subsetneq N_i \cap M_{m-1} = N_{i+1} \cap M_{m-1}$$

$$\subsetneq \cdots \subsetneq N_n \cap M_{m-1} = M_{m-1}.$$

is a composition series for M_{m-1} of length $n - 1$. The induction hypothesis gives $m-1 = n-1$ (hence $m = n$) and a bijection $\sigma : \{0, \cdots, i-1, i+1, \cdots, n\} \to \{0, \cdots, m-1\}$ such that $N_{j+1}/N_j \cong M_{\sigma(j)+1}/M_{\sigma(j)}$, for any $j \neq i$. Finally, we set $\sigma(i) = m - 1$.   □

VII.4.10 Example. The converse of this theorem does not hold true in general: two modules may have the same composition factors, appearing in the same order, and yet be nonisomorphic. An obvious example is that of the abelian groups $\mathbb{Z}/p\mathbb{Z} \oplus \mathbb{Z}/p\mathbb{Z}$ and $\mathbb{Z}/p^2\mathbb{Z}$ with p a prime. We give another one. Let K be a field and A the path algebra of the Kronecker quiver

$$\underset{1}{\circ} \overset{\alpha}{\underset{\beta}{\rightleftarrows}} \underset{2}{\circ}$$

Consider the representations H_α

$$K \overset{1}{\underset{0}{\rightleftarrows}} K$$

and H_β

$$K \overset{0}{\underset{1}{\rightleftarrows}} K$$

It is easily seen, by trying to construct a morphism as in Example II.3.3(h), that the only morphism from H_α to H_β, or from H_β to H_α, is the zero morphism. Thus, these representations (and the corresponding modules) cannot be isomorphic. However, both H_α and H_β admit the simple S_1 corresponding to the point 1 as unique proper nonzero submodule, and each has a unique composition series given respectively by:

$$0 \subsetneq S_1 \subsetneq H_\alpha \text{ and } 0 \subsetneq S_1 \subsetneq H_\beta.$$

Finally, we have short exact sequences of representations $0 \to S_1 \to H_\alpha \to S_2 \to 0$ and $0 \to S_1 \to H_\beta \to S_2 \to 0$, constructed as in Example II.4.16(d). Therefore,

$H_\alpha/S_1 \cong S_2 \cong H_\beta/S_1$, so that both modules H_α, H_β have the same composition factors appearing in the same order, but are not isomorphic.

What differs here is the embedding of S_1 as a submodule of H_α and H_β. Indeed, as K-vector spaces, $H_\alpha = Ke_2 \oplus K\alpha$ and $H_\beta = Ke_2 \oplus K\beta$: that is, S_1 embeds in the first as the subspace $K\alpha$ and in the second as the subspace $K\beta$.

The Jordan–Hölder theorem asserts that for any module having a composition series, that is, which is both noetherian and artinian, all series have the same length.

VII.4.11 Definition. A module M is of **finite length** if it has a composition series. Its **length**, or **composition length**, denoted as $\ell(M)$, is 0 if $M = 0$, and ℓ if M has a composition series of length ℓ. If M is not of finite length, then it is of **infinite length**, and we write $\ell(M) = \infty$.

VII.4.12 Corollary. *A module of finite length is finitely generated.*

Proof. Because of Theorem VII.2.9, any noetherian module is finitely generated. □

We prove later Theorem VIII.3.22 saying that any artinian algebra is also noetherian. Therefore, a finitely generated module over an artinian algebra is noetherian and artinian, because of Theorem VII.2.21, that is, it is of finite length. Hence, over an artinian algebra, a module is of finite length if and only if it is finitely generated.

VII.4.13 Examples. Lengths of modules are read on the Hasse quiver as the common length of any path from 0 to the module. For instance, in Example VII.4.8, we have:
- (a) Let K be a field and E an n-dimensional K-vector space. The length of E as a K-module is n. Thus, length generalises dimension of a vector space (as does the rank, see Definition V.2.3, but the latter only makes sense for free modules over a commutative ring).
- (b) Let S be a simple module, then $\ell(S) = 1$.
- (c) Let A be a principal ideal domain, $p \in A$ an irreducible element and $t > 0$ an integer, then $\ell(A/p^t A) = t$.
- (d) $\ell(\mathbb{Z}/2\mathbb{Z} \oplus \mathbb{Z}/2\mathbb{Z}) = 2$.
- (e) $\ell(\mathbb{Z}/12\mathbb{Z}) = 3$.
- (f) Let M be as in Example VII.4.8(e), then $\ell(M) = 5$.

The following results show that length behaves very much like the dimension of a vector space.

VII.4.14 Proposition. *Let $0 \to L \to M \to N \to 0$ be a short exact sequence of A-modules. Then M has finite length if and only if L and N have finite length. In this case, the composition factors of M (counting multiplicities) are the same as those of L and those of N (also counting multiplicities). Moreover, $\ell(M) = \ell(L) + \ell(N)$.*

Proof. The first statement follows from Theorem VII.2.11. We may assume, without loss of generality, that $L \subseteq M$ and that $N = M/L$. If

$$0 = L_0 \subsetneq \cdots \subsetneq L_t = L$$

is a composition series for L, and

$$0 = \frac{L_t}{L} \subsetneq \frac{M_{t+1}}{L} \subsetneq \cdots \subsetneq \frac{M_s}{L} = \frac{M}{L} \cong N$$

is a composition series for N, then the sequence

$$0 = L_0 \subsetneq \cdots \subsetneq L_t \subsetneq M_{t+1} \subsetneq \cdots \subsetneq M_s = M$$

is easily seen to be a composition series for M. This implies that $\ell(M) = \ell(L) + \ell(N)$ and composition factors are as asserted. $\qquad\square$

In particular, for two submodules L, N of M with zero intersection, we have $\ell(L \oplus N) = \ell(L) + \ell(N)$. We also deduce the well-known **Grassmann's formula**.

VII.4.15 Corollary. Grassmann's Formula. *Let M be a module and L, N submodules of M. Then*

$$\ell(L + N) = \ell(L) + \ell(N) - \ell(L \cap N).$$

Proof. There exist short exact sequences

$$0 \to L \cap N \to L \to \frac{L}{L \cap N} \to 0 \quad \text{and} \quad 0 \to N \to L + N \to \frac{L+N}{N} \to 0.$$

Because of Proposition VII.4.14, we have

$$\ell\left(\frac{L}{L \cap N}\right) = \ell(L) - \ell(L \cap N) \quad \text{and} \quad \ell\left(\frac{L+N}{N}\right) = \ell(L + N) - \ell(N).$$

The Isomorphism Theorem II.4.24 gives $(L + N)/L \cong N/(L \cap N)$, hence the result . $\qquad\square$

Exercises of Section VII.4

1. Let A be an algebra. Prove that an A-module S is simple if and only if there exists a maximal right ideal I in A such that $S \cong A_A/I$.

2. Let A be an algebra, M an A-module and S, L submodules M such that S is simple and $S \cap L \neq 0$. Prove that $S \subseteq L$.

3. Let A be an algebra and $f : M \to N$ a nonzero morphism between A-modules. Prove that:

(a) If N is simple and projective, then it is a direct summand of M.

(b) If M is simple and injective, then it is a direct summand of N.

4. Let A, B be algebras. Prove that:

(a) A right ideal of $A \times B$ is of the form $I \times J$, where I is a right ideal of A, and J a right ideal of B.

(b) A right ideal of $A \times B$ is maximal if and only if it is either of the form $I \times B$, with I a maximal right ideal in A, or of the form $A \times J$, with J a maximal right ideal in B.

(c) The set of simple $A \times B$-modules coincides with the union of the sets of simple A-modules and simple B-modules (considered as $A \times B$-modules by means of the projections).

5. Let I be a minimal right ideal of an algebra A. Prove that either $I^2 = 0$, or there exists an idempotent $e \in A$ such that $I = eA$ (so that I is a direct summand of A_A).

6. Give an example of a module which is not simple and whose endomorphism algebra is a division ring (so, the converse of Schur's lemma is false).

7. Let K be a field and E a nonzero finite-dimensional vector space viewed as left $\mathrm{End}_K E$-module under the action $(f, x) \mapsto f(x)$, for $f \in \mathrm{End}_K E, x \in E$. Prove that E is simple.

8. Let A be an algebra and S_1, S_2 simple A-modules. Prove that the submodule lattice of $S_1 \oplus S_2$ is distributive if and only if S_1 and S_2 are not isomorphic

9. Let K be a field. Compute the simple modules, up to isomorphism, over each of the algebras:

(a) $K[t]$.

(b) $T_2(K) = \begin{pmatrix} K & 0 \\ K & K \end{pmatrix}$.

10. Let A be the triangular matrix algebra

$$\begin{pmatrix} K & 0 & 0 \\ K & K & 0 \\ K & 0 & K \end{pmatrix} = \left\{ \begin{pmatrix} \alpha & 0 & 0 \\ \beta & \gamma & 0 \\ \delta & 0 & \epsilon \end{pmatrix} : \alpha, \cdots, \epsilon \in K \right\}$$

with the usual matrix operations. Consider each of the three rows $\begin{pmatrix} K & 0 & 0 \end{pmatrix}$, $\begin{pmatrix} K & K & 0 \end{pmatrix}$ and $\begin{pmatrix} K & 0 & K \end{pmatrix}$ as a right A-module as in Example II.2.4(d), compute a composition series of each, and deduce a composition series of A_A.

11. (a) Let p be a prime integer. Prove that the $\mathbb{Z}$-module $\mathbb{Z}/p\mathbb{Z} \oplus \mathbb{Z}/p\mathbb{Z}$ has exactly $p + 1$ composition series.

(b) Give an example of a module M with composition length 2 and infinitely many composition series.

(c) Let m be a positive integer. Exhibit a composition series of $\mathbb{Z}/m\mathbb{Z}$ and find its length.

(d) Characterise the integers $m > 0$ such that $\mathbb{Z}/m\mathbb{Z}$ is uniserial.

12. Prove that a noetherian and artinian module is uniserial if and only if it has a composition series involving all its submodules.

13. Prove that any submodule, and any quotient, of a uniserial module is also uniserial.

14. Let K be a field, $m > 1$ an integer and Q_m the linearly oriented quiver

$$1 \xleftarrow{\alpha_1} 2 \xleftarrow{\alpha_2} 3 \longleftarrow \cdots \longleftarrow m-1 \xleftarrow{\alpha_{m-1}} m$$

Prove that the projective module P_m associated to the point m is uniserial of length m and its composition factors are exactly all simple modules of the form S_i, with $1 \leqslant i \leqslant m$.

15. Let K be a field and Q the quiver

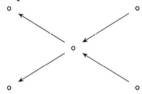

Exhibit three distinct composition series for the KQ-module

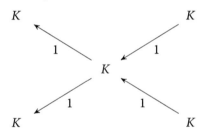

16. Let K be a field and $A = KQ$ be the path algebra of the quiver Q

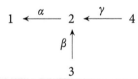

(a) Prove that the projective module P_4 corresponding to the point 4 is uniserial and give its composition series.

(b) Prove that the injective module I_2 corresponding to the point 2 has exactly two composition series.

17. Let K be a field. Exhibit a composition series for the projective module P_x over the algebra given by the one-loop quiver

bound by $I = <\alpha^5 >$.

18. Let A be an algebra and L, M modules of finite length. Show that each pair of composition series of L and M induces two composition series of $L \oplus M$, but there may be composition series of $L \oplus M$ which are not induced from composition series of L and M.

19. Let A be an algebra and $f : M \to N$ a nonzero linear map between noetherian and artinian A-modules. Prove that there exist composition series for M, N respectively

$$0 = M_0 \subsetneq M_1 \subsetneq \cdots \subsetneq M_m = M$$
$$0 = N_0 \subsetneq N_1 \subsetneq \cdots \subsetneq N_n = N$$

such that there exist i, j satisfying $M_{i+1}/M_i \cong N_{j+1}/N_j$.

20. Let A be an algebra, M a noetherian and artinian module and L a submodule of M. Prove that $\ell(L) \leqslant \ell(M)$ and equality holds if and only if $L = M$.

21. Let A be an algebra and $f : M \to N$ a linear map between A-modules, with M noetherian and artinian. Prove that

$$\ell(\operatorname{Im} f) + \ell(\operatorname{Ker} f) = \ell(M).$$

22. Let A be an algebra, M an A-module and L, N submodules of M such that $M/(L \cap N)$ is artinian and noetherian. Prove that

$$\ell(M/(L \cap N)) + \ell(M/(L + N)) = \ell(M/L) + \ell(M/N).$$

This is the analogue of the ***codimension formula*** of linear algebra.

23. Let A be an algebra, M a noetherian and artinian module and $M_1, \cdots, M_t$ maximal submodules of M. Find a formula for the length of $M/(\cap_{i=1}^t M_i)$.

24. Let A be an algebra. Prove the ***Schreier refinement theorem***: Let $0 = M_0 \subsetneq \cdots \subsetneq M_{t-1} \subsetneq M_t = M$ be a chain of submodules of a noetherian and artinian module M, then there exists a composition series whose terms include $M_0, \cdots, M_t$.

VII.5 Semisimple modules and algebras

VII.5.1 Semisimple modules

Let A be a K-algebra. A particularly nice situation occurs when a module is a sum of simple modules. Indeed, let K be a field and E a K-vector space with basis $(e_\lambda)_{\lambda \in \Lambda}$, say, then $E = \bigoplus_{\lambda \in \Lambda} K e_\lambda$, and each $K e_\lambda$ is a simple K-module, isomorphic to K. This property does not hold true for arbitrary modules (take, for instance, a nonsimple uniserial module), hence our interest in characterising modules having this property. In this section, modules are A-modules.

VII.5.1 Definition. A module is *semisimple* if it is a sum of simple modules.

In fact, we shall prove that a semisimple module is, even more, a *direct* sum of simple modules. For instance, a vector space over a field K is a semisimple K-module. We give more examples.

Example VII.5.2.
(a) Simple modules are semisimple. So, any algebra has semisimple modules.
(b) Any sum of semisimple modules is semisimple. Because finite products are direct sums, a finite product of semisimple modules is semisimple.
(c) Let S_1, S_2 be nonisomorphic simple modules, then the Hasse quiver of the submodule lattice of the semisimple module $S_1 \oplus S_2$ has the form

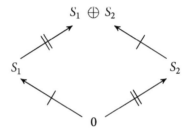

If the two simple modules are isomorphic, however, there exist nonzero submodules of the direct sum which are distinct from S_1 and S_2, see Example II.2.25(a).

(d) The $\mathbb{Z}$-module $\mathbb{Z}/105\mathbb{Z} \cong \mathbb{Z}/3\mathbb{Z} \oplus \mathbb{Z}/5\mathbb{Z} \oplus \mathbb{Z}/7\mathbb{Z}$ is semisimple, because, see Example VII.4.3(c), a $\mathbb{Z}$-module of the form $\mathbb{Z}/p\mathbb{Z}$, with p a prime integer, is simple. We show the Hasse quiver of its submodule lattice.

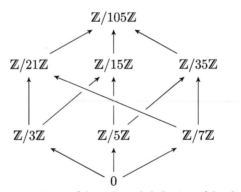

In general, the Hasse quiver of the submodule lattice of the direct sum of n simples has the shape of an n-dimensional cube.

(e) Let A be a principal ideal domain, p a prime integer and $t > 0$ an integer. The uniserial A-module A/p^tA of length t is semisimple if and only if $t = 1$, that is, if and only if it is simple.

(f) Let K be a field and Q the quiver $1 \xleftarrow{\alpha} 2$ and $A = KQ$. The A-module corresponding to the representation $(K \xleftarrow{0} K)$ is semisimple. Indeed, because the linear map between the two copies of K equals zero, we have

$$(K \xleftarrow{0} K) \cong (K \xleftarrow{0} 0) \oplus (0 \xleftarrow{0} K) \cong S_1 \oplus S_2$$

Actually, we prove in Example IX.2.8(c) below that a bound quiver representation is semisimple if and only if the linear maps between its component vector spaces are all zero.

Let S_1, S_2 be distinct simple submodules of a module M. Then $S_1 \cap S_2 = 0$ for, otherwise, $S_1 = S_1 \cap S_2 = S_2$. Hence $S_1 + S_2 = S_1 \oplus S_2$. Thus, a finite sum of simples is direct. We now show that the same statement holds true for arbitrary sums.

VII.5.3 Lemma. *Let $M = \sum_{\lambda \in \Lambda} S_\lambda$, where each S_λ is simple, and N a submodule of M. Then there exists $\Gamma \subseteq \Lambda$ such that $M = N \oplus (\oplus_{\gamma \in \Gamma} S_\gamma)$.*

Proof. Let $\mathcal{E}$ be the set of those subsets $\Gamma \subseteq \Lambda$ such that the sum $N + (\sum_{\gamma \in \Gamma} S_\sigma)$ is direct. Then $\mathcal{E}$ is nonempty because it contains the empty set. Moreover, $\mathcal{E}$, ordered by inclusion, is an inductive set. Because of Zorn's lemma, it has a maximal element Γ such that the sum $M' = N + (\sum_{\gamma \in \Gamma} S_\gamma)$ is direct. We claim that $M' = M$. Because $M = \sum_{\lambda \in \Lambda} S_\lambda$, this amounts to saying that each S_λ is contained in M'. But, if $S_\lambda \nsubseteq M'$, for some λ, then certainly $\lambda \notin \Gamma$. Because S_λ is simple, we get $S_\lambda \cap M' = 0$, see Exercise VII.4.2. Therefore, the sum $M' + S_\lambda$ is direct, which contradicts maximality of Γ. □

Semisimple modules can be characterised in several ways.

VII.5.4 Theorem. *The following conditions are equivalent for an A-module M :*
  (a) *M is semisimple.*
  (b) *M is a direct sum of simple A-modules.*
  (c) *Any submodule of M is a direct summand of M.*
  (d) *Any quotient of M is a direct summand of M.*
  (e) *Any short exact sequence of the form $0 \to L \to M \to N \to 0$ splits.*

Proof. (a) is equivalent to (b). Indeed, Lemma VII.5.3, with $N = 0$, says that (a) implies (b), while the converse is obvious.

(c), (d) and (e) are equivalent. Let L be a submodule of M. It is a direct summand if and only if the inclusion $L \to M$ is a section, or if and only if the short exact sequence $0 \to L \to M \to M/L \to 0$ splits. Thus, (c) is equivalent to (e). Because (d) is the dual of (c), while (e) is selfdual, this proves the equivalence of these three conditions.

Lemma VII.5.3 asserts that (a) implies (c). There remains to prove that (c) implies (a). We claim that any nonzero submodule L of M contains a simple submodule. Indeed, because $L \neq 0$, there exists a nonzero element $x \in L$. Because we are interested in simple *submodules* of L, we may assume that $L = xA \neq 0$. Because of Proposition II.2.23, L has a maximal submodule L'. Now, (c) implies that $M = L' \oplus L''$, for some L''. Because $L' \subseteq L$, the modular law gives:

$$L = (L' \oplus L'') \cap L = L' \oplus (L'' \cap L).$$

Then $L'' \cap L \cong L/L'$ is a simple submodule of L, due to Lemma VII.4.5. This proves our claim.

Let M' be the sum of all simple submodules of M. In particular, M' is semisimple. Because of (c), we have $M = M' \oplus M''$, for some M''. If $M'' \neq 0$, then it should contain a simple submodule S. But then $S \subseteq M'$, due to the definition of the latter, and we get the contradiction $0 = M' \cap M'' \supseteq S \neq 0$. Therefore, $M'' = 0$ and $M = M'$ is semisimple. □

In the course of the proof, we established that a module is semisimple if and only if it is the sum of *its* simple submodules.

VII.5.5 Corollary. *Let* $0 \to L \to M \to N \to 0$ *be a short exact sequence. If M is semisimple, then so are L and N.*

Proof. One can identify L to a submodule of M. Because of Theorem VII.5.4, there exists L' such that $M = L \oplus L'$. Because of Lemma VII.5.3, there exists a set Γ such that $M = L' \oplus (\oplus_{\gamma \in \Gamma} S_\gamma)$. Hence, $L \cong M/L' \cong (\oplus_{\gamma \in \Gamma} S_\gamma)$ is semisimple. The proof is similar for N. □

The converse of this corollary is false, as is seen from the sequence $0 \to \mathbb{Z}/p\mathbb{Z} \to \mathbb{Z}/p^2\mathbb{Z} \to \mathbb{Z}/p\mathbb{Z} \to 0$, with p a prime integer.

Finitely generated semisimple modules coincide with semisimple modules of finite length.

VII.5.6 Lemma. *The following conditions are equivalent for a semisimple module:*
   (a) *It is a finite direct sum of simple modules.*
   (b) *It is of finite length.*
   (c) *It is finitely generated.*

Proof. (a) implies (b). Assume that $M = \oplus_{i=1}^{t} S_i$. Then

$$0 \subsetneq S_1 \subsetneq S_1 \oplus S_2 \subsetneq \cdots \subsetneq \oplus_{i=1}^{t-1} S_i \subsetneq M$$

is a composition series for M. Thus, the module M has finite length t.

(b) implies (c). This follows from Corollary VII.4.12.

(c) implies (a). Assume that M is generated, say, by $\{x_1, \cdots, x_t\}$. Because M is semisimple, there exists a set Λ such that $M = \oplus_{\lambda \in \Lambda} S_\lambda$. For any i, there exists a finite set Λ_i such that $x_i \in \oplus_{\lambda \in \Lambda_i} S_\lambda$. The set $\Lambda' = \cup_{i=1}^{t} \Lambda_i$ is also finite and, for any i, we have $x_i \in \oplus_{\lambda \in \Lambda'} S_\lambda$. Therefore, $M = \oplus_{\lambda \in \Lambda'} S_\lambda$. □

VII.5.2 Semisimple algebras

Following a now familiar pattern, after defining semisimple modules, we consider algebras which are semisimple when viewed as modules over themselves.

VII.5.7 Definition. A K-algebra A is **semisimple** provided A_A is a semisimple A-module.

For instance, if K is a field, the K-module (vector space) K_K is one-dimensional, hence simple, hence semisimple. Thus, fields are semisimple algebras.

Because A_A is finitely generated (actually, it is even cyclic), semisimple algebras are artinian and noetherian, due to Lemma VII.5.6 .

As we shall prove, if A_A is semisimple, then so is $_AA$, so the previous definition is really left-right symmetric. Our objective in this subsection is to characterise semisimple algebras. The following example will turn out to be rather typical.

VII.5.8 Example. Let D be a division ring and $n > 0$ an integer, then $A = M_n(D)$ is a semisimple algebra. Indeed, let $\mathbf{e}_{ij}$ denote as usual the matrix having as coefficient 1 in position (i,j) and 0 elsewhere. Then $\{\mathbf{e}_{ii} : 1 \leqslant i \leqslant n\}$ is a complete set of pairwise orthogonal idempotents for A. Hence, $A = \oplus_{i=1}^n \mathbf{e}_{ii}A$, due to Lemma VII.3.5. It suffices to prove that each $I_i = \mathbf{e}_{ii}A$ is a simple A-module. As D-vector space, we have $I_i \cong D^{(n)}$, where the right action of the matrices of A can be considered as that of the endomorphisms of I_i. Let $x \in I_i$ be nonzero. There exists a D-basis of I_i containing x. For each $y \in I_i$, this basis allows to construct a D-linear map $f : I_i \to I_i$ such that $y = f(x)$. Thus, $xA = I_i$, for any nonzero $x \in I_i$. Because of Lemma VII.4.2, I_i is simple. Compare with Exercise VII.4.7.

VII.5.9 Lemma. *Let $A_1, \cdots, A_t$ be semisimple algebras. Then so is their product* $A = \prod_{i=1}^t A_i$.

Proof. Because of Exercise VII.4.4, the set of simple A-modules is the union of the sets of simple A_i-modules, when the latter are considered as A-modules. The statement follows. $\qquad\square$

Recall that, due to Schur's Lemma VII.4.4, the endomorphism algebra of a simple module is a division ring.

Lemma VII.5.10. *Let A be an algebra, $S_1, \cdots, S_t$ pairwise nonisomorphic simple modules and consider $M = \oplus_{i=1}^t S_i^{n_i}$, where $n_i \geq 1$ are integers. Setting $D_i = \mathrm{End}\,S_i$, we have*

$$\mathrm{End}\,M \cong M_{n_1}(D_1) \times \cdots \times M_{n_t}(D_t)$$

Proof. Due to Lemma VII.4.4, $\mathrm{Hom}_A(S_i, S_j) = 0$, for $i \neq j$. Also, due to Proposition III.3.14,

$$\mathrm{End}(S_i^{n_i}) \cong M_{n_i}(\mathrm{End}\,S_i) = M_{n_i}(D_i).$$

Hence,

$$\mathrm{End}M \;=\; [\mathrm{Hom}_A(S_i^{n_i}, S_j^{n_j})]_{i,j} \;\cong\; \prod_{i=1}^{t} \mathrm{End}(S_i^{n_i}) \;=\; \prod_{i=1}^{t} M_{n_i}(D_i). \qquad \square$$

VII.5.11 Example. Let M be the semisimple $\mathbb{Z}$-module $(\mathbb{Z}/2\mathbb{Z})^3 \oplus (\mathbb{Z}/3\mathbb{Z}) \oplus (\mathbb{Z}/5\mathbb{Z})^2$. Then $\mathrm{End}M \cong M_3(\mathbb{Z}/2\mathbb{Z}) \times (\mathbb{Z}/3\mathbb{Z}) \times M_2(\mathbb{Z}/5\mathbb{Z})$.

The following theorem, due to Wedderburn, characterises semisimple algebras.

VII.5.12 Theorem. *The following conditions are equivalent for an algebra A :*
  (a) *A is semisimple.*
  (b) *Any A-module is semisimple.*
  (c) *Any short exact sequence of A-modules splits.*
  (d) *Any A-module is projective.*
  (e) *Any A-module is injective.*
  (f) *$A \cong \prod_{i=1}^{t} M_{n_i}(D_i)$, where the $t, n_i \geq 0$ are integers and the D_i are division rings, which are extensions of K.*

Proof. (a) is equivalent to (b). If A is semisimple, then so is any free A-module. Due to Corollary VII.5.5 and Proposition III.4.8, so is any A-module. The converse is obvious.

(b) implies (c) because of Theorem VII.5.4, and (c) implies (d) because of Theorem VI.4.4. Conversely, for the same reason, (d) implies (b). Thus (b), (c) and (d) are equivalent. Dually, they are equivalent to (e). There remains to prove the equivalence of (a) and (f).

Assume that A is semisimple. Because A_A is finitely generated, it is a finite direct sum of simple modules $A_A = \bigoplus_{i=1}^{t} S_i$, due to Lemma VII.5.6. Moreover, $D_i = \mathrm{End}S_i$ is, because of Lemma VII.4.4, a division ring containing K. The conclusion (f) follows from Lemma VII.5.10.

Conversely, assume that (f) holds true. Because of Lemma VII.5.9, it is sufficient to prove that each $A_i = M_{n_i}(D_i)$ is semisimple. But this follows from Example VII.5.8. $\qquad \square$

For example, any field K is a semisimple algebra, therefore any K-vector space is at the same time projective and injective, something that we had already observed before.

Because condition (f) in Theorem VII.5.12 is selfdual, the left A-module $_AA$ is also semisimple, hence Definition VII.5.7 is left-right symmetric, as we claimed. When K is algebraically closed and A is finite-dimensional over K, condition (f) of the theorem takes a particularly nice form.

VII.5.13 Corollary. *Let A be a finite-dimensional algebra over an algebraically closed field K. Then A is semisimple if and only if $A \cong \prod_{i=1}^{t} M_{n_i}(K)$, for some integers $t, n_i \geq 0$.*

Proof. As seen in Example I.5.10, K has no proper division ring extension which is finite-dimensional. Therefore, it suffices to prove that the endomorphism algebra of a simple A-module S is finite-dimensional over K. Because of Lemma VII.4.2, S is cyclic, hence, because of Corollary III.4.9, finite-dimensional over K. Therefore, so is its endomorphism algebra. $\square$

The representation theory of finite groups is essentially the module theory of group algebras, see Example I.3.11(f). Hence the interest of the following class of semisimple algebras.

VII.5.14 Theorem. Maschke's Theorem. *Let G be a finite group of order n and K a field. If the characteristic of K does not divide n, then the group algebra KG is semisimple.*

Proof. We show that any submodule L of a KG-module M is a direct summand of M. This implies, because of Theorem VII.5.4, that M is semisimple and gives the desired result because of Theorem VII.5.12.

We claim that the inclusion $q: L \to M$ is a section, and for this, we construct a KG-linear map $p: M \to L$ such that $pq = 1_L$. Because q is K-linear, there exists a K-linear map $p_0: M \to L$ such that $p_0 q = 1_L$. The hypothesis on the characteristic of K implies that $n \cdot 1 \neq 0$ and so it has an inverse in K which we denote by $\frac{1}{n}$. Define p as follows:

$$p: \quad M \longrightarrow L$$
$$x \mapsto \frac{1}{n} \sum_{g \in G} p_0(xg^{-1})g$$

for $x \in M$. It is clear that p is K-linear. To see that p is actually KG-linear, we take $x \in M$ and $g' \in G$ and calculate $p(xg')$:

$$p(xg') = \frac{1}{n} \sum_{g \in G} p_0((xg')g^{-1})g.$$

Writing $g = hg'$, we get $h^{-1} = g'g^{-1}$ and so

$$p(xg') = \frac{1}{n} \sum_{h \in G} p_0(xh^{-1})hg' = \left(\frac{1}{n} \sum_{h \in G} p_0(xh^{-1})h \right) g' = p(x)g'$$

which establishes our claim. Now, if $x \in L$, then

$$pq(x) = \frac{1}{n} \sum_{g \in G} p_0(q(x)g^{-1})g = \frac{1}{n} \sum_{g \in G} (p_0 q)(xg^{-1})g = \frac{1}{n} \sum_{g \in G} (xg^{-1})g = x$$

because the inclusion $q: L \to M$ is KG-linear and $p_0 q = 1_L$. Hence $pq = 1_L$, as required. $\square$

Thus, if K is a field of characteristic 0 and G be a finite group, then KG is semisimple. A consequence of Maschke's theorem is that the representation theory of finite groups splits into two distinct areas: the classical where group algebras are semisimple, and the modular, where the characteristic of the field divides the order of the group.

VII.5.3 Simple algebras

VII.5.15 Definition. A nonzero algebra is **simple** if its only ideals are 0 and itself.

Because of Example I.3.11(b), if D is a division ring containing K, then $M_n(D)$ is simple. We now prove that, conversely, up to isomorphism, any simple artinian K-algebra is of this form.

VII.5.16 Theorem. Wedderbrun–Artin's Theorem. *Let A be a nonzero artinian K-algebra. Then A is simple if and only if there exist a division ring D extension of K and an integer $n > 0$ such that $A \cong M_n(D)$.*

Proof. We only need to prove necessity. Assume that A is simple. Because it is artinian, it contains minimal ideals, hence simple A-submodules. Let S be a simple submodule of A, then $AS = \sum_{a \in A} aS$ is a nonzero ideal of A. Because A is simple, $AS = A$. For any $a \in A$, there exists an epimorphism $S_A \to aS_A$ given by $x \mapsto ax$, for $x \in S$. Because of Lemma VII.4.4, its image aS is either zero or simple, and, in the latter case, isomorphic to S. Therefore, A_A is a semisimple module. Theorem VII.5.12 yields $A \cong \prod_{i=1}^{t} M_{n_i}(D_i)$, where each D_i is a division ring containing K, and t, n_i nonnegative integers. The fact that there are no nonzero ideals in A gives $t = 1$ and completes the proof. $\qquad\square$

A nice consequence of the preceding theorem and Theorem VII.5.12 is that an algebra is semisimple if and only if it is a finite product of simple artinian algebras.

VII.5.17 Corollary. *Let K be an algebraically closed field and A a finite-dimensional K-algebra. Then A is simple artinian if and only if there exists an integer $n > 0$ such that $A \cong M_n(K)$.*

Proof. This follows from Theorem VII.5.16 above and Example I.5.10. $\qquad\square$

We are now in position to give a relatively good description of the module category over a semisimple algebra.

VII.5.18 Proposition. *Let D be a division ring extension of K and $A = M_n(D)$. For any i such that $1 \leqslant i \leqslant n$, set $S_i = \mathbf{e}_{ii}A$. Then we have:*
   (a) *$A_A = \bigoplus_{i=1}^{n} S_i$.*
   (b) *Each S_i is a simple submodule of A.*
   (c) *$S_i \cong S_j$, for any i, j.*

Proof. (a) The $\mathbf{e}_{ii}$ form a complete set of pairwise orthogonal idempotents of $A = M_n(D)$, hence the statement.

(b) Let $\mathbf{x} \in S_i$ be nonzero, then $\mathbf{x} = \sum_{j=1}^n \mathbf{e}_{ij}x_j$, for some $x_j \in D$, and there exists j_0 such that $x_{j_0} \neq 0$. Hence, for any k, we have $\mathbf{x} \cdot \mathbf{e}_{j_0 k} = \mathbf{e}_{ik}x_{j_0}$ or, equivalently, $\mathbf{e}_{ik} = \mathbf{x} \cdot \mathbf{e}_{j_0 k} \cdot x_{j_0}^{-1}$. Therefore, if $\mathbf{y} = \sum_{k=1}^n \mathbf{e}_{ik}y_k \in S_i$ is an arbitrary element, then $\mathbf{y} = \mathbf{x}(\sum_{k=1}^n \mathbf{e}_{j_0 k}x_{j_0}^{-1}y_k) \in \mathbf{x}A$. Thus, $S_i = \mathbf{x}A$ and so S_i is simple because of Lemma VII.4.2.

(c) The morphism $f : S_i \to S_j$ defined by $\mathbf{x} \mapsto \mathbf{e}_{ji}\mathbf{x}$, for $\mathbf{x} \in S_i$, is nonzero, because $f(\mathbf{e}_{ii}) = \mathbf{e}_{ji}\mathbf{e}_{ii} = \mathbf{e}_{ji}$. The statement follows from Lemma VII.4.4. $\qquad\square$

That is, a simple artinian algebra admits, up to isomorphism, only one simple module.

VII.5.19 Corollary. *A semisimple algebra has only finitely many isomorphism classes of simple modules.*

Proof. Indeed, a semisimple algebra is a finite product of simple artinian algebras, each having, up to isomorphism, a unique simple module. $\qquad\square$

Let A be semisimple and $\{S_1, \cdots, S_t\}$ a complete set of representatives of isomorphism classes of simple modules. Due to Lemma VII.5.6, a finitely generated module M is a finite direct sum of simple modules, so we have $M \cong \oplus_{i=1}^t S_i^{n_i}$, where $S_i \not\cong S_j$ if $i \neq j$. Morphisms between two such modules are governed by Lemma VII.4.4: the image of a simple direct summand of the source is zero, or an isomorphic simple direct summand of the target.

Exercises of Section VII.5

1. Find necessary and sufficient conditions on m so that $\mathbb{Z}/m\mathbb{Z}$ is a semisimple $\mathbb{Z}$-module.

2. Prove that a noetherian and artinian module M is semisimple if and only if, for any proper submodule L of M, there exists a submodule $L' \neq 0$ such that $L \cap L' = 0$.

3. Let A be an algebra. Prove that the following conditions are equivalent for a module M:
   (a) M is semisimple.
   (b) For any monomorphism $j : L \to M$ and any morphism $f : L \to L'$, there exists a morphism $f' : M \to L'$ such that $f'j = f$.
   (c) For any epimorphism $p : M \to N$ and any morphism $g : N' \to N$, there exists a morphism $g' : N' \to M$ such that $pg' = g$.

4. Prove that a semisimple module is artinian if and only if it is noetherian.

5. Let A be an algebra, M an A-module and $M_1, \cdots, M_t$ maximal submodules of M. Prove that the quotient $M/(\cap_{i=1}^t M_i)$ is semisimple.

6. Let A be an algebra. Prove that the following conditions are equivalent for a module M:

   (a) M is semisimple of finite length.

   (b) M is noetherian and any maximal submodule of M is a direct summand.

   (c) M is noetherian and any simple submodule of M is a direct summand.

7. Let A be an algebra and M a semisimple module with decompositions $M \cong \oplus_{i=1}^{m} S_i \cong \oplus_{j=1}^{n} S'_j$, where the S_i, S'_j are simple modules, and $m, n \geq 0$ integers. Prove that $m = n$ and there exists a permutation σ of $\{1, \cdots, m\}$ such that $S_i \cong S'_{\sigma(i)}$.

8. Let A be a semisimple algebra and I a proper ideal of A. Show that A/I is semisimple.

9. Give an example showing that subalgebras of semisimple algebras are generally not semisimple.

10. Prove that a semisimple algebra has the same number of isomorphism classes of simple right modules as it has of simple left modules.

11. Prove that the centre of a semisimple algebra A is a product of fields. Prove next that if the centre of A is a field, then A is simple.

12. Let D be a division ring. Prove that two $M_n(D)$-modules which are finite-dimensional are isomorphic if and only if they have the same dimension.

13. Let A be an algebra and $M_n(A)$ the algebra of $n \times n$-matrices over A.

   (a) Prove that any ideal $\mathcal{I}$ of $M_n(A)$ is of the form $M_n(I)$, for some ideal I of A. (Hint: let I be the set of all $(1, 1)$-coefficients of matrices appearing in $\mathcal{I}$, then use the fact that for any matrix $\mathbf{x} = [x_{ij}]$, we have $\mathbf{e}_{ij}\mathbf{x}\mathbf{e}_{k\ell} = x_{jk}\mathbf{e}_{i\ell}$.)

   (b) Prove that the ideal I of (a) is uniquely determined by $\mathcal{I}$.

   (c) Deduce that, if A is simple, then so is $M_n(A)$.

14. Let $(A_\lambda)_{\lambda \in \Lambda}$ be a family of K-algebras. Show that $\prod_{\lambda \in \Lambda} A_\lambda$ is semisimple if and only if Λ is finite and each A_λ is semisimple.

15. Let K be a field. Find necessary and sufficient conditions for a path algebra KQ to be:

   (a) simple.

   (b) semisimple.

16. Let K be a field and Q the quiver $3 \longrightarrow 2 \longrightarrow 1$. Prove that the KQ-module given by the representation $K \xrightarrow{0} K^2 \xrightarrow{0} K$ is semisimple and decompose it as a direct sum of simple representations.

17. Let A be a commutative K-algebra. Show that A is semisimple if and only if it is a finite product of field extensions of K. Deduce that, if K is algebraically closed, any commutative semisimple K-algebra is isomorphic to K^n, for some integer $n \geq 0$.

Chapter VIII
Radicals

VIII.1 Introduction

The Wedderburn–Artin Theorems VII.5.12 and VII.5.16 characterise semisimple algebras and allow to describe their categories of finitely generated modules. However, there are relatively few semisimple algebras, because, as we have seen, only (finite products of) full matrix algebras over division rings are so. For instance, quotients of polynomial algebras, lower triangular matrix algebras and bound quiver algebras are generally not semisimple.

   If we wish to apply results about semisimple algebras to understand modules over an arbitrary algebra A, it seems reasonable to construct, starting from A, a semisimple algebra $\overline{A}$, that is 'closest' to A. Thus, we consider the ideal I of A such that $\overline{A} = A/I$ is the 'largest' semisimple quotient of A. This ideal is called the radical of the algebra. The same reasoning applied to modules leads us to the submodule L of a module M, called the radical of the module, such that M/L is the 'largest' semisimple quotient of M. In other words, the radical measures the defect of semisimplicity. Because algebras are modules over themselves, and K-categories can be thought of as algebras with many identities, there exists a corresponding notion of radical of a K-category. The study of the radical is particularly fruitful when dealing with finitely generated modules over artinian algebras, we shall therefore pay particular attention to this case. Again, the letter K stands for a commutative ring, and our algebras are K-algebras.

VIII.2 Radical and socle of a module

VIII.2.1 The radical

In this section, A denotes a K-algebra. Given an A-module M, we are looking for a submodule L of M such that M/L is the largest semisimple quotient of M. Because of Lemma VII.4.5, the quotient of M by some submodule is simple if and only if that submodule is maximal. But a module may have several (or no) maximal submodules. We look at Example II.2.25(b). The Hasse quiver of the submodule lattice of the abelian group $M = \mathbb{Z}/12\mathbb{Z}$ was computed as

An Introduction to Module Theory. Ibrahim Assem and Flávio U. Coelho, Oxford University Press.
© Ibrahim Assem and Flávio U. Coelho (2024). DOI: 10.1093/9780198904939.003.0009

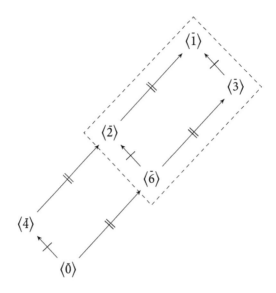

where similar notches indicate isomorphic simple composition factors. One sees that the largest semisimple quotient of M is the one enclosed by dotted lines, namely $M/\langle\bar{6}\rangle \cong (\mathbb{Z}/12\mathbb{Z})/(6\mathbb{Z}/12\mathbb{Z}) \cong \mathbb{Z}/6\mathbb{Z} \cong \mathbb{Z}/2\mathbb{Z} \oplus \mathbb{Z}/3\mathbb{Z}$. The sought submodule L is therefore $\langle\bar{6}\rangle = \langle\bar{2}\rangle \cap \langle\bar{3}\rangle$, that is, the intersection of the (only two) maximal submodules $\langle\bar{2}\rangle$ and $\langle\bar{3}\rangle$ of M. This suggests the definition.

VIII.2.1 Definition. Let A be a K-algebra and M an A-module. The **radical** of M, denoted by $\operatorname{rad} M$, is the intersection of all maximal submodules of M.

Because the radical of a module is the intersection of submodules, it is itself a submodule. If a module M has no maximal submodule, then its radical is the empty intersection, that is, $\operatorname{rad} M = M$. We give some easy examples (more will appear later).

VIII.2.2 Examples.
(a) If a module M is simple, then its only proper submodule (necessarily maximal) is zero, hence $\operatorname{rad} M = 0$.
(b) If a module has a unique maximal submodule, then the latter equals the radical. For instance, if A is a principal ideal domain, $p \in A$ an irreducible element and $t \geq 1$ an integer, then $\operatorname{rad}(A/p^tA) = pA/p^tA \cong A/p^{t-1}A$, because of Lemma V.3.21.
(c) The $\mathbb{Z}$-module $\mathbb{Q}$ has no maximal submodule (so $\operatorname{rad} \mathbb{Q} = \mathbb{Q}$). Indeed, if M is maximal, then $\mathbb{Q}/M$ is simple. So it is isomorphic to $\mathbb{Z}/p\mathbb{Z}$, for some prime integer p. Therefore, there exists a surjective group morphism $f \colon \mathbb{Q} \to \mathbb{Z}/p\mathbb{Z}$. But then, for any $a \in \mathbb{Q}$, we have $f(a) = f(\frac{pa}{p}) = pf(\frac{a}{p}) = 0$, hence $f = 0$, a contradiction which proves our claim.
(d) Let K be a field and Q the quiver $1 \leftarrow 2 \rightarrow 3 \rightarrow 4$. Consider the representation ($= KQ$-module) given by

$$M = e_2A = (K \xleftarrow{1} K \xrightarrow{1} K \xrightarrow{1} K).$$

Because $\dim_K M = 4$, nonzero proper submodules have possible dimensions 1, 2 or 3. An easy calculation shows that, up to isomorphism, there are only five nonzero proper submodules, namely $M_1 = (K \leftarrow 0 \rightarrow 0 \rightarrow 0)$, $M_2 = (0 \leftarrow 0 \rightarrow K \overset{1}{\rightarrow} K), M_3 = (0 \leftarrow 0 \rightarrow 0 \rightarrow K)$ and their direct sums $M_1 \oplus M_3$ and $M_1 \oplus M_2$. Now, $M_1 \oplus M_2$ contains the other submodules, because $M_3 \subseteq M_2$ implies $M_1 \oplus M_3 \subseteq M_1 \oplus M_2$. Because $\dim_K(M_1 \oplus M_2) = 3$, we get that $M_1 \oplus M_2$ is the unique maximal submodule of M. Hence, $\operatorname{rad} M = M_1 \oplus M_2$.

One may reformulate the definition.

VIII.2.3 Lemma. *Let M be an A-module. Then $\operatorname{rad} M$ is the intersection of the kernels of all nonzero morphisms $M \rightarrow S$, with S ranging through the class of simple A-modules.*

Proof. Because of Lemma VII.4.4, any nonzero morphism from M to a simple module is an epimorphism and, because of Lemma VII.4.5, its kernel is maximal. Conversely, the same lemma says that the quotient of M by a maximal submodule is simple. $\qquad\square$

We summarise below the main properties of the radical.

VIII.2.4 Proposition. *Let A be a K-algebra.*
   (a) *Let L, M be A-modules and $f : L \rightarrow M$ an A-linear map, then $f(\operatorname{rad} L) \subseteq \operatorname{rad} M$.*
   (b) *Let M be an A-module and N a submodule of M. Then $\operatorname{rad}(M/N) \supseteq (\operatorname{rad} M + N)/N$.*
   (c) *Let M be an A-module and N a submodule such that $N \subseteq \operatorname{rad} M$. Then $\operatorname{rad}(M/N) = (\operatorname{rad} M)/N$.*
   (d) *Let M be an A-module. Then $\operatorname{rad} M$ is the smallest submodule N of M such that $\operatorname{rad}(M/N) = 0$. In particular, $\operatorname{rad}(M/\operatorname{rad} M) = 0$.*
   (e) *Let $(M_\lambda)_\lambda$ be a family of A-modules, then $\operatorname{rad}(\oplus_\lambda M_\lambda) = \oplus_\lambda \operatorname{rad} M_\lambda$.*

Proof. (a) Let S be a simple module and $g : M \rightarrow S$ an A-linear map. Then $gf : L \rightarrow S$ vanishes over $\operatorname{rad} L$. Therefore, g vanishes over $f(\operatorname{rad} L)$.
   (b) Apply (a) to the projection $p : M \rightarrow M/N$ and use the fact that, because of Exercise II.4.24, $p(\operatorname{rad} M) = (\operatorname{rad} M + N)/N$.
   (c) Due to Theorem II.4.27, the correspondence $L \mapsto L/N$ induces a bijection between the set of maximal submodules of M containing N and the set of maximal submodules of M/N.
   (d) If $\operatorname{rad}(M/N) = 0$, then (b) implies that $\operatorname{rad} M + N = N$, hence $\operatorname{rad} M \subseteq N$. On the other hand, $N = \operatorname{rad} M$ implies, because of (c), that $\operatorname{rad}(M/\operatorname{rad} M) = (\operatorname{rad} M)/(\operatorname{rad} M) = 0$.
   (e) For any λ, the injection $M_\lambda \rightarrow \oplus_\mu M_\mu$ restricts to an injection $\operatorname{rad} M_\lambda \rightarrow \operatorname{rad}(\oplus_\mu M_\mu)$, because of (a). Hence, $\operatorname{rad} M_\lambda \subseteq \operatorname{rad}(\oplus_\mu M_\mu)$. Therefore, $\oplus_\lambda \operatorname{rad} M_\lambda \subseteq \operatorname{rad}(\oplus_\lambda M_\lambda)$, as submodules of the direct sum $\oplus_\lambda M_\lambda$.
   To prove the reverse inclusion, let $x \in \operatorname{rad}(\oplus_\lambda M_\lambda)$ and N_λ a maximal submodule of M_λ. Then $N_\lambda \oplus (\oplus_{\mu \neq \lambda} M_\mu)$ is a maximal submodule of $\oplus_\lambda M_\lambda$, because

the quotient of the latter by the former is the simple module M_λ/N_λ. Thus $x \in N_\lambda \oplus (\oplus_{\mu\neq\lambda}M_\mu)$. This being true for any N_λ, we get $x \in \operatorname{rad}M_\lambda \oplus (\oplus_{\mu\neq\lambda}M_\mu)$. If $x = (x_\lambda)_\lambda$, then $x_\lambda \in \operatorname{rad}M_\lambda$, for any λ, hence $x \in \oplus_\lambda\operatorname{rad}M_\lambda$. □

Part (e) of the proposition and Example VIII.2.2(a) imply that semisimple modules have zero radical.

On the other hand, part (a) means that any morphism $f : L \to M$ gives rise to a commutative square

$$
\begin{array}{ccc}
\operatorname{rad}L & \xrightarrow{\ f'\ } & \operatorname{rad}M \\
{\scriptstyle j_L}\downarrow & & \downarrow{\scriptstyle j_M} \\
L & \xrightarrow{\ f\ } & M
\end{array}
$$

where j_L, j_M are the injections and f' the restriction of f. This is expressed by saying that there exists a functor rad : Mod $A \to$ Mod A which is a **subfunctor of the identity** functor $1_{\mathrm{Mod}\,A}$, see Exercise VIII.2.12 below. This functor preserves monomorphisms: if f is injective, so is $f j_L = j_M f'$, hence so is f'. It does not always preserve epimorphisms, see Exercise VIII.2.3 below. We give a sufficient condition for an epimorphism f to restrict to an epimorphism f'.

VIII.2.5 Corollary. *Let* $0 \to L \to M \xrightarrow{f} N \to 0$ *be a short exact sequence with* $L \subseteq \operatorname{rad}M$. *Then* $f(\operatorname{rad}M) = \operatorname{rad}N$.

Proof. Due to Proposition VIII.2.4, rad N = rad (M/L) = $(\operatorname{rad}M)/L = f(\operatorname{rad}M)$. □

Because of Example VIII.2.2(c), $\mathbb{Q}$ is an abelian group without maximal subgroup and hence $\operatorname{rad}\mathbb{Q} = \mathbb{Q}$. But $\mathbb{Q}$ is infinitely generated. This suggests the following lemma.

VIII.2.6 Lemma. *If M is a nonzero finitely generated A-module, then* $\operatorname{rad}M \neq M$.

Proof. Because of Proposition II.2.23, a nonzero finitely generated module has maximal submodules. □

So, a nonzero finitely generated module A-module M fits as middle term of an exact sequence

$$0 \to \operatorname{rad}M \to M \to M/\operatorname{rad}M \to 0.$$

The quotient $M/\operatorname{rad}M$, is called the **top** of M, and denoted as top$M$. Thus, top$M$ is nonzero, finitely generated and has zero radical (because of Proposition VIII.2.4(d)). We prove in Corollary VIII.3.11 below that if M is artinian, then its top is semisimple.

VIII.2.2 Nakayama's lemma

The main result on radicals of finitely generated modules is known as **Nakayama's lemma**, of which several formulations are known. The one below is perhaps the most popular one.

VIII.2.7 Theorem. Nakayama's lemma. *Let A be a K-algebra, M a finitely generated A-module and L a submodule of M. Then $L \subseteq \text{rad}\, M$ if and only if, for any submodule N of M such that $L + N = M$, we have $N = M$.*

Proof. Necessity. Assume that $L \subseteq \text{rad}\, M$ and $L + N = M$. If $N \neq M$, then, because M is finitely generated, Proposition II.2.23 yields a maximal submodule N' containing N. Because of definition of radical, $L \subseteq N'$. Hence, $M = L + N \subseteq L + N' = N' \subsetneq M$, a contradiction.

Sufficiency. If $L \nsubseteq \text{rad}\, M$, then the definition of radical implies the existence of a maximal submodule M' of M such that $L \nsubseteq M'$. But then $L + M' = M$ and the hypothesis yield $M' = M$, a contradiction. $\qquad\square$

In particular, $N + \text{rad}\, M = M$ implies $N = M$.

This result is expressed by saying that the radical of M is a *superfluous* submodule of M, in the sense that it is 'not needed' to generate M. In order to express this property in terms of morphisms, we dualise the statement of the lemma '$L + N = M$ implies $N = M$' inside the submodule lattice of M. This gives '$L \cap N = 0$ implies $N = 0$', or, equivalently '$N \neq 0$ implies $L \cap N \neq 0$'. As seen in Lemma VI.4.32, this dual statement says that the embedding $L \to M$ is an essential monomorphism (or that L is an essential submodule of M). This leads us to dualise Definition VI.4.30 of essential monomorphism.

VIII.2.8 Definition. Let L, M be A-modules. An epimorphism $f : L \to M$ is *superfluous* if, whenever fg is an epimorphism, then g itself is an epimorphism.

For instance, the identity is a superfluous epimorphism. We can easily dualise Lemma VI.4.31 (and its proof).

VIII.2.9 Lemma. *Let $f : L \to M, g : M \to N$ be epimorphisms of A-modules.*
   (a) *If f, g are superfluous, then so is gf.*
   (b) *If gf is superfluous, then so is f.*

Proof. (a) Assume that $(gf)h$ is an epimorphism. Because g is superfluous and $g(fh) = (gf)h$ then fh is an epimorphism. But f is itself superfluous, hence h is an epimorphism.

(b) Assume that fh is an epimorphism. Because g is an epimorphism, so is $g(fh) = (gf)h$. But gf is superfluous. Hence, h is an epimorphism. $\qquad\square$

As expected from the heuristic argument preceding Definition VIII.2.8, we get an equivalent form of Nakayama's Lemma VIII.2.7.

VIII.2.10 Corollary. *Let M be a nonzero finitely generated A-module. Then $L \subseteq \operatorname{rad} M$ if and only if the projection $p : M \to M/L$ is a superfluous epimorphism.*

Proof. Necessity. Assume that $L \subseteq \operatorname{rad} M$ and that pg is an epimorphism. Therefore, for any $m \in M$, there exists x in the domain of g such that $p(m) = pg(x)$. Then $m - g(x) \in \operatorname{Ker} p = L$ so that $M = \operatorname{Im} g + L$. Theorem VIII.2.7 gives $M = \operatorname{Im} g$, that is, g is an epimorphism.

Sufficiency. Assume that p is a superfluous epimorphism and $L + N = M$ for some submodule N of M. Let $j : N \hookrightarrow M$ be the inclusion. Then, for any $m \in M$, there exist $n \in N, \ell \in L$ such that $m = j(n) + \ell$. Hence, $p(m) = pj(n) + p(\ell) = pj(n)$. Because p is surjective, this shows that also pj is surjective. Superfluousness of p implies that j is surjective, so $N = M$. Then Theorem VIII.2.7 implies that $L \subseteq \operatorname{rad} M$ and the proof is complete. $\qquad\square$

Accordingly, a submodule L of M is superfluous if and only if the projection $p : M \to M/L$ is a superfluous epimorphism (this is the dual of Lemma VI.4.32). In particular, the projection $p_M : M \to M/\operatorname{rad} M$ is a superfluous epimorphism. Therefore, the radical of a module is its largest superfluous submodule.

Let $f : L \to M$ be a morphism between A-modules. Because $f(\operatorname{rad} L) \subseteq \operatorname{rad} M$, due to Proposition VIII.2.4(a), there exists a morphism $\bar{f} : L/\operatorname{rad} L \to M/\operatorname{rad} M$ from the top of L to that of M, deduced by passing to cokernels in the commutative diagram with exact rows

$$
\begin{array}{ccccccccc}
0 & \longrightarrow & \operatorname{rad} L & \overset{i}{\longrightarrow} & L & \overset{p_L}{\longrightarrow} & L/\operatorname{rad} L & \longrightarrow & 0 \\
 & & \downarrow f & & \downarrow f & & \downarrow \bar{f} & & \\
0 & \longrightarrow & \operatorname{rad} M & \overset{j}{\longrightarrow} & M & \overset{p_M}{\longrightarrow} & M/\operatorname{rad} M & \longrightarrow & 0
\end{array}
$$

where i, j are the inclusions, see Proposition II.4.13. This morphism is explicitly given by

$$ \bar{f}(x + \operatorname{rad} L) = \bar{f}p_L(x) = p_M f(x) = f(x) + \operatorname{rad} M $$

for $x \in L$. The following corollary says that if one wants to check whether a morphism with finitely generated codomain is surjective or not, it suffices to check it on the level of the tops.

VIII.2.11 Corollary. *Let $f : L \to M$ be an A-linear map between A-modules with M finitely generated. Then f is an epimorphism if and only if so is $\bar{f}$.*

Proof. If f is an epimorphism, then so is $\bar{f}$, because of Proposition II.4.13. Conversely, if $\bar{f}$ is an epimorphism, then so is $\bar{f}p_L = p_M f$. Because of Corollary VIII.2.10, p_M is superfluous, hence f is an epimorphism. $\qquad\square$

Actually, an epimorphism with finitely generated source is superfluous if and only if it induces an isomorphism of the tops.

VIII.2.12 Corollary. *Let L be a finitely generated module. An epimorphism $f : L \to M$ is superfluous if and only if $\bar{f}$ is an isomorphism.*

Proof. *Necessity.* Because L is finitely generated and f an epimorphism, M is also finitely generated. Corollary VIII.2.11 gives that $\bar{f}$ is an epimorphism. We prove that it is a monomorphism. Let $x \in L$ be such that $\bar{f}(x + \operatorname{rad} L) = 0$. The explicit expression of $\bar{f}$, see the comments before Corollary VIII.2.11, gives $f(x) \in \operatorname{rad} M$. Because of superfluousness of f, Corollary VIII.2.5 gives $f(\operatorname{rad} L) = \operatorname{rad} M$ so there exists $y \in \operatorname{rad} L$ such that $f(x) = f(y)$, that is, $x - y \in \operatorname{Ker} f$. Again, because f is superfluous, $\operatorname{Ker} f \subseteq \operatorname{rad} L$. Therefore, $x = (x - y) + y$ lies in $\operatorname{rad} L$.

Sufficiency. Assume that g is such that fg is an epimorphism. Then $\bar{f}p_{Lg} = p_M(fg)$ is an epimorphism. But $\bar{f}$ is an isomorphism, hence p_{Lg} is an epimorphism. Because of Corollary VIII.2.10, p_L is superfluous. Hence, g is an epimorphism, as required. □

VIII.2.3 The socle of a module

The radical of a module M is the submodule $\operatorname{rad} M$ of M such that the quotient, namely the top, $M/\operatorname{rad} M$ is the largest semisimple quotient of M. The socle of M is the dual notion to the top: it is the largest semisimple submodule of M.

VIII.2.13 Definition. Let M be an A-module. The **socle** of M, denoted by $\operatorname{soc} M$, is the sum of all simple (minimal) submodules of M.

Thus, a module is semisimple if and only if it equals its socle. If a module M has no minimal submodule, then its socle is the empty sum, that is, $\operatorname{soc} M = 0$. Actually, $\operatorname{soc} M \neq 0$ if and only if M admits a minimal submodule.

VIII.2.14 Examples.
   (a) If A is a principal ideal domain, then the module A_A has no minimal submodule, see Example VII.4.3(d). Hence, the socle is the empty sum, that is, $\operatorname{soc} A = 0$. In particular, $\operatorname{soc} \mathbb{Z} = 0$. Also, $\operatorname{soc} K[t] = 0$.
   (b) If M is the abelian group $\mathbb{Z}/12\mathbb{Z}$, then it follows from its submodule lattice, see the beginning of Subsection VIII.2.1, that its socle equals $\langle \bar{4} \rangle \oplus \langle \bar{6} \rangle \cong 2\mathbb{Z}/12\mathbb{Z} \cong \mathbb{Z}/6\mathbb{Z}$.
   (c) If a module has a unique simple submodule, then the latter equals the socle. For instance, if A is a principal ideal domain, p an irreducible element and $t > 0$, then $\operatorname{soc}(A/p^t A) = p^{t-1}A/p^t A \cong A/pA$, due to Lemma V.3.21.
   (d) In Example VIII.2.2(d), each of M_1 and M_3 is simple. Therefore, $\operatorname{soc} M = M_1 \oplus M_3$.

Many properties of the socle are dual to those of the radical. We start by dualising Lemma VIII.2.3, thus reformulating the definition.

VIII.2.15 Lemma. *Let M be an A-module. Its socle* soc *M is the sum of the images of all morphisms S → M, with S ranging through the class of simple A-modules.*

Proof. Indeed, the images of such morphisms coincide with the simple submodules of *M*.   □

We summarise the main properties of the socle.

VIII.2.16 Proposition.
  (a) *Let L, M be A-modules and f : L → M an A-linear map, then f(soc L) ⊆ soc M.*
  (b) *Let M be an A-module and L a submodule of M. Then* soc *L = L ∩ soc M. In particular,* soc (soc *M*) = soc *M.*
  (c) *Let M be an A-module, then* soc *M is the intersection of all submodules L of M such that the inclusion L → M is an essential monomorphism.*
  (d) *Let $(M_\lambda)_\lambda$ be a family of A-modules, then* soc $(\oplus_\lambda M_\lambda) = \oplus_\lambda$ soc M_λ.

Proof. (a) The image under *f* of a simple submodule of *L* is either zero or a simple submodule of *M* (actually, an isomorphic simple).

(b) Applying (a) to the injection *L* ↪ *M* yields soc *L* ⊆ soc *M*. Hence, we have soc *L* ⊆ *L* ∩ soc *M*. Conversely, *L* ∩ soc *M* is a submodule of the semisimple module soc *M*, hence is semisimple. But *L* ∩ soc *M* is also a submodule of *L*. Hence, we have *L* ∩ soc *M* ⊆ soc *L*. Equality follows.

(c) Let *N* denote this intersection. For any submodule *L* of *M* such that the inclusion *L* ↪ *M* is an essential monomorphism, and any simple submodule *S* of *M*, we have *S* ∩ *L* ≠ 0, due to Lemma VI.4.32. Hence, *S* ⊆ *L*. Therefore, soc *M* ⊆ *L* for any *L*, and thus soc *M* ⊆ *N*.

In order to prove the reverse inclusion, it suffices to prove that *N* is semisimple. For this purpose, we prove that an arbitrary submodule *N'* of *N* is a direct summand of *N*, then we apply Theorem VII.5.4. Using Zorn's lemma, there exists a submodule *N''* of *N* maximal for the property *N'* ∩ *N''* = 0. Then *N'* + *N''* = *N'* ⊕ *N''* is a submodule of *N*.

We claim that the inclusion of *N'* ⊕ *N''* into *M* is essential. Let *U* be a nonzero submodule of *N* such that *U* ∩ (*N'* ⊕ *N''*) = 0. If *N'* ∩ (*N''* + *U*) ≠ 0, then this intersection contains a nonzero element *x'* = *x''* + *u*, with *x'* ∈ *N'*, *x''* ∈ *N''*, *u* ∈ *U*. Thus, *x'* − *x''* = *u* ∈ *U* ∩ (*N'* ⊕ *N''*). Hence, *u* = 0. So *x'* = *x''* ∈ *N'* ⊕ *N''*, that is, *x'* = *x''* = 0, a contradiction. Hence, *N'* ∩ (*N''* + *U*) = 0, which contradicts the maximality of *N''*. Our claim is established.

Because *N* is the smallest essential submodule of *M*, we have *N'* ⊆ *N* ⊆ *N'* ⊕ *N''*. The modular law, Proposition II.2.18, gives

$$N = N \cap (N' \oplus N'') = N' \oplus (N \cap N'')$$

and *N'* is indeed a direct summand of *N*, which is therefore semisimple, due to Theorem VII.5.4. Hence, *N* ⊆ soc *M* and equality follows.

(d) This is an easy exercise left to the reader.   □

Essentiality of an injection $L \hookrightarrow M$ being dual to superfluousness of a projection $M \to M/L$, the next statement corresponds to Nakayama's lemma for the radical. While in Nakayama's lemma, we needed to assume M finitely generated to ensure the existence of maximal modules, here, we need to assume M artinian to ensure existence of minimal, that is, simple, submodules.

VIII.2.17 Theorem. *Let M be a nonzero artinian module. Then soc M is the smallest submodule L of M such that the inclusion of L into M is an essential monomorphism.*

Proof. Due to Proposition VIII.2.16(c), it suffices to prove that soc $M \ne 0$ and the inclusion $j_M :$ soc $M \to M$ is essential. Because M is artinian, it has simple submodules, so soc $M \ne 0$. On the other hand, let L be a nonzero submodule of M. Because of Theorem VII.2.11, L itself is artinian, hence soc $L \ne 0$. Proposition VIII.2.16(b) implies that $L \cap$ soc $M \ne 0$. Therefore, j_M is essential and the proof is complete. □

Therefore, if M is a finitely generated module over an artinian algebra, then:
(a) the injection $j_M :$ soc $M \to M$ is an essential monomorphism, and
(b) the projection $p_M : M \to M/\mathrm{rad}\,M = \mathrm{top}M$ is a superfluous epimorphism.
Let $f : L \to M$ be an A-linear map between A-modules. Because of Proposition VIII.2.16(a), $f(\mathrm{soc}\,L) \subseteq$ soc M so we have a commutative square:

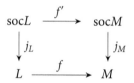

where the vertical maps are injections and f' is the restriction of f. In other words, the socle defines a subfunctor of the identity. We deduce the dual of Corollary VIII.2.11: a morphism with artinian domain is injective if and only if it remains so when restricted to the socle. Also, the proof is dual.

VIII.2.18 Corollary. *Let $f : L \to M$ be a morphism with L artinian. Then f is a monomorphism if and only if its restriction f' to soc L is a monomorphism.*

Proof. If f is a monomorphism, then so is $fj_L = j_M f'$. Hence, so is f'. Conversely, if f' is a monomorphism, then so is $j_M f' = fj_L$. But j_L is essential, because L is artinian. Hence, f is a monomorphism. □

In the artinian case, the socle of a module determines its injective envelope.

VIII.2.19 Corollary. *Let M be a nonzero artinian module. Then the injective envelopes of M and soc M are isomorphic.*

Proof. We claim that the injective envelope (E, e) of soc M is isomorphic to the injective envelope of M. It suffices to construct an essential monomorphism from M to E. We consider the diagram with exact row

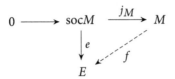

where j_M is the injection. Because E is injective, there exists $f : M \to E$ such that $fj_M = e$. Now, e is a monomorphism and j_M is an essential monomorphism. Therefore, f is a monomorphism. But e is also essential. Lemma VI.4.31(b) gives that f is essential.   □

VIII.2.20 Example. Let I be an artinian injective module having simple socle S. Because the inclusion $S \hookrightarrow I$ is essential, I is an injective envelope of S.

Exercises of Section VIII.2

1. Let A be an algebra and $M = xA$ a cyclic module. Prove that $y \in \operatorname{rad} M$ if and only if $M = (x + ya)A$, for any $a \in A$.

2. Let A be an algebra. We wish to prove that a module has zero radical if and only if it is a submodule of a product of simple modules. Let $(M_\lambda)_\lambda$ be a family of A-modules.
   (a) Prove that $\operatorname{rad}(\prod_\lambda M_\lambda) \subseteq \prod_\lambda \operatorname{rad} M_\lambda$.
   (b) Let all M_λ be simple. Prove that any submodule M of $\prod_\lambda M_\lambda$ has zero radical.
   (c) Conversely, assume that M has zero radical. Prove that M is a submodule of the product of the M/N_λ, where the N_λ are all maximal submodules of M.

3. Give an example of an epimorphism $f : L \to M$ such that $f(\operatorname{rad} L) \neq \operatorname{rad} M$. Prove that equality holds if $L/\operatorname{rad} L$ is semisimple.

4. Let A be an algebra and $(M_\lambda)_\lambda$ a family of modules. Prove that $\operatorname{top}(\oplus_\lambda M_\lambda) = \oplus_\lambda \operatorname{top} M_\lambda$.

5. Let A be an algebra. Prove that the socle of a module M is essential in M if and only if any nonzero submodule of M contains a minimal submodule.

6. Let M be a nonzero finitely generated abelian group. Prove that:
   (a) soc M is the subgroup generated by the elements of prime order.
   (b) If M is torsion-free, then soc $M = 0$.
   (c) If M is torsion, then soc M is essential in M.
   (d) If M is divisible, then $\operatorname{rad} M = M$.

7. Compute the socle, the radical and the top of each of the following abelian groups: $\mathbb{Z}, \mathbb{Z}/p^2q^3\mathbb{Z}$, where p, q are distinct prime integers, then $\mathbb{Z}/n\mathbb{Z}$, for an integer $n > 0$.

8. Let p be a prime integer. Prove the following properties of the Prüfer p-group $\mathbb{Z}(p^\infty)$.
    (a) The socle of $\mathbb{Z}(p^\infty)$ is cyclic of order p.
    (b) $\mathbb{Z}(p^\infty)$ is the injective envelope of $\mathbb{Z}/p\mathbb{Z}$ (Hint: see Exercise VII.2.6).
    (c) In the notation of Example VII.2.8(d), $\mathbb{Z}(p^\infty)$ is an essential extension of M_k, for any k.
    (d) Consider the subgroup $\mathbb{Z}_{(p)} = \{\frac{a}{b} \in \mathbb{Q} : p \nmid b\}$ of $\mathbb{Q}$ (actually, it is a subalgebra). The inclusion of $\mathbb{Z}[1/p]$ into $\mathbb{Q}$ induces a monomorphism $f : \mathbb{Z}(p^\infty) \to \mathbb{Q}/\mathbb{Z}_{(p)}$, and, actually, f is an isomorphism.

9. Let K be a field and A the path algebra of the Kronecker quiver

Compute the socle, the radical and the top of each of the following modules $P_1 = e_1A, P_2 = e_2A, I_1 = D(Ae_1)$ and $I_2 = D(Ae_2)$, in the notation of Examples VI.4.6(e) and VI.4.16(b). Deduce the Hasse quiver of the submodule lattice of each.

10. Let A be an algebra, M a module and L a submodule of M. Prove that:
    (a) $L = \operatorname{rad} M$ if and only if $L \subseteq \operatorname{rad} M$ and $\operatorname{rad}(M/L) = 0$.
    (b) $L = \operatorname{soc} M$ if and only if $L \supseteq \operatorname{soc} M$ and $\operatorname{soc} L = L$.
    (c) If L is superfluous in M and $\operatorname{rad}(M/L) = 0$, then $L = \operatorname{rad} M$.
    (d) If L is essential in M and $\operatorname{soc} L = L$, then $L = \operatorname{soc} M$.

11. Let A be an algebra and C a class of A-modules. We define its **trace** on a module M by
$$\operatorname{tr}_C M = \sum \{\operatorname{Im} f : f \in \operatorname{Hom}_A(X, M), X \in C\}.$$

Prove that:
    (a) $\operatorname{soc} M = \operatorname{tr}_S M$, where S denotes the class of simple A-modules.
    (b) $\operatorname{tr}_C M$ is the unique largest submodule L of M such that there exists an epimorphism $X \to L$, with X a direct sum of modules of C. We say that C **generates** L.
    (c) C generates M if and only if $\operatorname{tr}_C M = M$.
    (d) Let $f : L \to M$ be an A-linear map between A-modules, then we have $f(\operatorname{tr}_C L) \subseteq \operatorname{tr}_C M$.
    (e) If, in (d), f is a monomorphism and $\operatorname{tr}_C M \subseteq \operatorname{Im} f$, then $f(\operatorname{tr}_C L) = \operatorname{tr}_C M$.
    (f) If $(M_\lambda)_\lambda$ is a family of modules, then $\operatorname{tr}_C(\oplus_\lambda M_\lambda) = \oplus_\lambda \operatorname{tr}_C M_\lambda$.
    (g) If $A = \mathbb{Z}$ and $C = \{\mathbb{Z}/n\mathbb{Z} : n > 1\}$, then, for any abelian group M, the trace $\operatorname{tr}_C M$ equals the torsion subgroup $T(M)$ of M.

12. Let $F : \text{Mod } A \rightarrow \text{Mod } A$ be a subfunctor of the identity, that is, a functor which associates to each module M a submodule $FM \subseteq M$ and to each morphism $f : L \rightarrow M$ its restriction $Ff : FL \rightarrow FM$. Prove that:
   (a) The class $\mathcal{T}_F = \{M : FM = M\}$ is stable for quotients and direct sums.
   (b) The class $\mathcal{F}_F = \{M : FM = 0\}$ is stable for submodules and products.
   (c) If $L \in \mathcal{T}_F$ and $M \in \mathcal{F}_F$, then $\text{Hom}_A(L, M) = 0$.
   (d) If F is such that $F(M/FM) = 0$, for any M and $L \subseteq M$, then $F(M/L) \cong FM/L$.
   (e) The following conditions are equivalent for F :
       (i) F is left exact.
       (ii) If $L \subseteq M$, then $FL = (FM) \cap L$.
       (iii) $F \circ F = F$ and $\mathcal{T}_F$ is stable for submodules.
   (f) $\text{soc} : \text{Mod } A \rightarrow \text{Mod } A$ is a left exact subfunctor of the identity.

VIII.3 Radicals of algebras

VIII.3.1 The radical of an algebra

Let A be a K-algebra. The module A_A has a radical in the sense of Section VII.2, which we call the radical of the algebra.

VIII.3.1 Definition. Let A be a K-algebra. The *radical* or *Jacobson radical* $J = \text{rad}A$ of A is the intersection of its maximal right ideals.

The letter J stands for Jacobson who studied it systematically. The radical as defined should be called the *right* radical of A, the left radical being that of the left A-module $_AA$. But we prove below that this definition is left-right symmetric, that is, $\text{rad}(A_A) = \text{rad}(_AA)$, so ambiguity will disappear.

VIII.3.2 Lemma. *The radical of an algebra is a proper ideal.*

Proof. Let A be a K-algebra with radical J. Because J is the intersection of right ideals, it is itself a right ideal. In order to prove that it is also a left ideal, let $a \in A$. The map $f_a : A_A \rightarrow A_A$ defined by $x \mapsto ax$, for $x \in A$, is A-linear. Proposition VIII.2.4(a) yields

$$aJ = f_a(J) \subseteq J.$$

Therefore, J is an ideal of A.
   Because A_A is finitely generated (even cyclic), then $J \subsetneq A_A$, due to Lemma VIII.2.6.   □

The lemma implies that no invertible element of A belongs to J : indeed, if $x \in J$ is invertible, then $1 = xx^{-1} \in J$, and $J = A$, a contradiction.

VIII.3.3 Examples.

(a) Assume that $A = \mathbb{Z}$. Because of Theorem I.5.7, an ideal of $\mathbb{Z}$ is maximal if and only if it is of the form $p\mathbb{Z}$, with p prime (=irreducible). Then $\mathrm{rad}\,\mathbb{Z}$ is the intersection of all $p\mathbb{Z}$, with p running through all primes. An element in this intersection is divisible by all prime integers, hence can only be zero. Thus, $\mathrm{rad}\,\mathbb{Z} = 0$.

(b) Let n be a positive integer. In Example I.3.11(b), we proved that the full matrix algebra $M_n(D)$ over a division ring D (or a field K) has no nonzero proper ideal. Hence, $\mathrm{rad}\,M_n(D) = 0$.

(c) Let K be a field and A the lower triangular matrix algebra

$$A = \begin{pmatrix} K & 0 \\ K & K \end{pmatrix} = \left\{ \begin{pmatrix} \alpha & 0 \\ \beta & \gamma \end{pmatrix} : \alpha, \beta, \gamma \in K \right\}.$$

Because A is a three-dimensional vector space, its proper submodules (=right ideals) have dimension two or one. It is then easy to show, for instance, using Exercise II.2.11, that the maximal right ideals of A are

$$I_1 = \begin{pmatrix} K & 0 \\ K & 0 \end{pmatrix} \text{ and } I_2 = \begin{pmatrix} 0 & 0 \\ K & K \end{pmatrix}.$$

Therefore,

$$\mathrm{rad}\,A = I_1 \cap I_2 = \begin{pmatrix} 0 & 0 \\ K & 0 \end{pmatrix} = \left\{ \begin{pmatrix} 0 & 0 \\ \beta & 0 \end{pmatrix} : \beta \in K \right\}.$$

that is, it consists of the off-diagonal elements of A.

We set out to prove that the definition of the radical is left-right symmetric. This requires characterising the radical by means of its elements.

VIII.3.4 Theorem. *Let A be a K-algebra of radical J. Then $x \in J$ if and only if $1 - xa$ has a right inverse for any $a \in A$.*

Proof. *Necessity.* Assume that $x \in J$. If, for some $a \in A$, the element $1 - xa$ has no right inverse, then the submodule $(1 - xa)A$ of A is proper. Because A is finitely generated, there exists a maximal right ideal R such that $(1 - xa)A \subseteq R \subsetneq A$, due to Proposition II.2.23. On the other hand, $x \in J$ implies $x \in R$ and we get the contradiction $1 = xa + (1 - xa) \in R$.

Sufficiency. Assume that $x \in A$ is such that $1 - xa$ has a right inverse, for any $a \in A$. If $x \notin J$, then there exists a maximal right ideal R of A such that $x \notin R$. Because R is maximal, we have $R + xA = A$. Hence, there exists $a \in A$ such that $1 - xa \in R$, and so $(1 - xa)A \subseteq R \subsetneq A$. But then $1 - xa$ is not right invertible, a contradiction. Therefore, $x \in J$. $\qquad\square$

VIII.3.5 Examples.

(a) Let K be a field, then rad $K[t] = 0$. For, if $p \neq 0$ lies in the radical, then p is not constant (otherwise, it would be invertible), hence $\deg(1 - tp) > 1$. The previous theorem says that $1 - tp$ is (right) invertible in $K[t]$, hence a constant, a contradiction. The conclusion rad $K[t] = 0$ also follows from the fact that the intersection of the ideals of the form $\langle t - \lambda \rangle$, for some $\lambda \in K$, all of which are maximal, is zero.

(b) Let A be an algebra. The only idempotent in its radical J is zero. Indeed, let $e \in J$ be idempotent. Then $1 - e$ has a right inverse x, say, and $(1 - e)x = 1$ gives $x = 1 + ex$. Left-multiplication by e yields $ex = e + ex$ because e is idempotent. Hence, $e = 0$.

VIII.3.6 Corollary. *Let A be a K-algebra. Its radical J is the largest ideal I such that $1 - x$ is invertible, for any $x \in I$.*

Proof. We prove first that $x \in J$ implies $1 - x$ invertible. Because of Theorem VIII.3.4, $1 - x$ has a right inverse y, say. Then $(1 - x)y = 1$ implies $z = 1 - y = -xy \in J$, so that $1 - z$ has a right inverse y'. But $yy' = (1 - z)y' = 1$ shows that y has both a left and a right inverse. Therefore, it is invertible. Hence, so is $y^{-1} = 1 - x$.

We prove next that J is the largest ideal satisfying this condition. Let I be an ideal satisfying this condition and $x \in I$. Then $xa \in I$, for any $a \in A$. Hence, $1 - xa$ is invertible. Theorem VIII.3.4 gives $x \in J$. Hence, $I \subseteq J$. ☐

The condition of Corollary VIII.3.6 is left-right symmetric, Therefore, $\mathrm{rad}(A_A) = \mathrm{rad}(_AA)$, as claimed. We deduce further characterisations of the radical.

VIII.3.7 Corollary. *Let A be a K-algebra. Its radical J equals:*

(a) *The intersection of all maximal left ideals of A.*

(b) *The set of all $x \in A$ such that $1 - ax$ is left invertible, for any $a \in A$.* ☐

Another consequence of Corollary VIII.3.6 is the following relation between radicals of two algebras.

VIII.3.8 Corollary. *Let $\varphi : A \to B$ be a surjective algebra morphism. Then $\varphi(\mathrm{rad}\, A) \subseteq \mathrm{rad}\, B$.*

Proof. Because φ is surjective, the image $\varphi(\mathrm{rad}\, A)$ is an ideal in B, see Exercise I.4.4. Let $x \in \varphi(\mathrm{rad}\, A)$. There exists $a \in \mathrm{rad}\, A$ such that $x = \varphi(a)$. Due to Corollary VIII.3.6, $1 - a$ is invertible in A. Hence, $1 - x = \varphi(1 - a)$ is invertible in B. Another application of Corollary VIII.3.6 shows that the ideal $\varphi(\mathrm{rad}\, A)$ is contained in rad B. ☐

The statement of the corollary is false if φ is not surjective (think of the inclusion of $T_2(K)$ in $M_2(K)$, for K a field).

An element $x \in A$ is called **nilpotent** if there exists a positive integer m such that $x^m = 0$. The smallest such m is called the **nilpotency index** of x. An ideal of A is said to be a **nil ideal** if any element in it is nilpotent.

VIII.3.9 Corollary. *Let A be a K-algebra. Any nil ideal of A is contained in the radical.*

Proof. Let I be a nil ideal and $x \in I$. There exists a positive integer m such that $x^m = 0$. Then $1 - x$ is invertible because we have $(1 - x)(1 + x + \cdots + x^{m-1}) = (1 + x + \cdots + x^{m-1})(1 - x) = 1$. Because of Corollary VIII.3.6, x belongs to the radical J of A. Hence, $I \subseteq J$. $\qquad\square$

It does not follow from this corollary that the radical itself is nil. However, we prove below, in Corollary VIII.3.19, that this holds true for artinian algebras. The corollary does not imply either that all nilpotent elements lie in the radical: there may be nilpotent elements belonging to no nil ideal. For example, if K is a field and $n > 1$ an integer, then the matrix algebra $M_n(K)$ contains nonzero nilpotent elements but has zero radical, because it has no nonzero ideal, see Example I.3.11(b).

Examples of nil ideals are nilpotent ideals. An ideal I is called **nilpotent** if there exists an integer $m > 0$ such that $I^m = 0$, that is, any product of m (or more) elements of I vanishes. This implies that, for any $x \in I$, we have $x^m = 0$. Therefore, any nilpotent ideal is nil and hence contained in the radical. In Example VIII.3.3(c), the radical

$$J = \begin{pmatrix} 0 & 0 \\ K & 0 \end{pmatrix}$$

is nilpotent: we have $J^2 = 0$. On the other hand, nil ideals are generally not nilpotent, see Exercise VIII.3.12, but also Corollary VIII.3.19.

VIII.3.2 Artinian modules and algebras

We pointed out in Section VIII.2 that the radical and the socle of a module behave better when we deal with artinian and/or finitely generated modules. This subsection is devoted to these cases. To start with, recall that it follows from Proposition VIII.2.4(e) that semisimple modules have zero radical. The converse holds true in the artinian case.

VIII.3.10 Theorem. *A module is semisimple of finite length if and only if it is artinian with zero radical.*

Proof. Necessity. Semisimple modules have zero radical and, because of Lemma VII.4.7, any module of finite length is artinian.

Sufficiency. Let M be an artinian module with $\operatorname{rad} M = 0$. We may assume that $M \neq 0$. There exists a nonempty set $(N_\lambda)_\lambda$ of maximal submodules of M such that $\cap_\lambda N_\lambda = 0$. Because M is artinian, there exists a submodule $N = N_1 \cap \ldots \cap N_t$ minimal in the family of finite intersections of the N_λ.

We claim that $N = 0$. If not, then there exists an index μ such that $N \not\subseteq N_\mu$: for, if $N \subseteq N_\lambda$, for any λ, then $N \subseteq \cap_\lambda N_\lambda = 0$, a contradiction. For such an index μ, we have $N \cap N_\mu \subsetneq N$, which contradicts minimality of N. This proves our claim.

Define $f : M \to \oplus_{i=1}^t (M/N_i)$ by $x \mapsto (x + N_i)_{i=1}^t$, for $x \in M$. Then $\mathrm{Ker}\, f = \cap_{i=2}^t N_i = N = 0$, so f is a monomorphism. Therefore, M is isomorphic to a submodule of the semisimple module of finite length $\oplus_{i=1}^t (M/N_i)$. It is therefore itself semisimple of finite length, due to Corollary VII.5.5 and Proposition VII.4.14. $\square$

VIII.3.11 Corollary. *Let M be an artinian module, then* $\mathrm{top}\, M = M/\mathrm{rad}\, M$ *is semisimple of finite length.*

Proof. The top of an artinian module is artinian, because of Theorem VII.2.11, and Proposition VIII.2.4(d) says that it has zero radical. $\square$

VIII.3.12 Corollary. *An algebra is semisimple if and only if it is artinian with zero radical.*

Proof. Sufficiency follows from Theorem VIII.3.10, so we prove necessity. Let A be a semisimple algebra. Because A_A is a cyclic module, it is of finite length, due to Lemma VII.5.6. On the other hand, A_A is a semisimple module. Theorem VIII.3.10 gives A artinian with zero radical. $\square$

In particular, if K is a field, a finite-dimensional K-algebra is artinian. Hence, it is semisimple if and only if it has zero radical. This statement does not hold true for infinite-dimensional algebras (think of $K[t]$!).

VIII.3.13 Corollary. *Let A be an artinian algebra with radical J, then A/J is semisimple.*

Proof. Indeed, A/J is artinian and its radical is zero because of Proposition VIII.2.4(d). $\square$

It turns out that simple A/J-modules 'coincide' with simple A-modules.

VIII.3.14 Lemma. *Let A be an artinian algebra with radical J. Any simple A-module has a natural structure of simple A/J-module and conversely.*

Proof. Let S be a simple A-module, $x \in S$ and $f_x : A_A \to S_A$ the map defined by $a \mapsto xa$, for $a \in A$. Then f_x is A-linear. Lemma VIII.2.3 gives $J \subseteq \mathrm{Ker}\, f_x$, that is, $xJ = f_x(J) = 0$. Hence, the rule $x(a + J) = xa$, for $x \in S$, $a \in A$ defines unambiguously an A/J-module structure on S.

Conversely, if S is a simple A/J-module, then the projection of A onto A/J defines an A-module structure on S by setting $xa = x(a + J)$, for $x \in S$, $a \in A$. This is the change of scalars functor of Example III.2.7(g).

Assume that S is a simple A-module and $S' \subseteq S$ an A/J-submodule of S. Then it is an A-submodule for the structure we just defined. Hence, $S' = 0$ or $S' = S$ and S is a simple A/J-module. Conversely, if S is a simple A/J-module, then it is a simple A-module because of Exercise IV.4.9 (or using the same reasoning as above).   □

VIII.3.15 Corollary. *An artinian algebra has only finitely many isomorphism classes of simple modules.*

Proof. This follows from Lemma VIII.3.14 and Corollary VII.5.19.   □

Corollary VIII.3.13 also yields an explicit expression for the radical of a module over an artinian algebra.

VIII.3.16 Theorem. *Let A be an artinian algebra with radical J and M an A-module. Then $\operatorname{rad} M = MJ$.*

Proof. For any $x \in M$, the map $f_x : A \to M$ defined by $a \mapsto xa$, for $a \in A$, is A-linear. Hence, $xJ = f_x(J) \subseteq \operatorname{rad} M$, because of Proposition VIII.2.4(a). Therefore, $MJ \subseteq \operatorname{rad} M$.

Conversely, assume that $x \notin MJ$. The A-module M/MJ is annihilated by J, hence is an A/J-module. Because A/J is a semisimple algebra, due to Corollary VIII.3.13, M/MJ is a semisimple A/J-module, hence a semisimple A-module, due to Corollary VIII.3.14. Now $\bar{x} = x + MJ$ is nonzero in M/MJ, therefore there exist a simple module S and a morphism $f : M/MJ \to S$ such that $f(\bar{x}) \neq 0$. Let $p : M \to M/MJ$ be the projection, then $fp(x) = f(\bar{x}) \neq 0$. Due to Lemma VIII.2.3, $x \notin \operatorname{rad} M$. Hence, $\operatorname{rad} M \subseteq MJ$, and equality holds true.   □

We deduce that, in the artinian case, any epimorphism preserves the radical, compare with Corollary VIII.2.5.

VIII.3.17 Corollary. *Let A be an artinian algebra and $f : M \to N$ an epimorphism of A-modules. Then $f(\operatorname{rad} M) = \operatorname{rad} N$*

Proof. Let J be the radical of A. Then $f(\operatorname{rad} M) = f(MJ) = f(M)J = NJ = \operatorname{rad} N$.   □

Theorem VIII.3.16 entails characterisations of the radical of an artinian algebra in terms of nilpotency, see Subsection VIII.3.1.

VIII.3.18 Theorem. *The radical of an artinian algebra is its largest nilpotent ideal.*

Proof. Because any nilpotent ideal is nil, Corollary VIII.3.9 gives that any nilpotent ideal of an algebra A is contained in its radical J. It suffices to prove that J itself is nilpotent.

Because A is artinian, the descending chain of ideals $J \supseteq J^2 \supseteq J^3 \supseteq \cdots$ stabilises. That is, there exists an integer $n > 0$ such that $J^n = J^{n+1}$. Suppose that $J^n \neq 0$. Then the set $\mathcal{M}$ of nonzero right ideals M of A such that $MJ = M$ is nonempty because $J^n \in \mathcal{M}$. Hence, $\mathcal{M}$ has a least element M. Because $M = MJ = \cdots = MJ^n$, there exists $x \in M$ such that $xJ^n \neq 0$. So xJ^n is a nonzero right ideal of A, contained in M (because $x \in M$) and such that $(xJ^n)J = xJ^{n+1} = xJ^n$, due to definition of n. Minimality of M yields $M = xJ^n$ and the inclusions $M = xJ^n \subseteq xA \subseteq M$ yield $M = xA$, so M is cyclic (hence finitely generated). Lemma VIII.2.6 and Theorem VIII.3.16 yield the contradiction $M \neq \operatorname{rad} M = MJ = M$. The proof is complete. □

A consequence of the theorem is that any element of the radical of an artinian algebra is nilpotent, hence the radical is a nil ideal. But we prove actually a stronger statement: in an artinian algebra, nil and nilpotent ideals coincide.

VIII.3.19 Corollary. *In an artinian algebra, an ideal is a nil ideal if and only if it is nilpotent.*

Proof. Because any nilpotent ideal is clearly nil, we prove the converse. Let I be a nil ideal in an artinian algebra A and J its radical. Corollary VIII.3.9 gives $I \subseteq J$. Induction gives $I^n \subseteq J^n$, for any integer $n > 0$. Nilpotency of J implies that of I. □

The following characterisation of the radical is particularly useful in practical calculations.

VIII.3.20 Corollary. *The radical of an artinian algebra A is the unique nil ideal I such that A/I is semisimple.*

Proof. The radical J of A satisfies these conditions. Conversely, let I be a nil ideal such that A/I is semisimple. Because of Corollary VIII.3.9, $I \subseteq J$. Now, J is nilpotent in A, hence J/I is nilpotent in A/I, so that $J/I \subseteq \operatorname{rad}(A/I)$. But, due to the hypothesis, A/I is semisimple. Hence, $\operatorname{rad}(A/I) = 0$. Therefore, $I = J$. □

VIII.3.21 Corollary. *Let I be an ideal in an artinian algebra A, then $\operatorname{rad}(A/I) = (I + J)/I$.*

Proof. Because J is nilpotent in A, so is $(J + I)/I$ in A/I. On the other hand, we have algebra isomorphisms

$$\frac{A/I}{(J+I)/I} \cong \frac{A}{J+I} \cong \frac{A/J}{(J+I)/J}.$$

Because A/J is a semisimple algebra, so is the quotient on the right, hence so is the quotient on the left. We then apply Corollary VIII.3.20. □

A remarkable consequence of the theory developed so far is that any artinian algebra is also noetherian. This is a corollary of the following theorem, due to Hopkins and Levitski.

VIII.3.22 Theorem. Hopkins–Levitski's theorem. *An artinian module over an artinian algebra is also noetherian.*

Proof. Let J be the radical of an artinian algebra A and M_A an artinian module. Because J is nilpotent, there exists a least integer $n \geq 0$ such that $MJ^n = 0$. We prove our statement by induction on n. If $n = 0$, then $M = MA = MJ^0 = 0$ and the zero module is noetherian. If $n = 1$, the condition $MJ = 0$ says that $\operatorname{rad} M = 0$, because of Theorem VIII.3.16. Theorem VIII.3.10 gives that M has finite length. In particular, it is noetherian, due to Lemma VII.4.7. Assume that $n > 1$ and set $N = MJ^{n-1} \subseteq M$. Then N is artinian and $NJ = MJ^n = 0$. The case $n = 1$ just done yields N noetherian. On the other hand, M/N is artinian and $N = MJ^{n-1}$ gives $(M/N)J^{n-1} = 0$. Due to the induction hypothesis, M/N is noetherian. Then, the short exact sequence $0 \to N \to M \to M/N \to 0$ and Theorem VII.2.11 yield that M is noetherian. $\qquad\square$

VIII.3.23 Corollary. *An artinian algebra is also noetherian.* $\qquad\square$

There are several examples of noetherian algebras which are not artinian, notably the algebras $\mathbb{Z}$ of integers and $K[t]$ of polynomials in one indeterminate with coefficients in a field K.

VIII.3.24 Corollary. *The following conditions are equivalent for a module over an artinian algebra:*
    (a) *It is artinian.*
    (b) *It is noetherian.*
    (c) *It is finitely generated.*
    (d) *It is of finite length.*

Proof. Theorem VIII.3.22 shows that (a) implies (b) and Theorem VII.2.9 that (b) implies (c). In order to prove that (c) implies (a), let M be a finitely generated module over an artinian algebra A. Because of Exercise II.4.2, there exist an integer $m \geq 0$ and an epimorphism $A_A^{(m)} \to M$. Because A is artinian, so is the A-module $A_A^{(m)}$. Hence, so is M. This proves the equivalence of (a), (b), (c). Now, (d) is the conjunction of (a) and (b), due to Lemma VII.4.7. $\qquad\square$

If, for instance, A is a finite-dimensional algebra over a field K, then it is in particular artinian, and the objects of the category $\operatorname{mod}A$ of finitely generated A-modules coincide with artinian modules, noetherian modules and modules of finite length. Because of Corollary III.4.9, they also coincide with the modules which are finite-dimensional as K-vector spaces. Another nice consequence is that any finitely generated module M over an artinian algebra has a projective resolution of the form $\ldots \to P_i \to P_{i-1} \to \ldots \to P_1 \to P_0 \to M \to 0$ with all the P_i finitely generated projective. This indeed follows from Theorem VIII.3.22 and Corollary VII.2.16.

VIII.3.25 Examples.

(a) Let K be a field and $A = T_3(K)$ the algebra of 3×3-lower triangular matrices with coefficients in K, see Example I.3.11(c)

$$A = \begin{pmatrix} K & 0 & 0 \\ K & K & 0 \\ K & K & K \end{pmatrix} = \left\{ \begin{pmatrix} \alpha & 0 & 0 \\ \beta & \gamma & 0 \\ \delta & \epsilon & \lambda \end{pmatrix} : \alpha, ..., \lambda \in K \right\}.$$

We compute its radical. For this purpose, consider the K-algebra K^3, product of three copies of K, see Example I.3.11(d) and the map $\varphi : A \to K^3$ given by

$$\begin{pmatrix} \alpha & 0 & 0 \\ \beta & \gamma & 0 \\ \delta & \epsilon & \lambda \end{pmatrix} \mapsto (\alpha \ \ \gamma \ \ \lambda).$$

It is easily seen that φ is a surjective algebra morphism. Its kernel

$$I = \begin{pmatrix} 0 & 0 & 0 \\ K & 0 & 0 \\ K & K & 0 \end{pmatrix} = \left\{ \begin{pmatrix} 0 & 0 & 0 \\ \beta & 0 & 0 \\ \delta & \epsilon & 0 \end{pmatrix} : \beta, \delta, \epsilon \in K \right\}$$

is a nilpotent ideal of A, because $I^3 = 0$. Now $A/I \cong K^3$ is semisimple. We deduce from Corollary VIII.3.20 that $I = \operatorname{rad} A$ (that is, the radical of a lower triangular matrix algebra consists of its off-diagonal elements).

Consider the idempotent

$$e = \mathbf{e}_{33} = \begin{pmatrix} 0 & 0 & 0 \\ 0 & 0 & 0 \\ 0 & 0 & 1 \end{pmatrix}$$

in A. To it corresponds the projective module

$$eA = \begin{pmatrix} 0 & 0 & 0 \\ 0 & 0 & 0 \\ K & K & K \end{pmatrix},$$

see Corollary VI.4.5. Because of Theorem VIII.3.16, its radical is given by:

$$\operatorname{rad}(eA) = (eA)I = eI = \begin{pmatrix} 0 & 0 & 0 \\ 0 & 0 & 0 \\ K & K & 0 \end{pmatrix} = \left\{ \begin{pmatrix} 0 & 0 & 0 \\ 0 & 0 & 0 \\ \delta & \epsilon & 0 \end{pmatrix} : \delta, \epsilon \in K \right\}.$$

The computation is similar for other lower triangular matrix algebras over a field such as, for instance, incidence algebras of partially ordered sets.

(b) Let K be a field and $A = KQ/I$ a bound quiver algebra. Denote, as in Example I.3.11(h), by KQ^+ the ideal of KQ generated by the arrows of Q. We claim that $\operatorname{rad} A = KQ^+/I$.

Because I is admissible, there exists an integer $m \geq 2$ such that we have $KQ^{+m} \subseteq I \subseteq KQ^{+2}$. Therefore, $(KQ^+/I)^m = 0$, so KQ^+/I is a nilpotent ideal of $A = KQ/I$. On the other hand, we have an algebra isomorphism

$$\frac{A}{KQ^+/I} = \frac{KQ/I}{KQ^+/I} \cong \frac{KQ}{KQ^+}.$$

In order to compute the latter quotient, let $K^{|Q_0|} = \{(\lambda_x)_{x \in Q_0} : \lambda_x \in K\}$ be the product of $|Q_0|$ copies of K. Then, $K^{|Q_0|}$ is a semisimple algebra. Paths form a basis of KQ as a K-vector space. Hence, any $w \in KQ$ has a unique expression of the form $w = (\sum_{x \in Q_0} \lambda_x \epsilon_x) + w'$, where ϵ_x is the stationary path at x and w' is a linear combination of paths of length at least one. Define $\varphi : KQ \to K^{|Q_0|}$ by $w \mapsto (\lambda_x)_{x \in Q_0}$: this is easily seen to be a surjective algebra morphism with kernel equal to KQ^+. Thus $KQ/KQ^+ \cong K^{|Q_0|}$ which is semisimple. This completes the proof of our claim, which implies that KQ^+/I is the radical of $A = KQ/I$.

(c) Let, as in (b), $A = KQ/I$ be a bound quiver algebra and M an A-module identified with the corresponding bound representation of (Q, I), see Example IV.4.6(c). We compute its radical $R = \operatorname{rad} M$. Applying Theorem VIII.3.16 and (b) above,

$$R = M(KQ^+/I) = \sum_{\alpha \in Q_1} M\bar{\alpha}$$

where $\bar{\alpha} = \alpha + I$. For a point $x \in Q_0$, we have $R(x) = Re_x$, where $e_x = \epsilon_x + I$. Hence,

$$R(x) = \sum_{\alpha : y \to x} M\bar{\alpha}$$

where the sum is taken over all arrows of target x. For any such arrow, the definition of the equivalence functors in Examples IV.4.6(c) says that right multiplication by $\bar{\alpha}$ equals the action of the corresponding K-linear map $M(\alpha) : M(y) \to M(x)$, that is,

$$M\bar{\alpha} = Me_y\bar{\alpha} = M(y)\bar{\alpha} = M(\alpha)(M(y)) = \operatorname{Im} M(\alpha).$$

Therefore,

$$R(x) = \sum_{\alpha : y \to x} \operatorname{Im}(M(\alpha) : M(y) \to M(x)).$$

Now R is a submodule of M, so the map $R(\alpha)$, for an arrow $\alpha : y \to x$, is necessarily the restriction of $M(\alpha)$ to $R(y)$. This completely describes R as a bound quiver representation.

The top of M is then computed as the cokernel of the inclusion of $R = \operatorname{rad} M$ into M.

Let, for instance, $x \in Q_0$. We compute the radical R of the projective module $P_x = e_x A$, see Example VI.4.6(e). The formula above gives that, if $y \neq x$, then $R(y) = P_x(y)$, while, if $y = x$, then $R(x)$ is the K-vector space spanned by all $\bar{w} = w + I$, with w a path of positive length from x to x, that is, a nontrivial cycle. The top of P_x is the one-dimensional representation whose only nonzero component stands at the point x : this is the simple module S_x of Example VII.4.3(e).

For example, let Q be the quiver

bound by the (admissible) ideal $\langle \alpha^2 \rangle$. Then $P_1 = e_1 A$ is spanned, as a vector space, by $\{e_1, \bar{\alpha}, \bar{\beta}, \bar{\alpha}\bar{\beta}\}$, that is, it is given by the representation

and the previous calculation gives that the radical is spanned, as a vector space, by $\{\bar{\alpha}, \bar{\beta}, \bar{\alpha}\bar{\beta}\}$, thus, it is given by the representation R :

that is,

$$R \cong (K \xrightarrow{1} K) \oplus (0 \xrightarrow{0} K).$$

In particular, the top of P_1 is the simple module S_1

On the other hand, the projective module $P_2 = e_2 A$ equals the simple module S_2.

Exercises of Section VIII.3

1. Let A be an algebra with radical J. Prove that $x \in J$ if and only if, for any simple module S_A, we have $Sx = 0$ (so, the radical is the intersection of the annihilators of simple modules).

2. Prove that a principal ideal domain has zero radical if and only if it is either a field or it has infinitely many maximal ideals.

3. Let A be an algebra. Prove that $\operatorname{rad} A$ is the set of all $x \in A$ such that $1 - axb$ is invertible, for any $a, b \in A$.

4. Let A be an algebra and $a, b \in A$. Prove that, if $1 - ab$ is right invertible, then so is $1 - ba$.

5. Let A be an algebra and I a right ideal. Prove that $I \subseteq \mathrm{rad}\,A$ if and only if $MI = M$ for a finitely generated module M implies $M = 0$.

6. Let I be an ideal of an algebra A such that $\mathrm{rad}(A/I) = 0$. Prove that $\mathrm{rad}A \subseteq I$.

7. Let I be an ideal of an algebra A. Prove that the following conditions are equivalent:
   (a) $I = \mathrm{rad}A$.
   (b) I_A is the unique largest superfluous submodule of A.
   (c) I_A is superfluous in A and $\mathrm{rad}(A/I) = 0$.
   (d) I_A is superfluous in A and $\mathrm{rad}A \subseteq I$.

8. Let A, B be K-algebras. Prove that $\mathrm{rad}(A \times B) = \mathrm{rad}A \times \mathrm{rad}B$.

9. Let A, B be algebras and $\varphi : A \to B$ a surjective algebra morphism such that $\mathrm{Ker}\,\varphi \subseteq \mathrm{rad}A$. Prove that $\varphi(\mathrm{rad}A) = \mathrm{rad}B$.

10. Compute the radical of each of the following algebras:
    (a) $\mathbb{Z}/9\mathbb{Z}$.
    (b) $\mathbb{Z}/10\mathbb{Z}$.

11. Let K be a field. Compute the radical of each of the following K-algebras:
    (a) $K[t]/\langle t^n \rangle$, where $n > 1$ is an integer.
    (b) $K[t]/\langle p \rangle$, where p is a nonconstant polynomial.
    (c) $K[s, t]$.
    (d) $\begin{pmatrix} K & 0 & 0 \\ K & K & 0 \\ K & 0 & K \end{pmatrix}$.
    (e) $\begin{pmatrix} K & 0 & 0 & 0 \\ K & K & 0 & 0 \\ K & 0 & K & 0 \\ K & K & K & K \end{pmatrix}$.

12. Let K be a field and $TF(K)$ the algebra of row and column-finite lower triangular matrices of Exercise I.3.4. Prove that:
    (a) Its radical is $J = \{\mathbf{a} = (a_{ij}) \in TF(K) : a_{ii} = 0 \text{ for any } i\}$.
    (b) J is a nil ideal, but is not nilpotent.

13. Let K be a field, R the algebra of Exercise II.2.11, with A, B and M finite-dimensional over K. Prove that

$$\mathrm{rad}R = \begin{pmatrix} \mathrm{rad}A & 0 \\ M & \mathrm{rad}B \end{pmatrix}.$$

14. Let A be a K-algebra, $_A M_A$ an $A - A$-bimodule and define a new algebra $A \ltimes M$ as follows: its K-module structure is that of $A \oplus M$ and the product is defined by

$$(a, x)(b, y) = (ab, ay + xb),$$

for $a, b \in A$ and $x, y \in M$.

(a) Prove that $A \ltimes M$ is an algebra (called the **trivial extension** of A by M), isomorphic to the matrix algebra $\left\{ \begin{pmatrix} a & 0 \\ x & a \end{pmatrix} : a \in A, x \in M \right\}$, with the usual matrix operations (taking into account the bimodule structure of M).

(b) Prove that, if K is a field, and A, M are finite-dimensional over K, then $\mathrm{rad}(A \ltimes M) = (\mathrm{rad}A) \oplus M$.

(c) Let C be the category whose objects are pairs (U, u), where U is an A-module and $u : U \otimes_A M \to U$ an A-linear map such that $u(u \otimes 1_M) = 0$. A morphism in C from (U, u) to (V, v) is an A-linear map $f : U \to V$ such that $fu = v(f \otimes 1_M)$. Prove that C is equivalent to $\mathrm{Mod}\,(A \ltimes M)$.

15. Let K be a field and $A = \begin{pmatrix} K & 0 \\ K & K \end{pmatrix}$.

(a) Compute the socles of A_A and $_AA$.

(b) Prove that A has two nonisomorphic simple modules but that the direct summands of $\mathrm{soc}\,(A_A)$ are isomorphic.

16. Let A be a K-algebra and $n > 1$ an integer. Prove that $\mathrm{rad}\,M_n(A)$ is the set of matrices $(a_{ij})_{i,j}$ such that $a_{ij} \in \mathrm{rad}A$, for any i, j. (Hint: if $\mathbf{a} = (a_{ij})_{i,j}$ then $\sum_k \mathbf{e}_{ki}\mathbf{a}\mathbf{e}_{jk} = I \cdot a_{ij}$ where I is the identity matrix. Prove that if $I \cdot b \in \mathrm{rad}\,M_n(A)$, then $b \in \mathrm{rad}A$ and, if $b \in \mathrm{rad}A$, then $\mathbf{e}_{ij}b \in \mathrm{rad}\,M_n(A)$ for any i, j.)

17. Let A be an artinian algebra and $e \in A$ an idempotent. Prove that $\mathrm{rad}(eA) = e(\mathrm{rad}\,A)$.

18. Let K be a field. In each case, compute the radical and the top of the shown bound quiver representation.

(a)

$$
\begin{array}{ccccc}
1 & \xrightarrow{\beta} & 2 & \xrightarrow{\alpha} & 3 \\
\circ & \underset{\beta'}{\overset{\beta}{\rightleftarrows}} & \circ & \underset{\alpha'}{\overset{\alpha}{\rightleftarrows}} & \circ
\end{array}
$$

bound by $\langle \alpha\beta, \alpha'\beta' \rangle$

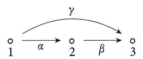

(b)

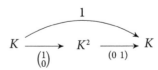

bound by $\langle \alpha\beta \rangle$

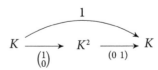

19. Let A be an artinian algebra. Prove that:
    (a) If J is an ideal such that A/J is semisimple, then $J = \cap_{i=1}^{t} I_i$, with the I_i maximal ideals containing J.
    (b) The radical of A is the smallest ideal J such that A/J is semisimple.
    (c) The radical of A is the intersection of its maximal ideals.

20. If M is a module over an artinian algebra A, prove that soc $M = \{x \in M : x(\text{rad}A) = 0\}$.

21. Let A be an algebra with radical J. Prove that $a \in A$ is invertible in A if and only if $a + J$ is invertible in $\bar{A} = A/J$.

22. Prove that an algebra A is noetherian if and only if it is artinian, its radical J is nilpotent and A/J is semisimple (Hint: For sufficiency, try induction on the nilpotency index of J).

VIII.4 Indecomposability

VIII.4.1 Direct sum decompositions

Because of Theorem V.3.18 , a finitely generated module over a principal ideal domain decomposes uniquely, up to isomorphism, as a direct sum of indecomposable modules. An analogous result holds for finitely generated semisimple modules, see Exercise VII.5.7. We ask whether similar, or more general, results hold true in other situations. As a first step, we explore the possibility of decomposing objects in a K-category. There are two reasons for working in this general context. The first is that the existence of a decomposition depends on a categorical condition that may go unnoticed if one works with elements. The second is that many of our results are valid not only in module categories, but also in subcategories satisfying reasonable conditions. Let K be a commutative ring, we define direct summands in a K-category C, generalising in an obvious way the notion for modules.

VIII.4.1 Definition. Let C be a K-category and X, Y objects in C. Then Y is a ***direct summand*** of X if there exist an object Y' and four morphisms $p : X \to Y$, $p' : X \to Y', q : Y \to X$ and $q' : Y' \to X$ in C making X a direct sum of Y and Y'.

As seen in Definition IV.2.11, this means that $pq = 1_Y, p'q' = 1_{Y'}, pq' = 0, p'q = 0$ and $qp + q'p' = 1_X$. As is the case for modules, this notion is closely connected with those of sections and retractions.

VIII.4.2 Lemma. Let C be a K-category and X, Y objects in C.
    (a) If $p : X \to Y$ has a kernel in C and $pq = 1_Y$ for some $q : Y \to X$, then Y is a direct summand of X.
    (b) If $q : Y \to X$ has a cokernel in C and $pq = 1_Y$ for some $p : X \to Y$, then Y is a direct summand of X.

Proof. We only prove (a), because the proof of (b) is similar. Let (Y', q') be a kernel of p. We claim that there exists $p' : X \to Y'$ such that $p'q' = 1_{Y'}, p'q = 0$ and $qp + q'p' = 1_X$. This proves the lemma, because we already have $pq = 1_Y$ and $pq' = 0$ (because $q' = \ker p$).

Consider the morphism $1_X - qp : X \to X$. We have $p(1_X - qp) = p - pqp = p - p = 0$, because $pq = 1_Y$. Therefore, $1_X - qp$ factors through the kernel Y' of p : there exists $p' : X \to Y'$ such that $1_X - qp = q'p'$, that is, $1_X = qp + q'p'$.

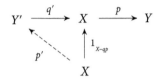

Now $q'p'q' = (1_X - qp)q' = q'$, because $pq' = 0$. But q' is a kernel, therefore a monomorphism. Hence, $p'q' = 1_Y$. For the same reason, $q'p'q = (1_X - qp)q = q - q = 0$ implies $p'q = 0$. $\qquad\square$

Because $pq = 1_Y$, the morphism p is a retraction and q an associated section. The proof of the lemma highlights the role played by the morphism $e = qp \in \mathrm{End}_C X$, which appears in the sum $qp + q'p' = 1_X$. In fact, the condition $pq = 1_Y$ implies that $(qp)^2 = (qp)(qp) = q(pq)p = qp$, so that $e = qp$ is an idempotent morphism. This motivates the next definition.

VIII.4.3 Definition. Let C be a K-category and X an object in C. A morphism $e \in \mathrm{End}_C X$ is a ***split idempotent*** if there exist an object Y and morphisms $p : X \to Y, q : Y \to X$ in C such that $e = qp$ and $pq = 1_Y$.

In the situation of Lemma VIII.4.2, the morphisms $e = qp$ and $e' = q'p'$ are split idempotents in $\mathrm{End}_C X$. On the other hand, e and e' are orthogonal because $pq' = 0$ implies $ee' = (qp)(q'p') = 0$. Similarly, $e'e = 0$. Finally, $1_X = e + e'$. It turns out that any finite direct sum decomposition of an object X induces a decomposition of $1_X \in \mathrm{End}_C X$ of this form.

VIII.4.4 Proposition. *Let C be a K-category. Then there exists a bijection between finite direct sum decompositions of an object and decompositions of the identity morphism on this object into a sum of pairwise orthogonal split idempotents.*

Proof. Let X be an object in C. Because of Theorems IV.2.13 and IV.2.12, to a finite direct sum decomposition $X = X_1 \oplus \cdots \oplus X_m$ correspond projections $p_i : X \to X_i$ and injections $q_i : X_i \to X$ such that $p_i q_i = 1_{X_i}$, for any i, that $p_i q_j = 0$ whenever $i \neq j$ and $q_1 p_1 + \cdots + q_m p_m = 1_X$. Setting $e_i = q_i p_i$, for any i, we see, as above, that the e_i are pairwise orthogonal split idempotents which add up to 1_X.

Conversely, if $1_X = e_1 + \cdots + e_m$, with the e_i pairwise orthogonal split idempotents, then, for any i, we have a factorisation $e_i = q_i p_i$, with $p_i : X \to X_i, q_i : X_i \to X$ morphisms such that $p_i q_i = 1_{X_i}$ for any i. We claim that $X = X_1 \oplus \cdots \oplus X_m$. Because of Theorems IV.2.13 and IV.2.12, it suffices to prove that $p_i q_j = 0$ whenever $i \neq j$.

Now, $i \neq j$ implies $e_i e_j = 0$, that is, $q_i p_i q_j p_j = 0$. But q_i is a section hence a monomorphism, and p_j is a retraction hence an epimorphism. Therefore $p_i q_j = 0$, as required. $\qquad\square$

In this proof, we have implicitly shown that, if C is abelian, then $X_i = \mathrm{Im}\, e_i$, for any i : this indeed follows from the fact that q_i is a monomorphism and p_i an epimorphism (and Proposition IV.2.27).

Indecomposable objects in K-categories are defined exactly as indecomposable modules, see Definition V.3.19.

VIII.4.5 Definition. Let C be a K-category. A nonzero object X in C is ***indecomposable*** if $X = Y \oplus Y'$ implies $X = Y$ or $X = Y'$. Otherwise, the object X is ***decomposable***.

If a module is simple, then it is certainly indecomposable, because it has no nonzero proper submodule. In fact, a semisimple module is indecomposable if and only if it is simple. We refer to Corollary V.3.22 for indecomposable modules over principal ideal domains. In general, indecomposability of an object is detected by its endomorphism algebra.

VIII.4.6 Corollary. *Let C be a K-category in which any idempotent splits. An object X of C is indecomposable if and only if the only idempotents of $\mathrm{End}_C X$ are 0 and 1_X.*

Proof. If $e \in \mathrm{End}_C X$ is an idempotent, then so is $1_X - e$. Moreover, e and $1_X - e$ are orthogonal because $e(1_X - e) = e - e^2 = 0$ and similarly $(1_X - e)e = 0$. Therefore, $1_X = e + (1_X - e)$ is a decomposition of 1_X as sum of orthogonal idempotents. These idempotents split due to the hypothesis on C. Because of Proposition VIII.4.4, X is indecomposable if and only if any such decomposition is trivial. This is the case if and only if $e = 0$ or $e = 1_X$. $\qquad\square$

Because of Lemma VII.3.4, this amounts to saying that X is indecomposable in C if and only if 1_X is a primitive idempotent of $\mathrm{End}_C X$.

We now verify that the hypothesis of Corollary VIII.4.6 holds true in a module category. The following lemma (and its proof) actually hold true in any abelian category.

VIII.4.7 Lemma. *Let A be an algebra and M an A-module. Any idempotent in $\mathrm{End}_A M$ splits.*

Proof. Let $e \in \mathrm{End}_A M$ be an idempotent, and denote by $e = qp$ its canonical factorisation through its image.

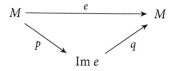

Because $e^2 = e$, we have $qpqp = qp$. But q is a monomorphism, while p is an epimorphism. Hence, $pq = 1_{\mathrm{Im}\, e}$ and e splits.   □

Therefore, Proposition VIII.4.4 takes on a simpler form when we deal with modules: there exists a bijection between finite direct sum decompositions of a module M and decompositions of 1_M as sum of pairwise orthogonal idempotents. The absence of the word 'split' should not mislead the reader: it is an important and nontrivial property of a module category (or a K-category) that idempotents split in it.

VIII.4.8 Examples.
(a) Let A be an algebra. Because of Lemma VIII.4.7, Corollary VIII.4.6 applies in Mod$A$. Combined with Lemma VII.3.4, it implies that a projective module of the form eA, with e an idempotent, is indecomposable if and only if e is primitive. Indeed, this follows from the fact that End $eA \cong eAe$.

For instance, let K be a field and $A = KQ/I$ a bound quiver algebra. The projective modules of the form $e_x A$, with x a point in Q, are indecomposable, because, due to Example VII.3.6(f), any idempotent of the form e_x is primitive.

(b) We continue here Example VII.3.6(e). Here K is a field, $A = M_2(K)$ and we want to prove that, for any $\lambda \in K$, the idempotent

$$e = \begin{pmatrix} 1 & \lambda \\ 0 & 0 \end{pmatrix}$$

is primitive. Because of Corollary VIII.4.19, it suffices to look at the projective module eA. A straightforward calculation shows that, for any λ, we have

$$eA = \begin{pmatrix} K & K \\ 0 & 0 \end{pmatrix} = \mathbf{e}_{11}A$$

where $\mathbf{e}_{11} = \begin{pmatrix} 1 & 0 \\ 0 & 0 \end{pmatrix}$. A direct calculation shows that $\mathbf{e}_{11}A\mathbf{e}_{11} \cong K$ whose only idempotents are 0 and the identity $\mathbf{e}_{11}$. Hence, $\mathbf{e}_{11}A$ is an indecomposable A-module. Therefore, so is eA, and e is primitive. The same reasoning shows that the other idempotents listed at the end of Example VII.3.6(e) are primitive.

We study existence of finite direct sum decompositions of modules.

VIII.4.9 Proposition. *Let M be an artinian or a noetherian module. Then M decomposes into a finite direct sum of indecomposable submodules.*

Proof. If $M = 0$, it equals the empty sum, so there is nothing to prove. If M is indecomposable, there is nothing to prove either. Otherwise, M has a nonzero proper direct summand. If M is artinian, it has a nonzero minimal proper direct summand N_1, which is therefore indecomposable. If M is noetherian, then it has a nonzero maximal proper direct summand M_1, whose direct complement N_1 is nonzero and indecomposable.

Induction gives

$$M = N_1 \oplus M_1 = \cdots = (N_1 \oplus \cdots \oplus N_m) \oplus M_m$$

with all N_i indecomposable and $M_i = N_{i+1} \oplus M_{i+1}$, for any $i < m$. If M is artinian, the descending chain $M \supsetneq M_1 \supsetneq M_2 \supsetneq \cdots$ yields $m > 0$ such that $M_m = 0$. If M is noetherian, the ascending chain $0 \subsetneq N_1 \subsetneq N_1 \oplus N_2 \subsetneq \cdots$ gives $m > 0$ such that $M = \oplus_{i=1}^{m} N_i$. □

VIII.4.10 Corollary. *Let A be an artinian algebra. Any finitely generated A-module decomposes into a finite direct sum of indecomposable modules.*

Proof. This follows from Corollary VIII.3.24 and Proposition VIII.4.9. □

Proposition VIII.4.9 gives a sufficient condition for existence of a finite direct sum decomposition into indecomposable modules. But given a direct sum decomposition, we should be able to decide whether the summands are indecomposable or not, and, if they are, whether such a decomposition is (essentially) unique. This is the subject of the next two subsections.

VIII.4.2 Local algebras

We first look for a reasonable criterion of indecomposability for a module. Corollary VIII.4.6 suggests to find a condition so that the endomorphism algebra has zero and the identity as its only idempotents. We describe a class of algebras having this property.

VIII.4.11 Definition. An algebra is *local* if it has a unique maximal right ideal.

If this is the case, then this unique maximal right ideal is necessarily the radical of the algebra (which is actually an ideal, not only a right ideal). Therefore, an algebra is local if and only if its radical is maximal as a right ideal. The definition implies that, in a local algebra, $0 \neq 1$, so local algebras are always nonzero. Before giving examples, it is convenient to prove characterisations of local algebras.

VIII.4.12 Theorem. *Let A be a K-algebra with radical J. The following are equivalent:*
  (a) *A is a local algebra.*
  (b) *J equals the set of noninvertible elements of A.*
  (c) *If $x \in A$, then x or $1 - x$ is invertible.*
  (d) *A/J is a division ring.*

Proof. (a) implies (b). Let $x \in J$. Then $xA \subseteq J \subsetneq A$ implies that x is not invertible. Conversely, if x is not invertible, then $xA \neq A$ or $Ax \neq A$. Assume the former, then, xA is contained in some maximal right ideal. But A is local, hence its only maximal right ideal is J. Therefore, $xA \subseteq J$. So $x \in J$. The proof is similar if we assume that Ax is contained in a maximal left ideal

(b) implies (c). If both x and $1 - x$ are not invertible, then, because they both belong to the ideal J, so does their sum $1 = x + (1 - x)$, a contradiction to the fact that J is a proper ideal, see Lemma VIII.3.2.

(c) implies (d). Because J is an ideal in A, see Lemma VIII.3.2, A/J is an algebra. We claim that any $x \notin J$ is invertible. Because of Theorem VIII.3.4 and Corollary VIII.3.7 (b), there exist $a, b \in A$ such that $1 - xa, 1 - bx$ are not invertible. Hypothesis (c) implies that xa and bx are invertible, that is, there exist $a', b' \in A$ such that $(xa)a' = 1$ and $b'(bx) = 1$. The element x has both a right and a left inverse, hence it is invertible. This establishes our claim, which amounts to saying that any nonzero element $x + J$ of A/J is invertible.

(d) implies (a). Because A/J is a division ring, it has no nonzero proper right ideal. Because of Theorem II.4.27, J is a maximal right ideal.   □

VIII.4.13 Corollary. *Any element of a local artinian algebra is either invertible or nilpotent.*

Proof. If A is artinian, then J is a nil ideal. If A is moreover local, then J is the set of noninvertible elements. The statement follows.   □

We deduce that modules with local endomorphism algebras are indecomposable.

VIII.4.14 Corollary. *Let C be a K-category and X an object in C. If $\mathrm{End}_C X$ is a local algebra, then X is indecomposable.*

Proof. Because of Corollary VIII.4.6, it suffices to prove that the only idempotents in $\mathrm{End}_C X$ are 0 and 1_X. Let e be an idempotent endomorphism. Because $\mathrm{End}_C X$ is local, e or $1_X - e$ is invertible. Now $e^2 = e$ implies $e(1_X - e) = 0$. Therefore, $e = 0$ or $e = 1_X$.   □

VIII.4.15 Examples.
  (a) Any division algebra, in particular any field, is local, because it has no nonzero proper ideals.
  (b) Let A be a principal ideal domain, p an irreducible element and $n \geq 1$ an integer. Because of Lemma V.3.21, the quotient algebra $A/p^n A$ is local: its unique maximal (right) ideal is $pA/p^n A$. For instance, if K is a field, the truncated polynomial algebra $K[t]/\langle t^n \rangle$ is local.

(c) Let K be a field, $n \geq 1$ an integer and A the matrix algebra

$$A = \left\{ \mathbf{a} = \begin{pmatrix} \alpha_1 & 0 & 0 & \cdots & 0 & 0 \\ \alpha_2 & \alpha_1 & 0 & \cdots & 0 & 0 \\ \alpha_3 & \alpha_2 & \alpha_1 & \cdots & 0 & 0 \\ \vdots & \vdots & & \ddots & \vdots & \vdots \\ \alpha_{n-1} & \alpha_{n-2} & \alpha_{n-3} & \cdots & \alpha_1 & 0 \\ \alpha_n & \alpha_{n-1} & \alpha_{n-2} & \cdots & \alpha_2 & \alpha_1 \end{pmatrix} : \alpha_i \in K \right\}$$

with the usual matrix operations. Then A is a subalgebra of the full lower triangular matrix algebra $T_n(K)$, see Exercise I.3.2(b). In order to prove that A is local, we compute its radical. Consider the map $\varphi : A \to K$, given by

$$\mathbf{a} = \begin{pmatrix} \alpha_1 & 0 & 0 & \cdots & 0 & 0 \\ \alpha_2 & \alpha_1 & 0 & \cdots & 0 & 0 \\ \alpha_3 & \alpha_2 & \alpha_1 & \cdots & 0 & 0 \\ \vdots & \vdots & & \ddots & \vdots & \vdots \\ \alpha_{n-1} & \alpha_{n-2} & \alpha_{n-3} & \cdots & \alpha_1 & 0 \\ \alpha_n & \alpha_{n-1} & \alpha_{n-2} & \cdots & \alpha_2 & \alpha_1 \end{pmatrix} \mapsto \alpha_1.$$

Then φ is a surjective algebra morphism with kernel being the ideal $I = \{\mathbf{a} \in A : \alpha_1 = 0\}$. Moreover, $A/I \cong K$ is a field. Because I is nilpotent, A is local with radical I.

(d) Let K be a field and Q the Kronecker quiver

$$\circ \rightleftarrows \circ$$

We claim that the KQ-module H given by the representation:

$$K^2 \xleftarrow[\substack{\binom{1\ 0}{0\ 1} \\ \binom{0\ 0}{1\ 0}}]{} K^2$$

is indecomposable. Because of Corollary VIII.4.14, it suffices to prove that $\mathrm{End}_{KQ}H$ is local. An endomorphism of H is given by a pair of K-linear maps (matrices) making the following two squares commute:

$$\begin{array}{ccc}
K^2 & \xleftarrow[\binom{0\ 0}{1\ 0}]{\binom{1\ 0}{0\ 1}} & K^2 \\
\binom{a\ c}{b\ d} \downarrow & & \downarrow \binom{a'\ c'}{b'\ d'} \\
K^2 & \xleftarrow[\binom{0\ 0}{1\ 0}]{\binom{1\ 0}{0\ 1}} & K^2
\end{array}$$

Commutativity of the square consisting of the vertical arrows and the two upper horizontal arrows yields

$$\begin{pmatrix} a & b \\ c & d \end{pmatrix} = \begin{pmatrix} a' & b' \\ c' & d' \end{pmatrix}$$

whereas commutativity of the square consisting of the vertical arrows and the two lower horizontal arrows gives

$$\begin{pmatrix} b & 0 \\ d & 0 \end{pmatrix} = \begin{pmatrix} a & b \\ c & d \end{pmatrix}\begin{pmatrix} 0 & 0 \\ 1 & 0 \end{pmatrix} = \begin{pmatrix} 0 & 0 \\ 1 & 0 \end{pmatrix}\begin{pmatrix} a & b \\ c & d \end{pmatrix} = \begin{pmatrix} 0 & 0 \\ a & b \end{pmatrix}$$

from which we get $a = d$ and $b = 0$. Thus the endomorphism algebra of H is isomorphic to the algebra of 2×2-matrices

$$R = \left\{ \begin{pmatrix} a & 0 \\ c & a \end{pmatrix} : a, c \in K \right\}.$$

Because of Example (c), $R \cong \mathrm{End}_{KQ}H$ is local. Therefore, H is indecomposable.

We claim that the converse of Corollary VIII.4.14 (for modules) holds true under reasonable hypotheses. We first generalise a well-known result of linear algebra. As usual, A denotes a K-algebra.

VIII.4.16 Lemma. *Let M be an A-module and f an endomorphism of M.*
 (a) *If M is noetherian and f an epimorphism, then f is an isomorphism.*
 (b) *If M is artinian and f a monomorphism, then f is an isomorphism.*

Proof. (a) The increasing chain $0 \subseteq \mathrm{Ker} f \subseteq \mathrm{Ker} f^2 \subseteq \cdots$ of submodules of M stabilises, because M is noetherian. So, there exists an integer $n > 0$ such that $\mathrm{Ker} f^n = \mathrm{Ker} f^{n+1}$. Because f is an epimorphism, so if f^n. Using Exercise II.3.11, we have $\mathrm{Ker} f = f^n(f^n)^{-1}(\mathrm{Ker} f) = f^n(f^{n+1})^{-1}(0) = f^n(\mathrm{Ker} f^{n+1}) = f^n(\mathrm{Ker} f^n) = 0$. Therefore, f is a monomorphism. Hence, it is an isomorphism.
 (b) The decreasing chain $M \supseteq \mathrm{Im} f \supseteq \mathrm{Im} f^2 \supseteq \cdots$ of submodules of M stabilises, because M is artinian. So there exists an integer $n > 0$ such that $\mathrm{Im} f^n = \mathrm{Im} f^{n+1}$. Assume that $y \in M$. There exists $x \in M$ such that $f^n(y) = f^{n+1}(x)$, that is, $f^n(y - f(x)) = 0$. Because f is a monomorphism, so is f^n. Therefore, $y = f(x)$. Thus f is an epimorphism, hence an isomorphism. $\qquad\square$

In other words, an endomorphism of an artinian and noetherian module, like that of a finite-dimensional vector space, is a monomorphism if and only if it is an epimorphism if and only it is an isomorphism. One deduces the following result, known as **Fitting's lemma.**

VIII.4.17 Lemma. Fitting's lemma. *Let M be an artinian and noetherian module and f an endomorphism of M. There exists an integer n > 0 such that*

$$M = \operatorname{Im} f^n \oplus \operatorname{Ker} f^n.$$

Proof. Because M is artinian, the decreasing chain $M \supseteq \operatorname{Im} f \supseteq \operatorname{Im} f^2 \supseteq \cdots$ stabilises. Hence, there exists an integer $n > 0$ such that $\operatorname{Im} f^n = \operatorname{Im} f^{2n}$. Then $f^n : \operatorname{Im} f^n \to \operatorname{Im} f^n$ is a surjective endomorphism of $\operatorname{Im} f^n$. But $\operatorname{Im} f^n$, being a submodule of M, is noetherian. Therefore, we may apply Lemma VIII.4.16(a) which gives that f^n is an automorphism of $\operatorname{Im} f^n$. So $\operatorname{Ker} f^n \cap \operatorname{Im} f^n = 0$. On the other hand, given $y \in M$, there exists $x \in M$ such that $f^n(y) = f^{2n}(x)$. Then we have, $y - f^n(x) \in \operatorname{Ker} f^n$. Consequently, $y = f^n(x) + (y - f^n(x)) \in \operatorname{Im} f^n + \operatorname{Ker} f^n$. The proof is complete. □

We deduce the sought criterion of indecomposability.

VIII.4.18 Corollary. *Let M be an artinian and noetherian module. The following are equivalent:*
 (a) *M is indecomposable.*
 (b) *Any endomorphism of M is nilpotent or invertible.*
 (c) *End$_A$M is local.*

Proof. (a) implies (b). If M is indecomposable, Fitting's lemma VIII.4.17 gives an integer $n > 0$ such that $\operatorname{Im} f^n = 0$ or $\operatorname{Ker} f^n = 0$. In the first case, $f^n = 0$, so f is nilpotent. In the second, f^n is a monomorphism, hence so is f. Lemma VIII.4.16(b) gives that f is an isomorphism.

(b) is equivalent to (c). The hypothesis (b) says that any endomorphism of M either lies in the radical or is invertible. Then (c) follows from Theorem VIII.4.12. The converse follows from Corollary VIII.4.13.

(c) implies (a). This is Corollary VIII.4.14. □

This corollary applies to finitely generated modules over artinian algebras, examples of which are projective modules generated by idempotents, see Examples VIII.4.8.

VIII.4.19 Corollary. *Let A be an artinian algebra and e $\in$ A an idempotent. The following conditions are equivalent:*
 (a) *eA is indecomposable.*
 (b) *eAe is a local algebra.*
 (c) *e is a primitive idempotent.*

Proof. Corollary VII.3.2 gives $\operatorname{End}(eA) \cong eAe$. The equivalence of (a) and (b) follows from Corollary VIII.4.18, and that of (a) and (c) from Example VIII.4.8(a). □

For instance, taking $e = 1$, we get that an artinian algebra A is local if and only if the module A_A is indecomposable (or if and only if 1 is a primitive idempotent).

VIII.4.3 The Unique Decomposition Theorem

This theorem, also called the **Krull–Schmidt theorem**, is the module-theoretical analogue of the Unique Factorisation Theorem I.5.12 for integers and polynomials. It
asserts that two finite direct sum decompositions of a module into summands with
local endomorphism algebras are essentially the same.

VIII.4.20 Theorem. *Let C be a K-category in which any idempotent splits. If*

$$\oplus_{i=1}^m X_i = \oplus_{j=1}^n Y_j$$

*where the objects X_i, Y_j have local endomorphism algebras, then $m = n$ and there
exists a permutation σ of $\{1, \cdots, m\}$ such that $X_i \cong Y_{\sigma(i)}$, for any i.*

Proof. Because the X_i, Y_j have local endomorphism algebras, they are indecomposable.
We prove the theorem by induction on m. If $m = 1$, then X_1 is indecomposable
and there is nothing to prove. Assume that $m \geq 2$ and set $X_1' = \oplus_{i>1} X_i$. Denote
respectively by p, p', q, q' the projections and injections associated to the direct sum
$X_1 \oplus X_1'$ and by p_j, q_j those associated to the direct sum $\oplus_{j=1}^n Y_j$. We have

$$1_{X_1} = pq = p\left(\sum_j q_j p_j\right)q = \sum_j (pq_j p_j q).$$

Because $\mathrm{End}_C X_1$ is local, Theorem VIII.4.12 gives an index j such that the morphism $v = pq_j p_j q : X_1 \to X_1$ is invertible. We may assume, up to a permutation
of indices, that $j = 1$. So $w = v^{-1}pq_1 : Y_1 \to X_1$ satisfies $wp_1 q = v^{-1}pq_1 p_1 q = v^{-1}v = 1_{X_1}$.
Hence, $(p_1 qw)^2 = p_1 q(wp_1 q)w = p_1 qw$ so $p_1 qw$ is an idempotent endomorphism of
Y_1. Because $\mathrm{End}_C Y_1$ is local and idempotents split in C, Corollary VIII.4.6 implies
that $p_1 qw = 0$ or $p_1 qw = 1_{Y_1}$.

The first case leads to an absurdity. Indeed, w is an epimorphism, because
$wp_1 q = 1_{X_1}$, therefore $p_1 qw = 0$ yields $p_1 q = 0$, which is impossible because
$v = pq_1 p_1 q$ is an isomorphism. Therefore, $p_1 qw = 1_{X_1}$ and we also have $wp_1 q = 1_{Y_1}$.
Thus, $f_{11} = p_1 q : X_1 \to Y_1$ is an isomorphism (with inverse w). Setting $Y_1' = \oplus_{j>1} Y_j$,
one can look at the identity as an isomorphism $X_1 \oplus X_1' = Y_1 \oplus Y_1'$, written in matrix
form

$$f = \begin{pmatrix} f_{11} & f_{12} \\ f_{21} & f_{22} \end{pmatrix}$$

where $f_{11} : X_1 \to Y_1$ is the previous isomorphism, $f_{12} : X_1' \to Y_1, f_{21} : X_1 \to Y_1'$ and
$f_{22} : X_1' \to Y_1'$ are morphisms. The required result will follow from the induction
hypothesis provided we can prove that $X_1' \cong Y_1'$. Now, f is an isomorphism, hence

so is the composite

$$\begin{pmatrix} 1_{Y_1} & 0 \\ -f_{21}f_{11}^{-1} & 1_{Y_1'} \end{pmatrix} \begin{pmatrix} f_{11} & f_{12} \\ f_{21} & f_{22} \end{pmatrix} \begin{pmatrix} 1_{X_1} & -f_{11}^{-1}f_{12} \\ 0 & 1_{X_1'} \end{pmatrix} = \begin{pmatrix} f_{11} & 0 \\ 0 & f_{22} - f_{21}f_{11}^{-1}f_{12} \end{pmatrix}.$$

In particular, $f_{22} - f_{21}f_{11}^{-1}f_{12} : X_1' \to Y_1'$ is an isomorphism. This completes the proof. □

Because of Proposition VIII.4.9, an artinian or noetherian module decomposes into finitely many indecomposable summands. If a module is both artinian and noetherian, then, due to Corollary VIII.4.18, it is indecomposable if and only if it has a local endomorphism algebra. This proves the following corollary.

VIII.4.21 Corollary. *Let A be a K-algebra. Any artinian and noetherian A-module decomposes into a finite direct sum of indecomposable summands, and this decomposition is unique up to isomorphism.* □

Because finitely generated modules over artinian algebras are both artinian and noetherian, we also have the next result.

VIII.4.22 Corollary. *Let A be an artinian algebra. Any finitely generated A-module decomposes into a finite direct sum of indecomposable summands, and this decomposition is unique up to isomorphism.* □

If A is artinian, then the results of this subsection say that the study of the category mod$A$ of finitely generated A-modules reduces to the study of the (isomorphism classes of) indecomposable finitely generated modules and morphisms between them.

In practice, however, it is very difficult to decompose a given module into indecomposable summands. No general algorithm, or computational technique, is known. We give an easy example.

VIII.4.23 Example. Let A be the path algebra of the quiver $3 \xrightarrow{\alpha} 2 \xrightarrow{\beta} 1$. We would like to decompose into indecomposable summands the module corresponding to the representation

$$V = (K \xrightarrow{\binom{1}{1}} K^2 \xrightarrow{(1\ 0)} K).$$

One knows that $P_3 = e_3 A = (K \xrightarrow{1} K \xrightarrow{1} K)$. Now, in V, the morphisms $V(\alpha), V(\beta)$ and $V(\alpha\beta) = V(\beta)V(\alpha)$ are nonzero. One may then 'suspect' that P_3 is a direct summand of V. In order to show it, it suffices to construct a retraction of V onto P_3. A quick calculation yields a morphism $f = (1, (1\ 0), 1) : V \to P_3$:

$$
\begin{array}{ccccc}
K & \xrightarrow{\binom{1}{1}} & K^2 & \xrightarrow{(1\ 0)} & K \\
\downarrow{1} & & \downarrow{(1\ 0)} & & \downarrow{1} \\
K & \xrightarrow{\ 1\ } & K & \xrightarrow{\ 1\ } & K
\end{array}
$$

Now, each of the K-linear maps $1 : K \to K, (1\,0) : K^2 \to K$ and $1 : K \to K$ is surjective. Hence, $f = (1, (1\ 0), 1)$ is an epimorphism of V onto P_3, due to Exercise II.3.24. Because the latter is projective, this epimorphism is a retraction. Thus P_3 is a direct summand of V. Another calculation shows that the kernel of f is $S_2 = (0 \to K \to 0)$. Therefore, $V = P_3 \oplus S_2$.

Exercises of Section VIII.4

1. Let $f : L \to M$ be a morphism of A-modules. Prove that:
   (a) If f is a section and M is indecomposable, then f is an isomorphism.
   (b) If f is a retraction and L is indecomposable, then f is an isomorphism.

2. Give an example of an indecomposable module that has:
   (a) a decomposable submodule,
   (b) a decomposable quotient.

3. Let K be a field. Prove the algebra of formal power series $K[[t]]$, see Example I.3.11(e), is local having $\langle t \rangle$ as maximal ideal.

4. Let K be a field, $n > 1$ an integer and A the set of matrices $[a_{ij}]$ for $1 \leqslant i, j \leqslant n$ such that $a_{11} = \cdots = a_{nn}$. Prove that A is local with maximal ideal $I = \{[a_{ij}] \in A : a_{11} = \cdots = a_{nn} = 0\}$.

5. (a) Let M be a finitely generated module over an artinian algebra. Prove that if $\mathrm{soc}\,M$ or $\mathrm{top}\,M$ is simple, then M is indecomposable.
   (b) Let K be a field and $A = T_n(K)$ the lower triangular matrix algebra of size $n \times n$. Denote by D the duality functor $\mathrm{Hom}_K(-, K)$. Prove that, for any i such that $1 \leqslant i \leqslant n$, each of the modules $e_{ii}A$ and $D(Ae_{ii})$ is indecomposable.

6. Let A be an algebra. Prove that the following are equivalent.
   (a) A is local.
   (b) If $x, y \in A$ are noninvertible, then $x + y$ is noninvertible.
   (c) $x \in A$ belongs to the radical if and only if $xA \neq A$.

7. Let K be a field, Q a finite connected quiver and I an admissible ideal of KQ. Prove that KQ/I is local if and only if $|Q_0| = 1$.

8. Give an example of a finitely generated $\mathbb{Z}$-module whose endomorphism algebra is not local.

9. Let A be an algebra and I_A an injective A-module.
   (a) Let M', M'' be nonzero submodules of I with respective injective envelopes I', I'' and such that $M' \cap M'' = 0$. Prove that $I' \oplus I''$ is a direct summand of I.

(b) Deduce that, if I is an indecomposable injective module, then any two nonzero submodules of I have a nonzero intersection.

(c) Assume that I is an indecomposable injective module and $f : I \to I$ is a monomorphism. Prove that f is an isomorphism.

(d) Deduce that indecomposable injective modules have local endomorphism algebras.

10. Prove that the algebra of Example VIII.4.15(b) is isomorphic to the truncated polynomial algebra $K[t]/\langle t^n \rangle$ and deduce another computation of its radical.

11. Let A be a noetherian algebra and L a free module having a finite basis X. Prove that all bases of L are finite and have same cardinality (Hint: if X' is a basis of L with cardinality larger than that of X, a surjection $X' \to X$ extends to an epimorphism $L \to L$. Apply Lemma VIII.4.16).

12. Using Fitting's lemma, prove that, if X, Y are square matrices of the same size such that XY equals the identity, then so does YX.

13. Let A be an algebra and M_A an artinian and noetherian module. Prove that, for any endomorphism f of M, there exists a direct sum decomposition $M = M' \oplus M''$ such that the restriction of f to M' is nilpotent and the restriction of f to M'' is invertible.

14. The criterion of Corollary VIII.4.18 does not hold true for modules which are not both artinian and noetherian. Let $A = K[t]$ with K a field. Prove that A_A is indecomposable but there exists an endomorphism of A as an A-module which is neither invertible nor nilpotent.

15. Let A be an algebra and M_A a module whose endomorphism algebra is local with maximal ideal I. Prove that, for any module N_A and morphisms $f : M \to N$, $g : N \to M$, one has $gf \in I$ or N is a direct summand of M.

16. Let $F : C \to D$ be a faithful functor between linear categories. Prove that, if X is indecomposable in C, then FX is indecomposable in D. If, in particular, F is an equivalence (or a duality), then there exists a bijection between the isomorphism classes of indecomposable objects in C and in D.

17. Let $M = \oplus_{i=1}^{m} M_i$ be a finitely generated module over an artinian algebra and N an indecomposable direct summand of M. Prove that there exists i, with $1 \leqslant i \leqslant m$, such that N is a direct summand of M_i.

18. Let K be a field. Prove that each of the following bound quiver representations is indecomposable.

(a) The following quiver bound by $I = 0$.

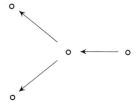

The representation

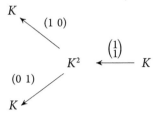

(b) The following quiver bound by $I = \langle \alpha\beta \rangle$.

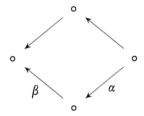

The representation

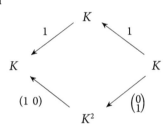

VIII.5 The radical of a module category

VIII.5.1 The radical of a linear category

A K-algebra can be considered as a K-category having just one object, identified to the identity, and whose elements are identified to the morphisms, see Example IV.2.16(c). Conversely, a K-category may be viewed as a K-algebra 'with many identities' whose morphisms behave like elements of an algebra. We may thus define ideals in categories as we do in algebras.

VIII.5.1 Definition. Let C be a K-category. An *ideal* $\mathcal{I}$ in C is the data, for any pair X, Y of objects in C, of a K-submodule $\mathcal{I}(X, Y)$ of $C(X, Y)$ such that:
 (a) $f \in \mathcal{I}(X, Y)$ and $g \in C(Y, Z)$ imply $gf \in \mathcal{I}(X, Z)$, and
 (b) $f \in \mathcal{I}(X, Y)$ and $h \in C(W, X)$ imply $fh \in \mathcal{I}(W, Y)$.

In other words, $\mathcal{I}$ is a family of submodules $(\mathcal{I}(X, Y))_{XY}$ of the K-modules of morphisms which is left and right stable under composition of morphisms.

VIII.5.2 Examples.

(a) Any K-category C has at least as ideals $\mathcal{I} = C$ defined by $\mathcal{I}(X, Y) = C(X, Y)$, for any pair X, Y, and $\mathcal{I} = 0$ defined by $\mathcal{I}(X, Y) = 0$, for any pair X, Y. The latter is called the **zero ideal**.

(b) Let A be a K-algebra viewed as K-category with a single object, see Example IV.2.16(c). Then ideals of the algebra coincide with those of the associated category.

(c) Let C, D be K-categories and $F : C \to D$ a linear functor. Define, for a pair X, Y of objects of C, the set $\mathcal{K}(X, Y) = \{f \in C(X, Y) : Ff = 0\}$. We leave to the reader the (easy) verification that this defines an ideal $\mathcal{K}$, called the **kernel** of F.

  A functor is faithful if and only if its kernel is zero. Indeed, if F is faithful and $f \neq 0$ then $Ff \neq F0 = 0$. Conversely, let g, h be morphisms such that $Fg = Fh$. Then $F(g - h) = 0$ implies $g - h = 0$ and hence $g = h$. Therefore, F is faithful.

(d) Let A be a K-algebra and, for a pair of A-modules M, N, let $P(M, N)$ be the set of morphisms from M to N which factor through a projective module. We prove that this data defines an ideal P of Mod$A$. Indeed, let $f_1, f_2 \in P(M, N)$, there exist projective modules P_1, P_2 and morphisms $h_1 : M \to P_1$, $h_2 : M \to P_2, g_1 : P_1 \to N, g_2 : P_2 \to N$ such that $f_1 = g_1 h_1, f_2 = g_2 h_2$. Then $P_1 \oplus P_2$ is projective and

$$f_1 + f_2 = g_1 h_1 + g_2 h_2 = \begin{pmatrix} g_1 & g_2 \end{pmatrix} \begin{pmatrix} h_1 \\ h_2 \end{pmatrix}.$$

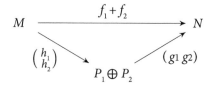

Therefore, $f_1 + f_2 \in P(M, N)$. If $\alpha \in K$ and $f \in P(M, N)$, then, clearly, $\alpha f \in P(M, N)$. Hence, each $P(M, N)$ is a K-submodule of $\text{Hom}_A(M, N)$.

  Let $f \in P(L, M)$ and $u \in \text{Hom}_A(M, N)$. There exist a projective module P and morphisms $h : L \to P, g : P \to M$ such that $f = gh$. Then $uf = u(gh) = (ug)h$ factors through P and thus belongs to $P(L, N)$. Hence, P is stable under left composition. Similarly, it is stable under right composition. Hence, it is an ideal of Mod A.

  One defines in the same way an ideal $\mathcal{I}$ of Mod$A$ consisting of the morphisms which factor through injective modules.

(e) When dealing with an ideal, one expects to encounter a quotient. Let $\mathcal{I}$ be an ideal in a K-category C. One defines the objects of the **quotient category** $C/\mathcal{I}$ to be those of C, and morphisms from an object X to an object Y are the quotient modules

$$C/\mathcal{I}(X,Y) = C(X,Y)/\mathcal{I}(X,Y).$$

The composition of morphisms in $C/\mathcal{I}$ is defined as follows. Let $f: X \to Y, g: Y \to Z$ be morphisms in C. Then set

$$(g + \mathcal{I}(Y,Z))(f + \mathcal{I}(X,Y)) = gf + \mathcal{I}(X,Z).$$

The definition of ideal implies that this composition is unambiguously defined. For instance, the quotient of $\mathrm{Mod}A$ by the ideal $\mathcal{P}$ of Example (d) above is called the **projectively stable category** $\underline{\mathrm{Mod}A} = (\mathrm{Mod}A)/\mathcal{P}$. The **injectively stable category** $\overline{\mathrm{Mod}\,A} = (\mathrm{Mod}\,A)/\mathcal{I}$ is defined in the same way.

We come to the radical. We define the radical of a K-category to consist of the morphisms that can be 'made into' elements of the radical of the endomorphism algebra of the target. Let $f: X \to Y$ be a morphism in a K-category C. In order to 'make it' an endomorphism of the target Y, we compose with an arbitrary morphism $g: Y \to X$, then $fg \in \mathrm{End}_C Y$. We would like that f belongs to the radical of C if and only if fg belongs to the radical of the algebra $\mathrm{End}_C Y$, for any $g: Y \to X$. If one assumes that $X = Y$, then, because of Theorem VIII.3.4, the latter occurs if and only if $1_X - fg$ has a right inverse, that is, is a retraction. This leads to the definition.

VIII.5.3 Definition. Let C be a K-category. Its **radical** rad_C is defined, for any pair of objects X, Y, by setting

$$\mathrm{rad}_C(X,Y) = \{f \in C(X,Y) : 1_Y - fg \text{ is a retraction for any } g: Y \to X\}.$$

It follows from this definition that, for any object X, we have

$$\mathrm{rad}_C(X,X) = \mathrm{rad}(\mathrm{End}_C X)$$

indeed, both sets consist of the $f \in \mathrm{End}_C X$ such that, for any $g: X \to X$, the morphism $1_X - fg$ has a right inverse.

We could have defined the radical by bringing back a morphism to its source rather than its target. We shall see in Theorem VIII.5.5 below that both definitions yield the same ideal in C.

VIII.5.4 Lemma. *The radical of a K-category is an ideal.*

Proof. Let C be a K-category. We prove first that, for any X, Y in C, the set $\mathrm{rad}_C(X,Y)$ is a submodule de $C(X,Y)$. We have $0 \in \mathrm{rad}_C(X,Y)$ and, if $\alpha \in K$ and $f \in \mathrm{rad}_C(X,Y)$, then $\alpha f \in \mathrm{rad}_C(X,Y)$. We must prove that, if $f_1, f_2 \in \mathrm{rad}_C(X,Y)$, then $f_1 + f_2$ also belongs to this set, that is, for any $g: Y \to X$, the morphism $1_Y - (f_1 + f_2)g$ is a retraction.

We know that $1_Y - f_1 g$ has a right inverse h_1, and similarly $1_Y - f_2 g h_1$ has a right inverse h_2. We claim that $(1_Y - (f_1 + f_2)g)h_1 h_2 = 1_Y$.

Indeed, $(1_Y - f_1 g)h_1 - 1_Y$ implies $(1_Y \ f_1 g)h_1 h_2 = h_2$, whence $h_1 h_2 - h_2 = f_1 g h_1 h_2$. Similarly, $(1_Y - f_2 g h_1)h_2 = 1_Y$ gives $h_2 - 1_Y = f_2 g h_1 h_2$. Therefore,

$$(1_Y - (f_1 + f_2)g)h_1 h_2 = h_1 h_2 - f_1 g h_1 h_2 - f_2 g h_1 h_2$$
$$= h_1 h_2 - (h_1 h_2 - h_2) - (h_2 - 1_Y) = 1_Y$$

which establishes our claim.

There remains to prove that the family rad_C is stable under left and right composition. Let $f \in \mathrm{rad}_C(X, Y)$ and $u : W \to X$. For any $g : Y \to W$, the morphism $1_Y - (fu)g = 1_Y - f(ug)$ is a retraction. Hence, $fu \in \mathrm{rad}_C(W, Y)$. Let $v : Y \to Z$ be a morphism. We claim that $vf \in \mathrm{rad}_C(X, Z)$. Given $g : Z \to X$, the morphism $1_Y - f(gv)$ is a retraction. Hence, there exists h such that $(1_Y - fgv)h = 1_Y$. This implies $fgvh = h - 1_Y$. We deduce that

$$(1_Z - vfg)(1_Z + vhfg) = 1_Z + vhfg - vfg - vfgvhfg$$
$$= 1_Z + vhfg - vfg - v(h - 1_Y)fg = 1_Z.$$

Thus, $1_Z - vfg$ is a retraction, and the proof is complete. $\square$

We give reformulations of the definition of radical, which respect a strict 'left-right' symmetry.

VIII.5.5 Theorem. *Let $f : X \to Y$ be a morphism in a K-category C. Then f belongs to the radical of C if and only if it satisfies the following equivalent conditions:*
(a) $1_Y - fg$ *is a retraction, for any* $g : Y \to X$.
(b) $1_X - gf$ *is a section, for any* $g : Y \to X$.
(c) $1_Y - fg$ *is an isomorphism, for any* $g : Y \to X$.
(d) $1_X - gf$ *is an isomorphism, for any* $g : Y \to X$.
(e) $fg \in \mathrm{rad}(End_C Y)$, *for any* $g : Y \to X$.
(f) $gf \in \mathrm{rad}(End_C X)$, *for any* $g : Y \to X$.

Proof. In view of the definition, it suffices to prove the equivalence of the stated conditions.

(a) is equivalent to (c). Because (c) implies (a) trivially, we prove the converse. Let $f : X \to Y$ be such that $1_Y - fg$ is a retraction for any $g : Y \to X$. There exists $h : Y \to Y$ such that $(1_Y - fg)h = 1_Y$. Then $h = 1_Y + fgh = 1_Y - f(-gh)$ is also a retraction, that is, h has a left inverse *and* a right inverse. Therefore, h is invertible and so is $1_Y - fg = h^{-1}$. This proves (c).

(b) is equivalent to (d). Follows from symmetry.

(c) is equivalent to (d). Let $f : X \to Y$ be such that $1_Y - fg$ is invertible for any $g : Y \to X$. Let $h : Y \to Y$ be such that $(1_Y - fg)h = 1_Y = h(1_Y - fg)$. The first equality gives $1_Y - h + fgh = 0$ and the second $1_Y - h + hfg = 0$, whence

$$(1_X - gf)(1_X + ghf) = 1_X - gf + ghf - gfghf = 1_X - g(1_Y - h + fgh)f = 1_X$$

and

$$(1_X + ghf)(1_X - gf) = 1_X - gf + ghf - ghfgf = 1_X - g(1_Y - h + hfg)f = 1_X.$$

Thus (c) implies (d). The converse follows from symmetry.

(c) is equivalent to (e). Assume that (e) holds true and $fg \in \text{End}_C Y$. Because of Corollary VIII.3.6, $1_Y - fg$ is invertible and we get (c). Conversely, if (c) holds true and $1_Y - fg$ is invertible, then it is a retraction so $f \in \text{rad}_C(X, Y)$. Let $h : Y \to Y$ be arbitrary. Then $hf \in \text{rad}_C(X, Y)$ due to Lemma VIII.5.4. Hence, because (a) implies (c), $1_Y - hfg$ is invertible and, in particular, is a section. This is true for any $h : Y \to Y$, so Corollary VIII.3.7(b) gives $fg \in \text{End}_C Y$.

(d) is equivalent to (e). This follows from symmetry.   □

This theorem does not offer good characterisations of the radical. The proof itself shows the superficiality of the stated conditions. Moreover, these equivalent conditions are difficult to verify: given $f : X \to Y$, one should test *all* morphisms $g : Y \to X$ in order to decide whether f belongs to the radical or not. We need handier criteria, and, for this, additional hypotheses.

VIII.5.6 Corollary. *Let $f : X \to Y$ be a morphism in a K-category C. If $f \in \text{rad}_C(X, Y)$, then f is not an isomorphism. The converse holds true if idempotents split in C and $\text{End}_C X, \text{End}_C Y$ are local algebras.*

Proof. Assume that $f \in \text{rad}_C(X, Y)$ is an isomorphism. Then $1_Y - ff^{-1} = 0$ is a retraction, an absurdity. So, f is not an isomorphism.

For the converse, let $f : X \to Y$ be a nonisomorphism. We claim that, under the stated hypotheses, fg is a nonisomorphism for any $g : Y \to X$. Indeed, assume that there exists g such that fg is an isomorphism. Let $h = (fg)^{-1}$ and $e = ghf : X \to X$. Then $e^2 = (ghf)(ghf) = ghf = e$, that is, $e \in \text{End}_C X$ is idempotent. Because idempotents split in C, Corollary VIII.4.6 yields $e = 0$ or $e = 1_X$. If $e = 0$, we get the contradiction $1_Y = 1_Y^2 = (fgh)(fgh) = fegh = 0$. Hence, $e = 1_X$. Then $ghf = 1_X$ and $fgh = 1_Y$ imply that f is an isomorphism, a contradiction which establishes our claim.

Because $\text{End}_C Y$ is local, it follows from Theorem VIII.4.12 that $1_Y - fg$ is invertible. Hence, $f \in \text{rad}_C(X, Y)$, due to Theorem VIII.5.5.   □

Assume that $\text{End}_C X, \text{End}_C Y$ are local algebras. Because of Corollary VIII.4.14, X and Y are indecomposable. If, moreover, idempotents split in C, Corollary VIII.5.6 says that, in this situation, a morphism from X to Y lies in the radical if and only if it is not an isomorphism (that is, is not invertible).

VIII.5.7 Example. The 'converse' part of Corollary VIII.5.6 does not hold true if we omit the hypothesis that endomorphism algebras are local. Let K be a field and $\text{mod} K$ the category of finite-dimensional K-vector spaces. Let E be a vector space of dimension $n \geq 2$ and $f : E \to E$ a nonzero nonisomorphism (thus, given in a basis by an $n \times n$-matrix, nonzero but with zero determinant). Then $\text{End} E \cong M_n(K)$, whose only proper ideal is the zero ideal. In particular $\text{rad}(E, E) \cong \text{radEnd} E = 0$. Hence, $f \notin \text{rad}(E, E)$.

VIII.5.2 The radical of modA.

Corollary VIII.5.6 suggests to look at finitely generated modules over artinian algebras. We first prove that radical commutes with finite direct sums if our K-category
admits them.

VIII.5.8 Lemma. *Let C be a linear category and $X = \oplus_{i=1}^{m} X_i, Y = \oplus_{j=1}^{n} Y_j$ objects in C.
Then we have*

$$\mathrm{rad}_C(X, Y) \cong \bigoplus_{i=1}^{m} \bigoplus_{j=1}^{n} \mathrm{rad}_C(X_i, Y_j).$$

Proof. Denote by p_i, q_i the projections and injections associated with the direct sum
X, and by p'_j, q'_j those associated with the direct sum Y. Define a morphism

$$\varphi : \mathrm{rad}_C(X, Y) \to \bigoplus_{i=1}^{m} \bigoplus_{j=1}^{n} \mathrm{rad}_C(X_i, Y_j)$$

by $f \mapsto (f_{ji})_{i,j}$, for $f \in \mathrm{rad}_C(X, Y)$, where $f_{ji} = p'_j f q_i$ for any i, j. Now
$f \in \mathrm{rad}_C(X, Y)$, gives $f_{ji} \in \mathrm{rad}_C(X_i, Y_j)$ because the radical is an ideal of C. Thus
φ maps into $\oplus_{i,j} \mathrm{rad}_C(X_i, Y_j)$. It is clear that φ is a K-linear map. We claim that the
morphism

$$\psi : \bigoplus_{i,j} \mathrm{rad}_C(X_i, Y_j) \to \mathrm{rad}_C(X, Y)$$

defined by $(f_{ji})_{i,j} \mapsto \sum_{i,j} q'_j f_{ji} p_i$, for $(f_{ji})_{i,j} \in \oplus_{i,j} \mathrm{rad}_C(X_i, Y_j)$, is the inverse of φ.
Again $f_{ji} \in \mathrm{rad}_C(X_i, Y_j)$ implies $q'_j f_{ji} p_i \in \mathrm{rad}_C(X, Y)$, for any i, j, so ψ maps indeed
into $\mathrm{rad}_C(X, Y)$. We have:

$$\psi\varphi(f) = \sum_{i,j} q'_j (p'_j f q_i) p_i = \left(\sum_j q'_j p'_j \right) f \left(\sum_i q_i p_i \right) = 1_Y f 1_X = f$$

for $f \in \mathrm{rad}_C(X, Y)$, and

$$\varphi\psi(f_{ji})_{ij} = \left(p'_j \left(\sum_{k,\ell} q'_\ell f_{\ell k} p_k \right) q_i \right)_{i,j} = \left(\sum_{k,\ell} \delta_{j\ell} f_{\ell k} \delta_{ki} \right)_{i,j} = (f_{ji})_{i,j}$$

for $(f_{ji})_{i,j} \in \oplus_{i,j} \mathrm{rad}_C(X_i, Y_j)$, where the δ are the Kronecker deltas. This completes
the proof. $\square$

From now on, we assume that A is artinian and study the radical of the cate-
gory modA of finitely generated right A-modules. We abbreviate $\mathrm{rad}_{\mathrm{mod}A}$ to rad_A.

Because of Corollary VIII.4.21, given finitely generated modules M, N, we have finite direct sum decompositions into indecomposable summands $M = \oplus_{i=1}^{m} M_i$, $N = \oplus_{j=1}^{n} N_j$. The previous lemma says that $\text{rad}_A(M, N) \cong \oplus_{i,j} \text{rad}_A(M_i, N_j)$, that is, a morphism from M to N belongs to the radical if and only if each of the induced morphisms between their indecomposable summands belongs to the radical. Thus, when studying morphisms lying in the radical, we may restrict ourselves to morphisms between indecomposable modules. The next corollary follows from Corollary VIII.5.6.

VIII.5.9 Corollary. *Let $f : M \to N$ be a linear map between finitely generated indecomposable modules over an artinian algebra A. Then $f \in \text{rad}_A(M, N)$ if and only if f is not an isomorphism.*

Proof. Because of Corollary VIII.4.18, a finitely generated module over an artinian algebra is indecomposable if and only if its endomorphism algebra is local. □

So, if M, N are indecomposable modules such that M is not isomorphic to N, then $\text{rad}_A(M, N) = \text{Hom}_A(M, N)$ while, if they are isomorphic, then $\text{rad}_A(M, N) = \text{rad}(\text{End}_A M)$.

VIII.5.10 Corollary. *Let $f : M \to N$ be a linear map between finitely generated indecomposable modules over an artinian algebra A. Then:*
(a) *If M is indecomposable, then $f \in \text{rad}_A(M, N)$ if and only if it is not a section.*
(b) *If N is indecomposable, then $f \in \text{rad}_A(M, N)$ if and only if it is not a retraction.*

Proof. We only prove (a), because the proof of (b) is similar. If f is a section, then there exists f' such that $f'f = 1_M$ and hence $f \notin \text{rad}_A(M, N)$. Conversely, if $f \notin \text{rad}_A(M, N)$, then, due to Lemma VIII.5.8, there exist indecomposable direct summands M', N' of M, N, respectively, such that the composition of f with the section $q : M' \to M$ and the retraction $p : N \to N'$ yields a morphism pfq which is not in $\text{rad}_A(M', N')$. Corollary VIII.5.9 implies that pfq is an isomorphism. Moreover, M is indecomposable, hence q also is an isomorphism. Therefore, $pf = pfqq^{-1}$ is an isomorphism and so f is a section. □

VIII.5.11 Example. Let K be a field and $A = K[t]/\langle t^2 \rangle$. Because A is a two-dimensional K-vector space, it is artinian. Moreover, see Example VIII.4.15(b), A is local. Because of Corollary VIII.4.19, the module A_A is indecomposable. Let $f : A_A \to A_A$ be a nonzero morphism. The image of f is a K-subspace of A, hence it has dimension 1 or 2. If $\dim_K \text{Im} f = 2$, then f is surjective. Because of Lemma VIII.4.16, it is an isomorphism, hence not in the radical. If $\dim_K \text{Im} f = 1$, then the image equals the unique one-dimensional (hence simple) submodule of A, given by $S_A = \langle t \rangle / \langle t^2 \rangle$. Hence, $\text{rad}_A(A, A) \cong \text{Hom}_A(A, S) \cong S$, because of Theorem II.3.9. Therefore, $\text{rad}_A(A, A)$ is a one-dimensional K-vector space. A nonzero morphism $f : A \to A$ with image S is the multiplication by t, defined by $p + \langle t^2 \rangle \mapsto tp + \langle t^2 \rangle$, for a polynomial p, and all other maps in $\text{rad}_A(A, A)$ are its multiples by elements of K.

Exercises of Section VIII.5

1. Let C be a K-category. Prove that the definition of quotient category given in Exercise III.2.1 is equivalent to that given in Example VIII.5.2.

2. Here, we generalise Example VIII.5.2(d). Let A be a K-algebra, M an A-module and Add$M$ the full subcategory of Mod$A$ consisting of the direct summands of direct sums of M.
   (a) Prove that Add$M$ is a linear subcategory of Mod$A$.
   (b) Given modules U, V, let $\mathcal{M}(U, V)$ be the subset of $\mathrm{Hom}_A(U, V)$ consisting of the maps which factor through an object of Add$M$. Prove that this data defines an ideal $\mathcal{M}$ in Mod$A$.

3. The following is a 'finitely generated' version of the previous Exercise VIII.5.2. Let A be a K-algebra, M a finitely generated A-module and add$M$ the full subcategory of mod$A$ consisting of the direct summands of *finite* direct sums of M.
   (a) Prove that add$M$ is a linear subcategory of mod$A$.
   (b) Given finitely generated modules U, V, let $\mathcal{M}(U, V)$ be the subset of $\mathrm{Hom}_A(U, V)$ consisting of the maps which factor through an object of add$M$. Prove that this data defines an ideal $\mathcal{M}$ in mod$A$.
   (c) The following generalises Lemma VI.4.11. Let A, B be K-algebras, $F, G : \mathrm{mod}A \to \mathrm{mod}B$ linear functors and $\varphi : F \to G$ a functorial morphism. If $L \cong L_1 \oplus L_2$ in add$M$, prove that φ_L is injective, or surjective, or an isomorphism if and only if so are φ_{L_1} and φ_{L_2}.

4. Let C be a K-category and $\mathcal{I}, \mathcal{J}$ ideals in C. Prove that:
   (a) Given objects X, Y in C, let $(\mathcal{I} \cdot \mathcal{J})(X, Y)$ be the set of sums $\sum_i g_i f_i$, with $g_i \in \mathcal{I}(Z_i, Y), f_i \in \mathcal{J}(X, Z_i)$, for some objects Z_i in C. Then this data defines an ideal $\mathcal{I} \cdot \mathcal{J}$ in C.
   (b) Given objects X, Y in C, define $(\mathcal{I} \cap \mathcal{J})(X, Y) = \mathcal{I}(X, Y) \cap \mathcal{J}(X, Y)$. Then this data defines an ideal $\mathcal{I} \cap \mathcal{J}$ in C, containing $\mathcal{I} \cdot \mathcal{J}$.
   (c) Define inductively $\mathcal{I}^n$, for any integer $n \geq 2$, by $\mathcal{I}^n = \mathcal{I}^{n-1} \cdot \mathcal{I}$ and $\mathcal{I}^\infty = \cap_{n \geq 1} \mathcal{I}^n$. Then the $\mathcal{I}^n$ and $\mathcal{I}^\infty$ are ideals in C.
   (d) For any integer $n \geq 1$, and objects X, Y in C, $\mathcal{I}^{n+1}(X, Y)$ and $\mathcal{I}^\infty(X, Y)$ are K-submodules of $\mathcal{I}^n(X, Y)$.

5. Let C be a K-category, $\mathcal{I}$ an ideal in C and $X = \oplus_{i=1}^m X_i, Y = \oplus_{j=1}^n Y_j$ objects in C. Prove that $\mathcal{I}(X, Y) \cong \oplus_{i=1}^m \oplus_{j=1}^n \mathcal{I}(X_i, Y_j)$.

6. Let C be a K-category. Prove that the assignment $(X,Y) \mapsto \mathrm{rad}_C(X,Y)$ defines a bifunctor from $C \times C$ to $\mathrm{Mod}K$, contravariant in the first variable, covariant in the second and linear in both.

7. Let C be a K-category. Prove that rad_C is the only ideal in C such that $\mathrm{rad}_C(X,X) = \mathrm{rad}(\mathrm{End}_C X)$ for any object X in C.

8. Let C be a K-category and X,Y objects in C. Prove that the following are equivalent for a morphism $f : X \to Y$:
   (a) $f \in \mathrm{rad}_C(X,Y)$.
   (b) $ufv \in \mathrm{rad}(\mathrm{End}_C Z)$, for any morphisms $u : Y \to Z, v : Z \to X$.
   (c) $1_Z - ufv$ is invertible, for any morphisms $u : Y \to Z, v : Z \to X$.

9. Let A be an artinian K-algebra and $\mathrm{ind}A$ denote a full subcategory of $\mathrm{mod}A$ consisting of exactly one representative from each isomorphism class of indecomposable finitely generated A-modules. Prove that:
   (a) $\mathrm{ind}A$ is a K-category.
   (b) The radical of $\mathrm{ind}A$ consists of the nonisomorphisms of the category.

10. Let A be an artinian algebra, M,N indecomposable finitely generated A-modules and $f : M \to N$ a morphism which is neither a monomorphism nor an epimorphism. Prove that $f \in \mathrm{rad}_A^2(M,N)$.

11. Let $f : M \to N$ be a linear map between finitely generated modules over an artinian algebra A. Prove that, if f lies in the radical, then, for every morphism $g : N \to M$, the morphisms gf and fg are nilpotent.

12. Let A be an artinian K-algebra, with radical J. Prove that:
    (a) $f \in \mathrm{rad}_A(A,A)$ if and only if $f(1) \in J$.
    (b) $\mathrm{rad}_A^n(M,N) = 0$, for some integer $n \geq 1$, and for any M,N in $\mathrm{mod}A$, if and only if $\mathrm{rad}_A^n(M',N') = 0$, for any M',N' indecomposables in $\mathrm{mod}A$.
    (c) If $\mathrm{rad}_A^n(M,N) = 0$ for all M,N in $\mathrm{mod}A$ and some integer $n \geq 0$, then $J^n = 0$.

13. Let A be an artinian K-algebra and S a simple A-module. Prove that:
    (a) If S is projective, then, for any M in $\mathrm{mod}A$, we have $\mathrm{rad}_A(M,S) = 0$.
    (b) If S is injective, then, for any M in $\mathrm{mod}A$, we have $\mathrm{rad}_A(S,M) = 0$.

14. Let A be an artinian K-algebra. Prove that A is semisimple if and only if, for any M,N in $\mathrm{mod}A$, one has $\mathrm{rad}(M,N) = 0$ (Hint: For necessity, use the previous Exercise VIII.5.13).

15. Let K be a field, A a finite-dimensional K-algebra and M,N finitely generated A-modules. Prove that the duality functor $D = \mathrm{Hom}_K(-,K) : \mathrm{mod}A \to \mathrm{mod}A^{op}$, see Example IV.4.8(b), induces isomorphisms of K-vector spaces:

$$\mathrm{rad}_A^n(M,N) \cong \mathrm{rad}_{A^{op}}^n(DN,DM)$$

for any integer $n \geq 1$, and

$$\mathrm{rad}_A^\infty(M,N) \cong \mathrm{rad}_{A^{op}}^\infty(DN,DM).$$

16. Let K be a field, A a finite-dimensional K-algebra and M, N finitely generated A-modules. Prove that we have inclusions of K-vector spaces:

$$\text{Hom}_A(M, N) \supseteq \text{rad}_A(M, N) \supseteq \text{rad}_A^2(M, N) \supseteq \ldots \supseteq \text{rad}_A^n(M, N) \supseteq \ldots \supseteq \text{rad}_A^\infty(M, N).$$

Deduce that there exists an integer $n \geq 1$ (depending on M and N) such that $\text{rad}_A^\infty(M, N) = \text{rad}_A^n(M, N)$.

17. Let A be an artinian algebra and $f : M \to N$ a morphism between finitely generated A-modules. Prove that:
   (a) If M is indecomposable, then f belongs to $\text{rad}_A(M, N)$ if and only if $\text{Im}\,\text{Hom}_A(f, M) \subseteq \text{rad}_A(M, M)$.
   (b) If N is indecomposable, then f belongs to $\text{rad}_A(M, N)$ if and only if $\text{Im}\,\text{Hom}_A(N, f) \subseteq \text{rad}_A(N, N)$.

18. Let A be an artinian algebra and M, N finitely generated A-modules. An A-linear map $f : M \to N$ is called an **irreducible morphism** if it is neither a section nor a retraction, and, whenever $f = f_1 f_2$ with $f_2 : M \to U, f_1 : U \to N$ morphisms in $\text{mod}A$, then f_2 is a section or f_1 is a retraction. Prove that:
   (a) If P is an indecomposable projective A-module, then the inclusion $\text{rad}P \hookrightarrow P$ is irreducible.
   (b) If f is irreducible, then it is either a monomorphism, or an epimorphism.
   (c) If f is irreducible and M or N is indecomposable, then $f \in \text{rad}_A(M, N)$.
   (d) If both M and N are indecomposable, then f is irreducible if and only if $f \in \text{rad}_A(M, N) \setminus \text{rad}_A^2(M, N)$.

19. Let K be a field, $n \geq 1$ an integer and $A = K[t]/\langle t^n \rangle$. Let $R = \langle t \rangle / \langle t^n \rangle$. Compute each of $\text{rad}_A(A, A), \text{rad}_A(A, R), \text{rad}_A(R, A), \text{rad}_A(R, R)$.

Chapter IX
Projectives and quivers

IX.1 Introduction

In the present chapter, we apply the results of Chapters VII and VIII to describe finitely generated indecomposable projective modules over artinian algebras. Their isomorphism classes turn out to be finitely many because they are in bijection with those of simple modules. If we deal with a finite–dimensional algebra over a field, then we are also able to describe finitely generated indecomposable injective modules. We then prove the Morita theorem, which gives a criterion allowing to verify whether two module categories are equivalent or not. This theorem holds true without the artinian hypothesis, but, when applied to an artinian algebra, it reduces the study of its module category to that of a 'smaller' algebra. In the second part of the chapter, we study a special type of finite–dimensional algebras over a field, called elementary, and prove that these coincide with bound quiver algebras. This allows us to give a nice visual description of the finitely generated modules over a class of such algebras, called Nakayama algebras.

IX.2 Projective modules over artinian algebras

IX.2.1 Idempotents elements and projective modules

Let K be a commutative ring and A a K-algebra, without additional hypothesis for the moment. Because of Lemma VII.3.5, to a complete set $\{e_1, \cdots, e_n\}$ of pairwise orthogonal idempotents corresponds a direct sum decomposition of the module A_A :

$$A_A = e_1 A \oplus \cdots \oplus e_n A.$$

We now prove the converse, thus the existence of a bijection between complete sets of pairwise orthogonal idempotents and direct sum decompositions of A into right ideals.

IX.2.1 Proposition. *Let A be a K-algebra. Then there exists a direct sum decomposition of A into right ideals*

$$A_A = I_1 \oplus \cdots \oplus I_n.$$

if and only if there exists a complete set $\{e_1, \cdots, e_n\}$ of pairwise orthogonal idempotents of A such that $I_i = e_i A$, for any i satisfying $1 \leqslant i \leqslant n$.

An Introduction to Module Theory. Ibrahim Assem and Flávio U. Coelho, Oxford University Press.
© Ibrahim Assem and Flávio U. Coelho (2024). DOI: 10.1093/9780198904939.003.0010

Proof. In view of Lemma VII.3.5, we just need to prove necessity. Assume that $A_A = I_1 \oplus \cdots \oplus I_n$. Because of Proposition VIII.4.4, the identity morphism 1_A on A decomposes as sum of a finite set of pairwise orthogonal idempotents $f_i \in \operatorname{End} A$ such that $I_i = \operatorname{Im} f_i$. Under the isomorphism of algebras $\operatorname{End}_A A \cong A$ of Corollary II.3.12, this says that $1 \in A$ decomposes as the sum of the pairwise orthogonal idempotents $e_i = f_i(1) \in I_i$. Therefore, $e_i A \subseteq I_i$, for each i. Conversely, let $x \in I_i$, then $x = 1 \cdot x = e_1 x + \cdots + e_n x$. Because the sum $A_A = \oplus_i I_i$ is direct, we get $e_j x = 0$ for $j \neq i$ and $x = e_i x \in e_i A$. Therefore, $I_i = e_i A$. $\square$

IX.2.2 Corollary. *Let A be an artinian K-algebra. Then there exists a direct sum decomposition of A into indecomposable right ideals*

$$A_A = I_1 \oplus \cdots \oplus I_n.$$

if and only if there exists a complete set $\{e_1, \cdots, e_n\}$ of primitive pairwise orthogonal idempotents of A such that $I_i = e_i A$, for any i satisfying $1 \leqslant i \leqslant n$.

Proof. If A is artinian, then, due to Corollary VIII.4.19, $e_i A$ is indecomposable if and only if e_i is a primitive idempotent. $\square$

Such a direct sum decomposition of A_A into a direct sum of indecomposable right ideals, which are also indecomposable projective A-modules, is called a ***Peirce decomposition*** of A. Our objective is to prove that, in the artinian case, any indecomposable projective finitely generated A-module is isomorphic to one of the $e_i A$, that is, a Peirce decomposition yields a complete list of representatives of the isomorphism classes of indecomposable projective finitely generated A-modules.

Assume that A is artinian with radical J. The strategy consists in reducing the problem to the semisimple case. Because of Corollary VIII.3.13, the quotient algebra $\bar{A} = A/J$ is semisimple. Therefore, its identity $\bar{1}$ decomposes as the sum of a complete set of primitive pairwise orthogonal idempotents in $\bar{A}$. We say that an idempotent $\bar{f} \in \bar{A}$ can be ***lifted modulo J*** if there exists an idempotent $e \in A$ such that $e + J = \bar{f}$. It is an important property of artinian algebras that any decomposition of the identity of $\bar{A}$ as sum of a complete set of primitive pairwise orthogonal idempotents lifts modulo the radical to a decomposition of the identity of A into a complete set of primitive pairwise orthogonal idempotents. We need a lemma. Given $x \in A$, we denote by $\bar{x}$ its residual class $x + J$ in $\bar{A} = A/J$.

IX.2.3 Lemma. *Let A be an artinian algebra, and $e \in A$ an idempotent. Then*

$$\operatorname{rad}(eAe) = e(\operatorname{rad} A)e.$$

Proof. Set $J = \operatorname{rad} A$. The map $eAe \to \bar{e}\bar{A}\bar{e}$ defined by $eae \mapsto \bar{e}\bar{A}\bar{e}$, for $a \in A$, is a surjective algebra morphism with kernel eJe. So, eJe is an ideal of eAe. It is nilpotent in it, because so is J in A, and $(eAe)/(eJe) \cong \bar{e}\bar{A}\bar{e}$ is semisimple. Corollary VIII.3.20 gives $eJe = \operatorname{rad}(eAe)$. $\square$

IX.2.4 Proposition. Lifting idempotents. *Let A be an artinian algebra.*
  (a) *For any idempotent $\bar{f} \in \bar{A}$, there exists an idempotent $e \in A$ such that $\bar{e} = \bar{f}$.*
  (b) *For any complete set $\{\bar{f}_1, \cdots, \bar{f}_n\}$ of pairwise orthogonal idempotents in $\bar{A}$, there exists a complete set $\{e_1, \cdots, e_n\}$ of pairwise orthogonal idempotents in A such that $\bar{e}_i = \bar{f}_i$, for any i satisfying $1 \leqslant i \leqslant n$.*

Proof. (a) Because of the hypothesis, $f - f^2 \in J$. Because J is a nil ideal, there exists an integer $m \geq 1$ such that $(f - f^2)^m = 0$, that is,

$$0 = (f - f^2)^m = f^m(1 - f)^m = f^m - f^{m+1}g$$

where g is a polynomial in f (and therefore commutes with f). Set $e = f^m g^m$, then

$$e^2 = f^{2m}g^{2m} = f^{m-1}(f^{m+1}g)g^{2m-1} = f^{m-1}f^m g^{2m-1} = f^{2m-1}g^{2m-1}.$$

Applying this procedure repeatedly, we end up with $e^2 = f^m g^m = e$, that is, e is idempotent in A. Moreover, because f is idempotent modulo J,

$$f + J = f^m + J = f^{m+1}g + J = (f^{m+1} + J)(g + J) = (f + J)(g + J) = fg + J.$$

Now, f and g commute, hence

$$f + J = (f + J)^m = (fg + J)^m = f^m g^m + J = e + J$$

as required.
  (b) We proceed by induction on n. Because of (a), there exists $e_1 \in A$ such that $\bar{e}_1 = \bar{f}_1$. Now, $\{\bar{f}_2, \cdots, \bar{f}_n\}$ is a complete set of pairwise orthogonal idempotents in

$$(1 - \bar{e}_1)\bar{A}(1 - \bar{e}_1) = (1 - \bar{e}_1)(A/J)(1 - \bar{e}_1) \cong (1 - e_1)A(1 - e_1)/(1 - e_1)J(1 - e_1).$$

Because the algebra $(1 - e_1)A(1 - e_1)$ admits $1 - e_1$ as its identity, and also $(1 - e_1)J(1 - e_1)$ as its radical, due to Lemma IX.2.3, the induction hypothesis yields pairwise orthogonal idempotents $\{e_2, \cdots, e_n\}$ of $(1 - e_1)A(1 - e_1)$ such that e_i lifts $\bar{f}_i$, that $\sum_{i>1} e_i = 1 - e_1$ and finally $e_1 e_i = 0 = e_i e_1$, for $i > 1$. This completes the proof.   □

Lifting idempotents preserves indecomposability of the associated projective modules and their isomorphism classes.

IX.2.5 Proposition. *Let A be an artinian algebra.*
  (a) *An idempotent $e \in A$ is primitive if and only if $\bar{e} \in \bar{A}$ is primitive.*
  (b) *Let $e_1, e_2 \in A$ be primitive idempotents. Then $e_1 A \cong e_2 A$ if and only if $\bar{e}_1 \bar{A} \cong \bar{e}_2 \bar{A}$.*

Proof. (a) Because of Corollary VIII.4.19, $\bar{e}$ is primitive if and only if $\bar{e}\bar{A}$ is indecomposable which happens if and only if it is simple, because $\bar{A}$ is semisimple. Lemma

VII.4.4 yields that $\mathrm{End}(\bar{e}\bar{A}) \cong \bar{e}\bar{A}\bar{e}$ is a division ring, hence a local algebra. Because $\bar{e}\bar{A}\bar{e} \cong (eAe)/(eJe)$, due to Lemma IX.2.3, we get that eAe is local if and only if so is $\bar{e}\bar{A}\bar{e}$.

(b) Assume that $e_1 A \cong e_2 A$. Because of Theorem VIII.3.16, $e_1 J = (e_1 A)J \cong (e_2 A)J = e_2 J$. Hence, $\bar{e}_1 \bar{A} \cong (e_1 A)/(e_1 J) \cong (e_2 A)/(e_2 J) \cong \bar{e}_2 \bar{A}$. This proves necessity. For sufficiency, let $f : \bar{e}_1 \bar{A} \to \bar{e}_2 \bar{A}$ be an isomorphism and $p_i : e_i A \to \bar{e}_i \bar{A}$, for $i = 1, 2$, the surjective morphisms defined by $e_i\, a \mapsto \bar{e}_i \bar{A}$, for $a \in A$.

$$
\begin{array}{ccccc}
e_1 A & \xrightarrow{\;p_1\;} & \bar{e}_1 \bar{A} & \longrightarrow & 0 \\[4pt]
\scriptstyle g \big\downarrow & & \scriptstyle f \big\downarrow \scriptstyle \cong & & \\[4pt]
e_2 A & \xrightarrow{\;p_2\;} & \bar{e}_2 \bar{A} & \longrightarrow & 0
\end{array}
$$

Projectivity of $e_1 A$ implies the existence of $g : e_1 A \to e_2 A$ such that $p_2 g = f p_1$. Because of Corollary VIII.2.11, g is surjective. But $e_2 A$ is projective, hence g is a retraction. Now, $e_1 A$ is indecomposable, because e_1 is primitive. Hence, g is an isomorphism.   □

We prove that, in the artinian case, there are only finitely many nonisomorphic indecomposable finitely generated projective modules. Moreover, there exists a natural bijection between them and the finitely many isomorphism classes of simple modules of Corollary VIII.3.15.

IX.2.6 Theorem. *Let A be an artinian algebra and $\{e_1, \cdots, e_n\}$ a complete set of primitive pairwise orthogonal idempotents of A. Then:*

  (a) *$e_1 A, \cdots, e_n A$ is a complete list of representatives of the isomorphism classes of finitely generated indecomposable projective A-modules.*
  (b) *$\bar{e}_1 \bar{A}, \cdots, \bar{e}_n \bar{A}$ is a complete list of representatives of the isomorphism classes of simple A-modules.*

Proof. (a) The set $\{e_1 A, \cdots, e_n A\}$ consists of indecomposable projective A-modules, which are finitely generated (even cyclic). Let P be an indecomposable finitely generated projective A-module. Because it is finitely generated, there exist an integer $m > 0$ and an epimorphism $p : A^{(m)} \to P$. Projectivity of P implies that p is a retraction, so P is a direct summand of $A^{(m)}$. Indecomposability of P and Theorem VIII.4.20 imply that it is in fact a direct summand of A_A. Hence, there exists i such that $P \cong e_i A$.

(b) Proposition IX.2.5(b) implies that the map φ from the set of isomorphism classes of indecomposable finitely generated projective A-modules to the set of isomorphism classes of simple $\bar{A}$-modules given by $\varphi : eA \mapsto \bar{e}\bar{A}$, for e an idempotent in A, is injective. Because of Proposition VIII.2.4(e), $A = \bigoplus_{i=1}^{n} e_i A$ implies $J = (\bigoplus_{i=1}^{n} e_i A)J = \bigoplus_{i=1}^{n} (e_i AJ) = \bigoplus_{i=1}^{n} (e_i J)$. Now $e_i J \subseteq e_i A$, for any i, hence $\bar{A} = A/J = (\bigoplus_{i=1}^{n} e_i A)/(\bigoplus_{i=1}^{n} e_i J) \cong \bigoplus_{i=1}^{n} (e_i A)/(e_i J)$, which is a decomposition of $\bar{A}$ into simple $\bar{A}$-modules. Because any simple $\bar{A}$-module is isomorphic to a simple summand of $\bar{A}$, the map φ is also surjective. The statement follows from Lemma VIII.3.14.   □

Combining Theorem IX.2.6 with the Unique Decomposition Theorem VIII.4.20, we get that any finitely generated projective module P over an artinian algebra A decomposes uniquely, up to isomorphism, as a direct sum of finitely many indecomposable projective modules of the form $e_i A$, with the e_i as above. In other words, there exist integers $t_1, \cdots t_n \geq 0$ such that

$$P_A \cong (e_1A)^{(t_1)} \oplus \cdots \oplus (e_nA)^{(t_n)},$$

with $e_i A \ncong e_j A$ for $i \neq i$. We reformulate the preceding results in terms of tops. Recall that the top of an A-module M is the module top $M = M/\text{rad } M$.

IX.2.7 Corollary. *Let A be an artinian algebra. Then:*
   (a) *A finitely generated projective module is indecomposable if and only if its top is simple.*
   (b) *There exists a bijection $eA \mapsto \text{top}(eA)$ between the isomorphism classes of indecomposable finitely generated projective A-modules and of simple A-modules.*

Proof. Part (a) is Proposition IX.2.5(a), and (b) follows from the proof of Theorem IX.2.6.   □

IX.2.8 Examples.
   (a) Let K be a field, $A = \begin{pmatrix} K & 0 \\ K & K \end{pmatrix}$, and $\mathbf{e}_{11} = \begin{pmatrix} 1 & 0 \\ 0 & 0 \end{pmatrix}, \mathbf{e}_{22} = \begin{pmatrix} 0 & 0 \\ 0 & 1 \end{pmatrix}$ the matrix idempotents. The projective modules $P_1 = \mathbf{e}_{11} A = \begin{pmatrix} K & 0 \\ 0 & 0 \end{pmatrix}$ and $P_2 = \mathbf{e}_{22}A = \begin{pmatrix} 0 & 0 \\ K & K \end{pmatrix}$ are indecomposable. The first one is simple (because it is one-dimensional), hence indecomposable. For the second, a direct calculation gives $\text{End}_A(\mathbf{e}_{22}A) \cong \mathbf{e}_{22}A\mathbf{e}_{22} \cong K$, which is local. In this example, one can describe the structure of P_2. Indeed, because of Proposition VII.3.1, $\text{Hom}_A(\mathbf{e}_{11}A, \mathbf{e}_{22}A) \cong \mathbf{e}_{22}A\mathbf{e}_{11} \cong K$, so there exists, up to scalars, a unique nonzero morphism from $S_1 = P_1 = \mathbf{e}_{11}A$ to P_2. Because S_1 is simple, this is a monomorphism embedding S_1 as a direct summand of $\text{soc}P_2$. Because P_2 has simple top and is two-dimensional, we infer that P_2 is uniserial of length two, having S_2 as its top and S_1 as its socle.
      The projective indecomposable *left* A-modules are, similarly, $A\mathbf{e}_{11} = \begin{pmatrix} K & 0 \\ K & 0 \end{pmatrix}$ and $A\mathbf{e}_{22} = \begin{pmatrix} 0 & 0 \\ 0 & K \end{pmatrix}$. While projective indecomposable right modules correspond to the rows of the matrix algebra, projective indecomposable left modules correspond to its columns.
   (b) In Proposition VI.4.12, we outlined a proof, showing that, over any algebra, one can reduce the calculation of a functorial isomorphism involving a projective module to the case where the latter equals the algebra. Here is

a shorter proof valid for finitely generated projective modules over artinian algebras. We were to prove that, if P is projective, then

$$\varphi : P \otimes_A \text{Hom}_B(M, N) \to \text{Hom}_B(\text{Hom}_A(P, M), N)$$

defined by $x \otimes f \mapsto (g \mapsto f(g(x)))$, for $x \in P, f \in \text{Hom}_B(M, N)$ and $g \in \text{Hom}_A(P, M)$, is a functorial isomorphism. Any finitely generated projective module is a direct sum of indecomposable projectives of the form eA, with e a primitive idempotent of A. Thus, setting $P = eA$ yields

$$eA \otimes_A \text{Hom}_B(M, N) \cong e\text{Hom}_B(M, N) \cong \text{Hom}_B(Me, N)$$

$$\cong \text{Hom}_B(\text{Hom}_A(eA, M), N)$$

because of the bimodule structure of M, see Lemma II.3.8, and using Proposition VII.3.1. Moreover, it is easily seen that the composed isomorphism equals φ_{eA}. The result then follows from linearity of the functors.

(c) Let K be a field and $A = KQ/I$ a bound quiver algebra. We have proved in Example VII.3.11(d) that the residual classes of the stationary paths $\{e_i = \epsilon_i + I : i \in Q_0\}$ form a complete set of primitive pairwise orthogonal idempotents in A.

Applying Theorem IX.2.6, a complete list of representatives of isomorphism classes of finitely generated indecomposable projective A-modules is furnished by the modules $\{P_i = e_i A : i \in Q_0\}$ constructed in Example VI.4.6(e). Also, a complete list of representatives of isomorphism classes of simple A-modules is given by the modules $\{S_i = \text{top}(e_i A) : i \in Q_0\}$ constructed in Example VII.4.3(e), see also Example VIII.3.25.

Warning: This statement holds true because A is finite–dimensional over K, hence artinian: for, if $A = K[t]$, which is infinite–dimensional, then, for $\lambda \in K$, the module $S_\lambda = K[t]/\langle t - \lambda \rangle$ is simple, see Example V.3.20(b). Now, $K[t]$ is the path algebra of the one-loop quiver

Applying the equivalence functor of Example IV.4.6(c), we get that S_λ is given by the representation

$$K \circ \overset{\curvearrowright}{} \cdot \lambda$$

where $\cdot\lambda$ is multiplication by λ. Clearly, if $\lambda \neq 0$, then S_λ is not of the form S_i, with i a point.

(d) As first application of the previous Example (b), we prove that a finitely generated bound representation (module) M over a bound quiver algebra $A = KQ/I$ is semisimple if and only if $M(\alpha) = 0$, for any arrow α in Q, see Example VII.5.2(e) and Exercise VII.5.16. In fact, we prove that $M(\alpha) = 0$, for any arrow α, if and only if $M = \bigoplus_{x \in Q_0} S_x^{(\dim_K M(x))}$. Indeed, finitely generated A-modules coincide with finite–dimensional, due to Corollary III.4.9, so the dimensions of the $M(x)$ are finite.

One implication is easy: assume M to be given by the previous formula. For a point x, the simple module S_x has only zero morphisms between coordinate vector spaces, hence the morphisms $M(\alpha)$ obtained by passing to the direct sums are also zero.

Conversely, assume that $M(\alpha) = 0$, for any arrow α. We prove by induction on $\dim_K M$ that M is given by the previous formula. If $\dim_K M = 1$, then there exists a point x such that $M(x) = K$, while $M(y) = 0$ for $y \neq x$. Then $M = S_x$ and we are done in this case. If $\dim_K M > 1$, let x be a point such that $\dim_K M(x) \geq 1$. Define $q : S_x \to M$ as follows: the map $S_x(x) = K \hookrightarrow M(x) = K^{(\dim_K M(x))}$ is the inclusion into the first component of $M(x)$ and, for $y \neq x$, the map $S_x(y) \to M(y)$ is (necessarily) zero. Then q is a morphism of representations, because of our hypothesis that $M(\alpha) = 0$, for any arrow α. Similarly, define $p : M \to S_x$ as follows: the map $M(x) = K^{(\dim_K M(x))} \to S_x(x)$ is the projection onto the first component of $M(x)$ while, for $y \neq x$, the map $M(y) \to S_x(y)$ is zero. Again our hypothesis that M vanishes on the arrows implies that p is a morphism of representations. Then $pq = 1_{S_x}$ implies $M \cong S_x \oplus M'$, with $\dim_K M' = \dim_K M - 1$. The result follows by induction.

(e) As a further application, we compute the socle of a bound representation M over a bound quiver algebra $A = KQ/I$. Because of Example (c) above, we need to find the largest subrepresentation M' of M such that $M'(\alpha) = 0$, for any arrow α. We claim that, for any point $x \in Q_0$,

$$
M'(x) = \begin{cases} M(x), & \text{if } x \text{ is a sink.} \\ \cap_{\alpha:x\to y}\text{Ker } M(\alpha) : M(x) \to M(y) & \text{otherwise.} \end{cases}
$$

With this definition, the restriction of any $M(\alpha) : M(x) \to M(y)$ to $M'(x)$ is zero. We thus set $M'(\alpha) = 0x$, for any arrow α.

Indeed, M' is a subrepresentation of M, because the $M'(\alpha)$ are the restrictions of the $M(\alpha)$. It is semisimple because the $M'(\alpha)$ are zero. There remains to prove that any simple subrepresentation S of M is a direct summand of M'. Example (b) says that there exists $x \in Q_0$ such that $S \cong S_x$. We thus have, for any arrow $\alpha : x \to y$, a commutative square

$$
\begin{array}{ccc}
K = S_x(x) & \xrightarrow{} & S_x(y) = 0 \\
\downarrow & & \downarrow \\
M(x) & \xrightarrow{M(\alpha)} & M(y)
\end{array}
$$

where the vertical arrows are inclusions. But then $S_x(x) \subseteq \text{Ker } M(\alpha)$, for any arrow α with source x, and thus $S_x(x) \subseteq M'(x)$, for any x. This shows that $S_x \subseteq M'$. Therefore, $S = S_x$ is a subrepresentation, and hence a direct summand (because of Theorem VII.5.4), of the semisimple representation M'. This establishes our claim that $M' = \text{soc} M$.

(f) Let A be an artinian K-algebra. The close connection between idempotents and finitely generated indecomposable projectives allows to define a category $S(A)$, which encodes much of the information on A. This category is sometimes called the **spectroid** of A. Let $\{e_1, \cdots, e_n\}$ be a complete set of primitive pairwise orthogonal idempotents. The objects $\{1, \cdots, n\}$ of $S(A)$ are in bijective correspondence with the e_i. Given objects i, j in $S(A)$, the morphism set from i to j is the K-module $S(A)(i, j) = e_i A e_j$, and composition is induced from the multiplication $e_i A e_j \times e_j A e_k \to e_i A e_k$ inside A.

Because $e_i A e_j \cong \mathrm{Hom}_A(e_j A, e_i A)$ and the proof of Proposition VII.3.1, multiplication corresponds to composition of morphisms, this yields indeed a category. Because of the Unique Decomposition Theorem VIII.4.20, this construction does not depend on the particular complete set of primitive pairwise orthogonal idempotents considered. Indeed, if $\{f_1, \cdots, f_m\}$ is another such set of idempotents, then $\oplus_{i=1}^n e_i A \cong \oplus_{j=1}^m f_j A$ yields $m = n$ and, up to order, $e_i A \cong f_i A$, for any i. Therefore, $e_i A e_j \cong \mathrm{Hom}_A(e_j A, e_i A) \cong \mathrm{Hom}_A(f_j A, f_i A) \cong f_i A f_j$, for any i, j.

If K is a field, and $A = KQ/I$ a bound quiver algebra, then $S(A)$ has a concrete description. Because of Example VIII.4.8(a), its objects are (in bijection with) the points of Q, while, if $i, j \in Q_0$, then

$$S(A)(i, j) = e_i A e_j \cong e_i(KQ/I)e_j \cong (\varepsilon_i KQ\varepsilon_j)/(\varepsilon_i I \varepsilon_j)$$

is the K-vector space spanned by the classes modulo I of the paths from i to j, namely the set of linear combinations of these paths with coefficients in K.

IX.2.2 Projective covers

Not only can any module be embedded into an injective one, but also there is a smallest such injective, unique up to isomorphism, namely the injective envelope, see Theorem VI.4.35. The dual statement would say that, given a module M, there exists a smallest projective module P, unique up to isomorphism, covering it, that is, such that there is an epimorphism from P to M.

In this subsection, A denotes a K-algebra. If M is an A-module and $f : P \to M$ an epimorphism with P projective, then $f(\mathrm{rad}\, P) \subseteq \mathrm{rad}\, M$, due to Proposition VIII.2.4(a). By passing to cokernels, we get an epimorphism $\bar{f} : \mathrm{top}\, P \to \mathrm{top}\, M$, see Corollary VIII.2.11. If A is artinian, then, due to Corollary VIII.3.11, $\mathrm{top}\, P$ and $\mathrm{top}\, M$ are semisimple. Because of Corollary IX.2.7(a), it is intuitively clear that P is 'smallest' when $\bar{f}$ is an isomorphism (in other words, all simple summands in the top of P appear in the top of M and there are no 'unused' simples in the top of P).

IX.2.9 Definition. Let M be an A-module. A **projective cover** of M is a pair (P, p), where P is projective and $p : P \to M$ an epimorphism which induces an isomorphism $\bar{p} : \mathrm{top}\, P \to \mathrm{top}\, M$.

For brevity, one sometimes calls P, or the morphism p, the projective cover of M, if no ambiguity may arise. We reformulate results from Chapter VIII.

IX.2.10 Lemma. *Let P, M be finitely generated A-modules, with P projective, and $p : P \to M$ an epimorphism. The following conditions are equivalent:*

(a) *(P, p) is a projective cover.*

(b) *p is a superfluous epimorphism.*

(c) *$\operatorname{Ker} p \subseteq \operatorname{rad} P$.*

Proof. These are Corollaries VIII.2.10 and VIII.2.12.   □

Accordingly, projective covers can equivalently be defined to be superfluous epimorphisms with projective source (this is the exact dual of the definition of injective envelope VI.4.33).

IX.2.11 Examples.

(a) As observed in Chapter VI, not all modules have projective covers. Consider the abelian group $\mathbb{Z}/2\mathbb{Z}$. Assume that $p : P \to \mathbb{Z}/2\mathbb{Z}$ is a projective cover. Let $f : \mathbb{Z} \to \mathbb{Z}/2\mathbb{Z}$ be the projection. Projectivity of $\mathbb{Z}$ gives $g : \mathbb{Z} \to P$ such that $f = pg$.

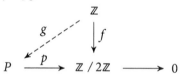

Because f is surjective and p is superfluous, g is surjective. But P is projective, hence g is a retraction. Because $\mathbb{Z}$ is an indecomposable abelian group, g is an isomorphism. We may thus assume that $(P, p) = (\mathbb{Z}, f)$ and so any g making the diagram commute must be an isomorphism. Let $k \geq 3$ be any odd integer and define $g : \mathbb{Z} \to \mathbb{Z}$ by $n \mapsto kn$, for $n \in \mathbb{Z}$. Then g makes the diagram commute but it is not surjective.

(b) Let A be artinian, with radical J, and M an A-module with simple top S. We construct the projective cover of M. There exists a primitive idempotent $e \in A$ such that $S \cong (eA)/(eJ)$. Let $P = eA$ and $p : P \to S$ the projection. Because M has simple top, it has a unique maximal submodule, namely, its radical. Hence the projection $f : M \to S \cong M/(\operatorname{rad} M)$ is a superfluous epimorphism due to Corollary VIII.2.10. Projectivity of P implies the existence of a morphism $g : P \to M$ such that $fg = p$. Surjectivity of P, superfluousness of f and Lemma VIII.2.9(b) yield that g is an epimorphism. Because $\bar{g}$ is the identity morphism 1_S, the morphism g is a projective cover.

(c) Let K be a field and A the algebra given by the quiver

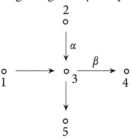

bound by $\langle \alpha\beta \rangle$. Consider the module M given by

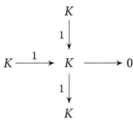

Applying, for instance, the technique of Example VIII.3.25(b), we find that rad M is

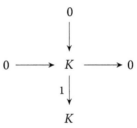

So $M/\mathrm{rad}\, M$ is the sum of the simple modules associated to the points 1 and 2, that is, $M/\mathrm{rad}\, M \cong S_1 \oplus S_2$. Let $P = e_1A \oplus e_2A$ be the direct sum of the indecomposable projective modules associated to 1 and 2. Then $P/\mathrm{rad}\, P \cong S_1 \oplus S_2$ because $e_iA/\mathrm{rad}(e_iA) \cong S_i$, for $i = 1, 2$. A quick calculation shows that, up to scalars, there exists a unique morphism $f_1 : e_1A \to M$ which is injective with cokernel S_2, and, similarly, a unique morphism $f_2 : e_2A \to M$ which is injective with cokernel S_1. One sees that $f = (f_1\, f_2) : P \to M$ is surjective with kernel rad M. Hence, (P, f) is a projective cover of M.

(d) Let A be an artinian K-algebra and P an indecomposable finitely generated projective A-module. An epimorphism $f : P \to L$ induces, because of Corollary VIII.2.11, an epimorphism from the top of P to that of L. The top of P is simple, due to Corollary IX.2.7, and, because L is finitely generated (for, so is P), we have top$L \neq 0$ and thus top$L$ is simple and isomorphic to top$P$. Therefore, P is a projective cover of L.

More generally, if P is as above and M_A is arbitrary, then P is a projective cover of the image of any morphism $P \to M$.

The proof of our main theorem is inspired from Example IX.2.11(b) above.

IX.2.12 Theorem. *Let A be an artinian algebra. Any finitely generated A-module admits a finitely generated projective cover, unique up to isomorphism.*

Proof. *Existence.* Let M be a finitely generated A-module. Because of Corollary VIII.3.11, its top is semisimple of finite length, say $\mathrm{top}M = \oplus_{i=1}^{t} S_i$, where each S_i is simple. For any i, with $1 \leqslant i \leqslant t$, there exists an indecomposable finitely generated projective module P_i such that $\mathrm{top}\, P_i \cong S_i$. The direct sum $P = \oplus_{i=1}^{t} P_i$ is finitely generated and projective. Moreover,

$$\mathrm{top}\, P \cong \bigoplus_{i=1}^{t} \mathrm{top}\, P_i \cong \bigoplus_{i=1}^{t} S_i \cong \mathrm{top}\, M$$

using Exercise VIII.2.4. Let $f : M \to \mathrm{top}\, M$ and $p : P \to \mathrm{top}\, P$ be the projections. Because P is projective, there exists $g : P \to M$ such that $p = fg$. Because f is superfluous, g is an epimorphism. But g induces an isomorphism between the tops, hence it is a projective cover.

Uniqueness. Assume that (P, p) and (P', p') are projective covers of M. It follows from projectivity of P and P' and the diagram

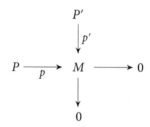

that there exist $h: P' \to P, h': P \to P'$ such that $ph = p'$ and $p'h' = p$. Because p and p' are superfluous, h and h' are epimorphisms. Because of Lemma VIII.4.16, both $hh' : P \to P$ and $h'h : P' \to P'$ are isomorphisms. Hence, so are h and h'. $\square$

The last part of the proof says that, if $p : P \to M$ is a projective cover and $p' : P' \to M$ an epimorphism with P' projective, then any morphism $h : P' \to P$ such that $ph = p'$ is an epimorphism, because p is superfluous. Projectivity of P implies that h is a retraction. Therefore, the projective cover P is (isomorphic to) a direct summand of any other projective 'covering' M. In that sense, P is indeed the smallest projective covering M. Another consequence of the proof is the next corollary, which is dual to Corollary VIII.2.19 and is useful in practical computations.

IX.2.13 Corollary. *Let A be artinian and M a finitely generated A-module. The projective cover of M is isomorphic to the projective cover of top M.* $\square$

Let A be artinian and S_i a simple module. As we saw in Example IX.2.11(b), its projective cover is an indecomposable projective module $P_i = e_i A$ of which S_i is the top. The kernel of the projective cover morphism $P_i \to S_i$ is the radical, because we have a short exact sequence

$$0 \to \operatorname{rad} P_i \to P_i \to \operatorname{top} P_i = S_i \to 0$$

and thus $\operatorname{rad} P_i$ is the unique maximal submodule of P_i.

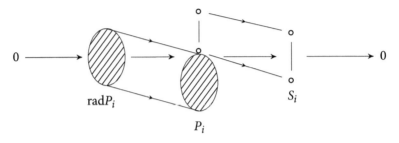

IX.2.3 Duality and injective modules

Theorem IX.2.6 characterises indecomposable finitely generated projective modules over an artinian algebra, and exhibits a bijection between them and simple modules. In order to obtain similar results for injectives, we need more hypotheses on the algebra.

In this subsection, we assume that K is a field, and A a finite–dimensional K-algebra. In this context, we proved in Example IV.4.8(b) that the functor $D = \operatorname{Hom}_K(-, K)$: $\operatorname{mod} A \to \operatorname{mod} A^{op}$ is a duality sending finitely generated right A-modules to finitely generated left A-modules. Its quasi-inverse is $D = \operatorname{Hom}_K(-, K)$: $\operatorname{mod} A^{op} \to \operatorname{mod} A$. Because the dual of a projective module is injective, we use this duality in order to describe injective modules.

IX.2.14 Lemma. *Let A be a finite–dimensional K-algebra and M an A-module. Then there exists an anti-isomorphism between the lattices of submodules of M and DM.*

Proof. Let L be a submodule of M. The projection $p : M \to M/L$ induces a monomorphism $Dp : D(M/L) \to DM$. We define $\varphi : S(M) \to S(DM)$ by mapping the submodule L of M to the submodule $D(M/L)$ of DM. The latter consists of the linear forms $f : M \to K$ such that $f(L) = 0$. That is, $D(M/L)$ equals the orthogonal complement $L^{\perp} = \{ f \in DM : f(L) = 0 \}$ of L in M. The rest of the proof is the straightforward verification that $\varphi : L \mapsto L^{\perp}$, for $L \in S(M)$, is an anti-isomorphism between the corresponding lattices. $\qquad \square$

From a graphical point of view, having an anti-isomorphism between lattices means turning the Hasse quiver upside-down.

Given a submodule L of M, the short exact sequence $0 \to L \to M \to M/L \to 0$ induces another short exact sequence $0 \to D(M/L) \to DM \to DL \to 0$, so we have $DM/L^\perp \cong DL$.

Let $\{e_1, \cdots, e_t\}$ be a complete set of primitive pairwise orthogonal idempotents of A, let J be the radical of A and $\bar A = A/J$. For any i, let $\bar e_i = e_i + J$. A full set of representatives of the isomorphism classes of simple right modules is $\{\bar e_1 \bar A, \cdots, \bar e_t \bar A\}$ and one of the simple left modules is $\{\bar A \bar e_1, \cdots, \bar A \bar e_t\}$. We show that they are dual to each other.

IX.2.15 Lemma. *Let A be a finite–dimensional K-algebra of radical J and set $\bar A = A/J$.*
  (a) *Let $e \in A$ be a primitive idempotent, then $D(\bar e \bar A) = \bar A \bar e$.*
  (b) *For a nonzero finitely generated module M, we have $\mathrm{soc}\, DM = D(\mathrm{top}\, M)$ and $\mathrm{top}\, DM = D(\mathrm{soc}\, M)$.*

Proof. (a) If $S = \bar e \bar A$ is a simple right A-module, then, because of Lemma IX.2.14, DS is a simple left A-module. Now $S\bar e = \bar e \bar A \bar e \neq 0$. Hence, there exists a linear form $f: S \to K$ such that $f(S\bar e) \neq 0$. Because $f(x\bar e) = (\bar e f)(x)$, for $x \in S$, we get $\bar e(DS) \neq 0$. Thus, $DS = \bar A \bar e$.

(b) The radical of M is the intersection of its maximal submodules, hence its orthogonal complement $(\mathrm{rad}\, M)^\perp$ is the sum of the minimal (= simple) submodules of DM, that is, the socle of the latter. Therefore, $\mathrm{soc}\, DM = (\mathrm{rad}\, M)^\perp = D(M/\mathrm{rad}\, M) = D(\mathrm{top}\, M)$. The second statement follows by duality (write $DM = N$ and apply the first statement to N). $\qquad\square$

IX.2.16 Theorem. *Let A be a finite–dimensional K-algebra and $\{e_1, \cdots, e_n\}$ a complete set of primitive pairwise orthogonal idempotents of A. Then $\{D(Ae_1), \cdots, D(Ae_n)\}$ is a complete set of representatives of the isomorphism classes of indecomposable finitely generated injective A-modules.*

Proof. Because of Theorem IX.2.6, $\{Ae_1, \cdots, Ae_n\}$ is a complete set of representatives of isomorphism classes of indecomposable finitely generated projective *left* A-modules. Let I be an indecomposable injective right A-module. Then DI is an indecomposable projective left A-module, so there exists i such that $DI \cong Ae_i$. Therefore, $I \cong D(Ae_i)$. Indecomposable finitely generated projective modules are identified by their simple tops, see Corollary IX.2.7. Lemma IX.2.15(b) says that the top of an indecomposable finitely generated projective module is isomorphic to the socle of its dual. This finishes the proof. $\qquad\square$

Therefore, besides the bijection $eA \mapsto \mathrm{top}(eA)$ of Corollary IX.2.7 between isomorphism classes of indecomposable finitely generated projective modules and simple modules, there is also a bijection $D(Ae) \mapsto \mathrm{soc}\, D(Ae)$ between isomorphism classes of indecomposable finitely generated injectives and simple modules. Moreover, $\mathrm{soc}\, D(Ae) \cong \mathrm{top}(eA)$, because of Lemma IX.2.15(b)

IX.2.17 Corollary. *Let A be a finite–dimensional K-algebra. The following conditions are equivalent for an A-module I.*
  (a) *I is an indecomposable injective module.*

(b) *I is isomorphic to an indecomposable direct summand of $D(_A A)$.*
(c) *There exists a primitive idempotent $e \in A$ such that $I \cong D(Ae)$.*
(d) *I is an injective module with a simple socle.*
(e) *I is the injective envelope of a simple module.*

Proof. The equivalence of (a), (b), (c) follows from Theorem IX.2.16. The duals of the statements that a projective module is indecomposable if and only if its top is simple, and if and only if it is the projective cover of a simple module give the equivalence of (a), (d), (e).   □

In particular, $D(_A A) = \bigoplus_{i=1}^n D(Ae_i)$ is an injective cogenerator for the category mod A (for, any module M_A is embedded in an injective module I, therefore, due to (b), in a direct sum of copies of $D(_A A)$). Moreover, the injective envelope of a simple module S_i is an indecomposable injective module $I_i = D(Ae_i)$ having S_i as its socle. The cokernel of the injective envelope morphism $S_i = \text{soc} I_i \to I_i$ is the unique maximal quotient module I_i/S_i of I_i. We thus get a short exact sequence dual to the one at the end of Subsection IX.2.2.

$$0 \to S_i \to I_i \to I_i/S_i \to 0.$$

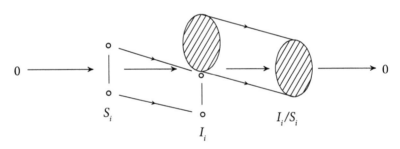

Comparing this picture with the one at the end of Subsection IX.2.2 shows clearly how the duality functor D acts on submodule lattices.

IX.2.18 Examples.

(a) We continue Example IX.2.8(a). Let K be a field, $A = \begin{pmatrix} K & 0 \\ K & K \end{pmatrix}$ and $e_{11} = \begin{pmatrix} 1 & 0 \\ 0 & 0 \end{pmatrix}, e_{22} = \begin{pmatrix} 0 & 0 \\ 0 & 1 \end{pmatrix}$ the matrix idempotents. We compute the indecomposable injective modules. We have

$$I_1 = D(Ae_{11}) = D\begin{pmatrix} K & 0 \\ K & 0 \end{pmatrix} \text{ and } I_2 = D(Ae_{22}) = D\begin{pmatrix} 0 & 0 \\ 0 & K \end{pmatrix}.$$

Thus, the one-dimensional module I_2 equals the simple module S_2, while I_1 is two-dimensional with simple socle S_1. We now compute I_1/S_1. The isomorphisms

$$\text{Hom}_A(I_1, I_2) \cong \text{Hom}_A(D(Ae_{11}), D(Ae_{22})) \cong \text{Hom}_{A^{op}}(Ae_{22}, Ae_{11})$$
$$\cong e_{22}Ae_{11} \cong K$$

where we used duality and (the left module version of) Proposition VII.3.1, show the existence of a nonzero morphism $I_1 \rightarrow S_2 = I_2$, unique up to scalars, and surjective because S_2 is simple. Because its kernel is S_1, we get that $I_2 = S_2 \cong I_1/S_1$ and I_1 is uniserial of length two, having S_2 as top and S_1 as socle. Thus, $I_1 \cong P_2$.

(b) Let K be a field and $A = KQ/I$ a bound quiver algebra. We proved in Theorem IX.2.16(b) that the injective modules of Example VI.4.16 form a complete set of representatives of the isomorphism classes of indecomposable finitely generated injective A-modules, each one attached to a point of the quiver.

A remarkable fact is that the bijection we constructed between indecomposable projective and injective modules extends to a functor. We define the **Nakayama functor** to be

$$\nu_A = D\,\text{Hom}_A(-, A) \colon \text{mod}\,A \rightarrow \text{mod}\,A.$$

Because ν_A is the composite of a duality and a left exact contravariant functor, it is covariant and right exact. In order to understand how this functor acts, we need a further homological lemma, which, besides its intrinsic interest, shows how one can use efficiently finite presentations.

IX.2.19 Lemma. *Let A, B be K-algebras, with A noetherian, and $L_{A}, {}_{A,B}M_{A,B}\, I$ modules with L_A finitely generated and ${}_B I$ injective. Then there exists a functorial isomorphism*

$$\varphi_L : L \otimes_A \text{Hom}_B(M, I) \rightarrow \text{Hom}_B(\text{Hom}_A(L, M), I)$$

given by $x \otimes f \mapsto (g \mapsto f(g(x)))$, for $x \in L, f \in \text{Hom}_B(M, I)$ and $g \in \text{Hom}_A(L, M)$.

Proof. Assume first that $L = A$. Then it is easily seen from their explicit expressions that

$$\varphi_A : A \otimes_A \text{Hom}_B(M, I) \rightarrow \text{Hom}_B(\text{Hom}_A(A, M), I)$$

is the composite of the isomorphisms

$$A \otimes_A \text{Hom}_B(M, I) \cong \text{Hom}_B(M, I) \cong \text{Hom}_B(\text{Hom}_A(A, M), I)$$

of Theorems II.3.9 and VI.2.8. In particular, φ_A is an isomorphism. Because all functors shown are K-linear, we infer that, if F is a finitely generated free module, then φ_F is an isomorphism.

Let L be a finitely generated A-module. Due to Corollary VII.2.15, it is finitely presented, that is, there exist finitely generated free modules F_1, F_0 and an exact sequence

$$F_1 \to F_0 \to L \to 0$$

Because tensor product functors are right exact, we have an exact sequence

$$F_1 \otimes_A \mathrm{Hom}_B(M, I) \to F_0 \otimes_A \mathrm{Hom}_B(M, I) \to L \otimes_A \mathrm{Hom}_B(M, I) \to 0.$$

The left exact functor $\mathrm{Hom}_A(-, M)$ yields an exact sequence

$$0 \to \mathrm{Hom}_A(L, M) \to \mathrm{Hom}_A(F_0, M) \to \mathrm{Hom}_A(F_1, M)$$

Because $_B I$ is injective, the functor $\mathrm{Hom}_B(-, I)$ is exact, and we get an exact sequence

$$\mathrm{Hom}_B(\mathrm{Hom}_A(F_1, M), I) \to \mathrm{Hom}_B(\mathrm{Hom}_A(F_0, M), I) \to \mathrm{Hom}_B(\mathrm{Hom}_A(L, M), I) \to 0$$

We deduce a commutative diagram with exact rows

$$
\begin{array}{ccccccc}
F_1 \otimes_A \mathrm{Hom}_B(M, I) & \longrightarrow & F_0 \otimes_A \mathrm{Hom}_B(M, I) & \longrightarrow & L \otimes_A \mathrm{Hom}_B(M, I) & \longrightarrow & 0 \\
\downarrow{\varphi_{F_1}} & & \downarrow{\varphi_{F_0}} & & \downarrow{\varphi_L} & & \\
\mathrm{Hom}_B(\mathrm{Hom}_A(F_1, M), I) & \longrightarrow & \mathrm{Hom}_B(\mathrm{Hom}_A(F_0, M), I) & \longrightarrow & \mathrm{Hom}_B(\mathrm{Hom}_A(L, M), I) & \longrightarrow & 0
\end{array}
$$

Because φ_{F_1} and φ_{F_0} are isomorphisms, so is φ_L, see Exercise II.4.8. $\qquad\square$

IX.2.20 Corollary. *Let K be a field and A a finite–dimensional K-algebra. Then, for any finitely generated A-modules L, M, one has a functorial isomorphism*

$$L \otimes_A DM \cong D\mathrm{Hom}_A(L, M).$$

Proof. Here, $B = K$ and $_B I = {}_K K$: indeed, $_K K$ is injective, because of Example VI.4.16(a). Our statement follows immediately from Lemma IX.2.19. $\qquad\square$

If we set $M = A$ in the previous corollary, we get that $\nu_A = D\mathrm{Hom}_A(-, A)$ is isomorphic to the functor $- \otimes_A DA$: the (right exact) Nakayama functor may be expressed as a tensor product.

IX.2.21 Theorem. *Let K be a field and A a finite–dimensional K-algebra. The functor ν_A induces an equivalence between the full subcategory of $\mathrm{mod}\, A$ formed by the projective modules and that formed by the injective modules. Its quasi-inverse is $\nu_A^{-1} = \mathrm{Hom}_A(DA, -)$.*

Proof. Let e be an idempotent in A. The functorial isomorphism

$$\nu_A(eA) = D\operatorname{Hom}_A(eA, A) \cong D(Ae)$$

shows that the restriction of the Nakayama functor ν_A to the projectives has its image among the injectives. The statement then follows from the functorial isomorphisms

$$\nu_A^{-1}(D(Ae)) = \operatorname{Hom}_A(DA, D(Ae)) \cong \operatorname{Hom}_{A^{op}}(Ae, A) \cong eA. \qquad \square$$

The following result is very useful for computing morphisms.

IX.2.22 Proposition. *Let K be a field, A a finite–dimensional K-algebra, $e \in A$ an idempotent and M an A-module. There exists a functorial isomorphism of K-vector spaces*

$$\operatorname{Hom}_A(eA, M) \cong D\operatorname{Hom}_A(M, D(Ae)) \cong Me.$$

Proof. The result follows from the sequence of functorial isomorphisms

$$\begin{aligned} D\operatorname{Hom}_A(M, D(Ae)) &\cong D\operatorname{Hom}_{A^{op}}(Ae, DM) \cong D(eDM) \\ &\cong D(D(Me)) \cong Me \\ &\cong \operatorname{Hom}_A(eA, M) \end{aligned}$$

where the second and the last isomorphism come from Proposition VII.3.1. $\square$

IX.2.23 Example. Let K be a field and $A = KQ/I$ a bound quiver algebra. Let M be an A-module. Under the equivalence between modules and bound representations, see Example IV.4.6(c), the representation V attached to M is such that, for $x \in Q_0$, we have $V(x) = Me_x$. Hence, $\dim_K V(x)$, which is the dimension of Me_x, equals the number of occurrences of the simple module S_x as composition factor of M. But, on the other hand, $V(x) = \operatorname{Hom}_A(e_x A, M) = D\operatorname{Hom}_A(M, D(Ae_x))$. Hence, a basis of each of these vector spaces is in bijection with the appearances of S_x as composition factor of M.

Exercises of Section IX.2

1. Let A be artinian and $e \in A$ a primitive idempotent. Prove that there exists a complete set of primitive pairwise orthogonal idempotents $\{e, e_2, \cdots, e_n\}$.

2. Prove that any finitely generated projective module over a local artinian algebra is free.

3. Let A be artinian and e_1, e_2 primitive idempotents such that $P_1 = e_1 A$ is not isomorphic to $P_2 = e_2 A$. Prove that $\operatorname{Hom}_A(P_1, P_2) \neq 0$ if and only if $e_2(\operatorname{rad} A)e_1 \neq 0$.

4. Let A be an artinian algebra and $\{e_1, \cdots, e_n\}, \{e_1', \cdots, e_m'\}$ complete sets of primitive pairwise orthogonal idempotents of A. Prove that $m = n$ and there exists $a \in A$ invertible such that, for any i, we have $e_i' = a e_i a^{-1}$, up to order.

5. Let A be artinian and M a finitely generated A-module. Prove that a pair (P, p) is a projective cover of M if and only if, for any epimorphism $p': P' \to M$, with P' projective, there exists an epimorphism $h: P' \to P$ such that $ph = p'$.

6. Let A be artinian and M a finitely generated module with simple top. Let P be a projective cover of M and $f : P' \to M$ a nonzero morphism with P' indecomposable projective not isomorphic to P. Prove that $\operatorname{Im} f \subseteq \operatorname{rad} M$.

7. Let A be artinian and $M_1, \cdots, M_m$ finitely generated A-modules. Assume that (P_i, p_i) is a projective cover of M_i, for any i, let also $P = \oplus_{i=1}^m P_i$ be their direct sum and $p: P \to M$ the morphism induced from the p_i by passing to direct sums. Prove that (P, p) is a projective cover of $\oplus_{i=1}^m M_i$.

8. Let A be artinian, I an ideal of A and M an A/I-module. If P is a projective cover of M considered as an A-module, prove that P/PI is a projective cover of M as an A/I-module.

9. Let $f : M \to N$ be a superfluous epimorphism and $p : P \to M$ a morphism, with P projective. Prove that p is a projective cover if and only if $fp : P \to N$ is a projective cover.

10. Let A be artinian and M a finitely generated A-module. Prove that if a projective cover of M is indecomposable, then so is M. Is the converse true?

11. Let A be an artinian K-algebra and M an indecomposable A-module. Prove that:
    (a) If M is nonprojective, then a projective cover $p : P \to M$ belongs to $\operatorname{rad}_A(P, M)$.
    (b) If M is noninjective, then an injective envelope $j : M \to E$ belongs to $\operatorname{rad}_A(M, E)$.

12. Let A be artinian and M a nonzero finitely generated A-module such that all its nonzero quotients are indecomposable. Prove that top M is simple.

13. Let K be a field, A a finite–dimensional K-algebra, M a finitely generated A-module and L, L_1, L_2 submodules of M. Prove the following properties of orthogonal complements:
    (a) $L_1 \subseteq L_2$ implies $L_2^{\perp} \subseteq L_1^{\perp}$.
    (b) $L^{\perp\perp} = L$.
    (c) $(L_1 + L_2)^{\perp} = L_1^{\perp} \cap L_2^{\perp}$.
    (d) $(L_1 \cap L_2)^{\perp} = L_1^{\perp} + L_2^{\perp}$.
    (e) L is semisimple if and only if so is $L^{\perp}$.

14. Let K be a field and A a finite–dimensional K-algebra. Prove that one has an isomorphism of functors $D \cong \operatorname{Hom}_A(-, DA)$.

15. Let K be a field, A a finite–dimensional K-algebra and I a finitely generated nonzero injective A-module. Prove that the following conditions are equivalent:

(a) I is indecomposable.

(b) I is an injective envelope of each of its submodules.

(c) Any pair of nonzero submodules of I has a nonzero intersection.

16. Let K be a field, A a finite–dimensional K-algebra and M a finitely generated A-module. Prove that the following conditions are equivalent:

(a) M is a faithful module.

(b) There exist an integer $m > 0$ and a monomorphism $A_A \to M^{(m)}$.

(c) There exist an integer $n > 0$ and an epimorphism $M^{(n)} \to D(_A A)$.

17. Let K be a field. Describe the isomorphism classes of finitely generated indecomposable projective and indecomposable injective modules over each of the algebras below.

(a) $A = \frac{K[t]}{\langle t^n \rangle}$, for an integer $n \geq 1$.

(b) $A = T_n(K) = \begin{pmatrix} K & & 0 \\ \vdots & \ddots & \\ K & \cdots & K \end{pmatrix}$, for an integer $n \geq 2$.

(c) $A = \begin{pmatrix} K & 0 & 0 \\ K & K & 0 \\ K & 0 & K \end{pmatrix}$.

(d) $A = \begin{pmatrix} K & 0 & 0 & 0 \\ K & K & 0 & 0 \\ K & 0 & K & 0 \\ K & K & K & K \end{pmatrix}$.

(e) A is the path algebra of the quiver

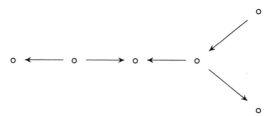

(f) A is the path algebra of the quiver

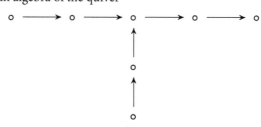

(g) A is given by the quiver $5 \xrightarrow{\alpha} 4 \xrightarrow{\beta} 3 \xrightarrow{\gamma} 2 \xrightarrow{\delta} 1$ bound by the ideal $\langle \alpha\beta, \beta\gamma\delta \rangle$.

18. Let K be a field, Q the quiver $1 \to 2 \to 3$ and $A = KQ$ its path algebra. Prove that the free module A_A is given by the representation

$$K \xrightarrow{\begin{pmatrix} 1 \\ 1 \end{pmatrix}} K^2 \xrightarrow{\begin{pmatrix} 1 & 1 \\ 1 & 1 \\ 1 & 1 \end{pmatrix}} K^3.$$

19. Let K be a field and $A = KQ/I$ a bound quiver algebra and $x, y \in Q_0$. Prove that:
   (a) S_y is a composition factor of P_x if and only if there exists a path w from x to y in Q such that $w \notin I$.
   (b) S_y is a composition factor of I_x if and only if there exists a path w from y to x in Q such that $w \notin I$.

IX.3 Morita equivalence

IX.3.1 The Morita equivalence theorem

It is standard practice in mathematics to replace the study of a difficult problem by that of a supposedly easier one, which helps solving the first. In this subsection, we give a criterion allowing to verify whether two module categories are equivalent or not. We then see in the next subsection how it allows to simplify the general problem of describing a module category.

Here, A and B are algebras over the same commutative ring K. We need to consider at the same time right and left module structures. Let T_A be a right A-module and $B = \text{End}_A T$ its endomorphism algebra. Then T has a natural left B-module structure given by $b \cdot t = b(t)$ for $b \in B, t \in T$. As we have seen in Subsection II.3.2, this endows T with a B-A-bimodule structure. We wish to compare the categories $\text{Mod}\,A$ and $\text{Mod}\,B$. There exists a pair of functors linking them. Indeed, for an A-module M, because of Lemma II.3.8, $\text{Hom}_A(T, M)$ has a natural B-module structure, while, for a B-module X, because of Lemma VI.2.5, $X \otimes_B T$ has a natural A-module structure. Due to Theorem VI.2.17, the functors $\text{Hom}_A(T, -)$ and $- \otimes_B T$ are adjoint. The module categories of A and B are equivalent if these functors are quasi-inverse. Now, adjoint functors are usually not quasi-inverse, but one may ask what condition(s) to impose on T so that they become so. We now show that an adjunction induces natural functorial morphisms.

Let C, D be categories and $F : C \to D, G : D \to C$ functors. If (G, F) is an adjoint pair, then, due to the definition, for any objects X in C and M in D, we have a functorial bijection $\psi_{X,M} : D(GX, M) \to C(X, FM)$. If we let $M = GX$, we get a morphism

$$\eta_X = \psi_{X, GX}(1_{GX}) : X \to FGX$$

associated to each object X in C. Similarly, if we set $X = FM$, we get a morphism

$$\varepsilon_M = \psi^{-1}_{FM, M}(1_{FM}) : GFM \to M$$

associated to each object M in $\mathcal{D}$. It is easily shown that this data defines a functorial endomorphism η in $\mathcal{C}$ and a functorial endomorphism ε in $\mathcal{D}$, see Exercise IX.3.6. The morphisms η and ε are respectively called the **co-unit** and the **unit** of the adjunction.

IX.3.1 Lemma. *Let A, B be K-algebras, ${}_BT_A$ a bimodule, M an A-module and X a B-module.*

(a) *The co-unit of the adjunction $(- \otimes_B T, \mathrm{Hom}_A(T, -))$ is $\eta_X : X \to \mathrm{Hom}_A(T, X \otimes_B T)$ defined by $\eta_X(x)(t) = x \otimes t$, for $x \in X, t \in T$.*

(b) *The unit of the adjunction $(-\otimes_B T, \mathrm{Hom}_A(T, -))$ is $\varepsilon_M : \mathrm{Hom}_A(T, M) \otimes_B T \to M$ defined by $\varepsilon_M(f \otimes t) = f(t)$, for $f \in \mathrm{Hom}_A(T, M), t \in T$.*

Proof. Theorem VI.2.17 exhibits inverse isomorphisms

$$\varphi_{X,M} : \mathrm{Hom}_B(X, \mathrm{Hom}_A(T, M)) \to \mathrm{Hom}_A(X \otimes_B T, M),$$

given by $f \mapsto (x \otimes t \mapsto f(x)(t))$, for $f \in \mathrm{Hom}_B(X, \mathrm{Hom}_A(T, M)), x \in X, t \in T$, and

$$\psi_{X,M} : \mathrm{Hom}_A(X \otimes_B T, M) \to \mathrm{Hom}_B(X, \mathrm{Hom}_A(T, M)),$$

given by $g \mapsto (x \mapsto (t \mapsto g(x \otimes t)))$, for $g \in \mathrm{Hom}_A(X \otimes_B T, M)$, $x \in X, t \in T$. If one sets in the first isomorphism $M_A = X \otimes_B T_A$ and $f = 1_{X \otimes_B T}$, one gets the formula for η_X. And setting in the second $X = \mathrm{Hom}_A(T, M)$ and $g = 1_{\mathrm{Hom}_A(T,M)}$ yields the one for ε_M. $\qquad\square$

Because of the formula expressing it, the unit of the adjunction may be viewed as an evaluation morphism.

We ask when $\mathcal{C} = \mathrm{Mod}\, A$ and $\mathcal{D} = \mathrm{Mod}\, B$ are equivalent categories. Let $G : \mathrm{Mod}\, B \to \mathrm{Mod}\, A, F : \mathrm{Mod}\, A \to \mathrm{Mod}\, B$ be an adjoint pair. Suppose that, for any A-module M and any B-module X, the maps $\varepsilon_M : GF(M) \to M$ and $\eta_X : X \to FG(X)$ are isomorphisms. Because they are functorial, this gives that F, G are quasi-inverse, as desired. We have the following lemma.

IX.3.2 Lemma. *Let A be a K-algebra, T_A an A-module and $B = \mathrm{End}\, T_A$. Then:*

(a) *η_B is an isomorphism.*

(b) *ε_T is an isomorphism.*

Proof. We only prove (a), because the proof of (b) is similar. The morphism $\eta_B : B \to \mathrm{Hom}_A(T, B \otimes_B T) \cong \mathrm{Hom}_A(T, T)$ is given by $b \mapsto (t \mapsto bt)$, for $b \in B, t \in T$, that is, it is induced from left multiplication of ${}_BT$. On the other hand, the right B-module structure of $B = \mathrm{Hom}_A(T, T)$ is given by $(fb)(t) = f(bt)$, for an A-linear map $f : T \to T$, in other words, it is determined by the left B-module structure of T. Thus, η_B is a B-module isomorphism. $\qquad\square$

We thus need to set conditions on T so that η and ε become isomorphisms on their respective categories. If the functors $\mathrm{Hom}_A(T, -)$ and $- \otimes_B T_A$ are equivalences, they send a projective module onto another, see Exercise IX.3.1(e). Hence, for a start, $T_A \cong B_B \otimes_B T_A$ should be projective. In this case, $\mathrm{Hom}_A(T, -)$ is an exact functor.

We recall a definition: an A-module T is said to **generate** another module M if there exist a set Λ and an epimorphism $T^{(\Lambda)} \to M$. It is called a **generator** of the module category $\mathrm{Mod}\,A$ if it generates any A-module, see Exercise VI.4.22. Because any module is quotient of a free module, the typical example of a generator of $\mathrm{Mod}\,A$ is the module $T_A = A_A$. In fact, any generator of the module category has A_A as direct summand of a finite power.

IX.3.3 Lemma. *An A-module T is a generator of $\mathrm{Mod}\,A$ if and only if there exists an integer $m > 0$ such that A_A is a direct summand of $T^{(m)}$.*

Proof. Because A_A is a generator of $\mathrm{Mod}\,A$, a module T generates this category if and only if there exist a set Λ and an epimorphism $f = (f_\lambda)_{\lambda \in \Lambda} : T^{(\Lambda)} \to A_A$. The module A_A is generated by 1. Hence, there exists a family $(x_\lambda)_{\lambda \in \Lambda} \in T^{(\Lambda)}$ such that $1 = f((x_\lambda)_\lambda) = \sum_\lambda f_\lambda(x_\lambda)$. Because this sum is finite, there exists a finite subset $\{\lambda_1, \cdots, \lambda_m\} \subseteq \Lambda$ such that $f_{\lambda_i}(x_{\lambda_i}) \neq 0$ so that $1 = \sum_{i=1}^{m} f_{\lambda_i}(x_{\lambda_i})$. Therefore, the image of $(f_{\lambda_1}, \cdots, f_{\lambda_m}) : T^{(m)} \to A$ contains 1, hence it is an epimorphism. Thus, T is a generator if and only if there exists an integer $m > 0$ and an epimorphism $T^{(m)} \to A$. Now, A_A is projective, hence any such epimorphism is a retraction, and A_A is a direct summand of $T^{(m)}$. $\square$

Because B_B is a generator of $\mathrm{Mod}\,B$, and equivalences preserve epimorphisms, $B_B \otimes_B T_A \cong T_A$ should be a generator of $\mathrm{Mod}\,A$. This, and the previous remark that T_A should be projective, motivate the following definition.

IX.3.4 Definition. Let A be an algebra. A finitely generated projective A-module which is also a generator of $\mathrm{Mod}\,A$ is a **progenerator** of $\mathrm{Mod}\,A$.

The hypothesis that a progenerator P is finitely generated will prove crucial, because it implies that the functor $\mathrm{Hom}_A(P, -)$ commutes with arbitrary direct sums, see Example IV.2.8(b).

IX.3.5 Example. Let A be an artinian algebra and $\{e_1, \cdots, e_n\}$ a complete set of primitive pairwise orthogonal idempotents. Then a module P is a progenerator if and only if there exist integers $t_i > 0$ such that $P = (e_1 A)^{(t_1)} \oplus \cdots \oplus (e_n A)^{(t_n)}$. For, assume that P is a progenerator. Because of Lemma IX.3.3, there exists an integer $m > 0$ and an A-module P' such that $P^{(m)} \cong A_A \oplus P'$. Because P is finitely generated projective, so is P'. Due to Theorem IX.2.6, P' is a direct sum of indecomposable finitely generated projective modules. Then P takes the required form, with the t_i strictly positive (because $P^{(m)} \cong A \oplus P'$). The converse is evident.

IX.3.6 Theorem. Morita's theorem. *Let A, B be K-algebras. Then there exists a K-linear equivalence $F: \mathrm{Mod}\,A \to \mathrm{Mod}\,B$ if and only if there exists a progenerator P_A such that $\mathrm{End}_A P \cong B$. In this case, the equivalence is given by the functors $F = \mathrm{Hom}_A(P, -)$ and $G = - \otimes_B P$.*

Proof. Sufficiency. Let P_A be a progenerator of $\mathrm{Mod}\,A$ and $B = \mathrm{End}_A P$. Let X be an arbitrary B-module, we claim that $\eta_X : X \to \mathrm{Hom}_A(P, X \otimes_B P)$ is an isomorphism.

Because of Corollary III.4.10, there exist sets Δ, Λ and a free presentation

$$B_B^{(\Delta)} \to B_B^{(\Lambda)} \to X_B \to 0.$$

The functor $- \otimes_B P$ is right exact and commutes with arbitrary direct sums, due to Proposition VI.2.7(c). On the other hand, because P_A is finitely generated projective, $\mathrm{Hom}_A(P, -)$ is exact and commutes with *arbitrary* direct sums. We deduce an exact sequence:

$$\mathrm{Hom}_A(P, B \otimes_B P)^{(\Delta)} \to \mathrm{Hom}_A(P, B \otimes_B P)^{(\Lambda)} \to \mathrm{Hom}_A(P, X \otimes_B P) \to 0.$$

Functoriality of η yields a commutative diagram with exact rows

$$
\begin{array}{ccccccccc}
B_B^{(\Delta)} & \longrightarrow & B_B^{(\Lambda)} & \longrightarrow & N & \longrightarrow & 0 \\
\downarrow{\scriptstyle \eta_{B^{(\Delta)}}} & & \downarrow{\scriptstyle \eta_{B^{(\Lambda)}}} & & \vdots\, {\scriptstyle \eta_X} & & \\
\mathrm{Hom}_A(P, B \otimes_B P)^{(\Delta)} & \longrightarrow & \mathrm{Hom}_A(P, B \otimes_B P)^{(\Lambda)} & \longrightarrow & \mathrm{Hom}_A(P, X \otimes_B P) & \longrightarrow & 0
\end{array}
$$

Because η_B is an isomorphism, due to Lemma IX.3.2(a), so are $\eta_{B^{(\Delta)}}$ and $\eta_{B^{(\Lambda)}}$. Passing to cokernels, we get that η_X is an isomorphism. Therefore, η is a functorial isomorphism.

The same argument works for ε. Because P is a progenerator of Mod A, any A-module M has a projective presentation of the form

$$P^{(\Delta)} \to P^{(\Lambda)} \to M \to 0$$

for some sets Δ, Λ. Because $\mathrm{Hom}_A(P, -)$ and $- \otimes_B P$ are both right exact and commute with arbitrary direct sums, we deduce the upper sequence in the diagram with exact rows

$$
\begin{array}{ccccccc}
(\mathrm{Hom}_A(P, P) \otimes_B P)^{(\Delta)} & \longrightarrow & (\mathrm{Hom}_A(P, P) \otimes_B P)^{(\Lambda)} & \longrightarrow & \mathrm{Hom}_A(P, M) \otimes_B P & \longrightarrow & 0 \\
\downarrow{\scriptstyle \varepsilon_{P^{(\Delta)}}} & & \downarrow{\scriptstyle \varepsilon_{P^{(\Lambda)}}} & & \downarrow{\scriptstyle \varepsilon_M} & & \\
P^{(\Delta)} & \longrightarrow & P^{(\Lambda)} & \longrightarrow & M & \longrightarrow & 0
\end{array}
$$

The diagram is commutative due to functoriality of ε. Because ε_P is an isomorphism, so are $\varepsilon_{P^{(\Delta)}}$ and $\varepsilon_{P^{(\Lambda)}}$. Hence, so is ε_M.

Necessity. Suppose that we have quasi-inverse equivalences $F : \mathrm{Mod}\, A \to \mathrm{Mod}\, B$ and $G : \mathrm{Mod}\, B \to \mathrm{Mod}\, A$. Set $P_A = G(B_B)$. Then $\mathrm{End}_A P = \mathrm{End}_A(GB) \cong \mathrm{End}_B B \cong B$. The module P_A is projective, because an equivalence sends a projective module onto another, see Exercise IX.3.1(e) below.

First, we prove that P is a generator of Mod A. Let M be a nonzero A-module. There exists a nonzero B-module X such that $M = GX$. But then $\mathrm{Hom}_A(P, M) = \mathrm{Hom}_A(GB, GX) \cong \mathrm{Hom}_B(B, X) \cong X \neq 0$. According to Example VI.3.4(c), the functor $\mathrm{Hom}_A(P, -)$ is faithful. Because of Exercise VI.4.22(b), P is a generator.

Next, we prove that P is finitely generated. There exists a module Y_B such that $GY \cong A_A$. Then, Y is a generator of $\operatorname{Mod} B$ because A is a generator of $\operatorname{Mod} A$ and G is an equivalence. Lemma IX.3.3 gives an integer $m > 0$ and an epimorphism $Y^{(m)} \to B$. We deduce an epimorphism $A^{(m)} = GY^{(m)} \to GB = P$. Hence, P is finitely generated.

Thus, P_A is a progenerator and $\operatorname{End} P_A \cong B$. There remains to show that the equivalence is realised by the functors $F = \operatorname{Hom}_A(P, -)$ and $G = - \otimes_B P$. Let M be an A-module. There exist functorial isomorphisms

$$FM \cong \operatorname{Hom}_B(B, FM) \cong \operatorname{Hom}_A(GB, GFM) \cong \operatorname{Hom}_A(P, M).$$

Hence, $F \cong \operatorname{Hom}_A(P, -)$. Let X_B be a B-module. Then, as above, there exist sets Δ, Λ and an exact sequence

$$B_B^{(\Delta)} \to B_B^{(\Lambda)} \to N_B \to 0.$$

Because of Theorem VI.2.8, there exists an isomorphism $\varphi_B : B \otimes_B P \to P$ defined by $\varphi_B(b \otimes x) = bx$, for $b \in B, x \in P$. Because G is exact and $- \otimes_B P$ is right exact and commutes with arbitrary direct sums, we get a commutative diagram with exact rows

$$
\begin{array}{ccccccc}
(B \otimes_B P)^{(\Delta)} & \longrightarrow & (B \otimes_B P)^{(\Lambda)} & \longrightarrow & X \otimes P & \longrightarrow & 0 \\
\downarrow{\varphi_B^{(\Delta)}} & & \downarrow{\varphi_B^{(\Lambda)}} & & \downarrow & & \\
GB^{(\Delta)} & \longrightarrow & GB^{(\Lambda)} & \longrightarrow & GX & \longrightarrow & 0
\end{array}
$$

Passing to cokernels, we get a functorial isomorphism $\varphi_X : X \otimes_B P \to GX$, whence the isomorphism of functors $- \otimes_B P \cong G$. $\qquad\square$

Algebras A, B for which there exist equivalences like those of Theorem IX.3.6 are called **Morita equivalent**.

IX.3.7 Example. Let $n > 0$ be an integer. Any algebra A is Morita equivalent to the full matrix algebra $M_n(A)$. This follows from the isomorphism $M_n(A) \cong \operatorname{End}_A(A^{(n)})$ and the observation that $A_A^{(n)}$ is a progenerator of $\operatorname{Mod} A$, because of Lemma IX.3.3.

Because progenerators are finitely generated modules, a Morita equivalence restricts to the level of finitely generated modules. More precisely, we show that if A, B are Morita equivalent algebras, then we have a commutative square of categories and functors

$$
\begin{array}{ccc}
\operatorname{mod}A & \xrightarrow{\ F'\ } & \operatorname{mod}B \\
\downarrow & & \downarrow \\
\operatorname{Mod}A & \xrightarrow{\ F\ } & \operatorname{Mod}B
\end{array}
$$

where the vertical arrows are inclusions, $F : \operatorname{Mod} A \to \operatorname{Mod} B$ is an equivalence functor, and $F' : \operatorname{mod} A \to \operatorname{mod} B$ its restriction, which is also an equivalence.

IX.3.8 Corollary. *Let A, B be K-algebras. The following conditions are equivalent:*
   (a) *A and B are Morita equivalent.*
   (b) *There exists a progenerator P such that $\operatorname{End}_A P \cong B$.*
   (c) *There exists a K-linear equivalence between $\operatorname{mod} A$ and $\operatorname{mod} B$.*

Proof. The equivalence between (a) and (b) was established in Theorem IX.3.6, hence we prove that between (a) and (c). Let A, B be Morita equivalent and M_A a finitely generated A-module. There exist an integer $n > 0$ and an epimorphism $f: A^{(n)} \to M$. If P is a progenerator of $\operatorname{mod} A$, there exist, because of Lemma IX.3.3, an integer $m > 0$ and an epimorphism $g: P^{(m)} \to A$. Hence, we have an epimorphism $g': P^{(mn)} \to A^{(n)}$. Now, fg' is an epimorphism and, because P is projective, so is also $\operatorname{Hom}_A(P, fg'): B^{(mn)} \to \operatorname{Hom}_A(P, M)$. Therefore, $\operatorname{Hom}_A(P, M)$ is finitely generated. We have established (c).

Conversely, if there exist quasi-inverse K-linear equivalences $F: \operatorname{mod} A \to \operatorname{mod} B$ and $G: \operatorname{mod} B \to \operatorname{mod} A$, set $P_A = G(B_B)$. Then P_A is a finitely generated projective module and $\operatorname{End} P_A \cong \operatorname{End} B_B \cong B$. Moreover, $F(A_A)$ is finitely generated, because so is A_A. Hence, there exist an integer $n > 0$ and an epimorphism $B_B^{(n)} \to F(A_A)$. We deduce an epimorphism

$$P_A^{(n)} = G(B_B^{(n)}) \to GF(A_A) \cong A_A.$$

Therefore, P is a progenerator of $\operatorname{Mod} A$ and the proof is complete. $\qquad\square$

The proof also shows that, for an algebra A, the progenerators of $\operatorname{Mod} A$ and $\operatorname{mod} A$ coincide.

We can say more if we decide to work with finite–dimensional algebras. Let K be a field and A a finite–dimensional K-algebra, T_A a finitely generated A-module and $B = \operatorname{End} T_A$. Because T is finitely generated, it is a finite–dimensional K-vector space, see Corollary III.4.9. Therefore, B is also a finite–dimensional K-algebra. For the same reason, given a finitely generated A-module M, the B-module $\operatorname{Hom}_A(T, M)$ is finitely generated. Similarly, if X_B is a finitely generated B-module, so is the A-module $X \otimes_B T$.

We need a notation. For a finitely generated A-module M, let add M be the full subcategory of $\operatorname{mod} A$ consisting of the finite direct sums of direct summands of M, see Exercise VIII.5.2. For example, the full subcategory of $\operatorname{mod} A$ consisting of the finitely generated projective A-modules becomes, in this notation, add A_A.

In our situation, the functor $\operatorname{Hom}_A(T, -)$ sends the subcategory add T to add B, that is, to the finitely generated projective B-modules. The following proposition says that, when restricted to add T, the functor $\operatorname{Hom}_A(T, -)$ is an equivalence.

IX.3.9 Proposition. *Let K be a field, A a finite–dimensional K-algebra, T_A a finitely generated A-module and $B = \operatorname{End} T_A$. Then:*

(a) *For T_0 in* $\mathrm{add}T$ *and M in* $\mathrm{mod}A$, *the map* $f \mapsto \mathrm{Hom}_A(T,f)$, *for*
  $f \in \mathrm{Hom}_A(T_0, M)$, *induces a functorial isomorphism*

$$\mathrm{Hom}_A(T_0, M) \overset{\cong}{\to} \mathrm{Hom}_B(\mathrm{Hom}_A(T, T_0), \mathrm{Hom}_A(T, M)).$$

(b) *The functors* $\mathrm{Hom}_A(T, -)$ *and* $- \otimes_B T$ *are quasi-inverse equivalences between*
  $\mathrm{add}T$ *and* $\mathrm{add}B$.

Proof. (a) Thanks to Exercise VIII.5.2(c) (or Lemma VI.4.11), it suffices to prove the
statement when $T_0 = T$. In this case, the well-known functorial isomorphisms

$$\mathrm{Hom}_B(\mathrm{Hom}_A(T, T), \mathrm{Hom}_A(T, M)) \cong \mathrm{Hom}_B(B, \mathrm{Hom}_A(T, M)) \cong \mathrm{Hom}_A(T, M)$$

send $\mathrm{Hom}_A(T,f)$ to $\mathrm{Hom}_A(T,f)(1_T) = f1_T = f$.

(b) Because of (a), the functor $\mathrm{Hom}_A(T, -) : \mathrm{add}T \to \mathrm{add}B$ is full and faithful.
Because of Theorem IV.4.5, it suffices to prove its essential surjectivity. Let P be
a finitely generated projective B-module, that is, an object in $\mathrm{add}\, B$. Because of
Lemmata IX.3.2 and VI.4.11, we have an isomorphism $\varepsilon_P : P_B \cong \mathrm{Hom}_A(T, P \otimes_B T)$.
Thus, P belongs to the essential image of $\mathrm{Hom}_A(T, -)$. This finishes the proof. $\square$

The equivalence $\mathrm{add}\, T \cong \mathrm{add}\, B$ can be read in two ways. If one has a good knowledge of the module T_A, it allows to construct an algebra B having preassigned finitely
generated projective modules $\mathrm{Hom}_A(T, T_0)$, with T_0 in $\mathrm{add}\, T$. If, on the other hand,
one needs information on the module T, this equivalence reduces the study of T to
that of finitely generated projective B-modules (and projectives are easier to handle).
This explains why this procedure is called ***projectivisation***.

IX.3.10 Example. With the hypotheses of the proposition, the finitely generated
injective B-modules are of the form $\mathrm{DHom}_A(T_0, T)$, with T_0 in $\mathrm{add}\, T$, where
D denotes the duality between $\mathrm{mod}\, B$ and $\mathrm{mod}\, B^{op}$, see Example IV.4.8(b).
Indeed, let I be a finitely generated indecomposable injective B-module. Because
of Theorem IX.2.16, there exists a primitive idempotent $e \in B$ such that $I \cong D(Be)$.
But then

$$I \cong D(Be) = D(\mathrm{Hom}_A(T, T)e) \cong \mathrm{DHom}_A(eT, T).$$

On the other hand, $T \cong eT \oplus (1 - e)T$, see Exercise VII.3.3, so that $eT \in \mathrm{add}T_A$.
This completes the proof.

We now try to identify the essential image $\mathrm{mod}\, A$ of the functor $(- \otimes_B T)$. Let
$\mathrm{pres}T$ denote the full subcategory of $\mathrm{mod}\, A$ consisting of the modules M such that
there exists an exact sequence

$$T_1 \to T_0 \to M \to 0$$

in $\mathrm{mod}\, A$, with T_0, T_1 in $\mathrm{add}\, T$. Such modules are called ***T-presented***, and the sequence
is a ***T-presentation***. For instance, T itself is T-presented, as well as all objects of $\mathrm{add}\, T$.

IX.3.11 Lemma. *For any finitely generated B-module X, the A-module $X \otimes_B T$ is T-presented.*

Proof. Let $P_1 \to P_0 \to X \to 0$ be a projective presentation of X in mod B, with P_0, P_1 finitely generated. Because tensor functors are right exact, we have an exact sequence in mod A

$$P_1 \otimes_B T \to P_0 \otimes_B T \to X \otimes_B T \to 0.$$

Now P_0, P_1 are finitely generated projectives, hence belong to add B_B and $B \otimes_B T \cong T_A$. Therefore, $P_1 \otimes_B T, P_0 \otimes_B T$ belong to add T. $\square$

This lemma shows that we have functors $- \otimes_B T : \text{mod}B \to \text{pres}T$, and, of course, $\text{Hom}_A(T, -) : \text{pres}T \to \text{mod } B$. We prove that if, moreover, the restriction of $\text{Hom}_A(T, -)$ to add T is exact (for instance, if T is a projective A-module), then these functors induce an equivalence of categories $\text{pres}T \cong \text{mod } B$.

IX.3.12 Proposition. *Let K be a field, A a finite–dimensional K-algebra, T_A a finitely generated A-module and $B = \text{End}T_A$. If the functor $\text{Hom}_A(T, -)$ is exact on $\text{pres}T$, then it induces an equivalence $\text{pres}T \cong \text{mod } B$, with quasi-inverse $- \otimes_B T$.*

Proof. It suffices to show that, for any T-presented A-module M and any finitely generated B-module X, the morphisms ε_M, η_X are isomorphisms. Take M in pres T. There exists an exact sequence $T_1 \to T_0 \to M \to 0$, with T_0, T_1 in add T. The hypothesis on T implies the existence of an exact sequence

$$\text{Hom}_A(T, T_1) \to \text{Hom}_A(T, T_0) \to \text{Hom}_A(T, M) \to 0.$$

If we apply the right exact functor $- \otimes_B T$ and we compare with the first sequence by means of the functorial morphism ε, we get a commutative diagram with exact rows

$$
\begin{array}{ccccccccc}
\text{Hom}_A(T, T_1) \otimes_B T & \longrightarrow & \text{Hom}_A(T, T_0) \otimes_B T & \longrightarrow & \text{Hom}_A(T, M) \otimes_B T & \longrightarrow & 0 \\
\downarrow{\varepsilon_{T_1}} & & \downarrow{\varepsilon_{T_0}} & & \downarrow{\varepsilon_M} & & \\
T_1 & \longrightarrow & T_0 & \longrightarrow & M & \longrightarrow & 0
\end{array}
$$

Because of Lemmata IX.3.2 and VI.4.11, $\varepsilon_{T_1}, \varepsilon_{T_0}$ are isomorphisms. Hence, so is ε_M.

Similarly, let X be a finitely generated B-module and $P_1 \to P_0 \to X \to 0$ a projective presentation, with P_0, P_1 finitely generated. Applying the right exact functor $- \otimes_B T$ yields an exact sequence in pres T

$$P_1 \otimes_B T \to P_0 \otimes_B T \to X \otimes_B T \to 0$$

due to Lemma IX.3.11. The functor $\text{Hom}_A(T, -)$ is thus exact on this sequence, and we get a commutative diagram with exact rows

$$
\begin{array}{ccccccc}
P_1 & \longrightarrow & P_0 & \longrightarrow & X & \longrightarrow & 0 \\
\downarrow{\scriptstyle \eta_{P_1}} & & \downarrow{\scriptstyle \eta_{P_0}} & & \downarrow{\scriptstyle \eta_X} & & \\
\mathrm{Hom}_A(T, P_1 \otimes_B T) & \longrightarrow & \mathrm{Hom}_A(T, P_0 \otimes_B T) & \longrightarrow & \mathrm{Hom}_A(T, X \otimes_B T) & \longrightarrow & 0
\end{array}
$$

Due to Lemmata IX.3.2 and VI.4.11, η_{P_0}, η_{P_1} are isomorphisms. Hence, so is η_X. This completes the proof. $\qquad \square$

If T is projective, there exists an idempotent $e \in A$ such that $T = eA$. Then $B = \mathrm{End}\, T = eAe$, because of Corollary VII.3.2. Hence an immediate corollary.

IX.3.13 Corollary. *Let A be a finite–dimensional algebra over a field and $e \in A$ an idempotent, then the adjoint pair of functors $(- \otimes_{eAe} eA, \mathrm{Hom}_A(eA, -))$ induces an equivalence* $\mathrm{pres}(eA) \cong \mathrm{mod}(eAe)$. $\qquad \square$

IX.3.14 Example. Let K be a field and $A = T_3(K)$ the algebra of lower triangular matrices of order 3. Let $e_i = e_{ii}$ denote the diagonal matrix idempotent having 1 in position (i, i) and 0 elsewhere. Consider the projective module $T_A = e_1 A \oplus e_3 A$, that is, $T = eA$, with $e = e_1 + e_3$. Compute $B = \mathrm{End}\, T_A = eAe$, with the help of Propositions III.3.14 and VII.3.1,

$$
B = \mathrm{End}\, T = \begin{pmatrix} \mathrm{Hom}_A(e_1 A, e_1 A) & \mathrm{Hom}_A(e_3 A, e_1 A) \\ \mathrm{Hom}_A(e_1 A, e_3 A) & \mathrm{Hom}_A(e_3 A, e_3 A) \end{pmatrix} = \begin{pmatrix} e_1 A e_1 & e_1 A e_3 \\ e_3 A e_1 & e_3 A e_3 \end{pmatrix} = \begin{pmatrix} K & 0 \\ K & K \end{pmatrix}.
$$

Thus, $B \cong T_2(K)$.

IX.3.2 Basic algebras

We apply Morita's theorem to the study of artinian algebras. Let K be a commutative ring and A an artinian K-algebra. If one decomposes A_A into indecomposable projective summands, some of the latter may be isomorphic, so one can write the decomposition as

$$
A_A = P_1^{(m_1)} \oplus \cdots \oplus P_t^{(m_t)},
$$

where the P_i are indecomposable, $P_i \not\cong P_j$ for $i \neq j$ and $m_i \geq 1$ for any i.

Set $P = \bigoplus_{i=1}^t P_i$, that is, P is the direct sum of a complete set of representatives of the isomorphism classes of indecomposable finitely generated projective A-modules. Because of Lemma IX.3.3, or of Example IX.3.5, P is a progenerator of Mod A. Therefore, A is Morita equivalent to $B = \mathrm{End}_A P$. But B is more 'economic' than A. For instance, if K is a field and A is finite–dimensional over K, then $\dim_K B \leqslant \dim_K A$, and equality holds if and only if $m_i = 1$, for all i, that is, if and only if $B = A$. This explains our interest in artinian algebras such that distinct indecomposable finitely generated projective modules are nonisomorphic.

IX.3.15 Lemma. *Let A be an artinian algebra of radical J and P a finitely generated projective module. Then*

$$\frac{\mathrm{End}_A P}{\mathrm{rad}(\mathrm{End}_A P)} \cong \mathrm{End}\left(\frac{P}{PJ}\right)_{A/J}.$$

Proof. For any morphism $f : P \to P$, one has $f(PJ) \subseteq PJ$, because of Proposition VIII.2.4(a). Therefore, f induces a morphism $\bar{f} : P/PJ \to P/PJ$ by passing to cokernels. Denoting by $p : P \to P/PJ$ the projection, $\bar{f}$ is the unique morphism such that $\bar{f}p = pf$.

$$
\begin{array}{ccc}
P & \xrightarrow{\ f\ } & P \\
\downarrow{\scriptstyle p} & & \downarrow{\scriptstyle p} \\
P/PJ & \xrightarrow{\ \bar{f}\ } & P/PJ
\end{array}
$$

Define $\varphi : \mathrm{End}\, P \to \mathrm{End}(P/PJ)$ by $\varphi : f \mapsto \bar{f}$, for $f \in \mathrm{End}\, P$. Then φ is K-linear, preserves identity and composition, so is an algebra morphism. Moreover, projectivity of P implies surjectivity of φ. There remains to prove that its kernel is $\mathrm{rad}(\mathrm{End}_A P)$.

Let $f \in \mathrm{Ker}\, \varphi$, then $pf = \varphi(f)p = 0$ gives $f(P) \subseteq \mathrm{Ker}\, p = PJ$. Inductively, $f^m(P) \subseteq PJ^m$, for any integer $m \geq 1$. Now, J is nilpotent, due to Theorem VIII.3.18. Hence, f is nilpotent. Therefore, $\mathrm{Ker}\, \varphi$ is a nil ideal of $\mathrm{End}_A P$. On the other hand, P/PJ is a semisimple A-module, because $PJ = \mathrm{rad}\, P$. Hence, $\mathrm{End}_A P/PJ$ is a semisimple algebra, because of Lemma VII.5.10. Applying Corollary VIII.3.20 to $(\mathrm{End}_A P)/\mathrm{Ker}\, \varphi \cong \mathrm{End}_A P/PJ$ yields the result. $\square$

IX.3.16 Theorem. *Let A be an artinian algebra of radical J. Distinct indecomposable finitely generated projective A-modules are nonisomorphic if and only if A/J is a product of division rings.*

Proof. Let $A_A = \bigoplus_{i=1}^n P_i$ be a decomposition of A into indecomposable direct summands. Because of Theorem IX.2.6, any indecomposable finitely generated projective A-module is isomorphic to one of the P_i. Moreover, $A/J = \bigoplus_{i=1}^n (P_i/P_iJ)$. Applying Lemma IX.3.15 yields:

$$\frac{A}{J} \cong \mathrm{End}\left(\frac{A}{J}\right)_{A/J} \cong \mathrm{End}\left(\bigoplus_{i=1}^n \left(\frac{P_i}{P_iJ}\right)\right)_{A/J}$$

where each P_i/P_iJ is a simple module, due to Theorem IX.2.6. Lemma VII.5.10 gives that A/J is a product of division rings if and only if $P_i/P_iJ \not\cong P_j/P_jJ$ for $i \neq j$. Because of Proposition IX.2.5(b), this happens if and only if $P_i \not\cong P_j$ for $i \neq j$. $\square$

IX.3.17 Definition. An artinian algebra A with radical J is a ***basic algebra*** if A/J is a product of division rings.

Because of the comments at the beginning of this subsection, given any artinian algebra A, there exists a basic algebra B which is Morita equivalent to A, and is the endomorphism algebra of the direct sum of exactly one representative from each isomorphism class of indecomposable projective finitely generated A-modules. It is called a **basic algebra of A**.

IX.3.18 Example.

(a) Let K be a field and $A = \begin{pmatrix} K & 0 \\ K & K \end{pmatrix}$ the K-algebra of 2×2 lower triangular matrices with coefficients in K. Because its radical is $J = \begin{pmatrix} 0 & 0 \\ K & 0 \end{pmatrix}$, we get $A/J \cong K \times K$. Therefore, A is basic. The same argument shows that the algebra $T_n(K)$ of lower triangular matrices is basic.

(b) Given a field K, a finite–dimensional K-algebra A and a finitely generated A-module T_A, it follows from Proposition IX.3.9 that $B = \operatorname{End} T_A$ is basic if and only the distinct indecomposable direct summands of T are pairwise nonisomorphic.

(c) Any bound quiver algebra $A = KQ/I$ is basic. Indeed, denoting by KQ^+ the ideal of KQ generated by the arrows, we proved in Example VIII.3.25(b) that $A/\operatorname{rad} A \cong K^{|Q_0|}$ is the product of $|Q_0|$ copies of K.

(d) Let D be a division ring. For any integer $n > 1$, the matrix algebra $M_n(D)$ is not basic. Indeed, Lemma VII.5.18 shows that a decomposition into indecomposable summands yields the direct sum of n isomorphic summands. The corresponding basic algebra of $M_n(D)$ is D, as is seen from the construction at the beginning of this subsection.

IX.3.3 The Eilenberg–Watts Theorems

Tensor functors are right exact, and Hom functors are left exact. So it is reasonable to ask whether, conversely, a right exact functor is a tensor functor and a left exact one a Hom functor. The answer is positive, provided we add a reasonable condition. The following theorems, proved independently by Eilenberg and Watts, explain why we give so much importance to the Hom and tensor functors. We start with an application of Yoneda's Lemma III.2.11.

IX.3.19 Lemma. *Let A, B be K-algebras and $F : \operatorname{Mod} A \to \operatorname{Mod} B$ a functor.*

(a) *If F has a right adjoint, then it preserves direct sums.*

(b) *If F has a left adjoint, then it preserves products.*

Proof. We prove (a), because the proof of (b) is similar. Let $G : \operatorname{Mod} B \to \operatorname{Mod} A$ be a right adjoint to F, and $(M_\lambda)_\lambda$ a family of A-modules. Then, for any B-module X, we have functorial isomorphisms:

$$\mathrm{Hom}_A(F(\oplus_\lambda M_\lambda), X) \cong \mathrm{Hom}_B(\oplus_\lambda M_\lambda, GX) \cong \prod_\lambda \mathrm{Hom}_B(M_\lambda, GX)$$

$$\cong \prod \mathrm{Hom}_A(F(M_\lambda), X) \cong \mathrm{Hom}_A(\oplus_\lambda F(M_\lambda), X)$$

where we applied Corollary III.3.12 at the second and the fourth isomorphisms. Corollary VIII.2.12(b), gives $F(\oplus_\lambda M_\lambda) \cong \oplus_\lambda F(M_\lambda)$, as required. $\square$

We prove the first Eilenberg–Watts theorem.

IX.3.20 Theorem. *Let A, B be K-algebras and F : Mod $A \to$ Mod B a K-linear functor. The following conditions are equivalent:*
  (a) *F has a right adjoint.*
  (b) *F is right exact and preserves direct sums.*
  (c) *There exists an A-B-bimodule M, unique up to a functorial isomorphism, such that $F \cong - \otimes_A M$.*

Proof. It is obvious that (c) implies (a). Because of Theorem VI.3.10 and Lemma IX.3.19 above, we also have that (a) implies (b). We only need to prove that (b) implies (c).

Let $M = F(A_A)$. In particular, M is a B-module. The functor F induces an algebra morphism

$$\varphi : A \cong \mathrm{End}\, A \to \mathrm{End}\, FA \cong \mathrm{End}\, M$$

which makes M_B a left A-module by change of scalars, that is, by setting $a \cdot x = \varphi(a)(x)$, for $a \in A, x \in M$. The K-linearity of the functor F implies that the action of K on the right and on the left of M is the same, so that the latter is indeed an A-B-bimodule.

Let L be an arbitrary A-module. Because A_A is a generator of Mod A, there exist sets Λ, Σ and an exact sequence $A^{(\Lambda)} \to A^{(\Sigma)} \to L \to 0$. Evaluating the right exact functors $- \otimes_A M$ and F on this sequence yields a commutative diagram with exact rows

$$
\begin{array}{ccccccc}
M^{(\Lambda)} & \longrightarrow & M^{(\Sigma)} & \longrightarrow & L \otimes_A M & \longrightarrow & 0 \\
\downarrow{\scriptstyle 1} & & \downarrow{\scriptstyle 1} & & \downarrow{\scriptstyle \psi_L} & & \\
FA^{(\Lambda)} & \longrightarrow & FA^{(\Sigma)} & \longrightarrow & FL & \longrightarrow & 0
\end{array}
$$

using Corollary VI.2.9 in the upper row. Passing to cokernels yields an isomorphism ψ_L. Thus, $FL \cong L \otimes_A M$, that is, $F \cong - \otimes_A M$. Moreover, ψ_L is functorial, so M is uniquely determined up to a functorial isomorphism. $\square$

The situation is more pleasant if we restrict ourselves to functors from mod A to mod B. In this case, we only have to handle finite direct sums and the latter are preserved by any linear functor, because of Exercise IV.2.11. On the other hand, we assume that we deal with noetherian algebras, in order to ensure that finitely generated modules are also finitely presented.

IX.3.21 Corollary. *Let A, B be noetherian K-algebras and $F : \operatorname{mod} A \to \operatorname{mod} B$ a K-linear functor. The following conditions are equivalent:*
  (a) *F has a right adjoint.*
  (b) *F is right exact.*
  (c) *There exists an A-B-bimodule M, unique up to a functorial isomorphism, such that $F \cong - \otimes_A M$.*

Proof. As above, we set $M = F(A_A)$ which belongs to mod B. The functor F induces a morphism $\varphi : A \to \operatorname{End} M$ which makes M_B an A-B-bimodule as above. Let L be a finitely generated A-module. Because of Proposition VII.2.15(b), there exist integers $m, n \geq 0$ and an exact sequence $A_A^{(m)} \to A_A^{(n)} \to L_A \to 0$. The proof continues as in Theorem IX.3.20 above, taking into account that F preserves finite direct sums. $\square$

IX.3.22 Examples.
  (a) Let A be a commutative algebra and $a \in A$. Consider the functor $F : \operatorname{Mod} A \to \operatorname{Mod} A$ of Exercise VI.3.5 given by $M \mapsto M/Ma$, for an A-module M. We saw there that it is right exact (but generally not left exact). It is easy to see that it preserves direct sums. Therefore, due to Theorem IX.3.20, it is a tensor product functor. Moreover, because $F(A_A) = A/Aa$, it follows from the proof that $F \cong - \otimes_A A/Aa$. In fact, we have recovered Corollary VI.3.17 which says $FM = M/Ma \cong M \otimes_A A/Aa$.
  (b) Let K be a field, A, B finite–dimensional K-algebras and $D = \operatorname{Hom}_K(-, K)$ the duality between right and left modules. Fix a bimodule $_B M_A$, finite–dimensional as K-vector space. The functor $F = \operatorname{DHom}_A(-, M) : \operatorname{mod} A \to \operatorname{mod} B$ is covariant and right exact. Therefore, $\operatorname{DHom}_A(-, M) \cong - \otimes_A F(A_A) \cong - \otimes_A \operatorname{DHom}_A(A, M) \cong - \otimes_A DM$. We have recovered the formula $\operatorname{DHom}_A(L, M) \cong L \otimes_A DM$ of Corollary IX.2.20. In particular, we observed after this corollary that the Nakayama functor ν_A, which is right exact, can be expressed as a tensor product $- \otimes_A DA$.

Contravariant left exact functors, such as contravariant Hom functors, are treated likewise. One proves easily, as in Lemma IX.3.19, that if a contravariant functor has a left adjoint, then it transforms direct sums into products, see Exercise IX.3.18.

IX.3.23 Theorem. *Let A, B be K-algebras and $F : \operatorname{Mod} A \to \operatorname{Mod} B$ a contravariant K-linear functor. The following conditions are equivalent:*
  (a) *F has a left adjoint.*
  (b) *F is left exact and transforms direct sums into products.*
  (c) *There exists an A-B-bimodule M, unique up to a functorial isomorphism, such that $F \cong \operatorname{Hom}_A(-, M)$.*

Proof. In view of the remark just before the theorem, (c) implies (a) which implies (b). We prove that (b) implies (c). Again the B-module $M_B = FA$ is made into an A-B-bimodule using the change of scalars arising from the algebra morphism $A \cong \operatorname{End} A \to \operatorname{End} FA$. Given any A-module L, there exist sets Λ, Σ and an exact sequence $A^{(\Lambda)} \to A^{(\Sigma)} \to L \to 0$. Evaluating on the latter the left exact functors $\operatorname{Hom}_A(-, M)$ and F yields a commutative diagram with exact rows

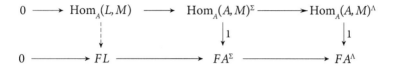

where we used Theorem II.3.9. The rest of the proof proceeds as in Theorem IX.3.20 above.   □

IX.3.24 Corollary. *Let A, B be noetherian K-algebras and $F : \operatorname{mod} A \to \operatorname{mod} B$ a K-linear contravariant functor. The following conditions are equivalent:*
  (a) *F has a left adjoint.*
  (b) *F is left exact.*
  (c) *There exists an A-B-bimodule M, unique up to a functorial isomorphism, such that $F \cong \operatorname{Hom}_A(-, M)$.*

Proof. Similar to that of Corollary IX.3.21 and left as an exercise.   □

The situation is different for covariant left exact functors, because, instead of using free presentations with the obvious generator A_A, one has to use a copresentation in terms of an injective cogenerator, and this implies technical complications when constructing the bimodule M which represents the functor.

IX.3.25 Theorem. *Let A, B be K-algebras and $F : \operatorname{Mod} A \to \operatorname{Mod} B$ a K-linear functor. The following conditions are equivalent:*
  (a) *F has a left adjoint.*
  (b) *F is left exact and preserves products.*
  (c) *There exists a B-A-bimodule M, unique up to a functorial isomorphism, such that $F \cong \operatorname{Hom}_A(M, -)$.*

Proof. As in the two previous theorems, it suffices to prove that (b) implies (c). Let Q be an injective cogenerator of $\operatorname{Mod} A$ and set $R = Q_A^{FQ}$. For $x \in FQ$, denote by $p_x : R \to Q$ the corresponding projection. Because F preserves products, $Fp_x : FR \to FQ$ is the projection associated to $x \in FQ$. Now $FR = FQ^{FQ}$ is the set of all maps from FQ to itself. Denote by $y \in FR$ the element corresponding to the identity $1_{FQ} : FQ \to FQ$. Then, for any $x \in FQ$, we have $F(p_x)(y) = x$.

Define $u : \operatorname{Hom}_A(R, Q) \to FQ$ by $f \mapsto F(f)(y)$, for $f \in \operatorname{Hom}_A(R, Q)$. Then u is surjective: we have $x = F(p_x)(y) = u(p_x)$, for any $x \in FQ$. We wish to compute the kernel of u.

An observation: let L_A be a submodule of R_A, with $i : L \to R$ the injection, then left exactness of F yields an injection $Fi : FL \to FR$ in Mod B, that is, FL is a submodule of FR.

Define M to be the intersection of all submodules L of R such that $y \in FL$. Thus, M is a submodule of R, and we have a short exact sequence

$$0 \to M \xrightarrow{j} R \to R/M \to 0.$$

We claim that Ker $u =$ Ker Hom$_A(j, Q)$.

Indeed, we have $f \in$ Ker u if and only if $F(f)(y) = u(f) = 0$, or, equivalently, $y \in$ Ker $F(f)$. Because F is left exact, this amounts to saying that $y \in F(\text{Ker} f)$. The definition of M implies that this happens if and only if $M \subseteq$ Ker f, which can be reformulated as Hom$_A(j, Q)(f) = fj = 0$, that is, $f \in$ Ker Hom$_A(j, Q)$, establishing our claim.

The functor Hom$_A(-, Q)$ is exact, because Q is injective, hence Hom$_A(j, Q)$ is an epimorphism. Evaluating this functor on the short exact sequence above yields a commutative diagram

$$
\begin{array}{ccc}
\text{Hom}_A(R, Q) & \xrightarrow{\ \text{Hom}_A(j,Q)\ } & \text{Hom}_A(M, Q) \\
\downarrow{\scriptstyle 1} & & \downarrow{\scriptstyle \psi} \\
\text{Hom}_A(R, Q) & \xrightarrow{\ \ u\ \ } & FQ
\end{array}
$$

Because Hom$_A(j, Q)$ and u are two epimorphisms with the same kernel, there exists a unique isomorphism $\psi :$ Hom$_A(M, Q) \to FQ$ such that ψHom$_A(j, Q) = u$.

We extend ψ to a functorial morphism $\varphi :$ Hom$_A(M, -) \to F$. For an A-module L, define $\varphi_L :$ Hom$_A(M, L) \to FL$ by $g \mapsto F(g)(y)$, for $g \in$ Hom$_A(M, L)$. Then φ is easily verified to be a functorial morphism. We claim that $\psi = \varphi_Q$. Let $g : M \to Q$ be a morphism. Because Q is injective, there exists $g' : R \to Q$ such that $g = g'j$. Then $\psi(g) = \psi(g'j) = u(g') = F(g')(y)$. But, in the latter expression, y is considered as an element of FR under the embedding $Fj : FM \to FR$. That is, $\psi(g) = F(g')(y) = F(g')F(j)(y) = F(g'j)(y) = F(g)(y) = \varphi_Q(g)$, and our claim is established.

Let now L be an A-module. Because Q is an injective cogenerator, it follows from Exercise VI.4.23 that there exist sets Λ, Σ and an exact sequence

$$0 \to L \to Q^\Lambda \to Q^\Sigma.$$

Evaluating on the latter the left exact functors Hom$_A(M, -)$ and F yields a commutative diagram with exact rows

$$0 \longrightarrow \mathrm{Hom}_A(M, L) \longrightarrow \mathrm{Hom}_A(M, Q)^\Lambda \longrightarrow \mathrm{Hom}_A(M, Q)^\Sigma$$

with vertical maps φ_L, φ_{Q^Λ}, φ_{Q^Σ} to

$$0 \longrightarrow FL \longrightarrow FQ^\Lambda \longrightarrow FA^\Sigma$$

Because $\varphi_Q = \psi$ is an isomorphism, so are φ_{Q^Λ} and φ_{Q^Σ}, hence so is φ_L. Consequently, $F \cong \mathrm{Hom}_A(M, -)$.

There remains to give a left B-module structure to M, making it a B-A-bimodule. Applying $\varphi_M : \mathrm{End}\, M \to FM$ to the identity morphism $1_M : M \to M$, we get $\varphi_M(1_M) \in FM$. But FM is a right B-module, hence, for any $b \in B$, we have $\varphi_M(1_M) \cdot b \in FM$. Moreover, φ_M is an isomorphism, hence $\varphi_M(1_M) \cdot b$ can be considered as an endomorphism of M and so we may define, for $b \in B, x \in M$,

$$bx = (\varphi_M(1_M) \cdot b)(x)$$

The rest is an immediate verification. □

This theorem is less handy than the two previous ones, because we do not have an explicit expression for the bimodule M. The situation is better in the case of finite–dimensional algebras over a field. Indeed, injective coresolutions in mod A are in terms of the injective cogenerator $D(_AA)$. If one wishes to work as in Theorems IX.3.20 and IX.3.23, one should set $M = FD(_AA)$. But the latter is a *right* B-module, while M should have a *left* B-module structure in order that $\mathrm{Hom}_A(M, -)$ sends mod A to mod B. The reasonable choice is thus $M = DFD(_AA)$, which is indeed a B-A-bimodule. This leads to the following result.

IX.3.26 Proposition. *Let K be a field, A, B finite–dimensional K-algebras and $F : \mathrm{mod}\, A \to \mathrm{mod}\, B$ a K-linear functor. The following conditions are equivalent:*
  (a) *F has a left adjoint.*
  (b) *F is left exact.*
  (c) *There exists a B-A-bimodule M, unique up to a functorial isomorphism, such that $F \cong \mathrm{Hom}_A(M, -)$.*

Proof. It suffices to prove that (b) implies (c). Let $D = \mathrm{Hom}_K(-, K)$ denote the duality between right and left modules (in mod A and in mod B). The functor $DFD : \mathrm{mod}\, A^{op} \to \mathrm{mod}\, B^{op}$ is right exact. The left-module version of Eilenberg–Watts Theorem IX.3.20 yields $DFD \cong DFD(_AA) \otimes_A - = M \otimes_A -$, where $M = DFD(_AA)$. Therefore, we have $DF \cong M \otimes_A D(-) = D\mathrm{Hom}_A(M, -)$, because of Corollary IX.2.20. Hence, $F \cong \mathrm{Hom}_A(M, -)$. □

IX.3.27 Example. Let K be a field, A a finite–dimensional K-algebra and $e \in A$ an idempotent. Define a functor $F : \mathrm{mod}\, A \to \mathrm{mod}\, K$ as follows. To an object M in mod A, we associate the finite–dimensional K-vector space Me, and to a morphism $f : L \to M$, we associate its restriction $Ff = f : Le \to Me$, which is K-linear. It is easily verified that this functor is left exact (and even exact). Applying the recipe of the previous Proposition IX.3.26, we get $DFD(A_A) = D(D(A_A)e) \cong D(D(eA_A)) \cong eA_A$, so that $F \cong \mathrm{Hom}_A(eA, -)$. We have recovered the result of Proposition VII.3.1.

Exercises of Section IX.3

1. Let A, B be K-algebras and $F: \text{Mod}\, A \to \text{Mod}\, B$ a K-linear equivalence. Prove that:

   (a) If M, N are A-modules, then there exists an isomorphism of K-modules
   $$\text{Hom}_A(M, N) \cong \text{Hom}_B(FM, FN).$$

   (b) If M is an A-module, then there exists an isomorphism of K-algebras
   $$\text{End}_A M \cong \text{End}_B FM.$$

   (c) A sequence $0 \to L_A \overset{f}{\to} M_A \overset{g}{\to} N_A \to 0$ is exact in $\text{Mod}\, A$ if and only if
   $0 \to FL \overset{Ff}{\to} FM \overset{Fg}{\to} FN \to 0$ is exact in $\text{Mod}\, B$. Moreover, the first splits if
   and only if so does the second.

   (d) Given a family $(M_\lambda)_{\lambda \in \Lambda}$ of A-modules, we have $F\left(\bigoplus_{\lambda \in \Lambda} M_\lambda\right) \cong \bigoplus_{\lambda \in \Lambda} FM_\lambda$ and $F\left(\prod_{\lambda \in \Lambda} M_\lambda\right) \cong \prod_{\lambda \in \Lambda} FM_\lambda$.

   (e) A module M satisfies one of the following properties if and only if so does
   FM: (i) artinian; (ii) noetherian; (iii) projective; (iv) injective; (v) simple;
   (vi) semisimple; (vii) indecomposable.

   (f) A pair (P, f) is a projective cover of a module M if and only if (FP, Ff) is
   a projective cover of FM. Also, a pair (I, j) is an injective envelope of M if
   and only if (FI, Fj) is an injective envelope of FM.

   (g) For any module M, there exists a lattice isomorphism between its submodule lattice and that of the B-module FM.

   (h) A module M has finite length if and only if FM has finite length and, in
   this case, $\ell(M) = \ell(FM)$.

   (i) The algebra A satisfies one of the following properties if and only if so does
   B: (i) artinian; (ii) noetherian; (iii) simple; (iv) semisimple.

2. Let A be an algebra and M a generator or a cogenerator (in the sense of Exercise
VI.4.23) of $\text{Mod}\, A$. Prove that M is a faithful A-module.

3. Let A be an algebra and $f: L \to M$ an epimorphism of A-modules. Prove that, if
M is a generator of $\text{Mod}\, A$, then so is L.

4. Let A be an algebra, T an A-module and $B = \text{End}_A T$. Prove that:
   (a) If T_A is a generator, then $_B T$ is finitely generated projective.
   (b) If T_A is finitely generated projective, then $_B T$ is a generator.
   (c) If T_A is a finitely generated projective generator, then so is $_B T$.

5. Prove that the unit and the co-unit of the adjunction between Hom and the tensor
product are functorial morphisms.

6. Let K be a field, A a finite–dimensional K-algebra, T_A a finitely generated A-module and $B = \text{End}_A T$. Prove that, for any finitely generated
projective B-module P, and any finitely generated B-module X, the map
$g \mapsto g \otimes T$, for $g \in \text{Hom}_B(X, P)$, induces an isomorphism between $\text{Hom}_B(X, P)$
and $\text{Hom}_A(X \otimes_B T, P \otimes_B T)$.

7. Let K be a field, A a finite–dimensional K-algebra, T_A a finitely generated A-module and $B = \operatorname{End}_A T$. Given an A-module M, let $\{f_1, \cdots, f_d\}$ be a generating set of the B-module $\operatorname{Hom}_A(T, M)$ and $f = (f_1, \cdots, f_d) : T^{(d)} \to M$.

   (a) Prove that, for any morphism $f_0 : T_0 \to M$, with T_0 in add T, there exists a morphism $g : T_0 \to T^{(d)}$ such that $f_0 = fg$.

   (b) Deduce that f is an epimorphism if and only M is generated by T.

8. Let K be a field, A a finite–dimensional K-algebra, T_A a finitely generated A-module and $B = \operatorname{End}_A T$. Prove that, if M_A is finitely generated, then the A-module $\operatorname{Hom}_A(T, M) \otimes_B T$ is generated by T_A. Deduce that the unit $\varepsilon_M : \operatorname{Hom}_A(T, M) \otimes_B T \to M$ is an epimorphism if and only if M is generated by T.

9. Let A, B be algebras and T an A-B-bimodule, assume that $_A T$ is finitely generated projective and set $T^t = \operatorname{Hom}_A(T, A)$. Prove that:

   (a) The functors from $\operatorname{Mod} A$ to $\operatorname{Mod} B$ given by $\operatorname{Hom}_A(T^t, -)$ and $- \otimes_A T$ are isomorphic.

   (b) The functor $- \otimes_A T$ is right adjoint to the functor $- \otimes_B T^t : \operatorname{Mod} B \to \operatorname{Mod} A$.

   (c) $I \otimes_A T$ is an injective B-module for any injective module I_A if and only if T^t is a flat B-module.

10. In this exercise, we give a more symmetric formulation of Morita's theorem. Let A, B be K-algebras. Prove that:

   (a) A K-linear functor $F : \operatorname{Mod} A \to \operatorname{Mod} B$ is an equivalence if and only if there exist bimodules $_A P_{B,B} Q_A$ with bimodule isomorphisms $P \otimes_B Q \cong A$, $Q \otimes_A P \cong B$ and $F \cong - \otimes_A P$.

   (b) Prove that, in this situation, $_A P, P_{B,B} Q, Q_A$ are progenerators.

   (c) Also, in this situation, we have the following module isomorphisms $\operatorname{Hom}_A(P, A) \cong_B Q$, $\operatorname{Hom}_B(P, B) \cong Q_A$, $\operatorname{Hom}_A(Q, A) \cong P_B$, $\operatorname{Hom}_B(Q, B) \cong_A P$.

   (d) In this situation, we have algebra isomorphisms $A \cong \operatorname{End}_B P \cong \operatorname{End}_B Q$ and $B \cong \operatorname{End}_A P \cong \operatorname{End}_A Q$.

11. Prove that the basic algebra of an artinian algebra is unique up to isomorphism.

12. Let K be a field, consider the matrix algebra

$$A = \begin{pmatrix} K & K & 0 \\ K & K & 0 \\ K & K & K \end{pmatrix} = \left\{ \begin{pmatrix} \alpha_1 & \alpha_2 & 0 \\ \alpha_3 & \alpha_4 & 0 \\ \alpha_5 & \alpha_6 & \alpha_7 \end{pmatrix} : \alpha_i \in K \right\}$$

with the usual matrix operations. Prove that A is not basic and compute its basic algebra.

13. Prove that an artinian algebra A is basic if and only if there exist an algebra B and a B-module T such that $A \cong \mathrm{End}_B T$ and T decomposes as direct sum of pairwise nonisomorphic indecomposable summands.

14. Let A, B be K-algebras. Prove that the following conditions are equivalent:
    (a) The basic algebras of A and B are isomorphic.
    (b) $B \cong \mathrm{End}_A P$, where P_A is a progenerator of $\mathrm{Mod}\,A$.
    (c) $B \cong \mathrm{End}_A P$ and $A \cong \mathrm{End}_B Q$, where P_A, Q_B are projective modules.

15. Let K be a field and A a finite–dimensional K-algebra. Prove that:
    (a) A is basic if and only if two distinct direct summands of $D(_A A)$ are not isomorphic.
    (b) If A is basic, then $D(_A A)$ is a direct summand of any cogenerator in $\mathrm{Mod}\,A$.

16. Prove that an artinian K-algebra is semisimple if and only if its basic algebra is a direct product of division rings containing K.

17. Let A, B be K-algebras and $F : \mathrm{Mod}\,A \to \mathrm{Mod}\,B$ be a contravariant K-linear functor. Prove that, if F has a left adjoint, then it transforms direct sums into products.

18. Let A be an artinian K-algebra and P a finitely generated projective right module. Prove that there exists a finitely generated projective left module $_A U$ such that $\mathrm{Hom}_A(P, -) \cong - \otimes_A U$.

19. Let K be a field, A, B finite–dimensional K-algebras, M a B-A-bimodule and D the duality between right and left modules. Prove that the functor $D(M \otimes_A D-) : \mathrm{mod}\,A \to \mathrm{mod}\,B$ is left exact and covariant, then express it as a Hom functor.

IX.4 Bound quiver algebras

IX.4.1 Representing elementary algebras

Bound quivers not only allow to visualise algebras as diagrams with relations between the paths, but also to visualise modules as representations. This allowed us to illustrate the theory on numerous examples. The applicability of quiver-theoretical techniques depends on how large is the class of algebras that can be represented in this way. The objective of this section is to determine this class. Certainly, as seen in Example I.3.11(h), they are finite–dimensional algebras over a field. Accordingly, in this section, we let K be a field.

IX.4.1 Definition. A finite–dimensional K-algebra A is an ***elementary algebra*** if the quotient of A by its radical is a product of copies of K.

So, elementary algebras are basic algebras. There is a partial converse.

IX.4.2 Lemma. *Let K be an algebraically closed field. A finite–dimensional K-algebra is elementary if and only if it is basic.*

Proof. We just need to prove sufficiency. Because of Corollary VII.5.13 , the (semisimple) quotient of the algebra over its radical is isomorphic to a product of the form $\prod_{i=1}^{t} M_{n_i}(K)$, for some integers $n_i \geq 1$. Because the algebra is basic, we have $n_i = 1$, for any i, as seen in Example IX.3.18(c). This completes the proof. $\qquad\square$

IX.4.3 Examples.

    (a) The lower triangular matrix algebra $A = T_n(K)$ is elementary. Indeed, according to Example VIII.3.25(a), the off-diagonal elements of A form a nilpotent ideal which is actually the radical of A. Hence, the quotient $A/\mathrm{rad}\, A$ is isomorphic to K^n.

    (b) Any bound quiver algebra $A = KQ/I$ is elementary. Let indeed KQ^+ denote the ideal of KQ generated by the arrows. We have proved in Example VIII.3.25(b) that $\mathrm{rad}\, A = KQ^+/I$ and $A/\mathrm{rad}\, A$ is isomorphic to the product $K^{|Q_0|}$ of $|Q_0|$ copies of K.

    (c) The five-dimensional $\mathbb{R}$-algebra

$$A = \begin{pmatrix} \mathbb{C} & 0 \\ \mathbb{C} & \mathbb{R} \end{pmatrix} = \left\{ \begin{pmatrix} u & 0 \\ v & x \end{pmatrix} : u, v \in \mathbb{C}, x \in \mathbb{R} \right\}$$

with the ordinary matrix operations, is basic but not elementary: indeed, the ideal

$$I = \begin{pmatrix} 0 & 0 \\ \mathbb{C} & 0 \end{pmatrix}$$

of A is nilpotent, while $A/I \cong \mathbb{C} \times \mathbb{R}$ is semisimple. Corollary VIII.3.20 yields that $\mathrm{rad}\, A = I$ and so $A/\mathrm{rad}\, A \cong \mathbb{C} \times \mathbb{R}$ is not a product of copies of the ground field $\mathbb{R}$.

We shall prove that elementary algebras are bound quiver algebras. For this purpose, we attach a quiver to an elementary algebra. We recall some features of bound quiver algebras. Let $A = KQ/I$ be a bound quiver algebra. Due to Example VII.3.6(e), a complete set of primitive pairwise orthogonal idempotents of A is $\{e_x : x \in Q_0\}$, where e_x is the class modulo I of the stationary path ε_x attached to a point x. Thus, points are in bijection with elements of a complete set of primitive pairwise orthogonal idempotents of A. We next look at arrows. As K-vector space, the ideal KQ^+ of KQ generated by the arrows has as basis all paths of (strictly) positive length. Similarly, KQ^{+2} has as basis all paths of length at least two. Thus, arrows correspond to vectors in a basis of the quotient space KQ^+/KQ^{+2}, and arrows from a given point x to another point

y, to a basis of the vector space $(\varepsilon_x KQ^+ \varepsilon_y)/(\varepsilon_x KQ^{+2}\varepsilon_y)$. Example VIII.3.25(c) says that rad $A \cong KQ^+/I$. Hence, rad$^2 A \cong KQ^{+2}/I$, which makes sense because I, being admissible, is contained in KQ^{+2}. Therefore, arrows from x to y correspond to a basis of the K-vector space $e_x((KQ^+/I)/(KQ^{+2}/I))e_y \cong e_x(\text{rad} A/\text{rad}^2 A)e_y$. This translation of combinatorial data into algebraic language motivates the definition of the quiver of an algebra.

IX.4.4 Definition. Let A be an elementary finite–dimensional K-algebra and $\{e_1, \cdots, e_n\}$ a complete set of primitive pairwise orthogonal idempotents of A. The **quiver** Q_A of A is defined as follows:

(a) The points $\{1, \cdots, n\}$ of Q_A are in bijection with the idempotents $\{e_1, \cdots, e_n\}$.
(b) If i, j are points of Q_A, then the arrows from i to j are in bijection with the vectors in a basis of the K-vector space $e_i(\text{rad} A/\text{rad}^2 A)e_j$.

The quiver of an algebra is sometimes referred to as its **ordinary quiver** or **Gabriel quiver**. Because A is finite–dimensional, Q_A has finitely many points and finitely many arrows, that is, is a finite quiver. We must however verify that Q_A is unambiguously defined.

IX.4.5 Lemma. *The quiver of an elementary algebra is independent of the choice of a complete set of primitive pairwise orthogonal idempotents.*

Proof. Let $\{e_1, \cdots, e_m\}$ and $\{f_1, \cdots, f_n\}$ be complete sets of primitive pairwise orthogonal idempotents for an elementary algebra A. Each determines a decomposition of A_A into indecomposable projective modules:

$$A_A = \bigoplus_{i=1}^{m} e_i A = \bigoplus_{j=1}^{n} f_j A.$$

Because of the Unique Decomposition Theorem VIII.4.20, $m = n$, so the number of points of the quiver Q_A is uniquely determined. Moreover, any $e_i A$ is isomorphic to some $f_j A$ and conversely. Up to a permutation of the indices, we may assume that $e_i A \cong f_i A$, for any i such that $1 \leqslant i \leqslant n$. We have isomorphisms of K-vector spaces:

$$e_i \left(\frac{\text{rad} A}{\text{rad}^2 A} \right) e_j \cong \left(\frac{\text{rad}(e_i A)}{\text{rad}^2(e_i A)} \right) e_j \cong \text{Hom}_A \left(e_j A, \frac{\text{rad}(e_i A)}{\text{rad}^2(e_i A)} \right)$$

$$\cong \text{Hom}_A \left(f_j A, \frac{\text{rad}(f_i A)}{\text{rad}^2(f_i A)} \right) \cong f_i \left(\frac{\text{rad} A}{\text{rad}^2 A} \right) f_j.$$

Thus, for any pair of points i, j, the number of arrows from i to j is uniquely determined. $\square$

The comments preceding Definition IX.4.4 show that, if $A \cong KQ/I$ is already given by a bound quiver, then $Q_A = Q$.

IX.4.6 Example. Let K be a field and A the matrix algebra

$$A = \begin{pmatrix} K & 0 & 0 \\ K & K & 0 \\ K & 0 & K \end{pmatrix} = \left\{ \begin{pmatrix} \alpha_1 & 0 & 0 \\ \alpha_2 & \alpha_3 & 0 \\ \alpha_4 & 0 & \alpha_5 \end{pmatrix} : \alpha_i \in K \right\}$$

with the usual matrix operations. As seen in Example IX.4.3(a), A is elementary with radical the off-diagonal elements, that is,

$$\operatorname{rad} A = \begin{pmatrix} 0 & 0 & 0 \\ K & 0 & 0 \\ K & 0 & 0 \end{pmatrix}.$$

Direct calculation gives $\operatorname{rad}^2 A = 0$. A complete set of primitive pairwise orthogonal idempotents is given by the matrix idempotents $e_i = \mathbf{e}_{ii}$, for $i = 1, 2, 3$. We get

$$e_2 \left(\frac{\operatorname{rad} A}{\operatorname{rad}^2 A} \right) e_1 \cong K \quad \text{and also} \quad e_3 \left(\frac{\operatorname{rad} A}{\operatorname{rad}^2 A} \right) e_1 \cong K$$

while all other $e_i(\operatorname{rad} A/\operatorname{rad}^2 A)e_j$ are zero. Thus, there is a single arrow from 2 to 1 and another one from 3 to 1. Hence, the quiver of A is:

$$3 \to 1 \leftarrow 2.$$

The proof of the main theorem needs some preparation. Let i, j be points in Q_A. The arrows $\{\alpha_1, \cdots, \alpha_k\}$ from i to j are in bijection with elements of a basis of the vector space $e_i(\operatorname{rad} A/\operatorname{rad}^2 A)e_j$. Then there exist $\{x_{\alpha_1}, \cdots, x_{\alpha_k}\}$ in $e_i(\operatorname{rad} A)e_j$ such that a basis of $e_i(\operatorname{rad} A/\operatorname{rad}^2 A)e_j$ is given by the classes $\{x_{\alpha_1} + \operatorname{rad}^2 A, \cdots, x_{\alpha_k} + \operatorname{rad}^2 A\}$.

IX.4.7 Lemma. *With this notation, the K-vector space $e_i(\operatorname{rad} A)e_j$ is generated by all products of the form $x_{\alpha_1} \cdots x_{\alpha_\ell}$, with $\alpha_1 \cdots \alpha_\ell$ a path from i to j*

Proof. We have a direct sum of K-vector subspaces of A:

$$e_i(\operatorname{rad} A)e_j = e_i \left(\frac{\operatorname{rad} A}{\operatorname{rad}^2 A} \right) e_j \oplus e_i(\operatorname{rad}^2 A)e_j.$$

Let $x \in e_i(\operatorname{rad} A)e_j$. Then there exists a linear combination $\sum_{\alpha:i \to j} \lambda_\alpha x_\alpha$, with $\lambda_\alpha \in K$, such that the difference $x - \sum_{\alpha:i \to j} \lambda_\alpha x_\alpha$ belongs to $e_i(\operatorname{rad}^2 A)e_j$. However, the definition of the square of an ideal implies that $e_i(\operatorname{rad}^2 A)e_j = \sum_{k \in (Q_A)_0} (e_i(\operatorname{rad} A)e_k)(e_k(\operatorname{rad} A)e_j)$. Hence, the same argument yields linear combinations $\sum_{\beta:i \to k} \lambda_\beta x_\beta$ and $\sum_{\gamma:k \to j} \lambda_\gamma x_\gamma$ such that:

$$x - \sum_{\alpha:i \to j} \lambda_\alpha x_\alpha - \left(\sum_{\beta:i \to k} \lambda_\beta x_\beta \right) \left(\sum_{\gamma:k \to j} \lambda_\gamma x_\gamma \right) = x - \sum_{\alpha:i \to j} \lambda_\alpha x_\alpha - \left(\sum_{\beta\gamma:i \to k \to j} \lambda_\beta \lambda_\gamma x_\beta x_\gamma \right)$$

belongs to $e_i(\operatorname{rad}^3 A)e_j$. We finish using induction and nilpotency of the radical of A. $\square$

IX.4.8 Theorem. *Let K be a field and A an elementary finite–dimensional K-algebra. Then there exists a surjective algebra morphism* $\varphi : KQ_A \to A$ *with admissible kernel I so that* $A \cong KQ_A/I$ *is a bound quiver algebra.*

Proof. We first define φ on points and arrows of Q_A. We use the same notation as before. Let:

$$\varphi(\epsilon_i) = e_i, \text{ for any point } i \in (Q_A)_0$$

$$\varphi(\alpha) = x_\alpha, \text{ for any arrow } \alpha \in (Q_A)_1$$

The K-vector space KQ_A has as basis the set of paths. Let $\alpha_1 \cdots \alpha_\ell$ be a path. Define

$$\varphi(\alpha_1 \cdots \alpha_\ell) = \varphi(\alpha_1) \cdots \varphi(\alpha_\ell) = x_{\alpha_1} \cdots x_{\alpha_\ell}$$

then extend this definition to KQ_A by linearity. This definition ensures that φ becomes a K-linear map which preserves the product of basis vectors, hence of any vectors. Finally,

$$\varphi(1) = \varphi\left(\sum_{i \in (Q_A)_0} \epsilon_i \right) = \sum_{i \in (Q_A)_0} e_i = 1$$

so φ is an algebra morphism.

This morphism is surjective: because A is elementary, the e_i generate the K-vector space $A/\mathrm{rad}\, A$ and, because of Lemma IX.4.7, products of the x_α generate $\mathrm{rad}\, A$. The e_i and the products of the x_α lie in the image of φ, hence φ is surjective.

There remains to prove that $I = \mathrm{Ker}\varphi$ is an admissible ideal. The construction of φ gives $\varphi(KQ_A^+) \subseteq \mathrm{rad}\, A$. Induction gives $\varphi(KQ_A^{+k}) \subseteq \mathrm{rad}^k A$, for any integer $k \geq 1$. Nilpotency of $\mathrm{rad}\, A$ and the fact that the image of an arrow is nonzero give $\varphi(KQ_A^{+m}) = 0$, for some $m \geq 2$. Therefore, $KQ_A^{+m} \subseteq \mathrm{Ker}\, \varphi = I$.

Let $x \in I$, that is, such that $\varphi(x) = 0$. Because there exists a decomposition $KQ_A = K^{|(Q_A)_0|} \oplus (KQ_A^+/KQ_A^{+2}) \oplus KQ_A^{+2}$ as vector spaces, there exist linear combinations $\sum_{i \in (Q_A)_0} \lambda_i \epsilon_i$ and $\sum_{\alpha \in (Q_A)_1} \lambda'_\alpha \alpha$, with $\lambda_i, \lambda'_\alpha$ scalars such that the difference $x - \sum_{i \in (Q_A)_0} \lambda_i \epsilon_i - \sum_{\alpha \in (Q_A)_1} \lambda'_\alpha \alpha$ lies in KQ_A^{+2}. Because $\varphi(x) = 0$, applying φ to the latter expression yields

$$\sum_i \lambda_i e_i + \sum_\alpha \lambda'_\alpha x_\alpha \in \varphi(KQ_A^{+2}) \subseteq \mathrm{rad}^2 A.$$

Because the e_i are orthogonal idempotents while $\mathrm{rad}\, A$ is nilpotent, $\lambda_i = 0$, for any i. Therefore, $\sum_\alpha \lambda'_\alpha x_\alpha \in \mathrm{rad}^2 A$. Equivalently, $\sum_\alpha \lambda'_\alpha(x_\alpha + \mathrm{rad}^2 A) = 0$. Because the $x_\alpha + \mathrm{rad}^2 A$ constitute a basis of the K-space $\mathrm{rad}\, A/\mathrm{rad}^2 A$, they are linearly independent. Hence, $\lambda'_\alpha = 0$, for all arrows α. Replacing, we get $x \in KQ_A^{+2}$. Thus, $I \subseteq KQ_A^{+2}$. Therefore, I is admissible and the proof is complete. $\square$

Because bound quiver algebras are themselves elementary, see Example IX.4.3(h), the theorem may be restated as: an algebra is a bound quiver algebra if and only if it is elementary.

IX.4.9 Corollary. *Any basic algebra over an algebraically closed field is a bound quiver algebra.*

Proof. This follows from the Theorem and Lemma IX.4.2.   □

Before giving examples, we use the notion of spectroid, introduced in Example IX.2.8(f), to interpret differently the equivalence between the module category of a bound quiver algebra and the category of representations of the bound quiver, see Example IV.4.6(c).

Let $A = KQ/I$ be a bound quiver algebra, then denote by $\mathrm{Fun}(S(A), \mathrm{Mod}\, K)$, and by $\mathrm{Fun}(S(A), \mathrm{mod}\, K)$ the categories of K-linear functors from the spectroid $S(A)$ to the category $\mathrm{Mod}\, K$ of K-vector spaces, and to the category $\mathrm{mod}\, K$ of finite–dimensional K-vector spaces, respectively. A functor F belonging to one of these categories associates to each object of $S(A)$, that is, to each point of Q, a K-vector space and, given objects x, y, to a morphism in $S(A)(x, y)$ which is the class modulo I of a linear combination $w = \sum_i \lambda_i w_i$ of paths w_i from x to y in Q, it associates a K-linear map $F(w) = \sum_i \lambda_i F(w_i)$ from $F(x)$ to $F(y)$. Moreover, $w \in I$ implies $F(w) = 0$. These observations suggest the following lemma.

IX.4.10 Lemma. *Let K be a field and $A = KQ/I$ a bound quiver algebra. Then there exists a commutative square of categories and functors*

$$
\begin{array}{ccc}
\mathrm{Rep}_K(Q, I) & \xrightarrow{\;\cong\;} & \mathrm{Fun}(S(A), \mathrm{Mod}K) \\[2pt]
\big\uparrow & & \big\uparrow \\[2pt]
\mathrm{rep}_K(Q, I) & \xrightarrow{\;\cong\;} & \mathrm{Fun}(S(A), \mathrm{mod}K)
\end{array}
$$

where the vertical functors are inclusions and the horizontal are isomorphisms of categories.

Proof. We construct a bijection between bound representations and functors. Let V be a bound representation of (Q, I). To an arrow $\alpha : x \to y$ is associated a morphism $V(\alpha) : V(x) \to V(y)$, hence, to a linear combination of paths $w = \sum_i \lambda_i w_i$ from x to y, say, is associated a unique K-linear map $V(w) = \sum_i \lambda_i V(w_i) : V(x) \to V(y)$, such that, if $w_i = \alpha_{i1} \cdots \alpha_{in_i}$ with the α_{ij} arrows, then $V(w_i) = V(\alpha_{in_i}) \cdots V(\alpha_{i1})$ (because maps compose opposite to arrows).

Conversely, let F be a K-linear functor from $S(A)$ to vector spaces. Because $S(A)$ is a small category, there exists a set embedding j from the disjoint union $Q_0 \coprod Q_1$ into $S(A)$, sending each $x \in Q_0$ to itself considered as an object in $S(A)$ and each arrow $\alpha \in Q_1$ to itself considered as a morphism in $S(A)$. Then the composition

$V = F \circ j$ is a K-linear representation of Q. Furthermore, because $F(w) = 0$, for any $w \in I$, this is a bound representation.   □

Combining this lemma with the equivalences of Example IV.4.6(c), we get an interpretation of modules over a bound quiver algebra as functors from its spectroid to vector spaces. Namely, there exists a commutative square of categories and functors:

$$
\begin{array}{ccc}
\mathrm{Mod}A & \xrightarrow{\;\cong\;} & \mathrm{Fun}(S(A), \mathrm{Mod}K) \\
\uparrow & & \uparrow \\
\mathrm{mod}A & \xrightarrow{\;\cong\;} & \mathrm{Fun}(S(A), \mathrm{mod}K)
\end{array}
$$

where the vertical functors are the inclusions and the horizontal functors are equivalences.

IX.4.11 Examples.
  (a) In Example IX.4.6, $\dim_K A = 5$ and the path algebra KQ_A is also five–dimensional (having as basis the three stationary paths plus the two arrows). Hence, $A \cong KQ_A$, thus, is a path algebra. We return to path algebras in Chapter XII, where we give other characterisations of this class.
  (b) Here is an historically important example. In the nineteenth century, Leopold Kronecker worked on the classification of simultaneously equivalent pairs of matrices, a problem stated as follows: given two pairs of $m \times n$ matrices (M, N) and (M', N'), with coefficients in a field K, do there exist invertible matrices P and Q such that we have both $M = Q^{-1}M'P$ and $N = Q^{-1}N'P$? In terms of representations of quivers, any such pair (M, N), of size (m, n), say, can be seen as a representation of the quiver $\circ \rightleftarrows \circ$ as follows

$$
K^m \underset{N}{\overset{M}{\rightleftarrows}} K^n
$$

where M and N represent the linear maps corresponding to the arrows in terms of fixed bases of K^n and K^m. In other words, Kronecker's problem was to classify, up to isomorphism, the representations of the Kronecker quiver

$$
\circ \rightleftarrows \circ
$$

As seen in Exercise IV.4.13, the path algebra of this quiver is isomorphic to the (four-dimensional) matrix algebra

$$
\begin{pmatrix} K & 0 \\ K^2 & K \end{pmatrix} = \left\{ \begin{pmatrix} a & 0 \\ (b, c) & d \end{pmatrix} : a, b, c, d \in K \right\}.
$$

(c) Let $A = K[t]/\langle t^5 \rangle$. Then Q_A has only one point, because A is local, see Example VIII.4.15(b), and its only nonzero idempotent is the identity. Also, $\operatorname{rad} A = \langle t \rangle / \langle t^5 \rangle$. Hence, $\operatorname{rad}^2 A = \langle t^2 \rangle / \langle t^5 \rangle$ and $\operatorname{rad} A/\operatorname{rad}^2 A \cong \langle t \rangle / \langle t^2 \rangle$ is one-dimensional. Therefore, there is a unique arrow α from the unique point to itself. The quiver Q_A is the single loop

$$\circ \; \rotatebox{0}{$\circlearrowleft$} \; \alpha$$

Here $KQ_A = K[t]$, which is infinite–dimensional, and the arrow α represents the indeterminate t. Because $(t + \langle t^5 \rangle))^n = 0$ in A, for any integer $n \geq 5$, the kernel of the surjection $\varphi : KQ_A \to A$ is the ideal $\langle \alpha^5 \rangle$ of KQ_A generated by α^5.

(d) Let $n > 1$ be an integer and $B = T_n(K)$ the algebra of lower triangular $n \times n$ matrices. Consider the ideal

$$I = \{[\alpha_{ij}] \in B : \alpha_{ij} = 0 \text{ whenever } (i, j) \neq (n, 1)\}$$

and set $A = B/I$. Because $n > 1$, the ideal I is contained in the radical of B (even in its square radical!), so a complete set of primitive pairwise orthogonal idempotents is provided by the matrix idempotents $e_i = \mathbf{e}_{ii}$. Because i ranges from 1 to n, the quiver has n points. The radical of B consists of the off-diagonal elements, that is,

$$\operatorname{rad} B = \{[\beta_{ij}] \in B : \beta_{ij} = 0 \text{ whenever } i \leq j\}.$$

Because of Corollary VIII.3.21, $\operatorname{rad} A = (\operatorname{rad} B)/I$. Therefore, we get that $\operatorname{rad} A/\operatorname{rad}^2 A \cong \operatorname{rad} B/\operatorname{rad}^2 B$ consists of the classes modulo I of the elements lying just below the main diagonal

$$\operatorname{rad} A/\operatorname{rad}^2 A = \{[\alpha_{ij}] + I : i = j + 1\}.$$

Left multiplying by e_i and right multiplying by e_j, for i, j with $1 \leq i, j \leq n$, gives $e_i(\operatorname{rad} A/\operatorname{rad}^2 A)e_{i-1} \cong K$ while, if $j \neq i - 1$, then $e_i(\operatorname{rad} A/\operatorname{rad}^2 A)e_j = 0$. Thus, the quiver of A is

$$n \xrightarrow{\alpha_n} n - 1 \xrightarrow{\alpha_{n-1}} \cdots \longrightarrow 2 \xrightarrow{\alpha_2} 1.$$

We compute the kernel of the surjection $\varphi : KQ_A \to A$ of Theorem IX.4.8. Because an arrow α_i represents a basis vector x_{α_i} of $e_i(\operatorname{rad} A/\operatorname{rad}^2 A)e_{i-1}$, for $i > 1$, and the product $x_{\alpha_n} \cdots x_{\alpha_2}$ is zero, the composite $\alpha_n \cdots \alpha_2$ lies in $\operatorname{Ker} \varphi$. On the other hand, $KQ_A \cong T_n(K)$, see Exercise I.4.10(d), hence $\dim_K A = \dim_K B - \dim_K I = \dim_K T_n(K) - 1 = \dim_K KQ_A - 1$. Thus, the kernel of φ is one-dimensional and in fact, $\operatorname{Ker} \varphi = \langle \alpha_n \cdots \alpha_2 \rangle$.

IX.4.2 Nakayama algebras

This subsection is devoted to a class of finite–dimensional algebras over a field K for which one can compute explicitly, up to isomorphism, all indecomposable finitely generated modules and morphisms between them. We start with the matrix algebra of Example IX.4.6

$$A = \begin{pmatrix} K & 0 & 0 \\ K & K & 0 \\ K & 0 & K \end{pmatrix}$$

which is the path algebra of the quiver $3 \rightarrow 1 \leftarrow 2$. A complete set of primitive pairwise orthogonal idempotents is $\{e_i = e_{ii} : i = 1, 2, 3\}$. So A_A decomposes into indecomposable projective modules as

$$A_A = e_1 A \oplus e_2 A \oplus e_3 A.$$

The module $e_1 A$ is simple, because it is one-dimensional. It corresponds to the representation

$$S_1 : 0 \rightarrow K \leftarrow 0$$

in the notation of Example VII.4.3(e). On the other hand, $e_2 A$ corresponds to the representation

$$P_2 : 0 \rightarrow K \xleftarrow{1} K.$$

It is uniserial: it has a unique nonzero proper submodule, isomorphic to S_1, which is maximal in P_2 and P_2/S_1 is isomorphic to the simple module S_2. The unique composition series of P_2 is thus $0 \subsetneq S_1 \subsetneq P_2$. In the notation of Subsection VII.4.2, it can be depicted as $\begin{smallmatrix} 2 \\ 1 \end{smallmatrix}$. Similarly, $P_3 = e_3 A$ is uniserial and given as $\begin{smallmatrix} 3 \\ 1 \end{smallmatrix}$. Therefore, $A_A = 1 \oplus \begin{smallmatrix} 2 \\ 1 \end{smallmatrix} \oplus \begin{smallmatrix} 3 \\ 1 \end{smallmatrix}$, that is, any indecomposable finitely generated projective A-module is uniserial. Before translating this property into quiver-theoretical terms, we characterise uniserial modules.

Let M be a finitely generated A-module. For any integer $i \geq 0$, one has $\mathrm{rad}^i M \neq \mathrm{rad}^{i+1} M$, if the former is nonzero, see Lemma VIII.2.6. Moreover, because the radical of the finite–dimensional algebra A is nilpotent, and because of Theorem VIII.3.16, there exists a least integer $t \geq 0$ such that $\mathrm{rad}^t M = M \cdot \mathrm{rad}^t A = 0$. Thus, we have a finite sequence as follows

$$0 = \mathrm{rad}^t M \subsetneq \cdots \subsetneq \mathrm{rad}^2 M \subsetneq \mathrm{rad}\, M \subsetneq M,$$

called the **radical filtration** of M.

In this subsection, let A be a finite dimensional K algebra. Uniserial modules have necessarily finite length, thus are finitely generated, due to Corollary VIII.3.24.

IX.4.12 Lemma. *A finitely generated A-module M is uniserial if and only if its radical filtration*

$$0 = \operatorname{rad}^t M \subsetneq \cdots \subsetneq \operatorname{rad}^2 M \subsetneq \operatorname{rad} M \subsetneq M$$

is a composition series.

Proof. If M is uniserial, then it has a unique maximal submodule, necessarily equal to $\operatorname{rad} M$. Induction on composition length, using Exercise VII.4.13, shows that its radical filtration is a composition series.

Conversely, assume that the radical filtration is a composition series. Then, for any i, the quotient $\operatorname{rad}^i M/\operatorname{rad}^{i+1} M$ is simple. Therefore, $\operatorname{rad}^{i+1} M$ is the unique maximal submodule of $\operatorname{rad}^i M$. Consequently, M is uniserial. □

The lemma can be reformulated to say that a finitely generated module is uniserial if and only if each of its nonzero submodules has a unique maximal submodule. Therefore, the composition length of a uniserial module M equals the least integer $t \geq 0$ such that $\operatorname{rad}^t M = 0$. This integer is the length of the radical filtration, called the **Loewy length** of M, denoted as $\ell\ell(M)$.

IX.4.13 Corollary.
 (a) *Any uniserial module is indecomposable.*
 (b) *If M is a uniserial module, then $\ell\ell(M) = \ell(M)$.*
 (c) *If $M = \bigoplus_{i=1}^m M_i$, then $\ell\ell(M) = max_{1 \leq i \leq m}\ell\ell(M_i)$.*
 (d) *Any submodule, and any quotient, of a uniserial module are uniserial.*

Proof. (a) Uniserial modules have simple top, so the statement follows from Corollary IX.2.7.
 (b) This follows directly from Lemma IX.4.12.
 (c) This follows from the definition of the Loewy length and the fact that $\operatorname{rad}^k(\bigoplus_i M_i) \cong \bigoplus_i \operatorname{rad}^k M_i$, for any integer $k \geq 0$, as follows from Proposition VIII.2.4(e) and induction.
 (d) This is Exercise VII.4.13. □

There are many examples of indecomposable modules which are not uniserial. An example is given after the following characterisation of algebras whose indecomposable finitely generated projective modules are uniserial.

IX.4.14 Proposition. *Let $A = KQ/I$ be a bound quiver algebra. Any indecomposable finitely generated projective A-module is uniserial if and only if, for any point $x \in Q_0$, at most one arrow starts at x.*

Proof. Necessity. Assume that $x \in Q_0$. Because the indecomposable projective module $P_x = e_x A$ is uniserial, its unique composition series equals its radical filtration, due to Lemma IX.4.12. Hence, the quotient $\mathrm{rad}P_x/\mathrm{rad}^2 P_x$ is simple or zero. Because simple modules over bound quiver algebras are one-dimensional, we have

$$\dim_K e_x\left(\frac{\mathrm{rad}\,A}{\mathrm{rad}^2 A}\right) = \dim_K \frac{e_x(\mathrm{rad}\,A)}{e_x(\mathrm{rad}^2 A)} = \dim_K\left(\frac{\mathrm{rad}P_x}{\mathrm{rad}^2 P_x}\right) \leqslant 1.$$

Thus, there exists at most one point $y \in Q_0$ such that $e_x(\mathrm{rad}\,A/\mathrm{rad}^2 A)e_y \neq 0$ and, if it is nonzero, then this space is one-dimensional. This amounts to saying that there exists at most one arrow starting from x and, if such an arrow exists, then its target is y.

Sufficiency. Assume that P_x is indecomposable projective but not uniserial. Then there exists an integer $i \geq 1$ such that $\mathrm{rad}^i P_x$ has more than one maximal submodule, that is $(\mathrm{rad}^i P_x)/(\mathrm{rad}^{i+1} P_x)$ is neither simple nor zero. Taking i minimal for this property, we have

$$\dim_K\left(\frac{\mathrm{rad}^i P_x}{\mathrm{rad}^{i+1} P_x}\right) \geq 2 \quad\text{and}\quad \dim_K\left(\frac{\mathrm{rad}^{i-1} P_x}{\mathrm{rad}^i P_x}\right) \leqslant 1.$$

Actually, $\dim_K(\mathrm{rad}^{i-1}P_x/\mathrm{rad}^i P_x) = 1$, because otherwise $\mathrm{rad}^{i-1}P_x = 0$, contrary to the definition of i. Let $p : P \to \mathrm{rad}^{i-1}P_x$ be a projective cover. Because $\mathrm{rad}^{i-1}P_x$ has a simple top equal to $\mathrm{rad}^{i-1}P_x/\mathrm{rad}^i P_x$, the projective module P is indecomposable, that is, $P = P_y$, for some $y \in Q_0$. Applying Corollary VIII.3.17, we get that p induces epimorphisms $\mathrm{rad}P_y \to \mathrm{rad}^i P_x$ and $\mathrm{rad}^2 P_y \to \mathrm{rad}^{i+1}P_x$. Passing to cokernels, we get an epimorphism $\mathrm{rad}P_y/\mathrm{rad}^2 P_y \to \mathrm{rad}^i P_x/\mathrm{rad}^{i+1}P_x$.

$$
\begin{array}{ccccccccc}
0 & \longrightarrow & \mathrm{rad}^2 P_y & \longrightarrow & \mathrm{rad}P_y & \longrightarrow & \mathrm{rad}P_y/\mathrm{rad}^2 P_y & \longrightarrow & 0 \\
 & & \downarrow & & \downarrow & & \downarrow & & \\
0 & \longrightarrow & \mathrm{rad}^{i+1} P_x & \longrightarrow & \mathrm{rad}^i P_x & \longrightarrow & \mathrm{rad}^i P_x/\mathrm{rad}^{i+1} P_x & \longrightarrow & 0
\end{array}
$$

Therefore,

$$\dim_K\left(\frac{\mathrm{rad}P_y}{\mathrm{rad}^2 P_y}\right) \geq \left(\dim_K \frac{\mathrm{rad}^i P_x}{\mathrm{rad}^{i+1} P_x}\right) \geq 2.$$

That is, the point y has at least two arrows starting in it. This completes the proof. $\square$

In the example computed above, all three indecomposable finitely generated projectives over the path algebra of the quiver $3 \to 1 \leftarrow 2$ are uniserial. However, the indecomposable finitely generated injective module I_1 is given by the representation $K \xrightarrow{1} K \xleftarrow{1} K$. It has two composition series, namely $0 \subsetneq S_1 \subsetneq P_2 \subsetneq I_1$ and $0 \subsetneq S_1 \subsetneq P_3 \subsetneq I_1$. Thus the fact that all indecomposable finitely generated projective modules are uniserial does not imply that all indecomposable finitely generated injective modules are uniserial. The surprising thing is that if we assume all indecomposable finitely generated projective *and* injective modules to be uniserial, then so is any indecomposable finitely generated module.

IX.4.15 Definition. An algebra is a *Nakayama algebra* if any indecomposable finitely generated projective module and any indecomposable finitely generated injective module is uniserial.

Combining Proposition IX.4.14 and its dual (which holds true thanks to the duality D existing for finitely generated modules over finite–dimensional algebras), we get the following theorem.

IX.4.16 Theorem. *A bound quiver algebra A = KQ/I is a Nakayama algebra if and only if Q is one of the two quivers:*

$$\circ \longleftarrow \circ \longleftarrow \cdots \longleftarrow \circ$$

or

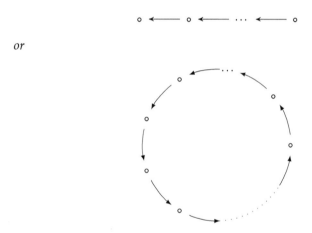

Proof. A bound quiver algebra is Nakayama if and only if any indecomposable finitely generated projective module and any indecomposable finitely generated injective module are uniserial. Because of Proposition IX.4.14 and its dual, this is the case if and only if, for any point $x \in Q_0$, at most one arrow starts at x and at most one arrow ends in x. Easy combinatorics show that the only quivers satisfying both conditions are those in the statement.   □

The theorem imposes no restriction on the binding ideal I. However, because of the shapes of the quivers in the theorem, this ideal has to be generated by paths. We next describe all indecomposable finitely generated modules, up to isomorphism.

IX.4.17 Theorem. *Let A be a Nakayama algebra and M an indecomposable finitely generated A-module. Then there exist an integer $t \geq 0$ and an indecomposable finitely generated projective A-module P such that $M \cong P/\mathrm{rad}^t P$.*

Proof. Let $t = \ell\ell(M)$ be the Loewy length of M, that is, the least integer $t \geq 0$ such that $\text{rad}^t M = 0$, and set $A' = A/\text{rad}^t A$. The proof is done in eight steps.

(1) Because $0 = \text{rad}^t M = M \cdot \text{rad}^t A$, the module M is annihilated by $\text{rad}^t A$, hence has a natural A'-module structure.

(2) The Loewy length of $A'_{A'}$ equals t. Indeed, we have $\text{rad}^t A' = 0$. On the other hand, because $\text{rad}^{t-1} M \neq 0$, we have $\text{rad}^{t-1} A \neq 0$, and therefore $\text{rad}^{t-1} A' \neq 0$.

(3) A' is a Nakayama algebra. Indeed, let $A_A = \bigoplus_{i=1}^n P_i$ be a decomposition into indecomposable projective A-modules. Because of Theorem VIII.3.16, we have $\text{rad}\, P_i = P_i(\text{rad}\, A)$, for any i, and therefore, inductively, $\text{rad}^t P_i = P_i(\text{rad}^t A)$. Hence,

$$\frac{A}{\text{rad}^t A} \cong \bigoplus_{i=1}^n \left(\frac{P_i}{P_i\text{rad}^t A}\right) \cong \bigoplus_{i=1}^n \left(\frac{P_i}{\text{rad}^t P_i}\right).$$

Each summand $P'_i = P_i/\text{rad}^t P_i$ is indecomposable because it has simple top, due to Corollary IX.2.7(a) and the fact that $\text{rad}^t P_i \subseteq \text{rad}\, P_i$. Consequently, any indecomposable finitely generated projective A'-module is isomorphic to some P'_i, because of Theorem IX.2.6. Finally, each P'_i is uniserial because so is the corresponding P_i, and because of Corollary IX.4.13. This statement and its dual imply that A' is indeed a Nakayama algebra. This also follows from the shapes of the quivers in Theorem IX.4.16.

(4) Because of Corollary IX.4.13(c), in the decomposition of A' into indecomposable projective modules $A'_{A'} = \bigoplus_{i=1}^n P'_i$, the integer $t = \ell\ell(A')$ equals the maximum of the $\ell\ell(P'_i)$.

(5) Let (P, f) be a projective cover of M in mod A'. Because $t = \ell\ell(A') \geq \ell\ell(P) \geq \ell\ell(M) = t$, there exists an indecomposable summand P' of P such that $\ell\ell(P') = t$.

(6) We claim that there exists an indecomposable summand P' of P such that $\ell\ell(P') = t$ and the restriction f' of f to P' is injective. For, assume that this is not the case for any P' having Loewy length t. Then each restriction $f' : P' \to \text{Im}\, f'$ is a proper epimorphism. Because P' is uniserial, so is its quotient $\text{Im}\, f'$, due to Corollary IX.4.13(c). Hence, $\ell\ell(\text{Im}\, f') = \ell(\text{Im}\, f') < \ell(P') = \ell\ell(P') = t$. On the other hand, if f'' is the restriction of f to another indecomposable summand P'', then $\ell\ell(\text{Im}\, f'') \leqslant \ell\ell(P'') < t$. Because f is an epimorphism, we get $\ell\ell(M) < t$, a contradiction which proves our claim.

(7) P' is an injective A'-module. Indeed, let $P' \to I$ be an injective envelope in mod A'. Because P' is uniserial, its socle is simple. Because of Corollary VIII.2.19, so is socI $= \text{soc}P'$. Therefore, I is indecomposable. Because both P' and I are uniserial, we have $t = \ell\ell(P') = \ell(P') \leqslant \ell(I) = \ell\ell(I) \leqslant \ell\ell(A') = t$. Hence, $P' \cong I$ and our claim is established.

(8) Because P' is injective, f' is a section and, because M is indecomposable, $M \cong P'$. The fact that $P' \cong P/\text{rad}^t P$, for some indecomposable projective A-module P yields the statement. $\qquad\square$

Thus, indecomposable finitely generated A-modules are obtained from indecomposable finitely generated projectives by looking at the composition series of each of the latter, and 'cutting' from below one simple composition factor at a time. This implies that any indecomposable finitely generated module over a Nakayama algebra is uniserial, because it is a quotient of a uniserial module. Therefore, an algebra is Nakayama if and only if any indecomposable finitely generated module is uniserial.

The next corollary collects a few properties of Nakayama algebras, all following from the theorem. It gives a good description of the category of finitely generated modules.

IX.4.18 Corollary. *Let A be a Nakayama algebra and $A_A = \oplus_{i=1}^{n} P_i$ a decomposition into indecomposable projective A-modules.*

(a) *There exist at most $n \cdot \ell\ell(A)$ (that is, finitely many) isomorphism classes of indecomposable finitely generated A-modules.*

(b) *If $M \cong P_i/rad^t P_i$, for some i and some integer $t \geq 0$, then the projection $P_i \to M$ is a projective cover.*

(c) *An indecomposable finitely generated module is uniquely determined by its simple top (or its simple socle) and its composition length.*

(d) *Let M, N be indecomposable finitely generated A-modules. The dimension of $Hom_A(M, N)$ equals the number of appearances of top M as a simple composition factor of N, or, equivalently, the number of appearances of soc$N$ as a simple composition factor of M.*

Proof. (a) Any indecomposable finitely generated projective A-module P_i has finite Loewy length, thus the number of isomorphism classes of indecomposable finitely generated A-modules is equal to $\sum_{i=1}^{n} \ell\ell(P_i)$, which does not exceed $n \cdot \ell\ell(A)$.

(b) Indeed, the kernel $rad^t P_i$ of this projection is zero or a superfluous submodule of M.

(c) Let S_i be the simple top of an indecomposable finitely generated module M having composition length t. Because of Lemma IX.4.12, $t = \ell\ell(M)$. Let P_i be a projective cover of S_i. The theorem gives $M \cong P_i/rad^t P_i$. The statement about the socle is dual.

(d) This follows from Example IX.2.23(b) and the comments preceding it.   □

Because of duality, one may equivalently describe the indecomposable finitely generated modules over a Nakayama algebra starting from indecomposable finitely generated injectives instead of projectives. In this case, the indecomposable finitely generated modules are obtained from the composition series of the indecomposable finitely generated injectives by 'cutting' from above one composition factor at a time, that is, they are submodules of the injectives.

IX.4.19 Examples.

(a) Let, as in Example IX.4.11(c), $A = K[t]/\langle t^5 \rangle$. It is given by the quiver

bound by the ideal $\langle \alpha^5 \rangle$. Then A is a Nakayama algebra, because its quiver is one of the two quivers depicted in Theorem IX.4.16. It has, up to isomorphism, a unique indecomposable finitely generated projective module which is also its unique indecomposable finitely generated injective module. Denoting by 1 the unique simple module, this projective is

$$P = \begin{matrix} 1 \\ 1 \\ 1 \\ 1 \\ 1 \end{matrix} \; .$$

Therefore, the indecomposable finitely generated modules, up to isomorphism, are

$$P = \begin{matrix} 1 \\ 1 \\ 1 \\ 1 \\ 1 \end{matrix} \;, \quad \frac{P}{\mathrm{rad}^4 P} = \begin{matrix} 1 \\ 1 \\ 1 \\ 1 \end{matrix} \;, \quad \frac{P}{\mathrm{rad}^3 P} = \begin{matrix} 1 \\ 1 \\ 1 \end{matrix} \;, \quad \frac{P}{\mathrm{rad}^2 P} = \begin{matrix} 1 \\ 1 \end{matrix} \quad \text{and} \quad \frac{P}{\mathrm{rad} P} = 1.$$

In fact, all algebras of the form $K[t]/\langle t^n \rangle$, for some integer $n \geq 2$, are Nakayama algebras, and one can calculate just as easily their indecomposable finitely generated modules.

(b) Let $n > 0$ be an integer and $A = T_n(K)$ the algebra of lower triangular matrices. As shown in Exercise I.4.10(d), it is the path algebra of the quiver

$$n \to n-1 \to \cdots \to 2 \to 1.$$

Its indecomposable finitely generated projective modules are, up to isomorphism,

$$P_n = \begin{matrix} n \\ n-1 \\ \vdots \\ 2 \\ 1 \end{matrix} \;, \quad P_{n-1} = \begin{matrix} n-1 \\ \vdots \\ 2 \\ 1 \end{matrix} \;, \quad \cdots, \quad P_2 = \begin{matrix} 2 \\ 1 \end{matrix} \;, \quad \text{and} \quad P_1 = 1.$$

It is readily seen that the isomorphism classes of indecomposable finitely generated A-modules are of the form

$$\begin{matrix} i \\ i-1 \\ \vdots \\ j \end{matrix}$$

for any pair (j, i) with $j < i \leq n$. They are in bijection with the nonempty segments $[j, i]$ in the quiver of A. In particular, there are exactly $n(n+1)/2$ isomorphism classes of indecomposable finitely generated modules. We also describe the isomorphism classes of indecomposable finitely generated injective modules:

$$I_1 = \begin{matrix} n \\ n-1 \\ \vdots \\ 2 \\ 1 \end{matrix} \quad , \quad I_2 = \begin{matrix} n \\ n-1 \\ \vdots \\ 2 \end{matrix} \quad , \quad \cdots, \quad I_{n-1} = \begin{matrix} n \\ n-1 \end{matrix} \quad , \quad \text{and} \quad I_n = n.$$

One observes that $P_n = I_1$: this indecomposable module is projective-injective.

For instance, assume $n = 3$, then we have exactly six nonisomorphic indecomposable modules, given respectively by

$$P_1 = 1, P_2 = \begin{matrix} 2 \\ 1 \end{matrix} , P_3 = I_1 = \begin{matrix} 3 \\ 2 \\ 1 \end{matrix} , I_2 = \begin{matrix} 3 \\ 2 \end{matrix} , I_3 = 3 \text{ and } S_2 = 2.$$

In Example IX.3.14, we considered the A-module $T = P_1 \oplus P_3$ and its endomorphism algebra $B \cong T_2(K)$. The same calculation as for A shows that there are exactly three nonisomorphic indecomposable $T_2(K)$-modules. Because of Proposition IX.3.12, we have $\mathrm{mod}B \cong \mathrm{pres}T$, and it is easily seen that, up to isomorphism, the only indecomposable T-presented A-modules are the direct summands of T, namely P_1, P_3, and also the module I_2. Actually, a T-presentation of I_2 is provided by the short exact sequence $0 \to P_1 \to P_3 \to I_2 \to 0$. Thus, $\mathrm{pres}T$ contains only three nonisomorphic indecomposable modules.

(c) Consider the algebra A given by the quiver

$$1 \circ \underset{\beta}{\overset{\alpha}{\rightleftarrows}} \circ 2$$

bound by the ideal $\langle \alpha\beta\alpha \rangle$. The indecomposable finitely generated projective A-modules are, up to isomorphism

$$P_1 = \begin{matrix} 1 \\ 2 \\ 1 \end{matrix} \quad \text{and} \quad P_2 = \begin{matrix} 2 \\ 1 \\ 2 \\ 1 \end{matrix} .$$

The remaining indecomposable finitely generated A-modules (again up to isomorphism) are

$$\frac{P_1}{\mathrm{rad}^2 P_1} = \begin{matrix} 1 \\ 2 \end{matrix} , \quad \frac{P_1}{\mathrm{rad}P_1} = 1, \quad \frac{P_2}{\mathrm{rad}^3 P_2} = \begin{matrix} 2 \\ 1 \\ 2 \end{matrix} , \quad \frac{P_2}{\mathrm{rad}^2 P_2} = \begin{matrix} 2 \\ 1 \end{matrix} , \quad \frac{P_2}{\mathrm{rad}P_2} = 2.$$

Using Example IX.2.23(b), one sees that $\mathrm{Hom}_A(P_1, P_2/\mathrm{rad}^3 P_2)$ is one-dimensional, while $\mathrm{Hom}_A(P_1, P_2)$ is two-dimensional. Bases of these vector spaces are computed as follows. A nonzero morphism from P_1 to $P_2/\mathrm{rad}^3 P_2$ maps the simple top 1 of P_1 onto the unique isomorphic composition factor of $P_2/\mathrm{rad}^3 P_2$. Hence, its image must be the submodule $\begin{smallmatrix}1\\2\end{smallmatrix}$ of the codomain, and the kernel is the simple socle 1 of P_1. This can be visualised using the following figure:

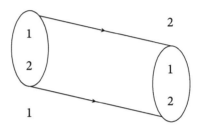

Similarly, the K-vector space $\mathrm{Hom}_A(P_1, P_2)$ has as basis two morphisms, one being the inclusion of P_1 as the radical of P_2 and the other the morphism with image the simple socle 1 of P_2 and kernel $\mathrm{rad}P_1 = \begin{smallmatrix}2\\1\end{smallmatrix}$. This second morphism is depicted as follows:

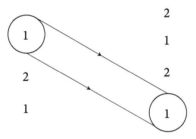

It is equally easy to compute the Hom-sets with nonprojective source. Assume we wish to compute $\mathrm{Hom}_A(L, M)$ with L nonprojective. We may assume that L, M are indecomposable. Let $p : P \to L$ be a projective cover. Because L is uniserial, $\mathrm{top}L$ is simple. Hence, so is $\mathrm{top}P$ (which equals $\mathrm{top}L$). Therefore, P is indecomposable. Now, a nonzero morphism $f : L \to M$ extends to a morphism $fp : P \to M$, nonzero because p is surjective, and fp is determined by the appearance of $\mathrm{top}P$ as a simple composition factor of M. Because p is an epimorphism, the map $f \mapsto fp$, for $f \in \mathrm{Hom}_A(L, M)$, is injective. Hence, f is determined by the appearance of $\mathrm{top}L = \mathrm{top}P$ as a simple composition factor of M. For instance, $\mathrm{End}_A(P_2/\mathrm{rad}^3 P_2)$ is two-dimensional, spanned by the identity morphism, and the morphism whose image is the simple socle 2 of $P_2/\mathrm{rad}^3 P_2$ and the kernel is $\mathrm{rad}(P_2/\mathrm{rad}^3 P_2) = \begin{smallmatrix}1\\2\end{smallmatrix}$.

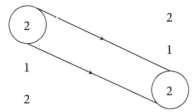

Now we compute projective resolutions of the two simple A-modules. We have a short exact sequence of the form $0 \to P_1 \overset{j}{\to} P_2 \to S_2 \to 0$ with j the inclusion of P_1 into P_2. This is, therefore, a projective resolution of S_2. The projective cover $P_1 \to S_1$ has as kernel the module $P_2/\mathrm{rad}^2 P_2$. A projective cover of the latter is P_2, so we have an exact sequence $P_2 \to P_1 \to S_1 \to 0$. The kernel of the morphism from P_2 to P_1 is $P_2/\mathrm{rad}^2 P_2$. So we have a longer exact sequence $P_2 \to P_2 \to P_1 \to S_1 \to 0$, with same kernel. Repeating this procedure, we deduce that S_1 has an infinite projective resolution

$$\cdots \to P_2 \to P_2 \to P_2 \to P_1 \to S_1 \to 0.$$

(d) The notation used for uniserial modules (in terms of the radical series) can be extended to nonuniserial ones. Let A be the algebra given by the quiver

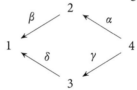

bound by the ideal $\langle \alpha\beta - \gamma\delta \rangle$, and consider the indecomposable finitely generated projective module P_4 corresponding to the point 4. As vector space, it has basis $\{e_4, \bar{\alpha}, \bar{\gamma}, \bar{\alpha}\bar{\beta} = \bar{\gamma}\bar{\delta}\}$, see Example VI.4.6(e). The corresponding bound quiver representation is

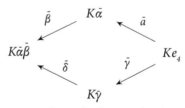

where the morphisms are right multiplication by the residual classes of the corresponding arrows. One can represent it graphically as

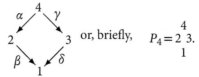

This diagram shows the radical series of P_4, namely $\mathrm{rad}P_4 = \begin{smallmatrix}2\,3\\1\end{smallmatrix}$ and $\mathrm{rad}^2 P_4 = \mathrm{rad}(\mathrm{rad}P_4) = 1$. It stops there because $\mathrm{rad}^3 P_4 = 0$. The diagram

also shows the submodules and quotients of P_4 : if M is a submodule of P_4, then it corresponds to a subdiagram of the diagram for P_4 in which all incident arrows enter and none leaves. Dually, quotients correspond to subdiagrams in which all arrows leave and none enters. Thus, for instance, the submodules of P_4, up to isomorphism, are

$$\begin{matrix} 2\,3 \\ 1 \end{matrix} \, , \quad \begin{matrix} 2 \\ 1 \end{matrix} \, , \quad \begin{matrix} 3 \\ 1 \end{matrix} \, , \quad 1$$

and its quotients are

$$\begin{matrix} 4 \\ 2\,3 \end{matrix} \, , \quad \begin{matrix} 4 \\ 2 \end{matrix} \, , \quad \begin{matrix} 4 \\ 3 \end{matrix} \, , \quad 4.$$

One composition series of P_4 is

$$1 \subsetneq \begin{matrix} 3 \\ 1 \end{matrix} \subsetneq \begin{matrix} 2\,3 \\ 1 \end{matrix} \subsetneq \begin{matrix} 4 \\ 2\,3 \\ 1 \end{matrix}$$

and the other is obtained by replacing $\begin{matrix} 3 \\ 1 \end{matrix}$ by $\begin{matrix} 2 \\ 1 \end{matrix}$. In particular, $\ell(P_4) = 4 \neq 3 = \ell\ell(P_4)$. One also has a short exact sequence $0 \rightarrow P_2 \rightarrow P_4 \rightarrow I_3 \rightarrow 0$

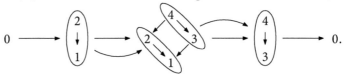

We compute a projective resolution for the simple module S_4. A projective cover is clearly P_4 and the kernel of the projective cover morphism is the radical of P_4, namely

$$\mathrm{rad}P_4 \;=\; \begin{matrix} 2\,3 \\ 1 \end{matrix} \, .$$

The top of the latter is the semisimple module $S_2 \oplus S_3$, therefore a projective cover of $\mathrm{rad}P_4$ is $P_2 \oplus P_3$. Finally, the kernel of the composite morphism $P_2 \oplus P_3 \rightarrow \mathrm{rad}P_4 \hookrightarrow P_4$ is the simple projective module $S_1 = P_1$. Thus, the projective resolution we constructed is:

$$0 \rightarrow P_1 \rightarrow P_2 \oplus P_3 \rightarrow P_4 \rightarrow S_4 \rightarrow 0.$$

(e) This way of visualising modules is generally not very precise (except, of course, for uniserial modules) and in large examples quickly becomes unpractical. However, we encourage the reader to draw his own pictures, keeping in mind that pictures are just a help to intuition, and do not replace rigorous proofs. For instance, we proved in Example VIII.4.15(d) that, if Q is the quiver

$$1 \rightleftharpoons 2$$

then the KQ-module H given by the representation:

$$K^2 \; \underset{\begin{pmatrix} 0 & 0 \\ 1 & 0 \end{pmatrix}}{\overset{\begin{pmatrix} 1 & 0 \\ 0 & 1 \end{pmatrix}}{\rightleftharpoons}} \; K^2$$

is indecomposable. Let $\{f_1, f_2\}$ and $\{e_1, e_2\}$ be respectively the canonical bases of the vector spaces $H(1)$ and $H(2)$. One can visualise the linear maps from the latter to the former as

The two vertical arrows represent the identity matrix, which maps Ke_i to Kf_i for $i = 1,2$ and the oblique, the nilpotent, which maps Ke_1 to Kf_2. This picture suggests that soc $H = S_1 \oplus S_1$ and top $H = S_2 \oplus S_2$ (respectively, the 'lower' and the 'upper' part of the diagram). Indeed, there exist two linearly independent morphisms from S_1 to H, one having as image

$$Kf_1 \rightleftharpoons 0$$

and the other with image

$$Kf_2 \rightleftharpoons 0$$

Because S_1 is simple, both morphisms are injective. Therefore, soc $H = S_1 \oplus S_1$. One proves similarly that top H is the semisimple module $S_2 \oplus S_2$ given by the representation

$$0 \rightleftharpoons K^2$$

We return to indecomposability: in the previous picture, one can look at H as given by a 'string' as follows:

$$Kf_1 \overset{1}{\leftarrow} Ke_1 \overset{1}{\rightarrow} Kf_2 \overset{1}{\leftarrow} Ke_2.$$

one does not see how and where to 'cut' the module into direct summands, therefore the mere fact that a module can be represented as a string (that is, a sequence of one-dimensional vector spaces related by identity morphisms) suggests indecomposablity. Needless to say, not all indecomposable modules can be represented as strings in this fashion, see Exercise VIII.4.17(a).

Consider now the module M given by the representation

$$K^2 \underset{\begin{pmatrix} 0 & 0 \\ 1 & 0 \end{pmatrix}}{\overset{\begin{pmatrix} 1 & 0 \\ 0 & 0 \end{pmatrix}}{\rightleftarrows}} K^2$$

Drawing a picture similar to the one for H yields

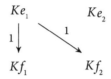

This picture suggests that we have a direct sum decomposition $M \cong P_2 \oplus S_2$: indeed, we have a 'string':

$$Kf_1 \xleftarrow{1} Ke_1 \xrightarrow{1} Kf_2$$

which is isomorphic to the indecomposable projective module P_2 and an 'isolated' copy of a one-dimensional space at the point 2, that is, S_2. We leave the proof as an exercise to the reader.

Exercises of Section IX.4

1. Let K be a field. Write each of the following lower triangular matrix algebras as a bound quiver algebra:

(a) $\begin{pmatrix} K & 0 & 0 & 0 \\ 0 & K & 0 & 0 \\ 0 & 0 & K & 0 \\ K & K & K & K \end{pmatrix}$
(b) $\begin{pmatrix} K & 0 & 0 & 0 & 0 \\ K & K & 0 & 0 & 0 \\ K & 0 & K & 0 & 0 \\ K & 0 & 0 & K & 0 \\ K & K & K & K & K \end{pmatrix}$
(c) $\begin{pmatrix} K & 0 & 0 & 0 & 0 \\ 0 & K & 0 & 0 & 0 \\ K & K & K & 0 & 0 \\ K & 0 & 0 & K & 0 \\ K & K & K & K & K \end{pmatrix}$

(d) The matrix algebra of Exercise I.4.11.

2. Let K be a field and Q a finite, connected and acyclic quiver with n points. Prove that:

(a) The points of Q can be numbered as $\{1, \cdots, n\}$ so that $j \leqslant i$ whenever there exists a path from i to j in Q.

(b) With such a numbering, the path algebra KQ is isomorphic to the triangular matrix algebra

$$\begin{pmatrix} \varepsilon_1(KQ)\varepsilon_1 & 0 & \cdots & 0 \\ \varepsilon_2(KQ)\varepsilon_1 & \varepsilon_2(KQ)\varepsilon_2 & \cdots & 0 \\ \vdots & \vdots & \ddots & \vdots \\ \varepsilon_n(KQ)\varepsilon_1 & \varepsilon_n(KQ)\varepsilon_2 & \cdots & \varepsilon_n(KQ)\varepsilon_n \end{pmatrix}$$

with the multiplication induced from the multiplication of KQ.

(c) Compute the lower triangular matrix algebra of (b) if the quiver Q is:
$$1 \leftarrow 3 \rightarrow 2.$$

3. Let $Q' = (Q'_0, Q'_1)$ be a full subquiver of a finite quiver $Q = (Q_0, Q_1)$ such that, if $\alpha : i \rightarrow j$ is an arrow in Q with $i \in Q'_0$, then $j \in Q'_0$ and $\alpha \in Q'_1$. Let I be an admissible ideal of KQ and $\epsilon = \sum_{i \in Q'_0} \epsilon_i$. Prove the following:

(a) $KQ' = \epsilon(KQ)\epsilon$.

(b) $I' = \epsilon I \epsilon$ is an admissible ideal in KQ'.

(c) KQ'/I' is isomorphic to the quotient of KQ/I by $J = \langle \epsilon_i + I : i \notin Q'_0 \rangle$.

4. Let A be a bound quiver algebra. Prove that:

(a) If $\operatorname{rad}^2 A = 0$ and $\{e_1, \cdots, e_n\}$ is a complete set of primitive pairwise orthogonal idempotents of A, then $e_i A e_j \neq 0$, for $i \neq j$, if and only if there exists an arrow $i \rightarrow j$ in the quiver of A.

(b) A is connected if and only if $A/\operatorname{rad}^2 A$ is connected.

5. Let K be a field, A a finite–dimensional K-algebra and M a finitely generated A-module. Consider the algebra

$$B = \begin{pmatrix} A & 0 \\ M & K \end{pmatrix} = \left\{ \begin{pmatrix} a & 0 \\ m & \lambda \end{pmatrix} : a \in A, m \in M, \lambda \in K \right\}$$

with the usual matrix addition and the multiplication induced from the K-A-bimodule structure of M. The algebra B is called a **one-point extension** of A by M. Prove that:

(a) $\operatorname{rad} B = \operatorname{rad} A \oplus M$, as K-vector spaces.

(b) If A is an elementary K-algebra, then so is B.

(c) In this case, the quiver Q_B of B contains the quiver Q_A of A as full subquiver, and there is exactly one additional point x, which is a source. Moreover, there is an additional arrow $x \rightarrow y$ each time the simple module S_y appears as a direct summand in $\operatorname{top} M$, and these are all the arrows in Q_B which are not in Q_A.

(d) Any indecomposable finitely generated projective A-module remains so when considered as a B-module, and there exists exactly one additional indecomposable finitely generated projective B-module whose radical is M.

(e) Compute the bound quiver of B when A is the path algebra of the quiver

$$1 \leftarrow 2 \leftarrow 3 \leftarrow 4 \quad \text{and} \quad M = 4 \oplus \begin{smallmatrix} 3 \\ 2 \end{smallmatrix} .$$

6. The present exercise shows that generators of an admissible ideal are generally not uniquely determined. Let K be a field of characteristic different from 2 and Q the quiver

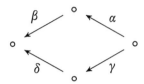

Set $I = \langle \alpha\beta + \gamma\delta \rangle$, $I' = \langle \alpha\beta - \gamma\delta \rangle$. Prove that I, I' are distinct admissible ideals such that $KQ/I \cong KQ/I'$.

7. Let K be a field, A the path algebra of the quiver $1 \to 2$ and M the A-module given as a representation by:

$$K \xrightarrow{\binom{1}{1}} K^2$$

Prove that, in the notation of Examples VI.4.6(d) and VII.4.3(d), we have $M \cong S_2 \oplus P_1$.

8. Let A be a bound quiver algebra. Prove that any indecomposable finitely generated projective A-module is uniserial if and only if, for any indecomposable finitely generated projective module P, the quotient $\mathrm{rad}P/\mathrm{rad}^2 P$ is simple or zero.

9. Let K be a field and KQ/I a bound quiver algebra. Prove that:
   (a) If any indecomposable finitely generated projective module is uniserial, then, given two points $i, j \in Q_0$, there is at most one path from i to j not passing twice through the same point.
   (b) The converse of (a) is not true.

10. Consider the indecomposable projective module P_4 in Example IX.4.19(d). Construct the Hasse quiver of the submodule lattice of P_4.

11. Let A be a finite–dimensional algebra over a field and $0 \to L \to M \to N \to 0$ a short exact sequence of finitely generated A-modules. Prove that:
   (a) $\ell\ell(M) \geq \max\{\ell\ell(L), \ell\ell(N)\}$.
   (b) $\ell\ell(M) \leq \ell\ell(L) + \ell\ell(N)$.

12. Prove that an elementary algebra A is a Nakayama algebra if and only if $A/\mathrm{rad}^2 A$ is a Nakayama algebra.

13. Let A be a finite–dimensional algebra over a field and I a proper ideal of A, contained in the radical.
   (a) If any indecomposable finitely generated projective A-module is uniserial, then so is any indecomposable finitely generated projective A/I-module.
   (b) If A is a Nakayama algebra, then so is A/I.

14. Let K be a field. For each of the following bound quivers (Q, I), compute all isomorphism classes of indecomposable finitely generated KQ/I-modules.
   (a) $\underset{1}{\circ} \xleftarrow{\ \delta\ } \underset{2}{\circ} \xleftarrow{\ \gamma\ } \underset{3}{\circ} \xleftarrow{\ \beta\ } \underset{4}{\circ} \xleftarrow{\ \alpha\ } \underset{5}{\circ}$
   bound by the ideal $\langle \alpha\beta, \beta\gamma\delta \rangle$.

(b) $\underset{1}{\circ} \xleftarrow{\ \varepsilon\ } \underset{2}{\circ} \xleftarrow{\ \delta\ } \underset{3}{\circ} \xleftarrow{\ \gamma\ } \underset{4}{\circ} \xleftarrow{\ \beta\ } \underset{5}{\circ} \xleftarrow{\ \alpha\ } \underset{6}{\circ}$

bound by the ideal $\langle \alpha\beta, \gamma\delta\varepsilon \rangle$.

(c)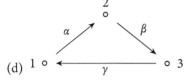

bound by the ideal $\langle \alpha\beta, \beta\alpha \rangle$.

(d)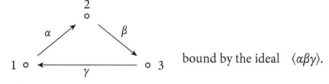

bound by the ideal $\langle \alpha\beta\gamma, \beta\gamma\alpha \rangle$.

15. Let K be a field and A be given by the quiver

$$\underset{1}{\circ} \xleftarrow{\ \delta\ } \underset{2}{\circ} \xleftarrow{\ \gamma\ } \underset{3}{\circ} \xleftarrow{\ \beta\ } \underset{4}{\circ} \xleftarrow{\ \alpha\ } \underset{5}{\circ}$$

bound by the ideal $\langle \alpha\beta\gamma, \gamma\delta \rangle$. For the idempotent $e = e_1 + e_2 + e_4 + e_5$, compute the bound quiver of eAe and show explicitly the equivalence between mod (eAe) and the subcategory pres (eA), consisting of the eA-presented modules, see Proposition IX.3.12.

16. Let K be a field and A be given by the quiver

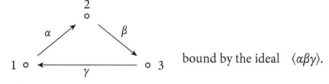

bound by the ideal $\langle \alpha\beta\gamma \rangle$.

(a) Prove that $\mathrm{End}_A\, P_2$ is two-dimensional as a K-vector space, and exhibit a basis of the space.

(b) Compute projective resolutions for each of the simple modules S_1, S_2 and S_3.

17. Let A be the algebra of Example IX.4.19(c). Prove that the endomorphism algebra of each of the indecomposable projective modules P_1 and P_2 is isomorphic to $K[t]/\langle t^2 \rangle$, as algebras.

18. Let K be a field and Q the Kronecker quiver $1 \rightleftarrows 2$. Prove that the KQ-module given by the representation:

$$K^2 \underset{\begin{pmatrix} 0 & 0 \\ 0 & 1 \end{pmatrix}}{\overset{\begin{pmatrix} 1 & 0 \\ 0 & 0 \end{pmatrix}}{\rightleftarrows}} K^2$$

is, in the notation of Example VII.4.10, isomorphic to $H_\alpha \oplus H_\beta$.

Chapter X

Homology

X.1 Introduction

Originally, homological algebra was meant to be a language, designed to describe in algebraic terms topological properties of geometrical objects. It is now a well-developed branch of mathematics in its own right, with connections to several other areas. In module theory, homological techniques are mainly used to derive invariants for modules and algebras.

Perhaps the easiest way to approach homology is to think of the Hom and tensor functors. They are just one-sided exact: evaluating a Hom functor, or a tensor functor, on a short exact sequence, yields a sequence which is only left exact, or right exact, respectively. Homology is (among others) a tool to complete a one-sided exact sequence on the other side so that it becomes exact. The resulting sequence is usually longer than the original one, and sometimes infinite. We start by showing how these long exact sequences are constructed. In several important cases, concrete computations may actually be performed. The second half of the chapter is devoted to the formalism of derived functors which offers a framework allowing to repair the lack of exactness of a one-sided exact functor. This formalism is the one we shall use in the next chapter for constructing the derived functors of the Hom and tensor functors. As usual, K denotes a commutative ring and our algebras are K-algebras.

X.2 Homology and cohomology

X.2.1 Basic definitions

Homology arises when we have a not necessarily exact sequence such that the composite of two consecutive morphisms is zero.

X.2.1 Definition. Let A be a K-algebra.

(a) A ***chain complex*** of A-modules $C_{\bullet} = (C_n, d_n)_{n \in \mathbb{Z}}$ is the data of a sequence $(C_n)_{n \in \mathbb{Z}}$ of A-modules and a sequence $(d_n : C_n \to C_{n-1})_{n \in \mathbb{Z}}$ of A-linear maps

$$C_{\bullet} : \cdots \to C_{n+1} \xrightarrow{d_{n+1}} C_n \xrightarrow{d_n} C_{n-1} \to \cdots$$

such that $d_n d_{n+1} = 0$, for any $n \in \mathbb{Z}$. The module C_n is said to be of ***degree*** n. The morphism d_n is the ***differential*** of degree n.

(b) Dually, a ***cochain complex*** $C^{\bullet} = (C^n, d^n)_{n \in \mathbb{Z}}$ is the data of a sequence $(C^n)_{n \in \mathbb{Z}}$ of A-modules and a sequence $(d^n : C^n \to C^{n+1})_{n \in \mathbb{Z}}$ of A-linear

An Introduction to Module Theory. Ibrahim Assem and Flávio U. Coelho, Oxford University Press.
© Ibrahim Assem and Flávio U. Coelho (2024). DOI: 10.1093/9780198904939.003.0011

maps

$$C^\bullet : \cdots \to C^{n-1} \xrightarrow{d^{n-1}} C^n \xrightarrow{d^n} C^{n+1} \to \cdots$$

such that $d^n d^{n-1} = 0$, for any $n \in \mathbb{Z}$. Here as well, n is the **degree** of the module C^n and d^n is the **differential** of degree n.

Thus, the difference between chain and cochain complexes only depends on indexing. More precisely, there is a bijection between chain and cochain complexes given by $n \mapsto -n$. The condition on the differentials is written succinctly as $d^2 = 0$.

The following picture illustrates a chain complex. Comparing this picture with the one following Definition II.4.8, one sees that, because $d_n d_{n+1} = 0$, the image of d_{n+1} (the hatched circle) is contained in the kernel of d_n, usually properly (the difference being the dotted area of $\mathrm{Ker} d_n$).

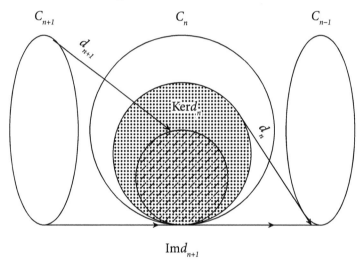

When talking about complexes, it is not necessary to deal with modules. One defines in the same way complexes of objects and morphisms taken inside any abelian category. Another remark is that complexes need not be infinite on both sides as it may seem from the definition, it may indeed happen that all terms, from some degree onward and/or backward, are zero.

We make a category out of complexes. If $C_\bullet = (C_n, d_n)_{n \in \mathbb{Z}}$ and $C'_\bullet = (C'_n, d'_n)_{n \in \mathbb{Z}}$ are chain complexes, we define a **morphism of complexes** from the first to the second to be a family of A-linear maps $f_\bullet = (f_n : C_n \to C'_n)_{n \in \mathbb{Z}}$ compatible with the differentials, that is, such that the following diagram commutes

$$
\begin{array}{ccccccccc}
\cdots & \longrightarrow & C_{n+1} & \xrightarrow{d_{n+1}} & C_n & \xrightarrow{d_n} & C_{n-1} & \longrightarrow & \cdots \\
& & \downarrow{f_{n+1}} & & \downarrow{f_n} & & \downarrow{f_{n-1}} & & \\
\cdots & \longrightarrow & C'_{n+1} & \xrightarrow{d'_{n+1}} & C'_n & \xrightarrow{d'_n} & C'_{n-1} & \longrightarrow & \cdots
\end{array}
$$

for any n, or, in other words, such that $f_{n-1}d_n = d'_n f_n$, for any n. This condition can be written briefly as $fd = d'f$.

This definition of morphisms implies that they are composable by degrees. If $f_{\cdot} : C_{\cdot} \to C'_{\cdot}$ and $g_{\cdot} : C'_{\cdot} \to C''_{\cdot}$ are morphisms of chain complexes, then so is $g_{\cdot}f_{\cdot} = (g_n f_n : C_n \to C''_n)_{n \in \mathbb{Z}}$, which we define to be their composite. This gives a category of chain complexes of A-modules.

We define subcomplexes and quotient complexes in the obvious way. Let $C_{\cdot} = (C_n, d_n)_{n \in \mathbb{Z}}$ and $C'_{\cdot} = (C'_n, d'_n)_{n \in \mathbb{Z}}$ be chain complexes of A-modules. We say that $C'_{\cdot}$ is a **subcomplex** of $C_{\cdot}$ if, for any $n \in \mathbb{Z}$, we have $C'_n \subseteq C_n$ and d'_n equals the restriction of d_n to C'_n. The latter condition amounts to saying that the inclusions $j_n : C'_n \to C_n$, with $n \in \mathbb{Z}$, constitute a morphism of complexes $j_{\cdot} : C'_{\cdot} \to C_{\cdot}$, called the **injection**.

If $C'_{\cdot}$ is a subcomplex of $C_{\cdot}$, then we define the **quotient complex** $C''_{\cdot} = C_{\cdot}/C'_{\cdot}$ as

$$C''_{\cdot} : \cdots \to C''_n = C_n/C'_n \xrightarrow{d''_n} C''_{n-1} = C_{n-1}/C'_{n-1} \to \cdots$$

where $d''_n : C''_n \to C''_{n-1}$ is deduced from d_n, d'_n by passing to cokernels, that is,

$$d''_n(x_n + C'_n) = d_n(x_n) + C'_{n-1}$$

for $x_n \in C_n$. This defines indeed a family of differentials, because

$$d''_n d''_{n+1}(x_{n+1} + C'_{n+1}) = d''_n(d_{n+1}(x_{n+1}) + C'_n) = d_n d_{n+1}(x_{n+1}) + C'_{n-1} = 0 + C'_{n-1}$$

for $x_{n+1} \in C_{n+1}$ and $n \in \mathbb{Z}$. In other words, $d''_n d''_{n+1} = 0$.

The A-linear maps $p_n : C_n \to C''_n$ defined by $x_n \mapsto x_n + C'_n$, for $x_n \in C_n$, are obviously surjective and constitute a morphism of complexes because

$$p_n d_{n+1}(x_{n+1}) = d_{n+1}(x_{n+1}) + C'_n = d''_{n+1}(x_{n+1} + C'_{n+1}) = d''_{n+1}p_{n+1}(x_{n+1})$$

for $x_{n+1} \in C_{n+1}$ and $n \in \mathbb{Z}$. This morphism of complexes $p_{\cdot} : C_{\cdot} \to C''_{\cdot}$ is called the **projection**.

With these definitions, it is easily seen that the category of chain complexes of A-modules is abelian. Furthermore, its abelian structure is constructed degree by degree, see Exercise X.2.1. This implies that a sequence $0 \to C'_{\cdot} \to C_{\cdot} \to C''_{\cdot} \to 0$ in the category of complexes is short exact if and only if each of the component sequences of A-modules $0 \to C'_n \to C_n \to C''_n \to 0$ is short exact.

One defines, in the same way, a category of cochain complexes, which is also abelian.

X.2.2 Examples.

(a) It follows from the definition that any exact sequence is a complex, but the converse is generally not true. Because $d_n d_{n+1} = 0$ amounts to saying that $\mathrm{Im}\, d_{n+1} \subseteq \mathrm{Ker}\, d_n$, a complex is an exact sequence if and only if $\mathrm{Im}\, d_{n+1} \supseteq \mathrm{Ker}\, d_n$ for any n. A complex which is also exact is called an **exact complex** or **acyclic complex**. For instance, resolutions (free, projective, injective or flat) of modules are exact complexes.

(b) A complex $C. = (C_n, d_n)_{n \in \mathbb{Z}}$ is called a **concentrated complex** in $m \in \mathbb{Z}$, or a **stalk complex**, if $C_n = 0$ for $n \neq m$. Thus, a complex concentrated in $m \in \mathbb{Z}$ is nonzero if and only if $C_m \neq 0$, and it is exact if and only if it is zero.

Let $C. = (C_n, d_n)_{n \in \mathbb{Z}}$ and $C'. = (C'_n, d'_n)_{n \in \mathbb{Z}}$ be complexes concentrated in m, m' respectively. If $m \neq m'$, then the only morphism of complexes from $C.$ to $C'.$ is the zero morphism (the one with all component maps equal to zero). If $m = m'$, the morphisms of complexes from $C.$ to $C'.$ are in bijection with the A-linear maps from C_m to C'_m.

(c) Let A be a principal ideal domain, $p \in A$ an irreducible element and $k \geq 1$ an integer. For $n \in \mathbb{Z}$, set $C_n = A/p^k A$ and $d_n = d : \bar{x} \mapsto p^2 \bar{x}$, for $\bar{x} \in C_n$. We thus have

$$\cdots \to A/p^k A \xrightarrow{d} A/p^k A \xrightarrow{d} A/p^k A \xrightarrow{d} A/p^k A \to \cdots$$

If $k = 1$ or $k = 2$, then this data defines a complex with all differentials equal to zero (in particular, it is not exact). If $k = 3$, it defines a complex with nonzero differentials which is not exact. Indeed, $d_n d_{n+1} = 0$, for any n, but also Im $d_{n+1} = p^2 A/p^3 A \cong A/pA$, while Ker $d_n = pA/p^3 A \cong A/p^2 A$, for any n. If $k = 4$, it defines an exact complex because in this case, Im $d_{n+1} \cong A/p^2 A \cong$ Ker d_n, for any n. Finally, if $k > 4$, it is not a complex because $d_n d_{n+1} \neq 0$.

(d) Let A, B be K-algebras and $F : \mathrm{Mod}\, A \to \mathrm{Mod}\, B$ a covariant K-linear functor. To any complex $C. = (C_n, d_n)_{n \in \mathbb{Z}}$ of A-modules is associated a complex $FC. = (FC_n, Fd_n)_{n \in \mathbb{Z}}$ of B-modules: indeed, $d_n d_{n+1} = 0$ implies $(Fd_n)(Fd_{n+1}) = F(d_n d_{n+1}) = F(0) = 0$. In the same way, to a morphism of complexes $f. : C. \to C'.$ of A-modules corresponds a morphism of complexes $Ff. : FC. \to FC'.$ of B-modules.

The situation is similar if F is contravariant, but then a chain complex is transformed into a cochain complex and vice-versa.

(e) In the notation of Example (d) above, even if the complex $C.$ is exact, the complex $FC.$ is generally *not* exact, see Examples VI.3.9 and VI.3.16. Here is a further example. Let A be an integral domain, $a \subset A$ a nonzero noninvertible element and consider the short exact sequence of A-modules

$$0 \longrightarrow A \xrightarrow{\mu_a} A \longrightarrow A/aA \longrightarrow 0$$

where μ_a is multiplication by a, that is, the map $x \mapsto ax$, for $x \in A$. This map is injective because A is integral and $a \neq 0$. Apply to this sequence the functor $\mathrm{Hom}_A(A/aA, -)$. Then $\mathrm{Hom}_A(A/aA, A) = 0$, because A/aA is a torsion module, while A is torsion-free. On the other hand, $\mathrm{End}_A(A/aA) = \mathrm{Hom}_A(A/aA, A/aA) \neq 0$, because it contains the identity morphism, and $A/aA \neq 0$, because a is noninvertible. Hence, the image of the short exact sequence under the functor $\mathrm{Hom}_A(A/aA, -)$ is the concentrated complex

$$0 \to 0 \to 0 \to \mathrm{End}_A(A/aA) \to 0$$

which is certainly not exact.

Homology measures the lack of exactness of a complex. Let $C_. = (C_n, d_n)_{n \in \mathbb{Z}}$ be a chain complex. Because $d_n d_{n+1} = 0$, we have $\mathrm{Im}\, d_{n+1} \subseteq \mathrm{Ker}\, d_n$, and the complex is exact at C_n, that is, in degree n, if and only if we have equality. Lack of exactness is measured by the 'difference' between $\mathrm{Ker}\, d_n$ and $\mathrm{Im}\, d_{n+1}$. This leads us to consider the following submodules of C_n:

(a) $Z_n(C_.) = \mathrm{Ker}\, d_n$, and
(b) $B_n(C_.) = \mathrm{Im}\, d_{n+1}$.

The elements of $Z_n(C_.)$ are called **n-cycles** and those of $B_n(C_.)$ are **n-boundaries**. This terminology comes from algebraic topology. We sometimes abbreviate $Z_n(C_.)$ and $B_n(C_.)$ as Z_n and B_n respectively. Because $B_n \subseteq Z_n$, there exists, for any n, an inclusion $j_n : B_n \to Z_n$. Composing it with the inclusion $z_n : Z_n \to C_n$ yields the inclusion $i_{n+1} = z_n j_n : B_n \to C_n$.

Homology represents the difference between the sets B_n and Z_n. In algebraic terms, this means the quotient module of the second by the first.

X.2.3 Definition. Let $C_. = (C_n, d_n)_{n \in \mathbb{Z}}$ be a chain complex of A-modules. Its nth-**homology module** is the quotient module $H_n(C_.) = Z_n(C_.)/B_n(C_.)$.

Returning to the picture following Definition X.2.1, $H_n(C_.)$ is represented by the dotted area between $Z_n(C_.) = \mathrm{Ker}\, d_n$ and $B_n(C_.) = \mathrm{Im}\, d_{n+1}$. Thus, $C_.$ is exact in degree n if and only if $H_n(C_.) = 0$ and it is exact if and only if all its homology modules vanish.

We show how homology behaves with respect to morphisms. Let $f_. : C_. \to C'_.$ be a morphism of complexes. Observe that, for any n, there exists a unique morphism $Z_n(f_.) : Z_n(C_.) \to Z_n(C'_.)$ making commutative the left-hand square in the following diagram with exact rows

$$
\begin{array}{ccccccc}
0 & \longrightarrow & Z_n(C_.) & \overset{z_n}{\longrightarrow} & C_n & \overset{d_n}{\longrightarrow} & C_{n-1} \\
 & & \big\downarrow{\scriptstyle Z_n(f_.)} & & \big\downarrow{\scriptstyle f_n} & & \big\downarrow{\scriptstyle f_{n-1}} \\
0 & \longrightarrow & Z_n(C'_.) & \overset{z'_n}{\longrightarrow} & C'_n & \overset{d'_n}{\longrightarrow} & C'_{n-1}
\end{array}
$$

Indeed, it is obtained from the inclusions z_n, z'_n by passing to kernels. Similarly, there exists a unique morphism $B_n(f_.) : B_n(C_.) \to B_n(C'_.)$ making the following diagram commute:

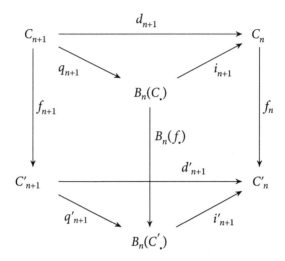

where $d_{n+1} = i_{n+1}q_{n+1}$ and $d'_{n+1} = i'_{n+1}q'_{n+1}$ are the canonical factorisations through the image. In fact, both morphisms $Z_n(f_{\bullet})$ and $B_n(f_{\bullet})$ are the restrictions of f_n to the corresponding submodules.

X.2.4 Lemma. *Let $f_{\bullet} : C_{\bullet} \to C'_{\bullet}$ be a morphism of complexes. For any n, there exists a unique morphism $H_n(f_{\bullet}) : H_n(C_{\bullet}) \to H_n(C'_{\bullet})$ making the following diagram with exact rows commute:*

$$
\begin{array}{ccccccccc}
0 & \longrightarrow & B_n(C_{\bullet}) & \overset{j_n}{\longrightarrow} & Z_n(C_{\bullet}) & \overset{p_n}{\longrightarrow} & H_n(C_{\bullet}) & \longrightarrow & 0 \\
& & \downarrow{\scriptstyle B_n(f_{\bullet})} & & \downarrow{\scriptstyle Z_n(f_{\bullet})} & & \downarrow{\scriptstyle H_n(f_{\bullet})} & & \\
0 & \longrightarrow & B_n(C'_{\bullet}) & \overset{j'_n}{\longrightarrow} & Z_n(C'_{\bullet}) & \overset{p'_n}{\longrightarrow} & H_n(C'_{\bullet}) & \longrightarrow & 0
\end{array}
$$

where p_n, p'_n are the respective projections.

Proof. It suffices to prove commutativity of the left-hand square, because then existence and uniqueness of $H_n(f_{\bullet})$ follow by passing to cokernels. We have

$$z'_n Z_n(f_{\bullet})j_n = f_n z_n j_n = f_n i_{n+1} = i'_{n+1}B_n(f_{\bullet}) = z'_n j'_n B_n(f_{\bullet}).$$

But z'_n is a monomorphim, hence $Z_n(f_{\bullet})j_n = j'_n B_n(f_{\bullet})$, as required. □

The commutative diagram of the previous Lemma X.2.4 shows that the morphism $H_n(f_{\bullet}) : H_n(C_{\bullet}) \to H_n(C'_{\bullet})$ is given explicitly by the formula:

$$H_n(f_{\bullet})(z_n + B_n(C_{\bullet})) = f_n(z_n) + B_n(C'_{\bullet})$$

for $z_n \in Z_n(C_{\bullet})$. One proves easily that each H_n is a covariant linear functor from complexes to modules: this is the nth-**homology functor**.

The definitions of Z_n, B_n and H_n imply the existence of exact sequences involving them.

X.2.5 Lemma. *Let $C_\bullet$ be a chain complex. For any $n \in \mathbb{Z}$, we have exact sequences*
(a) $0 \to H_n(C_\bullet) \to C_n/B_n(C_\bullet) \to B_{n-1}(C_\bullet) \to 0.$
(b) $0 \to H_n(C_\bullet) \to C_n/B_n(C_\bullet) \to Z_{n-1}(C_\bullet) \to H_{n-1}(C_\bullet) \to 0.$

Proof. (a) Due to the Isomorphism Theorem II.4.25, we have, for any n, a short exact sequence

$$0 \to Z_n(C_\bullet)/B_n(C_\bullet) \to C_n/B_n(C_\bullet) \to C_n/Z_n(C_\bullet) \to 0.$$

Now, $C_n/Z_n(C_\bullet) = C_n/\mathrm{Ker}\, d_n \cong \mathrm{Im}\, d_n = B_{n-1}(C_\bullet)$, due to the Isomorphism Theorem II.4.21, and $H_n(C_\bullet) = Z_n(C_\bullet)/B_n(C_\bullet)$, because of its very definition. We deduce the sequence in the statement.
(b) We splice the short exact sequence of (a) with the sequence

$$0 \to B_{n-1}(C_\bullet) \to Z_{n-1}(C_\bullet) \to H_{n-1}(C_\bullet) \to 0$$

thus obtaining the required sequence

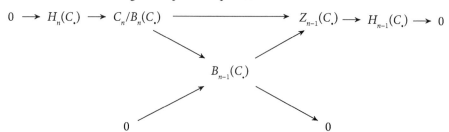

where the morphism from $C_n/B_n(C_\bullet)$ to $Z_{n-1}(C_\bullet)$ is the composite of the epimorphism from $C_n/B_n(C_\bullet)$ to $B_{n-1}(C_\bullet)$ with the injection of the latter into $Z_{n-1}(C_\bullet)$. $\qquad\square$

X.2.6 Examples.
(a) Let $C_\bullet = (C_n, d_n)_{n \in \mathbb{Z}}$ be a chain complex with all its differentials d_n equal to zero. Then $Z_n = C_n$ while $B_n = 0$. Hence, $H_n(C_\bullet) = C_n$, for any n.
(b) Let $C_\bullet = (C_n, d_n)_{n \in \mathbb{Z}}$ be a chain complex concentrated in degree m. Then $H_m(C_\bullet) = C_m$ while $H_n(C_\bullet) = 0$ for $n \neq m$.
(c) Let A be a principal ideal domain, $p \in A$ an irreducible element and $C_\bullet = (C_n, d_n)_{n \in \mathbb{Z}}$ the complex of A-modules defined by $C_n = A/p^3 A$ and $d_n : \bar{x} \mapsto p^2 \bar{x}$, for any n and $\bar{x} \in A/p^3 A$, see Example X.2.2(c). We have computed $B_n \cong A/pA$ and $Z_n \cong A/p^2 A$. Therefore, $H_n(C_\bullet) \cong (A/p^2 A)/(A/pA) \cong A/pA$, for any n.
(d) Let $\cdots \to P_1 \to P_0 \to M \to 0$ be a projective resolution of a module M. The complex $P_\bullet : \cdots \to P_1 \to P_0 \to 0$ obtained by deleting M is exact in any degree except in degree 0, and we have $H_0(P_\bullet) \cong M$.

(e) Let A be an integral domain and $a \in A$ a nonzero noninvertible element. The short exact sequence $0 \longrightarrow A \overset{\mu_a}{\longrightarrow} A \longrightarrow A/aA \longrightarrow 0$ is an exact complex, hence its homology vanishes everywhere. On the other hand, the two-terms complex $0 \to C_1 = A \overset{\mu_a}{\to} C_0 = A \to 0$ is not exact, and its only nonzero homology is $H_0(C_{\textbf{.}}) = \mathrm{Coker}\,\mu_a = A/aA$.

(f) Here is a variation on the previous Example (e). Let A be an integral domain and $a \in A$ a nonzero noninvertible element. Define $C_{\textbf{.}}$ by setting $C_n = A$, for any $n \in \mathbb{Z}$, while

$$d_n = \begin{cases} \mu_a & \text{if } n \text{ is even,} \\ 0 & \text{if } n \text{ is odd.} \end{cases}$$

Because one of two consecutive morphisms is zero, this is a complex

$$C_{\textbf{.}} : \cdots \longrightarrow A \overset{0}{\longrightarrow} A \overset{\mu_a}{\longrightarrow} A \overset{0}{\longrightarrow} A \overset{\mu_a}{\longrightarrow} A \overset{0}{\longrightarrow} A \longrightarrow \cdots$$

We compute its homology. We have two cases:

(i) If n is even, then $d_{n+1} = 0$ and $d_n = \mu_a$. Hence, $Z_n(C_{\textbf{.}}) = 0$, because μ_a is injective, while $B_n(C_{\textbf{.}}) = 0$. Therefore, $H_n(C_{\textbf{.}}) \cong 0$.

(ii) If n is odd, then $d_{n+1} = \mu_a$ and $d_n = 0$. Hence, $Z_n(C_{\textbf{.}}) = A$, while $B_n(C_{\textbf{.}}) = aA$. Therefore, $H_n(C_{\textbf{.}}) \cong A/aA$.

The dual considerations apply to cochain complexes. Let $C^{\textbf{.}} = (C^n, d^n)_{n \in \mathbb{Z}}$ be a cochain complex. For a given degree n, we define

(a) $Z^n(C^{\textbf{.}}) = \mathrm{Ker}\, d^n$, and

(b) $B^n(C^{\textbf{.}}) = \mathrm{Im}\, d^{n-1}$.

Elements of $Z^n(C^{\textbf{.}})$ are **n-cocycles** and those of $B^n(C^{\textbf{.}})$ are **n-coboundaries**.

X.2.7 Definition. Let $C^{\textbf{.}} = (C^n, d^n)_{n \in \mathbb{Z}}$ be a cochain complex of A-modules. Its nth-**cohomology module** is the quotient module $H^n(C^{\textbf{.}}) = Z^n(C^{\textbf{.}})/B^n(C^{\textbf{.}})$.

A morphism of cochain complexes $f^{\textbf{.}} : C^{\textbf{.}} \to C'^{\textbf{.}}$ induces, for any $n \in \mathbb{Z}$, a morphism between the cohomology modules $H^n(f^{\textbf{.}}) : H^n(C^{\textbf{.}}) \to H^n(C'^{\textbf{.}})$ and the statements corresponding to Lemmata X.2.4 and X.2.5 hold true.

X.2.8 Example. Let A be an integral domain, $a \in A$ a nonzero noninvertible element and consider the short exact sequence of A-modules

$$0 \longrightarrow A \overset{\mu_a}{\longrightarrow} A \longrightarrow A/aA \longrightarrow 0$$

where μ_a is multiplication by a, that is, the map $x \mapsto ax$, for $x \in A$. Evaluating the contravariant functor $\mathrm{Hom}_A(-, A)$ yields the left exact cochain complex

$$0 \to \mathrm{Hom}_A(A/aA, A) \to \mathrm{Hom}_A(A, A) \xrightarrow{\mathrm{Hom}_A(\mu_a, A)} \mathrm{Hom}_A(A, A).$$

Because a is nonzero and noninvertible, A/aA is a nonzero torsion module, while A is torsion-free, hence $\mathrm{Hom}_A(A/aA, A) = 0$. Applying Theorem II.3.9 yields a commutative square

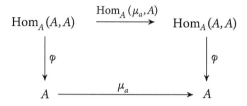

where the vertical maps are the isomorphisms $\varphi : u \mapsto u(1)$, with $u \in \mathrm{Hom}_A(A, A)$, of Theorem II.3.9. An easy calculation shows that the map below is indeed μ_a, which is injective. Because μ_a is not surjective, the left exact complex is not exact. Its only nonzero cohomology module is thus $\mathrm{Coker}\, \mu_a \cong A/aA \neq 0$.

X.2.2 The long exact (co)homology sequence

In the Snake Lemma II.4.19, we constructed a morphism which we called the connecting morphism. It links a sequence consisting of three kernels with one consisting of three cokernels. We need to use it in the proof of our main result about (co)homology, and for this we need another diagram-chasing lemma. In this subsection, all modules are over a K-algebra A.

X.2.9 Lemma. *The connecting morphism is functorial.*

Proof. To say that the connecting morphism δ is functorial amounts to saying that, if we have a commutative diagram with exact rows

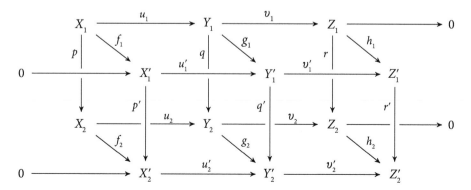

then there exists a commutative square

$$
\begin{array}{ccc}
\operatorname{Ker} h_1 & \xrightarrow{\;\delta_1\;} & \operatorname{Coker} f_1 \\
\downarrow{\scriptstyle\varphi} & & \downarrow{\scriptstyle\psi} \\
\operatorname{Ker} h_2 & \xrightarrow{\;\delta_2\;} & \operatorname{Coker} f_2
\end{array}
$$

where δ_1, δ_2 are the respective connecting morphisms and the vertical morphisms φ, ψ are obtained by passing to kernels for the first, and to cokernels for the second. Thus, letting $k_i : \operatorname{Ker} h_i \to Z_i$, for $i = 1, 2$, denote the injections, we have $rk_1 = k_2\varphi$, and, letting $p_i : X'_i \to \operatorname{Coker} f_i$, for $i = 1, 2$, denote the projections, we have $\psi p_1 = p_2 p'$.

The explicit formula giving the connecting morphism was:

$$
\delta_i = p_i u_i'^{-1} g_i v_i^{-1} k_i
$$

for $i = 1, 2$, where v_i^{-1} corresponds to taking an *arbitrary* preimage, while $u_i'^{-1}$ corresponds to taking a *unique* one, see the proof of Lemma II.4.19.

Let $x \in \operatorname{Ker} h_1$. Because v_1 is surjective, there exists $y \in Y_1$ such that $v_1(y) = k_1(x)$. Then

$$
\delta_1(x) = p_1 u_1'^{-1} g_1 v_1^{-1} k_1(x) = p_1 u_1'^{-1} g_1(y).
$$

Because $g_1(y) \in \operatorname{Ker} u'_2$, there exists $x' \in X'_1$ such that $u'_1(x') = g_1(y)$. Moreover, x' is unique because u'_1 is a monomorphism. We have $\delta_1(x) = p_1(x')$. Hence, $\psi\delta_1(x) = \psi p_1(x') = p_2 p'(x')$. On the other hand, $u'_2 p'(x') = q' u'_1(x')$. Therefore, $p'(x') = u_2'^{-1} q' u'_1(x')$ and so

$$
\psi\delta_1(x) = p_2 u_2'^{-1} q' u'_1(x') = p_2 u_2'^{-1} q' g_1(y) = p_2 u_2'^{-1} g_2 q(y).
$$

But $v_2 q = r v_1$, hence $q(y) = v_2^{-1} r v_1(y)$ and we have

$$
\psi\delta_1(x) = p_2 u_2'^{-1} g_2 v_2^{-1} r v_1(y) = p_2 u_2'^{-1} g_2 v_2^{-1} r k_1(x) = p_2 u_2'^{-1} g_2 v_2^{-1} k_2 \varphi(x) = \delta_2 \varphi(x).
$$

This completes the proof.   □

Though we have already seen applications of the Snake lemma, its real *raison d'être*, and in particular, that of the connecting morphism, is to prove our main result about (co)homology.

X.2.10 Theorem. *Let* $0 \to C'_\cdot \xrightarrow{f_\cdot} C_\cdot \xrightarrow{g_\cdot} C''_\cdot \to 0$ *be a short exact sequence of chain complexes. Then there exists an exact sequence*

$$\cdots \to H_n(C'') \xrightarrow{H_n(f')} H_n(C''') \xrightarrow{H_n(g')} H_n(C') \xrightarrow{\delta_n} H_{n-1}(C'') \xrightarrow{H_{n-1}(f')} H_{n-1}(C') \to \cdots$$

with all morphisms functorial.

Proof. Set $C'_\bullet = (C'_n, d'_n)_{n \in \mathbb{Z}}$, $C_\bullet = (C_n, d_n)_{n \in \mathbb{Z}}$ and $C''_\bullet = (C''_n, d''_n)_{n \in \mathbb{Z}}$ then apply the Snake Lemma II.4.19 to the commutative diagram with exact rows

$$
\begin{array}{ccccccccc}
0 & \longrightarrow & C'_n & \xrightarrow{f_n} & C_n & \xrightarrow{g_n} & C''_n & \longrightarrow & 0 \\
& & \downarrow{d'_n} & & \downarrow{d'_n} & & \downarrow{d''_n} & & \\
0 & \longrightarrow & C'_{\acute{n}} & \xrightarrow{f_{n-1}} & C_{n-1} & \xrightarrow{g_{n-1}} & C''_{n-1} & \longrightarrow & 0
\end{array}
$$

In each degree n, we deduce first, by passing to kernels, a left exact sequence

$$0 \to Z_n(C'_\bullet) \to Z_n(C_\bullet) \to Z_n(C''_\bullet)$$

then, by passing to cokernels, a right exact sequence

$$C'_n/B_n(C'_\bullet) \to C_n/B_n(C_\bullet) \to C''_n/B_n(C''_\bullet) \to 0.$$

The exact sequence of Lemma X.2.5(b) yields a commutative diagram where the two middle rows and the three columns are exact, and the δ_n stand for the connecting morphisms. The exactness of the required sequence follows from another application of the Snake lemma.

We prove functoriality. Because each H_n is a functor, each of the morphisms $H_n(f_\bullet)$ and $H_n(g_\bullet)$ is functorial. Finally, it follows from Lemma X.2.9 that so is each δ_n. $\square$

The exact sequence in the theorem is referred to as the **long exact homology sequence** associated to the short exact sequence of chain complexes $0 \to C'_\bullet \to C_\bullet \to C''_\bullet \to 0$.

We hasten to point out that Theorem X.2.10 holds true in any abelian category. Indeed, the Snake lemma, which we proved using diagram chasing, has also a purely categorical proof which, for brevity, we shall not present here. All other morphisms in the previous diagram were constructed using universal properties (that is, categorically).

Theorem X.2.10 implies that:

(a) If two of the complexes $C'_\bullet, C_\bullet, C''_\bullet$ are exact (that is, have vanishing homology), then so is the third.

(b) If one of the complexes $C'_\bullet, C_\bullet, C''_\bullet$ is exact, then the two others have isomorphic homology modules.

More generally, we have the following sufficient condition.

X.2.11 Corollary. *Let $f_\bullet : C_\bullet \to C'_\bullet$ be a morphism of complexes. If $\mathrm{Ker}\,(f_\bullet)$ and $\mathrm{Coker}\,(f_\bullet)$ are exact, then $H_n(C_\bullet) \cong H_n(C'_\bullet)$, for any n.*

Proof. Let $f_\bullet = j_\bullet p_\bullet$ be the canonical factorisation of $f_\bullet$ through its image. Because each H_n is a functor, $H_n(f_\bullet) = H_n(j_\bullet)H_n(p_\bullet)$, for any n. Now, we have short exact sequences of complexes:

$$0 \longrightarrow \mathrm{Ker}\,(f_\bullet) \longrightarrow C_\bullet \xrightarrow{p_\bullet} \mathrm{Im}\,(f_\bullet) \longrightarrow 0$$

$$0 \longrightarrow \mathrm{Im}\,(f_\bullet) \xrightarrow{j_\bullet} C'_\bullet \longrightarrow \mathrm{Coker}\,(f_\bullet) \longrightarrow 0.$$

Applying remark (b) above, each of $H_n(j_\bullet)$ and $H_n(p_\bullet)$ is an isomorphism, for any n. Therefore, so is their composite $H_n(f_\bullet)$. $\square$

X.2.12 Example. We construct an (easy) example of a long homology sequence. Let A be an integral domain and $a \in A$ a nonzero noninvertible element. Consider the short exact sequence of complexes $0 \to C'_\bullet \xrightarrow{j_\bullet} C_\bullet \xrightarrow{p_\bullet} C''_\bullet \to 0$, where $C'_\bullet = C''_\bullet$ is the complex of Example X.2.6(f), namely, for any $n \in \mathbb{Z}$, we set $C'_n = C''_n = A$ and

$$d'_n = d''_n = \begin{cases} \mu_a & \text{if } n \text{ is even} \\ 0 & \text{if } n \text{ is odd} \end{cases}$$

and then we let $C_n = C'_n \oplus C''_n = \Lambda^2$, while

$$d_n = \begin{cases} \begin{pmatrix} \mu_a & 0 \\ 0 & \mu_a \end{pmatrix} & \text{if } n \text{ is even} \\ 0 & \text{if } n \text{ is odd} \end{cases}$$

is the morphism obtained by passing to coproducts. The morphisms of complexes $j_., p_.$ are defined by $j_n = \binom{1}{0}, p_n = (0\ 1)$, for any $n \in \mathbb{Z}$. It is easy to see that this data yields indeed a short exact sequence of complexes. In view of the results of Example X.2.6(f), the long exact homology sequence is of the form

$$\cdots \to 0 \to A/aA \xrightarrow{\binom{1}{0}} (A/aA)^2 \xrightarrow{(0\ 1)} A/aA \to 0 \to 0 \to 0 \to A/aA \xrightarrow{\binom{1}{0}} (A/aA)^2 \xrightarrow{(0\ 1)}$$

$$A/aA \to 0 \to 0 \to \cdots$$

In particular, it is the union of split short exact sequences.

In view of the importance of Theorem X.2.10, we state the corresponding result for cochains.

X.2.13 Theorem. *Let* $0 \to C'^\bullet \xrightarrow{f^\bullet} C^\bullet \xrightarrow{g^\bullet} C''^\bullet \to 0$ *be a short exact sequence of cochain complexes. Then there exists an exact sequence*

$$\cdots \to H^n(C'^\bullet) \xrightarrow{H^n(f^\bullet)} H^n(C^\bullet) \xrightarrow{H^n(g^\bullet)} H^n(C''^\bullet) \xrightarrow{\delta^n} H^{n+1}(C'^\bullet) \xrightarrow{H^{n+1}(f^\bullet)} H^{n+1}(C^\bullet) \to \cdots$$

with all morphisms functorial. □

The exact sequence of the theorem is called the **long exact cohomology sequence** associated to the short exact sequence of cochain complexes $0 \to C'^\bullet \to C^\bullet \to C''^\bullet \to 0$. Again, Theorem X.2.13 holds true in any abelian category. We leave as an exercise the formulation of a corollary similar to X.2.11 in the case of cochains.

X.2.3 Homotopy

Two complexes may have the same (co)homology modules without necessarily being isomorphic, see Exercise X.2.6. In the same way, distinct morphisms of complexes may induce the same morphism in (co)homology. In this subsection, all our complexes are complexes of modules over a K-algebra A.

X.2.14 Definition. The morphisms of complexes $f_., g_. : C_. \to C'_.$ are **homotopic** if there exists, for any $n \in \mathbb{Z}$, a morphism $s_n : C_n \to C'_{n+1}$ such that we have $f_n - g_n = d'_{n+1}s_n + s_{n-1}d_n$, for any n. The sequence $s_. = (s_n)_{n\in\mathbb{Z}}$ is then a **homotopy** between $f_.$ and $g_.$.

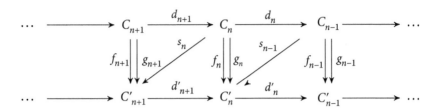

The relation $f_n - g_n = d'_{n+1}s_n + s_{n-1}d_n$, for any n, can be written briefly as $f - g = sd + d's$. Homotopy clearly defines a relation on the set of morphisms of complexes from $C_\bullet$ to $C'_\bullet$.

X.2.15 Lemma.

(a) *Homotopy is an equivalence relation on each morphism set and is compatible with composition.*

(b) *Homotopic morphisms induce the same morphism in homology.*

(c) *Evaluating a K-linear functor on homotopic morphisms yields homotopic morphisms.*

Proof. (a) Reflexivity and symmetry of the homotopy relation are evident. For transitivity, let $s_\bullet$ be a homotopy between $f_\bullet$ and $g_\bullet$, and $t_\bullet$ a homotopy between $g_\bullet$ and $h_\bullet$, then the relations $f - g = sd + d's$ and $g - h = td + d't$ give $f - h = (s+t)d + d'(s+t)$. So, homotopy is an equivalence in each morphism set.

In order to verify compatibility with composition, let, as above, $s_\bullet$ be a homotopy between $f_\bullet$ and $g_\bullet$, and $h_\bullet : C'_\bullet \to C''_\bullet$ a morphism which is left composable with $f_\bullet$ and $g_\bullet$. Then $f - g = sd + d's$ implies $hf - hg = hsd + hd's$. But h is a morphism of complexes, hence $hd' = d''h$, where d'' is the differential of $C''_\bullet$. We thus have $hf - hg = (hs)d + d''(hs)$. Therefore, $h_\bullet f_\bullet$ and $h_\bullet g_\bullet$ are homotopic. We proceed similarly for right composition.

(b) Let $s_\bullet$ be a homotopy between $f_\bullet$ and $g_\bullet$. Set $h_\bullet = f_\bullet - g_\bullet$. We must prove that $H_n(h_\bullet) = 0$, for any n. But $Z_n(h_\bullet)$ is the only morphism which makes the left hand square commutative in the diagram with exact rows

$$
\begin{array}{ccccccc}
0 & \longrightarrow & Z_n(C_\bullet) & \xrightarrow{\ z_n\ } & C_n & \xrightarrow{\ d_n\ } & C_{n-1} \\
& & {\scriptstyle Z_n(h_\bullet)}\big\downarrow & & {\scriptstyle h_n}\big\downarrow & & \big\downarrow{\scriptstyle h_{n-1}} \\
0 & \longrightarrow & Z_n(C'_\bullet) & \xrightarrow{\ z_n\ } & C'_n & \xrightarrow{\ d_n\ } & C'_{n-1}
\end{array}
$$

where z_n, z'_n are the injections. Thus,

$$z'_n Z_n(h_\bullet) = h_n z_n = s_{n-1} d_n z_n + d'_{n+1} s_n z_n = d'_{n+1} s_n z_n,$$

because $d_n z_n = 0$.

Hence, $Z_n(h.)$ factors through $B_n(C'.)$. The commutative diagram with exact rows

$$
\begin{array}{ccccccccc}
0 & \longrightarrow & B_n(C.) & \xrightarrow{\ j_n\ } & Z_n(C.) & \xrightarrow{\ p_n\ } & H_n(C.) & \longrightarrow & 0 \\
& & \downarrow{\scriptstyle B_n(h.)} & & \downarrow{\scriptstyle Z_n(h.)} & & \downarrow{\scriptstyle H_n(h.)} & & \\
0 & \longrightarrow & B_n(C'.) & \xrightarrow{\ j'_n\ } & Z_n(C'.) & \xrightarrow{\ p'_n\ } & H_n(C'.) & \longrightarrow & 0
\end{array}
$$

implies that $H_n(h.)p_n = 0$, whence $H_n(h.) = 0$, because $p.$ is an epimorphism.

(c) Assume that F is a covariant K-linear functor. Then $f - g = sd + d's$ implies $Ff - Fg = (Fs)(Fd) + (Fd')(Fs)$. The proof is similar if F is contravariant. $\square$

The most obvious applications of homotopy come from part (b) of the lemma: when a given construction does not result in a unique morphism, one tries to prove that two morphisms obtained using this construction are homotopic, and thus induce the same morphism in homology. It is therefore a tool for proving that certain morphisms (between homology modules) are unambiguously defined, see Theorem X.3.1 below and its consequences.

X.2.16 Examples.

(a) Let $L \xrightarrow{f} M \xrightarrow{g} N$ be modules and linear maps such that $gf = 0$. Define a complex $C.$ by setting $C_2 = L, C_1 = M, C_0 = N$ and $C_n = 0$, for $n \neq 0, 1, 2$ with differential $d_2 = f, d_1 = g$ and $d_n = 0$, for $n \neq 1, 2$. The complex $C.$ is exact if and only if so is the sequence $0 \to L \xrightarrow{f} M \xrightarrow{g} N \to 0$. We claim that the identity on $C.$ is homotopic to the zero morphism if and only if the sequence splits.

Indeed, $1_{C.}$ is homotopic to 0 if and only if there exist morphisms $s : N \to M$ and $t : M \to L$ such that $gs = 1, tf = 1$ and $sg + ft = 1$, and this is the case if and only if the short exact sequence splits, see Exercise IV.2.24.

(b) The converse of Lemma X.2.15(b) is not true: two morphisms of complexes may induce the same morphism in homology without being homotopic. Let A be an integral domain and $a \in A$ nonzero and noninvertible. The multiplication map $\mu_a : A \to A$ given by $x \mapsto ax$, for $x \in A$, is a monomorphism with nonzero cokernel equal to A/aA. Define two complexes as follows:

(a) $C.$ is given by $C_1 = C_0 = A$ and $C_n = 0$, for $n \neq 0, 1$, with differential $d_1 = \mu_a$ and $d_n = 0$, for $n \neq 1$.

(b) $C'.$ is given by $C'_1 = A$ and $C'_n = 0$, for $n \neq 1$, and the differential zero everywhere.

We claim that that the morphism $f_{\bullet} : C_{\bullet} \to C'_{\bullet}$, given by $f_1 = 1_A$ and $f_n = 0$, for $n \neq 1$, induces in homology the same morphism as the zero morphism, but is not homotopic to it.

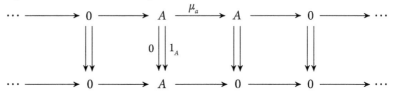

Indeed, passing to homology, we have $H_0(C_{\bullet}) = A/aA$ and $H_n(C_{\bullet}) = 0$, for $n \neq 0$, and $H_1(C'_{\bullet}) = A$. Therefore, $H_n(f_{\bullet}) = 0$, so $f_{\bullet}$ vanishes in homology. Now, assume that it is homotopic to zero. Because of Lemma X.2.15(c), it remains so after evaluation by the K-linear functor $- \otimes_A (A/aA)$. Then $\mu_a \otimes_A (A/aA) = 0$, so that the complex $C_{\bullet} \otimes_A (A/aA)$ is given by $C_1 \otimes_A (A/aA) = C_0 \otimes_A (A/aA) = A \otimes_A (A/aA) \cong A/aA$ with differential zero everywhere, while $C'_{\bullet} \otimes_A (A/aA)$ is given by $C'_1 \otimes_A (A/aA) \cong A/aA$ with differential also zero everywhere. Moreover, $H_1(f_{\bullet} \otimes_A A/aA) = 1_{A/aA}$ is nonzero.

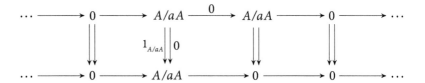

That is, $f_{\bullet} \otimes_A (A/aA)$ does not induce in homology the same morphism as the zero morphism, a contradiction which completes our proof.

We define in the same way homotopy between morphisms of cochain complexes. Two homotopic morphisms between cochain complexes induce the same morphism in cohomology.

Exercises of Section X.2

1. Let A be a K-algebra and $C_{\bullet}(\mathrm{Mod}\,A)$ the category of chain complexes of A-modules.
   (a) Let $f_{\bullet} : C_{\bullet} \to C'_{\bullet}$ be a morphism of complexes. Compute the kernel and the cokernel of $f_{\bullet}$ inside $C_{\bullet}(\mathrm{Mod}\,A)$.
   (b) Prove that a morphism of complexes $f_{\bullet}$ is a monomorphism, or an epimorphism, in $C_{\bullet}(\mathrm{Mod}\,A)$, if and only if each component map f_n is a monomorphism, or an epimorphism, respectively, in $\mathrm{Mod}\,A$.
   (c) Prove that the category $C_{\bullet}(\mathrm{Mod}\,A)$ is abelian.

2. Repeat the previous exercise for the category $C^{\bullet}(\text{Mod}\,A)$ of cochain complexes of A-modules.

3. Let A be an algebra and $f_{\bullet} : C_{\bullet} \to C'_{\bullet}$ a morphism of complexes. Prove that the following conditions are equivalent:
   (a) $f_{\bullet}$ is an isomorphism of complexes.
   (b) The induced morphism $\text{Hom}_A(M, f_{\bullet}) : \text{Hom}_A(M, C_{\bullet}) \to \text{Hom}_A(M, C'_{\bullet})$ is an isomorphism of complexes for any A-module M.
   (c) The induced morphism $\text{Hom}_A(f_{\bullet}, M) : \text{Hom}_A(C'_{\bullet}, M) \to \text{Hom}_A(C_{\bullet}, M)$ is an isomorphism of complexes for any A-module M.

4. In each case, verify that the sequence is a complex, then compute its homology modules.
   (a) Let p be a prime integer and consider the sequence of abelian groups
   $$\cdots \to 0 \to \mathbb{Z}/p^2\mathbb{Z} \xrightarrow{d} \mathbb{Z}/p^2\mathbb{Z} \to 0 \to \cdots \text{ where } d(x + p^2\mathbb{Z}) = px + p^2\mathbb{Z}, \text{ for } x \in \mathbb{Z}.$$
   (b) Let p be a prime integer and consider the sequence of abelian groups
   $$\cdots \to \mathbb{Z}/p^5\mathbb{Z} \xrightarrow{d} \mathbb{Z}/p^5\mathbb{Z} \xrightarrow{d} \mathbb{Z}/p^5\mathbb{Z} \to \cdots \text{ where } d(x + p^5\mathbb{Z}) = p^3x + p^5\mathbb{Z}, \text{ for } x \in \mathbb{Z}.$$
   (c) Let K be a field, $A = K[t]/\langle t^4 \rangle$ and consider the sequence of A-modules
   $$\cdots \to A_A \xrightarrow{d} A_A \xrightarrow{d} A_A \to \cdots \text{ where } d(p + \langle t^4 \rangle) = t^3 p + \langle t^4 \rangle, \text{ for } p \in K[t].$$

5. Let K be a field and A the algebra given by the quiver
   $$1 \circ \underset{\beta}{\overset{\alpha}{\underset{\longleftarrow}{\longrightarrow}}} \circ 2$$
   bound by the ideal $\langle \beta\alpha\beta\alpha \rangle$.
   (a) Compute the indecomposable projective modules P_1, P_2.
   (b) Construct an endomorphism $f \in \text{rad}_A(P_1, P_1)$ such that $f \circ f = f^2 \neq 0$.
   (c) Prove that, for any endomorphism $g \in \text{rad}_A(P_2, P_2)$, we have $g^2 = 0$.
   (d) Deduce that $\text{End}\, P_1 \cong K[t]/\langle t^3 \rangle$, while $\text{End}\, P_2 \cong K[t]/\langle t^2 \rangle$.
   (e) Using the endomorphism f of (b), prove that we have a complex
   $$\cdots \to P_1 \xrightarrow{f^2} P_1 \xrightarrow{f^2} P_1 \to \cdots \text{ and compute its homology modules.}$$

6. Construct an example of two nonisomorphic and nonexact complexes with the same homology modules.

7. Let $n > 1$ be an integer, G an abelian group and consider the short exact sequence of abelian groups $0 \longrightarrow \mathbb{Z} \xrightarrow{\mu_n} \mathbb{Z} \longrightarrow \mathbb{Z}/n\mathbb{Z} \longrightarrow 0$ where μ_n is multiplication by n. Evaluate the functor $G \otimes_{\mathbb{Z}} -$ on this sequence, then compute the homology groups of the resulting complex.

8. Let $C_{\bullet}$ be a complex consisting of free abelian groups. Prove that, for any integer n:
   (a) Z_n and B_n are free, and
   (b) Z_n is a direct summand of C_n.

9. Let $C_{\bullet} = (C_n, d_n)_{n \in \mathbb{Z}}$ be a chain complex of vector spaces over a field K. Prove that, for any integer n, we have $C_n \cong B_n(C_{\bullet}) \oplus B_{n-1}(C_{\bullet}) \oplus H_n(C_{\bullet})$.

10. Let A, B be K-algebras, $C.$ a chain complex of A-modules, ${}_A M_B$ a bimodule and N_B a B-module. Prove that we have an isomorphism of cochain complexes

$$\mathrm{Hom}_B(C. \otimes_A M, N) \cong \mathrm{Hom}_A(C., \mathrm{Hom}_B(M, N))$$

functorial in M and N.

11. Prove that the homology of a family of complexes commutes with the product and the direct sum of this family: let $(C_\cdot^\lambda)_{\lambda \in \Lambda}$ be a family of chain complexes of A-modules, prove that, for any $n \in \mathbb{Z}$, we have isomorphisms:
    (a) $\oplus_{\lambda \in \Lambda} H_n(C_\cdot^\lambda) \cong H_n(\oplus_{\lambda \in \Lambda} C_\cdot^\lambda)$, and
    (b) $\prod_{\lambda \in \Lambda} H_n(C_\cdot^\lambda) \cong H_n(\prod_{\lambda \in \Lambda} C_\cdot^\lambda)$.

12. Prove that a complex $C.$ such that the identity $1_{C.}$ is homotopic to the zero morphism is exact.

13. A morphism of chain complexes $f. : C. \to C'.$ is called a **homotopy equivalence** if there exists a morphism $f'. : C'. \to C.$ such that $f.f'. \sim 1_{C'.}$ and $f'.f. \sim 1_{C.}$, where $\sim$ denotes the homotopy relation. Prove that:
    (a) If $f.$ is a homotopy equivalence and $f'., f''.$ are such that $f'.f. \sim 1_{C.}$ and $f.f''. \sim 1_{C.}$, then $f'. \sim f''.$.
    (b) If $f.$ is a homotopy equivalence, then $H_n(f.)$ is an isomorphism, for any integer n.
    (c) If $f., g.$ are homotopy equivalences, then so is $g.f.$, if the composite exists.
    (d) If $f.$ is a homotopy equivalence and $g. \sim f.$, then $g.$ is a homotopy equivalence.

14. Let A, B be algebras, $F : \mathrm{Mod}\,A \to \mathrm{Mod}\,B$ an exact functor and $C.$ a chain complex of A-modules. Prove that:
    (a) If F is covariant, then $H_n(FC.) \cong FH_n(C.)$, for any n.
    (b) If F is contravariant, then $H^n(FC.) \cong FH_n(C.)$, for any n.

15. Let $C.$ be a chain complex. Prove that:
    (a) There exist complexes $Z.$ and $B_\cdot^-$ where $Z_n = Z_n(C.)$ and $B_n^- = B_{n-1}(C.)$, for any n (the exponent $-$ means that the complex is shifted by minus one) such that we have a short exact sequence of complexes $0 \to Z. \to C. \to B_\cdot^- \to 0$.
    (b) The long exact homology sequence resulting from the short exact sequence of complexes in (a) is a union of short exact sequences of modules.

16. Let $f. : C. \to C'.$ be a morphism of chain complexes. The **cone** of $f.$ is $K. = (K_n, \delta_n)_{n \in \mathbb{Z}}$ defined by $K_n = C_{n-1} \oplus C'_n$ and $\delta_n = \begin{pmatrix} -d_{n-1} & 0 \\ f_{n-1} & d_n \end{pmatrix}$, for any n.
    (a) Prove that $K.$ is a chain complex.
    (b) Let $C_\cdot^-$ be the complex obtained from $C.$ by shifting the indices by minus one: $C_n^- = C_{n-1}$ with the obvious differentials. Prove that $H_n(C_\cdot^-) = H_{n-1}(C.)$, for any n.
    (c) Prove that the injection of $C'.$ in $K.$ and the projection of $K.$ onto $C_\cdot^-$ induce a short exact sequence of complexes $0 \to C'. \to K. \to C_\cdot^- \to 0$.

(d) Prove that the long exact homology sequence associated to the short exact sequence of complexes in (c) is:

$$\cdots \to H_{n+1}(C'_{\bullet}) \to H_{n+1}(K_{\bullet}) \to H_n(C_{\bullet}) \xrightarrow{H_n(f_{\bullet})} H_n(C'_{\bullet}) \to H_n(K_{\bullet}) \to \cdots$$

(e) Deduce that $H_n(f_{\bullet})$ is an isomorphism, for any n, if and only if the cone of $f_{\bullet}$ is exact.

X.3 Derived functors

X.3.1 Left derived functors

In this section, we show how, given a one-sided exact functor (such as a Hom or a tensor functor) taking (co)homology can be used to repair its lack of exactness.

Let A, B be K-algebras and $F : \mathrm{Mod}A \to \mathrm{Mod}B$ a right exact K-linear and covariant functor, such as a tensor product functor. Evaluating it on a short exact sequence of A-modules $0 \to M' \to M \to M'' \to 0$ yields a sequence of B-modules which is only right exact $FM' \to FM \to FM'' \to 0$. By repairing its lack of exactness, we mean continuing the sequence on the left, getting a longer exact sequence, involving new functors $L_n F$ which we call the left derived functors of F. The resulting "long" exact sequence would be of the form:

$$\cdots \to L_2 F M'' \to L_1 F M' \to L_1 F M \to L_1 F M'' \to F M' \to F M \to F M'' \to 0$$

The construction of the $L_n F$ is done using the long exact homology sequence of Theorem X.2.10. It is achieved in three steps: first, construct a projective resolution of the module considered, second, evaluate F on the complex obtained by deleting the module from this resolution, and third, take homology of the resulting complex. We need a notation. Given a chain complex $C_{\bullet}$ of A-modules of the form

$$C_{\bullet} : \cdots \to C_1 \to C_0 \to X \to 0,$$

the complex obtained by deleting X from $C_{\bullet}$ is denoted by $(C_X)_{\bullet}$:

$$(C_X)_{\bullet} : \cdots \to C_1 \to C_0 \to 0.$$

We use the same notation for deleted cochain complexes. The proof of the next theorem illustrates how one can construct homotopies between morphisms of complexes.

X.3.1 Theorem. The Comparison Theorem. *Let A be an algebra and consider the diagram*

$$P_{\bullet} \quad \cdots \longrightarrow P_2 \xrightarrow{d_2} P_1 \xrightarrow{d_1} P_0 \xrightarrow{d_0} M \longrightarrow 0$$

with vertical maps f_2, f_1, f_0, f and lower row

$$P'_{\bullet} \quad \cdots \longrightarrow P'_2 \xrightarrow{d'_2} P'_1 \xrightarrow{d'_1} P'_0 \xrightarrow{d'_0} M' \longrightarrow 0$$

in $\mathrm{Mod}\,A$, where $P_{\bullet}$ is a complex with each P_n projective, while $P'_{\bullet}$ is exact. There exists a morphism of complexes $f_{\bullet} : (P_M)_{\bullet} \to (P'_{M'})_{\bullet}$, unique up to homotopy, such that $f d_0 = d'_0 f_0$.

Proof. Both existence and uniqueness up to homotopy are proved by induction on degree.

(a) *Existence.* For $n = 0$, we have a diagram with exact row

Projectivity of P_0 implies the existence of a morphism $f_0 : P_0 \to P'_0$ such that $f d_0 = d'_0 f_0$. Let $n \geq 0$. Assume that $f_k : P_k \to P'_k$ is constructed for any k such that $k \leq n$. We have, in particular, a commutative diagram with exact lower row

where we agree to set $P_{-1} = M, P'_{-1} = M'$ and $f_{-1} = f$. It then follows that $d'_n f_n d_{n+1} = f_{n-1} d_n d_{n+1} = 0$, because the upper row is a complex. Therefore, $\mathrm{Im}\,(f_n d_{n+1}) \subseteq \mathrm{Ker}\,d'_n = \mathrm{Im}\,d'_{n+1}$ and we get a diagram with exact row

Again, projectivity of P_{n+1} yields a morphism $f_{n+1} : P_{n+1} \to P_n$ such that $f_n d_{n+1} = d'_{n+1} f_{n+1}$. This completes the construction of $f_{\bullet} : (P_M)_{\bullet} \to (P'_{M'})_{\bullet}$. However, each of the f_n was constructed using the lifting property of projective modules which does *not* ensure uniqueness. Therefore, $f_{\bullet}$ is not unique in general.

(b) *Uniqueness up to homotopy.* Assume that $f_{\bullet}, g_{\bullet} : (P_M)_{\bullet} \to (P'_{M'})_{\bullet}$ are morphisms of complexes such that $f d_0 = d'_0 f_0 = d'_0 g_0$. We construct a homotopy $s_{\bullet} = (s_n)_{n \in \mathbb{Z}}$ between $f_{\bullet}$ and $g_{\bullet}$. We start by setting $s_n = 0$ for $n < 0$. Because

$d'_0 f_0 = d'_0 g_0$, we have $d'_0(f_0 \quad g_0) = 0$ hence Im $(f_0 - g_0) \subseteq$ Ker $d'_0 -$ Im d'_1. Because P_0 is projective, there exists $s_0 : P_0 \to P'_1$ such that $f_0 - g_0 = d'_1 s_0$.

$$P'_1 \xrightarrow{\ d'_1\ } \text{Im } d'_1 \longrightarrow 0$$

(with s_0 and $f_0 - g_0$ arrows from P_0)

Let $n \geq 0$. Assume that $s_k : P_k \to P'_{k+1}$ is constructed for any k such that $k \leqslant n$. Then $f_n - g_n = d'_{n+1} s_n + s_{n-1} d_n$, due to the induction hypothesis. We have

$$d'_{n+1}(f_{n+1} - g_{n+1} - s_n d_{n+1}) = d'_{n+1} f_{n+1} - d'_{n+1} g_{n+1} - d'_{n+1} s_n d_{n+1}$$
$$= f_n d_{n+1} - g_n d_{n+1} - (f_n - g_n - s_{n-1} d_n) d_{n+1}$$
$$= s_{n-1} d_n d_{n+1} = 0.$$

Hence, Im $(f_{n+1} - g_{n+1} - s_n d_{n+1}) \subseteq$ Ker $d'_{n+1} =$ Im d'_{n+2} and the projectivity of P_{n+1} yields a morphism $s_{n+1} : P_{n+1} \to P'_{n+2}$ such that

$$f_{n+1} - g_{n+1} - s_n d_{n+1} = d'_{n+2} s_{n+1}.$$

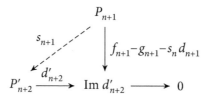

This completes the proof.   □

The morphism $f_\bullet$ is said to **lift** or to be a **lifting** of f. Because $f_\bullet$ is unique up to homotopy, the morphism it induces in homology is unique, due to Lemma X.2.15. The Comparison Theorem implies that certain modules, or morphisms, constructed with the help of a projective resolution, become independent of the choice of the latter once we pass to (co)homology.

X.3.2 Corollary. *Let A, B be K-algebras, $F :$ Mod $A \to$ Mod B a K-linear covariant functor and n an integer.*

(a) For any A-module M and projective resolution $P_\bullet$ of M, the module $H_n(F(P_M)_\bullet)$ is uniquely determined by M, up to isomorphism.

(b) For any A-linear map $f : M \to N$, projective resolutions $P_\bullet, Q_\bullet$ of M, N, respectively, and lifting $f_\bullet$ of f, the morphism $H_n(Ff_\bullet)$ is uniquely determined by f, up to isomorphism.

Proof. (a) Let

$$P_{\bullet} : \cdot \to P_1 \to P_0 \to M \to 0$$

and

$$P'_{\bullet} : \cdot \to P'_1 \to P'_0 \to M \to 0$$

be projective resolutions of M. We claim that $H_n(F(P_M)_{\bullet}) \cong H_n(F(P'_M)_{\bullet})$, for any n. The Comparison Theorem yields two morphisms $u_{\bullet} : (P_M)_{\bullet} \to (P'_M)_{\bullet}$ and $v_{\bullet} : (P'_M)_{\bullet} \to (P_M)_{\bullet}$ lifting the identity $1_M : M \to M$ and such that the composites $v_{\bullet}u_{\bullet} : (P_M)_{\bullet} \to (P_M)_{\bullet}$ and $u_{\bullet}v_{\bullet}; (P'_M)_{\bullet} \to (P'_M)_{\bullet}$ are homotopic to the respective identities. Because of Lemma X.2.15(c), $F(v_{\bullet}u_{\bullet})$ and $F(u_{\bullet}v_{\bullet})$ are also homotopic to the respective identity morphisms. The same lemma, part (b), gives that $H_nF(v_{\bullet}u_{\bullet}) = H_nF(v_{\bullet})H_nF(u_{\bullet})$ and $H_nF(u_{\bullet}v_{\bullet}) = H_nF(u_{\bullet})H_nF(v_{\bullet})$ are equal to the identities. That is, $H_nF(v_{\bullet})$ and $H_nF(u_{\bullet})$ are inverse isomorphisms, for any n.

(b) We claim that $H_n(Ff_{\bullet})$ depends neither on the choice of a projective resolution nor on the choice of a lifting $f_{\bullet}$ of f. Let

$$P_{\bullet} : \cdots \to P_1 \to P_0 \to M \to 0$$

and

$$P'_{\bullet} : \cdots \to P'_1 \to P'_0 \to M \to 0$$

be projective resolutions of M, and

$$Q_{\bullet} : \cdots \to Q_1 \to Q_0 \to N \to 0$$

and

$$Q'_{\bullet} : \cdots \to Q'_1 \to Q'_0 \to N \to 0$$

projective resolutions of N. Let also $g_{\bullet} : (P_M)_{\bullet} \to (Q_N)_{\bullet}, h_{\bullet} : (P'_M)_{\bullet} \to (Q'_N)_{\bullet}$ be liftings of $f : M \to N$. There exists a commutative diagram with exact rows:

$$
\begin{array}{ccccccccc}
\cdots & \longrightarrow & P_1 & \longrightarrow & P_0 & \longrightarrow & M & \longrightarrow & 0 \\
& & \downarrow{\scriptstyle g_1} & & \downarrow{\scriptstyle g_0} & & \downarrow{\scriptstyle f} & & \\
\cdots & \longrightarrow & Q_1 & \longrightarrow & Q_0 & \longrightarrow & N & \longrightarrow & 0 \\
& & & & & & \downarrow{\scriptstyle 1_N} & & \\
\cdots & \longrightarrow & Q'_1 & \longrightarrow & Q'_0 & \longrightarrow & N & \longrightarrow & 0
\end{array}
$$

The Comparison Theorem X.3.1 gives $u_{\bullet} : (Q_N)_{\bullet} \to (Q'_N)_{\bullet}$ lifting the identity 1_N. Hence, $u_{\bullet}g_{\bullet}$ lifts f.

Similarly, the commutative diagram with exact rows

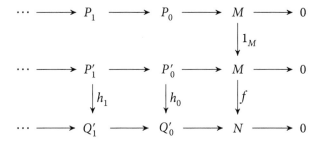

yields $v_\cdot : (P_M)_\cdot \to (P'_M)_\cdot$ lifting 1_M. Consequently, $h_\cdot v_\cdot$ lifts f. But lifts are unique up to homotopy. Therefore, $u_\cdot g_\cdot$ and $h_\cdot v_\cdot$ are homotopic. Because of Lemma X.2.15(c), $F(u_\cdot g_\cdot)$ and $F(h_\cdot v_\cdot)$ are also homotopic and we have $H_n F(u_\cdot) H_n F(g_\cdot) = H_n F(u_\cdot g_\cdot) = H_n F(h_\cdot v_\cdot) = H_n F(h_\cdot) H_n F(v_\cdot)$, for any n. Now, because of part (a) above, $H_n F(u_\cdot)$ and $H_n F(v_\cdot)$ are isomorphisms, for any n. This completes the proof. $\qquad\square$

The next definition is implicitly contained in the previous corollary.

X.3.3 Definition. Let A, B be K-algebras and $F : \operatorname{Mod} A \to \operatorname{Mod} B$ a K-linear covariant functor. For any integer $n \geq 0$, its ***nth-left derived functor*** $L_n F$ is defined as follows:

(a) Given an A-module M, we set $(L_n F)(M) = H_n(F(P_M)_\cdot)$, where $P_\cdot$ is an arbitrary projective resolution of M.

(b) Given an A-linear map $f : M \to N$, we set $(L_n F)(f) = H_n(Ff_\cdot)$, where $f_\cdot : (P_M)_\cdot \to (Q_N)_\cdot$ is any lifting of f using projective resolutions $P_\cdot, Q_\cdot$ of M, N, respectively.

Each $L_n F$, which is unambiguously defined thanks to Corollary X.3.2, is easily shown to be a K-linear covariant functor from $\operatorname{Mod} A$ to $\operatorname{Mod} B$. Indeed, let $P_\cdot, Q_\cdot$ be projective resolutions of M, N respectively, then $P_\cdot \oplus Q_\cdot$ is a projective resolution of $M \oplus N$. The statement follows from the fact that the functors F and H_n are K-linear.

An interesting remark is that, if we want to compute the left derived functors $L_n F$, we do not need to know how the functor F evaluates on *any* module: we just need to know how it evaluates on *projective* modules.

In order to construct a long exact homology sequence, we need a short exact sequence of complexes. The following lemma, which is Exercise VI.4.2(a), asserts that, starting from a short exact sequence of modules, we may construct a short exact sequence of projective resolutions of these modules. This construction carries the name of Horseshoe Lemma, due to the shape of the diagram one has to complete.

X.3.4 Lemma. The Horseshoe Lemma. *Let A be an algebra and assume that we have a diagram with exact row*

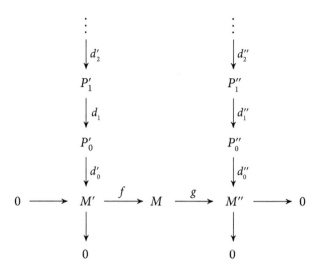

in Mod A, where the columns are projective resolutions $P'_\bullet, P''_\bullet$ of M', M'', respectively. There exists a projective resolution

$$P_\bullet : \cdots \to P_1 \to P_0 \to M \to 0$$

of M and morphisms $f_\bullet : (P'_{M'})_\bullet \to (P_M)_\bullet, g_\bullet : (P_M)_\bullet \to (P''_{M''})_\bullet$ lifting f, g, respectively, such that we have a short exact sequence of complexes

$$0 \to (P'_{M'})_\bullet \xrightarrow{f_\bullet} (P_M)_\bullet \xrightarrow{g_\bullet} (P''_{M''})_\bullet \to 0.$$

Proof. Using induction, it suffices to complete the diagram with exact rows and columns

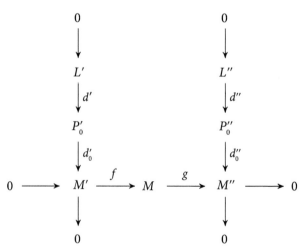

where P_0' and P_0'' are projective. Set $P_0 = P_0' \oplus P_0''$ and consider the morphisms
$f_0 = \begin{pmatrix} 1 \\ 0 \end{pmatrix} : P_0' \to P_0, g_0 = \begin{pmatrix} 0 & 1 \end{pmatrix} : P_0 \to P_0''$. Then P_0 is projective and the sequence

$$0 \to P_0' \xrightarrow{f_0} P_0 \xrightarrow{g_0} P_0'' \to 0$$

is exact (and split, because P_0'' is projective, see Theorem VI.4.4). Because P_0'' is projective, there exists $h : P_0'' \to M$ such that $d_0'' = gh$. We claim that $d_0 = f d_0' \begin{pmatrix} 1 & 0 \end{pmatrix} + h \begin{pmatrix} 0 & 1 \end{pmatrix}$ is surjective. Indeed, let $x \in M$, there exists $y'' \in P_0''$ such that $g(x) = d_0''(y'') = gh(y'')$. Then $x - h(y'') \in \text{Ker } g = \text{Im } f$ and there exists $y' \in P_0'$ such that $x - h(y'') = f d_0'(y')$. Hence,

$$x = f d_0' \begin{pmatrix} 1 & 0 \end{pmatrix} \begin{pmatrix} y' \\ y'' \end{pmatrix} + h \begin{pmatrix} 0 & 1 \end{pmatrix} \begin{pmatrix} y' \\ y'' \end{pmatrix} = d_0 \begin{pmatrix} y' \\ y'' \end{pmatrix}.$$

This establishes our claim. We have obtained a diagram with exact rows and columns

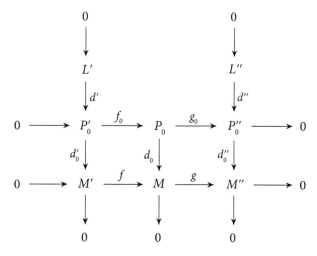

Furthermore, this diagram is commutative, because $g d_0 = gh \begin{pmatrix} 0 & 1 \end{pmatrix} = d_0'' \begin{pmatrix} 0 & 1 \end{pmatrix}$
$= d_0'' g_0$ and $d_0 f_0 = f d_0' \begin{pmatrix} 1 & 0 \end{pmatrix} \begin{pmatrix} 1 \\ 0 \end{pmatrix} + h \begin{pmatrix} 0 & 1 \end{pmatrix} \begin{pmatrix} 1 \\ 0 \end{pmatrix} = f d_0'$. Set $L = \text{Ker } d_0$. Applying the Snake Lemma II.4.19, we complete the diagram with the short exact sequence $0 \to L' \to L \to L'' \to 0$ obtained by passing to the kernels. The proof is complete. $\square$

In the previous proof, f_0 is a section, and g_0 a retraction (as they should be). If $g_0' : P_0'' \to P_0$ is a section such that $g_0 g_0' = 1$, then commutativity of the right square $g d_0 = d_0'' g_0$, upon right-multiplying by g_0', yields $g d_0 g_0' = d_0''$. No such identity results from the left square by considering a retraction to f_0.

X.3.5 Theorem. *Let A, B be K-algebras, $F : \operatorname{Mod} A \to \operatorname{Mod} B$ a K-linear covariant right exact functor and $0 \to M' \xrightarrow{f} M \xrightarrow{g} M'' \to 0$ a short exact sequence of A-modules. Then there exists an exact sequence of B-modules*

$$\cdots \to L_2 FM'' \xrightarrow{\delta} L_1 FM' \to L_1 FM \to L_1 FM'' \xrightarrow{\delta} FM' \to FM \to FM'' \to 0$$

with all morphisms functorial.

Proof. Because of the Horseshoe Lemma X.3.4, there exist projective resolutions $P'_\bullet, P_\bullet, P''_\bullet$ of M', M, M'', respectively, and a short exact sequence of complexes

$$0 \to (P'_{M'})_\bullet \xrightarrow{f_\bullet} (P_M)_\bullet \xrightarrow{g_\bullet} (P''_{M''})_\bullet \to 0$$

with $f_\bullet, g_\bullet$ liftings of f, g respectively. A component sequence of this short exact sequence of complexes is of the form $0 \to P'_n \xrightarrow{f_n} P_n \xrightarrow{g_n} P''_n \to 0$ and splits, because P''_n is projective. Therefore, for each n, we have a (split) short exact sequence $0 \to FP'_n \xrightarrow{Ff_n} FP_n \xrightarrow{Fg_n} FP''_n \to 0$. Applying Theorem X.2.10 yields the long exact homology sequence

$$\cdots \to L_2 FM'' \xrightarrow{\delta} L_1 FM' \to L_1 FM \to L_1 FM'' \xrightarrow{\delta} L_0 FM' \to L_0 FM \to L_0 FM'' \to 0$$

where the δ are the connecting morphisms. The sequence stops at $L_0 FM''$ because $P'_{-1} = 0$ gives $H_{-1}(F(P'_{M'})_\bullet) = 0$.

There remains to prove that $L_0 F \cong F$ as functors. This is a consequence of right exactness (which was not used so far). Let M be an A-module and $P_1 \to P_0 \to M \to 0$ an exact sequence with the P_i projective. Right exactness of F gives an exact sequence $FP_1 \to FP_0 \to FM \to 0$. Taking homology, we get $H_0(F(P_M)_\bullet) = FM$, so $L_0 F$ and F coincide on objects. Let $f : M \to M'$ be A-linear. Taking projective resolutions and applying the Comparison Theorem X.3.1 yields a commutative diagram with exact rows

$$
\begin{array}{ccccccccc}
\cdots & \longrightarrow & P_1 & \longrightarrow & P_0 & \longrightarrow & M & \longrightarrow & 0 \\
& & \downarrow{\scriptstyle f_1} & & \downarrow{\scriptstyle f_0} & & \downarrow{\scriptstyle f} & & \\
\cdots & \longrightarrow & P'_1 & \longrightarrow & P'_0 & \longrightarrow & M' & \longrightarrow & 0
\end{array}
$$

Evaluating the right exact functor F, we get a commutative diagram with exact rows

$$
\begin{array}{ccccccccc}
\cdots & \longrightarrow & FP_1 & \longrightarrow & FP_0 & \longrightarrow & FM & \longrightarrow & 0 \\
& & \downarrow{\scriptstyle Ff_1} & & \downarrow{\scriptstyle Ff_0} & & \downarrow{\scriptstyle Ff} & & \\
\cdots & \longrightarrow & FP'_1 & \longrightarrow & FP'_0 & \longrightarrow & FM' & \longrightarrow & 0
\end{array}
$$

so that $H_0(Ff.) = Ff$, as required. □

As we have observed in the proof, the existence of the long exact homology sequence

$$\cdots \to L_2FM'' \xrightarrow{\delta} L_1FM' \to L_1FM \to L_1FM'' \xrightarrow{\delta} L_0FM' \to L_0FM \to L_0FM'' \to 0$$

holds true for *any* linear covariant functor, not necessarily a right exact one. Right exactness is needed only to ensure that $L_0F \cong F$.

We postpone concrete examples to the next chapter where they will appear more naturally.

X.3.2 Right derived functors

The construction of right derived functors is analogous. But here, one should consider a left exact functor, such as the Hom functor. The latter is covariant in its second variable and contravariant in the first. This forces us to look at two cases: the right derived functors of a covariant functor and those of a contravariant functor. Let A, B be K-algebras.

(a) Assume that $F : \operatorname{Mod} A \to \operatorname{Mod} B$ is a K-linear covariant functor. In order to proceed as for left derived functors, we start by stating an 'injective version' of the Comparison Theorem X.3.1. The proof is dual to that of the latter and can be safely left as an exercise.

X.3.6 Theorem. The Comparison Theorem for injectives. *Let A be an algebra and consider the following diagram*

$$
\begin{array}{ccccccccccc}
I'^{\bullet}: & 0 & \longrightarrow & N' & \xrightarrow{d'^{\,0}} & I'^{\,0} & \xrightarrow{d'^{\,1}} & I'^{\,1} & \xrightarrow{d'^{\,2}} & I'^{\,2} & \longrightarrow & \cdots \\
& & & {\scriptstyle f}\downarrow & & {\scriptstyle f^0}\downarrow & & {\scriptstyle f^1}\downarrow & & {\scriptstyle f^2}\downarrow & & \\
I^{\bullet}: & 0 & \longrightarrow & N & \xrightarrow{d^0} & I^0 & \xrightarrow{d^1} & I^1 & \xrightarrow{d^2} & I^2 & \longrightarrow & \cdots
\end{array}
$$

in $\operatorname{Mod} A$, where $I^{\bullet}$ is a complex with each I^n injective, while $I'^{\bullet}$ is exact. There exists a morphism of complexes $f^{\bullet} : (I'_{N'})^{\bullet} \to (I_N)^{\bullet}$, unique up to homotopy, such that $d^0 f = f^0 d'^0$. □

The morphism $f^{\bullet}$ is said to **extend** or to be an **extension** of f.

X.3.7 Definition. Let A, B be K-algebras and $F : \operatorname{Mod} A \to \operatorname{Mod} B$ a K-linear covariant functor. For any integer $n \geq 0$, the nth**right derived functor** R_nF of F is defined by:

(a) Given an A-module N, we let $R_nF(N) = H^n(F(I_N)^{\bullet})$, where $I^{\bullet}$ is an arbitrary injective coresolution of N.

(b) Given an A-linear map $f : N \rightarrow N'$, we let $(R_n F)(f) = H^n(Ff^{\cdot})$, where $f^{\cdot} : (I_N)^{\cdot} \rightarrow (J_{N'})^{\cdot}$ is any extension of f to injective coresolutions $I^{\cdot}, J^{\cdot}$ of N, N', respectively.

The Comparison Theorem for injectives X.3.6 shows that this definition is unambiguous. Clearly, each $R_n F$ is a K-linear covariant functor. The injective version of the Horseshoe lemma, which is Exercise VI.4.2(b), gives the long exact cohomology sequence.

X.3.8 Theorem. *Let A, B be K-algebras, $F : \operatorname{Mod} A \rightarrow \operatorname{Mod} B$ a K-linear covariant left exact functor and $0 \rightarrow N' \rightarrow N \rightarrow N'' \rightarrow 0$ a short exact sequence of A-modules. Then there exists an exact sequence of B-modules*

$$0 \rightarrow FN' \rightarrow FN \rightarrow FN'' \xrightarrow{\delta} R_1 FN' \rightarrow R_1 FN \rightarrow R_1 FN'' \xrightarrow{\delta} R_2 FN' \rightarrow \cdots$$

with all morphisms functorial.   □

Again, right derived functors exist and give rise to a long exact cohomology sequence for any linear covariant functor F. Left exactness is here to ensure that $R_0 F = F$.
  (b) If $F : \operatorname{Mod} A \rightarrow \operatorname{Mod} B$ is contravariant, then we use projective resolutions.

X.3.9 Definition. Let A, B be K-algebras and $F : \operatorname{Mod} A \rightarrow \operatorname{Mod} B$ a K-linear contravariant functor. For any integer $n \geq 0$, the nth **right derived functor** $R_n F$ of F is defined by:
  (a) Given an A-module M, we let $R_n F(M) = H^n(F(P_M).)$, where $P.$ is an arbitrary projective resolution of M.
  (b) Given an A-linear map $f : M \rightarrow M'$, we let $(R_n F)(f) = H^n(Ff.)$, where $f. : (P_M)^{\cdot} \rightarrow (Q_{M'})^{\cdot}$ is any lifting of f to projective resolutions $P., Q.$ of M, M', respectively.

Because F is a K-linear contravariant functor from $\operatorname{Mod} A$ to $\operatorname{Mod} B$, so is each $R_n F$.

X.3.10 Theorem. *Let A, B be K-algebras, $F : \operatorname{Mod} A \rightarrow \operatorname{Mod} B$ a K-linear contravariant left exact functor and $0 \rightarrow M' \rightarrow M \rightarrow M'' \rightarrow 0$ a short exact sequence of A-modules. Then there exists an exact sequence of B-modules*

$$0 \rightarrow FM'' \rightarrow FM \rightarrow FM' \xrightarrow{\delta} R_1 FM'' \rightarrow R_1 FM \rightarrow R_1 FM' \xrightarrow{\delta} R_2 FM'' \rightarrow \cdots$$

with all morphisms functorial.   □

Arrived at this point, one may ask why we constructed the right derived functors of left exact functors which are both covariant and contravariant, while we restricted our study of left derived functors to right exact covariant functors. Indeed, one may

define and study in the same manner the left derived functors of a right exact contravariant functor. We refrain from doing it, because our intention is to apply this
construction to the bifunctors Hom and tensor product. While the former, which is
left exact, is contravariant in its first variable and covariant in the second, the tensor product, which is the prototype of a right exact functor, is covariant in each
variable.

Exercises of Section X.3

1. Let A, B be K-algebras and $F : \mathrm{Mod}A \to \mathrm{Mod}B$ a K-linear functor. Prove that:
   (a) If F is covariant, then $L_n F(P) = 0$, for any projective module P_A and any
       integer $n > 0$.
   (b) If F is covariant, then $R_n F(I) = 0$, for any injective module I_A and any
       integer $n > 0$.
   (c) If F is contravariant, then $R_n F(P) = 0$, for any projective module P_A and
       any integer $n > 0$.

2. Let A, B be algebras and $F : \mathrm{Mod}A \to \mathrm{Mod}B$ a K-linear covariant right exact
   functor. Prove that, for any integers m, n, we have

$$L_n L_m F = \begin{cases} L_n F & \text{if } m = 0, \\ 0 & \text{otherwise.} \end{cases}$$

3. Let G be an exact K-linear functor.
   (a) If F is a K-linear right exact functor then $G(L_n F) = L_n(GF)$, for any
       integer $n \geq 0$.
   (b) State and prove similar results for left exact (covariant and contravariant)
       functors.

4. Let p be a prime integer. Consider the short exact sequence $0 \to \mathbb{Z}/p\mathbb{Z} \to
   \mathbb{Z}/p^2\mathbb{Z} \to \mathbb{Z}/p\mathbb{Z} \to 0$, with the obvious morphisms, and the projective resolu-
   tion $0 \longrightarrow \mathbb{Z} \xrightarrow{\mu_p} \mathbb{Z} \longrightarrow \mathbb{Z}/p\mathbb{Z} \longrightarrow 0$, where μ_p denotes multiplication by p.
   Construct a morphism of complexes lifting the identity morphism of $\mathbb{Z}/p\mathbb{Z}$ as in
   the Comparison Theorem X.3.1, computing explicitly the morphisms.

$$
\begin{array}{ccccccccc}
0 & \longrightarrow & \mathbb{Z} & \xrightarrow{\mu_p} & \mathbb{Z} & \longrightarrow & \mathbb{Z}/p\mathbb{Z} & \longrightarrow & 0 \\
 & & \Big\downarrow & & \Big\downarrow & & \Big\downarrow{\scriptstyle 1_{\mathbb{Z}/p\mathbb{Z}}} & & \\
0 & \longrightarrow & \mathbb{Z}/p\mathbb{Z} & \longrightarrow & \mathbb{Z}/p^2\mathbb{Z} & \longrightarrow & \mathbb{Z}/p\mathbb{Z} & \longrightarrow & 0
\end{array}
$$

5. With the same notation as in the previous Exercise X.3.4, complete the Horseshoe
   below computing explicitly the morphisms.

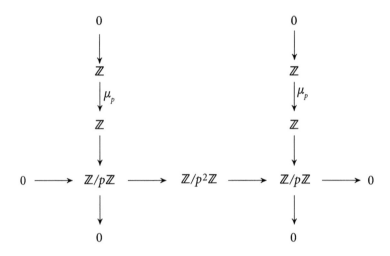

6. In the notation of the Horseshoe Lemma X.3.4, prove that there exists, for any integer $n > 0$, a morphism $u_n : P''_n \to P'_{n-1}$ such that $d_n = \begin{pmatrix} d'_n & u_n \\ 0 & d''_n \end{pmatrix}$.

7. State and prove a similar result for the Horseshoe lemma for injectives.

8. Let A be the Nakayama algebra given by the quiver

bound by the ideal $\langle \alpha\beta \rangle$, and the short exact sequence $0 \to 3 \to \begin{smallmatrix} 2 \\ 3 \end{smallmatrix} \to 2 \to 0$.
   (a) Write projective resolutions for the simple A-modules S_2 and S_3.
   (b) Repeat Exercise X.3.4 above with the given short exact sequence.
   (c) Repeat Exercise X.3.5 above with the given short exact sequence and the projective resolutions of (a).

9. This exercise generalises part of the proof of Theorem X.3.5. Let A, B be K-algebras and $F : \mathrm{Mod}A \to \mathrm{Mod}B$ a K-linear functor. Prove that:
   (a) If F is covariant, then $L_0F \cong F$ if and only if F is right exact.
   (b) If F is covariant, then $R_0F \cong F$ if and only if F is left exact.
   (c) If F is contravariant, then $R_0F \cong F$ if and only if F is left exact.

Chapter XI
Extension and torsion

XI.1 Introduction

In the previous chapter, we developed a considerable machinery but did not explain its use. Let us sum up: given a half-exact functor, and a module, we construct a projective resolution or an injective coresolution of this module, delete the module from it, then evaluate on the deleted complex the given functor, the (co)homology of the complex thus obtained gives (left or right) derived functors. Applying this procedure to a short exact sequence of modules, using compatible projective resolutions or injective coresolutions (by compatible, we mean constructed with the help of the Horseshoe Lemma X.3.4 or its dual) yields a long exact (co)homology sequence.

Typical examples of half-exact functors are the Hom and tensor functors. Applying the theory of derived functors to these, we obtain what are respectively called extension and torsion functors. Now, each of the Hom and tensor functor is actually a bifunctor, and one can construct derived functors either with respect to the first, or to the second variable. It turns out that both constructions are compatible, so that extension and torsion functors become bifunctors, which, in degree zero, are isomorphic to the Hom and tensor bifunctor, respectively.

Besides studying these functors, we relate the extension functor to extensions as defined in Subsection VII.4.2, namely short exact sequences. In the second part of the chapter, we construct a bijection between the elements of the nth-extension module and equivalence classes of exact sequences of length n, thus giving a concrete interpretation of the former. Throughout this chapter, unless otherwise specified, K is a commutative ring and algebras are K-algebras.

XI.2 The extension and torsion functors

XI.2.1 The extension functors

Consider the right derived functors of the Hom functors (which are left exact). Because Hom is a bifunctor, there are two types of right derived functors: those associated to the first variable (for which the Hom functor is contravariant) and the other associated to the second variable (for which the functor is covariant). Let A be a K-algebra.

(a) Let $n \geq 0$ be an integer and M an A-module. Consider the covariant Hom-functor $\mathrm{Hom}_A(M, -) : \mathrm{Mod}\,A \to \mathrm{Mod}\,K$. Its nth right derived functor is denoted (provisionally) by $E_n(M, -)$. Applying Definition X.3.7 to $F = \mathrm{Hom}_A(M, -)$ (and thus $R_n F = E_n(M, -)$), we have, for any module N,

An Introduction to Module Theory. Ibrahim Assem and Flávio U. Coelho, Oxford University Press.
© Ibrahim Assem and Flávio U. Coelho (2024). DOI: 10.1093/9780198904939.003.0012

$$E_n(M, -)(N) = E_n(M, N) = H^n(M, (I_N)^\bullet).$$

where $I^\bullet$ is an injective coresolution of N and $(I_N)^\bullet$ the complex obtained after deleting N. In particular, each of the $E_n(M, N)$ is a K-module. Due to Theorem X.2.13, to a short exact sequence $0 \to N' \to N \to N'' \to 0$ in $\mathrm{Mod}\,A$ corresponds a long exact cohomology sequence

$$0 \to \mathrm{Hom}_A(M, N') \to \mathrm{Hom}_A(M, N) \to \mathrm{Hom}_A(M, N'') \xrightarrow{\delta} E_1(M, N') \to \cdots$$

$$\cdots \to E_{n-1}(M, N'') \xrightarrow{\delta} E_n(M, N') \to E_n(M, N) \to E_n(M, N'') \xrightarrow{\delta} E_{n+1}(M, N') \to \cdots$$

with all morphisms functorial.

(b) Let $n \geq 0$ be an integer and N an A-module. Consider the contravariant Hom-functor $\mathrm{Hom}_A(-, N) : \mathrm{Mod}\,A \to \mathrm{Mod}\,K$. Its nth right derived functor is denoted (provisionally) by $E^n(-, N)$. It follows from Definition X.3.9 applied to $F = \mathrm{Hom}_A(-, N)$ that, for any module M, we have

$$E^n(-, N)(M) = E^n(M, N) = H^n((P_M)_\bullet, N).$$

where $P_\bullet$ is a projective resolution of M and $(P_M)_\bullet$ the complex obtained after deleting M. Thus, each of the $E^n(M, N)$ is a K-module. Due to Theorem X.2.13, to a short exact sequence $0 \to M' \to M \to M'' \to 0$ in $\mathrm{Mod}\,A$ corresponds a long exact cohomology sequence

$$0 \to \mathrm{Hom}_A(M'', N) \to \mathrm{Hom}_A(M, N) \to \mathrm{Hom}_A(M', N) \xrightarrow{\delta} E^1(M'', N) \to \cdots$$

$$\cdots \to E^{n-1}(M', N) \xrightarrow{\delta} E^n(M'', N) \to E^n(M, N) \to E^n(M', N) \xrightarrow{\delta} E^{n+1}(M'', N) \to \cdots$$

with all morphisms functorial.

We claim that these constructions are compatible, that is, for any pair of modules M, N, and integer $n \geq 1$, there exists an isomorphism $E_n(M, N) \cong E^n(M, N)$, functorial in each variable, thus defining a bifunctor in each degree n. The strategy consists in first reducing the problem to the case of low degrees, then solving the latter. Reduction to a lower degree is done through a device called **dimension shifting**.

XI.2.1 Lemma. Dimension shifting. *Let M, N be A-modules.*

(a) *Assume that $I^\bullet$:*

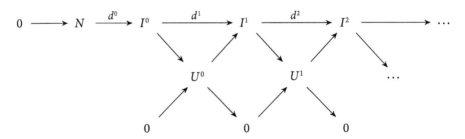

is an injective coresolution of N, with $U^n = \text{Im } d^{n+1}$, for $n \geq 0$ an integer. Then we have functorial isomorphisms:

$$E_{n+1}(M, N) \cong E_n(M, U^0) \cong \cdots \cong E_1(M, U^{n-1}).$$

(b) *Assume that* $P_\bullet$:

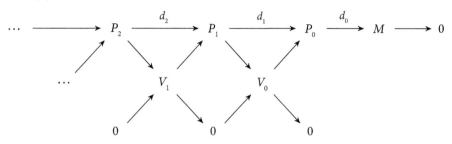

is a projective resolution of M, with $V_n = \text{Im } d_{n+1}$, for $n \geq 0$ an integer. Then we have functorial isomorphisms:

$$E^{n+1}(M, N) \cong E^n(V_0, N) \cong \cdots \cong E^1(V_{n-1}, N).$$

Proof. We prove (a), because the proof of (b) is similar. Inductively, it suffices to prove the first isomorphism $E_{n+1}(M, N) \cong E_n(M, U^0)$. We have an injective coresolution of U^0:

$$I'^\bullet : 0 \to U^0 \to I^1 \xrightarrow{d^2} I^2 \xrightarrow{d^3} \cdots$$

Evaluating $\text{Hom}_A(M, -)$ on $(I'_{U^0})^\bullet$, we obtain a complex

$$\text{Hom}_A(M, (I'_{U^0})^\bullet) : 0 \to \text{Hom}_A(M, I^1) \xrightarrow{\text{Hom}_A(M, d^2)} \text{Hom}_A(M, I^2) \xrightarrow{\text{Hom}_A(M, d^3)} \cdots$$

Therefore,

$$E_n(M, U^0) = H^n(\text{Hom}_A(M, (I'_{U^0})^\bullet)) = \text{Ker } \text{Hom}_A(M, d^{n+2})/\text{Im } \text{Hom}_A(M, d^{n+1})$$
$$\cong H^{n+1}(\text{Hom}_A(M, (I'_N)^\bullet)) = E^{n+1}(M, N).$$

It is clear that this isomorphism is functorial. □

In the statement of this lemma, we represented on the same diagram a (co)resolution together with the images of the morphisms on it, so that the diagram becomes commutative. Images of the morphisms lying on a projective resolution of M are called the **syzygies** of M while those of the morphisms lying on an injective coresolution are its **cosyzygies**. The word *syzygy*, which comes from greek, is an astronomical term expressing that three celestial bodies are in conjunction (or in opposition).

The following lemma simplifies considerably our calculations, because we are dealing with projective resolutions and/or injective coresolutions.

XI.2.2 Lemma. (a) *If N is injective, then $E_n(M, N) = 0$, for any module M and integer $n \geq 1$.*

(b) *If M is projective, then $E^n(M, N) = 0$, for any module N and integer $n \geq 1$.*

Proof. We prove (a), because the proof of (b) is similar. Because N is injective, the sequence $0 \to N \xrightarrow{1} N \to 0$ is an injective coresolution (with $I^0 = N$ and $I^n = 0$, for $n \geq 1$). Therefore, for any module M and any $n \geq 1$, we have $E_n(M, N) = H^n(\mathrm{Hom}_A(M, (I_N)^\bullet)) = 0$. $\qquad\qquad\square$

We establish our earlier claim.

XI.2.3 Theorem. *Let M, N be A-modules and $n \geq 1$ an integer. Then we have an isomorphism $E_n(M, N) \cong E^n(M, N)$, functorial in both variables.*

Proof. Let

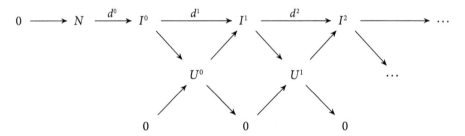

be an injective coresolution of N, with $U^n = \mathrm{Im}\, d^{n+1}$, and

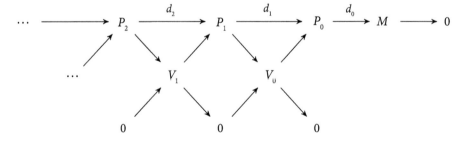

a projective resolution of M, with $V_n = \mathrm{Im}\, d_{n+1}$. We use induction on n. For any integers $i, j \geq 0$, we have short exact sequences

$$0 \to V_i \to P_i \to V_{i-1} \to 0$$
$$0 \to U^{j-1} \to I^j \to U^j \to 0$$

where we set $U^{-1} = N$ and $V_{-1} = M$. Taking the long exact cohomology sequences, in which, because of Lemma XI.2.2, several terms vanish, we get, because of functoriality, a commutative diagram with exact rows and columns

$$
\begin{array}{ccccccccc}
0 & \to & \mathrm{Hom}_A(V_{i-1}, U^{j-1}) & \to & \mathrm{Hom}_A(V_{i-1}, I^j) & \xrightarrow{u} & \mathrm{Hom}_A(V_{i-1}, U^j) & \to & E_1(V_{i-1}, U^{j-1}) \to 0 \\
& & \downarrow & & \downarrow & & \downarrow & & \\
0 & \to & \mathrm{Hom}_A(P_i, U^{j-1}) & \to & \mathrm{Hom}_A(P_i, I^j) & \xrightarrow{v} & \mathrm{Hom}_A(P_i, U^j) & \to & 0 \\
& & f\downarrow & & g\downarrow & & h\downarrow & & \\
0 & \to & \mathrm{Hom}_A(V_i, U^{j-1}) & \to & \mathrm{Hom}_A(V_i, I^j) & \xrightarrow{w} & \mathrm{Hom}_A(V_i, U^j) & \to & E_1(V_i, U^{j-1}) \to 0 \\
& & \downarrow & & \downarrow & & \downarrow & & \\
& & E^1(V_{i-1}, U^{j-1}) & & 0 & & E^1(V_{i-1}, U^j) & & \\
& & \downarrow & & & & \downarrow & & \\
& & 0 & & & & 0 & &
\end{array}
$$

Applying the Snake Lemma II.4.19 to the morphisms f, g, h yields an exact sequence

$$0 \to \mathrm{Hom}_A(V_{i-1}, U^{j-1}) \to \mathrm{Hom}_A(V_{i-1}, I^j) \xrightarrow{u} \mathrm{Hom}_A(V_{i-1}, U^j) \to E^1(V_{i-1}, U^{j-1}) \to 0$$

because the columns are left exacts. Hence,

$$E^1(V_{i-1}, U^{j-1}) \cong \mathrm{Coker}\, u \cong E_1(V_{i-1}, U^{j-1}).$$

Setting $i = j = 0$, we get $E^1(M, N) = E_1(M, N)$, so our statement is proved for $n = 1$. On the other hand, g, v are epimorphisms, therefore

$$E^1(V_{i-1}, U^j) = \mathrm{Coker}\, h = \mathrm{Coker}\, hv = \mathrm{Coker}\, wg = \mathrm{Coker}\, w = E_1(V_i, U^{j-1}).$$

These formulae imply that

$$E^1(V_{n-1}, U^{-1}) \cong E_1(V_{n-1}, U^{-1}) \cong \cdots \cong E_1(V_{-1}, U^{n-1}).$$

On the other hand, because of dimension shifting, see Lemma XI.2.1, we have

$$E^{n+1}(M, N) \cong \cdots \cong E^1(V_{n-1}, N) = E^1(V_{n-1}, U^{-1})$$

and

$$E_{n+1}(M, N) \cong \cdots \cong E_1(M, U^{n-1}) = E_1(V_{-1}, U^{n-1}).$$

Therefore, $E^{n+1}(M, N) \cong E_{n+1}(M, N)$, as required. Because this isomorphism is the composite of a sequence of functorial isomorphisms, it is itself functorial. $\square$

XI.2.4 Definition. Let M, N be A-modules. For any integer $n \geq 1$, the ***n^{th}-extension module*** of N by M, denoted by $\mathrm{Ext}_A^n(M, N)$, is the common value of $E^n(M, N)$ and $E_n(M, N)$.

It follows from our discussion on derived functors that, for any integer $n > 0$, $\text{Ext}_A^n(-, ?)$ is, like Hom, a K-linear bifunctor, covariant in its second variable and contravariant in the first, taking values in Mod K.

If either M or N, or both, have bimodule structures, then $\text{Ext}_A^n(M, N)$ inherits a module structure, in the same way as $\text{Hom}_A(M, N)$ does, see Lemma II.3.8. For instance, if $_BM_A$ is a B-A-bimodule, and $I^\bullet$ an injective coresolution of N, then each of the K-modules $\text{Hom}_A(M, (I_N)^n)$ has a B-module structure, each of the differentials $\text{Hom}_A(M, (I_N)^n) \rightarrow \text{Hom}_A(M, (I_N)^{n+1})$ is B-linear, hence $\text{Ext}_A^n(M, N) \cong H^n(\text{Hom}_A(M, (I_N)^\bullet))$ has an induced B-module structure. If, in particular, A is commutative, then $\text{Ext}_A^n(M, N)$ has a natural A-module structure.

We reformulate in our context Theorems X.3.8 and X.3.10.

XI.2.5 Theorem. (a) *Let M be an A-module and $0 \rightarrow N' \rightarrow N \rightarrow N'' \rightarrow 0$ a short exact sequence of A-modules. Then we have a long exact cohomology sequence:*

$$0 \rightarrow \text{Hom}_A(M, N') \rightarrow \text{Hom}_A(M, N) \rightarrow \text{Hom}_A(M, N'') \xrightarrow{\delta} \text{Ext}_A^1(M, N') \rightarrow \cdots$$

$$\cdots \rightarrow \text{Ext}_A^{n-1}(M, N'') \xrightarrow{\delta} \text{Ext}_A^n(M, N') \rightarrow \text{Ext}_A^n(M, N) \rightarrow \text{Ext}_A^n(M, N'') \xrightarrow{\delta} \text{Ext}_A^{n+1}(M, N') \rightarrow \cdots$$

with all morphisms functorial.

(b) *Let N be an A-module and $0 \rightarrow M' \rightarrow M \rightarrow M'' \rightarrow 0$ a short exact sequence of A-modules. Then we have a long exact cohomology sequence:*

$$0 \rightarrow \text{Hom}_A(M'', N) \rightarrow \text{Hom}_A(M, N) \rightarrow \text{Hom}_A(M', N) \xrightarrow{\delta} \text{Ext}_A^1(M'', N) \rightarrow \cdots$$

$$\cdots \rightarrow \text{Ext}_A^{n-1}(M', N) \xrightarrow{\delta} \text{Ext}_A^n(M'', N) \rightarrow \text{Ext}_A^n(M, N) \rightarrow \text{Ext}_A^n(M', N) \xrightarrow{\delta} \text{Ext}_A^{n+1}(M'', N) \rightarrow \cdots$$

with all morphisms functorial. $\qquad\square$

We leave to the reader the straightforward reformulation of Lemma XI.2.1 on dimension shifting, using the notation Ext^n instead of E_n, E^n. In Lemma XI.2.2, we proved that, for any integer $n \geq 1$, we have $\text{Ext}_A^n(M, N) = 0$ if M is projective, or if N is injective. We now prove the converse. We need a useful (and easy) lemma.

XI.2.6 Lemma. *If $\text{Ext}_A^1(M, N) = 0$, then any short exact sequence of the form* $0 \rightarrow N \xrightarrow{f} E \xrightarrow{g} M \rightarrow 0$ *splits.*

Proof. Evaluating $\text{Hom}_A(M, -)$ on the given sequence yields an exact cohomology sequence:

$$0 \rightarrow \text{Hom}_A(M, N) \rightarrow \text{Hom}_A(M, E) \rightarrow \text{Hom}_A(M, M) \rightarrow \text{Ext}_A^1(M, N) = 0$$

where the last term vanishes due to the hypothesis. The statement follows from Example VI.3.9(c). $\qquad\square$

XI.2.7 Proposition. (a) *Let M be an A-module. The following conditions are equivalent:*
  (i) *M is projective.*
  (ii) $\mathrm{Ext}_A^1(M, -) = 0.$
  (iii) $\mathrm{Ext}_A^n(M, -) = 0,$ *for any integer* $n \geq 1.$
  (b) *Let N be an A-module. The following conditions are equivalent:*
  (i) *N is injective.*
  (ii) $\mathrm{Ext}_A^1(-, N) = 0.$
  (iii) $\mathrm{Ext}_A^n(-, N) = 0,$ *for any integer* $n \geq 1.$

Proof. We prove (a), because the proof of (b) is similar. Due to Lemma XI.2.2, (i) implies (iii). Because (iii) implies (ii) trivially, we need to prove that (ii) implies (i). Assume that $\mathrm{Ext}_A^1(M, -) = 0$. Due to Theorem VI.4.4, we have that M is projective if and only if any short exact sequence of the form $0 \to X \to Y \to M \to 0$ splits. But this is just Lemma XI.2.6.   □

Computing extensions starting from their definitions can be cumbersome and requires a good knowledge of the morphisms arising in a (co)resolution. Assume for instance that we wish to compute $\mathrm{Ext}_A^n(M, N)$ with the help of, say, a projective resolution of M.

$$P_\bullet : \cdots \longrightarrow P_{n+1} \xrightarrow{d_{n+1}} P_n \xrightarrow{d_n} \cdots \longrightarrow P_0 \xrightarrow{d_0} M \longrightarrow 0$$

We know that $\mathrm{Ker\,Hom}_A(d_{n+1}, N)$ consists of the morphisms $f : P_n \to N$ such that $f d_{n+1} = 0$, while $\mathrm{Im\,Hom}_A(d_n, N)$ consists of those morphisms $g : P_n \to N$ such that there exists $g' : P_{n-1} \to N$ satisfying $g = g' d_n$. So, $\mathrm{Ext}_A^n(M, N) = \mathrm{Ker\,Hom}_A(d_{n+1}, N)/\mathrm{Im\,Hom}_A(d_n, N)$ represents the morphisms which vanish when composed with d_{n+1} but do not factor through d_n. The situation is similar if one is using an injective coresolution of N.

Other information can be useful for computing extension modules, such as the following generalisation of Proposition XI.2.7.

XI.2.8 Lemma. (a) *Let M be a module having a finite projective resolution*

$$0 \to P_m \xrightarrow{d_m} P_{m-1} \to \cdots \to P_1 \xrightarrow{d_1} P_0 \xrightarrow{d_0} M \to 0$$

then $\mathrm{Ext}_A^n(M, -) = 0,$ *for any integer* $n > m,$ *and* $\mathrm{Ext}_A^m(M, N) = \mathrm{Coker\,Hom}_A(d_m, N),$ *for any module N.*
  (b) *Let N be a module having a finite injective coresolution*

$$0 \to N \xrightarrow{d^0} I^0 \xrightarrow{d^1} I^1 \to \cdots \to I^{m-1} \xrightarrow{d^m} I^m \to 0$$

then $\mathrm{Ext}_A^n(-, N) = 0,$ *for any integer* $n > m,$ *and* $\mathrm{Ext}_A^m(M, N) = \mathrm{Coker\,Hom}_A(M, d^m),$ *for any module M.*

Proof. We prove (a), because the proof of (b) is similar. The first statement follows from the fact that $P_n = 0$, for $n > m$. For the second, Ker $\mathrm{Hom}_A(d_{m+1}, N) = \mathrm{Hom}_A(P_m, N)$, yields $\mathrm{Ext}_A^m(M, N) = \mathrm{Ker}\ \mathrm{Hom}_A(d_{m+1}, N)/\mathrm{Im}\ \mathrm{Hom}_A(d_m, N) = \mathrm{Coker}\ \mathrm{Hom}_A(d_m, N)$. $\qquad\square$

XI.2.9 Examples.

(a) Let K be a field and $A = \begin{pmatrix} K & 0 \\ K & K \end{pmatrix}$ the 2×2 lower triangular matrix algebra. Let, as usual, $\mathbf{e}_{ij}$, for $(i, j) = (1, 1), (2, 1), (2, 2)$, be the canonical basis vectors of A. Consider the finitely generated indecomposable projective modules $P_1 = \mathbf{e}_{11}A, P_2 = \mathbf{e}_{22}A$, see Example IX.2.8(a). Left multiplication by $\mathbf{e}_{21}$ induces a nonzero morphism $f: P_1 \to P_2$ given by $\mathbf{e}_{11}a \mapsto \mathbf{e}_{21}\mathbf{e}_{11}a = \mathbf{e}_{22}a$, for $a \in A$. Because P_1 is simple, f is a monomorphism. Moreover, $S_2 = P_2/P_1$ has dimension $\dim_K S_2 = \dim_K P_2 - \dim_K P_1 = 2 - 1 = 1$, hence it is simple and we have a projective resolution $0 \to P_1 \overset{f}{\to} P_2 \to S_2 \to 0$. Evaluating on the latter the functor $\mathrm{Hom}_A(-, P_1)$ yields the exact cohomology sequence

$$0 \to \mathrm{Hom}_A(S_2, P_1) \to \mathrm{Hom}_A(P_2, P_1) \to \mathrm{Hom}_A(P_1, P_1) \to \mathrm{Ext}_A^1(S_2, P_1) \to$$
$$\to \mathrm{Ext}_A^1(P_2, P_1) = 0$$

where the last term vanishes because P_2 is projective. Now, $\mathrm{Hom}_A(P_2, P_1) = 0$: indeed, if there exists a nonzero morphism from P_2 to P_1, then it must be surjective, because P_1 is simple, but then it is a retraction, because P_1 is projective. This contradicts the indecomposability of P_2.

Therefore, $\mathrm{Ext}_A^1(S_2, P_1) \cong \mathrm{Hom}_A(P_1, P_1) = \mathrm{End}P_1$ and the latter is isomorphic to K, because P_1 is a one-dimensional vector space. Hence, $\mathrm{Ext}_A^1(S_2, P_1) \cong K$.

One proves just as easily that $\mathrm{Ext}_A^1(S_2, S_2) = 0$. Indeed, evaluating $\mathrm{Hom}_A(-, S_2)$ on the previous projective resolution, we get an exact sequence

$$\cdots \to \mathrm{Hom}_A(P_1, S_2) \to \mathrm{Ext}_A^1(S_2, S_2) \to \mathrm{Ext}_A^1(P_2, S_2) = 0$$

because P_2 is projective. Now, P_1, S_2 are nonisomorphic simple modules, hence Lemma VII.4.4 gives $\mathrm{Hom}_A(P_1, S_2) = 0$. Therefore, $\mathrm{Ext}_A^1(S_2, S_2) = 0$. For $n > 1$, we have $\mathrm{Ext}_A^n(S_2, -) = 0$, due to Lemma XI.2.8.

(b) Let A be an integral domain and $a \in A$ a nonzero element. The multiplication map $\mu_a : A_A \to A_A$ given by $x \mapsto ax$, for $x \in A$, is injective. We have a short exact sequence

$$0 \longrightarrow A \overset{\mu_a}{\longrightarrow} A \longrightarrow A/aA \longrightarrow 0.$$

Let M be an A-module. Evaluating $\mathrm{Hom}_A(-, M)$ on this sequence yields an exact cohomology sequence

$$0 \longrightarrow \operatorname{Hom}_A(A/aA, M) \longrightarrow \operatorname{Hom}_A(A, M) \overset{\operatorname{Hom}_A(\mu_a, M)}{\longrightarrow} \operatorname{Hom}_A(A, M) \longrightarrow$$

$$\longrightarrow \operatorname{Ext}_A^1(A/aA, M) \longrightarrow \operatorname{Ext}_A^1(A, M) = 0$$

because $\operatorname{Ext}_A^1(A, M) = 0$, due to projectivity of A_A. Therefore, $\operatorname{Ext}_A^1(A/aA, M) \cong \operatorname{Coker} \operatorname{Hom}_A(\mu_a, M)$. Because of Example III.2.10(b), there exists a functorial isomorphism $\varphi : \operatorname{Hom}_A(A, M) \to M$ given by $\varphi : u \mapsto u(1)$, for a morphism $u : A \to M$. We get a commutative square.

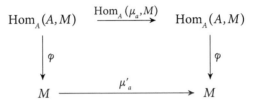

where $\mu'_a : M \to M$ is given by $m \mapsto ma$, for $m \in M$ (which is A-linear due to the commutativity of A). Hence, $\operatorname{Ext}_A^1(A/aA, M) \cong \operatorname{Coker} \mu'_a \cong M/Ma$, as K-modules. Because A is commutative, this is also an isomorphism of A-modules. A special case was done in Example X.2.8.

(c) Let A be a principal ideal domain and $p \in A$ an irreducible element. The algebra $R = A/p^2 A$ has a unique proper nonzero submodule $S = A/pA$ which is simple (because it is a simple abelian group or, alternatively, a one-dimensional A/pA-vector space). Therefore, R is a local algebra, and R_R is an indecomposable projective R-module, because of Corollary VIII.4.19. Multiplication by p defines an epimorphism $g : R \to S$. Moreover, $\operatorname{Hom}_R(R, S) \cong S = A/pA$, due to Theorem II.3.9, hence any morphism from R to S is a multiple of g. On the other hand, Lemma V.3.23 implies that $\operatorname{Hom}_R(R, R)$ is generated by the identity 1_R and the composite $f = ig : R \to R$ of g with the inclusion $i : S \to R$. Now, $i = \ker g$, hence $gi = 0$, and so $f^2 = (ig)^2 = 0$. We get the infinite projective resolution

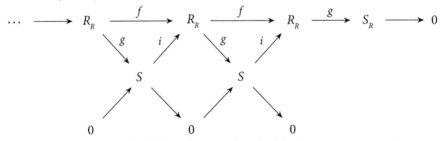

We claim that $\operatorname{Ext}_R^n(S, S) \cong A/pA$, for any integer $n > 0$. Indeed, the resolution and dimension shifting give that $\operatorname{Ext}_R^n(S, S) \cong \cdots \cong \operatorname{Ext}_R^1(S, S)$. So, it suffices to consider the case $n = 1$.

Due to its definition, $\operatorname{Ext}_R^1(S, S) = \operatorname{Ker} \operatorname{Hom}_R(f, S)/\operatorname{Im} \operatorname{Hom}_R(f, S)$. Now, $gf = 0$. Hence, $g \in \operatorname{Ker} \operatorname{Hom}_R(f, S)$, which is thus a nonzero submodule of the simple R-module $\operatorname{Hom}_R(R, S) \cong S$. Hence, $\operatorname{Ker} \operatorname{Hom}_R(f, S) =$

$\operatorname{Hom}_R(R, S)$. Therefore, $\operatorname{Im} \operatorname{Hom}_R(f, S) \cong \operatorname{Hom}_R(R, S)/\operatorname{Ker} \operatorname{Hom}_R(f, S) = 0$, using the Isomorphism Theorem II.4.21. Hence,

$$\operatorname{Ext}_R^n(S, S) \cong \cdots \cong \operatorname{Ext}_R^1(S, S) \cong \operatorname{Ker} \operatorname{Hom}_R(f, S) \cong \operatorname{Hom}_R(R, S) \cong S = A/pA.$$

One of the nicest properties of the ring R is that one may work with injectives exactly as one does with projectives. Indeed, because of Example VI.4.21, R_R is an injective module. Hence, the short exact sequence $0 \to S \overset{i}{\to} R \overset{g}{\to} S \to 0$, spliced with itself as many times as one wants also yields an infinite injective coresolution

$$0 \to S_R \overset{i}{\to} R_R \overset{f}{\to} R_R \to \cdots \to R_R \overset{f}{\to} R_R \to \cdots$$

starting from which one can reprove that $\operatorname{Ext}_R^n(S, S) \cong A/pA$, for any integer $n \geq 1$.

Typical examples of this situation are the ring $\mathbb{Z}/p^2\mathbb{Z}$, with p a prime (in this case, $A/pA \cong \mathbb{Z}/p\mathbb{Z}$), and, if K is a field, the algebra $K[t]/\langle t^2 \rangle$ (then, $A/pA \cong K$).

(d) Extensions offer an alternative interpretation of the character group $G^+ = \operatorname{Hom}_{\mathbb{Z}}(G, \mathbb{Q}/\mathbb{Z})$ of a torsion abelian group G, see Subsection V.4.5. For, evaluating the functor $\operatorname{Hom}_{\mathbb{Z}}(G, -)$ on the short exact sequence $0 \to \mathbb{Z} \to \mathbb{Q} \to \mathbb{Q}/\mathbb{Z} \to 0$ gives an exact cohomology sequence

$$0 \to \operatorname{Hom}_{\mathbb{Z}}(G, \mathbb{Z}) \to \operatorname{Hom}_{\mathbb{Z}}(G, \mathbb{Q}) \to \operatorname{Hom}_{\mathbb{Z}}(G, \mathbb{Q}/\mathbb{Z}) \to \operatorname{Ext}_{\mathbb{Z}}^1(G, \mathbb{Z}) \to$$
$$\to \operatorname{Ext}_{\mathbb{Z}}^1(G, \mathbb{Q}) = 0.$$

The last term vanishes because $\mathbb{Q}$ is a divisible, hence injective, abelian group. On the other hand, $\operatorname{Hom}_{\mathbb{Z}}(G, \mathbb{Q}) = 0$, because G is torsion, while $\mathbb{Q}$ is torsion-free. Therefore, $\operatorname{Ext}_{\mathbb{Z}}^1(G, \mathbb{Z}) \cong \operatorname{Hom}_{\mathbb{Z}}(G, \mathbb{Q}/\mathbb{Z}) = G^+$.

(e) Let K be a field and Q the Kronecker quiver

$$\underset{1}{\circ} \overset{\alpha}{\underset{\beta}{\rightleftarrows}} \underset{2}{\circ}$$

Set $A = KQ$ for short. Consider the A-module H_α given by the representation

$$K \overset{1}{\underset{0}{\rightleftarrows}} K$$

We claim that $\operatorname{Ext}_A^1(H_\alpha, H_\alpha) \cong K$. Indeed, we have constructed in Example VI.4.9(b) a projective resolution

$$0 \to P_1 \overset{j}{\to} P_2 \overset{p}{\to} H_\alpha \to 0.$$

where P_1, P_2 are the indecomposable projective modules corresponding to the points 1, 2, respectively. Evaluating $\operatorname{Hom}_A(-, H_\alpha)$ on this exact sequence yields a long exact sequence

$$0 \to \mathrm{Hom}_A(H_\alpha, H_\alpha) \xrightarrow{\mathrm{Hom}_A(p, H_\alpha)} \mathrm{Hom}_A(P_2, H_\alpha) \to \mathrm{Hom}_A(P_1, H_\alpha) \to$$
$$\to \mathrm{Ext}^1_A(H_\alpha, H_\alpha) \to \mathrm{Ext}^1_A(P_2, H_\alpha) = 0.$$

The last equality comes from projectivity of P_2. As pointed out in Example IX.2.23, the dimension of the vector space $\mathrm{Hom}_A(P_2, H_\alpha)$ equals that of $H_\alpha(2)$, that is, one.

On the other hand, the morphism $\mathrm{Hom}_A(p, H_\alpha)$ is nonzero (because, if it were zero, then $p = \mathrm{Hom}_A(p, H_\alpha)(1_{H_\alpha}) = 0$, an absurdity). Hence, it is surjective. Exactness implies the vector space isomorphisms $\mathrm{Ext}^1_A(H_\alpha, H_\alpha) \cong \mathrm{Hom}_A(P_1, H_\alpha) \cong H_\alpha(1) \cong K$. Our claim is established.

(f) Let K be a field and $A = KQ/I$ a bound quiver algebra. For distinct points $i \neq j$ in Q, the dimension of the first extension space between the simple modules S_i, S_j equals the number of arrows from i to j in Q. In order to prove it, we construct a projective resolution of S_i. Its projective cover is the indecomposable projective module $P_i = e_i A$, where e_i is the primitive idempotent corresponding to the point $i \in Q_0$. The short exact sequence $0 \to \mathrm{rad}\, P_i \to P_i \to S_i \to 0$ shows that we can take as next term in the resolution the projective cover P' of $\mathrm{rad}\, P_i$ which is defined by

$$\mathrm{top} P' \cong \mathrm{top}(\mathrm{rad} P_i) \cong (\mathrm{rad}\, P_i)/(\mathrm{rad}^2 P_i).$$

Similarly, the next projective P'' is the projective cover of the kernel of the morphism $P' \to \mathrm{rad} P_i$, and this kernel is certainly contained in $\mathrm{rad} P'$, because $\mathrm{top} P'$ is isomorphic to $\mathrm{top}(\mathrm{rad} P_i)$. Thus, the beginning of our projective resolution of S_i is:

$$\cdots \to P'' \xrightarrow{d_2} P' \xrightarrow{d_1} P_i \xrightarrow{d_0} S_i \to 0.$$

with $\mathrm{Im}\, d_2 = \mathrm{Ker}\, d_1 \subseteq \mathrm{rad} P'$. The first extension space is defined to be $\mathrm{Ext}^1_A(S_i, S_j) = \mathrm{Ker}(\mathrm{Hom}_A(d_2, S_j))/\mathrm{Im}(\mathrm{Hom}_A(d_1, S_j))$. Now, d_2 maps into the radical of P', whereas only the top of P' maps onto S_j, because the latter is simple. Hence, $\mathrm{Ker}(\mathrm{Hom}_A(d_2, S_j)) = \mathrm{Hom}_A(P', S_j)$. On the other hand, $i \neq j$ implies $\mathrm{Hom}_A(P_i, S_j) = 0$, thus the map $\mathrm{Hom}_A(d_1, S_j) : \mathrm{Hom}_A(P_i, S_j) \to \mathrm{Hom}_A(P', S_j)$ is zero. We get $\mathrm{Im}(\mathrm{Hom}_A(d_1, S_j)) = 0$. Therefore,

$$\mathrm{Ext}^1_A(S_i, S_j) \cong \mathrm{Hom}_A(P', S_j) \cong \mathrm{Hom}_A(\mathrm{top} P', S_j)$$

$$\cong \mathrm{Hom}_A((\mathrm{rad}\, P_i/\mathrm{rad}^2 P_i), S_j) \cong \mathrm{Hom}_A((\mathrm{rad}\, P_i/\mathrm{rad}^2 P_i), I_j)$$

$$\cong D\mathrm{Hom}_A(P_j, (\mathrm{rad}\, P_i/\mathrm{rad}^2 P_i))$$

$$\cong D\mathrm{Hom}_A(e_j A, (e_i \mathrm{rad}\, A)/(e_i \mathrm{rad}^2 A))$$

$$= D(e_i(\mathrm{rad}\, A)e_j/e_i(\mathrm{rad}^2 A)e_j).$$

At the fourth isomorphism, we used the fact that any morphism from a semisimple module to I_j maps into its simple socle $S_j = \mathrm{soc} I_j$. At the fifth, we used Proposition IX.2.22, and at the sixth, Proposition VII.3.1. Now, the number of arrows from i to j is equal to $\dim_K(e_i(\mathrm{rad}\, A)e_j/e_i(\mathrm{rad}^2 A)e_j)$, see Definition IX.4.4. Our claim follows.

We give a concrete example. Let Q be the quiver $3 \to 2 \to 1$ and $A = KQ$. We claim that $\mathrm{Ext}^1_A(S_3, P_2) \cong K$. Applying the functor $\mathrm{Hom}_A(S_3, -)$ to the short exact sequence $0 \to S_1 \to P_2 \to S_2 \to 0$ yields an exact cohomology sequence

$$\cdots \to \mathrm{Ext}^1_A(S_3, S_1) \to \mathrm{Ext}^1_A(S_3, P_2) \to \mathrm{Ext}^1_A(S_3, S_2) \to \mathrm{Ext}^2_A(S_3, S_1) \to \cdots$$

We have a projective resolution $0 \to P_2 \to P_3 \to S_3 \to 0$. Thus, applying Lemma XI.2.8, we get $\mathrm{Ext}^2_A(S_3, S_1) = 0$. On the other hand, $\mathrm{Ext}^1_A(S_3, S_1) = 0$, because there is no arrow from 3 to 1, while $\mathrm{Ext}^1(S_3, S_2) \cong K$, because there is exactly one arrow from 3 to 2. Hence, $\mathrm{Ext}^1_A(S_3, P_2) \cong \mathrm{Ext}^1_A(S_3, S_2) \cong K$, as required.

(g) The same technique can be used to compute extensions between arbitrary modules over bound quiver algebras. Let, for instance, A be the algebra of Example IX.4.19(a) and $M = \dfrac{P}{\mathrm{rad}^2 P} = \begin{smallmatrix} 1 \\ 1 \end{smallmatrix}$ uniserial of length two. We claim that $\mathrm{Ext}^3_A(M, M) \cong K$. Proceeding as above, we get an exact sequence:

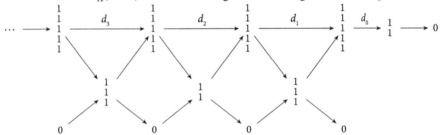

Namely, we get an infinite projective resolution:

$$\cdots \to P_4 \xrightarrow{d_4} P_3 \xrightarrow{d_3} P_2 \xrightarrow{d_2} P_1 \xrightarrow{d_1} P_0 \xrightarrow{d_0} M \to 0$$

where all P_i equal the unique indecomposable projective module $P = A_A$, all even syzygies equal $P/\mathrm{rad}^3 P$ while all odd syzygies equal $P/\mathrm{rad}^2 P = M$. Now $\mathrm{Ext}^3_A(M, M) = \mathrm{Ker}(\mathrm{Hom}_A(d_4, M))/\mathrm{Im}(\mathrm{Hom}_A(d_3, M))$. It is easily seen that the space $\mathrm{Hom}_A(P_4, M)$ is two-dimensional, that is, there exist only two nonzero linearly independent morphisms from P_4 to M, the first one sends the simple top of P_4 to that of M, and the second one sends it to the simple socle of M. Moreover, it is readily seen that the first does not vanish when composed with d_4, while the second one does. This proves that $\mathrm{Ker}(\mathrm{Hom}_A(d_4, M))$ is one-dimensional. Moreover, this unique morphism (up to scalars) does not factor through d_3. Hence, $\mathrm{Ext}^3_A(M, M) \cong \mathrm{Ker}(\mathrm{Hom}_A(d_4, M)) \cong K$.

Thus, computing an extension module is done generally in three steps. First, one reduces it to a first extension module, using dimension shifting. The latter is computed either using the definition, or a long exact cohomology sequence in which one or several terms vanish (due to projectivity or injectivity of certain modules). Finally, the resulting term is computed using appropriate isomorphisms, a computation that can be hard.

We end this subsection by proving that the Ext functors, which are derived from the Hom functor, behave like the latter with respect to direct sums and products.

XI.2.10 Proposition. (a) *Let* $M, (N_\lambda)_{\lambda \in \Lambda}$ *be A-modules and* $n \geq 0$ *an integer. There exists a functorial isomorphism*

$$\operatorname{Ext}_A^n \left(M, \prod_{\lambda \in \Lambda} N_\lambda \right) \cong \prod_{\lambda \in \Lambda} \operatorname{Ext}_A^n(M, N_\lambda).$$

(b) *Let* $N, (M_\sigma)_{\sigma \in \Sigma}$ *be A-modules and* $n \geq 0$ *an integer. There exists a functorial isomorphism*

$$\operatorname{Ext}_A^n \left(\bigoplus_{\sigma \in \Sigma} M_\sigma, N \right) \cong \prod_{\sigma \in \Sigma} \operatorname{Ext}_A^n(M_\lambda, N).$$

Proof. We prove (a), because the proof of (b) is similar. The case $n = 0$ is Theorem III.3.10, so assume that $n = 1$. For any $\lambda \in \Lambda$, there exists a short exact sequence $0 \to N_\lambda \to I_\lambda \to L_\lambda \to 0$, with I_λ injective. Using Exercise III.3.6, we deduce a short exact sequence $0 \to \prod_{\lambda \in \Lambda} N_\lambda \to \prod_{\lambda \in \Lambda} I_\lambda \to \prod_{\lambda \in \Lambda} L_\lambda \to 0$, with $\prod_{\lambda \in \Lambda} I_\lambda$ injective because of Proposition VI.4.17. Evaluating $\operatorname{Hom}_A(M, -)$ on these sequences, and applying again Exercise III.3.6, yields a commutative diagram with exact rows

$$
\begin{array}{ccccccc}
\operatorname{Hom}_A(M, \Pi_\lambda I_\lambda) & \longrightarrow & \operatorname{Hom}_A(M, \Pi_\lambda L_\lambda) & \longrightarrow & \operatorname{Ext}_A^1(M, \Pi_\lambda N_\lambda) & \longrightarrow & 0 \\
\downarrow{\scriptstyle\cong} & & \downarrow{\scriptstyle\cong} & & \downarrow & & \\
\Pi_\lambda \operatorname{Hom}_A(M, I_\lambda) & \longrightarrow & \Pi_\lambda \operatorname{Hom}_A(M, L_\lambda) & \longrightarrow & \Pi_\lambda \operatorname{Ext}_A^1(M, N_\lambda) & \longrightarrow & 0
\end{array}
$$

where we have used that $\operatorname{Ext}_A^1(M, \prod_\lambda I_\lambda) = 0$, and $\operatorname{Ext}_A^1(M, I_\lambda) = 0$, for any $\lambda \in \Lambda$, due to Proposition XI.2.7(b). Passing to cokernels, we get $\operatorname{Ext}_A^1(M, \prod_{\lambda \in \Lambda} N_\lambda) \cong \prod_{\lambda \in \Lambda} \operatorname{Ext}_A^1(M, N_\lambda)$. We have established the statement for $n = 1$.

If $n > 1$, then the statement follows from the diagram with exact rows

$$
\begin{array}{ccccccc}
0 & \longrightarrow & \operatorname{Ext}_A^{n-1}(M, \Pi_\lambda L_\lambda) & \longrightarrow & \operatorname{Ext}_A^n(M, \Pi_\lambda N_\lambda) & \longrightarrow & 0 \\
& & \downarrow{\scriptstyle\cong} & & \downarrow{\scriptstyle\cong} & & \\
0 & \longrightarrow & \Pi_\lambda \operatorname{Ext}_A^{n-1}(M, L_\lambda) & \longrightarrow & \Pi_\lambda \operatorname{Ext}_A^n(M, N_\lambda) & \longrightarrow & 0
\end{array}
$$

using again Proposition XI.2.7(b). □

Both statements of the proposition can be assembled in one: if $(M_\sigma)_{\sigma \in \Sigma}$, $(N_\lambda)_{\lambda \in \Lambda}$ are families of A-modules, then for any integer $n \geq 0$, we have a functorial isomorphism:

$$\operatorname{Ext}_A^n \left(\bigoplus_{\sigma \in \Sigma} M_\sigma, \prod_{\lambda \in \Lambda} N_\lambda \right) \cong \prod_{(\sigma, \lambda) \in \Sigma \times \Lambda} \operatorname{Ext}_A^n(M_\sigma, N_\lambda).$$

XI.2.2 The torsion functors

Torsion functors are the left derived functors of the tensor product. Our exposition follows closely the pattern of the previous subsection. Let A be a K-algebra. Left A-modules are considered as A^{op}-modules.

(a) For an integer $n \geq 0$ and an A-module M, the nth-left derived functor of $M \otimes_A - : \operatorname{Mod} A^{op} \to \operatorname{Mod} K$ is denoted (provisionally) by $T_n(M, -)$, thus, for a left A-module N, we have

$$T_n(M, -)(N) = T_n(M, N) = H_n(M \otimes_A (P_N).)$$

where $P.$ is a projective resolution of N in $\operatorname{Mod} A^{op}$. If $0 \to N' \to N \to N'' \to 0$ is a short exact sequence in $\operatorname{Mod} A^{op}$, then we have a long exact homology sequence

$$\cdots \to T_{n+1}(M, N'') \overset{\delta}{\to} T_n(M, N') \to T_n(M, N) \to T_n(M, N'') \overset{\delta}{\to} T_{n-1}(M, N') \to \cdots$$

$$\cdots \to T_1(M, N'') \overset{\delta}{\to} M \otimes_A N' \to M \otimes_A N \to M \otimes_A N'' \to 0$$

with all morphisms functorial.

(b) For an integer $n \geq 0$ and an A^{op}-module N, the nth-left derived functor of $- \otimes_A N : \operatorname{Mod} A \to \operatorname{Mod} K$ is denoted (provisionally) by $T^n(-, N)$, thus, for a right A-module M, we have

$$T^n(-, N)(M) = T^n(M, N) = H_n((P_M). \otimes_A N)$$

where $P.$ is a projective resolution of M in $\operatorname{Mod} A$. If $0 \to M' \to M \to M'' \to 0$ is a short exact sequence in $\operatorname{Mod} A$, then we have a long exact homology sequence

$$\cdots \to T^{n+1}(M'', N) \overset{\delta}{\to} T^n(M', N) \to T^n(M, N) \to T^n(M'', N) \overset{\delta}{\to} T^{n-1}(M', N) \to \cdots$$

$$\cdots \to T^1(M'', N) \overset{\delta}{\to} M' \otimes_A N \to M \otimes_A N \to M'' \otimes_A N \to 0$$

with all morphisms functorial.

We prove that these constructions are compatible, that is, we prove that, for any modules $M_A, {}_A N$ and integer $n \geq 0$, there exists an isomorphism $T_n(M, N) \cong T^n(M, N)$, functorial in both variables, thus defining a bifunctor. This is done as for the extension functors, see Subsection IX.3.3, so we outline the steps, leaving details to the reader.

XI.2.11 Lemma. Dimension shifting. *Let M be an Λ-module and N an A^{op} module.*
  (a) *If $P_{\bullet}$:*

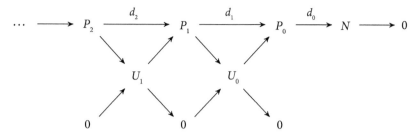

is a projective resolution of N in $\operatorname{Mod} A^{op}$ and $U_n = \operatorname{Im} d_{n+1}$, for $n \geq 0$ an integer, then we have functorial isomorphisms

$$T_{n+1}(M, N) \cong T_n(M, U_0) \cong \cdots \cong T_1(M, U_{n-1}).$$

  (b) *If $P_{\bullet}$:*

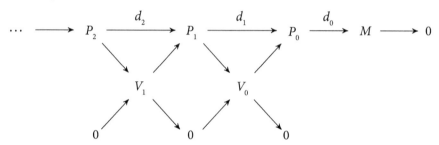

is a projective resolution of M in $\operatorname{Mod} A$ and $V_n = \operatorname{Im} d_{n+1}$, for $n \geq 0$ an integer, then we have functorial isomorphisms

$$T^{n+1}(M, N) \cong T^n(V_0, N) \cong \cdots \cong T^1(V_{n-1}, N).$$

Proof. Similar to the proof of Lemma XI.2.1 and left to the reader. ☐

XI.2.12 Lemma. (a) *If M_A is flat (for instance, projective), then $T_n(M, N) = 0$, for any module $_A N$ and any integer $n > 0$.*
  (b) *If $_A N$ is flat (for instance, projective), then $T^n(M, N) = 0$, for any module M_A smd any integer $n > 0$.*

Proof. We prove (a), because the proof of (b) is similar. Let $_A N$ be a module and $P_{\bullet}$ a projective resolution of N in $\operatorname{Mod} A^{op}$. Because M is flat, the complex $M \otimes_A P_{\bullet}$ is exact. Hence, so is $M \otimes_A (P_N)_{\bullet}$ in positive degrees. That is, we have an isomorphism $T_n(M, N) \cong H_n(M \otimes_A (P_N)_{\bullet}) = 0$, for $n > 0$. ☐

XI.2.13 Theorem. *Let $M_{A}, {}_A N$ be modules and $n \geq 0$ an integer, then we have an isomorphism $T_n(M, N) \cong T^n(M, N)$, functorial in each variable.*

Proof. Similar to the proof of Theorem XI.2.3 and left to the reader. $\square$

XI.2.14 Definition. Let M be an A-module and N an A^{op}-module. The nth-*torsion module* of M and N, denoted by $\text{Tor}_n^A(M, N)$, is the common value of $T^n(M, N)$ and $T_n(M, N)$.

Thus, $\text{Tor}_n^A(-, ?)$ is a bifunctor, covariant in each variable. Moreover, if M, or N, or both, has an additional B-module structure (so that it is a bimodule), then $\text{Tor}_n^A(M, N)$ has an induced B-module structure. We state below the 'torsion version' of Theorem X.3.5.

XI.2.15 Theorem. (a) *Let M be an A-module and $0 \to N' \to N \to N'' \to 0$ a short exact sequence in $\text{Mod} A^{op}$, then we have a long exact homology sequence*

$$\cdots \to \text{Tor}_{n+1}^A(M, N'') \xrightarrow{\delta} \text{Tor}_n^A(M, N') \to \text{Tor}_n^A(M, N) \to \text{Tor}_n^A(M, N'')$$

$$\xrightarrow{\delta} \text{Tor}_{n-1}^A(M, N') \cdots \to \text{Tor}_1^A(M, N'') \xrightarrow{\delta} M \otimes_A N' \to M \otimes_A N \to M \otimes_A N'' \to 0$$

with all morphisms functorial.

   (b) *Let N be an A^{op}-module and $0 \to M' \to M \to M'' \to 0$ a short exact sequence in $\text{Mod} A$, then we have a long exact homology sequence*

$$\cdots \to \text{Tor}_{n+1}^A(M'', N) \xrightarrow{\delta} \text{Tor}_n^A(M', N) \to \text{Tor}_n^A(M, N) \to \text{Tor}_n^A(M'', N)$$

$$\xrightarrow{\delta} \text{Tor}_{n-1}^A(M', N) \cdots \to \text{Tor}_1^A(M'', N) \xrightarrow{\delta} M' \otimes_A N \to M \otimes_A N \to M'' \otimes_A N \to 0$$

with all morphisms functorial. $\square$

XI.2.16 Proposition. (a) *Let M be an A-module. The following conditions are equivalent:*
   (i) *M is flat.*
   (ii) *$\text{Tor}_1^A(M, -) = 0$.*
   (iii) *$\text{Tor}_n^A(M, -) = 0$, for any integer $n \geq 1$.*
   (b) *Let N be an A^{op}-module. The following conditions are equivalent:*
   (i) *N is flat.*
   (ii) *$\text{Tor}_1^A(-, N) = 0$.*
   (iii) *$\text{Tor}_n^A(-, N) = 0$, for any integer $n \geq 1$.*

Proof. We prove (a), because the proof of (b) is similar. Because of Lemma XI.2.12, (i) implies (iii), and the latter implies (ii) trivially. Assume that (ii) holds true and let $0 \to N' \to N \to N'' \to 0$ be exact in $\text{Mod} A^{op}$. Evaluating $M \otimes_A -$ on the latter yields an exact sequence

$$0 = \text{Tor}_1^A(M, N'') \to M \otimes_A N' \to M \otimes_A N \to M \otimes_A N'' \to 0$$

where the first term vanishes due to the hypothesis. Hence, $M \otimes_A -$ is exact and M is flat. $\square$

The next lemma shows that, in the definition of torsion functors, one may use flat resolutions instead of projective ones.

XI.2.17 Lemma. *Let M be an A-module, N an A^{op}-module and $n \geq 1$.*

(a) *If* $F_\bullet : \cdots \to F_1 \xrightarrow{d_1} F_0 \xrightarrow{d_0} N \to 0$ *is a flat resolution of N in $\operatorname{Mod} A^{op}$, then we have a functorial isomorphism* $\operatorname{Tor}_n^A(M, N) \cong H_n(M \otimes_A (F_N)_\bullet)$.

(b) *If* $F_\bullet : \cdots \to F_1 \xrightarrow{d_1} F_0 \xrightarrow{d_0} M \to 0$ *is a flat resolution of M in $\operatorname{Mod} A$, then we have a functorial isomorphism* $\operatorname{Tor}_n^A(M, N) \cong H_n((F_M)_\bullet \otimes_A N)$.

Proof. We prove (a), because the proof of (b) is similar. We use induction on degree n. The statement holds true for $n = 0$, because the functor $M \otimes_A -$ is right exact. Let $L = \operatorname{Im} d_1$ and $f : L \to F_0$ be the inclusion. Evaluating the functor $M \otimes_A -$ on the short exact sequence $0 \to L \xrightarrow{f} F_0 \xrightarrow{d_0} N \to 0$ yields the exact homology sequence

$$0 = \operatorname{Tor}_1^A(M, F_0) \to \operatorname{Tor}_1^A(M, N) \to M \otimes_A L \xrightarrow{1_M \otimes f} M \otimes_A F_0 \xrightarrow{1_M \otimes d_0} M \otimes_A N \to 0$$

because F_0 is flat. So, $\operatorname{Tor}_1^A(M, N) \cong \operatorname{Ker}(1_M \otimes f)$. Also, we have an exact sequence $F_2 \xrightarrow{d_2} F_1 \xrightarrow{g} L \to 0$, where $d_1 = fg$ is the canonical factorisation through the image. Because $1_M \otimes d_2 = (1_M \otimes f)(1_M \otimes g)$ and $M \otimes_A -$ is a right exact functor, we get a commutative diagram with exact rows

$$
\begin{array}{ccccccc}
M \otimes_A F_2 & \xrightarrow{1_M \otimes d_2} & M \otimes_A F_1 & \xrightarrow{1_M \otimes g} & M \otimes_A L & \longrightarrow & 0 \\
\downarrow u & & \downarrow 1_M \otimes F_1 & & \downarrow 1_M \otimes f & & \\
0 \longrightarrow \operatorname{Ker}(1_M \otimes d_1) & \xrightarrow{i} & M \otimes_A F_1 & \xrightarrow{1_M \otimes d_1} & M \otimes_A F_0 & &
\end{array}
$$

where i is the inclusion and u the morphism induced from the universal property of the kernel and the equality $d_1 d_2 = 0$ which gives $(1_M \otimes d_1) 1_{M \otimes F_1} (1_M \otimes d_2) = 0$. We have $\operatorname{Im}(1_M \otimes d_2) \cong \operatorname{Im}(iu) \cong \operatorname{Im} u$, because i is a monomorphism. Applying the Snake Lemma II.4.19 gives

$$\operatorname{Ker}(1_M \otimes f) \cong \operatorname{Ker}(1_M \otimes d_1)/\operatorname{Im}(1_M \otimes d_2) \cong H_1(M \otimes_A (F_N)_\bullet).$$

This proves the statement for $n = 1$. For $n > 1$, consider the flat resolution $F'_\bullet : \cdots \to F_2 \xrightarrow{d_2} F_1 \xrightarrow{g} L \to 0$, then use dimension shifting and the induction hypothesis

$$\operatorname{Tor}_n^A(M, N) \cong \operatorname{Tor}_{n-1}^A(M, L) \cong H_{n-1}(M \otimes_A (F'_L)_\bullet) \cong H_n(M \otimes_A (F_N)_\bullet). \qquad \square$$

In order to apply induction, we had to consider separately the cases $n = 0$ and $n = 1$. This often happens with homological proofs, see also the proof of Proposition XI.2.10.

Recall that, if an algebra A is commutative and M, N are A-modules, then $M \otimes_A N \cong N \otimes_A M$, see Proposition VI.2.7(a). The torsion functors enjoy the same property.

XI.2.18 Proposition. *Let A be a commutative algebra, M, N modules and $n \geq 0$ an integer. Then we have an isomorphism $\mathrm{Tor}_n^A(M, N) \cong \mathrm{Tor}_n^A(N, M)$, functorial in both variables.*

Proof. Let $P_\bullet$ be a projective (or a flat) resolution of N. Because of Proposition VI.2.7(a), we have, for any $n \geq 0$, a functorial isomorphism $f_n : M \otimes_A P_n \to P_n \otimes_A M$ given by $x \otimes y_n \mapsto y_n \otimes x$, where $x \in M, y_n \in P_n$. The morphisms $(f_n)_{n \geq 0}$ induce an isomorphism of complexes $M \otimes_A (P_N)_\bullet \cong (P_N)_\bullet \otimes_A M$. These complexes thus have isomorphic homology. $\qquad\square$

Before turning to examples, we state a useful computational tool.

XI.2.19 Lemma. *Let M be an A-module and N an A^{op}-module.*
   (a) *Assume that M has a finite projective (or flat) resolution in $\mathrm{Mod}\, A$*

$$P_\bullet : 0 \to P_m \xrightarrow{d_m} P_{m-1} \to \cdots \to P_1 \xrightarrow{d_1} P_0 \xrightarrow{d_0} M \to 0$$

then $\mathrm{Tor}_n^A(M, -) = 0$, for any integer $n > m$, and $\mathrm{Tor}_m^A(M, N) = \mathrm{Ker}\,(d_m \otimes 1_N)$, for any A^{op}-module N.
   (b) *Assume that N has a finite projective (or flat) resolution in $\mathrm{Mod}\, A^{op}$*

$$P_\bullet : 0 \to P_m \xrightarrow{d_m} P_{m-1} \to \cdots \to P_1 \xrightarrow{d_1} P_0 \xrightarrow{d_0} M \to 0$$

then $\mathrm{Tor}_n^A(-, N) = 0$, for any integer $n > m$, and $\mathrm{Tor}_m^A(M, N) = \mathrm{Ker}\,(1_M \otimes d_m)$, for any A-module M.

Proof. Similar to that of Lemma XI.2.8 and left to the reader. $\qquad\square$

XI.2.20 Examples.
   (a) Let A be an algebra. For any left A-module N, there exists a short exact sequence

$$0 \to N' \xrightarrow{j} P \to N \to 0$$

with $_A P$ projective. Evaluating the functor $M \otimes_A -$ yields an exact homology sequence

$$0 = \mathrm{Tor}_1^A(M, P) \to \mathrm{Tor}_1^A(M, N) \to M \otimes_A N' \xrightarrow{1_M \otimes j} M \otimes_A P \to M \otimes_A N \to 0$$

where the first term vanishes because P is projective (hence flat). Hence, $\mathrm{Tor}_1^A(M, N) = \mathrm{Ker}\,(1_M \otimes j)$.

Moreover, for any integer $n \geq 1$, we have an exact sequence

$$0 = \operatorname{Tor}_{n+1}^A(M, P) \to \operatorname{Tor}_{n+1}^A(M, N') \to \operatorname{Tor}_n^A(M, N) \to \operatorname{Tor}_n^A(M, P) = 0$$

due to projectivity of P. Hence, $\operatorname{Tor}_{n+1}^A(M, N') \cong \operatorname{Tor}_n^A(M, N)$, for any $n \geq 1$. Assume that N is cyclic. Then $N = A/I$, for some left ideal I of A, and the exact sequence $0 \to I \to A \to A/I \to 0$ induces the exact homology sequence

$$0 = \operatorname{Tor}_1^A(M, A) \to \operatorname{Tor}_1^A(M, A/I) \to M \otimes_A I \to M \otimes_A A \to M \otimes_A A/I \to 0$$

Because of Theorem VI.2.8, the multiplication map induces an isomorphism $M \otimes_A A \cong M$, hence $\operatorname{Tor}_1^A(M, A/I)$ can be identified to the kernel of the multiplication morphism $M \otimes_A I \to M$ defined by $x \otimes c \mapsto xc$, for $x \in M, c \in I$. Also, for any $n \geq 1$, we have an isomorphism $\operatorname{Tor}_{n+1}^A(M, A/I) \cong \operatorname{Tor}_n^A(M, I)$.

Suppose now that A is an integral domain and $N = A/aA$, with $a \in A$ nonzero noninvertible. We have a free resolution

$$0 \to A \xrightarrow{\mu_a} A \to A/aA \to 0$$

where μ_a denotes multiplication by a. that is, the map $x \mapsto ax$, for $x \in A$. Given an A-module M, we have a commutative diagram with exact rows

$$
\begin{array}{ccccccccc}
0 & \longrightarrow & \operatorname{Tor}_1^A(M, A/aA) & \longrightarrow & M \otimes_A A & \xrightarrow{1_M \otimes \mu_a} & M \otimes_A A & \longrightarrow & M \otimes_A A/aA \longrightarrow 0 \\
& & & & \mu \downarrow \cong & & \mu \downarrow \cong & & \\
0 & \longrightarrow & \operatorname{Ker}(\mu_a) & \longrightarrow & M & \xrightarrow{\mu_a} & M & \longrightarrow & M/aM \longrightarrow 0
\end{array}
$$

where the multiplication map $\mu : x \otimes c \mapsto xc$, with $x \in M, c \in A$, is an isomorphism, and, in the lower row, μ_a is considered as a map from M to itself. Thus, $\operatorname{Tor}_1^A(M, A/aA) \cong \operatorname{Ker}(\mu_a) = \{x \in M : ax = 0\}$ is a submodule of the torsion submodule of M. We also obtain an isomorphism $M \otimes_A A/aA \cong M/aM$, recovering a result from Corollary VI.3.17. Moreover, because of Lemma XI.2.19, $\operatorname{Tor}_n^A(M, -) = 0$, for any integer $n > 1$. Finally, commutativity of A gives $\operatorname{Tor}_1^A(A/aA, M) \cong \operatorname{Tor}_1^A(M, A/aA) = \{x \in M : ax = 0\}$ and $A/aA \otimes_A M \cong M \otimes_A A/aA \cong M/aM$.

(b) Let, as in Example XI.2.9(c), A be a principal ideal domain, $p \in A$ an irreducible element, and $R = A/p^2A$. We have an infinite projective resolution of the simple module $S = A/pA$:

$$P_\bullet : \cdots \to R_R \xrightarrow{f} R_R \to \cdots \to R_R \xrightarrow{f} R_R \xrightarrow{g} S_R \to 0$$

where f and g are as in Example XI.2.20(c).

Evaluating the functor $- \otimes_R S$ yields $R \otimes_R S \cong S$, due to Theorem VI.2.8, and $f \otimes 1_S = 0$ because f is multiplication by p. Thus, the resolution becomes, after deleting its first term

$$(P_S). : \cdots \to S_R \xrightarrow{0} S_R \to \cdots \to S_R \xrightarrow{0} S_R \to 0$$

Therefore, $\text{Tor}_n^R(S, S) \cong S = A/pA$, for any integer $n > 0$.

(c) Actually, the torsion subgroup of an abelian group G, considered as a functor as in Exercise III.2.11, is a Tor functor in our sense (hence the name of torsion). We prove it in three steps.

(i) First, if G is a torsion group, then $\text{Tor}_1^\mathbb{Z}(G, \mathbb{Q}/\mathbb{Z}) \cong G$, functorially. Indeed, evaluating the functor $G \otimes_\mathbb{Z} -$ on the short exact sequence $0 \to \mathbb{Z} \to \mathbb{Q} \to \mathbb{Q}/\mathbb{Z} \to 0$, where the morphisms are inclusion and projection, yields an exact sequence

$$\text{Tor}_1^\mathbb{Z}(G, \mathbb{Q}) \to \text{Tor}_1^\mathbb{Z}(G, \mathbb{Q}/\mathbb{Z}) \xrightarrow{\delta} G \otimes_\mathbb{Z} \mathbb{Z} \to G \otimes_\mathbb{Z} \mathbb{Q}$$

Because $\mathbb{Q}$ is torsion-free, it is flat, due to Theorem VI.4.43. Hence, $\text{Tor}_1^\mathbb{Z}(G, \mathbb{Q}) = 0$. But also, $G \otimes_\mathbb{Z} \mathbb{Q} = 0$, for, if $x \in G$, then there exists an integer $m > 0$ such that $mx = 0$, hence, for any $y \in \mathbb{Q}$, we have $x \otimes y = x \otimes \frac{my}{m} = mx \otimes \frac{y}{m} = 0$ (this is Exercise VI.2.2(d)). So δ is an isomorphism. Because $G \otimes_\mathbb{Z} \mathbb{Z} \cong G$, due to Theorem VI.2.8, we infer that $\text{Tor}_1^\mathbb{Z}(G, \mathbb{Q}/\mathbb{Z}) \cong G$. This isomorphism is functorial, because it is the composite of two functorial isomorphisms.

(ii) Second, if G is torsion-free, then $\text{Tor}_1^\mathbb{Z}(G, \mathbb{Q}/\mathbb{Z}) = 0$. Indeed, because of Theorem VI.4.26, there exists an embedding $j : G \to I$, where I is an injective (that is, divisible) abelian group. Let $T(I)$ denote the torsion subgroup of I and consider the short exact sequence $0 \to T(I) \xrightarrow{i} I \xrightarrow{p} I/T(I) = E \to 0$, where i is the inclusion and p the projection. If, for $x \in G$, we have $pj(x) = 0$, then $j(x) \in T(I)$ and so $x = 0$, because G is torsion-free. Hence, $pj : G \to E$ is a monomorphism. But E is torsion-free, hence flat, due to Theorem VI.4.43. Evaluating $- \otimes_\mathbb{Z} \mathbb{Q}/\mathbb{Z}$ on the short exact sequence $0 \to G \xrightarrow{pj} E \to E/G \to 0$ yields an exact sequence

$$0 = \text{Tor}_2^\mathbb{Z}(E/G, \mathbb{Q}/\mathbb{Z}) \to \text{Tor}_1^\mathbb{Z}(G, \mathbb{Q}/\mathbb{Z}) \dashrightarrow \text{Tor}_1^\mathbb{Z}(E, \mathbb{Q}/\mathbb{Z}) = 0$$

where the first term vanishes because $\text{Tor}_2^\mathbb{Z}(-, ?) = 0$, as seen in Example (a) above, and the third vanishes because E is flat. The statement follows.

(iii) Lastly, let G be an arbitrary abelian group. The short exact sequence $0 \to T(G) \to G \to G/T(G) \to 0$ induces an exact sequence

$$0 = \text{Tor}_2^\mathbb{Z}(G/T(G), \mathbb{Q}/\mathbb{Z}) \to \text{Tor}_1^\mathbb{Z}(T(G), \mathbb{Q}/\mathbb{Z}) \to \text{Tor}_1^\mathbb{Z}(G, \mathbb{Q}/\mathbb{Z}) \to$$
$$\to \text{Tor}_1^\mathbb{Z}(G/T(G), \mathbb{Q}/\mathbb{Z}) = 0$$

where the first term vanishes due to Lemma XI.2.19 above and the last term vanishes because of (ii). Therefore,

$$\operatorname{Tor}_1^{\mathbb{Z}}(G, \mathbb{Q}/\mathbb{Z}) \cong \operatorname{Tor}_1^{\mathbb{Z}}(T(G), \mathbb{Q}/\mathbb{Z}) \cong T(G)$$

because of (i). This isomorphism is functorial. Hence, as functors, $\operatorname{Tor}_1^{\mathbb{Z}}(-, \mathbb{Q}/\mathbb{Z}) \cong T(-)$.

Tensor functors commute with direct sums, see Proposition VI.2.7(c)(d). The same holds true for torsion functors.

XI.2.21 Proposition. (a) *Let M be an A-module, $(N_\lambda)_{\lambda \in \Lambda}$ a family of A^{op}-modules and $n \geq 1$ an integer. There exists a functorial isomorphism*

$$\operatorname{Tor}_n^A\left(M, \bigoplus_{\lambda \in \Lambda} N_\lambda\right) \cong \bigoplus_{\lambda \in \Lambda} \operatorname{Tor}_n^A(M, N_\lambda).$$

(b) *Let N be an A^{op}-module, $(M_\sigma)_{\sigma \in \Sigma}$ a family of A-modules and $n \geq 1$ an integer. There exists a functorial isomorphism*

$$\operatorname{Tor}_n^A\left(\bigoplus_{\sigma \in \Sigma} M_\sigma, N\right) \cong \bigoplus_{\sigma \in \Sigma} \operatorname{Tor}_n^A(M_\sigma, N).$$

Proof. Similar to that of Proposition XI.2.10 and left to the reader. ☐

Both statements can be combined: if $(M_\sigma)_{\sigma \in \Sigma}$ is a family of right A-modules, and $(N_\lambda)_{\lambda \in \Lambda}$ a family of left A-modules, then, for any integer $n \geq 0$, there exists a functorial isomorphism

$$\operatorname{Tor}_n^A\left(\bigoplus_{\sigma \in \Sigma} M_\sigma, \bigoplus_{\lambda \in \Lambda} N_\lambda\right) \cong \bigoplus_{(\sigma, \lambda) \in \Sigma \times \Lambda} \operatorname{Tor}_n^A(M_\sigma, N_\lambda).$$

XI.2.3 Relation between extension and torsion

Computing an extension, or torsion, module is sometimes difficult. However, there exists (in the noetherian case) a relation between the two which may help. We start with a technical lemma.

XI.2.22 Lemma. *Let A be an algebra and I an injective A-module. For any cochain complex $C^\bullet$ and integer $n \geq 0$, we have a functorial isomorphism*

$$H_n(\operatorname{Hom}_A(C^\bullet, I)) \cong \operatorname{Hom}_A(H^n(C^\bullet), I).$$

Proof. Let $C^\bullet = (C^n, d^n)_{n \in \mathbb{Z}}$. For any $n \geq 0$, we have a short exact sequence $0 \to \operatorname{Im} d^n \to \operatorname{Ker} d^{n+1} \to H^n(C^\bullet) \to 0$ in Mod A. Evaluating the exact functor

$\text{Hom}_A(-, I)$ yields an exact sequence

$$0 \to \text{Hom}_A(H^n(C^\cdot), I) \to \text{Hom}_A(\text{Ker } d^{n+1}, I) \to \text{Hom}_A(\text{Im } d^n, I) \to 0.$$

Moreover, exactness of $\text{Hom}_A(-, I)$ also implies the isomorphisms

$$\text{Hom}_A(\text{Ker } d^{n+1}, I) \cong \text{Coker Hom}_A(d^{n+1}, I) = \text{Hom}_A(C^n, I)/\text{Im Hom}_A(d^{n+1}, I).$$

and

$$\text{Hom}_A(\text{Im } d^n, I) \cong \text{Coim Hom}_A(d^n, I) = \text{Hom}_A(C^n, I)/\text{Ker Hom}_A(d^n, I).$$

The previous short exact sequence becomes

$$0 \to \text{Hom}_A(H^n(C^\cdot), I) \to \text{Hom}_A(C^n, I)/\text{Im Hom}_A(d^{n+1}, I) \to$$
$$\to \text{Hom}_A(C^n, I)/\text{Ker Hom}_A(d^n, I) \to 0.$$

The Isomorphism Theorem II.4.21 gives

$$\text{Hom}_A(H^n(C^\cdot), I) \cong \text{Ker Hom}_A(d^n, I)/\text{Im Hom}_A(d^{n+1}, I) = H_n(\text{Hom}_A(C^\cdot, I)). \quad \square$$

XI.2.23 Theorem. *Let A be a noetherian algebra, B an algebra, L_A a finitely generated A-module, $_BM_A$ a B-A-bimodule and $_BI$ an injective B-module. For any integer $n \geq 0$, there exists a functorial isomorphism*

$$\text{Tor}_n^A(L, \text{Hom}_B(M, I)) \cong \text{Hom}_B(\text{Ext}_A^n(L, M), I).$$

Proof. Because A is noetherian and L_A finitely generated, there exists, due to Corollary VII.2.16, a projective resolution $P_\cdot : \cdots \to P_1 \to P_0 \to L \to 0$ with all P_i finitely generated. On the other hand, injectivity of $_BI$ and Lemma IX.2.19 give a functorial isomorphism

$$L \otimes_A \text{Hom}_B(M, I) \cong \text{Hom}_B(\text{Hom}_A(L, M), I).$$

Replacing L by each term of $(P_L)_\cdot$ yields an isomorphism of complexes

$$(P_L)_\cdot \otimes_A \text{Hom}_B(M, I) \cong \text{Hom}_B(\text{Hom}_A((P_L)_\cdot, M), I).$$

Thus, these complexes have isomorphic homology. Lemma XI.2.22 yields

$$H_n((P_L). \otimes_A \operatorname{Hom}_B(M, I)) \cong H_n(\operatorname{Hom}_B(\operatorname{Hom}_A((P_L)., M), I))$$
$$\cong \operatorname{Hom}_B(H^n(\operatorname{Hom}_A((P_L)., M), I),$$

that is,

$$\operatorname{Tor}_n^A(L, \operatorname{Hom}_B(M, I)) \cong \operatorname{Hom}_B(\operatorname{Ext}_A^n(L, M), I). \qquad \square$$

If, in particular, K is a field and A a finite–dimensional K-algebra, then we can use the duality $D = \operatorname{Hom}_K(-, K)$ between finitely generated right and left A-modules, see Example IV.4.8(b).

XI.2.24 Corollary. *Let K be a field, A a finite–dimensional K-algebra and L, M finitely generated A-modules. For any integer $n \geq 0$, there exists a functorial isomorphism*

$$\operatorname{Tor}_n^A(L, DM) \cong D\operatorname{Ext}_A^n(L, M).$$

In particular, $L \otimes_A DM \cong D\operatorname{Hom}_A(L, M)$.

Proof. Because K is an injective K-module, we can apply Theorem XI.2.23. $\qquad \square$

The case $n = 0$, that is, the isomorphism $L \otimes_A DM \cong D\operatorname{Hom}_A(L, M)$, was also derived previously as an application of Lemma IX.2.19.

XI.2.25 Example. Let A be the algebra of triangular matrices of Example XI.2.9(a). Denote by S_1' the simple left A-module $Ae_{11}/\operatorname{rad} Ae_{11}$. Because of Lemma IX.2.15(a), we have $DS_1' \cong P_1 = e_{11}A$ (which is also a simple right A-module). Hence,

$$\operatorname{Tor}_1^A(S_2, S_1') \cong D\operatorname{Ext}_A^1(S_2, DS_1') \cong D\operatorname{Ext}_A^1(S_2, P_1) \cong K$$

where we have used the result of Example XI.2.9(a). In particular, the simple module S_1' is not projective as a left module. Also, for $n \geq 2$, $\operatorname{Ext}_A^n(-, ?) = 0$ implies $\operatorname{Tor}_n^A(-, ?) = 0$.

Exercises of Section XI.2

1. Let K be a field and A a finite–dimensional K-algebra. Prove that:
   (a) If M_A, N_A are finitely generated A-modules, then $\operatorname{Ext}_A^1(M, N)$ is a finite–dimensional K-vector space.
   (b) If $M_A, {}_A N$ are finitely generated A-modules, then $\operatorname{Tor}_1^A(M, N)$ is a finite–dimensional K-vector space.

2. Let A be a K-algebra and $0 \to M' \to M \to M'' \to 0$ a short exact sequence of A-modules. Prove that:

(a) For any module N, a morphism from M' to N extends to a morphism $M \to N$ if and only if its image under the connecting morphism $\mathrm{Hom}_A(M'', N) \to \mathrm{Ext}^1_A(M', N)$ is zero.

(b) For any module L, a morphism from L to M'' lifts to a morphism $L \to M$ if and only if its image under the connecting morphism $\mathrm{Hom}_A(L, M'') \to \mathrm{Ext}^1_A(L, M')$ is zero.

3. Let A be a K-algebra. If $0 \to M' \to P \to M \to 0$ and $0 \to L \to I \to L' \to 0$ are short exact sequences of A-modules, with P projective and I injective, prove that $\mathrm{Ext}^1_A(M', L) \cong \mathrm{Ext}^1_A(M, L')$.

4. Let A be a K-algebra and $0 \to M \xrightarrow{f} P \to X \to 0$, $0 \to Y \to I \xrightarrow{g} N \to 0$ short exact sequences of A-modules, with P projective and I injective. Prove that:

   (a) $\mathrm{Ext}^1_A(X, Y) \cong \dfrac{\mathrm{Ker}\,\mathrm{Hom}_A(f, g)}{\mathrm{Ker}\,\mathrm{Hom}_A(f, I) + \mathrm{Ker}\,\mathrm{Hom}_A(P, g)}$.

   (b) $\mathrm{Ext}^2_A(X, Y) \cong \mathrm{Coker}\,\mathrm{Hom}_A(f, g)$.

   (c) $\mathrm{Ext}^n_A(X, Y) \cong \mathrm{Ext}^{n-2}_A(M, N)$, for any integer $n > 2$.

5. Let A be a noetherian K-algebra, M a finitely generated A-module and $(N_\lambda)_{\lambda \in \Lambda}$ an arbitrary family of modules. Prove that, for any integer $n \geq 0$, we have an isomorphism

$$\mathrm{Ext}^n_A(M, \oplus_{\lambda \in \Lambda} N_\lambda) \cong \oplus_{\lambda \in \Lambda} \mathrm{Ext}^n_A(M, N_\lambda).$$

6. Let A be the triangular matrix algebra of Example XI.2.9(a). Prove that $\mathrm{Ext}^1_A(S_2, P_2) = 0$.

7. Let G be an abelian group. Prove that:
   (a) If G has elements of finite order, then $\mathrm{Ext}^1_{\mathbb{Z}}(G, \mathbb{Z}) \neq 0$.
   (b) If G is finitely generated and such that $\mathrm{Hom}_{\mathbb{Z}}(G, \mathbb{Z}) = 0$, $\mathrm{Ext}^1_{\mathbb{Z}}(G, \mathbb{Z}) = 0$, then $G = 0$.
   (c) If G is finite, then $\mathrm{Ext}^1_{\mathbb{Z}}(G, \mathbb{Z}) \cong G$.
   (d) If G is torsion, then $\mathrm{Ext}^1_{\mathbb{Z}}(G, \mathbb{Z}) \cong \mathrm{Hom}_{\mathbb{Z}}(G, \mathbb{R}/\mathbb{Z})$.

8. (a) Let m, n be integers having d as gcd. Prove that $\mathrm{Ext}^1_{\mathbb{Z}}(\mathbb{Z}/m\mathbb{Z}, \mathbb{Z}/n\mathbb{Z}) \cong \mathbb{Z}/d\mathbb{Z}$.

   (b) Let G, H be finite abelian groups of coprime orders. Prove that $\mathrm{Ext}^1_{\mathbb{Z}}(G, H) = 0$.

9. Let $m > 0$ be an integer and G an abelian group.
   (a) If $G = mG$, prove that any short exact sequence $0 \to G \to E \to \mathbb{Z}/m\mathbb{Z} \to 0$ splits.
   (b) Set $_mG = \{x \in G : mx = 0\}$. Construct a short exact sequence $0 \to \mathrm{Ext}^1_{\mathbb{Z}}(mG, \mathbb{Z}) \to \mathrm{Ext}^1_{\mathbb{Z}}(G, \mathbb{Z}) \to \mathrm{Ext}^1_{\mathbb{Z}}(_mG, \mathbb{Z}) \to 0$.
   (c) Construct an exact sequence $0 \to \mathrm{Hom}_{\mathbb{Z}}(G, \mathbb{Z}) \to \mathrm{Hom}_{\mathbb{Z}}(mG, \mathbb{Z}) \to \mathrm{Ext}^1_{\mathbb{Z}}(G/mG, \mathbb{Z}) \to \mathrm{Ext}^1_{\mathbb{Z}}(G, \mathbb{Z}) \to \mathrm{Ext}^1_{\mathbb{Z}}(mG, \mathbb{Z}) \to 0$.
   (d) Prove that $\mathrm{Hom}_{\mathbb{Z}}(G, \mathbb{Z}) \cong \mathrm{Hom}_{\mathbb{Z}}(mG, \mathbb{Z})$.

10. Let F, G, H be finitely generated abelian groups. Prove that $\mathrm{Ext}^1_{\mathbb{Z}}(F, \mathrm{Ext}^1_{\mathbb{Z}}(G, H)) \cong \mathrm{Ext}^1_{\mathbb{Z}}(G, \mathrm{Ext}^1_{\mathbb{Z}}(F, H))$.

11. Let p be a prime integer. Consider the short exact sequence of abelian groups

$0 \to \mathbb{Z}/p\mathbb{Z} \xrightarrow{f} \mathbb{Z}/p^2\mathbb{Z} \xrightarrow{g} \mathbb{Z}/p\mathbb{Z} \to 0$, where f is the inclusion and $g = \mathrm{coker} f$. Compute the long exact (co)homology sequence obtained by evaluating on this short exact sequence:

    (a) the covariant functor $\mathrm{Hom}_{\mathbb{Z}}(\mathbb{Z}/p\mathbb{Z}, -)$.

    (b) the contravariant functor $\mathrm{Hom}_{\mathbb{Z}}(-, \mathbb{Z}/p\mathbb{Z})$.

    (c) the covariant functor $\mathbb{Z}/p\mathbb{Z} \otimes_{\mathbb{Z}} -$.

12. Let A be a K-algebra.

    (a) Given an exact sequence $0 \to L \xrightarrow{f} P_{n-1} \to \cdots \to P_0 \to M \to 0$ of modules, with $P_0, \cdots, P_{n-1}$ projective. Prove that, for any N_A, there is an exact sequence $\mathrm{Hom}_A(P_{n-1}, N) \xrightarrow{\mathrm{Hom}_A(f,N)} \mathrm{Hom}_A(L, N) \to \mathrm{Ext}^n_A(M, N) \to 0$.

    (b) Given an exact sequence $0 \to L \to I^0 \to \cdots \to I^{n-1} \xrightarrow{g} N \to 0$ of modules, with $I^0, \cdots, I^{n-1}$ injective. Prove that, for any M_A, there is an exact sequence $\mathrm{Hom}_A(M, I^{n-1}) \xrightarrow{\mathrm{Hom}_A(M,g)} \mathrm{Hom}_A(M, N) \to \mathrm{Ext}^n_A(M, N) \to 0$.

    (c) State and prove the corresponding statements for the torsion functors.

13. Let A be a principal ideal domain and M, M', N, N' be A-modules. Prove that:

    (a) An epimorphism $N \to N'$ induces an epimorphism $\mathrm{Ext}^1_A(M, N) \to \mathrm{Ext}^1_A(M, N')$.

    (b) A monomorphism $M' \to M$ induces a monomorphism $\mathrm{Tor}^A_1(M', N) \to \mathrm{Tor}^A_1(M, N)$.

14. Let K be a field, Q the quiver $1 \leftarrow 3 \to 2$ and $A = KQ$. Using a projective resolution of the simple module S_3, prove that $\mathrm{Ext}^1_A(S_3, P_3) \cong K$.

15. Let K be a field and A the algebra given by the quiver

$$1 \circ \underset{\beta}{\overset{\alpha}{\underset{\longleftarrow}{\longrightarrow}}} \circ 2$$

bound by the ideal $\langle \alpha\beta \rangle$. Using a projective resolution of the indecomposable injective module I_1, prove that $\mathrm{Ext}^1_A(I_1, P_1) \cong K$.

16. Prove that, for each of the two bound quiver algebras over a field K, the K-vector space $\mathrm{Ext}^2_A(I_x, P_y)$ is one-dimensional. In both cases, x is the unique source, and y the unique sink.

(a)

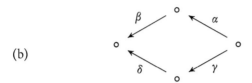

bound by the ideal $\langle \alpha\beta\gamma \rangle$.

(b)

bound by the ideal $\langle \alpha\beta - \gamma\delta \rangle$.

17. Let p be a prime integer and M a $\mathbb{Z}/p^2\mathbb{Z}$-module M. Set $_pM = \{x \in M : px = 0\}$. Prove that:

$$\mathrm{Tor}_n^{\mathbb{Z}/p^2\mathbb{Z}}(M, \mathbb{Z}/p^2\mathbb{Z}) = \begin{cases} M/pM & \text{if } n = 0, \\ _pM/pM & \text{if } n > 0. \end{cases}$$

18. Let A be a K-algebra and $0 \to M' \to M \to M'' \to 0$ a short exact sequence of A-modules. Prove that, if M', M'' are flat, then so is M.

19. Let A be a K-algebra and $0 \to M_A \xrightarrow{f} P_A \to X_A \to 0, 0 \to_A Y \to_A Q \xrightarrow{g}_A N \to 0$ short exact sequences of (right and left, respectively) A-modules, with P_A and $_AQ$ projective. Prove that:
   (a) $\mathrm{Tor}_1^A(X, Y) \cong \dfrac{\mathrm{Im}\,(f \otimes 1_Q) \cap \mathrm{Im}\,(1_P \otimes g)}{\mathrm{Im}\,(f \otimes_A g)}$.
   (b) $\mathrm{Tor}_2^A(X, Y) \cong \mathrm{Ker}\,(f \otimes_A g)$.
   (c) $\mathrm{Tor}_n^A(X, Y) \cong \mathrm{Tor}_{n-2}^A(M, N)$, for any integer $n > 2$.

20. Let I be a right ideal and J a left ideal of an algebra A, prove that $\mathrm{Tor}_1^A(A/I, A/J) \cong (I \cap J)/IJ$.

21. For an abelian group G, denote by $T(G)$ its torsion subgroup. Prove that:
   (a) There exists an exact sequence $0 \to T(G) \to G \to G \otimes_\mathbb{Z} \mathbb{Q} \to G \otimes_\mathbb{Z} \mathbb{Q}/\mathbb{Z} \to 0$.
   (b) There exists a short exact sequence $0 \to G/T(G) \to G \otimes_\mathbb{Z} \mathbb{Q} \to G \otimes_\mathbb{Z} \mathbb{Q}/\mathbb{Z} \to 0$.
   (c) The group G is torsion if and only if $G \otimes_\mathbb{Z} \mathbb{Q} = 0$.

22. Prove that an abelian group G is torsion-free if and only if $\mathrm{Tor}_1^\mathbb{Z}(G, -) = 0$ or, equivalently, if and only if $\mathrm{Tor}_1^\mathbb{Z}(-, G) = 0$.

23. If G, H are abelian groups such that $mG = 0$ and $nH = 0$, with m, n coprime integers, prove that $\mathrm{Tor}_1^\mathbb{Z}(G, H) = 0$.

24. If G, H are finite abelian groups, prove that $\mathrm{Tor}_1^\mathbb{Z}(G, H) \cong G \otimes_\mathbb{Z} H$.

25. Let G, H be abelian groups with respective torsion subgroups $T(G), T(H)$. Prove that:
   (a) $\mathrm{Tor}_1^\mathbb{Z}(G, H) \cong \mathrm{Tor}_1^\mathbb{Z}(T(G), T(H))$.

(b) If $mT(G) = 0$, for an integer m, then $m\mathrm{Tor}_1^{\mathbb{Z}}(G, H) = 0$.

XI.3 Exact sequences and extensions

XI.3.1 Short exact sequences and Ext^1

Let A be a K-algebra and M, N be A-modules. As seen in Subsection VII.4.2, the expression 'extension of N by M' means a short exact sequence of the form

$$\mathbf{e} : 0 \to N \to E \to M \to 0$$

We claim that the elements of the extension module $\mathrm{Ext}_A^1(M, N)$ correspond bijectively to equivalence classes of extensions of N by M.

Let $\mathbf{e} : 0 \to N \xrightarrow{f} E \xrightarrow{g} M \to 0$ and $\mathbf{e'} : 0 \to N \xrightarrow{f'} E' \xrightarrow{g'} M \to 0$ be extensions of N by M. We declare $\mathbf{e}$ **equivalent** to $\mathbf{e'}$ if there exists a morphism $h : E \to E'$ making commutative the following diagram with exact rows

$$
\begin{array}{ccccccccc}
0 & \longrightarrow & N & \xrightarrow{f} & E & \xrightarrow{g} & M & \longrightarrow & 0 \\
& & \downarrow{1_N} & & \downarrow{h} & & \downarrow{1_M} & & \\
0 & \longrightarrow & N & \xrightarrow{f'} & E' & \xrightarrow{g'} & M & \longrightarrow & 0
\end{array}
$$

that is, such that $hf = f'$ and $g'h = g$. Because of the Five Lemma II.4.18, h is necessarily an isomorphism. Therefore, equivalence of extensions is a reflexive, symmetric and transitive relation. The equivalence class of an extension $\mathbf{e}$ of N by M is denoted by $[\mathbf{e}]$ and their set, that is, the quotient set of the extensions of N by M modulo this equivalence, by $\mathcal{E}(M, N)$.

Conversely, the mere existence of an isomorphism $h : E \to E'$ does *not* suffice to make the extensions $\mathbf{e}$ and $\mathbf{e'}$ equivalent: we need a commutative diagram as above.

XI.3.1 Examples.

(a) Let M, N be A-modules. It follows from Definition IV.2.31 that any two split extensions of N by M are equivalent.

(b) Let $A = \mathbb{Z}$, $p > 2$ a prime integer and a short exact sequence

$$\mathbf{e} : 0 \to \mathbb{Z}/p\mathbb{Z} \to E \to \mathbb{Z}/p\mathbb{Z} \to 0.$$

Then the cardinality of E is p^2. Because of Theorem V.3.18, either $E \cong \mathbb{Z}/p\mathbb{Z} \oplus \mathbb{Z}/p\mathbb{Z}$, in which case the sequence splits, or else $E \cong \mathbb{Z}/p^2\mathbb{Z}$. Assume that the latter holds true and let a be an integer such that $1 \leqslant a < p$. Consider the short exact sequences

$$\mathbf{e}_a : 0 \to \mathbb{Z}/p\mathbb{Z} \xrightarrow{f_a} \mathbb{Z}/p^2\mathbb{Z} \xrightarrow{g_a} \mathbb{Z}/p\mathbb{Z} \to 0$$

with $f_a : 1 + p\mathbb{Z} \mapsto ap + p^2\mathbb{Z}$ and $g_a = \mathrm{coker} f_a$. We prove that, if $a \neq 1$, then $\mathbf{e}_1$ and $\mathbf{e}_a$ are not equivalent.

We claim that $\operatorname{Im} f_a = \operatorname{Im} f_1 = p\mathbb{Z}/p^2\mathbb{Z}$. First, $\operatorname{Im} f_a = ap\mathbb{Z}/p^2\mathbb{Z} \subseteq p\mathbb{Z}/p^2\mathbb{Z} = \operatorname{Im} f_1$. To prove the reverse inclusion, it suffices to show that the generator $p + p^2\mathbb{Z}$ of $\operatorname{Im} f_1 = p\mathbb{Z}/p^2\mathbb{Z}$ belongs to $ap\mathbb{Z}/p^2\mathbb{Z}$, that is, there exists an integer k such that $p + p^2\mathbb{Z} = akp + p^2\mathbb{Z}$ or, equivalently, an integer k such that p divides $ak - 1$. Because $1 \leqslant a < p$, we take for k an integer whose residual class modulo p is the inverse of that of a. Such an integer exists because p is prime and hence $\mathbb{Z}/p\mathbb{Z}$ is a field. This establishes our claim. The latter implies that the cokernels of f_1 and f_a are equal and hence, that $g_1 = g_a$. We may suppose that g_1, g_a are both given by $1 + p^2\mathbb{Z} \mapsto 1 + p\mathbb{Z}$, because the kernel of the latter morphism is $p\mathbb{Z}/p^2\mathbb{Z}$.

Assume that there exists $h : \mathbb{Z}/p^2\mathbb{Z} \to \mathbb{Z}/p^2\mathbb{Z}$ such that $f_a = hf_1$ and $g_a h = g_1$. The second relation gives $g_a h(1 + p^2\mathbb{Z}) = g_1(1 + p^2\mathbb{Z}) = 1 + p\mathbb{Z}$. Hence, $h(1 + p^2\mathbb{Z})$ is a preimage of $1 + p\mathbb{Z}$ under g_a. That is, there exists ℓ, with $1 \leqslant \ell < p$, such that $h(1 + p^2\mathbb{Z}) = (1 + \ell p) + p^2\mathbb{Z}$. Now, the first relation $f_a = hf_1$ gives $h(p + p^2\mathbb{Z}) = ap + p^2\mathbb{Z}$. Hence, $ap + p^2\mathbb{Z} = h(p + p^2\mathbb{Z}) = ph(1 + p^2\mathbb{Z}) = (p + \ell p^2) + p^2\mathbb{Z} = p + p^2\mathbb{Z}$, which gives $ap - p \in p^2\mathbb{Z}$, that is, a is congruent to 1 modulo p. But $1 \leqslant a < p$. Therefore, $a = 1$. This completes the proof.

The same proof shows that if a, b are distinct integers with $1 \leqslant a, b < p$, then $\mathbf{e}_a$ and $\mathbf{e}_b$ are not equivalent.

We now construct a bifunctor starting from the correspondence of objects $(M, N) \mapsto \mathcal{E}(M, N)$. Given an extension $\mathbf{e} : 0 \to N \xrightarrow{f} E \xrightarrow{g} M \to 0$ and a morphism $u : N \to N'$, it follows from Theorem IV.3.9 that we have a commutative diagram with exact rows

$$
\begin{array}{ccccccccc}
\mathbf{e} : & 0 & \longrightarrow & N & \xrightarrow{\ f\ } & E & \xrightarrow{\ g\ } & M & \longrightarrow 0 \\
& & & \downarrow{\scriptstyle u} & & \downarrow{\scriptstyle v} & & \downarrow{\scriptstyle 1_M} & \\
\mathbf{e'} : & 0 & \longrightarrow & N' & \xrightarrow{\ f'\ } & E' & \xrightarrow{\ g'\ } & M & \longrightarrow 0
\end{array}
$$

where E' is the amalgamated sum of f, u. Moreover, also because of Theorem IV.3.9, the equivalence class of the lower sequence $\mathbf{e'}$ is uniquely determined by that of $\mathbf{e}$. This defines a map $\mathcal{E}(M, u) : \mathcal{E}(M, N) \to \mathcal{E}(M, N')$. The image $\mathbf{e'}$ of $\mathbf{e}$ is denoted by $\mathcal{E}(M, u)([\mathbf{e}]) = [u\mathbf{e}]$.

Dually, to a morphism $v : M' \to M$ corresponds a commutative diagram with exact rows

$$
\begin{array}{ccccccccc}
\mathbf{e'} : & 0 & \longrightarrow & N & \xrightarrow{\ f'\ } & E' & \xrightarrow{\ g'\ } & M' & \longrightarrow 0 \\
& & & \downarrow{\scriptstyle 1_N} & & \downarrow{\scriptstyle v} & & \downarrow{\scriptstyle w} & \\
\mathbf{e} : & 0 & \longrightarrow & N & \xrightarrow{\ f\ } & E & \xrightarrow{\ g\ } & M & \longrightarrow 0
\end{array}
$$

where E' is the fibered product of g, w. This again defines a map $\mathcal{E}(w, N) : \mathcal{E}(M, N) \to \mathcal{E}(M', N)$, see Theorem IV.3.4. The image $\mathbf{e'}$ of $\mathbf{e}$ is denoted by $\mathcal{E}(w, N)([\mathbf{e}]) = [\mathbf{e}w]$.

XI.3.2 Lemma. *The correspondence* $(M, N) \mapsto \mathcal{E}(M, N)$ *induces a set-valued bifunctor* $\mathcal{E}(-, ?)$ *on pairs of modules, contravariant in the first variable and covariant in the second.*

Proof. Let $\mathbf{e} : 0 \to N \xrightarrow{f} E \xrightarrow{g} M \to 0$ be an extension and $u : N \to N'$, $w : M' \to M$ morphisms. We must prove that $\mathcal{E}(w, N)\mathcal{E}(M, u)[\mathbf{e}] = \mathcal{E}(M, u)\mathcal{E}(w, N)[\mathbf{e}]$ that is, the extensions $(u\mathbf{e})w$ and $u(\mathbf{e}w)$ are equivalent.

We construct these extensions by means of the diagrams

$$
\begin{array}{ccccccccc}
\mathbf{e} : & 0 & \longrightarrow & N & \xrightarrow{\ f\ } & E & \xrightarrow{\ g\ } & M & \longrightarrow 0 \\
 & & & \downarrow{\scriptstyle u} & & \downarrow{\scriptstyle h_-} & & \downarrow{\scriptstyle 1_M} & \\
u\mathbf{e} : & 0 & \longrightarrow & N' & \xrightarrow{\ f_-\ } & E_- & \xrightarrow{\ g_-\ } & M & \longrightarrow 0 \\
 & & & \uparrow{\scriptstyle 1_{N'}} & & \uparrow{\scriptstyle k_-} & & \uparrow{\scriptstyle w} & \\
(u\mathbf{e})w : & 0 & \longrightarrow & N' & \xrightarrow{\ f_{-+}\ } & E_{-+} & \xrightarrow{\ g_{-+}\ } & M' & \longrightarrow 0 \\
\end{array}
$$

and

$$
\begin{array}{ccccccccc}
\mathbf{e} : & 0 & \longrightarrow & N & \xrightarrow{\ f\ } & E & \xrightarrow{\ g\ } & M & \longrightarrow 0 \\
 & & & \uparrow{\scriptstyle 1_N} & & \uparrow{\scriptstyle h_+} & & \uparrow{\scriptstyle w} & \\
\mathbf{e}w : & 0 & \longrightarrow & N & \xrightarrow{\ f_+\ } & E_+ & \xrightarrow{\ g_+\ } & M' & \longrightarrow 0 \\
 & & & \downarrow{\scriptstyle u} & & \downarrow{\scriptstyle k_+} & & \downarrow{\scriptstyle 1_{M'}} & \\
u(\mathbf{e}w) : & 0 & \longrightarrow & N' & \xrightarrow{\ f_{+-}\ } & E_{+-} & \xrightarrow{\ g_{+-}\ } & M' & \longrightarrow 0 \\
\end{array}
$$

Our strategy consists in using the universal properties of fibered product and amalgamated sum in order to construct a morphism $\varphi : E_{+-} \to E_{-+}$ realising the equivalence of the considered extensions.

We first consider the morphisms $h_- h_+ : E_+ \to E_-$ and $g_+ : E_+ \to M'$. We have $g_-(h_- h_+) = (g_- h_-)h_+ = g h_+ = w g_+$. The universal property of the fibered product E_{-+} (of w and g_-) yields a unique $\ell : E_+ \to E_{-+}$ such that $g_{-+}\ell = g_+$ and $k_-\ell = h_- h_+$.

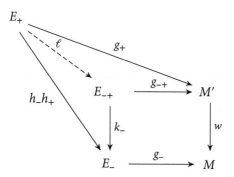

We next consider the morphisms $\ell : E_+ \to E_{-+}$ and $f_{-+} : N' \to E_{-+}$. We claim that $\ell f_+ = f_{-+}u$. Because of the definition of ℓ, it suffices to prove that $k_-\ell f_+ = k_- f_{-+}u$ and $g_{-+}\ell f_+ = g_{-+}f_{-+}u$. The second equality is immediate, because $g_{-+}f_{-+} = 0$ while $g_{-+}\ell f_+ = g_+ f_+ = 0$, so both sides equal zero. The first equality follows from $k_-\ell f_+ = h_- h_+ f_+ = h_- f = f_- u = k_- f_{-+}u$. The universal property of the amalgamated sum E_{+-} (of u and f_+) yields a unique morphism $\varphi : E_{+-} \to E_{-+}$ such that $\varphi\ell = k_+$ and $\varphi f_{+-} = f_{-+}$.

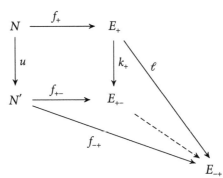

We claim that φ realises the sought equivalence of extensions

$$
\begin{array}{ccccccccc}
0 & \longrightarrow & N' & \xrightarrow{f_{+-}} & E_{+-} & \xrightarrow{g_{+-}} & M' & \longrightarrow & 0 \\
 & & \downarrow{\scriptstyle 1_{N'}} & & \downarrow{\scriptstyle \varphi} & & \downarrow{\scriptstyle 1_{M'}} & & \\
0 & \longrightarrow & N' & \xrightarrow{f_{-+}} & E_{-+} & \xrightarrow{g_{-+}} & M' & \longrightarrow & 0
\end{array}
$$

Because $\varphi f_{+-} = f_{-+}$, it suffices to prove that $g_{-+}\varphi = g_{+-}$. Due to definition of φ, this amounts to proving that $g_{-+}\varphi k_+ = g_{+-}k_+$ and $g_{-+}\varphi f_{+-} = g_{+-}f_{+-}$. The second equality is trivial, because both sides vanish (here, we use $\varphi f_{+-} = f_{-+}$), whereas the first follows from $g_{-+}\varphi k_+ = g_{-+}\ell = g_+ = g_{+-}k_+$. The proof is complete. $\qquad\square$

We now construct a map $\varphi : \mathcal{E}(M, N) \to \mathrm{Ext}^1_A(M, N)$. Let $\mathbf{e} : 0 \to N \xrightarrow{f} E \xrightarrow{g} M \to 0$ be an extension. Evaluating $\mathrm{Hom}_A(M, -)$ on this short exact sequence yields a long exact cohomology sequence

$$
\cdots \to \mathrm{Hom}_A(M, E) \xrightarrow{\mathrm{Hom}_A(M,g)} \mathrm{Hom}_A(M, M) \xrightarrow{\delta_{\mathbf{e}}} \mathrm{Ext}^1_A(M, N) \to \cdots
$$

where we denote the connecting morphism by $\delta_{\mathbf{e}}$ in order to emphasise dependence on $\mathbf{e}$. Now, $\mathrm{Hom}_A(M, M)$ has a distinguished element, namely the identity 1_M. Its image $\delta_{\mathbf{e}}(1_M)$ in $\mathrm{Ext}^1_A(M, N)$ is called the **obstruction** of $\mathbf{e}$. This term should be taken as meaning 'obstruction to being split': indeed, $\mathbf{e}$ is split if and only if g is a retraction, and this is the case if and only if 1_M lies in the image of $\mathrm{Hom}_A(M, g)$, that is, the obstruction $\delta_{\mathbf{e}}(1_M)$ vanishes.

We verify that the obstruction of $\mathbf{e}$ only depends on the equivalence class of $\mathbf{e}$. Let $\mathbf{e}$ be as above and $\mathbf{e}' : 0 \to N \xrightarrow{f'} E' \xrightarrow{g'} M \to 0$ an equivalent extension. There exists a commutative diagram with exact rows

$$
\begin{array}{ccccccccc}
0 & \longrightarrow & N & \xrightarrow{f} & E & \xrightarrow{g} & M & \longrightarrow & 0 \\
 & & \downarrow{1_N} & & \downarrow{h} & & \downarrow{1_M} & & \\
0 & \longrightarrow & N & \xrightarrow{f'} & E' & \xrightarrow{g'} & M & \longrightarrow & 0
\end{array}
$$

The functoriality of the connecting morphism, see Lemma X.2.9, yields a commutative square

$$
\begin{array}{ccc}
\mathrm{Hom}_A(M, M) & \xrightarrow{\delta e} & \mathrm{Ext}^1_A(M, N) \\
\downarrow{1_{\mathrm{Hom}_A(M,M)}} & & \downarrow{1_{\mathrm{Ext}^1_A(M,N)}} \\
\mathrm{Hom}_A(M, M) & \xrightarrow{\delta e'} & \mathrm{Ext}^1_A(M, N)
\end{array}
$$

so that $\delta_{\mathbf{e}}(1_M) = \delta_{\mathbf{e}'}(1_M)$. We have thus defined a map $\varphi : \mathcal{E}(M, N) \to \mathrm{Ext}^1_A(M, N)$ given by $[\mathbf{e}] \mapsto \delta_{\mathbf{e}}(1_M)$, for $[\mathbf{e}] \in \mathcal{E}(M, N)$.

We compute $\varphi[\mathbf{e}]$. Let

$$
\cdots \to P_2 \xrightarrow{d_2} P_1 \xrightarrow{d_1} P_0 \xrightarrow{d_0} M \to 0
$$

be a projective resolution of M. The Comparison Theorem X.3.1 gives a commutative diagram with exact rows

$$
\begin{array}{ccccccccc}
\cdots & \longrightarrow & P_2 & \xrightarrow{d_2} & P_1 & \xrightarrow{d_1} & P_0 & \xrightarrow{d_0} & M & \longrightarrow & 0 \\
 & & \downarrow & & \downarrow{h_1} & & \downarrow{h_0} & & \downarrow{1_M} & & \\
 & & & & 0 & \longrightarrow & N & \longrightarrow & E & \longrightarrow & M & \longrightarrow & 0
\end{array}
$$

In particular, $h_1 d_2 = 0$ says that $h_1 \in \mathrm{Ker}\,\mathrm{Hom}_A(d_2, N)$, so that h_1 determines an element $h_1 + \mathrm{Im}\,\mathrm{Hom}_A(d_1, N)$ of $\mathrm{Ext}^1_A(M, N) = \mathrm{Ker}\,\mathrm{Hom}_A(d_2, N)/\mathrm{Im}\,\mathrm{Hom}_A(d_1, N)$. The Comparison Theorem X.3.1 says that h_1 is unique up to homotopy, and hence its residual class in the cohomology module $\mathrm{Ext}^1_A(M, N)$ is uniquely determined. Moreover, h_1 does not change if we replace $\mathbf{e}$ by an equivalent extension, so its class in $\mathrm{Ext}^1_A(M, N)$ only depends on the equivalence class $[\mathbf{e}]$.

Let $d_1 = jp$ be the factorisation of d_1 through its image $L = \mathrm{Im}\, d_1$. Because $h_1 d_2 = 0$, there exists $h : L \to N$ such that $h_1 = hp$, that is, we have a commutative diagram with exact rows

$$\mathbf{e_0}:\quad 0 \longrightarrow L \xrightarrow{\ j\ } P_0 \xrightarrow{\ d_0\ } M \longrightarrow 0$$

with vertical maps h, h_0, 1_M

$$\mathbf{e}:\quad 0 \longrightarrow N \xrightarrow{\ f\ } E \xrightarrow{\ g\ } M \longrightarrow 0$$

and, moreover, $[\mathbf{e}] = [h\mathbf{e_0}]$. We are now able to prove that the obstruction of the extension $\mathbf{e}$ equals the class of the morphism h_1 in $\operatorname{Ext}_A^1(M, N)$.

XI.3.3 Lemma. *With the previous notation, $\varphi[\mathbf{e}] = h_1 + \operatorname{Im} \operatorname{Hom}_A(d_1, N)$.*

Proof. This is a direct computation. The short exact sequences $\mathbf{e}$ and $\mathbf{e_0}$ induce a commutative diagram with exact rows and columns

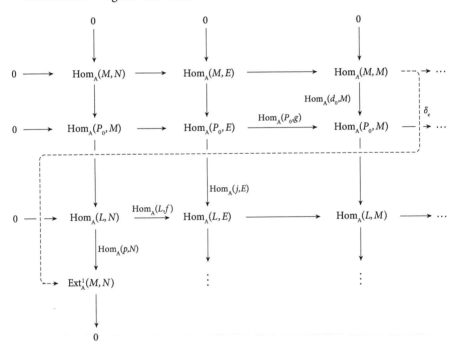

The connecting morphism $\delta_{\mathbf{e}}$ arises from the Snake Lemma II.4.19 applied to the two lower rows. Using its explicit expression, we get that $\delta_{\mathbf{e}}(1_M)$ equals

$$\operatorname{Hom}_A(p, N)\operatorname{Hom}_A(L, f)^{-1}\operatorname{Hom}_A(j, E)\operatorname{Hom}_A(P_0, g)^{-1}\operatorname{Hom}_A(d_0, M)(1_M)$$

$$+ \operatorname{ImHom}_A(d_1, N) = \operatorname{Hom}_A(p, N)\operatorname{Hom}_A(L, f)^{-1}\operatorname{Hom}_A(j, E)\operatorname{Hom}_A(P_0, g)^{-1}(d_0)$$

$$+ \operatorname{Im} \operatorname{Hom}_A(d_1, N).$$

Now $d_0 = gh_0 = \text{Hom}_A(P_0, g)(h_0)$. Hence, $\text{Hom}_A(P_0, g)^{-1}(d_0) = h_0$ and so

$$\delta_{\mathbf{e}}(1_M) = \text{Hom}_A(p, N)\text{Hom}_A(L, f)^{-1}\text{Hom}_A(j, E)(h_0) + \text{Im Hom}_A(d_1, N)$$
$$= \text{Hom}_A(p, N)\text{Hom}_A(L, f)^{-1}(h_0 j) + \text{Im Hom}_A(d_1, N)$$
$$= \text{Hom}_A(p, N)\text{Hom}_A(L, f)^{-1}(fh) + \text{Im Hom}_A(d_1, N)$$
$$= \text{Hom}_A(p, N)(h) + \text{Im Hom}_A(d_1, N)$$
$$= hp + \text{Im Hom}_A(d_1, N)$$
$$= h_1 + \text{Im Hom}_A(d_1, N).$$
□

XI.3.4 Theorem. *Let M, N be A-modules. The map $\varphi : \mathcal{E}(M, N) \to \text{Ext}_A^1(M, N)$ is a bijection functorial in M and N, in which the class of the split extension corresponds to $0 \in \text{Ext}_A^1(M, N)$.*

Proof. We first observe that φ applies a split extension to the zero of $\text{Ext}_A^1(M, N)$: indeed, given an extension $\mathbf{e} \in \mathcal{E}(M, N)$, we have $\varphi[\mathbf{e}] = \delta_{\mathbf{e}}(1_M)$ and the obstruction vanishes if and only if $\mathbf{e}$ splits.

In order to prove that φ is bijective, we construct the inverse map $\psi : \text{Ext}_A^1(M, N) \to \mathcal{E}(M, N)$. Take as above a projective resolution of M

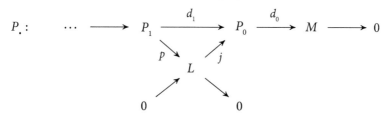

where $d_1 = jp$ is the canonical factorisation through $L = \text{Im } d_1$ and $x \in \text{Ext}_A^1(M, N)$. There exists $h_1 : P_1 \to N$ such that $h_1 d_2 = 0$ and $x = h_1 + \text{Im Hom}_A(d_1, N)$. The first condition amounts to existence of a morphism $h : L \to N$ such that $h_1 = hp$. We set $\psi(x) = [h\mathbf{e}_0]$, where $\mathbf{e}_0$ is the extension $0 \to L \xrightarrow{j} P_0 \xrightarrow{d_0} M \to 0$. Thus, $h\mathbf{e}_0$ is the lower sequence in the commutative diagram with exact rows

$$
\begin{array}{ccccccccc}
\mathbf{e}_0 : & 0 & \longrightarrow & L & \xrightarrow{\ j\ } & P_0 & \xrightarrow{\ d_0\ } & M & \longrightarrow & 0 \\
& & & \downarrow{\scriptstyle h} & & \downarrow{\scriptstyle h_0} & & \downarrow{\scriptstyle 1_M} & & \\
h\mathbf{e}_0 : & 0 & \longrightarrow & N & \xrightarrow{\ f\ } & E & \xrightarrow{\ g\ } & M & \longrightarrow & 0
\end{array}
$$

We prove that this definition does not depend on the choice of the representative h_1 of the class x. Let h_1, h_1' be representatives of x, that is, $h_1, h_1' \in \text{Ker Hom}_A(d_2, N)$ are such that $h_1' - h_1 \in \text{Im Hom}_A(d_1, N)$. Then there exists $s \in \text{Hom}_A(P_0, N)$ such that $h_1' - h_1 = \text{Hom}_A(d_1, N)(s) = sd_1$. Moreover, $h_1 d_2 = 0, h_1' d_2 = 0$ imply the existence of h, h' such that $h_1 = hp, h_1' = h'p$. Then $h'p - hp = sd_1 = sjp$ and p is an epimorphism, hence $h' - h = sj$. We claim that we have a commutative diagram with exact rows

$$
\begin{array}{ccccccccc}
0 & \longrightarrow & L & \overset{j}{\longrightarrow} & P_0 & \overset{d_0}{\longrightarrow} & M & \longrightarrow & 0 \\
 & & {\scriptstyle h'}\downarrow & & {\scriptstyle h_0+fs}\downarrow & & {\scriptstyle 1_M}\downarrow & & \\
0 & \longrightarrow & N & \underset{f}{\longrightarrow} & E & \underset{g}{\longrightarrow} & M & \longrightarrow & 0
\end{array}
$$

Indeed, $fh' = f(h + sj) = fh + fsj = h_0j + fsj = (h_0 + fs)j$ and $g(h_0 + fs) = gh_0 - d_0$ because $gf = 0$. Theorem IV.3.9 implies that $[he_0] - [h'e_0']$. Thus, ψ is unambiguously defined.

We prove that φ and ψ are mutually inverse. Let $x = h_1 + \mathrm{Im}\,\mathrm{Hom}_A(d_1, N)$. Then $\psi(x) = [he_0]$, where h and e_0 are as before. Lemma XI.3.3 gives

$$\varphi\psi(x) = \varphi[he_0] = hp + \mathrm{Im}\mathrm{Hom}_A(d_1, N) = h_1 + \mathrm{Im}\mathrm{Hom}_A(d_1, N) = x,$$

so that $\varphi\psi = 1$.

Conversely, let $[e] \in \mathcal{E}(M, N)$. Then $\varphi[e] = hp + \mathrm{Im}\,\mathrm{Hom}_A(d_1, N)$, where $hp : P_1 \to N$ comes, as we have seen, from an application of the Comparison Theorem X.3.1. Then

$$\psi\varphi[e] = \psi(hp + \mathrm{Im}\,\mathrm{Hom}_A(d_1, N)) = [he_0] = [e]$$

where the last equality follows from Theorem IV.3.9. Hence, $\psi\phi = 1$.

Functoriality of φ follows from that of the connecting morphism.   □

We have shown that φ defines an isomorphism of $\mathcal{E}(-, ?)$ and $\mathrm{Ext}_A^1(-, ?)$ as *set-valued* bifunctors. But, for any modules M, N, the set $\mathrm{Ext}_A^1(M, N)$ is actually a K-module. Therefore, using transportation of structure, see Exercise II.3.17, the bijection φ defines a K-module structure on $\mathcal{E}(M, N)$ such that φ becomes an isomorphism. For reasons of space, we do not describe here explicitly the module structure of $\mathcal{E}(M, N)$, and relegate it to the exercises.

Similarly, one can evaluate the contravariant functor $\mathrm{Hom}_A(-, N)$ on an extension $\mathbf{e} : 0 \to N \to E \to M \to 0$ and consider the image of the identity 1_N under the connecting morphism $\delta_\mathbf{e} : \mathrm{Hom}_A(N, N) \to \mathrm{Ext}_A^1(M, N)$. This gives a map from $\mathcal{E}(M, N)$ to $\mathrm{Ext}_A^1(M, N)$ defined by $[\mathbf{e}] \mapsto \delta_\mathbf{e}(1_N)$. One can prove, as before, but using this time an injective coresolution, that this map is a bijection, functorial in M, N, in which split exact sequences correspond to the zero of $\mathrm{Ext}_A^1(M, N)$.

Identifying $\mathcal{E}(M, N)$ with $\mathrm{Ext}_A^1(M, N)$ by means of the bijection φ, we deduce an explicit expression for the connecting morphism.

XI.3.5 Corollary. (a) *Let M be an A-module and $\mathbf{e} : 0 \to N' \to N \to N'' \to 0$ a short exact sequence. The connecting morphism $\delta : \mathrm{Hom}_A(M, N'') \to \mathrm{Ext}_A^1(M, N')$ is given by $w \mapsto [\mathbf{e}w]$, for $w \in \mathrm{Hom}_A(M, N'')$.*

(b) *Let N be an A-module and $\mathbf{e} : 0 \to M' \to M \to M'' \to 0$ a short exact sequence. The connecting morphism $\delta : \mathrm{Hom}_A(M', N) \to \mathrm{Ext}_A^1(M'', N)$ is given by $u \mapsto [u\mathbf{e}]$ for $u \in \mathrm{Hom}_A(M', N)$.*

Proof. We prove (a), because the proof of (b) is similar. To $w \in \mathrm{Hom}_A(M, N'')$ corresponds a commutative diagram with exact rows

$$
\begin{array}{ccccccccc}
\mathbf{ew}: & 0 & \longrightarrow & N' & \longrightarrow & E & \longrightarrow & M & \longrightarrow & 0 \\
 & & & \downarrow{\scriptstyle 1_{N'}} & & \downarrow & & \downarrow{\scriptstyle w} & & \\
\mathbf{e}: & 0 & \longrightarrow & N' & \longrightarrow & N & \longrightarrow & N'' & \longrightarrow & 0
\end{array}
$$

Evaluating the functor $\mathrm{Hom}_A(M, -)$, this diagram induces a commutative square

$$
\begin{array}{ccc}
\mathrm{Hom}_A(M, M) & \xrightarrow{\ \delta_{ew}\ } & \mathrm{Ext}^1_A(M, N') \\
\downarrow{\scriptstyle \mathrm{Hom}_A(M, w)} & & \downarrow{\scriptstyle 1_{\mathrm{Ext}^1_A(M, N')}} \\
\mathrm{Hom}_A(M, N'') & \xrightarrow{\ \delta_e\ } & \mathrm{Ext}^1_A(M, N')
\end{array}
$$

because of functoriality of the connecting morphism. Then we have

$$
\delta_e(w) = \delta_e \mathrm{Hom}_A(M, w)(1_M) = \delta_{ew}(1_M) = \varphi[\mathbf{ew}]
$$

which is identified to $[\mathbf{ew}]$ under the bijection φ. This completes the proof.   □

XI.3.6 Examples.

(a) Let $A = \mathbb{Z}$ and p a prime integer. It follows from Example XI.2.9(c) that $\mathrm{Ext}^1_A(\mathbb{Z}/p\mathbb{Z}, \mathbb{Z}/p\mathbb{Z}) \cong (\mathbb{Z}/p\mathbb{Z})/(p\mathbb{Z}/p\mathbb{Z}) \cong \mathbb{Z}/p\mathbb{Z}$. On the other hand, for any integer a such that $1 \leqslant a < p$, we have an extension

$$
\mathbf{e}_a : 0 \to \mathbb{Z}/p\mathbb{Z} \xrightarrow{f_a} \mathbb{Z}/p^2\mathbb{Z} \xrightarrow{g_a} \mathbb{Z}/p\mathbb{Z} \to 0
$$

see Example XI.3.1(a), and the $p - 1$ distinct extensions $\mathbf{e}_a$ are pairwise nonequivalent. Therefore, the $\mathbf{e}_a$ together with the split extension $0 \to \mathbb{Z}/p\mathbb{Z} \to \mathbb{Z}/p\mathbb{Z} \oplus \mathbb{Z}/p\mathbb{Z} \to \mathbb{Z}/p\mathbb{Z} \to 0$ furnish a complete set of representatives of residual classes in $\mathrm{Ext}^1_A(\mathbb{Z}/p\mathbb{Z}, \mathbb{Z}/p\mathbb{Z})$.

(b) Let K be a field, $A = T_2(K), P_1 = \mathbf{e}_{11}A, P_2 = \mathbf{e}_{22}A$ and $S_2 = P_2/P_1$ as in Example XI.2.9(a). We know that $\mathrm{Ext}^1_A(S_2, P_1)$ is a one-dimensional K-vector space. Hence, a basis of this space is represented by any nonsplit extension of P_1 by S. Now the short exact sequence $0 \to P_1 \to P_2 \to S \to 0$ is not split, because P_2 is indecomposable. Therefore, it represents a basis vector of $\mathrm{Ext}^1_A(S, P_1)$.

(c) Let A be an algebra and assume that we have a commutative square in $\mathrm{Mod}A$

$$
\begin{array}{ccc}
M & \xrightarrow{\ f\ } & N \\
\downarrow{\scriptstyle v} & & \downarrow{\scriptstyle w} \\
M' & \xrightarrow{\ f'\ } & N'
\end{array}
$$

If f, f' are epimorphisms having the same kernel U, then, thanks to Theorem IV.3.3, in the commutative diagram

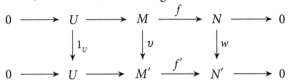

the right-hand square is cartesian, so, if the class of the lower sequence is $\mathbf{e} \in \operatorname{Ext}_A^1(N', U)$, then the class of the upper sequence is $\mathbf{e}w \in \operatorname{Ext}_A^1(N, U)$. The situation is completely dual with respect to cocartesian squares.

For instance, let A be the path algebra of the quiver $1 \leftarrow 3 \rightarrow 2$. One has short exact sequences $0 \rightarrow S_1 \rightarrow I_1 \rightarrow S_3 \rightarrow 0$ and $0 \rightarrow P_3 \rightarrow I_1 \oplus I_2 \rightarrow S_3 \rightarrow 0$ which are actually injective coresolutions of S_1, P_3, respectively (because S_3 is injective). Now, S_1 is a direct summand of $\operatorname{soc}P_3$, hence there exists an embedding $u : S_1 \rightarrow P_3$. On the other hand, $\operatorname{top}(I_1 \oplus I_2) = S_3^{(2)}$, it is a two-dimensional K-vector space having at least one simple (one-dimensional) submodule E distinct from $\operatorname{top}I_1$ and $\operatorname{top}I_2$. Therefore, mapping the top of I_1 onto E, as in Exercise IV.2.26, yields a commutative diagram

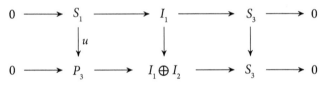

The left-hand square is cocartesian and, if the upper sequence is $\mathbf{e}$, then the lower one is $u\mathbf{e}$.

(d) Let $A = KQ/I$ be a bound quiver algebra. As seen in Example XI.2.9(e), given two points $i \neq j$ in Q, the dimension n of $\operatorname{Ext}_A^1(S_i, S_j)$ equals the number of arrows from i to j. Let these arrows be $\alpha_1, \cdots, \alpha_n$. Each α_k defines a uniserial module $U_k = \binom{S_i}{S_j}$ given as bound representation by $U_k(i) = U_k(j) = K$, and $U_k(\ell) = 0$, for any point $\ell \neq i, j$, also $U_k(\alpha_k) = 1_K$ and $U_k(\beta) = 0$, for any arrow $\beta \neq \alpha_k$. We get n short exact sequences $\mathbf{e}_k : 0 \rightarrow S_j \rightarrow U_k \rightarrow S_i \rightarrow 0$, with $1 \leqslant k \leqslant n$. It is easy to see that $k \neq k'$ implies $\operatorname{Hom}_A(U_k, U_{k'}) = 0$ and, therefore, $U_k \not\cong U_{k'}$. Hence, the n short exact sequences $\mathbf{e}_k$ are pairwise nonequivalent. That is, they form a basis of the extension space $\operatorname{Ext}_A^1(S_i, S_j)$.

(e) Let K be a field and A the path algebra of the Kronecker quiver

$$\underset{1}{\circ} \overset{\alpha}{\underset{\beta}{\leftleftarrows}} \underset{2}{\circ}$$

Consider the A-module H_α given by the representation

$$K \mathrel{\underset{0}{\overset{1}{\leftleftarrows}}} K$$

We have proved in Example VII.4.10 that H_α is uniserial of length two and, in particular, is indecomposable. Also, it follows from

Example XI.2.9(e) that $\mathrm{Ext}_A^1(H_\alpha, H_\alpha) \cong K$. Thus, a basis in this extension space contains only one vector. We exhibit a nonsplit short exact sequence representing it. Consider the A-module H given by the representation:

$$K^2 \;\; \underset{\begin{pmatrix} 0 & 0 \\ 1 & 0 \end{pmatrix}}{\overset{\begin{pmatrix} 1 & 0 \\ 0 & 1 \end{pmatrix}}{\rightleftarrows}} \;\; K^2$$

It is easily verified, for instance, using Exercise IV.2.29, that we have a short exact sequence

$$0 \to H_\alpha \xrightarrow{f} H \xrightarrow{g} H_\alpha \to 0$$

where $f = (f_1 = \binom{0}{1}, f_2 = \binom{0}{1}))$ and $g = (g_1 = (1\ 0), g_2 = (1\ 0))$. This sequence is not split because H is indecomposable, as seen in Example VIII.4.15(d).

XI.3.2 Longer exact sequences and Ext^n

The generalisation of Theorem XI.3.4 to higher extension modules, due to Yoneda, interprets the elements of Ext^n as equivalence classes of longer exact sequences. As in the previous subsection, A denotes an algebra over a commutative ring K, and our modules are A-modules.

XI.3.7 Definition. Let L, M be A-modules and $n > 0$ an integer, an **n-extension** of L by M is an exact sequence of the form

$$\mathbf{e} : 0 \longrightarrow L \xrightarrow{f_n} E_n \xrightarrow{f_{n-1}} \cdots \xrightarrow{f_1} E_1 \xrightarrow{f_0} M \longrightarrow 0.$$

Thus, a short exact sequence is a 1-extension.
Given an n-extension of L by X

$$\mathbf{e} : 0 \longrightarrow L \xrightarrow{f_n} E_n \xrightarrow{f_{n-1}} \cdots \xrightarrow{f_1} E_1 \xrightarrow{f_0} X \longrightarrow 0.$$

and an m-extension of X by M

$$\mathbf{e}' : 0 \longrightarrow X \xrightarrow{g_m} F_m \xrightarrow{g_{m-1}} \cdots \xrightarrow{g_1} F_1 \xrightarrow{g_0} M \longrightarrow 0.$$

we can form a composite exact sequence

which is an $(n + m)$-extension of L by M. This operation is called the **splicing** or **Yoneda product** of $\mathbf{e}$ and $\mathbf{e}'$, and the resulting sequence is denoted by $\mathbf{ee}'$.

Conversely, any n-extension, with $n > 1$, can be considered as the splicing of shorter extensions. In particular, it may be obtained as the splicing of n short exact sequences (=1-extensions). Splicing sequences is plainly an associative operation.

Following the approach of Subsection IX.4.1, we define an equivalence relation between n-extensions of L by M. Given two such extensions

$$\mathbf{e} : 0 \longrightarrow L \xrightarrow{f_n} E_n \xrightarrow{f_{n-1}} \cdots \xrightarrow{f_1} E_1 \xrightarrow{f_0} M \longrightarrow 0$$

and

$$\mathbf{e}' : 0 \longrightarrow L \xrightarrow{f'_n} E'_n \xrightarrow{f'_{n-1}} \cdots \xrightarrow{f'_1} E'_1 \xrightarrow{f'_0} M \longrightarrow 0,$$

we define $\mathbf{e} \curvearrowright \mathbf{e}'$ (or, equivalently, $\mathbf{e}' \curvearrowleft \mathbf{e}$) to mean that there exist n morphisms $u_i : E_i \to E'_i$, with i such that $1 \leqslant i \leqslant n$, making commutative the diagram with exact rows

$$
\begin{array}{ccccccccccc}
\mathbf{e} : & 0 & \longrightarrow & L & \xrightarrow{f_n} & E_n & \longrightarrow & \cdots & \xrightarrow{f_1} & E_1 & \xrightarrow{f_0} & M & \longrightarrow & 0 \\
 & & & \downarrow{\scriptstyle 1_L} & & \downarrow{\scriptstyle u_n} & & & & \downarrow{\scriptstyle u_1} & & \downarrow{\scriptstyle 1_M} & & \\
\mathbf{e}' : & 0 & \longrightarrow & L & \xrightarrow{f'_n} & E'_n & \longrightarrow & \cdots & \xrightarrow{f'_1} & E'_1 & \xrightarrow{f'_0} & M & \longrightarrow & 0
\end{array}
$$

If $n = 1$, this relation on 1-extensions, that is, on short exact sequences is the same as in the previous subsection and, in particular, is an equivalence (because u_1 is an isomorphism in this case).

If $n \geq 2$, there is no reason why the u_i should be isomorphisms. Hence, the relation $\curvearrowright$ is reflexive and transitive, but generally not symmetric. Let $\sim$ be the smallest equivalence relation containing $\curvearrowright$, that is, the intersection of all equivalence relations containing it. It is called its **symmetric closure**. Routine set-theoretical considerations imply that, given $\mathbf{e}, \mathbf{e}'$ as above, $\mathbf{e} \sim \mathbf{e}'$ if and only if there exist an integer $k \geq 0$ and n-extensions $\mathbf{e_i}$, with $0 \leqslant i \leqslant 2k$, of L by M such that

$$\mathbf{e} = \mathbf{e_0} \curvearrowright \mathbf{e_1} \curvearrowleft \mathbf{e_2} \curvearrowright \cdots \curvearrowright \mathbf{e_{2k-1}} \curvearrowleft \mathbf{e_{2k}} = \mathbf{e}'.$$

In contrast to the situation with short exact sequences, this definition of equivalence involves no direct relation between $\mathbf{e}$ and $\mathbf{e}'$. Consequently, it is hard, in practice, to verify whether two n-extensions are equivalent or not.

Let $\mathcal{E}^n(M, N)$ denote the set of equivalence classes of n-extensions of N by M. The equivalence class of an extension $\mathbf{e}$ is denoted by $[\mathbf{e}]$. We prove that equivalence is compatible with splicing.

XI.3.8 Lemma. *Given integers $n, m > 0$, let $\mathbf{e_1}, \mathbf{e_2}$ be n-extensions of L by X, and $\mathbf{e'_1}, \mathbf{e'_2}$ be m-extensions of X by M. If $\mathbf{e_1} \sim \mathbf{e_2}$ and $\mathbf{e'_1} \sim \mathbf{e'_2}$, then $\mathbf{e'_1 e_1} \sim \mathbf{e'_2 e_2}$.*

Proof. Assume first that $e_1' = e_2'$ is the extension

$$0 \longrightarrow L \xrightarrow{f_m} E_m \xrightarrow{f_{m-1}} \cdots \xrightarrow{f_1} E_1 \xrightarrow{f_0} X \longrightarrow 0$$

while e_1 and e_2 are given respectively by

$$0 \longrightarrow X \xrightarrow{g_n} F_n \xrightarrow{g_{n-1}} \cdots \xrightarrow{g_1} F_1 \xrightarrow{g_0} M \longrightarrow 0$$

and

$$0 \longrightarrow X \xrightarrow{h_n} G_n \xrightarrow{h_{n-1}} \cdots \xrightarrow{h_1} G_1 \xrightarrow{h_0} M \longrightarrow 0.$$

If $e_1 \curvearrowright e_2$, then there exists a commutative diagram with exact rows

The diagram obtained by splicing

is commutative with exact rows. Hence, $e_1' e_1 \curvearrowright e_2' e_2$ in this case. The definition of symmetric closure implies that if $e_1 \sim e_2$, then $e_1' e_1 \sim e_2' e_2$. This finishes the proof in case $e_1' = e_2'$.

One proves similarly the statement in case $e_1 = e_2$ and the general case follows from a transitivity argument: $e_1' e_1 \sim e_1' e_2 \sim e_2' e_2$. $\qquad\square$

We prove that the correspondence of objects $(M, N) \mapsto \mathcal{E}^n(M, N)$ extends to a set-valued bifunctor $\mathcal{E}(-, ?)$. Given an n-extension

$$e : 0 \longrightarrow L \xrightarrow{f_n} E_n \xrightarrow{f_{n-1}} \cdots \xrightarrow{f_1} E_1 \xrightarrow{f_0} M \longrightarrow 0$$

and a morphism $u : L \to L'$, we define an n-extension ue as follows. Let $f_{n-1} = jp$ be the canonical factorisation of f_{n-1} through its image X. Then $\mathbf{e}$ is the splicing of the short exact sequence $\mathbf{e}' : 0 \to L \xrightarrow{f_n} E_n \xrightarrow{p} X \to 0$ with the $(n-1)$-extension

$$\mathbf{e}'' : 0 \to X \xrightarrow{j} E_{n-1} \xrightarrow{f_{n-2}} \cdots \xrightarrow{f_1} E_1 \xrightarrow{f_0} M \to 0.$$

The amalgamated sum of f_n with u yields a commutative diagram with exact rows

Splicing the lower exact sequence with $\mathbf{e}''$ yields the n-extension

which we define to be ue. In other words, writing $\mathbf{e} = \mathbf{e}'\mathbf{e}''$, the definition says that $u\mathbf{e} = (u\mathbf{e}')\mathbf{e}''$. This definition of left multiplication is the obvious generalisation of the one in $\mathcal{E}(M, N)$, see Subsection XI.3.1. We must prove that the correspondence $\mathbf{e} \mapsto u\mathbf{e}$ induces a map $\mathcal{E}^n(M, L) \to \mathcal{E}^n(M, L')$ given by $[\mathbf{e}] \mapsto [u\mathbf{e}]$, for $[\mathbf{e}] \in \mathcal{E}^n(M, L)$.

XI.3.9 Lemma. *With this notation, if* $\mathbf{e}, \mathbf{f}$ *are n-extensions, then* $\mathbf{e} \sim \mathbf{f}$ *implies* $u\mathbf{e} \sim u\mathbf{f}$.

Proof. Assume first that $\mathbf{e} \curvearrowright \mathbf{f}$. We have a commutative diagram with exact rows

Let $f_{n-1} = jp$ and $g_{n-1} = iq$ be the canonical factorisations through the images X and Y respectively. Write $\mathbf{e} = \mathbf{e}'\mathbf{e}''$, where $\mathbf{e}' : 0 \to L \xrightarrow{f_n} E_n \xrightarrow{p} X \to 0$ and $\mathbf{e}'' : 0 \to X \xrightarrow{j} E_{n-1} \to \cdots \to E_0 \xrightarrow{f_0} M \to 0$. Similarly $\mathbf{f} = \mathbf{f}'\mathbf{f}''$, where $\mathbf{f}' : 0 \to L \xrightarrow{g_n} F_n \xrightarrow{q} Y \to 0$ and $\mathbf{f}'' : 0 \to Y \xrightarrow{i} F_{n-1} \to \cdots \to F_0 \xrightarrow{g_0} M \to 0$. Computing amalgamated sums yields a commutative diagram with exact rows

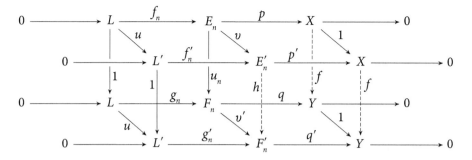

where $f : X \to Y$ is obtained by passing to cokernels in the two exact sequences on the back, that is, it is the unique morphism such that $fp = qu_n$.

We claim that there exists a morphism $h : E_n' \to F_n'$ making the whole diagram commute. We have $(v'u_n)f_n = v'(u_nf_n) = v'g_n = g_n'u$. But E_n' is the amalgamated sum of u and f_n. Hence, there exists a unique morphism $h : E_n' \to F_n'$ such that $hf_n' = g_n'$ and $hv = v'u_n$.

There remains to prove that $q'h = fp'$. Due to definition of h, this amounts to proving that $q'hv = fp'v$ and $q'hf_n' = fp'f_n'$. But $p'f_n' = 0$ and $q'hf_n' = q'g_n' = 0$, so the second equality is satisfied. On the other hand, $q'hv = q'v'u_n = qu_n = fp = fp'v$. This establishes our claim.

Splicing with $\mathbf{e}''$, $\mathbf{f}''$, respectively, gives $u\mathbf{e} \curvearrowright u\mathbf{f}$. The statement of the lemma then follows from the definition of the equivalence $\sim$. $\qquad\square$

We define similarly right multiplication by a morphism: given an n-extension

$$\mathbf{e} : 0 \longrightarrow L \xrightarrow{f_n} E_n \xrightarrow{f_{n-1}} \cdots \xrightarrow{f_1} E_1 \xrightarrow{f_0} M \longrightarrow 0$$

and a morphism $w : M' \to M$, let $f_1 = jp$ be the factorisation of f_1 through its image $Z = \mathrm{Im}\, f_1$. The fibered product of f_0 and w yields a commutative diagram with exact rows

$$
\begin{array}{ccccccccc}
0 & \longrightarrow & Z & \xrightarrow{j'} & E_1' & \xrightarrow{f_0} & M' & \longrightarrow & 0 \\
 & & \downarrow{\scriptstyle 1_Z} & & \downarrow{\scriptstyle v} & & \downarrow{\scriptstyle w} & & \\
0 & \longrightarrow & Z & \xrightarrow{j} & E_1 & \xrightarrow{f_0} & M & \longrightarrow & 0
\end{array}
$$

and finally splicing gives the extension $\mathbf{e}w$

$$0 \to L \xrightarrow{f_n} E_n \to \cdots \xrightarrow{f_2} E_2 \xrightarrow{f_1' = j'p} E_1' \xrightarrow{f_0'} M' \to 0$$

with p and j' factoring through Z, and 0's below.

Right multiplication induces a map from $\mathcal{E}^n(M, L)$ to $\mathcal{E}^n(M', L)$ given by $[e] \mapsto [ew]$, for $[e] \in \mathcal{E}^n(M, L)$.

XI.3.10 Lemma. *For any integer* $n > 0$, *the correspondence of objects* $(M, N) \mapsto \mathcal{E}^n(M, N)$ *induces a set-valued bifunctor, covariant in its second variable and contravariant in the first.*

Proof. We have to prove that, if e is an n-extension from N to M and $u : N \to N', w : M' \to M$ are morphisms, then $(ue)w \sim u(ew)$. If $n = 1$, this is Lemma XI.3.2. If $n \geq 2$, let $e = e_1 e' e_2$ where e_1, e_2 are short exact sequences and e' is an $(n-2)$-extension (empty if $n = 2$). Then we have $[(ue)w] = [(u(e_1 e' e_2))w] = [((ue_1)(e' e_2))w]$, because of Lemma XI.3.9, and the latter equals $[(ue_1)((e' e_2)w)] = [(ue_1)(e'(e_2 w))] = [u(e_1(e'(e_2 w)))] = [u(ew)]$. $\square$

We finally prove that equivalence classes of n-extensions are in bijection with elements of the nth-extension module.

XI.3.11 Theorem. *Let* M, N *be A-modules and* $n > 0$ *an integer. There exists a bijection* $\mathcal{E}^n(M, N) \to \operatorname{Ext}_A^n(M, N)$, *functorial in both variables.*

Proof. We use induction on n. In view of Theorem XI.3.4, we may assume that $n > 1$. There exists a short exact sequence

$$\mathbf{p} : 0 \to L \xrightarrow{f} P \xrightarrow{g} M \to 0$$

with projective middle term P. Dimension shifting gives a functorial isomorphism $\operatorname{Ext}_A^n(M, N) \cong \operatorname{Ext}_A^{n-1}(L, N)$ and the induction hypothesis a functorial bijection $\operatorname{Ext}_A^{n-1}(L, N) \cong \mathcal{E}^{n-1}(L, N)$, so we need to construct a functorial bijection $\varphi : \mathcal{E}^{n-1}(L, N) \to \mathcal{E}^n(M, N)$.

Let $e : 0 \to N \to E_n \to \cdots \to E_2 \to L \to 0$ be an $(n-1)$-extension of L by N. Splicing with $\mathbf{p}$ yields an n-extension $\mathbf{ep}$ of N by M :

Due to Lemma XI.3.8, if $e \sim e'$, then $\mathbf{ep} \sim \mathbf{e'p}$. We thus have a map $\varphi : \mathcal{E}^{n-1}(L, N) \to \mathcal{E}^n(M, N)$, given by $[e] \mapsto [\mathbf{ep}]$ for $[e] \in \mathcal{E}^{n-1}(L, N)$. We prove that φ is bijective.

(a) *φ is surjective.* Given an n-extension of N by L,

$$e : 0 \longrightarrow N \xrightarrow{f_n} E_n \xrightarrow{f_{n-1}} \cdots \xrightarrow{f_1} E_1 \xrightarrow{f_0} M \longrightarrow 0,$$

let $f_k = j_k p_k$ be the canonical factorisation of each f_k through its image L_k, with $0 \leqslant k \leqslant n, L_n = N, L_0 = M$. Then $\mathbf{e}$ is the splicing of short exact sequences $\mathbf{e_k} : 0 \to L_k \xrightarrow{j_k} E_k \xrightarrow{p_{k-1}} L_{k-1} \to 0$. We claim that there exists an $(n-1)$-extension of N by L, of the form

$$\mathbf{f} : 0 \to N \xrightarrow{f_n} E_n \xrightarrow{f_{n-1}} \cdots \to E_3 \xrightarrow{f_2'} F_2 \xrightarrow{f_1'} L \to 0$$

such that $\mathbf{e} = \mathbf{fp}$. We construct $\mathbf{f}$ inductively starting from the right part of $\mathbf{e}$.

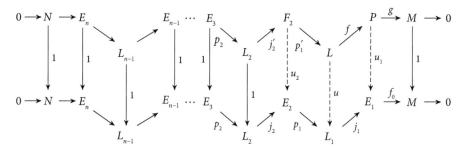

Projectivity of P gives $u_1 : P \to E_1$ such that $f_0 u_1 = g$. Passing to kernels, we get $u : L \to L_1$ such that $j_1 u = u_1 f$. Let F_2 be the fibered product of u and p_1. Then the morphism $p_1' : F_2 \to L$ arising by taking fibered product is an epimorphism. Because of Theorem IV.3.4, the kernel of p_1' is isomorphic to the kernel L_2 of p_1. Let $j_2' : L_2 \to F_2$ be the kernel morphism of p_1', then $u_2 j_2' = j_2$. Taking $1_{E_k} : E_k \to E_k$ and $1_{L_k} : L_k \to L_k$ for $k \geq 3$ completes the construction of the upper exact sequence in the diagram, and its truncation

$$\mathbf{f} : 0 \longrightarrow N \xrightarrow{f_n} E_n \xrightarrow{f_{n-1}} \cdots \longrightarrow E_3 \xrightarrow{f_2' = j_2' p_2} F_2 \xrightarrow{f_1' = p_1'} L \longrightarrow 0$$

satisfies $\mathbf{e} = \mathbf{fp}$. This completes the proof of surjectivity.

An important remark before proving injectivity. Assume that $u_1' : P \to E_1$ is a morphism such that $f_0 u_1' = g = f_0 u_1$. Then $f_0(u_1' - u_1) = 0$ implies the existence of $h : P \to L_1$ such that $u_1' = u_1 + j_1 h$. Projectivity of P gives $k : P \to E_2$ such that $h = p_1 k$ and then $u_1' = u_1 + j_1 p_1 k$. We claim that the morphism $u_2' = u_2 + kfp_1'$ satisfies $f_1 u_2' = u_1' fp_1'$ and $u_2' j_2' = j_2$, that is, the u_i' make the diagram commute.

Indeed, $u_2' j_2' = (u_2 + kfp_1')j_2' = u_2 j_2' = j_2$, because $p_1' j_2' = 0$. Also, $f_1 u_2' = f_1 u_2 + f_1 kfp_1' = j_1 p_1 u_2 + j_1 p_1 kfp_1' = j_1 up_1' + j_1 hfp_1' = (u_1 + j_1 h)fp_1' = u_1' fp_1'$, as required.

(b) φ *is injective.* Because of the definition of the equivalence, it suffices to prove that, if $\mathbf{fp} \curvearrowright \mathbf{e}, \mathbf{f'p} \curvearrowright \mathbf{e'}$ and $\mathbf{e} \curvearrowright \mathbf{e'}$ then $\mathbf{f} \curvearrowright \mathbf{f'}$. Let

$$\mathbf{e} : 0 \to N \to E_n \to \cdots \to E_3 \to E_2 \to E_1 \to M \to 0$$

$$\mathbf{f} : 0 \to N \to E_n \to \cdots \to E_3 \to F_2 \to L \to 0$$

$$\mathbf{e'} : 0 \to N \to E_n' \to \cdots \to E_3' \to E_2' \to E_1' \to M \to 0$$

$$\mathbf{f'} : 0 \to N \to E_n' \to \cdots \to E_3' \to F_2' \to L \to 0$$

be the extensions under consideration. We have commutative diagrams with exact rows

$$
\mathbf{fp}: \quad 0 \longrightarrow N \longrightarrow E_n \longrightarrow \cdots \longrightarrow E_3 \longrightarrow F_2 \longrightarrow P \longrightarrow M \longrightarrow 0
$$

with vertical maps 1, 1, 1, u_2, u_1, 1

$$
\mathbf{e}: \quad 0 \longrightarrow N \longrightarrow E_n \longrightarrow \cdots \longrightarrow E_3 \longrightarrow E_2 \longrightarrow E_1 \longrightarrow M \longrightarrow 0
$$

$$
\mathbf{f'p}: \quad 0 \longrightarrow N \longrightarrow E'_n \longrightarrow \cdots \longrightarrow E'_3 \longrightarrow F'_2 \longrightarrow P \longrightarrow M \longrightarrow 0
$$

with vertical maps 1, 1, 1, u'_2, u'_1, 1

$$
\mathbf{e'}: \quad 0 \longrightarrow N \longrightarrow E'_n \longrightarrow \cdots \longrightarrow E'_3 \longrightarrow E'_2 \longrightarrow E'_1 \longrightarrow M \longrightarrow 0
$$

$$
\mathbf{e}: \quad 0 \longrightarrow N \longrightarrow E_n \longrightarrow \cdots \longrightarrow E_3 \longrightarrow E_2 \longrightarrow E_1 \longrightarrow M \longrightarrow 0
$$

with vertical maps 1, v_n, v_3, v_2, v_1, 1

$$
\mathbf{e'}: \quad 0 \longrightarrow N \longrightarrow E'_n \longrightarrow \cdots \longrightarrow E'_3 \longrightarrow E'_2 \longrightarrow E'_1 \longrightarrow M \longrightarrow 0
$$

We claim that there exist a morphism $w : F_2 \to F'_2$ and a commutative diagram with exact rows

$$
\mathbf{f}: \quad 0 \longrightarrow N \longrightarrow E_n \longrightarrow \cdots \longrightarrow E_3 \longrightarrow F_2 \longrightarrow L \longrightarrow 0
$$

with vertical maps 1, v_n, v_3, w, 1

$$
\mathbf{f'}: \quad 0 \longrightarrow N \longrightarrow E'_n \longrightarrow \cdots \longrightarrow E'_3 \longrightarrow F'_2 \longrightarrow L \longrightarrow 0
$$

As before, consider the images of the morphisms in the right-hand portions of each of the first three commutative diagrams, using a similar notation

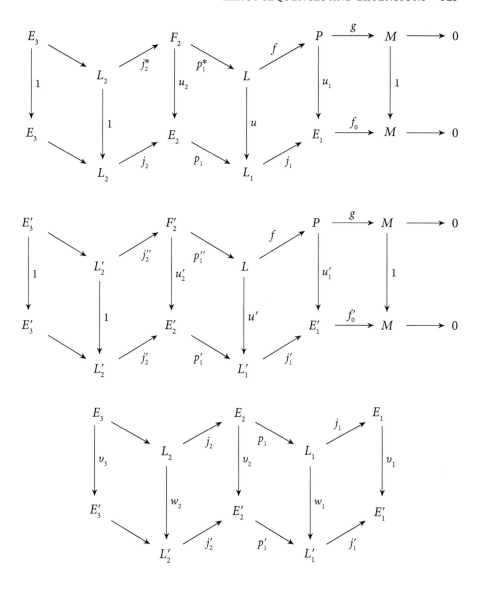

Because of the remark at the end of the surjectivity part, we may assume that $u_1' = v_1 u_1$. Hence, $j_1' w_1 u = v_1 j_1 u = v_1 u_1 f = u_1' f = j_1' u'$. Because j_1' is a monomorphism, we have $w_1 u = u'$. Therefore, $u' p_1^* = w_1 u p_1^* = w_1 p_1 u_2 = p_1' v_2 u_2$. Now, F_2' is the fibered product of u' and p_1', hence there exists a unique $w : F_2 \to F_2'$ such that $p_1'' w = p_1^*$ and $u_2' w = v_2 u_2$.

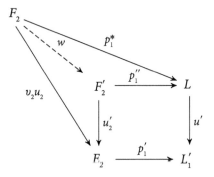

We must prove that the following diagram is commutative

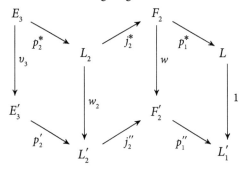

that is, $wj_2^* = j_2'' w_2$. Due to definition of w, this amounts to proving that $p_1'' wj_2^* = p_1'' j_2'' w_2$ and $u_2' wj_2^* = u_2' j_2'' w_2$. In the first equality, both sides vanish: $p_1' wj_2^* = p_1^* j_2^* = 0$ and $p_1'' j_2'' = 0$. For the second, $u_2' wj_2^* = v_2 u_2 j_2^* = v_2 j_2 = j_2' w_2 = u_2' j_2'' w_2$. The proof is complete. $\qquad\square$

Exactly as the first extensions, the sets $\mathcal{E}^n(M, N)$ of n-extensions can be given a K-module structure making it isomorphic to $\mathrm{Ext}_A^n(M, N)$. We refrain from describing it because of space considerations. As pointed out, if $n > 1$, it is hard to determine whether two n-extensions are equivalent or not, and so to compute full sets of representatives of equivalence classes. We give easy examples.

XI.3.12 Examples.

(a) Let A be a principal ideal domain, $p \in A$ an irreducible element, $R = A/p^2A$ and $S_R = A/pA$. Because of Example XI.2.9, there exists, for any integer $n > 0$, an n-extension

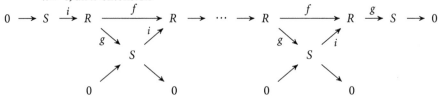

where $i : S \rightarrow R$ denotes the inclusion, $g : R \rightarrow S$ multiplication by p and $f = ig$. We also know that $\mathrm{Ext}_R^n(S, S) \cong A/pA$, which is a one-dimensional A/pA-vector space. Therefore, any n-extension which is not equivalent to zero is a basis of the space. Using the explicit description of dimension shifting, see Lemma XI.2.1, we see that, in the sequence of isomorphisms $\mathrm{Ext}_R^n(S, S) \cong \cdots \cong \mathrm{Ext}_R^1(S, S)$, the n-extension above corresponds to the short exact sequence $0 \rightarrow S \xrightarrow{i} R \xrightarrow{p} S \rightarrow 0$ which is nonsplit and hence nonzero as an extension. Therefore, the n-extension above is not equivalent to zero, and represents a basis vector of $\mathrm{Ext}_R^n(S, S)$.

(b) Let K be a field and A the Nakayama K-algebra given by the quiver

$$5 \xrightarrow{\alpha} 4 \xrightarrow{\beta} 3 \xrightarrow{\gamma} 2 \xrightarrow{\delta} 1$$

bound by the ideal $\langle \alpha\beta, \beta\gamma, \gamma\delta \rangle$. One easily computes a projective resolution of S_5:

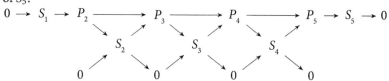

$$0 \rightarrow S_1 \rightarrow P_2 \longrightarrow P_3 \longrightarrow P_4 \longrightarrow P_5 \rightarrow S_5 \rightarrow 0$$

Using dimension shifting as in Example (a) above, we get:

$$\mathrm{Ext}_A^4(S_5, S_1) \cong \mathrm{Ext}_A^3(S_4, S_1) \cong \mathrm{Ext}_A^2(S_3, S_1) \cong \mathrm{Ext}_A^1(S_2, S_1).$$

The dimension of the latter equals the number of arrows from 2 to 1, that is, one, see Example XI.2.9(e). So, $\mathrm{Ext}_A^4(S_5, S_1)$ is one-dimensional, and, as in (a) above, the 4-extension

$$0 \rightarrow S_1 \rightarrow P_2 \rightarrow P_3 \rightarrow P_4 \rightarrow P_5 \rightarrow S_5 \rightarrow 0$$

represents a basis vector of this space.

(c) Here is a somewhat different example. Let A be the K-algebra given by the quiver Q

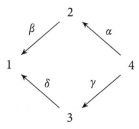

bound by the ideal $\langle \alpha\beta, \gamma\delta \rangle$. We wish to find long extensions representing a basis of the K-vector space $\mathrm{Ext}_A^2(S_4, S_1)$. Here, as usual, S_i, P_i, I_i represent respectively the simple module, the indecomposable projective and the indecomposable injective corresponding to a point $i \in Q_0$. We need additional notation: to $i \in Q_0$, we associate the projective cover morphism $q_i : P_i \to S_i$ (thus q_1 is the identity, because S_1 is projective), also, because $P_4 = {}^{4}_{23}, P_2 = {}^{2}_{1}$ and $P_3 = {}^{3}_{1}$, we denote by $i' : S_1 \to P_2, i'' : S_1 \to P_3$, $i_2 : S_2 \to P_4$ and $i_3 : S_3 \to P_4$ the inclusions of the corresponding simple modules as socle factors of the projectives. Computing a projective resolution of the simple module S_4, we get

$$\mathbf{p}: \quad 0 \to S_1 \oplus S_1 \xrightarrow{d_2} P_2 \oplus P_3 \xrightarrow{d_1} P_4 \xrightarrow{d_0} S_4 \to 0$$

because S_1 is projective. Here, $d_0 = q_4, d_1 = (i_2 q_2 \ i_3 q_3)$ equals the composition $(i_2 \ i_3) \begin{pmatrix} q_2 & 0 \\ 0 & q_3 \end{pmatrix} : P_2 \oplus P_3 \to S_2 \oplus S_3 \hookrightarrow P_4$ and finally $d_2 = \begin{pmatrix} i' & 0 \\ 0 & i'' \end{pmatrix}$. Therefore, using dimension shifting, we have isomorphisms of K-vector spaces

$$\mathrm{Ext}^2_A(S_4, S_1) \cong \mathrm{Ext}^1_A(S_2 \oplus S_3, S_1) \cong \mathrm{Ext}^1_A(S_2, S_1) \oplus \mathrm{Ext}^1_A(S_3, S_1) \cong K^2$$

because $\dim_K \mathrm{Ext}^1_A(S_i, S_j)$ equals the number of arrows from i to j in Q. Thus, $\mathrm{Ext}^2_A(S_4, S_1)$ is two-dimensional over K. Its basis consists of two linearly independent vectors of the cohomology space, thus, two nonequivalent 2-extensions of S_1 by S_4.

Consider the projection morphisms $p_1 = (1\ 0), p_2 = (0\ 1)$ of $S_1 \oplus S_1$ on its first and second coordinate, respectively. Because $\mathrm{Hom}_A(S_1 \oplus S_1, S_1) \cong K \oplus K$, due to Lemma VII.4.4, $\{p_1, p_2\}$ is a basis of this Hom-space. Now consider the short exact sequence

$$\mathbf{p}': \quad 0 \to S_1 \oplus S_1 \xrightarrow{d_2} P_2 \oplus P_3 \xrightarrow{u} S_2 \oplus S_3 \to 0,$$

extracted from $\mathbf{p}$, where $u = \begin{pmatrix} q_2 & 0 \\ 0 & q_3 \end{pmatrix} : P_2 \oplus P_3 \to S_2 \oplus S_3$ is the epimorphism from $P_2 \oplus P_3$ onto the image of d_1. Evaluating $\mathrm{Hom}_A(-, S_1)$ on $\mathbf{p}'$ yields an exact sequence

$$\cdots \to \mathrm{Hom}_A(P_2 \oplus P_3, S_1) \to \mathrm{Hom}_A(S_1 \oplus S_1, S_1) \to \mathrm{Ext}^1_A(S_2 \oplus S_3, S_1) \to$$

$$\to \mathrm{Ext}^1{}_A(P_2 \oplus P_3, S_1) = 0$$

due to projectivity of $P_2 \oplus P_3$. On the other hand, $\mathrm{Hom}_A(P_2 \oplus P_3, S_1) = 0$: for, any morphism $P_2 \oplus P_3 \to S_1$ is surjective, because S_1 is simple, and hence induces an isomorphism of a top summand of $P_2 \oplus P_3$ with S_1, but this is impossible because $\mathrm{top}(P_2 \oplus P_3) = S_2 \oplus S_3$. We deduce an isomorphism $\mathrm{Hom}_A(S_1 \oplus S_1, S_1) \cong \mathrm{Ext}^1_A(S_2 \oplus S_3, S_1)$, so that the cohomology classes of p_1, p_2 constitute a basis of $\mathrm{Ext}^1_A(S_2 \oplus S_3, S_1)$ and, in particular, are linearly independent.

We construct the corresponding 2-extensions. Note that the indecomposable injective module at the point 2 is $I_2 = \frac{4}{2}$. We claim that the 2-extension corresponding to the projection p_1 is the lower sequence in the commutative diagram with exact rows

$$
\begin{array}{ccccccccccc}
\mathbf{p} : 0 & \longrightarrow & S_1 \oplus S_1 & \xrightarrow{\ d_2\ } & P_2 \oplus P_3 & \xrightarrow{\ d_1\ } & P_4 & \xrightarrow{\ d_0\ } & S_4 & \longrightarrow & 0 \\
& & \downarrow{\scriptstyle p_1 = (1\ 0)} & & \downarrow{\scriptstyle (1\ 0)} & & \downarrow{\scriptstyle f} & & \downarrow{\scriptstyle 1} & & \\
\mathbf{e}_1 : 0 & \longrightarrow & S_1 & \xrightarrow{\ i'\ } & P_2 & \xrightarrow{\ iq_2\ } & I_2 & \xrightarrow{\ q\ } & S_4 & \longrightarrow & 0
\end{array}
$$

where $i : S_2 \to I_2$ is the injection, and $q : I_2 \to S_4$ the projection, while $f : P_4 \to I_2$ is an epimorphism with kernel S_3. Thus, we have $fi_2 = i, fi_3 = 0$. Moreover, $qf = q_4 = d_0$.

As seen in the proof of Theorem XI.3.11, the 2-extension corresponding to a cohomology class is constructed inductively using the commutative square

$$
\begin{array}{ccc}
\mathrm{Ext}^2_A(S_4, S_1) & \dashrightarrow & \mathcal{E}^2(S_4, S_1) \\
\cong \uparrow & & \uparrow \cong \\
\mathrm{Ext}^1_A(S_2 \oplus S_3, S_1) & \xrightarrow{\ \cong\ } & \mathcal{E}^1(S_2 \oplus S_3, S_1)
\end{array}
$$

where the left vertical arrow arises from dimension shifting, the right vertical arrow from splicing and the lower horizontal arrow is the bijection of Theorem XI.3.4. Applying Corollary XI.3.5, the image of p_1 under the composition $\mathrm{Hom}_A(S_1 \oplus S_1, S_1) \cong \mathrm{Ext}^1_A(S_2 \oplus S_3, S_1) \cong \mathcal{E}^1(S_2 \oplus S_3, S_1)$ is the equivalence class $[p_1 \mathbf{p}']$ of the lower sequence in the commutative diagram with exact rows

$$
\begin{array}{ccccccccccc}
\mathbf{p}' : 0 & \longrightarrow & S_1 \oplus S_1 & \xrightarrow{\ d_2\ } & P_2 \oplus P_3 & \xrightarrow{\ u\ } & S_2 \oplus S_3 & \longrightarrow & 0 \\
& & \downarrow{\scriptstyle p_1} & & \downarrow{\scriptstyle v} & & \downarrow{\scriptstyle 1} & & \\
p_1 \mathbf{p}' : 0 & \longrightarrow & S_1 & \xrightarrow{\binom{i'}{0}} & P_2 \oplus S_3 & \xrightarrow{\ w\ } & S_2 \oplus S_3 & \longrightarrow & 0
\end{array}
$$

where $v = \begin{pmatrix} 1 & 0 \\ 0 & q_3 \end{pmatrix} : P_2 \oplus P_3 \to P_2 \oplus S_3$, and $w = \begin{pmatrix} q_2 & 0 \\ 0 & 1 \end{pmatrix} : P_2 \oplus S_3 \to S_2 \oplus S_3$. The lower sequence is then transferred to $\mathcal{E}^2(S_4, S_1)$ by means of splicing with the short exact sequence $0 \to S_2 \oplus S_3 \xrightarrow{(i_2\ i_3)} P_4 \xrightarrow{d_0} S_4 \to 0$. This yields the 2-extension corresponding to the class of p_1, namely

$$
\mathbf{e}'_1 : 0 \to S_1 \xrightarrow{\binom{i'}{0}} P_2 \oplus S_3 \xrightarrow{(i_2 q_2\ i_3)} P_4 \xrightarrow{d_0} S_4 \to 0.
$$

But the latter sequence is equivalent to $\mathbf{e}_1$, because of the commutative diagram with exact rows

$$\mathbf{e}_1' : 0 \longrightarrow S_1 \xrightarrow{\binom{i}{0}} P_2 \oplus S_3 \xrightarrow{(i_2 q_2\ i_3)} P_4 \xrightarrow{d_0} S_4 \longrightarrow 0$$

with vertical maps 1, $(1\ 0)$, f, 1 down to

$$\mathbf{e}_1 : 0 \longrightarrow S_1 \xrightarrow{i'} P_2 \xrightarrow{iq_2} I_2 \xrightarrow{q} S_4 \longrightarrow 0$$

This establishes our claim that the 2-extension $\mathbf{e}_1$ corresponds to the cohomology class of the morphism p_1. In exactly the same way, one proves that the 2-extension corresponding to the cohomology class of p_2 is given by

$$\mathbf{e}_2 : 0 \to S_1 \to P_3 \to I_3 \to S_4 \to 0$$

with the obvious morphisms.

Exercises of Section XI.3

1. Give a full set of representatives for the equivalence classes in each of the extension sets:
   (a) $\mathrm{Ext}_{\mathbb{Z}}^1(\mathbb{Z}/p\mathbb{Z}, \mathbb{Z}/q\mathbb{Z})$, where p, q are distinct prime integers.
   (b) $\mathrm{Ext}_{\mathbb{Z}}^1(\mathbb{Z}/p^2\mathbb{Z}, \mathbb{Z}/p\mathbb{Z})$, where p is a prime integer.
   (c) $\mathrm{Ext}_{\mathbb{Z}}^1(\mathbb{Z}/p^2\mathbb{Z}, \mathbb{Z})$, where p is a prime integer.
   (d) $\mathrm{Ext}_{\mathbb{Z}}^1(\mathbb{Z}/p\mathbb{Z}, \mathbb{Z}/p^2\mathbb{Z})$, where p is a prime integer.
   (e) $\mathrm{Ext}_{\mathbb{Z}}^1(\mathbb{Z}/p^2\mathbb{Z}, \mathbb{Z}/p^2\mathbb{Z})$, where p is a prime integer.
   (f) $\mathrm{Ext}_{\mathbb{Z}}^1(\mathbb{Z}/p^2q\mathbb{Z}, \mathbb{Z}/pq^2\mathbb{Z})$, where p, q are distinct prime integers.

2. Let p be an odd prime integer, prove that the following two extensions

$$0 \to \mathbb{Z} \xrightarrow{\mu_p} \mathbb{Z} \xrightarrow{f} \mathbb{Z}/p\mathbb{Z} \to 0 \text{ and } 0 \to \mathbb{Z} \xrightarrow{\mu_p} \mathbb{Z} \xrightarrow{g} \mathbb{Z}/p\mathbb{Z} \to 0$$

where μ_p is multiplication by p, and f, g are defined by $f(1) = 1 + p\mathbb{Z}, g(1) = 2 + p\mathbb{Z}$, respectively, are not equivalent.

3. Let A be a K-algebra and $\mathbf{e} : 0 \to L \to M \to N \to 0$ a split short exact sequence of A-modules. Prove that, for any morphisms $f : L \to L'$ and $g : N' \to N$, the short exact sequences $f\mathbf{e}$ and $\mathbf{e}g$ split.

4. Let A be a K-algebra and N_A, N_A' modules. Prove that the following conditions are equivalent:
   (a) $\mathrm{Ext}_A^1(N, N') = 0$.
   (b) For any short exact sequence $0 \to L \xrightarrow{f} M \xrightarrow{g} N \to 0$, the morphism $\mathrm{Hom}_A(f, N')$ is surjective.
   (c) For any short exact sequence $0 \to L \xrightarrow{f} M \xrightarrow{g} N \to 0$, with M projective, the morphism $\mathrm{Hom}_A(f, N')$ is surjective.
   (d) For any short exact sequence $0 \to L' \xrightarrow{f'} M' \xrightarrow{g'} N' \to 0$, the morphism $\mathrm{Hom}_A(N, g')$ is surjective.

(e) For any short exact sequence $0 \to L' \xrightarrow{f'} M' \xrightarrow{g'} N' \to 0$, with M' injective, the morphism $\mathrm{Hom}_A(N, g')$ is surjective.

5. Let A be a K-algebra. Consider the diagram in $\mathrm{Mod}\,A$ with exact rows

$$
\begin{array}{ccccccccc}
\mathbf{e}_1 : & 0 & \longrightarrow & L_1 & \longrightarrow & M_1 & \longrightarrow & N_1 & \longrightarrow & 0 \\
& & & {\scriptstyle f}\downarrow & & & & {\scriptstyle h}\downarrow & & \\
\mathbf{e}_2 : & 0 & \longrightarrow & L_2 & \longrightarrow & M_2 & \longrightarrow & N_2 & \longrightarrow & 0
\end{array}
$$

Prove that there exists $g : M_1 \to M_2$ making the diagram commute if and only if $[f\mathbf{e}_1] = [\mathbf{e}_2 h]$.

6. Let $\varphi : \mathcal{E}(M, N) \to \mathrm{Ext}_A^1(M, N)$ be the bijection of Theorem XI.3.4 and $\mathbf{e} : 0 \to N \to E \to M \to 0$ an extension. Prove that:

(a) For any morphism $u : N \to N'$, we have $\varphi[u\mathbf{e}] = u\varphi[\mathbf{e}]$.

(b) For any morphisms $w : M' \to M$ and w_1 obtained by lifting w to a projective resolution of M'

$$
\begin{array}{ccccccc}
P_1 & \longrightarrow & P_0 & \longrightarrow & M' & \longrightarrow & 0 \\
{\scriptstyle w_1}\downarrow & & \downarrow & & {\scriptstyle w}\downarrow & & \\
0 \longrightarrow N & \longrightarrow & E & \longrightarrow & M & \longrightarrow & 0
\end{array}
$$

we have $\varphi[\mathbf{e}w] = \varphi[\mathbf{e}]w_1$.

(c) Deduce that, if u, w are as above, then $\varphi[(u\mathbf{e})w] = \varphi[u(\mathbf{e}w)]$.

7. Let A be a K-algebra, $\mathbf{e} : 0 \to N \xrightarrow{f} E \xrightarrow{g} M \to 0$, $\mathbf{e}' : 0 \to N \xrightarrow{f} E' \xrightarrow{g'} M \to 0$ extensions in $\mathrm{Mod}\,A$ of N by M and $\mathbf{e} \oplus \mathbf{e}'$ denote the extension

$$
0 \to N \oplus N \xrightarrow{\begin{pmatrix} f & 0 \\ 0 & f' \end{pmatrix}} E \oplus E' \xrightarrow{\begin{pmatrix} g & 0 \\ 0 & g' \end{pmatrix}} M \oplus M' \to 0
$$

Let again $\varphi : \mathcal{E}(M, N) \to \mathrm{Ext}_A^1(M, N)$ be the bijection of Theorem XI.3.4. Prove that

$$
\varphi[\mathbf{e} \oplus \mathbf{e}'] = \begin{pmatrix} \varphi[\mathbf{e}] & 0 \\ 0 & \varphi[\mathbf{e}'] \end{pmatrix}.
$$

8. Let A be a K-algebra. For an A-module X, let $\Delta = \begin{pmatrix} 1 \\ 1 \end{pmatrix} : X \to X \oplus X$ and $\nabla = (1 \quad 1) : X \oplus X \to X$. Let $\varphi : \mathcal{E}(M, N) \to \mathrm{Ext}_A^1(M, N)$ be the bijection of Theorem XI.3.4. Prove that:

(a) If $\mathbf{e}, \mathbf{e}'$ are extensions of N by M, then $\varphi[\nabla(\mathbf{e} \oplus \mathbf{e}')\Delta] = \varphi[\mathbf{e}] + \varphi[\mathbf{e}']$.

(b) Deduce that $\mathcal{E}(M, N)$ becomes an abelian group under the so-called **Baer sum**: $[\mathbf{e}] + [\mathbf{e}'] = [\nabla(\mathbf{e} \oplus \mathbf{e}')\Delta]$ and φ an isomorphism of groups.

(c) In this abelian group, $-[\mathbf{e}] = [(-1_N)\mathbf{e}] = [\mathbf{e}(-1_M)]$.

(d) Let p be a prime integer. Compute the negative in the abelian group $\mathcal{E}(\mathbb{Z}/p\mathbb{Z}, \mathbb{Z}/p\mathbb{Z})$ of the class of the exact sequence $0 \to \mathbb{Z}/p\mathbb{Z} \xrightarrow{f} \mathbb{Z}/p^2\mathbb{Z} \xrightarrow{g} \mathbb{Z}/p\mathbb{Z} \to 0$, where $f : 1 + p\mathbb{Z} \mapsto p + p^2\mathbb{Z}$ is the inclusion and $g = \operatorname{coker} f$.

9. Let K be a field, A given by the quiver

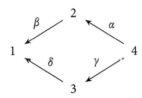

bound by the ideal $\langle \alpha\beta - \gamma\delta \rangle$, and the short exact sequence $\mathbf{e} : 0 \to P_1 \to P_2 \to S_2 \to 0$. Prove that there exists an epimorphism $v : \operatorname{rad} P_4 \to S_2$, then construct a short exact sequence representing $[\mathbf{e}v]$.

10. For each of the algebras of Exercise XI.2.16, give a representative for a basis vector of $\operatorname{Ext}_A^2(I_x, P_y)$.

11. Let E be a set, $\curvearrowright$ a relation on E that is reflexive and transitive and $\sim$ the smallest equivalence relation on E containing $\curvearrowright$. Prove that, for $x, y \in E$, the following conditions are equivalent:

(a) $x \sim y$.
(b) There exist an integer $k \geq 0$ and $x_1, x_2, \cdots, x_{2k-1} \in E$ such that

$$x = x_0 \curvearrowright x_1 \curvearrowleft x_2 \curvearrowright \cdots \curvearrowright x_{2k-1} \curvearrowleft x_{2k} = y.$$

(c) There exist an integer $\ell \geq 0$ and $y_1, y_2, \cdots, y_{2\ell-1} \in E$ such that

$$x = y_0 \curvearrowleft y_1 \curvearrowright y_2 \curvearrowleft \cdots \curvearrowleft y_{2k-1} \curvearrowright y_{2\ell} = y.$$

12. Let A be a K-algebra. Assume that $n \geq 2$ and consider the n-extensions $\mathbf{e} : 0 \to N \xrightarrow{f_n} E_n \to \cdots \to E_1 \xrightarrow{f_0} M \to 0$ and $\vartheta : 0 \to N \xrightarrow{1} N \to \cdots \to 0 \to M \xrightarrow{1} M \to 0$ (the class of ϑ plays the role of the zero in $\mathcal{E}^n(M, N)$) in $\operatorname{Mod} A$. Prove that:

(a) $\mathbf{e} \curvearrowright \vartheta$ if and only if f_n is a section.
(b) $\vartheta \curvearrowright \mathbf{e}$ if and only if f_0 is a retraction.

Chapter XII
Homological dimensions

XII.1 Introduction

Homological algebra allows to define invariants for modules or algebras. For instance, extension and torsion modules are algebraic invariants. We define another type of invariants, of a numerical nature. Homological dimensions measure how far a module, or an algebra, is far from a situation considered to be ideal (for the purpose of the definition, of course). For example, a module M is projective if and only if it admits a projective resolution of length zero $0 \to P_0 \to M \to 0$ (with P_0 projective) or, equivalently, if and only if $\mathrm{Ext}_A^1(M, -) = 0$. In order to measure how far a module M is from being projective, we define its so-called projective dimension, as the shortest length of a projective resolution of M, and we show that the latter equals the least integer n such that $\mathrm{Ext}_A^{n+1}(M, -) = 0$. If M has only infinitely long projective resolutions, then we say that its projective dimension is infinite. The injective and flat dimensions are defined similarly. It turns out that the supremum of the projective dimensions of all (right) modules equals the supremum of their injective dimension, and this common value is called the (right) global dimension of the algebra. Thus, the global dimension measures how far an algebra is from having all of its modules projective, or all of its modules injective, that is, of being semisimple. The left global dimension is defined similarly. In general, the right and the left global dimension of an algebra do not coincide, but they do if the algebra is right and left noetherian. Our final chapter is devoted to studying these dimensions and two classes of algebras which are extremal from this point of view, the first is the class of algebras of global dimension one, and the second consists of algebras of infinite global dimension. As always, K denotes a commutative ring and our algebras are K-algebras.

XII.2 Homological dimensions of modules

XII.2.1 Projective, injective and flat dimension

The projective, or injective, or flat, dimension of a module measure how far this module is from being projective, or injective, or flat, respectively.

XII.2.1 Definition. Let A be a K-algebra and M an A-module.
(a) The **projective dimension** $\mathrm{pd}M$ of M is the least nonnegative integer n, if it exists, such that there exists a finite projective resolution

$$0 \to P_n \to \cdots \to P_1 \to P_0 \to M \to 0.$$

An Introduction to Module Theory. Ibrahim Assem and Flávio U. Coelho, Oxford University Press.
© Ibrahim Assem and Flávio U. Coelho (2024). DOI: 10.1093/9780198904939.003.0013

(b) The **injective dimension** id M of M is the least nonnegative integer n, if it exists, such that there exists a finite injective coresolution

$$0 \to M \to I^0 \to I^1 \to \cdots \to I^n \to 0.$$

(c) The **flat dimension** fd$M$ of M is the least nonnegative integer n, if it exists, such that there exists a finite flat resolution

$$0 \to F_n \to \cdots \to F_1 \to F_0 \to M \to 0.$$

If there is no such finite (co)resolution, then we say that M has infinite projective, injective, or flat, dimension, according to the case, and we write pd$M = \infty$, id $M = \infty$, or fd$M = \infty$, respectively. By convention, the zero module has all three dimensions equal to $-\infty$.

In each case, the reason for defining the dimension as the least possible length of a (co)resolution is that, given a finite (co)resolution, it is always possible to construct a longer one for the same module. Let, for instance, an A-module M have a finite projective resolution:

$$0 \to P_n \to \cdots \to P_1 \to P_0 \to M \to 0.$$

Then, given any projective module P, one can construct another projective resolution

$$0 \to P \xrightarrow{\binom{1}{0}} P \oplus P_n \to \cdots \to P_1 \to P_0 \to M \to 0.$$

A remark on indices. The projective dimension n of a module of finite projective dimension is *not* the number of projective modules in a projective resolution of least length of this module (the latter, indeed, is $n + 1$). It is the number of nonzero *morphisms* between projective modules in such a resolution. The same remark can be made about injective and flat dimensions.

XII.2.2 Examples.

(a) A nonzero module is projective (or injective, or flat) if and only if its projective (or injective, or flat, respectively) dimension is 0.

(b) Assume that the algebra A is such that any submodule of a projective module is projective. This is the case, for instance, if A is a principal ideal domain, see Example VI.4.6(b). For any A-module M, there exists a short exact sequence $0 \to L \to P_0 \to M \to 0$, with P_0 projective. Because of the hypothesis, the submodule L of P_0 is projective. Therefore, this sequence is a projective resolution of M and pd$M \leqslant 1$.

Dually, if A is such that any quotient of an injective module is injective, then any module has injective dimension at most one. This also holds true for principal ideal domains: a module is injective if and only if it is divisible, see Exercise VI.4.24(a), and quotients of divisible modules are divisible. Thus, modules over principal ideal domains have both projective and injective dimension at most one.

(c) Let A be an integral domain and $a \in A$ a nonzero noninvertible element. We have a short exact sequence $0 \to A \xrightarrow{\mu_a} A \to A/aA \to 0$, where $\mu_a : A \to A$ is the multiplication morphism given by $x \mapsto ax$, for $x \in A$. We claim that $\mathrm{pd}(A/aA) = 1$. Indeed, the exact sequence gives $\mathrm{pd}(A/aA) \leqslant 1$. If $\mathrm{pd}(A/aA) = 0$, then A/aA is projective and the sequence splits but this is impossible, because A is an indecomposable A-module, due to Example V.3.20(c). This proves our claim.

(d) Let K be a field and $A = T_2(K)$ Set, as in Example XI.2.9(a), $P_1 = e_{11}A$, $P_2 = e_{22}A$ and $S_2 = P_2/P_1$. The short exact sequence $0 \to P_1 \to P_2 \to S_2 \to 0$ is a projective resolution of S_2, hence $\mathrm{pd}S_2 \leqslant 1$. If S_2 is projective, then the sequence splits, which, as in Example (c) above, contradicts indecomposability of P_2. Hence, $\mathrm{pd}S_2 \neq 0$ and so $\mathrm{pd}S_2 = 1$. Using Example IX.2.18, and that $S_2 = I_2$, we have a (nonsplit) injective coresolution $0 \to S_1 \to I_1 \to I_2 \to 0$. Because S_1 is not injective, we get $\mathrm{id}S_1 = 1$.

(e) The same module may have different homological dimensions according to the algebra on which it is considered. Let A be a principal ideal domain and $p \in A$ an irreducible element. Then $S = A/pA$, as an A-module, has projective and injective dimension at most 1 (actually, exactly equal to 1, because it is neither projective, nor injective), see Example (b) above. But S may also be viewed as an $R = A/p^2A$-module, and in this case, we have a short exact sequence of R-modules $0 \to S \xrightarrow{i} R \xrightarrow{g} S \to 0$ where i denotes the inclusion and g the projection. Then $f : R \to R$ defined by $f = ig$ is the morphism induced from multiplication by p. As in Example XI.2.9(c), we deduce an infinite projective resolution of S :

$$\cdots \to R \xrightarrow{f} R \to \cdots \to R \xrightarrow{f} R \xrightarrow{g} S \to 0.$$

There is no finite projective resolution of S_R, because any projective module P_0 such that there is an epimorphism $P_0 \to S$, must have as direct summand the unique indecomposable projective module R. Also, the unique morphism, up to scalars, from R to S is g, which has kernel S. Repeating inductively this argument, the process never stops. Hence, $\mathrm{pd}S = \infty$. We also have an infinite injective coresolution of S :

$$0 \to S \xrightarrow{i} R \xrightarrow{f} R \to \cdots \to R \xrightarrow{f} R \to \cdots$$

see Example XI.2.9(c). Therefore, $\mathrm{id}S = \infty$.

(f) In Example XI.3.12(b), one has $\mathrm{pd}S_5 \leqslant 4$.

XII.2.3 Theorem. *Let A be a K-algebra, M an A-module and $n \geq 0$ an integer. The following conditions are equivalent:*
(a) $\mathrm{pd}M \leqslant n$.
(b) $\mathrm{Ext}_A^k(M, -) = 0$, *for any* $k > n$.
(c) $\mathrm{Ext}_A^{n+1}(M, -) = 0$.

(d) *For any exact sequence* $0 \to L_{n-1} \to P_{n-1} \to \cdots \to P_0 \to M \to 0$, *with the* P_i *projective, we have* L_{n-1} *projective.*

Proof. (a) implies (b). Because $\mathrm{pd}M \leqslant n$, there exists a projective resolution $P_\bullet$ of M with $P_k = 0$ for $k > n$. Then, for any A-module X, we have $\mathrm{Ext}_A^k(M,X) \cong H^k((P_M)_\bullet,X) = 0$ for $k > n$.
(b) implies (c). This is evident.
(c) implies (d). Given an exact sequence as in (d) and an A-module X, dimension shifting gives $0 = \mathrm{Ext}_A^{n+1}(M,X) \cong \mathrm{Ext}_A^1(L_{n-1},X)$. The statement follows from Proposition XI.2.7(a).
(d) implies (a). Let

$$P_\bullet : \cdots \to P_{n-1} \xrightarrow{d_{n-1}} P_{n-2} \to \cdots \to P_0 \xrightarrow{d_0} M \to 0$$

be a projective resolution. The hypothesis says that $L_{n-1} = \mathrm{Ker}\, d_{n-1}$ is projective. Hence,

$$0 \to L_{n-1} \to P_{n-1} \to \cdots \to P_0 \to M \to 0$$

is a finite projective resolution of M. So $\mathrm{pd}M \leqslant n$. $\square$

The theorem can be read as saying that $\mathrm{pd}M$, if finite, equals the largest of the integers $n \geq 0$ such that $\mathrm{Ext}_A^n(M,-) \neq 0$. Here are some consequences.

XII.2.4 Corollary. *Let M be an A-module of finite projective dimension. Then, for any positive integer $m \leqslant \mathrm{pd}M$, we have $\mathrm{Ext}_A^m(M,-) \neq 0$.*

Proof. By descending induction: if, for some m, one has $\mathrm{Ext}_A^{m+1}(M,-) \neq 0$, then there exists an A-module X such that $\mathrm{Ext}_A^{m+1}(M,X) \neq 0$. Also, there exists a short exact sequence of the form $0 \to X \to I \to I/X \to 0$, with I injective. It induces an exact sequence

$$0 = \mathrm{Ext}_A^m(M,I) \to \mathrm{Ext}_A^m(M,I/X) \to \mathrm{Ext}_A^{m+1}(M,X) \to \mathrm{Ext}_A^{m+1}(M,I) = 0$$

where the end terms vanish because I is injective and $m > 0$. Therefore, $\mathrm{Ext}_A^m(M,I/X) \neq 0$. Hence, $\mathrm{Ext}_A^m(M,-) \neq 0$. $\square$

XII.2.5 Corollary. *Let $(M_\lambda)_\lambda$ be a family of A-modules, then*

$$\mathrm{pd}(\oplus_\lambda M_\lambda) = \sup_\lambda(\mathrm{pd}M_\lambda).$$

Proof. This is the equivalence of (a) and (c) in the theorem plus the fact that $\mathrm{Ext}_A^n(\oplus_\lambda M_\lambda,X) \cong \prod_\lambda \mathrm{Ext}_A^n(M_\lambda,X)$, for any module X, see Proposition XI.2.10. $\square$

XII.2.6 Corollary. *Let $0 \to L \to M \to N \to 0$ be a short exact sequence of A-modules. Then:*

(a) $\mathrm{pd}N \leqslant \sup\{\mathrm{pd}M, \mathrm{pd}L + 1\}$ *and equality holds if* $\mathrm{pd}M \neq \mathrm{pd}L$.
(b) $\mathrm{pd}L \leqslant \sup\{\mathrm{pd}M, \mathrm{pd}N - 1\}$ *and equality holds if* $\mathrm{pd}M \neq \mathrm{pd}N$.
(c) $\mathrm{pd}M \leqslant \sup\{\mathrm{pd}L, \mathrm{pd}N\}$ *and equality holds if* $\mathrm{pd}N \neq \mathrm{pd}L + 1$.

Proof. We only prove (a), because the proofs of (b) and (c) are similar. For an integer $n \geq 0$ and a module X, we have an exact sequence

$$\cdots \to \mathrm{Ext}_A^{n+1}(L, X) \to \mathrm{Ext}_A^{n+2}(N, X) \to \mathrm{Ext}_A^{n+2}(M, X) \to \cdots$$

Here $\mathrm{pd}L \leqslant n$ and $\mathrm{pd}M \leqslant n + 1$ imply $\mathrm{pd}N \leqslant n + 1$, as required.
If inequality is strict, that is, $\mathrm{pd}N < \sup\{\mathrm{pd}M, \mathrm{pd}L + 1\}$, then, in particular, $\mathrm{pd}N \leqslant n < \infty$. Let $k > n$ be an integer and X a module, the exact sequence above implies $\mathrm{Ext}_A^{k+1}(M, X) \cong \mathrm{Ext}_A^{k+1}(L, X)$, that is, $\mathrm{pd}M = \mathrm{pd}L$. □

In particular, if any two of L, M, N have finite projective dimension, then so does the third.

XII.2.7 Example. Let M be a module of projective dimension $n > 0$. A projective resolution $0 \to P_n \to \cdots \to P_0 \overset{f}{\to} M \to 0$ induces a short exact sequence $0 \to L \to P_0 \overset{f}{\to} M \to 0$ with $L = \mathrm{Ker}\,f$. Then Corollary XII.2.6 gives $\mathrm{pd}L \leqslant n - 1$. If inequality is strict, then splicing a projective resolution of L of length strictly smaller than $n - 1$ with the latter short exact sequence gives a shorter projective resolution for M, a contradiction. Therefore, $\mathrm{pd}L = n - 1$. Here, inequality (c) of Corollary XII.2.6 is strict, but neither (a) nor (b).

Given an A-module M, a projective resolution of M is not unique. However, the syzygies of two different projective resolutions, that is, the images of the morphisms on the resolutions, behave in a similar way.

XII.2.8 Lemma. *Let A be a K-algebra, M an A-module and*

$$P_\bullet : \cdots \to P_1 \overset{d_1}{\longrightarrow} P_0 \overset{d_0}{\longrightarrow} M \to 0 \ \text{and} \ P'_\bullet : \cdots \to P'_1 \overset{d'_1}{\longrightarrow} P'_0 \overset{d'_0}{\longrightarrow} M \to 0$$

projective resolutions with respective syzygies $U_n = \mathrm{Im}\, d_n$, $U'_n = \mathrm{Im}\, d'_n$, then, for any integer $n \geq 1$:
   (a) *There exist projective A-modules P^n, P'^n such that $U_n \oplus P^n \cong U'_n \oplus P'^n$.*
   (b) *For any module X, we have $\mathrm{Ext}_A^1(U_n, X) \cong \mathrm{Ext}_A^1(U'_n, X)$.*

Proof. The first statement follows from the generalisation of Schanuel's lemma, which is Exercise VI.4.5, the second follows from the first and additivity of Ext :

$$\mathrm{Ext}_A^1(U_n, X) \cong \mathrm{Ext}_A^1(U_n \oplus P_n, X) \cong \mathrm{Ext}_A^1(U'_n \oplus P'_n, X) \cong \mathrm{Ext}_A^1(U'_n, X).$$

using the fact that extensions by a projective module are always split. □

We are, of course, mostly interested in Ext^1_A because the calculation of Ext^n_A, for $n > 1$, can be reduced to $n = 1$ using dimension shifting. We finish this discussion of projective dimensions by proving a change of scalars formula, see Example II.2.4(e).

XII.2.9 Lemma. *Let A, B be K-algebras, $\varphi : A \to B$ an algebra morphism and M a B-module. Then we have*

$$\mathrm{pd}M_A \leqslant \mathrm{pd}M_B + \mathrm{pd}B_A.$$

Proof. If $\mathrm{pd}M_B$ is infinite, there is nothing to prove, so assume that $\mathrm{pd}M_B = n < \infty$ and use induction. If $n = 0$, then M_B is projective, therefore there exists a B-module N and a set Λ such that $M \oplus N \cong B^{(\Lambda)}$, due to Theorem VI.4.4. Because the change of scalars functor $\mathrm{Mod}B \to \mathrm{Mod}A$ preserves isomorphisms, we also have $(M \oplus N)_A \cong B^{(\Lambda)}_A$. Then, Corollary XII.2.5 gives $\mathrm{pd}M_A \leqslant \mathrm{pd}(M \oplus N)_A = \mathrm{pd}B^{(\Lambda)}_A = \mathrm{pd}B_A$, which proves the inequality in this case.

Suppose $n > 0$. Then M is the image of a free module, hence there exists a set Λ and a short exact sequence

$$0 \to L \to B^{(\Lambda)} \to M \to 0.$$

Because of Example XII.2.7, $\mathrm{pd}M_B = n$ implies $\mathrm{pd}L_B = n - 1$. Hence, the induction hypothesis gives $\mathrm{pd}L_A \leqslant n - 1 + \mathrm{pd}B_A$. But then the short exact sequence and Corollary XII.2.6 yield:

$$\mathrm{pd}M_A \leqslant 1 + \max(\mathrm{pd}L_A, \mathrm{pd}B_A) = 1 + \max(n - 1 + \mathrm{pd}B_A, \mathrm{pd}B_A) \leqslant n + \mathrm{pd}B_A. \quad \square$$

We state the corresponding results for injectives and flats, the proofs are left as exercises.

XII.2.10 Theorem. *Let A be a K-algebra, M an A-module and $n \geq 0$ an integer. The following conditions are equivalent:*
(a) $\mathrm{id}M \leqslant n$.
(b) $\mathrm{Ext}^k_A(-, M) = 0$, for any $k > n$.
(c) $\mathrm{Ext}^{n+1}_A(-, M) = 0$.
(d) For any exact sequence $0 \to M \to I^0 \to \cdots \to I^{n-1} \to L^{n-1} \to 0$ with the I^i injective, we have L^{n-1} injective. $\qquad\square$

Thus, $\mathrm{id}M = \sup\{n \geq 0 : \mathrm{Ext}^n_A(-, M) \neq 0\}$.

XII.2.11 Corollary. *Let M be an A-module of finite injective dimension. Then, for any positive integer $m \leqslant \mathrm{id}M$, we have $\mathrm{Ext}^m_A(-, M) \neq 0$.* $\qquad\square$

XII.2.12 Corollary. *Let $(M_\lambda)_\lambda$ be a family of A-modules. Then*

$$\mathrm{id}\left(\prod_\lambda M_\lambda\right) = \sup_\lambda(\mathrm{id}M_\lambda).$$ $\qquad\square$

If the algebra A is noetherian, then Theorem VII.2.22 implies that $\mathrm{id}(\oplus_\lambda M_\lambda) = \sup_\lambda(\mathrm{id}M_\lambda)$.

XII.2.13 Corollary. *Let* $0 \to L \to M \to N \to 0$ *be a short exact sequence of A-modules. Then:*
(a) $\mathrm{id}N \leqslant \sup\{\mathrm{id}M, \mathrm{id}L - 1\}$ *and equality holds if* $\mathrm{id}M \neq \mathrm{id}L$.
(b) $\mathrm{id}L \leqslant \sup\{\mathrm{id}M, \mathrm{id}N + 1\}$ *and equality holds if* $\mathrm{id}M \neq \mathrm{id}N$.
(c) $\mathrm{id}M \leqslant \sup\{\mathrm{id}L, \mathrm{id}N\}$ *and equality holds if* $\mathrm{id}N \neq \mathrm{id}L - 1$. $\square$

In particular, if any two of L, M, N have finite injective dimension, then so does the third.

XII.2.14 Example. Assume that a module M has injective dimension $n > 0$. Then there exists an injective coresolution $0 \to M \xrightarrow{g} I^0 \to \cdots \to I^n \to 0$. The injective dimension of $\mathrm{Coker}\, g$ equals $n - 1$. Again, inequality (c) of Corollary XII.2.13 is strict, but neither (a) nor (b).

XII.2.15 Lemma. *Let A be a K-algebra, M an A-module and*

$$I^\cdot : 0 \to M \xrightarrow{d^0} I^0 \xrightarrow{d^1} I^1 \to \cdots \ \text{and}\ I'^\cdot : 0 \to M \xrightarrow{d'^0} I'^0 \xrightarrow{d'^1} I'^1 \to \cdots$$

injective coresolutions with respective cosyzygies $U^n = \mathrm{Im}\, d^n$, $U'^n = \mathrm{Im}\, d'^n$, then, for any integer $n \geq 1$:
(a) *There exist injective A-modules I^n, I'^n such that $U^n \oplus I^n \cong U'^n \oplus I'^n$.*
(b) *For any module X, we have $\mathrm{Ext}_A^1(X, U^n) \cong \mathrm{Ext}_A^1(X, U'^n)$.* $\square$

XII.2.16 Theorem. *Let A be a K-algebra, M an A-module and $n \geq 0$. The following conditions are equivalent:*
(a) $\mathrm{fd}M \leqslant n$.
(b) $\mathrm{Tor}_k^A(M, -) = 0$, *for any $k > n$.*
(c) $\mathrm{Tor}_{n+1}^A(M, -) = 0$.
(d) *For any exact sequence $0 \to L_{n-1} \to F_{n-1} \to \cdots \to F - 0 \to M \to 0$ with the F_i flat, we have L_{n-1} flat.* $\square$

Thus, $\mathrm{fd}M = \sup\{n \geq 0 : \mathrm{Tor}_n^A(M, -) \neq 0\}$.

XII.2.17 Corollary. *Let M be an A-module of finite flat dimension. Then, for any positive integer $m \leqslant \mathrm{fd}M$, we have $\mathrm{Tor}_m^A(M, -) \neq 0$.* $\square$

Projective modules are flat, hence, for any module M, we have $\mathrm{fd}M \leqslant \mathrm{pd}M$. But in general, these two dimensions are *not* equal: for instance, the abelian group $\mathbb{Q}$ is flat, but nonprojective, see Example VI.4.44. Hence, $\mathrm{fd}\mathbb{Q} = 0$, but $\mathrm{pd}\mathbb{Q} = 1$, due to Example XII.2.2(b).

We leave to the reader the formulation and proofs of results similar to Corollaries XII.2.5 and XII.2.6 for flat modules, see Exercise XII.2.10.

XII.2.2 Minimal (co)resolutions

Homological dimensions are minimal lengths of (co)resolutions. So we ask when this minimum is attained, namely, which are the (co)resolutions having least length? Intuitively, that should happen whenever each term of a (co)resolution is as small as possible, that is, is a projective cover or an injective envelope. Throughout, let A be a K-algebra.

XII.2.18 Definition. Let M be an A-module.

(a) We say that $\cdots \to P_n \xrightarrow{d_n} \cdots \to P_1 \xrightarrow{d_1} P_0 \xrightarrow{d_0} M \to 0$ is a *minimal projective resolution* if, for any $n \geq 0$, the module P_n is a projective cover of Im d_n.

(b) We say that $0 \to M \xrightarrow{d^0} I^0 \xrightarrow{d^1} I^1 \to \cdots \xrightarrow{d^n} I^n \to \cdots$ is a *minimal injective coresolution* if, for any $n \geq 0$, the module I^n is an injective envelope of Im d^n.

Because of Theorem VI.4.35, any module (over any algebra) has an injective envelope, unique up to isomorphism. We prove below that it also admits a minimal injective coresolution, again unique up to isomorphism. On the other hand, projective covers do not always exist hence minimal projective resolutions do not necessarily exist.

Our objective is to prove that the minimal injective coresolutions, and the minimal projective resolutions (when the latter exist) are the shortest ones.

XII.2.19 Lemma. (a) *Any A-module admits a minimal injective coresolution.*
(b) *If*

$$I^{\cdot} : 0 \to M \xrightarrow{d^0} I^0 \xrightarrow{d^1} I^1 \to \cdots \xrightarrow{d^n} I^n \to \cdots$$

$$J^{\cdot} : 0 \to M \xrightarrow{f^0} J^0 \xrightarrow{f^1} J^1 \to \cdots \xrightarrow{f^n} J^n \to \cdots$$

are injective coresolutions, with $I^{\cdot}$ minimal, then there exists a morphism of complexes $u^{\cdot} : (I_M)^{\cdot} \to (J_M)^{\cdot}$ extending 1_M and such that each u^i is a section.
(c) *Two minimal injective coresolutions are isomorphic.*

Proof. (a) We use induction on degree. Let I^0 be an injective envelope of M and $d^0 : M \to I^0$ the inclusion. Assume that I^i and d^i are constructed for all $i \leq k$. Letting $p^k = \operatorname{coker} d^k$, we have an exact sequence

$$I^{k-1} \xrightarrow{d^k} I^k \xrightarrow{p^k} \operatorname{Coker} d^k \to 0.$$

Let I^{k+1} be the injective envelope of Coker d^k and $j^k : \operatorname{Coker} d^k \to I^{k+1}$ the inclusion. Then the exact sequence

$$0 \to M \xrightarrow{d^0} I^0 \to \cdots \to I^k \xrightarrow{d^{k+1} = j^k p^k} I^{k+1}$$

is (the beginning of) a minimal injective coresolution of M.

(b) Let $I^\bullet, J^\bullet$ be as in the statement. We may assume that M is not injective, oth erwise the statement is trivial. We construct $u^\bullet$ by induction on degree. We use the following notation: for any integer $i \geq 0$, let $L^i = \operatorname{Coker} d^i, p^i = \operatorname{coker} d^i$, $j^i = \ker d^{i+1}$ and $L'^i = \operatorname{Coker} f^i, p'^i = \operatorname{coker} f^i, j'^i = \ker f^{i+1}$.

Because I^0, J^0 are injective, there exist $u^0 : I^0 \to J^0, v^0 : J^0 \to I^0$ such that $u^0 d^0 = f^0$ and $v^0 f^0 = d^0$. We claim that $v^0 u^0$ is an isomorphism. Indeed, $v^0 u^0 d^0 = v^0 f^0 = d^0$ is an essential monomorphism, hence $v^0 u^0$ is a monomorphism. On the other hand, I^0 is an injective envelope of $d^0(M) \subseteq v^0 u^0(I^0) \subseteq I^0$. Because $M \cong d^0(M)$ is not injective, we have $d^0(M) \neq I^0$. Now, no injective can be properly inserted between a module and its injective envelope, see Exercise VI.4.26. Hence, $v^0 u^0(I^0) = I^0$. Therefore, $v^0 u^0$ is surjective and so is an isomorphism. In particular, u^0 is a section.

Passing to cokernels in the commutative diagram with exact rows

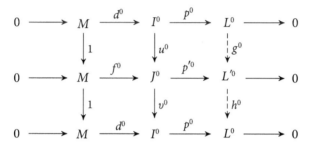

we get $g^0 : L^0 \to L'^0, h^0 : L'^0 \to L^0$ such that $g^0 p^0 = p'^0 u^0$ and $h^0 p'^0 = p^0 v^0$. Because 1_M and $v^0 u^0$ are isomorphisms, so is $h^0 g^0$. Moreover, L^0, L'^0 are nonzero because M is not injective.

Inductively, assume that $g^i : L^i \to L'^i, h^i : L'^i \to L^i$ are given, with $L^i, L'^i \neq 0$ and not injective, and such that $h^i g^i$ is an isomorphism. Injectivity of I^{i+1}, J^{i+1} implies the existence of morphisms $u^{i+1} : I^{i+1} \to J^{i+1}, v^{i+1} : J^{i+1} \to I^{i+1}$ such that $u^{i+1} j^i = j'^i g^i$ and $v^{i+1} j'^i = j^i h^i$.

$$
\begin{array}{ccccccccc}
0 & \longrightarrow & L^i & \longrightarrow & I^{i+1} & \longrightarrow & L^{i+1} & \longrightarrow & 0 \\
& & \downarrow g^i & & \downarrow u^{i+1} & & \downarrow g^{i+1} & & \\
0 & \longrightarrow & L^i & \longrightarrow & J^{i+1} & \longrightarrow & L'^{i+1} & \longrightarrow & 0 \\
& & \downarrow h^i & & \downarrow v^{i+1} & & \downarrow h^{i+1} & & \\
0 & \longrightarrow & L^i & \longrightarrow & I^{i+1} & \longrightarrow & L^{i+1} & \longrightarrow & 0
\end{array}
$$

We prove exactly as before that $v^{i+1} u^{i+1}$ is an isomorphism (hence u^{i+1} is a section). Passing to cokernels, we find g^{i+1}, h^{i+1} making the diagram commute and such that $h^{i+1} g^{i+1}$ is an isomorphism. This completes the proof of (b).

(c) Assume that in (b), $I^\bullet, J^\bullet$ are both minimal. Then any u^i is an isomorphism. If $i = 0$, this follows from uniqueness of injective envelopes, see Theorem VI.4.35. Passing to cokernels, g^0 is an isomorphism. Inductively, if g^i is an isomorphism, so is u^{i+1}, hence so is g^{i+1}. $\qquad\square$

A nice consequence is that the length of a minimal injective coresolution is at most equal to the length of one which is not, and two minimal injective coresolutions have same length.

XII.2.20 Theorem. *Let M be an A-module, $n \geq 0$ an integer and $0 \to M \to I^0 \to I^1 \to \cdots$ a minimal injective coresolution. Then $\mathrm{id}\, M = n$ if and only if $I^n \neq 0$ and $I^i = 0$ for $i > n$.*

Proof. Definition XII.2.18(b) implies that, if $I^i = 0$ for some i, then $I^j = 0$ for $j > i$. Assume that $\mathrm{id}\, M = n$, then there exists an injective coresolution of the form $0 \to M \to J^0 \to J^1 \to \cdots \to J^n \to 0$ and any other injective coresolution has length at least n. This is in particular the case for the minimal injective coresolution in the statement. Thus, $I^n \neq 0$. Because of Lemma XII.2.19, I^{n+1} is a direct summand of $J^{n+1} = 0$. Hence, $I^{n+1} = 0$. This proves necessity.

For sufficiency, if $\mathrm{id}\, M = m < n$, then the same argument gives $I^{m+1} = 0$ and hence $I^n = 0$, a contradiction. $\square$

While in general projective covers do not exist, see Example IX.2.11(a), they do when we deal with finitely generated modules over artinian algebras, due to Theorem IX.2.12. In this case, results dual to Lemma XII.2.19 and Theorem XII.2.20 hold true, with dual proofs. The only remark is that one has to construct resolutions into finitely generated projectives (to ensure that kernels are finitely generated), and this is possible because of Corollary VII.2.16. We state the results for ease of reference.

XII.2.21 Lemma. (a) *Any finitely generated module over an artinian algebra admits a minimal projective resolution.*
(b) *If M is a finitely generated module over an artinian algebra and*

$$P_\bullet : \cdots \to P_1 \to P_0 \to M \to 0$$
$$Q_\bullet : \cdots \to Q_1 \to Q_0 \to M \to 0$$

are projective resolutions of M, with the P_i, Q_i finitely generated and $P_\bullet$ minimal, then there exists a morphism $u_\bullet : (Q_M)_\bullet \to (P_M)_\bullet$, which lifts 1_M and such that each u_i is a retraction.
(c) *Two minimal projective resolutions are isomorphic.* $\square$

XII.2.22 Theorem. *Let M be a finitely generated module over an artinian algebra, $n \geq 0$ an integer and $\cdots \to P_1 \to P_0 \to M \to 0$ a minimal projective resolution. Then $\mathrm{pd}\, M = n$ if and only if $P_n \neq 0$ and $P_i = 0$ for $i > n$.* $\square$

It is natural to ask whether similar results hold true for flat modules. This requires a notion of 'flat cover' similar to projective cover. For a long time, existence of flat covers (for arbitrary modules over arbitrary algebras) remained an open conjecture. It was proved in 2001. This result is even more remarkable if one remembers that projective covers do not always exist.

We give computational examples in the case of finite–dimensional algebras. Recall that, because of the proof of Theorem IX.2.12, the projective cover of a module M equals the projective cover of its top, and the projective cover of a simple module is the corresponding projective under the bijection of Theorem IX.2.6. If $\mathrm{top}M = \bigoplus_{i=1}^{n} S_i$, and P_i denotes the finitely generated indecomposable projective having S_i as its top, then the projective cover of M is $P = \bigoplus_{i=1}^{n} P_i$. The situation is dual for injectives, we use the socle instead of the top (because duality turns the submodule lattice upside down), see Theorem IX.2.16.

XII.2.23 Examples.

(a) Let K be a field and A given by the quiver $1 \xrightarrow{\alpha} 2 \xrightarrow{\beta} 3$, bound by the ideal $\langle \alpha\beta \rangle$. Then A has three simple modules, up to isomorphism, namely S_1, S_2, S_3, in the notation of Example VII.4.3(e).

First, S_3 is projective, hence its minimal projective resolution is $0 \to P_3 \to S_3 \to 0$. Second, the projective cover of S_2 is P_2, and the kernel of the projection $P_2 \to S_2$ is $S_3 = P_3$, hence the minimal projective resolution is the short exact sequence $0 \to P_3 \to P_2 \to S_2 \to 0$. Thus, $\mathrm{pd}S_2 = 1$. Third, S_1 has P_1 as projective cover. The kernel of the projection $P_1 \to S_1$ is S_2 of which we have just computed a minimal projective resolution. Splicing gives a minimal projective resolution $0 \to P_3 \to P_2 \to P_1 \to S_1 \to 0$. So, $\mathrm{pd}S_1 = 2$.

Minimal injective coresolutions are constructed similarly. First, $S_1 = I_1$ is injective. Next, the injective envelope of S_2 is $I_2 = P_1$ and the cokernel of the injection is S_1. Therefore, $0 \to S_2 \to I_2 \to I_1 \to 0$ is a minimal injective coresolution and $\mathrm{id}S_2 = 1$. Finally, the injective envelope of S_3 is $I_3 = P_2$, the cokernel of the injection is S_2 of which we just computed a minimal injective coresolution. Therefore, a minimal injective coresolution of S_3 is $0 \to S_3 \to I_3 \to I_2 \to I_1 \to 0$. In particular, $\mathrm{id}S_3 = 2$.

(b) Over Nakayama algebras, minimal (co)resolutions are easy to compute: any indecomposable module is uniserial, hence has simple top and socle, therefore it has uniserial projective cover or injective envelope, and the (co)syzygy is itself uniserial, hence has a uniserial projective cover or injective envelope, etc. Let A be the Nakayama algebra of Example IX.4.19(c). The projective resolutions of the simple modules computed there are minimal:

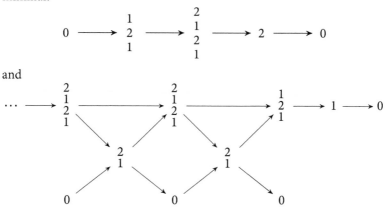

and

Thus, $\mathrm{pd}S_2 = 1$ while $\mathrm{pd}S_1 = \infty$. Here is a point: because a minimal projective resolution, or a minimal injective coresolution, of a given module is a shortest one, then, if it is infinite, all others (of the same module) are infinite as well, and the projective dimension, or injective dimension, respectively, of the module is infinite.

A similar calculation yields the minimal injective coresolutions of the simple modules:

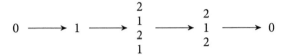

and

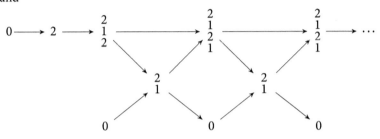

Therefore, $\mathrm{id}S_1 = 1$ while $\mathrm{id}S_2 = \infty$.

(c) Let A be the bound quiver algebra of Example IX.4.19(d). The simple module S_1 is projective (equal to P_1), so $\mathrm{pd}S_1 = 0$. The minimal projective resolutions of the other simple modules are:

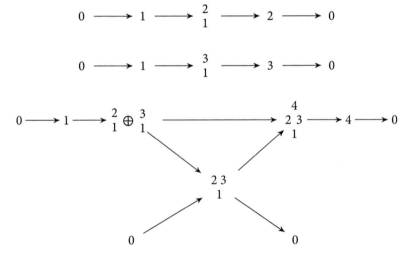

The kernel of the projection $P_4 \to S_4$ is the module $\mathrm{rad}P_4$ whose top equals $S_2 \oplus S_3$. Therefore, the projective cover of the latter is $P_2 \oplus P_3$. Thus, $\mathrm{pd}S_2 = 1 = \mathrm{pd}S_3$ while $\mathrm{pd}S_4 = 2$. One computes in the same way minimal injective coresolutions.

(d) In Example XI.3.12(b), one has $\mathrm{pd}S_5 = 4$, because the projective resolution exhibited is actually minimal.

We end with a lemma which will be used later.

XII.2.24 Lemma. *Let K be a field, A a finite–dimensional K-algebra and M a finitely generated A-module.*

(a) *If* $\cdots \to P_n \xrightarrow{d_n} P_{n-1} \to \cdots \to P_0 \xrightarrow{d_0} M \to 0$ *is a minimal projective resolution, then for any indecomposable direct summand L' of $L = \mathrm{Ker}\, d_n$, we have* $\mathrm{Ext}_A^{n+1}(M, L') \neq 0$.

(b) *If* $0 \to M \xrightarrow{d^0} I^0 \to \cdots \to I^{n-1} \xrightarrow{d^n} I^n \to \cdots$ *is a minimal injective coresolution, then for any indecomposable direct summand L' of $L = \mathrm{Coker}\, d^n$, we have* $\mathrm{Ext}_A^{n+1}(L', M) \neq 0$.

Proof. We prove (a), because the proof of (b) is similar. We have an exact sequence

$$0 \to L \xrightarrow{d} P_n \xrightarrow{d_n} \cdots \to P_0 \xrightarrow{d_0} M \to 0$$

and a retraction $f : L \to L'$, that is, there exists $f' : L' \to L$ such that $ff' = 1$. Set $N = \mathrm{Im}\, d_n$. We have a short exact sequence $0 \to L \xrightarrow{d} P_n \to N \to 0$. Dimension shifting gives $\mathrm{Ext}_A^{n+1}(M, L') \cong \mathrm{Ext}_A^1(N, L')$. It thus suffices to prove that $\mathrm{Ext}_A^1(N, L') \neq 0$. If not, then in the commutative diagram with exact rows

$$
\begin{array}{ccccccccc}
0 & \longrightarrow & L & \xrightarrow{d} & P_n & \longrightarrow & N & \longrightarrow & 0 \\
& & \downarrow{f} & & \downarrow{g} & & \downarrow{1_N} & & \\
0 & \longrightarrow & L' & \xrightarrow{u} & E & \longrightarrow & N & \longrightarrow & 0
\end{array}
$$

the lower sequence splits, that is, there exists $u' : E \to L'$ such that $u'u = 1$. But we know that $ff' = 1$. Hence,

$$(u'g)(df') = u'(gd)f' = u'(uf)f' = (u'u)(ff') = 1.$$

Therefore, $df' : L' \to P_n$ is a section, and one can write $L = L' \oplus L''$ and $P_n = P_n' \oplus P_n''$ with $df' : L' \to P_n'$ an isomorphism. We get an exact sequence

$$\cdots \to P_n'' \to P_{n-1} \to \cdots \to P_0 \xrightarrow{d_0} M \to 0.$$

Because L' is indecomposable, it is nonzero, therefore P_n'' is a *proper* direct summand of P_n, contradicting minimality of the original resolution. $\qquad\square$

Exercises of Section XII.2

1. Let A be a K-algebra and $0 \to L \to M \to N \to 0$ a short exact sequence of A-modules.
    (a) If M is projective, then either L, N are both projective or $\mathrm{pd}N = \mathrm{pd}L + 1$.
    (b) Assume that $\mathrm{pd}L > \mathrm{pd}M$ or $\mathrm{pd}N > 1 + \mathrm{pd}M$. Prove that $\mathrm{pd}N = 1 + \mathrm{pd}L$.
    (c) Prove that:
        (i) $\mathrm{pd}L < \mathrm{pd}M$ implies $\mathrm{pd}M = \mathrm{pd}N$.
        (ii) $\mathrm{pd}L > \mathrm{pd}M$ implies $\mathrm{pd}N = 1 + \mathrm{pd}M$.
        (iii) $\mathrm{pd}L = \mathrm{pd}M$ implies $\mathrm{pd}L \leqslant 1 + \mathrm{pd}M$.
        (iv) $\mathrm{pd}M \leqslant \max(\mathrm{pd}L, \mathrm{pd}N)$, with equality unless $\mathrm{pd}N = 1 + \mathrm{pd}L$.
    (d) State and prove similar results for injective dimensions.

2. Let A be a K-algebra and $0 = M_0 \subsetneq M_1 \subsetneq \cdots \subsetneq M_m = M$ a finite chain of submodules of an A-module M. Prove that $\mathrm{pd}M \leqslant \max_i(\mathrm{pd}(M_{i+1}/M_i))$ and $\mathrm{id}M \leqslant \max_i(\mathrm{id}(M_{i+1}/M_i))$

3. Let A be a K-algebra.
    (a) Assume that there exists a nonprojective A-module of finite projective dimension. Prove that there exists an A-module of projective dimension one.
    (b) Assume that there exists a noninjective A-module of finite injective dimension. Prove that there exists an A-module of injective dimension one.

4. Let A be a K-algebra and I an ideal in A. Prove that, if P is a projective A-module, then $\mathrm{pd}(PI_A) \leqslant \mathrm{pd}I_A$.

5. Let A be a K-algebra and M an A-module of projective dimension $n < \infty$.
    (a) Prove that there exists a free module L such that $\mathrm{Ext}_A^n(M, L) \neq 0$.
    (b) If A is noetherian and M finitely generated, prove that $\mathrm{Ext}_A^n(M, A) \neq 0$.

6. Let A be a K-algebra and M an A-module. Prove that:
    (a) If $\mathrm{pd}M = \infty$, then, for any integer $n > 0$, we have $\mathrm{Ext}_A^n(M, -) \neq 0$.
    (b) If $\mathrm{id}M = \infty$, then, for any integer $n > 0$, we have $\mathrm{Ext}_A^n(-, M) \neq 0$.

7. Consider the change of scalars functor induced by an algebra morphism $\varphi : A \to B$. Prove that, for any B-module M:
    (a) If Q_B is an injective cogenerator of $\mathrm{Mod}B$, then $\mathrm{id}M_A \leqslant \mathrm{id}M_B + \mathrm{id}Q_A$.
    (b) We have $\mathrm{fd}M_A \leqslant \mathrm{fd}M_B + \mathrm{fd}B_A$.
    (c) If, in the inequality of Lemma XII.2.9, $\mathrm{pd}M_A \leqslant \mathrm{pd}M_B + \mathrm{pd}B_A$, one has equality for any $M \neq 0$ such that $\mathrm{pd}M_B \leqslant 1$, then one has equality for any $M \neq 0$ such that $\mathrm{pd}M_B < \infty$.

8. Let A be a K-algebra and an exact sequence $0 \to L \to M_1 \to \cdots \to M_n \to N \to 0$ of A-modules. For a given integer $d \geq 0$, prove that:
    (a) If $\mathrm{pd}M_i \leqslant d$, for any i, then, for any $k > d$, we have $\mathrm{Ext}_A^k(L, -) \cong \mathrm{Ext}_A^{k+n}(N, -)$. Hence, if also $\mathrm{pd}L \leqslant d$, then $\mathrm{pd}N \leqslant d + n$.
    (b) If $\mathrm{id}M_i \leqslant d$, for any i, then, for any $k > d$, we have $\mathrm{Ext}_A^k(-, N) \cong \mathrm{Ext}_A^{k+n}(-, L)$. Hence, if also $\mathrm{id}N \leqslant d$, then $\mathrm{id}L \leqslant d + n$.

(c) If $\mathrm{fd}M_i \leqslant d$, for any i, then, for any $k > d$, we have $\mathrm{Tor}_k^A(L, -) \cong \mathrm{Tor}_{k+n}^A(N, -)$. Hence, if also $\mathrm{fd}L \leqslant d$, then $\mathrm{fd}N \leqslant d + n$.

9. Let A be a K-algebra. Prove that an A-module M is flat if and only if $\mathrm{Tor}_1^A(M, A/I) = 0$, for any left ideal I of A.

10. Let A be a K-algebra and $0 \to L \to M \to N \to 0$ a short exact sequence of A-modules. Prove that:
   (a) $\mathrm{fd}N \leqslant \sup\{\mathrm{fd}M, \mathrm{fd}L + 1\}$, and equality holds if and only if $\mathrm{fd}M \neq \mathrm{fd}L$.
   (b) $\mathrm{fd}L \leqslant \sup\{\mathrm{fd}M, \mathrm{fd}N - 1\}$, and equality holds if and only if $\mathrm{fd}M \neq \mathrm{fd}N$.
   (c) $\mathrm{fd}M \leqslant \sup\{\mathrm{fd}L, \mathrm{fd}N\}$, and equality holds if and only if $\mathrm{fd}N \neq \mathrm{fd}L + 1$.

11. Let A be a principal ideal domain and M a finitely generated A-module. Denote by $T(M)$ the torsion submodule of M. Prove that $\mathrm{Hom}_A(M, A) \cong \mathrm{Hom}_A(M/T(M), A)$ and $\mathrm{Ext}_A^1(M, A) \cong \mathrm{Ext}_A^1(T(M), A)$.

12. Let K be a field. For each of the following bound quiver K-algebras, compute a minimal injective coresolution and a minimal projective resolutions for each simple module.

(a) $\underset{5}{\circ} \xleftarrow{\ \delta\ } \underset{4}{\circ} \xleftarrow{\ \gamma\ } \underset{3}{\circ} \xleftarrow{\ \beta\ } \underset{2}{\circ} \xleftarrow{\ \alpha\ } \underset{1}{\circ}$ bound by the ideal $\langle \alpha\beta\gamma\delta \rangle$.

(b) $\underset{5}{\circ} \xleftarrow{\ \delta\ } \underset{4}{\circ} \xleftarrow{\ \gamma\ } \underset{3}{\circ} \xleftarrow{\ \beta\ } \underset{2}{\circ} \xleftarrow{\ \alpha\ } \underset{1}{\circ}$ bound by the ideal $\langle \alpha\beta, \beta\gamma\delta \rangle$.

(c) $\underset{1}{\circ} \xrightarrow{\ \alpha\ } \underset{2}{\circ} \overset{\beta}{\circlearrowright}$ bound by the ideal $\langle \alpha\beta, \beta^3 \rangle$.

(d) $\underset{1}{\circ} \underset{\delta}{\overset{\alpha}{\rightleftarrows}} \underset{2}{\circ} \underset{\beta}{\overset{\gamma}{\rightleftarrows}} \underset{3}{\circ}$ bound by the ideal $\langle \alpha\delta, \gamma\beta, \delta\alpha - \gamma\beta \rangle$.

13. Let K be a field and A the K-algebra given by the quiver

$$\underset{1}{\circ} \underset{\beta}{\overset{\alpha}{\rightleftarrows}} \underset{2}{\circ} \quad \text{bound by the ideal } \langle \alpha\beta, \beta\alpha \rangle$$

Write a minimal projective resolution of the simple module S_1 and use it to prove that

$$\mathrm{Ext}_A^n(S_1, S_1) = \begin{cases} K & n \text{ even,} \\ 0 & n \text{ odd.} \end{cases}$$

XII.3 Homological dimensions of algebras

XII.3.1 Global dimensions

Let A be a K-algebra. The projective dimension of an A-module M is the supremum of the integers n such that there exists a module X satisfying $\mathrm{Ext}_A^n(M, X) \neq 0$. Dually, the injective dimension of an A-module N is the supremum of the n such that there exists a module Y satisfying $\mathrm{Ext}_A^n(Y, N) \neq 0$. When we let both variables M and N run separately through the whole module category, we get an invariant of the category, called the (right) global dimension of A.

XII.3.1 Definition. Let A be a K-algebra. Its **right global dimension** r.gl.dim A is the largest integer n, if it exists, such that there exist A-modules M, N satisfying $\mathrm{Ext}_A^n(M, N) \neq 0$.

In other words, r.gl.dim $A = \sup\{n \geq 0 : \mathrm{Ext}_A^n(-,?) \neq 0\}$ if this supremum is finite. If it is not, then we say that the right global dimension is infinite, and we write r.gl.dim $A = \infty$. By convention, the zero algebra has right global dimension equal to $-\infty$.

We define similarly the **left global dimension** l.gl.dim A of A using left modules. There is no reason for the right and left global dimensions to be equal and this is generally not the case, see Example XII.4.5(c) below. When equality holds true, for instance, when the algebra is commutative, then the common value of the right and the left global dimensions is called the **global dimension** gl.dim.A of the algebra A. We have the following reformulation.

XII.3.2 Theorem. *Let A be a K-algebra, then*

$$\mathrm{r.gl.dim}\,A = \sup\{\mathrm{pd}M : M \text{ an } A\text{-module}\} = \sup\{\mathrm{id}N : N \text{ an } A\text{-module}\}.$$

Proof. We prove the first equality, the second being similar. Because of Theorem XII.2.3, to say that $\mathrm{pd}M \leq n$, for some n, means that $\mathrm{Ext}_A^{n+1}(M, -) = 0$ or, equivalently, $\mathrm{Ext}_A^{n+1}(M, X) = 0$ for any module X. Thus, to say that $\mathrm{pd}M \leq n$, for some n, and all modules M, amounts to saying that r.gl.dim $A \leq n$. □

XII.3.3 Corollary. *If $\varphi : A \to B$ is an algebra morphism, then*

$$\mathrm{r.gl.dim}.A \leq \mathrm{r.gl.dim}.B + \mathrm{pd}B_A.$$

Proof. This follows from the theorem and Lemma XII.2.9. □

Similarly, the left global dimension of an algebra equals the supremum of the projective dimensions of its left modules, and the supremum of the injective dimensions of its left modules. Also, if $\varphi : A \to B$ is an algebra morphism, then l.gl.dim.$A \leq$ l.gl.dim.B + $\mathrm{pd}_A B$.

XII.3.4 Examples.
  (a) An algebra has right global dimension zero if and only if $\mathrm{Ext}_A^1(M, N) = 0$, for all modules M, N, that is, if and only if any short exact sequence of A-modules splits. Because of Theorem VII.5.4, this happens if and only if A is semisimple. The latter property being left-right symmetric, this holds true if and only if l.gl.dim $A = 0$. An algebra is semisimple if and only if its (right or left) global dimension is zero.

(b) Let A be a principal ideal domain. Because of Example XII.2.2(b), any module has projective dimension at most one. Hence, r.gl.dim$A \leqslant 1$. Also, l.gl.dim$A \leqslant 1$, because A is commutative, that is, the global dimension of a principal ideal domain is at most one.

(c) It follows from Theorem XII.3.2 that an algebra has infinite (right or left) global dimension if and only if it admits a module of infinite projective (or injective) dimension. Let A be a principal ideal domain and $p \in A$ an irreducible element. The simple A/p^2A-module $S = A/pA$ has infinite projective and injective dimensions, see Example XII.2.2(e). Because A/p^2A is commutative, gl.dim.$(A/p^2A) = \infty$.

   For instance, $\mathbb{Z}/p^2\mathbb{Z}$ with p prime, and $K[t]/\langle t^2 \rangle$, with K a field, have infinite global dimensions.

(d) We (slightly) generalise the previous Example (c). Let A be an integral domain and $a, b \in A$. We are interested in the global dimension of the commutative algebra A/abA. We have a short exact sequence of A-modules, which is also a short exact sequence of A/abA-modules:

$$0 \to bA/abA \to A/abA \to aA/abA \to 0$$

where the first morphism is inclusion and the second multiplication by a. This sequence splits if and only if a, b are coprime, see Example IV.2.33(b). Thus, if a, b are coprime, $aA/abA \cong A/bA$ and $bA/abA \cong A/aA$ are direct summands of a free A/abA-module, so are projective. Otherwise, we have a nonsplit short exact sequence of A/abA-modules:

$$0 \to A/aA \to A/abA \to A/bA \to 0$$

but also another one

$$0 \to A/bA \to A/abA \to A/aA \to 0.$$

Splicing these sequences yields an infinite projective (actually, free) resolution

$$\cdots \to A/abA \to A/abA \to A/aA \to 0$$

that is, pd$(A/aA) = \infty$. Similarly, pd$(A/bA) = \infty$. Setting $n = ab$, we have proved that, if there exists a decomposition of n as a product of irreducible elements (for instance, if A is principal) $n = p_1^{t_1} \cdots p_m^{t_m}$, with at least one of the t_i greater than or equal to 2, then gl.dim.$(A/nA) = \infty$. On the other hand, if $t_i = 1$, for all i, then $A/nA \cong \bigoplus_{i=1}^m A/p_iA$. Now, each A/p_iA is semisimple, even simple. Hence, gl.dim.$(A/nA) = 0$.

Finding the supremum of the homological dimensions of all modules seems a difficult task, so we must devise practical ways to compute the global dimension. We first prove that it suffices to compute the projective dimensions of cyclic modules. Let A be an algebra, not necessarily commutative. Due to Proposition II.4.22, a module is cyclic if and only if it is of the form A/I, with I a right ideal in A. We have an unexpected reformulation of Baer's criterion VI.4.19.

XII.3.5 Lemma. *An A-module M is injective if and only if $\operatorname{Ext}^1_A(A/I, M) = 0$, for any right ideal I in A.*

Proof. Evaluating the functor $\operatorname{Hom}_A(-, M)$ on the short exact sequence $0 \to I \xrightarrow{i} A \to A/I \to 0$, where i is the inclusion, yields an exact sequence

$$\cdots \to \operatorname{Hom}_A(A, M) \xrightarrow{\operatorname{Hom}_A(i,M)} \operatorname{Hom}_A(I, M) \to \operatorname{Ext}^1_A(A/I, M) \to \operatorname{Ext}^1_A(A, M) = 0.$$

The last term vanishes because A is projective. Hence, $\operatorname{Ext}^1_A(A/I, M) = 0$ if and only if $\operatorname{Hom}_A(i, M)$ is surjective, that is, if and only if any morphism from I to M extends to a morphism from A to M. Baer's criterion says that this happens if and only if M is injective. $\qquad\square$

The next corollary is due to **Auslander**.

XII.3.6 Corollary. *For any algebra A, we have*

$$\operatorname{r.gl.dim} A = \sup\{\operatorname{pd}(A/I) : I \text{ right ideal in } A\} = \sup\{\operatorname{pd} M_A : M \text{ finitely generated}\}.$$

Proof. Both statements are clear if the suprema are infinite, so assume that they are finite. If $\operatorname{r.gl.dim} A \leqslant n$, for some integer n, then the projective dimension of any module, including the cyclic and the finitely generated ones, does not exceed n.

Assume that there exists an integer $n \geq 0$ such that $\operatorname{pd}(A/I) \leqslant n$, or, equivalently, such that $\operatorname{Ext}^{n+1}_A(A/I, -) = 0$, for any right ideal I. Let M be an arbitrary module and

$$0 \to M \to I^0 \to \cdots \to I^{n-1} \to L^{n-1} \to 0$$

an exact sequence with all I^i injective. Dimension shifting gives

$$\operatorname{Ext}^1_A(A/I, L^{n-1}) \cong \operatorname{Ext}^{n+1}_A(A/I, M) = 0.$$

Lemma XII.3.5 gives L^{n-1} injective, hence $\operatorname{id} M \leqslant n$. Therefore, $\operatorname{r.gl.dim} A \leqslant n$. This proves the first equality.

For the second, assume that $\operatorname{pd} M \leqslant n$, for M finitely generated. This holds true in particular for cyclic modules. Applying the now proven first equality yields $\operatorname{r.gl.dim} A \leqslant n$, as required. $\qquad\square$

If A is artinian, then it has only finitely many isomorphism classes of simple modules, see Theorem IX.2.6. For this reason, the following theorem is particularly useful.

XII.3.7 Theorem. *If A is an artinian algebra, then*

$$\text{r.gl.dim}\,A = \sup\{\text{pd}S_A : S \text{ a simple module}\}.$$

Proof. Because any simple module is cyclic, it suffices to consider the case where the right-hand term equals $n < \infty$. We claim that, in this case, any finitely generated module M has projective dimension at most n. Finitely generated modules over artinian algebras have finite composition length, so we use induction on the length ℓ of M. If $\ell = 1$, then M is simple and there is nothing to prove. Assume that $\ell > 1$ and M' is a maximal submodule of M. We have a short exact sequence $0 \to M' \to M \to M/M' \to 0$. The length of M' equals $\ell - 1$, hence the induction hypothesis gives $\text{pd}M' \leqslant n$. On the other hand, M/M' is simple, so $\text{pd}(M/M') \leqslant n$, because of the case $\ell = 1$. Applying Corollary XII.2.6 gives $\text{pd}M \leqslant n$. This establishes our claim. The theorem then follows from Corollary XII.3.6. $\qquad\square$

Similarly, if A is a left artinian algebra, then its left global dimension is the supremum of the projective dimensions of the simple left modules.

XII.3.8 Examples. Let K be a field.
 (a) Let $A = T_2(K)$. Then A has only two nonisomorphic simple modules, namely $P_1 = \mathbf{e}_{11}A$ and $S_2 = P_2/P_1$. Now $\text{pd}P_1 = 0$ and $\text{pd}S_2 = 1$, see Example XII.2.2(d). Hence, $\text{r.gl.dim}\,A = 1$
 (b) Let A be the bound quiver algebra of Example XII.2.23(a). We have computed the projective dimensions of its simple modules. It follows that the right global dimension of this algebra equals 2. Similarly, the bound quiver algebras of Examples XII.2.23(b) and (c) have respective right global dimensions ∞ and 2.
 (c) Let A be the bound quiver algebra of Example XI.3.12(b). We know that $\text{pd}S_5 = 4$, see Example XII.2.23(d). Computing the minimal projective resolutions of simple modules, we find that $\text{pd}S_4 = 3$, $\text{pd}S_3 = 2$, $\text{pd}S_2 = 1$ while S_1 is projective. Hence, $\text{r.gl.dim}.A = 4$.

XII.3.2 Equality of left and right global dimension

If an algebra is commutative, or semisimple, its left and right global dimensions coincide. If not, and one wants to compare these dimensions, one needs to handle at the same time right and left modules. For this purpose, tensor products, and their derived torsion functors, are obviously the appropriate tools. We define an analogue of the right and left global dimensions, using Tor instead of Ext.

XII.3.9 Definition. Let A be a K-algebra. Its **weak global dimension** w.gl.dim.A is the largest integer n, if it exists, such that there exist a right A-module M and a left A-module N satisfying $\text{Tor}_n^A(M, N) \neq 0$.

Thus, w.gl.dim.$A \leqslant n$ if and only if $\text{Tor}_{n+1}^A(-, ?) = 0$. Because $\text{fd}M \leqslant \text{pd}M$, for any A-module M, we get, taking suprema, w.gl.dim.$A \leqslant$ r.gl.dim.A. Now, the weak global dimension measures at the same time right and left modules, so we also have $\text{w.gl.dim.}A \leqslant \text{l.gl.dim.}A$.

If there is no integer n as in the definition, then we say that the weak global dimension of A is infinite, which we write as $\text{w.gl.dim.}A = \infty$. Of course, if this is the case, then $\text{r.gl.dim.}A = \infty$ and $\text{l.gl.dim.}A = \infty$. We agree that $\text{w.gl.dim.}0 = -\infty$.

XII.3.10 Theorem. *Let A be a K-algebra, then*

$$\text{w.gl.dim.}A = \sup\{\text{fd}M_A : M \text{ a right module}\} = \sup\{\text{fd}_A N : N \text{ a left module}\}.$$

Proof. We prove the first equality, the second being similar. To say that $\text{fd}M \leqslant n$, for some integer $n \geq 0$, means that $\text{Tor}^A_{n+1}(M, -) = 0$ or, equivalently, $\text{Tor}^A_{n+1}(M, X) = 0$, for any left A-module X. Thus, $\text{fd}M \leqslant n$, for any right module M, if and only if $\text{w.gl.dim}A \leqslant n$. $\qquad\square$

Our main result says that over a noetherian algebra, the weak global dimension equals the right global dimension. For this, we need the following theorem.

XII.3.11 Theorem. *Over a noetherian algebra A, any finitely generated flat module is projective.*

Proof. Let F_A be a finitely generated flat module. In order to prove that F is projective, we shall prove that an epimorphism $f : M \to N$ induces an epimorphism $\text{Hom}_A(F, f) : \text{Hom}_A(F, M) \to \text{Hom}_A(F, N)$. Consider each of M, N as a $\mathbb{Z}$-A-bimodule. Because of Proposition VI.4.29, it suffices to prove that, for any injective cogenerator Q of $\text{Mod}\mathbb{Z}$ (for instance, $I = \mathbb{Q}/\mathbb{Z}$), the induced map $\text{Hom}_\mathbb{Z}(\text{Hom}_A(F, f), Q) : \text{Hom}_\mathbb{Z}(\text{Hom}_A(F, N), Q) \to \text{Hom}_\mathbb{Z}(\text{Hom}_A(F, M), Q)$ is injective.

According to Lemma VI.4.29, we have a commutative square

$$
\begin{array}{ccc}
\text{Hom}_\mathbb{Z}(\text{Hom}_A(F, N), Q) & \longrightarrow & \text{Hom}_\mathbb{Z}(\text{Hom}_A(F, M), Q) \\
\downarrow{\scriptstyle\cong} & & \downarrow{\scriptstyle\cong} \\
F \otimes_A \text{Hom}_\mathbb{Z}(N, Q) & \longrightarrow & F \otimes_A \text{Hom}_\mathbb{Z}(M, Q)
\end{array}
$$

where the vertical arrows are isomorphisms and the horizontal arrows are induced from f. Because Q is $\mathbb{Z}$-injective, the morphism $\text{Hom}_\mathbb{Z}(f, Q) : \text{Hom}_\mathbb{Z}(N, Q) \to \text{Hom}_\mathbb{Z}(M, Q)$ is injective. Because F_A is flat, so is the lower horizontal arrow in the commutative square. Hence, so is the upper. $\qquad\square$

XII.3.12 Corollary. *Let A be noetherian and M_A a finitely generated module. Then $\text{fd}M = \text{pd}M$.*

Proof. We already know that $\text{fd}M \leqslant \text{pd}M$, see the comments after Theorem XII.2.16. So we must prove that $\text{pd}M \leqslant \text{fd}M$. Let n be an integer such that $\text{fd}M \leqslant n$.

Then $\mathrm{Tor}_{n+1}^{A}(M,-) = 0$. Because M is finitely generated, there exists an exact sequence

$$0 \to L_{n-1} \to P_{n-1} \to \cdots \to P_0 \to M \to 0,$$

with the P_i finitely generated projective, due to Corollary VII.2.16. Dimension shifting gives $\mathrm{Tor}_1^{A}(L_{n-1}, X) \cong \mathrm{Tor}_{n+1}^{A}(M, X) = 0$, for any left module X. Due to Proposition XI.2.16, L_{n-1} is flat. Because A is noetherian and P_{n-1} is finitely generated, then L_{n-1} is finitely generated. Thanks to Theorem XII.3.11, L_{n-1} is projective, and the previous exact sequence is a projective resolution. So $\mathrm{pd} M \leqslant n$. $\qquad\square$

The following theorem is also due to Auslander.

XII.3.13 Theorem. *Let A be noetherian, then* r.gl.dim.A = w.gl.dim.A.

Proof. One has $\mathrm{fd} M \leqslant \mathrm{pd} M$, for any A-module M. Hence, w.gl.dim.$A \leqslant$ r.gl.dim.A. Conversely, the right global dimension of A is the supremum of the projective dimensions of finitely generated modules, because of Corollary XII.3.6, which is the same as the supremum of the flat dimensions, because of Lemma XII.3.12. The latter does not exceed the supremum of the flat dimensions of all modules, namely the weak global dimension. Therefore, r.gl.dim.$A \leqslant$ w.gl.dim.A. Equality follows. $\qquad\square$

Similarly, if A is left noetherian, w.gl.dim.A = l.gl.dim.A. We deduce the following corollary.

XII.3.14 Corollary. *Let A be right and left noetherian, then* r.gl.dim.A = l.gl.dim.A
$\qquad\square$

We may thus speak about global dimension for this class of algebras.

Because of the Hopkins–Levitski Theorem VIII.3.22, artinian algebras are noetherian, so, if A is artinian, then r.gl.dim.A = w.gl.dim.A. If A is left and right artinian, for instance, if A is a finite–dimensional algebra over a field, then r.gl.dim.A = l.gl.dim.A.

XII.3.3 The global dimension of a polynomial algebra

Given a commutative ring K, it is reasonable to ask what is the global dimension of the polynomial algebra $K[t_1, \cdots, t_n]$ in n indeterminates t_i, with coefficients in the ring K. For instance, if $n = 1$ and K is a field, then, because of Example XII.3.4(b) and, because $K[t]$ is not semisimple, we have gl.dim.$K[t] = 1$. The answer to this problem is known as **Hilbert's Syzygy Theorem**, and goes back to 1890. This subsection is devoted to its proof.

Below, we deal with two rings, K and the ring $K[t]$ of polynomials in t. Both being commutative, left and right modules coincide. We need to consider the forgetful functor $|-| : \mathrm{Mod} K[t] \to \mathrm{Mod} K$ (for brevity, given a $K[t]$-module M, we write M instead of $|M|$) and the change of scalars functors $- \otimes_K K[t] : \mathrm{Mod}K \to \mathrm{Mod}K[t]$. If X_K is a K-module, then $X \otimes_K K[t]$ is a $K[t]$-module, whose elements can be written as sums $\Sigma_i(x_i \otimes t^i)$, with the $x_i \in X$ almost all zero, see Exercise VI.2.7. Moreover, as a K-module, $K[t] \cong K^{(\mathbb{N})}$, and in particular is projective (even free).

XII.3.15 Lemma. *For any K-module X, we have* $\mathrm{pd}(X \otimes_K K[t])_{K[t]} \leqslant \mathrm{pd}X_K$.

Proof. The statement is trivial if $\mathrm{pd}X_K = \infty$. So assume that $\mathrm{pd}X_K = n > 0$. We have a projective resolution in $\mathrm{Mod}K$

$$0 \to P_n \to \cdots \to P_1 \to P_0 \to X \to 0$$

with $P_n \neq 0$. Now $K[t]$ is projective, hence flat, over K. Therefore, evaluating the functor $- \otimes_K K[t]$ yields an exact sequence

$$0 \to P_n \otimes_K K[t] \to \cdots \to P_1 \otimes_K K[t] \to P_0 \otimes_K K[t] \to X \otimes_K K[t] \to 0.$$

in $\mathrm{Mod}K[t]$. Because of Exercise VI.4.15, each $P_i \otimes_K K[t]$ is projective in $\mathrm{Mod}K[t]$. Thus, this sequence is a projective resolution of $X \otimes_K K[t]$, and the statement follows. $\qquad\square$

Multiplication by t defines an injective $K[t]$-linear map $\mu_{t,K[t]}$ from $K[t]$ to itself, whose cokernel is $K[t]/tK[t] \cong K$, and yields our familiar short exact sequence of $K[t]$-modules from Example II.4.16(b)

$$0 \to K[t] \xrightarrow{\mu_{t,K[t]}} K[t] \to K \to 0.$$

This suggests us to look at the short exact sequence in the next lemma.

XII.3.16 Lemma. *For any $K[t]$-module M, there exists a short exact sequence in* $\mathrm{Mod}K[t]$:

$$0 \to M \otimes_K K[t] \to M \otimes_K K[t] \to M \to 0.$$

Proof. There exists an obvious epimorphism of $K[t]$-modules $f : M \otimes_K K[t] \to M$, given by $\Sigma_i(x_i \otimes t^i) \mapsto \Sigma_i(x_i t^i)$, for $x_i \in M$ almost all zero, and $i \geq 0$. It suffices to construct an isomorphism $g : M \otimes_K K[t] \to \mathrm{Ker}f$. Set $g : \Sigma_{i=0}^n(x_i \otimes t^i) \to \Sigma_{i=1}^n x_i(1 \otimes t - t \otimes 1)t^i$, for $\Sigma_{i=1}^n(x_i \otimes t^i) \in M \otimes_K K[t]$. Then g is $K[t]$-linear and maps into $\mathrm{Ker}f$. In order to prove that g is an isomorphism, it is convenient to write its image otherwise:

$$g(\Sigma_{i=0}^n x_i \otimes t^i) = x_0(1 \otimes t - t \otimes 1) + x_1(1 \otimes t - t \otimes 1)t + \cdots + x_n(1 \otimes t - t \otimes 1)t^n$$

$$= -x_0 t \otimes 1 + (x_0 - x_1 t) \otimes 1 \cdot t + \cdots + (x_{n-1} - x_n t) \otimes 1 \cdot t^n + x_n \otimes 1 \cdot t^{n+1}$$

$$= -x_0 t \otimes 1 + \Sigma_{i=1}^n(x_{i-1} - x_i t) \otimes 1 \cdot t^i + x_n \otimes 1 \cdot t^{n+1}.$$

The latter expression allows to prove easily bijectivity.

(a) g is injective: suppose that $g(\Sigma_{i=0}^n x_i \otimes t^i) = 0$, then we have

$$x_n t = x_{n-1} - x_n t = \cdots = x_0 - x_1 t = -x_0 t = 0.$$

Hence, $x_i = 0$, for all i, and $\Sigma_{i=0}^n (x_i \otimes t^i) = 0$.

(b) g is surjective: if $\Sigma_{i=0}^n (y_i \otimes t^i) \in \text{Ker} f$, then $\Sigma_{i=0}^n y_i t^i = 0$ in M. The following set of equations (in the x_i) has a solution which can be obtained by a straightforward induction

$$0 = x_n, \quad y_n = x_{n-1}, \quad y_{n-1} = x_{n-2} - x_{n-1} t, \cdots y_0 = -x_0 t.$$

The proof is now complete.   □

XII.3.17 Corollary. *For any $K[t]$-module M, we have $\text{pd} M_K \leqslant \text{pd} M_{K[t]} \leqslant 1 + \text{pd} M_K$.*

Proof. Because $K[t]$ is projective as a K-module, the first inequality follows from Lemma XII.2.9. The exact sequence of Lemma XII.3.16 and Corollary XII.2.6 imply that $\text{pd} M_{K[t]} \leqslant 1 + \text{pd}(M \otimes_K K[t])_{K[t]}$. The second inequality now follows from Lemma XII.3.15.   □

Multiplication by t actually yields a morphism when defined on an arbitrary nonzero $K[t]$-module M. The (functorial) endomorphism $\mu_{t,M} : M \to M$ defined by $x \mapsto xt$, for $x \in M$, has Mt as its image.

XII.3.18 Lemma. *Let M be a $K[t]$-module.*
(a) If $\mu_{t,M}$ is injective, then $\text{pd}(M/Mt)_K \leqslant \text{pd} M_{K[t]}$.
(b) If $Mt = 0$, then $\text{pd} M_{K[t]} = 1 + \text{pd} M_K$.

Proof. (a) The statement is trivial if $\text{pd} M_{K[t]} = \infty$. We use induction on $\text{pd} M_{K[t]} = n$. If $n = 0$, then M is a projective $K[t]$-module. Because of Corollary VI.3.17, we have $M/Mt \cong M \otimes_{K[t]} K[t]/K[t]t \cong M \otimes_{K[t]} K$, which is a projective K-module, because $\text{Hom}_K(M/Mt, -) \cong \text{Hom}_{K[t]}(M, \text{Hom}_K(K, -))$, due to the adjunction isomorphism.

Assume that $\text{pd} M_{K[t]} = n > 0$. Then there exists a short exact sequence $0 \to N \to L \to M \to 0$ of $K[t]$-modules with L free. Using the functorial endomorphism $\mu_{t,-}$, we deduce a commutative diagram with exact rows in $\text{Mod} K[t]$

$$
\begin{array}{ccccccccc}
0 & \longrightarrow & N & \longrightarrow & L & \longrightarrow & M & \longrightarrow & 0 \\
& & \downarrow{\scriptstyle \mu_{t,N}} & & \downarrow{\scriptstyle \mu_{t,L}} & & \downarrow{\scriptstyle \mu_{t,M}} & & \\
0 & \longrightarrow & N & \longrightarrow & L & \longrightarrow & M & \longrightarrow & 0
\end{array}
$$

Because L is free, $\mu_{t,L}$ is injective (because so is $\mu_{t,K[t]}$, as observed above). Hence, $\mu_{t,N}$ is also injective. The short exact sequence $0 \to N \to L \to M \to 0$ and Corollary XII.2.6 yield $\text{pd} N_{K[t]} = n - 1$. The induction hypothesis gives $\text{pd}(N/Nt)_K \leqslant n - 1$. On the other hand, the Snake Lemma II.4.19, or the 3×3-

Lemma II.4.20, applied to the shown commutative diagram yields a short exact sequence consisting of the cokernels

$$0 \to N/Nt \to L/Lt \to M/Mt \to 0$$

in $\mathrm{Mod}K[t]$, but also in $\mathrm{Mod}K$ (because the forgetful functor is exact). Because L is free over $K[t]$, then L/Lt is free over K. Hence, Corollary XII.2.6 gives $\mathrm{pd}(M/Mt)_K \leqslant 1 + \mathrm{pd}(N/Nt)_K \leqslant 1 + (n-1) = n$.

(b) Due to Corollary XII.3.17, the statement holds true if $\mathrm{pd}M_{K[t]} = \infty$. Because M is nonzero such that $Mt = 0$, it cannot be projective: indeed a projective module is a submodule of a free module, hence any nonzero $x \in M$ has coordinates in $K[t]$, thus there exists a nonzero polynomial p that is a coordinate of x, and then $pt \neq 0$, therefore, $xt \neq 0$, a contradiction to $Mt = 0$. We may thus suppose that $\mathrm{pd}M_{K[t]} = n > 0$. The hypothesis $Mt = 0$ says that $\mu_{t,M} = 0$, and hence $\mathrm{Ker}\,\mu_{t,M} = M$. Thus, the Snake Lemma II.4.19 applied to the commutative diagram in (a) yields a four-terms exact sequence

$$0 \to M \to N/Nt \to L/Lt \to M \to 0.$$

Let X denote the image of the morphism in the middle. The short exact sequence $0 \to X \to L/Lt \to M \to 0$ and the fact that L/Lt is a free K-module imply that $\mathrm{pd}X_K = \mathrm{pd}M_K - 1$. On the other hand, because $\mathrm{pd}N_{K[t]} = n - 1$ and $\mu_{t,N}$ is injective, we have $\mathrm{pd}(N/Nt) \leqslant n - 1$, due to (a). Corollary XII.2.6 and the short exact sequence $0 \to M \to N/Nt \to X \to 0$ yield $\mathrm{pd}M_K \leqslant n - 1$. Finally, Corollary XII.3.17 gives $\mathrm{pd}M_K \geq \mathrm{pd}M_{K[t]} - 1 = n - 1$. Therefore, $\mathrm{pd}M_K = n - 1$ and the statement is established.   □

The main result of this subsection follows easily.

XII.3.19 Theorem. *Let K be a commutative ring, then* $\mathrm{gl.dim.}K[t] = 1 + \mathrm{gl.dim.}K$.

Proof. Let M be a $K[t]$-module. Because of Corollary XII.3.17, we have

$$\mathrm{pd}M_{K[t]} \leqslant 1 + \mathrm{pd}M_K \leqslant 1 + \mathrm{gl.dim.}K.$$

Any K-module X can be considered as a $K[t]$-module, with the action defined by $xt = 0$ for any $x \in X$, that is, $Xt = 0$. Applying Lemma XII.3.18(b), we get

$$\mathrm{pd}X_K = \mathrm{pd}X_{K[t]} - 1 \leqslant \mathrm{gl.dim.}K[t] - 1.$$

Taking the supremum yields $\mathrm{gl.dim.}K \leqslant \mathrm{gl.dim.}K[t] - 1$. This completes the proof.   □

An immediate induction show that, for a commutative ring K, we have $\mathrm{gl.dim.}K[t_1, \cdots, t_n] = n + \mathrm{gl.dim.}K$. Observe that the proof did not really use the full hypothesis that K is commutative, we only needed that t is central in $K[t]$.

Taking K to be a field, we obtain Hilbert's Syzygy Theorem.

XII.3.20 Corollary. *Let K be a field. Then* gl.dim.$K[t_1, \cdots, t_n] = n$.

Proof. The global dimension of a field is zero, see Example XII.3.4(a). $\square$

In particular, for any positive integer n, there exists a commutative algebra having global dimension n. A noncommutative example is given in Exercise XII.4.5.

XII.3.21 Example. The global dimension of a principal ideal domain is at most 1, due to Example XII.3.4(b). Hence, if K is a principal ideal domain, then gl.dim.$K[t] \leqslant 2$ and in general gl.dim.$K[t_1, \cdots, t_n] \leqslant n + 1$. For instance, gl.dim.$\mathbb{Z}[t] = 2$.

XII.3.4 The global dimension of a triangular matrix algebra

This last subsection is devoted to computing bounds for the global dimension of a triangular matrix algebra. Let K be a commutative ring, A, B be K-algebras, ${}_B M_A$ a B - A-bimodule and

$$R = \begin{pmatrix} A & 0 \\ M & B \end{pmatrix}.$$

This algebra has been studied in Exercises II.2.11, VII.2.11 and others. We are interested in the global dimension of R. Besides the intrinsic interest of the problem, the proof shows how to handle change of scalar functors.

In order not to distinguish between right and left global dimension, we assume R to be right and left noetherian, see Corollary XII.3.14. This is in particular the case if K is a field and R a finite–dimensional K-algebra. According to Exercise VII.2.11, R is right and left noetherian if and only if A, B are right and left noetherian and the modules ${}_B M$ et M_A are noetherian. Consequently, the right and left global dimensions of A and B coincide as well.

Let I be an ideal in R, we shall compare R-modules with R/I-modules. Due to Example IV.4.6(a), the R/I-modules can be identified with the R-modules X such that $XI = 0$. On the other hand, if P is a projective R-module, then P/PI is a projective R/I-module: indeed, $P/PI \cong P \otimes_R (R/I)$, due to Corollary VI.3.17, hence the functor

$$\mathrm{Hom}_{R/I}(P/PI, -) \cong \mathrm{Hom}_{R/I}(P \otimes_R (R/I), -)$$
$$\cong \mathrm{Hom}_R(P, \mathrm{Hom}_{R/I}(R/I, -))$$
$$\cong \mathrm{Hom}_R(P, -)$$

is exact (this is Exercise VI.4.13). In particular, if P is an R/I-module which is projective when viewed as an R-module, then it is also projective as an R/I-module (because, in this case, $PI = 0$). The next two lemmata are valid without the assumption that R is a triangular matrix algebra.

XII.3.22 Lemma. *Let R be a right and left noetherian algebra and I an ideal of R such that $I^2 = I$ and I_R is a projective R-module. Then*

$$\mathrm{gl.dim.}(R/I) \leqslant \mathrm{gl.dim.}R.$$

Proof. We prove that, for any R/I-module X, we have $\mathrm{pd}X_{R/I} \leqslant \mathrm{pd}X_R$. If X is projective as an R-module, then it is also projective as an R/I-module, according to the previous comments. We may thus assume that X is not projective. There exists a short exact sequence of R-modules

$$0 \to Y \to L \to X \to 0$$

with L free, and we may assume that $Y \subseteq L$. Because $X \cong L/Y$ is an R/I-module, we have $(L/Y)I = XI = 0$, that is, $LI \subseteq Y$. Multiplying by I, we get $L \supseteq Y \supseteq LI \supseteq YI \supseteq LI^2 = LI$, which gives $LI = YI$. Now, because $LI \subseteq Y$, we have a short exact sequence of R-modules

$$0 \to Y/LI \to L/LI \to X \to 0.$$

due to the Isomorphism Theorem II.4.25. Because $LI = YI$, the sequence can be written as

$$0 \to Y/YI \to L/LI \to X \to 0$$

and it is also an exact sequence of R/I-modules. We use induction on $\mathrm{pd}X = n \geq 1$.

If $n = 1$, the sequence $0 \to Y \to L \to X \to 0$ and Theorem XII.2.3 yield Y projective as an R-module. Hence, Y/YI is projective as an R/I-module (according to the comments preceding the lemma). Consequently, $\mathrm{pd}X_{R/I} \leqslant 1$, and the statement is verified in this case.

Suppose that $n > 1$. We claim that LI is a projective R-module. Indeed, because L is a free R-module, we have $\mathrm{Tor}_1^R(L, -) = 0$, hence a commutative diagram with exact rows

$$
\begin{array}{ccccccccc}
0 = \mathrm{Tor}_1^R(L, R/I) & \longrightarrow & L \otimes_R I & \longrightarrow & L \otimes_R R & \longrightarrow & L \otimes_R R/I & \longrightarrow & 0 \\
& & \downarrow{\mu'} & & \downarrow{\mu} & & \downarrow{\mu''} & & \\
0 & \longrightarrow & LI & \longrightarrow & L & \longrightarrow & L/LI & \longrightarrow & 0
\end{array}
$$

where μ is the multiplication map $x \otimes a \mapsto xa$, for $x \in L, a \in R$, the morphism μ' is its restriction, and μ'' is induced by passing to cokernels. Because μ and μ'' are isomorphisms, so is μ'. Hence, $LI \cong L \otimes_R I$. Because I_R is projective, the functor

$$\mathrm{Hom}_R(LI, -) \cong \mathrm{Hom}_R(L \otimes_R I, -) \cong \mathrm{Hom}_R(L, \mathrm{Hom}_R(I, -))$$

is exact, which establishes our claim.

The short exact sequence $0 \to LI \to L \to L/LI \to 0$ gives $\mathrm{pd}(L/LI)_R \leqslant 1$. Then, Corollary XII.2.6(a) and the exact sequence $0 \to Y/YI \to L/LI \to X \to$

0 yield $\mathrm{pd}(Y/YI)_R \leqslant n - 1$. The induction hypothesis gives $\mathrm{pd}(Y/YI)_{R/I} \leqslant n - 1$. Because $\mathrm{pd}(L/LI)_{R/I} \leqslant 1$, according to the case $n = 1$ above, it follows that $\mathrm{pd}X_{R/I} \leqslant 1$. The proof is complete.   □

XII.3.23 Lemma. *Let R be a right and left noetherian algebra and I, J ideals in R such that JI = 0. Then*

$$\mathrm{gl.dim.}R \leqslant \sup\{\mathrm{gl.dim.}(R/I) + \mathrm{pd}(R/I)_R, \mathrm{gl.dim.}(R/J) + \mathrm{pd}(R/J)_R\}.$$

Proof. Let X be an R-module. Because XJ is a submodule of X, the quotient X/XJ is defined in Mod$R$. But it is also an R/J-module, because it is annihilated by J. Due to Lemma XII.2.9, we have

$$\mathrm{pd}(X/XJ)_R \leqslant \mathrm{pd}(X/XJ)_{R/J} + \mathrm{pd}(R/J)_R \leqslant \mathrm{gl.dim.}(R/J) + \mathrm{pd}(R/J)_R.$$

On the other hand, XJ is an R/I-module, because $(XJ)I = X(JI) = 0$. Hence,

$$\mathrm{pd}(XJ)_R \leqslant \mathrm{gl.dim.}(R/I) + \mathrm{pd}(R/I)_R.$$

Corollary XII.2.6(a) applied to the short exact sequence of R-modules $0 \to XJ \to X \to X/XJ \to 0$ gives the required inequality.   □

From now on, we assume again R to be a triangular matrix algebra. The following notation is useful. Set

$$R = \begin{pmatrix} A & 0 \\ M & B \end{pmatrix}.$$

as before. The identities 1_A and 1_B of A and B, respectively, correspond to idempotents in R,

$$e_A = \begin{pmatrix} 1_A & 0 \\ 0 & 0 \end{pmatrix} \text{ and } e_B = \begin{pmatrix} 0 & 0 \\ 0 & 1_B \end{pmatrix}.$$

We then have $A = e_A R e_A$ and $B = e_B R e_B$.

XII.3.24 Proposition. *Let*

$$R = \begin{pmatrix} A & 0 \\ M & B \end{pmatrix}$$

be a right and left noetherian algebra. Then we have

$$\sup\{\mathrm{gl.dim.}A, \mathrm{gl.dim.}B\} \leqslant \mathrm{gl.dim.}R \leqslant \sup\{\mathrm{gl.dim.}A, \mathrm{gl.dim.}B + \mathrm{pd}M_A + 1\}.$$

Proof. Consider the ideals $I = \begin{pmatrix} 0 & 0 \\ M & B \end{pmatrix}$ and $J = \begin{pmatrix} A & 0 \\ M & 0 \end{pmatrix}$ in R. We have $JI = 0, I^2 = I$ and $I_R = e_B R$ is a projective R-module. Moreover, $R/I \cong A$ and $R/J \cong B$. Because of Lemma XII.3.22, we have gl.dim.$A \leqslant$ gl.dim.R and gl.dim.$B \leqslant$ gl.dim.R, hence the first inequality.

Because of Lemma XII.3.23, we have

$$\text{gl.dim.}R \leqslant \sup\{\text{gl.dim.}A + \text{pd}A_R, \text{gl.dim.}B + \text{pd}B_R\}.$$

Now, $A_R \cong e_A R$ is projective. Hence, pd$A_R = 0$. On the other hand, we have a short exact sequence of R-modules $0 \rightarrow M \rightarrow I \rightarrow B \rightarrow 0$. Because I_R is projective, we have pd$B_R = 1 + \text{pd}M_R$. Because of Lemma XII.2.9, pd$M_R \leqslant \text{pd}M_A + \text{pd}A_R = \text{pd}M_A$, due to projectivity of A_R. Therefore, pd$B_R \leqslant 1 + \text{pd}M_A$, and we have indeed gl.dim.$R \leqslant \sup\{\text{gl.dim.}A, \text{gl.dim.}B + \text{pd}M_A + 1\}$, as required.   □

XII.3.25 Examples. We show that both bounds may be attained. Let K be a field.

(a) Let $R = T_2(K)$. In this case, $A = B = K$ which is semisimple, hence gl.dim.$A = $ gl.dim.$B = 0$. For the same reason, pd$M_A = 0$. The lower bound is thus 0 and the upper bound is 1. Now, gl.dim.$R = 1$ due to Example XII.3.8(a). Here, the upper bound is attained.

(b) Let

$$R = T_4(K) = \begin{pmatrix} K & 0 & 0 & 0 \\ K & K & 0 & 0 \\ K & K & K & 0 \\ K & K & K & K \end{pmatrix}$$

Here, $A = B = T_2(K)$ hence gl.dim.$A = $ gl.dim.$B = 1$. The module M_A is isomorphic to the direct sum of two copies of the indecomposable projective $T_2(K)$-module $P_2 = e_{22}A$, in the notation of Example XII.3.8(a) (why?). Hence, pd$M_A = 0$ and the upper bound is 2, while the lower is 1. Now the global dimension of R is equal to 1 : indeed, $S_1 = e_{11}R$ is projective, and the other simple R-modules have all projective dimension one, because of the short exact sequences:

$$0 \rightarrow e_{11}R \rightarrow e_{22}R \rightarrow S_2 \rightarrow 0 \,, 0 \rightarrow e_{22}R \rightarrow e_{33}R \rightarrow S_3 \rightarrow 0$$

$$\text{and } 0 \rightarrow e_{33}R \rightarrow e_{44}R \rightarrow S_4 \rightarrow 0.$$

Theorem XII.3.7 applies here, because R is a finite–dimensional K-algebra, and yields gl.dim.$R = 1$. In this case, the lower bound has been attained.

Exercises of Section XII.3

1. Prove that the following conditions are equivalent for an algebra A :
  (a) r.gl.dim$A \leqslant 2$.
  (b) The kernel of any morphism between projectives is projective.
  (c) The cokernel of any morphism between injectives is injective.
  (d) For any module M, the functor $\text{Ext}_A^2(M, -)$ is right exact.
  (e) For any module M, the functor $\text{Ext}_A^2(-, M)$ is right exact.

2. Let A be a nonsemisimple algebra. Prove that

$$\text{r.gl.dim.}A = 1 + \sup\{\text{pd}I : I \text{ right ideal of } A\}.$$

3. Let $0 \to L \to M \to N \to 0$ be a short exact sequence of A-modules.
   (a) If $\text{pd}L = \text{r.dim.gl.}A$, prove that $\text{pd}M = \text{r.dim.gl.}A$.
   (b) State and prove a similar result for the injective dimensions.

4. Let A be an algebra of right global dimension n. Prove that:
   (a) If M is a module of projective dimension $n - 1$, then any submodule of M has projective dimension at most $n - 1$.
   (b) If M is a module of injective dimension $n - 1$, then any quotient of M has injective dimension at most $n - 1$.

5. Let $A_1, \cdots, A_t$ be algebras. Prove that

$$\text{r.gl.dim}\left(\prod_{i=1}^{t} A_i\right) = \sup_i\{\text{r.gl.dim.}A_i : i = 1\cdots, t\}.$$

6. Let A be a noetherian algebra and $n > 0$ an integer. Prove that the following conditions are equivalent:
   (a) $\text{r.gl.dim}\,A \leqslant n$.
   (b) For any finitely generated modules M_A, N_A, we have $\text{Ext}_A^{n+1}(M, N) = 0$.
   (c) For any finitely generated modules $M_A, {}_A N$, we have $\text{Tor}_{n+1}^A(M, N) = 0$.

7. Let A be a noetherian algebra and P a finitely generated A-module. Prove that P is projective if and only if, for any left ideal I of A, the multiplication map $P \otimes_A I \to PI$ given by $x \otimes a \mapsto xa$, for $x \in P, a \in A$, is an isomorphism of K-modules.

8. Prove that the right global dimension of an artinian algebra equals the supremum of the projective dimensions of its indecomposable modules.

9. Prove that, for any algebra A, the weak global dimension equals the supremum of the flat dimensions of the A/I, with I a finitely generated right ideal of A.

10. Let K be a field and $A = K[s, t]/\langle st \rangle$. Prove that:
    (a) There exists a short exact sequence of A-modules $0 \to sA \to A \to tA \to 0$.
    (b) The right ideal sA contains no nonzero idempotent.
    (c) This exact sequence does not split, and $\text{gl.dim.}A = \infty$.

11. Let K be a field and A a finite–dimensional K-algebra. Let T be the direct sum of a complete set of representatives of the isomorphism classes of simple A-modules. Prove that:
    (a) For an A-module M, we have $\text{pd}M = \sup\{n : \text{Ext}_A^n(M, T) \neq 0\} = \sup\{n : \text{Tor}_n^A(M, T) \neq 0\}$ and $\text{id}M = \sup\{n : \text{Ext}_A^n(T, M) \neq 0\}$.
    (b) $\text{gl.dim.}A = \text{pd}T_A = \text{id}_A T = \sup\{n : \text{Ext}_A^n(T, T) \neq 0\} = \sup\{n : \text{Tor}_n^A(T, T) \neq 0\}$.

12. Prove that, in Lemma XII.3.15, one has actually equality.

13. This exercise generalises Lemma XII.3.22. Let A be a right and left noetherian algebra and I an ideal in A such that $I^m = I^{m+1}$ for some integer $m > 0$, and I_A is a projective A-module. Prove that $\mathrm{gl.dim.}(A/I) \leqslant \mathrm{gl.dim.}A + 2m - 2$.

14. Compute the global dimension of each of the following triangular matrix algebras:

(a) $A = \begin{pmatrix} K & 0 & 0 \\ K & K & 0 \\ K & 0 & K \end{pmatrix}$.

(b) $B = \begin{pmatrix} K & 0 & 0 & 0 \\ K & K & 0 & 0 \\ K & 0 & K & 0 \\ K & K & K & K \end{pmatrix}$.

15. Let K be a field. Compute the global dimension of the algebra given by the quiver

$$\underset{1}{\circ} \; \underset{\beta}{\overset{\alpha}{\rightleftarrows}} \; \underset{2}{\circ} \quad \text{bound by the ideals}$$

(a) $I_1 = \langle \alpha\beta \rangle$,
(b) $I_2 = \langle \alpha\beta\alpha \rangle$ and
(c) $I_3 = \langle \alpha\beta, \beta\alpha \rangle$.

16. Let K be a field and A the K-algebra given by the quiver

$$\underset{5}{\circ} \xleftarrow{\delta} \underset{4}{\circ} \xleftarrow{\gamma} \underset{3}{\circ} \xleftarrow{\beta} \underset{2}{\circ} \xleftarrow{\alpha} \underset{1}{\circ}$$

bound by the ideal $\langle \alpha\beta\gamma, \gamma\delta \rangle$.
   (a) Compute the global dimension of A.
   (b) Prove that $\mathrm{Ext}_A^3(S_5, S_1) = K$, and give a representative for a basis of this space.
   (c) Compute $\mathrm{Ext}_A^1(S_2 \oplus P_3, S_2 \oplus P_3)$.
   (d) Prove that $\mathrm{id}P_3 > 1$.

XII.4 Classes of algebras

XII.4.1 Hereditary algebras

We illustrate the previous discussion by looking at algebras with small global dimension. There is not much to say about algebras of global dimension zero: we have seen in Example XII.3.4(a) that they coincide with semisimple algebras. The next case is that of global dimension one. It is convenient, however, to define this class using another property.

XII.4.1 Definition. A K-algebra A is a **right hereditary algebra** if any right ideal in A is projective as an A-module.

The definition amounts to saying that A is right hereditary if and only if any cyclic right A-module has projective dimension one. Indeed, an A-module is cyclic if and only if it is the quotient of A_A by a right ideal, see Proposition II.4.22.

We already encountered right hereditary algebras in Subsection V.2.2, without naming them so, and proved in Theorem V.2.4 that if A is right hereditary, then any submodule of a free module is a direct sum of right ideals (hence is projective).

Similarly, A is **left hereditary** if any left ideal is projective as left A-module. Right hereditary algebras are generally not left hereditary, and conversely, see Example XII.4.5(d) below. Because of our comments above, one can also think of hereditary algebras as being those which are closest to the semisimple ones, from the global dimension point of view. In order to prove our first characterisation of hereditary algebras, we need a lemma, which is actually Exercise VI.4.6(b).

XII.4.2 Lemma. *Let A be a K-algebra. An A-module I is injective if and only if, for any monomorphism $j : P' \to P$, with P projective, and any morphism $u : P' \to I$, there exists $v : P \to I$ such that $vj = u$.*

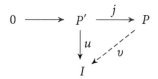

Proof. Let $j : M' \to M$ be a monomorphism and $u : M' \to I$ an A-linear map. We must construct an A-linear map $w : M \to I$ extending u. There exists a short exact sequence $0 \to L \xrightarrow{g} P \xrightarrow{f} M \to 0$ with P projective. Let P' be the fibered product of j and f. Because of Theorem IV.3.4, we have a commutative diagram with exact rows

$$
\begin{array}{ccccccccc}
0 & \longrightarrow & L & \xrightarrow{g'} & P' & \xrightarrow{f'} & M' & \longrightarrow & 0 \\
& & \downarrow{\scriptstyle 1_L} & & \downarrow{\scriptstyle j'} & & \downarrow{\scriptstyle j} & & \\
0 & \longrightarrow & L & \xrightarrow{g} & P & \xrightarrow{f} & M & \longrightarrow & 0
\end{array}
$$

Because of the Snake Lemma II.4.19, or Lemma IV.3.3, j' is a monomorphism. Consider the morphism $uf' : P' \to I$. The hypothesis yields $v : P \to I$ such that $vj' = uf'$. Now, $vg = vj'g' = uf'g' = 0$, because $f'g' = 0$. Hence, v factors through $M = \mathrm{Coker}\, g$, that is, there exists $w : M \to I$ such that $wf = v$. But then $uf' = vj' = wfj' = wjf'$ gives $u = wj$, because f' is an epimorphism. The proof is complete. $\qquad\square$

XII.4.3 Theorem. *Let A be a K-algebra, the following conditions are equivalent:*
(a) *A is right hereditary.*
(b) *r.gl.dim $A \leqslant 1$.*
(c) *Any submodule of a projective A-module is projective.*
(d) *Any quotient of an injective A-module is injective.*

Proof. (a) implies (c). Let A be right hereditary. A projective module P is a direct summand, hence a submodule, of a free module. Therefore, any submodule M of

P is a submodule of a free module. Because of Theorem V.2.4, *M* is a direct sum of right ideals, each of which is projective, because *A* is right hereditary. Proposition VI.4.3 gives that *M* is projective.

(c) implies (b). Let *M* be an *A*-module. There exists an epimorphism $f : P \to M$ with *P* projective. The hypothesis implies that $\mathrm{Ker}\, f$ is projective, and the short exact sequence $0 \to \mathrm{Ker}\, f \to P \xrightarrow{f} M \to 0$ gives $\mathrm{pd}\, M \leqslant 1$. Therefore, r.gl.dim.$A \leqslant 1$.

(b) implies (a). Let *I* be a right ideal in *A*. Because $\mathrm{pd}(A/I) \leqslant 1$, the short exact sequence $0 \to I \to A \to A/I \to 0$ gives that *I* is projective.

This proves the equivalence of (a), (b) and (c). There remains to prove the equivalence with (d). Consider the diagram with exact rows:

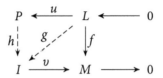

where *P* is projective and *I* injective. If (c) holds true, then there exists $g : L \to I$ such that $f = vg$. Injectivity of *I* yields $h : P \to I$ such that $g = hu$. Then $f = v(hu)$. Because of Lemma XII.4.2, *M* is injective. Thus, (c) implies (d). The proof of the converse is dual. ◻

Further conditions equivalent to those of the theorem are: an algebra *A* is right hereditary if and only if, for any *A*-module *M*, we have $\mathrm{pd}\, M \leqslant 1$, or if and only if, for any *A*-module *M*, we have $\mathrm{id}\, M \leqslant 1$, Thus, one may think of a right hereditary algebra as one in which any module is not too far from being projective, or, equivalently, not too far from being injective.

XII.4.4 Corollary. *Let A be a right and left noetherian algebra. Then it is right hereditary if and only if it is left hereditary.*

Proof. This follows from the theorem and Corollary XII.3.14. ◻

The previous corollary applies to right and left artinian algebras, and finite–dimensional algebras over a field.

XII.4.5 Examples.
  (a) Semisimple algebras are right and left hereditary.
  (b) Principal ideal domains have global dimension at most one, see Example XII.3.4(b). Hence, they are hereditary (right and left, because they are commutative).
  (c) Let *K* be a field and $A = T_2(K)$. Then gl.dim.$A = 1$, see Example XII.3.8(a). Hence, *A* is hereditary (right and left, because it is a finite–dimensional algebra). The algebra *A* is the path algebra of a quiver, see Exercise I.4.9(d). We prove in Example XII.4.11 below that path algebras of acyclic quivers are hereditary.

(d) The $\mathbb{Z}$-algebra of triangular matrices $A = \begin{pmatrix} \mathbb{Q} & 0 \\ \mathbb{Q} & \mathbb{Z} \end{pmatrix}$ is right hereditary but not left hereditary. We prove right heredity. The right ideals of A are of one of the following three types, see Example VII.3.15 and Exercise II.2.11,

$$I_1 = \begin{pmatrix} \mathbb{Q} & 0 \\ \mathbb{Q} & n\mathbb{Z} \end{pmatrix} \text{ for some } n \neq 0, I_2 = \begin{pmatrix} 0 & 0 \\ \mathbb{Q} & n\mathbb{Z} \end{pmatrix} \text{ for some } n \neq 0, \text{ or}$$

$$I_E = \left\{ \begin{pmatrix} c_1 & 0 \\ c_2 & 0 \end{pmatrix} : \begin{pmatrix} c_1 \\ c_2 \end{pmatrix} \in E \right\}$$

where E is a vector subspace of $\mathbb{Q}^{(2)}$.

First, we claim that I_1 is free, hence projective. Indeed, any element of I_1 can be written in the form

$$\begin{pmatrix} a & 0 \\ b & nk \end{pmatrix} = \begin{pmatrix} n(a/n) & 0 \\ n(b/n) & nk \end{pmatrix} = n \begin{pmatrix} a/n & 0 \\ b/n & k \end{pmatrix} \in nA$$

that is, $I_1 \cong nA \cong A_A$, which establishes our claim. But then $I_1 \cong I_2 \oplus I_0$, with $I_0 = \begin{pmatrix} \mathbb{Q} & 0 \\ 0 & 0 \end{pmatrix}$, implies that I_2 and I_0 are projective as well.

Second, we look at I_E as an A-module. We claim that I_E is isomorphic to $E_{\mathbb{Q}}$ considered as an A-module under the change of scalars functor induced by the surjective algebra morphism $A \to \mathbb{Q}$ defined by $\begin{pmatrix} a & 0 \\ b & k \end{pmatrix} \mapsto a$, for $\begin{pmatrix} a & 0 \\ b & k \end{pmatrix} \in A$. This indeed follows from the equalities

$$\begin{pmatrix} c_1 & 0 \\ c_2 & 0 \end{pmatrix} \begin{pmatrix} a & 0 \\ b & k \end{pmatrix} = \begin{pmatrix} c_1 a & 0 \\ c_2 a & 0 \end{pmatrix} = \begin{pmatrix} c_1 & 0 \\ c_2 & 0 \end{pmatrix} a.$$

This reduces the problem to E, considered as an A-module. Now, $E_{\mathbb{Q}}$ is a vector subspace of $\mathbb{Q}^{(2)}$, hence isomorphic to $0, \mathbb{Q}$ or $\mathbb{Q}^{(2)}$. If E is one-dimensional over $\mathbb{Q}$, then $I_E \cong I_0$, where I_0 is as above. But I_0 is projective, as a direct summand of I_1, hence I_E is projective as well. If $E \cong \mathbb{Q}^{(2)}$, then I_E is projective. This finishes the proof that A is right hereditary.

In order to prove that A is *not* left hereditary, we prove that the left ideal

$$N = \left\{ \begin{pmatrix} 0 & 0 \\ b & 0 \end{pmatrix} : b \in \mathbb{Q} \right\}$$

is not projective as a left A-module. Indeed, consider the surjective algebra morphism $A \to \mathbb{Z}$ given by $\begin{pmatrix} a & 0 \\ b & k \end{pmatrix} \mapsto k$, for $\begin{pmatrix} a & 0 \\ b & k \end{pmatrix} \in A$, and denote by J its kernel. The isomorphisms of functors

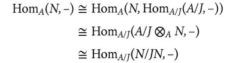

$$\text{Hom}_A(N, -) \cong \text{Hom}_A(N, \text{Hom}_{A/J}(A/J, -))$$
$$\cong \text{Hom}_{A/J}(A/J \otimes_A N, -)$$
$$\cong \text{Hom}_{A/J}(N/JN, -)$$

the isomorphism $A/J \cong \mathbb{Z}$ of algebras and the isomorphism $N/JN \cong \mathbb{Q}$ of modules yield that $\text{Hom}_A(N, -) \cong \text{Hom}_{\mathbb{Z}}(\mathbb{Q}, -)$. Thus, $_A N$ is projective if and only if $\mathbb{Q}$ is a projective $\mathbb{Z}$-module, and this is not true, see Example VI.4.44. Therefore, N is not projective.

In the noetherian case, being hereditary is decided on the level of finitely generated modules.

XII.4.6 Proposition. *A noetherian algebra is right hereditary if and only if any submodule of a finitely generated projective module is projective.*

Proof. Necessity is obvious, so we prove sufficiency. A right ideal I in a noetherian algebra A is a submodule of the finitely generated (even cyclic) module A_A. Hence, it is projective. ☐

Artinian algebras have only finitely many nonisomorphic finitely generated indecomposable projective modules, hence the interest of the following proposition.

XII.4.7 Proposition. *An artinian algebra is right hereditary if and only if the radical of any indecomposable finitely generated projective module is projective.*

Proof. Necessity being obvious, we prove sufficiency. Let A be an artinian algebra such that radicals of finitely generated projectives are projective. Thanks to Proposition XII.4.6, it suffices to prove that any submodule M of a finitely generated projective module P is projective. We use induction on the composition length $\ell(P)$ of P which, due to Corollary VIII.3.24, is finite. If $\ell(P) = 1$, then P is simple and there is nothing to prove. Assume that $\ell(P) > 1$, and that the statement holds true for any finitely generated projective module of lesser length. Then P can be written as $P = P' \oplus P''$, where P' is indecomposable and P'' may be zero. Let $p : P \to P'$ be the projection and consider the image $p(M)$ of M in P'. We have two cases.

If $p(M) = P'$, then the composite $pj : M \to P'$ of the inclusion $j : M \to P$ with p is an epimorphism. Because P' is projective, it is a retraction, so we can write $M \cong P' \oplus M'$, where $M' = M \cap P'' \subseteq P''$. Because $\ell(P'') < \ell(P)$, the induction hypothesis gives M' projective. Therefore, so is $M \cong P' \oplus M'$.

If $p(M) \neq P'$, then $p(M) \subseteq \text{rad}P'$, so that $M \subseteq \text{rad}P' \oplus P''$. The hypothesis implies that $\text{rad}P'$ is projective. Moreover, $\ell(\text{rad}P' \oplus P'') = \ell(\text{rad}P') + \ell(P'') = \ell(P')-1+\ell(P'') = \ell(P)-1$. Hence, M is projective due to the induction hypothesis. This finishes the proof. ☐

XII.4.8 Corollary. *Let A be an artinian algebra. The following conditions are equivalent:*

(a) *A is right hereditary.*
(b) *Any indecomposable finitely generated module has projective dimension at most 1.*
(c) *Any simple A-module has projective dimension at most 1.*

Proof. Clearly, (a) implies (b) which implies (c). Finally, (c) implies (a) because of Theorem XII.3.7. $\qquad\square$

XII.4.9 Corollary. *Let A be an artinian right hereditary algebra. Then:*
  (a) *Any nonzero morphism between finitely generated indecomposable projective modules is a monomorphism.*
  (b) *The endomorphism algebra of a finitely generated indecomposable projective module is a division ring.*

Proof. (a) Let $f : P \to P'$ be a nonzero morphism, with P, P' finitely generated indecomposable projective modules. Because A is hereditary, the submodule $\operatorname{Im} f$ of P' is projective. Hence, the short exact sequence $0 \to \operatorname{Ker} f \to P \to \operatorname{Im} f \to 0$ splits so $P \cong \operatorname{Ker} f \oplus \operatorname{Im} f$. Because $f \neq 0$, we have $\operatorname{Im} f \neq 0$. But P is indecomposable. Hence, $\operatorname{Ker} f = 0$.

(b) Because of (a), any nonzero endomorphism of P is a monomorphism, but then, because of Lemma VIII.4.16, it is an isomorphism. $\qquad\square$

For finite–dimensional algebras over a field, where we have a duality between finitely generated right and left modules, the preceding results may be interpreted in terms of injective modules, because of Theorem IX.2.16. The following corollary summarises our characterisations. Even more, we may replace right by left modules throughout.

XII.4.10 Corollary. *Let K be a field and A a finite–dimensional K-algebra. The following conditions are equivalent:*
  (a) *A is hereditary.*
  (b) *gl.dim.$A \leqslant 1$.*
  (c) *The projective dimension of any simple module is at most 1.*
  (d) *The injective dimension of any simple module is at most 1.*
  (e) *The projective dimension of any indecomposable module is at most 1.*
  (f) *The injective dimension of any indecomposable module is at most 1.*
  (g) *Any submodule of a projective module is projective.*
  (h) *Any quotient of an injective module is injective.*
  (i) *Any submodule of a projective finitely generated module is projective.*
  (j) *Any quotient of an injective finitely generated module is injective.*
  (k) *The radical of an indecomposable projective finitely generated module is projective.*
  (l) *The quotient of an indecomposable injective finitely generated module by its socle is injective.* $\qquad\square$

XII.4.11 Examples. Let K be a field.
  (a) We prove that the path algebra A of any finite acyclic quiver Q is hereditary. Because Q is acyclic, $A = KQ$ is finite–dimensional over K. For points y, z

in Q, let $w(y, z)$ be the number of paths from y to z. Because Q is finite and acyclic, $w(y, z)$ is finite for any y, z. Given a point x, let $y_1, \cdots, y_t$ be the distinct points such that there exists at least an arrow $x \to y_i$, and, for each i, let n_i be the number of arrows from x to y_i. Because of Example VIII.3.25(c), the top of radP_x equals $\oplus_{i=1}^{t} S_{y_i}^{n_i}$, so we have a projective cover morphism $p : \oplus_{i=1}^{t} P_{y_i}^{n_i} \to \mathrm{rad}P_x$. Let $y \neq x$ be arbitrary in Q, then

$$\dim_K(\mathrm{rad}P_x)e_y = \dim_K(P_x e_y) = \dim_K(e_x A e_y) = w(x, y) = \sum_{i=1}^{t} n_i w(y_i, y)$$

$$= \sum_{i=1}^{t} \dim_K(P_{y_i} e_y) = \dim_K \left(\bigoplus_{i=1}^{t} P_{y_i}^{n_i} \right) e_y.$$

So, p is an isomorphism and radP_x is projective. Due to Proposition XII.4.7, A is hereditary.

This statement does *not* extend to quotients of path algebras: the algebras of Examples XII.3.8(b)(c)(d) are not hereditary.

(b) Conversely, we prove that the quiver of an elementary hereditary algebra A is acyclic. For, assume that $x_0 \to x_1 \to \cdots \to x_n \to x_0$ is an oriented cycle in the quiver of A. Each arrow $\alpha : a \to b$ in a quiver induces a morphism between the corresponding indecomposable projective modules $e_a A \to e_b A$; indeed, $\alpha e_a = e_b \alpha$ in the bound quiver algebra, hence left multiplication by α maps $e_a A$ to $e_b A$. In particular, this morphism is nonzero, and it is a monomorphism due to Corollary XII.4.9(a). Therefore, the cycle $x_0 \to x_1 \to \cdots \to x_n \to x_0$ induces a nonzero composite monomorphism $e_{x_0} A \to e_{x_1} A \to \cdots \to e_{x_n} A \to e_{x_0} A$. Because of Corollary XII.4.9(b), it is an isomorphism. Hence, so is each of the individual morphisms $e_{x_i} A \to e_{x_{i+1}} A$, an absurdity because these two distinct indecomposable projective modules have distinct tops.

(c) Moreover, if a bound quiver algebra $A = KQ/I$ is hereditary, then $I = 0$. Indeed, if this is not the case, then there exist points $x, y \in Q_0$ such that $\epsilon_x I \epsilon_y \neq 0$. We may also assume x to be minimal in the sense that, for any $z \neq x, y$ lying on a path from x to y, we have $\epsilon_z I \epsilon_y \neq 0$. Then, using the notation of (a), we have

$$\dim_K(\mathrm{rad}P_x)e_y = \dim_K(e_x A e_y) = \dim_K(\epsilon_x K Q \epsilon_y) - \dim_K (\epsilon_x I \epsilon_y)$$

$$< w(x, y) = \sum_{i=1}^{t} n_i w(y_i, y) = \dim_K \left(\bigoplus_{i=1}^{t} P_{y_i}^{n_i} \right) e_y.$$

In other words, radP_x is not projective, and A is not hereditary, due to Proposition XII.4.7.

We have proved that an elementary finite–dimensional K-algebra is hereditary if and only if it is the path algebra of an acyclic quiver.

Hereditary finite–dimensional algebras are characterised by the property that, for any indecomposable finitely generated module M, one has $\mathrm{pd}M \leqslant 1$, or equivalently, for any indecomposable finitely generated module N, one has $\mathrm{id} N \leqslant 1$. Following this philosophy, one could think of characterising finite–dimensional algebras such that any indecomposable finitely generated A-module has bounded projective or injective dimension. Then the global dimension cannot be too large.

XII.4.12 Lemma. *Let $m, n \geqslant 1$ be integers, K a field and A a finite–dimensional K-algebra such that, for any indecomposable finitely generated A-module M, one has $\mathrm{pd} M \leqslant m$ or $\mathrm{id} M \leqslant n$. Then $\mathrm{gl.dim.} A \leqslant m + n + 1$.*

Proof. Let M be an indecomposable finitely generated A-module. We claim that $\mathrm{pd} M \leqslant m + n + 1$. If $\mathrm{pd} M < n + 1$, there is nothing to prove so assume that M is such that $\mathrm{pd} M \geq n + 1$. Then we have the start of a minimal projective resolution

$$0 \to L \to P_n \to P_{n-1} \to \cdots \to P_0 \to M \to 0.$$

with the P_i projective and $L \neq 0$. Moreover, because of Lemma XII.2.24, we have $\mathrm{Ext}_A^{n+1}(M, L') \neq 0$, for any indecomposable summand L' of L. Hence, $\mathrm{id} L' \geq n + 1$. The hypothesis implies $\mathrm{pd} L' \leqslant m$. This holds true for any indecomposable summand of L, hence $\mathrm{pd} L \leqslant m$. Thus, L has a projective resolution of length at most m. Splicing such a resolution with the previous projective resolution of M yields a projective resolution of length $m + n + 1$. Hence, $\mathrm{pd} M \leqslant m + n + 1$ and so $\mathrm{gl.dim.} A \leqslant m + n + 1$. $\qquad\square$

XII.4.13 Example. The converse of Lemma XII.4.12 does not hold true. Let A be given by the quiver

$$
\begin{array}{ccc}
4 & \xrightarrow{\alpha} & 3 \\
{\scriptstyle \delta}\downarrow & & \downarrow{\scriptstyle \beta} \\
1 & \xleftarrow{\gamma} & 2
\end{array}
$$

bound by the ideal $\langle \gamma\beta, \beta\alpha \rangle$. One has $\mathrm{pd} S_1 = 0, \mathrm{pd} S_2 = 1, \mathrm{pd} S_3 = 2$ and $\mathrm{pd} S_4 = 3$. Therefore, $\mathrm{gl.dim.} A = 3$. But there exists an indecomposable finitely generated A-module M having both projective and injective dimension three. Indeed, let M denote the uniserial A module whose top is S_4 and socle is S_1, that is, the module given by the representation

$$
\begin{array}{ccc}
K & \xrightarrow{0} & 0 \\
{\scriptstyle 1_K}\downarrow & & \downarrow{\scriptstyle 0} \\
K & \xleftarrow{0} & 0
\end{array}
$$

A minimal projective resolution is

$$0 \longrightarrow 1 \longrightarrow \begin{smallmatrix} 2 \\ 1 \end{smallmatrix} \longrightarrow \begin{smallmatrix} 3 \\ 2 \end{smallmatrix} \longrightarrow \begin{smallmatrix} 4 \\ 3\,1 \end{smallmatrix} \longrightarrow M \longrightarrow 0$$

and a minimal injective coresolution is

$$0 \longrightarrow M \longrightarrow \begin{smallmatrix} 2\,4 \\ 1 \end{smallmatrix} \longrightarrow \begin{smallmatrix} 3 \\ 2 \end{smallmatrix} \longrightarrow \begin{smallmatrix} 4 \\ 3 \end{smallmatrix} \longrightarrow 4 \longrightarrow 0$$

XII.4.2 Selfinjective algebras

Hereditary algebras have global dimension one. At the other extreme, from the homological point of view, lie algebras having infinite global dimension. For instance, let A be a principal ideal domain and $p \in A$ an irreducible element. Because of Example VI.4.21, the quotient algebra A/p^2A is injective when considered as a module over itself, and, because of Example XII.3.4(c), it has infinite global dimension. It turns out that the first condition implies the second. This suggests the definition.

XII.4.14 Definition. A right and left noetherian K-algebra A is **right selfinjective** or **right quasi-Frobenius** if A_A is an injective A-module.

Left selfinjective algebras are defined in the same way. As usual, when we say 'selfinjective' without adjective, we mean 'right selfinjective'. The two concepts coincide if the algebra is commutative. Because selfinjective algebras are right and left noetherian, one can speak of their global dimension, see Corollary XII.3.14.

The importance of selfinjective algebras comes from the fact that the algebras of finite groups are selfinjective, a fact proven below in Example XII.4.28(a).

XII.4.15 Examples.
    (a) Let A be a semisimple algebra. Due to Wedderburn's Theorem VII.5.12, any module is injective. Hence, A is (trivially) selfinjective.
    (b) Let A be a principal ideal domain and I a nonzero ideal in A. Then A/I is selfinjective, due to Example VI.4.21.
    (c) Consider the Nakayama algebra given by the quiver

bound by the ideal $\langle \alpha\beta, \beta\gamma, \gamma\alpha \rangle$. Then $A_A = P_1 \oplus P_2 \oplus P_3$. Because $P_1 = I_2 = \begin{smallmatrix}1\\2\end{smallmatrix}$, $P_2 = I_3 = \begin{smallmatrix}2\\3\end{smallmatrix}$ and $P_3 = I_1 = \begin{smallmatrix}3\\1\end{smallmatrix}$, each indecomposable finitely generated projective module (up to isomorphism) is injective, hence A_A is injective: the algebra is right selfinjective.

This last example suggests the following theorem.

XII.4.16 Theorem. *A right and left noetherian K-algebra A is selfinjective if and only if any projective A-module is injective.*

Proof. Sufficiency being obvious, we prove necessity. Let A be selfinjective and P a projective A-module. Due to Theorem VI.4.4, there exists a set Λ such that P is a direct summand of $A_A^{(\Lambda)}$. Because A is noetherian, it follows from Theorem VII.2.22 that $A_A^{(\Lambda)}$ is injective. Therefore, its submodule P_A is injective as well. $\qquad\square$

It follows almost immediately that selfinjective algebras have infinite (or zero) global dimension.

XII.4.17 Proposition. *Let A be selfinjective. Then any nonprojective module has infinite projective dimension. In particular, either A is semisimple, or* gl.dim.$A = \infty$.

Proof. Assume that M_A is a nonprojective module of finite projective dimension n. Then there exists a projective resolution of the form

$$0 \to P_n \xrightarrow{d_n} P_{n-1} \to \cdots \to P_1 \to P_0 \to M \to 0$$

and no shorter resolution. Because P_n is projective, it is injective, so the short exact sequence $0 \to P_n \xrightarrow{d_n} P_{n-1} \to \operatorname{Coker} d_n \to 0$ splits. Hence, $\operatorname{Coker} d_n$ is projective and the sequence

$$0 \to \operatorname{Coker} d_n \to \cdots \to P_1 \to P_0 \to M \to 0$$

is a projective resolution of length $n - 1$, a contradiction which establishes the first statement. The second comes from the fact that either any A-module is projective (and then A is semisimple) or else there exists a nonprojective A-module. $\qquad\square$

We say a few words about another homological invariant. Given an algebra A, we define its **finitistic dimension** to be

$$\text{Fin.dim.}A = \sup\{\text{pd}M: M \text{ an } A\text{-module}, \text{pd}M < \infty\}$$

that is, we take the supremum only over those modules which have finite projective dimension. If one restricts to finitely generated modules, this becomes

$$\text{fin.dim.}A = \sup\{\text{pd}M: M \text{ a finitely generated } A\text{-module}, \text{pd}M < \infty\}.$$

In particular, fin.dim.$A \leqslant$ Fin.dim.A. If the (right) global dimension of an algebra is finite, then it follows from the previous definition that it equals its finitistic dimension. But the converse is not true: there exist algebras with infinite global dimension but finite finitistic dimension. There is a well-known conjecture, going back to the 1960s, still unproved at the time of writing these lines, stating that for any algebra A, its finitistic dimension Fin.dim.A is finite. It is not even known if fin.dim.A is always finite.

In this context, Proposition XII.4.17 says that a nonsemisimple selfinjective algebra has infinite global dimension and zero finitistic dimension. Another example is the following one.

XII.4.18 Example. Consider the Nakayama algebra A given by the quiver

$$1 \circ \underset{\beta}{\overset{\alpha}{\rightleftarrows}} \circ 2$$

bound by the ideal $\langle \alpha\beta\alpha \rangle$.

In order to compute the global and the finitistic dimension, it suffices, because of Corollary XII.2.5, to compute the projective dimensions of the indecomposable modules. As seen in Example IX.4.19(c), the indecomposable A-modules are, up to isomorphism,

$$P_1 = \begin{matrix} 1 \\ 2 \\ 1 \end{matrix}, \ P_2 = \begin{matrix} 2 \\ 1 \\ 2 \\ 1 \end{matrix}, \ P_1/\mathrm{rad}^2 P_1 = \begin{matrix} 1 \\ 2 \end{matrix}, \ P_1/\mathrm{rad}P_1 = 1, \ P_2/\mathrm{rad}^3 P_2 = \begin{matrix} 2 \\ 1 \\ 2 \end{matrix},$$

$$P_2/\mathrm{rad}^2 P_2 = \begin{matrix} 2 \\ 1 \end{matrix}, \text{ and } P_2/\mathrm{rad}P_2 = 2$$

The modules P_1 and P_2 are projective, so $\mathrm{pd}P_1 = \mathrm{pd}P_2 = 0$. Computing minimal projective resolutions, as we did in Examples XII.2.23, we find that $\mathrm{pd}(P_2/\mathrm{rad}P_2) = 1$ and all other indecomposable A-modules have infinite projective dimension. Hence, A is not selfinjective, see Proposition XII.4.17, but we have $\mathrm{gl.dim}.A = \infty$ and $\mathrm{fin.dim}.\,A = 1$.

From now on, let A be a finite–dimensional algebra over a field K. In this case, right selfinjective coincides with left selfinjective. In order to prove it, we need the duality $D = \mathrm{Hom}_K(-, K)$ between right and left finitely generated A-modules and the results of Subsection IX.2.3.

XII.4.19 Theorem. *Let A be a finite–dimensional K-algebra. The following conditions are equivalent:*

(a) *A is right selfinjective.*
(b) *A is left selfinjective.*
(c) *$(DA)_A$ is a projective right module.*
(d) *$_A(DA)$ is a projective left module.*
(e) *Any finitely generated projective right A-module is injective.*
(f) *Any finitely generated projective left A-module is injective.*
(g) *Any finitely generated injective right A-module is projective.*
(h) *Any finitely generated injective left A-module is projective.*

Proof. (a), (d), (h) and (e) are equivalent: A is selfinjective if and only if A_A is injective hence if and only if $_A(DA) = D(A_A)$ is projective. Because any finitely generated injective left A-module is isomorphic to a finite direct sum of indecomposable summands of $_A(DA)$, due to Theorem IX.2.16, we get the equivalence with (h). Because of Theorem XII.4.16, (a) implies (e), and the converse follows from the fact that A_A is a cyclic module.

Dually, (b), (c), (f), (g) are equivalent.

Finally, using the duality functor D shows that (e) and (h) are equivalent. $\quad\square$

XII.4.20 Definition. A finite–dimensional K-algebra A is a ***Frobenius algebra*** if there exists an isomorphism of right A-modules $A_A \cong D({}_AA)$.

It follows from the preceding Theorem XII.4.19 that Frobenius algebras are selfinjective.

XII.4.21 Theorem. *The following conditions are equivalent for a finite–dimensional algebra A :*

   (a) *A is a Frobenius algebra.*
   (b) *There exists a nondegenerate K-bilinear form $\langle -, ? \rangle : A \times A \to K$ which is **associative**, that is, such that $\langle ab, c \rangle = \langle a, bc \rangle$, for any $a, b, c \in A$.*
   (c) *There exists a linear form $g \in DA$ whose kernel contains no nonzero right or left ideal.*

Proof. (a) implies (b). Because A is Frobenius, there exists an isomorphism $f : A_A \to D({}_AA)$ of right A-modules. For $a, b \in A$, we have $f(ab) = f(a)b$. Moreover, the definition of the right module structure on $D({}_AA)$ implies that $f(ab)(x) = (f(a)b)(x) = f(a)(bx)$, for $x \in A$. This suggests to define the form $\langle -, ? \rangle : A \times A \to K$ by:

$$\langle x, y \rangle = f(x)(y)$$

for $x, y \in A$. It is clear that $\langle -, ? \rangle$ is K-bilinear. It is nondegenerate : indeed, $\langle -, y \rangle = 0$ means that any linear form applied to y gives zero, hence $y = 0$, while $\langle x, - \rangle = 0$ gives $f(x) = 0$, but then $x = 0$ because f is an isomorphism. Associativity comes from:

$$\langle x, yz \rangle = f(x)(yz) = f(xy)(z) = \langle xy, z \rangle$$

for $x, y, z \in A$.

   (b) implies (a). Define $f : A_A \to D({}_AA)$ by setting $f(x) = \langle x, - \rangle$, for $x \in A$, that is, $f(x)(y) = \langle x, y \rangle$, for $x, y \in A$. Then f is a K-linear isomorphism, because the form is nondegenerate, and it is actually A-linear, because the form is associative.

   (b) implies (c). Given a bilinear form $\langle -, ? \rangle$, define a linear form $g \in DA$ by $g(x) = \langle x, 1 \rangle$, for $x \in A$. Assume that the kernel of g contains a nonzero right ideal I. If $x \in I$ is a nonzero element, then $g(xA) = \langle xA, 1 \rangle = 0$. Associativity yields $\langle x, A \rangle = 0$. Nondegeneracy yields $x = 0$, a contradiction. Similarly, the kernel of g contains no nonzero left ideal.

   (c) implies (b). If the form g is given, one defines $\langle -, ? \rangle : A \times A \to K$ by setting $\langle x, y \rangle = g(xy)$, for $x, y \in A$. There remains to prove that $\langle -, ? \rangle$ is bilinear, nondegenerate and associative, and this is an easy exercise left to the reader. $\square$

Let A be a basic selfinjective algebra, see Subsection IX.3.2, and $\{e_1, \cdots, e_n\}$ a complete set of primitive pairwise orthogonal idempotents. Because of Theorems IX.2.6 and IX.2.16, the sets $\{e_1A, \cdots, e_nA\}$ and $\{D(Ae_1), \cdots, D(Ae_n)\}$ are respectively complete lists of representatives of the isomorphism classes of finitely generated indecomposable projective and injective modules.

XII.4.22 Corollary. *A basic selfinjective algebra is a Frobenius algebra.*

Proof. It follows from Theorem XII.4.19 that there exists a bijection $v : \{1, \cdots, n\} \rightarrow \{1, \cdots, n\}$ such that $D(Ae_i) \cong e_{v(i)}A$, for any i. Therefore, we have an isomorphism $(DA)_A = \oplus_{i=1}^{n} D(Ae_i) \cong \oplus_{i=1}^{n} (e_{v(i)}A) = A_A$. □

In fact, not only do we have a permutation v of $\{1, \cdots, n\}$ as in the proof, but actually this permutation is induced from an automorphism of the algebra.

XII.4.23 Proposition. *Let A be a basic Frobenius algebra and $\{e_1, \cdots, e_n\}$ a complete set of primitive pairwise orthogonal idempotents. Then there exists an automorphism $v : A \rightarrow A$ such that $D(Ae_i) \cong e_{v(i)}A$, for any i with $1 \leqslant i \leqslant n$.*

Proof. Let $\langle -, ? \rangle : A \times A \rightarrow K$ be a nondegenerate associative bilinear form and define a mapping $v : A \rightarrow A$ by:

$$\langle v(a), - \rangle = \langle -, a \rangle$$

for $a \in A$. Then v is clearly K-linear. Moreover, for $a, b, x \in A$,

$$\langle x, ab \rangle = \langle xa, b \rangle = \langle v(b), xa \rangle = \langle v(b)x, a \rangle = \langle v(a), v(b)x \rangle = \langle v(a)v(b), x \rangle$$

therefore, $v(ab) = v(a)v(b)$. Finally,

$$\langle v(1), x \rangle = \langle x, 1 \rangle = \langle 1 \cdot x, 1 \rangle = \langle 1, x \cdot 1 \rangle = \langle 1, x \rangle$$

for any $x \in A$, which gives $v(1) = 1$. Because v is clearly invertible, it is an automorphism.

Therefore, $\{v(e_1), \cdots, v(e_n)\}$ is a complete set of primitive pairwise orthogonal idempotents. Hence, for any i, the module $v(e_i)A$ is indecomposable projective. Therefore, there exists an index j such that $v(e_i)A \cong e_jA$. That is, v induces a permutation of $\{1, \cdots, n\}$.

Let now I_A be an indecomposable injective module. Because of Theorem XII.4.19, there exists i such that $I = e_iA$. Let S be the (simple) socle of I. Then $S = e_iS$: indeed, any $s \in S$ can be written as $s = e_ia$, for some $a \in A$, hence $s = e_is$, which proves our claim. On the other hand, S is a submodule of A_A, so it is a right ideal of A. Let g be a linear form attached to A, according to Theorem XII.4.21(c). There exists $x \in S$ such that $g(e_ix) \neq 0$ and then:

$$g(xv^{-1}(e_i)) = \langle x, v^{-1}(e_i) \rangle = \langle e_i, x \rangle = \langle e_ix, 1 \rangle = g(e_ix) \neq 0.$$

Therefore, $\text{Hom}_A(v^{-1}(e_i)A, S) \cong Sv^{-1}(e_i) \neq 0$ and so the top of $v^{-1}(e_i)A$ is isomorphic to the socle S of e_iA. According to Theorem XII.4.19, this implies that $e_iA \cong D(Av^{-1}(e_i))$. □

The automorphism v and the induced permutation of $\{1, \cdots, n\}$ are called the **Nakayama automorphism** and the **Nakayama permutation** of A, respectively.

The next lemma, which is Exercise IX.2.4, shows that the Nakayama permutation is uniquely determined up to an inner automorphism of A.

XII.4.24 Lemma. *Let A be a finite–dimensional K-algebra and $\{e_1, \cdots, e_n\}, \{e'_1, \cdots, e'_m\}$ two complete sets of primitive pairwise orthogonal idempotents of A. Then $m = n$ and there exists $a \in A$ invertible such that $e'_i = ae_ia^{-1}$, up to order.*

Proof. Because $A_A = \bigoplus_{i=1}^n e_iA = \bigoplus_{j=1}^m e'_jA$, it follows from the Unique Decomposition Theorem VIII.4.20 that $m = n$ and that, up to order, $e_iA \cong e'_iA$. Because of the remark following Proposition VII.3.1, there exists $a_i \in e'_iAe_i$ such that $a_i = e'_ia_i = ae_i$. Therefore, $a = a_1 + \cdots + a_n$ satisfies $a = e'_ia = ae_i$, for any i.

There remains to prove that a is invertible. But the remark following Proposition VII.3.1 gives also $b_i \in e_iAe'_i$ such that $b_i = e_ib_i = b_ie'_i$ and again $a_ib_i = e_i, b_ia_i = e'_i$. Then one sees that $b = b_1 + \cdots + b_n$ satisfies $ab = 1$ and $ba = 1$. $\qquad\square$

When the bilinear form $\langle -, ? \rangle$ is symmetric, the algebra itself is called symmetric.

XII.4.25 Definition. A finite–dimensional K-algebra A is a **symmetric algebra** if there exists an isomorphism of $A - A$-bimodules ${}_AA_A \cong D({}_AA_A)$.

Obviously, any symmetric algebra is a Frobenius algebra.

XII.4.26 Corollary. *The following conditions are equivalent for a finite–dimensional algebra A :*
  (a) *A is a symmetric algebra.*
  (b) *There exists a nondegenerate associative symmetric K-bilinear form $\langle -, ? \rangle : A \times A \to K$.*
  (c) *There exists a linear form $g \in DA$ whose kernel contains no nonzero right or left ideal and, moreover, such that $g(ab) = g(ba)$, for any $a, b \in A$.*

Proof. This is an easy adaptation of the proof of Theorem XII.4.21, left to the reader. $\qquad\square$

Because of the symmetry of the bilinear form, it follows from the proof of Proposition XII.4.23 that the Nakayama automorphism of a symmetric algebra equals the identity. This proves the following corollary.

XII.4.27 Corollary. *Let A be a symmetric algebra and $e \in A$ a primitive idempotent, then $eA \cong D(Ae)$.* $\qquad\square$

XII.4.28 Examples.
  (a) Let K be a field and G a finite group. The group algebra KG is symmetric: indeed, let, for $g, h \in G$

$$\langle g, h \rangle = \begin{cases} 1 & \text{if } gh = 1 \\ 0 & \text{if } gh \neq 1 \end{cases}$$

and extend by bilinearity to all elements of KG. It is clear that $\langle -, ? \rangle$ is symmetric and associative. We prove that it is nondegenerate. Let $\sum_{g \in G} g\alpha_g \in KG$ be such that $\langle \sum_{g \in G} g\alpha_g, ? \rangle = 0$. Then, for any $h \in G$, one has

$$0 = \left\langle \sum_{g \in G} g\alpha_g, h^{-1} \right\rangle = \sum_{g \in G} \alpha_g \langle g, h^{-1} \rangle = \alpha_h.$$

Therefore, $\sum_{g \in G} g\alpha_g = 0$.

(b) Let A be a finite–dimensional K-algebra. Its trivial extension $A \ltimes DA$ (see Exercise VIII.3.14) by the injective cogenerator bimodule $_A(DA)_A$ is symmetric. Indeed, it is easily verified that the form $\langle -, ? \rangle$ defined, for $(a, f), (b, g) \in A \ltimes DA$, by:

$$\langle (a, f), (b, g) \rangle = g(a) + f(b)$$

satisfies all required properties.

(c) The algebra of Example XII.4.15(c) is selfinjective but not symmetric because its Nakayama automorphism is not equal to the identity (actually, the Nakayama permutation in this example is given by $\nu = \begin{pmatrix} 1 & 2 & 3 \\ 2 & 3 & 1 \end{pmatrix}$).
On the other hand, the algebra given by the same quiver bound by $\langle \alpha\beta\gamma, \beta\gamma\alpha, \gamma\alpha\beta \rangle$ is symmetric.

Exercises of Section XII.4

1. Let K be a field and A a finite–dimensional K-algebra. Prove that the following conditions are equivalent:
   (a) A is hereditary.
   (b) For any module M, the functor $\mathrm{Ext}_A^1(M, -)$ is right exact.
   (c) For any simple module S, the functor $\mathrm{Ext}_A^1(S, -)$ is right exact.
   (d) For any module N, the functor $\mathrm{Ext}_A^1(-, N)$ is right exact.
   (e) For any simple module S, the functor $\mathrm{Ext}_A^1(-, S)$ is right exact.

2. Let K be a field. Prove that the following lower triangular matrix algebra is hereditary:

$$\begin{pmatrix} K & 0 & 0 & 0 \\ K & K & 0 & 0 \\ K & 0 & K & 0 \\ K & 0 & 0 & K \end{pmatrix}.$$

3. Let K be a field and A a hereditary finite–dimensional K-algebra. Prove that, for any indecomposable injective A-module I_A, the endomorphism algebra $\operatorname{End} I_A$ is a division ring.

4. Let, as in Exercise IX.4.5, the algebra B be a one-point extension of a bound quiver algebra A by an A-module M. Prove that:
    (a) gl.dim.B = max{gl.dim.A, 1 + pd$M$}.
    (b) B is hereditary if and only if A is hereditary and M is projective.

5. Let K be a field, $n > 1$ an integer and $A = T_n(K)$ the lower triangular matrix algebra. Prove that A is hereditary but that gl.dim$(A/\operatorname{rad}^2 A) = n - 1$.

6. Prove that a noetherian algebra is hereditary if and only if any right ideal is flat.

7. Let A be an artinian algebra with radical J. Prove that the following conditions are equivalent:
    (a) A is right hereditary.
    (b) J_A is a projective module.
    (c) For any idempotent e, the module eJ_A is projective.
    (d) For any primitive idempotent e, the module eJ_A is projective.
    (e) Any maximal submodule of A is a projective module.

8. An algebra is called **right semihereditary** if any finitely generated right ideal is projective. Left semihereditary algebras are defined dually. Prove the following:
    (a) If an algebra is right semihereditary, then any finitely generated submodule of a free module is the finite direct sum of finitely generated right ideals.
    (b) An algebra is right semihereditary if and only if any finitely generated submodule of a projective module is projective.
    (c) The algebra of Example XII.4.5(c) is right semihereditary.

9. Let K be a field and Q the quiver $1 \xrightarrow{\alpha} 2 \xrightarrow{\beta} 3 \xrightarrow{\gamma} 4 \xrightarrow{\delta} 5$. Prove that:
    (a) If A is given by the quiver Q bound by the ideal $\langle \beta\gamma \rangle$, then gl.dim.$A = 2$, and for any indecomposable finitely generated module M, one has pd$M \leqslant 1$ or id$M \leqslant 1$.
    (b) If A is given by the quiver Q bound by the ideal $\langle \alpha\beta, \gamma\delta \rangle$ then gl.dim.$A = 2$ but there exists an indecomposable finitely generated module M having both projective and injective dimension equal to 2.
    (c) If A is given by the quiver Q bound by the ideal $\langle \alpha\beta\gamma, \gamma\delta \rangle$ then gl.dim.$A = 3$, and for any indecomposable finitely generated module M, one has pd$M \leqslant 1$ or id$M \leqslant 1$.

10. Let A be a Frobenius algebra. Prove that so is $M_n(A)$.

11. Let $A = K$. Prove that its trivial extension $A \ltimes DA$, see Exercise VIII.3.14, is isomorphic to $K[t]/\langle t^2 \rangle$.

12. Let K be a field. Prove that each of the following bound quiver K-algebras is selfinjective.

(a)

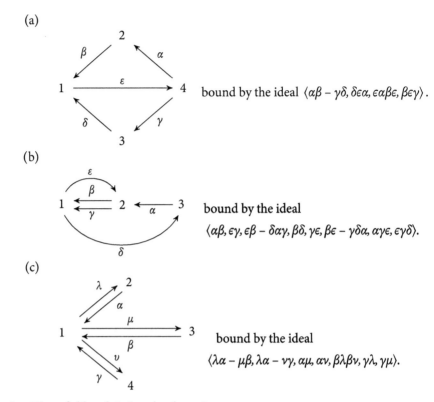

bound by the ideal $\langle \alpha\beta - \gamma\delta, \delta\epsilon\alpha, \epsilon\alpha\beta\epsilon, \beta\epsilon\gamma \rangle$.

(b)

bound by the ideal
$$\langle \alpha\beta, \epsilon\gamma, \epsilon\beta - \delta\alpha\gamma, \beta\delta, \gamma\epsilon, \beta\epsilon - \gamma\delta\alpha, \alpha\gamma\epsilon, \epsilon\gamma\delta \rangle.$$

(c)

bound by the ideal
$$\langle \lambda\alpha - \mu\beta, \lambda\alpha - \nu\gamma, \alpha\mu, \alpha\nu, \beta\lambda\beta\nu, \gamma\lambda, \gamma\mu \rangle.$$

13. Let K be a field and A given by the quiver

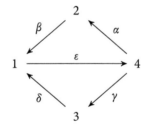

bound by the ideal $\langle \alpha\beta - \gamma\delta, \epsilon\alpha, \epsilon\gamma, \beta\epsilon, \delta\epsilon \rangle$. Prove that $\mathrm{id}A_A = 1$, $\mathrm{pd}(DA)_A = 1$ but gl.dim. $A = \infty$. Thus, A has infinite global dimension but is not selfinjective.

14. Let A be an elementary Nakayama K-algebra. Prove that A is selfinjective if and only if it is given by the quiver

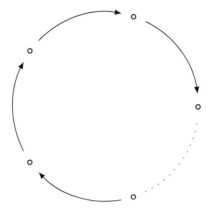

bound by the ideal $\mathrm{rad}^i A$, for some integer $i \geq 2$. This generalises Example XII.4.15(c).

Bibliography

ADAMSON, I. T., Rings Modules and Algebras, Oliver and Boyd, Edinburgh, 1971.

ANDERSON, F. W., and FULLER, K. R., Rings and Categories of Modules, Springer-Verlag, Berlin, Heidelberg, New York, 1974.

ASSEM, I., Algèbres et modules, Les Presses de l'Université d'Ottawa, Ottawa, et Masson, Paris, 1997.

ASSEM, I., Introduction au langage catégorique, Calvage et Mounet, Paris, 2022.

ASSEM, I., and COELHO, F. U., Basic Representation Theory of Algebras, GTM 283 Springer-Verlag, Berlin, Heidelberg, New York, 2020.

ASSEM, I., SIMSON, D., and SKOWRONSKI, A., Elements of the Representation Theory of Associative Algebras, London Mathematical Society Student Texts, Cambridge University Press, Cambridge, 2006.

BLYTH, T. S., Module Theory, An Approach to Linear Algebra, Oxford Science Publications, Oxford, 1990.

BOURBAKI, N., Algèbre: Chapitres 1 à 3, Hermann, Paris, 1970.

BOURBAKI, N., Algèbre: Chapitre 8, Modules et anneaux semisimples, Hermann, Paris, 1973.

BOURBAKI, N., Algèbre: Chapitre 10, Algèbre homologique, Masson, Paris, 1980.

BURTON, D. M., A First Course in Rings and Ideals, Addison-Wesley, Reading, Massachusetts, 1970.

CARTAN, H., and EILENBERG, S., Homological Algebra, Princeton University Press, Princeton, 1956.

COHN, P. M., Algebra, 3 Volumes, John Wiley and Sons, 2nd Edition, Cambridge, 1982.

CURTIS, C. W., and REINER, I., Representation Theory of Finite Groups and Associative Algebras. John Wiley and Sons, Cambridge, 1962.

ERDMANN, K. and HOLM, T., Algebras and Representation Theory, Springer-Verlag, Berlin, Heidelberg, New York, 2018.

GABRIEL, P., and ROITER, A. V., Representations of finite-dimensional algebras, Algebra VIII, Vol. Representations of finite-dimensional algebras, Algebra VIII, Vol. 73 of the Encyclopaedia of Mathamatical Sciences, Springer-Verlag, Berlin, Heidelberg, New York, 1992.

GELFAND, S. I., and MANIN, Yu. I., Methods of Homological Algebra, Springer-Verlag, Berlin, Heidelberg, New York, 1996.

HARTLEY, H., and HAWKES, T. O., Rings, Modules and Linear Algebra, Chapman and Hall, London, New York, 1983.

HILTON, P., and STAMMBACH, U., A Course in Homological Algebra, Springer-Verlag, Berlin, Heidelberg, New York, 2nd edition, 1997.

HILTON, P., and WU, Y.C., A Course in Modern Algebra, John Wiley and Sons, Cambridge, 1989.

JANS, J. P., Rings and Homology, Holt, Rinehart and Winston, New York, 1964.

LAFON, J.-P., Les formalismes fondamentaux de l'Algèbre commutative, Hermann, Paris, 1974.

LAM, T. Y., Lectures on Modules and Rings, Springer-Verlag, Berlin, Heidelberg, New York, 1999.

LAM, T. Y., A First Course in Noncommutative Rings, Springer-Verlag, Berlin, Heidelberg, New York, 2nd edition, 2001.

LAMBEK, J., Lectures on Rings and Modules, Blaisdell Publishing Compagny, Waltham, Massachusetts, 1966.

MAC LANE, S., and BIRKHOFF, G., Algèbre, 2 Volumes, Gauthier-Villars, Paris, 1971.

MAC LANE, S., Homology, Springer-Verlag, Berlin, Heidelberg, New York, 1963.

MAC LANE, S., Categories for the Working Mathematician, Springer-Verlag, Berlin, Heidelberg, New York, 1971.

MITCHELL, B., Theory of Categories, Academic Press, New York, 1965.

NORTHCOTT, D.G., A First Course of Homological Algebra, Cambridge University Press, Cambridge, 1973.

PAREIGIS, B., Categories and Functors, Academic Press, New York and London, 1970.

ROTMAN, J. J., An Introduction to Homological Algebra, Academic Press, New York, 1979.

SCHIFFLER, R., Quiver Representations, Springer-Verlag, Berlin, Heidelberg, New York, 2014.

SCOTT OSBORNE, M., Basic Homological Algebra, Springer-Verlag, Berlin, Heidelberg, New York, 2000.

STENSTRÖM, B. Rings of Quotients, Springer-Verlag, Berlin, Heidelberg, New York, 1975.

WEIBEL, C. A., An Introduction to Homological Algebra, Cambridge University Press, Cambridge, 1994.

Index

Ab-category, 145
A-B-bimodule, 54
A-bilinear, 231
A-linear map, 69
K-abelian, 158
K-algebra, 17
K-algebra of endomorphisms, 146
K-category, 145
K-linear, 147
K-linear category, 154
T-presentation, 412
T-presented, 412
f-invariant, 52
n-boundaries, 453
n-coboundaries, 456
n-cocycles, 456
n-cycles, 453
n-extension, 515
n^{th} right derived functor, 475, 476
n^{th}-cohomology module, 456
n^{th}-extension module, 483
n^{th}-homology functor, 454
n^{th}-homology module, 453
n^{th}-left derived functor, 471
n^{th}-torsion module, 494
p-group, 167
p-primary, 214

A-balanced, 231
abelian, 158
abelian group, 5
abelianisation, 244
acyclic, 26
acyclic complex, 452
addition, 5, 12
additive, 147
additive category, 154
adjoining an identity, 28
adjoint pair, 239
adjunction, 239
admissible, 27
algebra of formal power series, 25
algebra of polynomials, 18
algebra over *K*, 17
algebraically closed, 45
amalgamated sum, 172
annihilated, 182

annihilator, 66
arrows, 26
ascending chain, 287
ascending chain condition, 287
associate, 42
associate irreducible, 44
associative, 571
associative algebra, 18
atom, 313
atoms, 315
Auslander, 548
automorphism, 33, 72, 113
automorphism group, 33, 72

Bézout-Bachet theorem, 43
Baer sum, 529
Baer's criterion, 266
basic algebra, 415
basic algebra of *A*, 416
basis, 56
becomes stationary, 287, 288
bicartesian, 174
bifunctor, 117
bimodule morphism, 72
binomial theorem, 15
biproduct, 152
block, 305
block decomposition, 305
block diagonal matrix, 222
blocks, 222
boolean ring, 15
bound quiver, 27
bound quiver algebra, 27
bound representation, 183

cancellation property, 8
canonical basis, 57
canonical decomposition, 97
canonical factorisation, 158
canonical isomorphism, 97
canonical morphism, 158
cartesian square, 168
category, 112
category enriched in Mod*K*, 145
central, 304
centrally primitive, 304
chain, 29

chain complex, 449
change of scalars, 52
change of scalars functor, 118
character module, 277
characteristic, 15
characteristic polynomial, 227
Chinese Remainder Theorem, 312
closed, 14
co-unit, 407
cocartesian square, 172
cochain complex, 449
codimension formula, 329
codomain, 112
cogenerated, 284
coimage, 157
cokernel, 83, 156
column-finite, 29
comaximal, 310
commutative, 6
commutative algebra, 18
companion matrix, 229
complete, 301
complete lattice, 59
Complexification of a real vector space, 242
composable, 112
composition, 112
composition factor, 318
composition length, 325
composition series, 318
concentrated complex, 452
concrete categories, 114
cone, 466
connected, 305, 306
connecting morphism, 95
contravariant functor, 116
coprime, 43
coproduct, 129
correspondence of ideals theorem, 36
correspondence of submodules, 102
coset, 22, 58
cosyzygies, 481
covariant functor, 115
covers, 60
cyclic (sub)module, 56

decomposable, 216, 365
degree, 449, 450
dense, 180
descending chain, 288
descending chain condition, 288
diagonal, 201
diagram chasing, 92
differential, 449, 450
dimension shifting, 480, 493
direct sum, 62, 129, 154

direct summand, 62, 363
direct sums, 129
disconnected, 306
discrete category, 124
divide, 41
divisible, 146, 285
division algebra, 23
division ring, 9
domain, 112
dual, 114
dual categories, 186
dual category, 114
duality principle, 114

elementary algebra, 424
elementary operations, 199
embedding, 179
endofunctor, 116
endomorphism, 31, 70, 112
endomorphism algebra, 76
enough injectives, 270
epimorphism, 81, 147
equivalent, 179, 505
equivalent matrices, 201
essential, 273
essential extension, 274
essential image, 179
essential submodule, 274
essentially small category, 182
essentially surjective, 180
euclidean domains, 41
evaluation, 186
evaluation morphism, 235
exact, 85, 160, 245
exact at, 160
exact at M, 85
exact complex, 451
extend, 263, 475
extension, 10, 263, 317, 475
external multiplication, 12
external product, 17

faithful, 66, 179
fibered product, 168
field, 9
field of fractions, 11
final object, 154
finite dimensional algebra, 19
finite dimensional representation, 184
finite length, 325
finite order, 195
finite quiver, 26
finitely generated module, 56
finitely presented, 141
finitistic dimension, 569

fitting's Lemma, 371
flat, 276
flat dimension, 532
flat resolution, 277
for almost all, 21
Forgetful functors, 116
fraction, 10
free, 138
free algebra, 144
free module, 137, 138
free module functor, 142
free presentation, 141
free resolution, 141
Frobenius algebra, 571
full, 179
full and faithful, 179
full matrix algebra, 18
full subcategory, 115
full subquiver, 26
fully faithful, 179
functor in two variables, 117
functorial isomorphism, 119
functorial morphism, 118
functoriality, 118
functorisation, 142

Gabriel quiver, 426
Gaussian integers, 16
generate, 408
generated, 20, 21, 55, 284
generates, 349
generator, 408
generators, 55
global dimension, 546
Grassmann Formula, 326
greatest common divisor, 42
greatest lower bound, 59
group algebra, 25

Hasse quiver, 60
Hausdorff space, 149
Hilbert's Syzygy Theorem, 551
Hom-sets, 74
homeomorphisms, 113
homomorphism, 30, 69, 102
homotopic, 461
homotopy, 461
homotopy equivalence, 466
Hopkins-Levitski's Theorem, 295, 357

ideal, 20, 376
idempotent, 78, 257
identity, 6, 112, 118
identity functor, 116
image, 32, 157

incidence algebra, 25
inclusion, 70, 156
inclusion functor, 116
inclusion morphism, 31
indecomposable, 216, 365
indecomposable as an algebra, 305
infimum, 59
infinite dimensional algebra, 19
infinite length, 325
initial object, 154
injection, 70, 156, 451
injection morphism, 31
injections, 129
injective, 263
injective cogenerator, 268
injective copresentation, 271
injective coresolution, 271
injective dimension, 532
injective envelope, 274
injective hull, 274
injectively stable category, 378
integral domain, 8
internal multiplication, 17
internal operation, 5
invariant factors, 205
inverse, 9, 113, 119, 178
invertible, 9
irreducible elements, 44
irreducible morphism, 385
isomorphic, 33, 72, 113, 178
isomorphism, 33, 72, 102, 112

Jacobson radical, 350
Jordan block, 222
Jordan matrix, 222
Jordan-Hölder Theorem, 323

Ker-Coker sequence, 95
kernel, 32, 156, 377
Kronecker quiver, 190
Krull's theorem, 44
Krull-Schmidt theorem, 372

lattice, 59
least common multiple, 42
least upper bound, 59
left A-modules, 49
Left K-modules, 12
left adjoint, 239
left and right distributive, 6
left artinian, 290, 294
left exact, 86, 245
left global dimension, 546
left hereditary, 561
left ideal, 51

left noetherian, 290, 294
length, 46, 318, 325
lift, 255, 469
lifted modulo J, 388
lifting, 255, 469
lifting idempotents, 389
linear, 147
linear category, 154
linear combination, 14, 55
linear extension of maps theorem, 139
linearly dependent, 56
linearly independent, 56
local, 367
Loewy length, 433
long exact cohomology sequence, 461
long exact homology sequence, 460
lower triangular, 29

Maschke, 335
matrix, 193
maximal, 36, 63, 289
minimal injective coresolution, 538
minimal polynomial, 45, 228
minimal projective resolution, 538
minor, 205
mixed associativity, 12
modular, 59
modular law, 29, 59
monoid, 6
monomorphism, 80, 147
Morita equivalent, 410
Morita Theorem, 408
morphism, 30, 69, 102
morphism category, 124
morphism of complexes, 450
morphism of representations, 71
morphisms, 112
multiplication, 6

Nakayama algebra, 435
Nakayama automorphism, 572
Nakayama functor, 401
Nakayama permutation, 572
Nakayama's Lemma, 343
natural transformation, 118
negative, 6, 12, 17
nil ideal, 352
nilpotency index, 228, 352
nilpotent, 29, 78, 228, 352, 353
normalised, 42

objects, 112
obstruction, 508
of finite support, 21
one, 6

one-point extension, 445
operation, 5
opposite, 6, 12, 17
opposite algebra, 18
opposite category, 114
opposite quiver, 37
opposite ring, 7
order, 30
ordinary quiver, 426
orthogonal, 301

parallelogram law, 100
passing to cokernels, 89
passing to coproducts, 135
passing to kernels, 87
passing to products, 135
path, 26
path algebra, 26
Peirce decomposition, 388
points, 26
Pontrjagin dual, 277
powers, 22
Prüfer p-group, 291
preimage, 32
preserve exactness, 231
primary components, 214
primary decomposition, 214
prime ideal, 40
primitive, 301
principal ideal, 21
principal ideal domain, 40
product, 6, 12, 15, 17, 22, 24, 126, 128
product category, 124
product module, 51
progenerator, 408
projection, 31, 70, 156, 451
projections, 128
projective, 255
Projective Basis Theorem, 283
projective cover, 394
projective dimension, 531
projective functor, 261
projective module, 141
projective presentation, 260
projective resolution, 260
projective-injective module, 265
projectively stable category, 378
projectivisation, 412
projectivity, 141
projector, 228
proper value, 227
proper vector, 227
pullback, 168
pure subgroup, 108
pushout, 172

quasi-inverse dualities, 186
quasi-inverse equivalences, 179
quaternion, 9
quiver, 26, 426
quotient, 41, 58
quotient algebra, 23
quotient category, 377
quotient complex, 451

radical, 340, 350, 378
radical filtration, 432
rank, 192
refinement, 289
remainder, 41
representation, 52
representative, 22, 58
result, 5
retraction, 149
right A-module, 49
right K-action, 12
right adjoint, 239
right artinian, 290, 294
right exact, 88, 245
right global dimension, 546
right hereditary algebra, 560
right ideal, 50
right K-module, 12
right noetherian, 290, 294
right quasi-Frobenius, 568
right selfinjective, 568
right semihereditary, 575
ring, 5
ring homomorphism, 10
ring isomorphism, 10
root, 45
row-finite, 29

scalars, 49
Schanuel's Lemma, 281
Schreier refinement theorem, 329
Schur's Lemma, 316
section, 149
selfdual, 114
semisimple, 330, 333
separated, 148
sequence, 25
short exact sequence, 89, 160
simple, 314, 336
simple module, 78
skeleton, 178
skew field, 9
small category, 112
Smith normal form, 205
Smith normal matrix, 201
socle, 345

source, 26, 112
spectroid, 394
splicing, 515
split, 162
split idempotent, 364
splits, 162
stabilises, 287, 288
stable, 14
stalk complex, 452
stationary path, 26
structure constants, 37
subalgebra, 20
subcategory, 115
subcomplex, 451
subfunctor of the identity, 342
submodule, 14, 50
subquiver, 26
subrepresentation, 54
sum, 5, 12, 17, 21, 61
superfluous, 343
support, 21
supremum, 59
surjection, 70, 156
surjection morphism, 31
symmetric algebra, 573
symmetric closure, 516
syzygies, 481

target, 26, 112
tensor product, 232, 233
tensor product of the matrices, 243
tensors, 233
The 3×3 Lemma, 95
The abelianisation of a group, 244
The adjunction isomorphism, 240
The Comparison Theorem, 467
The Comparison Theorem for injectives, 475
The field of fractions of an integral domain, 243
The Five Lemma, 92
The Four Lemma, 91
The Hilbert Basis Theorem, 296
The Horseshoe Lemma, 471
The Invariant Factors Theorem, 207
The Snake Lemma, 93
top, 342
topological abelian group, 146
torsion, 196
torsion class, 200
torsion objects, 200
torsion pair, 200
torsion subgroup, 125
torsion-free, 196
torsion-free class, 200
torsion-free objects, 200
trace, 349

transition matrices, 193
transportation of structure., 78
trivial, 6, 13, 18
trivial extension, 362
two-sided ideal, 20

uniserial, 322
unit, 407
unitary algebra, 18
unitary rings, 6
universal property, 83, 84, 127

valuation, 41
vectors, 49

weak global dimension, 549
Wedderburn-Artin, 336
Well-Ordering Principle, 41

Yoneda embeddings, 179
Yoneda product, 515
Yoneda's Lemma, 121, 123

Zassenhaus' Butterfly Lemma, 108
zero, 6, 12, 13, 17, 18
zero divisor, 8
zero ideal, 377
zero map, 70
zero object, 154